MICROBIOLOGY
molecules, microbes, and man

MICROBIOLOGY

molecules, microbes, and man

Eugene W. Nester
C. Evans Roberts
Brian J. McCarthy
Nancy N. Pearsall
All of the University of Washington

HOLT, RINEHART and WINSTON, Inc.

New York   Chicago   San Francisco   Atlanta
Dallas   Montreal   Toronto   London   Sydney

Preface

Why yet another Microbiology text? The answer to this question comes from a number of years cumulative experience of the authors in teaching an introductory course at the University of Washington. This course, one quarter in length, attracts a large percentage of students preparing to enter health-related fields. Most of the students have had some exposure to chemistry and biology, but are far from expert in handling these subjects. The course aims to teach basic microbiological concepts and the role of microorganisms in health and other applied areas. In teaching this course, we have tried numerous texts or combinations of texts but have not found them satisfactory. They are either too detailed, presuppose too much chemical and biological knowledge, or neglect basic microbiology or the relationship of microbes to disease. This text is designed to eliminate these problems.

We have been guided by certain principles in writing this text. First and foremost, we have attempted to present the fundamentals of microbiology in such a way that their relevance to the applied material is clearly evident. In our experience, students inclined toward careers in the health sciences area have limited backgrounds in biology and chemistry and find discussions of such topics as microbial physiology, cell structure and function, and genetics not only difficult but not particularly important to their real interests. Thus, they try to memorize facts merely for exams rather than understand the material and apply it to the sections on applied microbiology. We have used several devices to interest the student and help him interrelate material presented throughout the text. First, we have included numerous examples, particularly from familiar areas to illustrate abstract concepts. Second, we have employed cross-referencing in the margins to help the student recall material previously covered and see relationships that may not have been obvious. Third, we have reduced the amount of minutiae by providing fewer facts than most texts. It is our belief that the understanding of a relatively few fundamentals is more important than the memorization of a host of facts, most of which will be forgotten soon after exams have been taken.

Another principle we have followed is to cover material to which every student has been exposed but which we often find is poorly understood or even forgotten. For this reason an early chapter covers the principles of biological chemistry. This chapter will serve as a review for some students and as reference material for others, but we predict it will provide some appreciation to the relationships of biology to chemistry for most students. We have also covered such topics as meiosis and mitosis, because we have found that these subjects are not well-understood by the average student.

We have reduced the number of new terms, for the student should not be expected to learn the language of the specialist. In particular, latinized names are used only in the genus and species names of microorganisms; diseases and anatomical features are given their common names wherever possible, with the corresponding medical designation provided in the glossary. We have attempted to refer to microorganisms by the names being used in the laboratories on the forefront of research which we expect will be used by all microbiologists within the next few years. For example, *Pneumococcus pneumoniae* is now *Streptococcus pneumoniae*, since the latter term is in common usage at the Center for Disease Control, Atlanta, Georgia.

We feel that the text is highly adaptable to different courses for students of varying interests. Thus, the text has been written to include a number of core chapters appropriate for all courses and a number of additional chapters which are particularly appropriate for other courses. The "core" chapters include 1–11, 16, 19, and 20. Students in the health sciences area would probably read Chapters 17–33. Students concerned primarily with applied aspects of microbiology other than health-related topics might read Chapters 34–39. For students interested in an overall view of microbiology, Chapters 12, 13, 14, 15, 16, 22, 32, 34–38 would be appropriate. One chapter (3) summarizes material included in greater detail in other chapters (12, 13, 14, and 15). Since all chapters have been cross-referenced, it is possible for the student to obtain relevant information on a particular subject without reading an entire chapter.

The chapters on microbial diseases represent a sufficiently radical departure from other textbooks to warrant some explanation. First, they are organized along an "organ-system" approach rather than by microbial classification—more interesting to students although more difficult to teach. Second, some elementary anatomical and physiological description of the area of the body involved is given to enhance understanding of host-microbial interactions. Third, diseases chosen for discussion are those which students are likely to have heard about through news media or other prior experience. We have rejected the idea that only diseases of present-day suburban United States are important, and have chosen examples according to their interest and the availability of sufficient data to illustrate biological principles. Finally, we have not attempted comprehensive coverage. These sections are not a handbook for paramedical personnel, but exemplary material for all introductory students, including those entering health-related fields.

In writing this text, we have had abundant help from many members of our own department, particularly Velma Chambers, Dorothy Cramer, Joe Dalmasso, Howard Douglas, Sam Eng, Charles Evans, Carol Laxson, Ramona Memmer, Erling Ordal, Helen Pollock, George Ray, Fritz Schoenknecht, John Sherris, James Staley, and Russell Weiser. The artwork for this text was designed and executed by Robert Anderson and Iris Nichols. Eugene Smith and Dan Serebrakian of Holt, Rinehart and Winston assumed most of the problems involved with seeing the rough type script reach a printed page. For their patience and forbearance we owe a great deal. A number of other distinguished teachers, both at the University of Washington and in universities and junior colleges around the country, have also made important suggestions regarding the manuscript. Investigators have been extremely generous in supplying illustrations. The final product is, of course, our own. We sincerely hope it will be of value to students and teachers of introductory microbiology and we solicit your comments and suggestions.

Seattle, Washington

EWN
CER
BJMc
NNP

Contents

PART III   Microbes and Man

PART IV   Microbes and the Environment

I

Foundations of Microbial Life

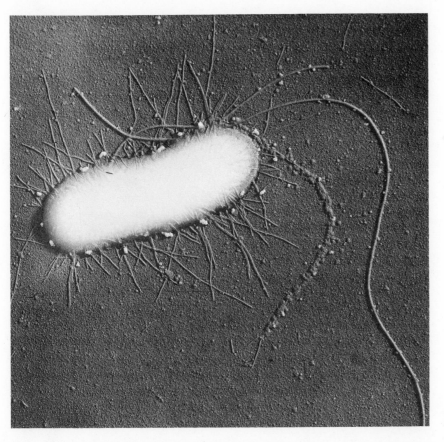

Pili of Escherichia coli (courtesy of D. C. Brinton, Jr.)

Chapter 1

Microbiology in the Biological World

DISCOVERY OF THE MICROBIAL WORLD

The birth of microbiology occurred in 1674 when Anton van Leeuwenhoek, an inquisitive Dutch drapery merchant, peered at a drop of lake water through a glass lens that he had carefully ground. What he observed through this simple magnifying glass was undoubtedly one of the most startling and amazing sights that man has ever beheld—the first glimpse of the world of microbes. As he recorded in a letter to the Royal Society of London, he saw:

> ". . . very many little animalcules, whereof some were roundish, while others a bit bigger consisted of an oval. On these last, I saw two little legs near the head, and two little fins at the hind most end of the body. Others were somewhat longer than an oval, and these were very slow a-moving, and few in number. These animalcules had divers colours, some being whitish and transparent; others with green and very glittering little scales, others again were green in the middle, and before and behind white; others yet were ashen grey. And the motion of most of these animalcules in the water was so swift, and so various, upwards, downwards, and round about, that 'twas wonderful to see: and I judge that some of these little creatures were about a thousand times smaller than the smallest ones I have ever yet seen . . . ."

FIRST DESCRIPTION OF THE MICROBIAL WORLD

This letter, one of his earliest to the Royal Society, became the first known description of *protozoa* and *algae*. It was followed by 150 more letters which spanned 50 years. Each of these remarkable letters was filled with descriptions of his important discoveries. Although van Leeuwenhoek had little formal education, his inquisitive nature led him to examine a wide variety of materials—river, well, and sea water as well as vinegar, pepper, and ginger water, to list only a few.

3

One of his more interesting observations resulted from the examination of his own spittle:

"... in the said matter there were many very little animalcules, very prettily a-moving. The biggest sort had the shape of Fig. A (Fig. 1-1): these had a very strong and swift motion, and shot through the water (or spittle) like a pike does through water. These were most always few in number. The second sort had the shape of Fig. B. These oft time spun around like a top, and every now and then took a course like that shown between C and D: and these were far more in number. To the third sort I could assign no figure: for at times they seemed to be oblong, while anon they looked perfectly round. These were so small that I could see them no bigger than Fig. E: yet therewithal they went ahead so nimbly, and hovered so together, that you might imagine them to be a big swarm of gnats or flies, flying in and out among one another ... one being bent crooked, another straight like Fig. F ..."

"While I was talking to an old man ... my eye fell upon his teeth which were all coated over ... so I took some spittle out of his mouth and examined it. The biggest sort bent their body into curves in going forward, as in Fig. G."

Thus did van Leeuwenhoek discover bacteria in the human mouth. It is even more remarkable that his descriptions of the organisms were so precise that it is now possible to assign them to specific genera with reasonable certainty.

Van Leeuwenhoek left no descriptions of the apparatus with which he observed protozoa and bacteria. He kept "for himself alone" his best microscopes and his manner of observing the animalcules. He never did divulge his method. His jealously prized lenses and guarded techniques enabled him to surpass all other microscopists for at least a century.

Van Leeuwenhoek left over 400 microscopes behind when he died; except for a few, all have long since disappeared mysteriously. We do, however, know

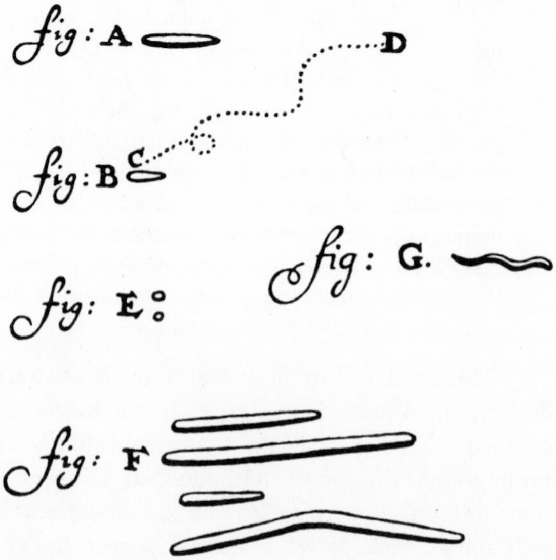

FIGURE 1-1

Van Leeuwenhoek's figures of bacteria from the human mouth enlarged 1½ times, from the engravings published in *Arc. Nat. Det.*, 1695. A. A motile *Bacillus;* B. *Selenomonas sputigena;* C. and D. the path of its motion; E. Micrococci; F. *Leptothrix buccalis;* G. a spirochaete, probably *"Spirochaeta buccalis,"* the largest form found in this situation.

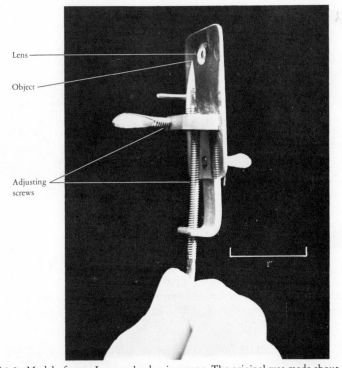

Lens

Object

Adjusting
screws

1"

FIGURE 1-2   Model of a van Leeuwenhoek microscope. The original was made about 1673 and magnified almost 300 times. Stained bacteria between 1 and 2 μm in size can be seen with the original model. The object being viewed is brought into focus with the adjusting screws. This replica was made according to the directions given in *The American Biology Teacher,* **30** (1968), 537. Note its small size. (Photograph courtesy of J. P. Dalmasso)

DESCRIPTION OF
VAN LEEUWENHOEK
MICROSCOPE

something about his apparatus. The microscopes that he fashioned consisted of a spherical lens, meticulously ground, and mounted between two metal plates (Fig. 1-2). The object to be examined was placed on a blunt pin on one plate and the specimen was brought into focus by moving the object relative to the lens with screws. Instead of removing his specimen once it was fixed on the pin and in focus, van Leeuwenhoek would make a new microscope for the next specimen he wished to study. This accounts for the fact that he made several hundred instruments.

These instruments were not much more than powerful magnifying glasses capable of enlarging objects up to 300 times. However, what incredible magnifying glasses they were! One complete original van Leeuwenhoek microscope exists at the University Museum at Utrecht, Netherlands. Although the lens is now badly scratched, it is still usable. Recently a photograph of a diatom as viewed through this original van Leeuwenhoek microscope was taken (Fig. 1-3a). The lens, although not in the best condition and probably not one of van Leeuwenhoek's best, can still separate two objects which are less than 1/17,500 of an inch apart. A picture of a similar diatom taken with a modern microscope is shown in Figure 1-3b.

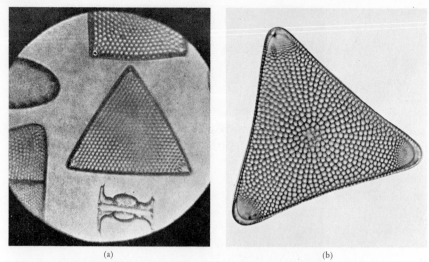

(a)   (b)

FIGURE 1-3   (a) Photograph of diatoms as seen through an original van Leeuwenhoek microscope. (Photograph taken by Professor Van Cittert and provided by Professor S. Tolansky.) (b)
Photograph of a similar diatom as viewed through a modern microscope. Magnification of the
microscope is ×210, total magnification, ×510. (Courtesy of Johns-Manville Research and Engi-
neering Center, Manville, N.J.)

Even though there is no question that van Leeuwenhoek was a master in
the art of lens making, it is highly unlikely that he could have observed the detail
and clarity in microorganisms that he reported unless he had unusually good
methods of illuminating the specimens. Unfortunately, he "kept for himself
alone his best microscopes and his particular manner for observing small creatures." It seems probable that he discovered the technique of darkfield microscopy, a technique in use today, in which faint objects appear as brightly lit against
a dark background.

Darkfield microscopy
Page 70

THE CONTROVERSY OVER
SPONTANEOUS GENERATION

One of the most intriguing questions that the discovery of microbes immediately
prompted concerned the origin of these microscopic forms. At the end of the
seventeenth century, the theory of spontaneous generation, namely that organisms visible to the naked eye can arise spontaneously in decomposing organic
matter, had just been completely debunked. Francesco Redi, an Italian biologist
and physician, showed conclusively that worms found on putrefying meat originated from the eggs of flies. If he surrounded the meat with a gauze fine enough
to prevent flies from depositing their eggs, the worms no longer appeared.

Despite these findings, the possibility that organic matter could give rise
to microscopic forms was quickly suggested. This hypothesis was not disproven
for almost 200 years. Because it was not possible to keep out microscopic forms

by means of gauze, the general approach to determining whether microbes could arise spontaneously consisted of boiling the organic material in a vessel to kill all forms of life and then sealing the vessel to the air. The appearance of a turbid solution signaled the presence of microorganisms and supported the theory of spontaneous generation. The results of this kind of experiment, however, varied with different investigators.

In 1749, Needham, a Catholic priest and a capable scientist, reported that many different types of broth (infusions) had given rise to microorganisms even though they had been boiled and then sealed with corks. However, these experiments in turn appeared to be refuted by the results of other similar experiments published by Father Spallanzani in 1765. His experiments differed from Needham's in two significant ways: Spallanzani boiled the infusions for longer periods of time and sealed the flasks by melting the neck of the flask. With this technique he demonstrated by hundreds of separate experiments that the infusions would remain free of microorganisms indefinitely. However, if the neck of the flask cracked, the broth rapidly became turbid. Spallanzani thus concluded that the microbes must have entered the infusion with the unsterilized air. The controversy, however, was far from over. "Spontaneous generation" continued to be reported for the next 100 years by investigators who either did not boil their infusions long enough or sealed the flasks imperfectly.

Experiments of Pasteur

Although Spallanzani's conclusions were correct, Needham raised some serious objections that were not convincingly answered for still another 100 years. Thus Needham remarked, "From the way he [Spallanzani] has treated and tortured his vegetable infusions, it is obvious that he has not only much weakened and maybe even destroyed the "vegetative force" of the infused substances, but also that he has completely degraded . . . the small amount of air which was left in his vials. It is not surprising, thus, that his infusions did not show any sign of life. . . ."

The two giants in science who did the most to lay the theory of spontaneous generation to rest once and for all were the French chemist, Louis Pasteur, considered by many to be the father of microbiology, and the English physicist, John Tyndall. The refutation published by Pasteur in 1861 was a masterpiece of logic. Indeed the force of his logic was probably as important in swaying the scientific community to his side as were the results of the experiments themselves, many of which had already been performed by other investigators. First, he demonstrated that air was filled with microorganisms by filtering air through a cotton filter and then examining microscopically the organisms which had been trapped by the cotton. Many of the organisms were indistinguishable from those which had previously been observed in many infusions. Furthermore, dropping the cotton plug into a sterilized infusion promptly resulted in an increase in the number of these organisms. Perhaps Pasteur's most unique and dramatic experiment consisted in demonstrating that infusions once sterilized would remain sterile even

PASTEUR REFUTES
SPONTANEOUS
GENERATION

in a flask open to the air if the path from the outside air to the infusion was tortuous. In these swan-necked flasks the organisms would settle out in the bends and never reach the fluid. This experiment finally disposed of all claims that unheated air contained in itself a "vital force" sufficient for spontaneous generation.

Experiments of Tyndall

Although a large proportion of the scientific population, especially in France, were convinced by Pasteur's experiments, many respected scientists still doggedly held to the concept of spontaneous generation. This persistence undoubtedly stemmed in part from the fact that many scientists were unable to verify his results. It remained for the English physicist, John Tyndall, to provide a logical explanation for the differences in experimental results among laboratories. Tyndall, an ardent admirer of Pasteur, was convinced of the validity of his experiments, even though he was unable to obtain consistent results when he repeated them. He then realized that different infusions required widely varying boiling times in order to be sterilized. Thus some materials could be sterilized by boiling 5 minutes, whereas others, most notably hay infusions, could be boiled for 5 hours and still contain living organisms. Furthermore, the presence of hay in the laboratory made it virtually impossible to sterilize even the infusions that had previously been sterilized by boiling for 5 minutes. What was present in hay that caused these effects? Tyndall finally realized that organisms were being brought into his laboratory on the hay which were very resistant to heat. These organisms infected all infusions, thereby making everything difficult to sterilize. He concluded that some bacteria must be able to exist in two forms: a heat-labile form (*vegetative cell*) readily killed by boiling and a heat-resistant form (*endospore*). In the same year a German botanist Ferdinand Cohn also described endospores and demonstrated their heat-resistant properties.

Tyndall then developed a procedure for destroying even the most resistant forms. This consisted of boiling the infusion intermittently; the endospores developed into vegetative cells in the intervals between heatings and were then readily killed in the next round of heating. With this technique, named *tyndallization,* Tyndall could sterilize a hay infusion by boiling it for one minute, five separate times, although continuous boiling for one hour would not kill all of the endospores. This discovery of the extreme heat resistance of spores explained the discrepancies between Pasteur's results and those of other investigators. Organisms which produce endospores are commonly found in the soil and so would be present in hay infusions. Since Pasteur, however, used only infusions prepared from sugar and yeast extract, this broth would not likely contain endospores.

Thus the concept of spontaneous generation, as envisioned by Needham and others, was disproved less than 100 years ago. This fact emphasizes the fantastic progress that has been made in our knowledge of microorganisms in these last 100 years.

Although scientists do not believe that, under present-day conditions, life can originate from nonliving matter, it undoubtedly did under the far different conditions which existed on earth when life first arose. However, today even the simplest microorganism is extremely complex and must have evolved over millions of years from far simpler forms of life. Indeed, electron photomicrographs taken from the inside of a chert (type of rock) two billion years old revealed what appear to be the fossil remains of bacteria.

The above studies demonstrated that microscopic microorganisms and visible organisms are similar in at least one respect: they both arise only from pre-existing organisms. In order to gain some insight into the role of microorganisms in the biological world, it will be enlightening to view other similarities and differences between the microscopic forms and plants and animals.

CELL THEORY

Since bacteria are single-celled organisms, an understanding of microbiology can be best gained by recognizing the basic significance of the *cell* to all life. About the time that van Leeuwenhoek was peering at "animalcules" through his simple microscopes, an English microscopist, Robert Hook, was making significant observations on the structure of cork. He perceived that cork consisted of a "great many little boxes, separated out of one continued long pore, by certain Diaphragms . . . ." The little boxes were cells. Despite this significant discovery of the structure of living forms, intensive studies of cells were not carried out until the early nineteenth century.

Before the 1830s most people viewed the entire organism as the fundamental unit of structure. This implied that no part of the total organism was capable of surviving unless the entire organism remained intact. In 1838 a German botanist, Matthias Schleiden, and in 1839 a German zoologist, Theodor Schwann, published statements to the effect that all organisms were composed of cells, and that cells were the fundamental units of life. They clearly recognized that these units were capable of carrying out all of the basic functions which were attributed to living organisms at that time. Their hypothesis, termed the *cell theory,* represents one of the significant conceptual advances in biological thinking. Although Schleiden and Schwann were not the first to proclaim this concept, the clarity and logic of their presentation made a great impact on the thinking of scientists. The attention of scientists which was previously focused on the apparent diversity of organisms now switched to their basic component parts, the cells, which shared obvious similarities even in the most diverse forms of life. In the intervening 125 years between the publication of this concept and today, the work of thousands of scientists in areas of zoology, botany, biochemistry, genetics, and microbiology has detailed just how basic these similarities are, and yet at the same time how much variety nature can tolerate. The study of bacteria and other unicellular organisms has been especially important to an understanding of cell structure and its functions.

THE CELL IS THE BASIC UNIT OF STRUCTURE AND FUNCTION

CONCEPT OF BIOCHEMICAL UNITY AND CELLULAR DIVERSITY

The concept of simultaneous unity and diversity can be best expressed as follows: cells display a *unity* of biochemical principle and a *diversity* of cellular apparatus and methods. To amplify this statement, all cells, whether a free-living bacterium or a brain cell, which is part of a larger structure, have to deal with similar problems in order to live. Superficially the basic mechanisms for solving these problems are the same in all cells. However, more precise analyses reveal that there are extensive variations both in the structures and in the detailed mechanisms which different cells use in carrying on life processes.

Unity of Cell Function

Virtually all living cells, whether in animals, plants, or unicellular forms, must carry out three basic life processes to endure. First, *cells must be capable of reproducing exact copies of themselves* but also must be capable of undergoing small changes in their genetic makeup. Next, they must be able to *synthesize the constituents of living matter* from the much simpler foodstuffs available to them. Finally, they must be capable of *producing and utilizing energy* through the breakdown of foodstuffs.

In order to replicate itself, a cell must contain genetic instructions for exact duplication as well as the complex apparatus required to synthesize vast arrays of cellular components. Since biosynthetic processes involve the expenditure of energy, all cells must have mechanisms for deriving, storing, and utilizing energy obtained from nutrients.

One of the most striking findings of investigations of many cells (animal, plant, and unicellular) is the remarkable similarity in the basic mechanisms whereby these life processes are carried out. Representative cells from each group of organisms duplicate their genetic material in the same way, break down foodstuffs to gain energy through the same general sequence of biochemical reactions, and synthesize each of their cellular components from the same starting materials, Chapters 7 and 8 using the same biochemical reactions. Several later chapters will be devoted to discussing these aspects of cell function.

LEVELS OF ORGANIZATION

The science of microbiology can be considered in terms of a *hierarchical order of organization*. Each unit in the organization is composed of units lower in the order, and in turn it comprises part of a larger unit. Thus the levels of organization can be viewed as horizontal slabs, one on top of another, which form a trapezoid. (Fig. 1-4). The foundation, for this course, are the *atoms*. The atoms interact to form *small molecules,* which in turn are joined together to form very large molecules, *macromolecules*. The macromolecules associate with other macro-

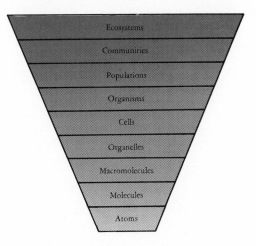

FIGURE 1-4
Levels of organization in the microbial world. Each level includes all levels beneath it. Neither tissues nor organs are present in members of the microbial world and are not included in this organizational scheme.

molecules to form structures which perform a specific function, the *organelles*. In turn, organelles and other cell structures form a part of the cell. Some cells may represent the entire organism, but in other cases the organism is composed of many cells. A *population* is composed of a number of similar organisms and many different populations interact with one another to form a *community*.

The discussion will begin with the least complex, the atoms and molecules. Much of the rest of the text will be a consideration of the more complex levels.

Atoms of Life

A chemical analysis of cells taken from a diversity of organisms reveals an obvious uniformity in the kind and quantity of elements present. Table 1-1 gives the elemental composition of a bacterial cell. Several interesting aspects stand out. First, cells are composed of only a small fraction of the total number of elements (over one hundred elements are known). Four constituents, carbon, hydrogen, oxygen, and nitrogen, comprise about 98.5 percent of the atoms by weight. Phosphorus and sulfur together make up an additional 1 percent, and all of the remaining elements together comprise less than 0.5 percent. This composition differs in some specialized cells in which certain elements are associated with a specific structure or function. Cells in bone, for example, have a very high calcium content; diatom cells have a high silicon content.

The carbon atom forms the backbone of most organic molecules in the cell. A significant feature of this atom is its ability to form stable, covalent bonds with four other atoms. In many biologically important molecules, three of these atoms are also carbon atoms, thereby producing a three-dimensional network of carbon atoms. The fourth bond may involve a variety of atoms, each conferring distinctive properties on the molecule. The unique properties of carbon can be compared to silicon, an element which shares many of the properties of the carbon atom. In particular, silicon also has the property of achieving a stable configuration by the addition of four electrons. Although silicon is 146 times more abun-

C, H, O, N COMPRISE 98% OF CELLS WEIGHT

Page 23

CARBON ATOM HAS UNIQUE PROPERTIES

TABLE 1-1
Elemental Composition of a Bacterial Cell

Element	Percent of Cell Weight	Derived Macromolecules
Carbon	13	All macromolecules
Hydrogen	10	All macromolecules
Oxygen	70	All macromolecules
Nitrogen	5.5	Proteins, nucleic acids
Phosphorus	0.8	Nucleic acids
Sulfur	0.3	Proteins
Potassium	0.3	
Magnesium	0.05	
Iron	0.01	
Calcium	0.01	
Manganese Cobalt Nickel Copper Zinc Molybdenum Vanadium Silicon Selenium Boron Sodium Chlorine	Total weight less than 0.01% of total	

dant than carbon in the earth's crust, it rarely appears in biological materials. This probably is a result of the fact that silicon-silicon bonds are considerably weaker than carbon-carbon bonds, and many silicon-hydrogen compounds are relatively unstable and react with oxygen. Thus silicon is not nearly as versatile as carbon in the number of compounds it is able to form.

INERT AND LARGE
ATOMS NOT FOUND IN
LIVING SYSTEMS
Several groups of elements are also excluded from living systems. These include the *inert* or noble elements, such as argon and neon which will not share electrons and therefore will not form bonds. A second, heterogenous group contains the elements which are very large, those which have a great number of electron shells. Apparently the smaller-sized elements provide sufficient versatility in the numbers of bonds that they can form that it is unnecessary to employ the larger-sized atoms. The latter atoms form less stable bonds than the smaller atoms because of the greater distance of the nucleus from the distant electron shells.

Molecular Composition

Carbon, hydrogen, oxygen, and nitrogen are found in all *important building blocks of cells* such as amino acids, purines, and pyrimidines as well as in other molecules that play important roles in energy metabolism. Sulfur is limited to certain amino acids as well as to vitamins and coenzymes.

Amino acids
Pages 33–35
Purines and pyrimidines
Page 37
The most abundant molecule in all living cells is water, which makes up about 70 percent of the cell by weight. The chemistry of the cell is based on water

TABLE 1-2
Approximate Chemical Composition of a Typical Bacterial Cell

Component	Percent of Total Cell Weight	Approximate Number of Molecules per Cell	Number of Different Kinds
H_2O	70	4×10^{10}	1
Inorganic ions	1	2.5×10^8	20
Carbohydrates and precursors	3	2×10^8	200
Precursors of proteins	0.4	3×10^7	100
Precursors of nucleic acids	0.4	1.2×10^7	200
Precursors of fats	2	2.5×10^7	50
Breakdown products of food molecules	0.2	1.5×10^7	200
Proteins	15	10^6	2,000 to 3,000
Nucleic acids			
DNA	1	1–4	1
RNA	6	5×10^5	1,000

as the solvent, and all the molecules within the cell would function quite differently if a material such as benzene were the solvent. The unique properties of water will be considered in Chapter 2.

Water
Pages 29–30

All of the molecules within the cell (Table 1-2) can be classified within a relatively few groups in terms of their broad functions within the cell. One group contributes to the *ionic environment*. These include water, other inorganic molecules, and some small organic molecules. Other molecules serve as fuel in the production of energy. Some of the small organic molecules such as sugars fit into this category. Yet another group of small molecules, such as amino acids, purines, and pyrimidines as well as some sugars, serves as the subunits for macromolecules.

Macromolecular Composition

All cells contain the same major macromolecules in approximately the same proportions which perform essentially the same functions (Table 1-2). For example, some proteins function as *enzymes,* which are biological catalysts that facilitate the myriad reactions involved in energy metabolism and biosynthesis. Other proteins also form an integral part of certain structures of the cell. Deoxyribonucleic acid (DNA) is the macromolecule which carries all the genetic information and specifies the structure and properties of the cell by determining the composition of its proteins. Another macromolecule, ribonucleic acid (RNA), plays a major role in translating this information coded in the DNA into the sequence of amino acids in the proteins.

Enzymes
Page 142

The fundamental unity of molecular composition in all forms of life is especially evident from an analysis of the detailed chemical makeup of the two most important groups of biological macromolecules, proteins and nucleic acids. The same twenty amino acids appear in all cells as constituents of cell proteins.

Amino acids—formulae
Page 34; Fig. 2-7

The same two purine and three pyrimidine nitrogenous bases make up the basic subunits of DNA and RNA in all cells. A few smaller organic compounds are also ubiquitous, functioning universally in various crucial metabolic reactions. Thus *adenosine triphosphate* (ATP) serves as the storage and transfer form of energy in all cells. The *coenzymes,* which are synthesized from vitamins, function universally as agents for the transfer of atoms and electrons.

Cell Organelles

Macromolecules of the same or different types associate with each other in very specific patterns to form a variety of cell structures. Some structures are termed *organelles* if they are inside a cell and perform a specific function. Each structure has one or more functions essential to the life of the cell. Not all structures are found in all cells, and it is at this level of cellular organization that significant differences between different cells can be identified. For example, not all cells have mitochondria, the complex organelles concerned with energy metabolism. However, since energy metabolism is crucial to all life, some structure which performs this function must be present in every cell.

Basic Cell Types

BASIC DICHOTOMY IN
CELL TYPES—
PROCARYOTIC AND
EUCARYOTIC

The assemblage of individual cellular structures into a smoothly functioning, autonomous unit, the *cell,* represents the next level of biological organization. It is at this level that basic differences between organisms emerge. These differences result from collective variations in cell structures. Two distinctly different cell types emerge: the *eucaryotic* and the *procaryotic.* All organisms, both macroscopic and microscopic, are composed of eucaryotic cells, except for bacteria and a close relative, the blue-greens, which comprise procaryotic cells. This division of cells into two types represents one of the most significant dichotomies in all of biology. Their differences are listed in Table 1-3. Although each of these structures are explored in more detail in Chapter 4, this table illustrates the major point to be emphasized at this time—that the two cell types differ from each other in a great many respects.

The Organism

The next level of organization in the biological world relates to the physical and functional relationships between cells, of multicellular beings, that is, *multicellular organisms.* If we consider these organisms as an assemblage of cells, we can categorize the living world into groups according to the way in which these cells are organized to form the total being.

The simplest type of cell organization is that in which the entire organism consists of a single cell. This organization predominates among the organisms we will study in this course. Single-celled organisms are almost invariably microscopic in size. Since such cells exist independently and are self-sufficient, each one

TABLE 1-3
Comparison of Eucaryotic and Procaryotic Cells

	Procaryotic	Eucaryotic
Size of cells (diameter)	0.3–2 μm	2–20 μm
Genetic structure		
Chromosome Number[a]	1–4 (all identical)	>1 (all different)
Nuclear division by mitosis	–	+
Protein bound to DNA	–	+
Nuclear membrane	–	+
Cytoplasmic structures		
Cell wall	Unique chemical components	If present, composition varies between organisms
Mitochondria	–	+
Ribosomes	70S[b]	80S
Cytoplasmic streaming	–	+
Photosynthetic pigments associated with	Chromatophores	Chloroplasts

[a] The DNA of procaryotic cells functions as the chromosome of eucaryotic cells. However, its structure is far less complex than a true chromosome and therefore it is often referred to as a nuclear body. In this text we employ the term chromosome, emphasizing that this defines its function and not its detailed structure.
[b] The size of the ribosome is indicated by the speed at which it sediments upon centrifugation (S value) the larger the S value, the heavier and larger the molecule.

must carry out all vital life functions. Although unicellular organisms are characterized by their smallness, there may be very profound differences among them. Thus some members, such as protozoa and many algae, possess the complex eucaryotic cell structure, whereas others, the bacteria and the blue-greens, for example, are of the simpler procaryotic type. As a result, there is a wide range in size as well as in internal organization among the unicellular organisms. Figure 1-5 shows the comparative sizes of some molecules, cells, and organisms, as well as the scales used for their measurement.

Multicellularity is another type of cellular organization in which organisms arise from single cells which increase in number and remain permanently attached to one another in a characteristic way. Multicellular organisms can differ from one another in a number of fundamental ways. The most outstanding variation involves the level of *structural complexity* achieved within the organism. Although an organism is multicellular, it may be nevertheless quite simple in terms of how the cells are organized. Many multicellular organisms are large in size, but microscopic examination reveals that most of the cells appear very similar. Although differentiation occurs, one does not observe the formation of structures composed of specialized cells which function as a unit. An example of a simple but seemingly complex form is the mushroom or toadstool. Mushrooms have several distinctive structures that are important in the life of the organism. The cells in all of these structures, however, are essentially similar (Fig. 1-6a and b). For example, spore bearing cells and spores, although specialized, function independently of one another rather than cooperatively. In higher organisms such as plants and animals there are distinct regions of cells, all of which are identical to each other but different in shape, structure, and function from cells in other

COMPLEXITY OF
ORGANISM IS
UNRELATED TO SIZE

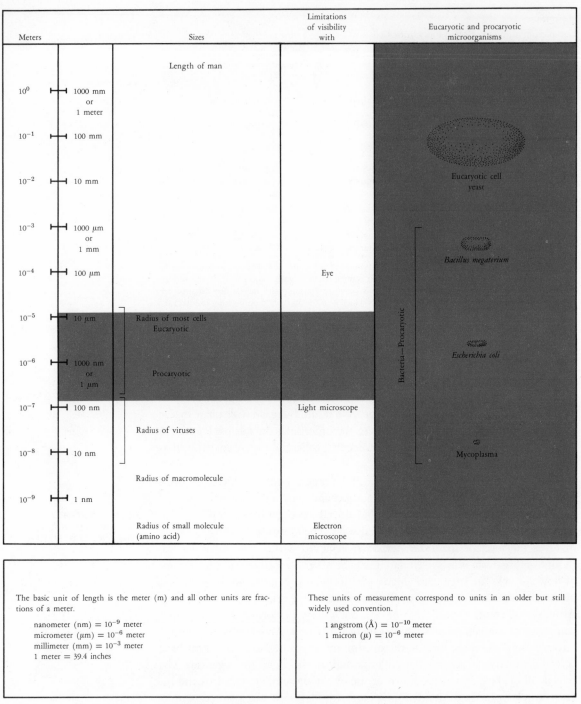

FIGURE 1-5 Measurements of molecules, cells, and organisms.

The basic unit of length is the meter (m) and all other units are fractions of a meter.

nanometer (nm) = 10^{-9} meter
micrometer (μm) = 10^{-6} meter
millimeter (mm) = 10^{-3} meter
1 meter = 39.4 inches

These units of measurement correspond to units in an older but still widely used convention.

1 angstrom (Å) = 10^{-10} meter
1 micron (μ) = 10^{-6} meter

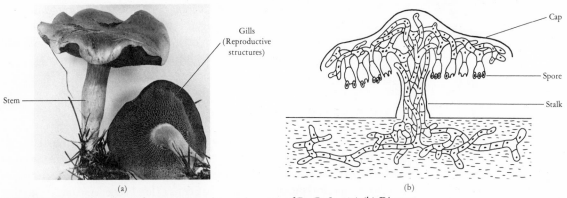

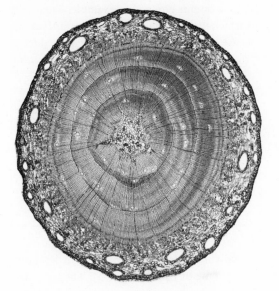

FIGURE 1-6 (a) Photograph of a common mushroom (courtesy of Dr. D. Stuntz) (b) Diagram of a section through a mushroom. The entire stalk (stem) and cap are composed of mycelia. These mycelia differentiate to form the club-shaped structures in which reproductive spores are produced.

regions. Each such specialized group of cells is termed a *tissue,* functions as a unit, and performs a specific function. Different tissues in turn may combine into an even more complex structure, an *organ.* These multicellular organisms represent the highest level of cell organization (Fig. 1-7).

This increasing order of complexity of cells, tissues, and organs is characteristic of many plants and animals; *it is not typical of members of the microbial world.* This complex type of organization results in a high degree of specialization. Each of the vital functions of the organism is performed by a different organ. Accordingly the life of the organism requires the proper functioning of each organ.

MEMBERS OF MICROBIAL WORLD HAVE NO TISSUES

In a table of increasing levels of organization several additional levels can be identified (Fig. 1-4). These include the association of organisms of similar type

FIGURE 1-7
An organ. Cross section of a pine stem composed of many types of identical cells, each type grouped together to form a specialized functional unit, a tissue. The assemblage of tissues constitutes an organ. Some tissues are concerned with the transport of nutrients, others with water transport and still others provide structural integrity to the organ. (Courtesy of Carolina Biological Supply Company)

to form a *population,* the association of different populations to form *communities,* and the association of communities, together with nonliving environment, to form *ecosystems.* These higher levels, from organisms to ecosystems, are discussed in increasing order of complexity in later chapters.

NOMENCLATURE OF ORGANISMS

Nomenclature refers to the system by which organisms are named. Virtually all organisms are named according to the *bionomial system* devised by the Swedish botanist, Carl Von Linné. In the nomenclature of bacteria, each distinct kind is given both a species (plural species) and a genus (plural genera) name. A number of different species are included within the same genus. The first word is the genus name and the first letter is always capitalized. The second word is the species name and is not capitalized. Both words are always italicized. For example, *Escherichia coli* is a member of the genus *Escherichia.* The genus name is generally abbreviated with the first letter capitalized, that is, *E. coli.* Some organisms may be virtually identical to other species except that they may differ in one or two ways. For example, one species may not be able to break down a foodstuff that the other organism can. This slight difference generally does not justify giving the organisms different species names, and so they are designated as *strains* or varieties of the same species. For example, two common strains of *E. coli* are K-12 (*E. coli* K-12) isolated from a patient at Stanford University Hospital and ML (*E. coli* ML) which is reputed to have been isolated from the bowel of a famous French scientist, whose initials are ML. They are identical in their gross characteristics, but are distinctly different in many details.

The precise identification of an organism requires both a genus and species name. However, many organisms are commonly referred to by less precise terms which serve to place the organism in a certain group. For example, bacilli are rod-shaped (cylindrical) organisms which include members of the genus *Bacillus* and also the genera *Clostridium, Lactobacillus,* and other genera. The general term bacillus is neither capitalized nor italicized and thus can be differentiated from the genus *Bacillus.*

SIGNIFICANCE OF THE SIZE OF MICROORGANISMS

The smallness of microorganisms has a number of consequences. These relate primarily to the simplicity of the cell, and its growth rate.

Simplicity of the Cell

The unicellular organisms range in cell volume from 0.01 to about 50,000 μm^3, with a typical bacterial cell having a volume of 1 μm^3 (Fig. 1-5). The large single-celled organisms, very large protozoans, can be seen without the aid of a microscope. The lower size limit of a microorganism is probably determined by the dimensions of the macromolecules it needs to carry out its life functions. If the organism were any smaller, it would not be physically large enough to contain the equipment necessary to sustain life. Indeed, there are physical entities, the *viruses,* which may be considerably smaller than 0.01 μm^3. Although they will be considered as microorganisms, they have a number of features which set them apart from the other groups of microorganisms which will be considered. They are too small to contain any of the organelles necessary for life, and therefore must utilize the machinery of the cell they invade in order to multiply. Outside the infected cell they are essentially inert.

CELLS MUST BE LARGE
ENOUGH TO CONTAIN
ORGANELLES NECESSARY
FOR LIFE

The size of the smaller microorganisms also restricts the complexity of the cellular apparatus that they can contain. Since a single mitochondrion in a liver cell may be as large as an entire bacterial cell, obviously the latter is too small to contain a liver mitochondrion. The same holds true for many other organelles found in eucaryotic cells. As a general rule, as organisms increase in size, they also increase in complexity. Thus eucaryotic cells are generally larger than procaryotic cells (Fig. 1-5; Table 1-3) and their increased size allows them to contain a number of organelles not present in procaryotic cells.

The upper size limit of a single-celled organism is probably related to the ratio of the surface area to cell volume. All the nutrients consumed by the cell must enter across the surface of the cell. Likewise, waste products must leave through the cell surface. As cell size increases, the volume increases much more rapidly than the surface area: the volume increases as the *cube* of the cell radius, whereas the surface area increases only as the *square* of the radius. Thus when the cell's radius has increased three times, the volume has increased 27 (3^3) times, whereas the surface area has increased 9 times (3^2) (Fig. 1-8). With such a disparity between surface area and volume, the cell faces the problem of not having enough surface area for the intake of nutrients and exit of waste materials to supply the increased demands of the enlarged cell volume.

Eucaryotic cells, being larger, have had to solve this problem in a number of ways. Many cells assume shapes which bring the internal parts of the cell close to the surface. Thus a long nerve cell has a large, threadlike shape. The cytoplasmic streaming evident in many eucaryotic cells (but not in procaryotic) serves to stir up the cytoplasm, moving nutrients and waste products from one area to another. In addition, many cells, both procaryotic and eucaryotic, have their surface areas markedly enlarged as a result of extensive invaginations or convolutions of the cytoplasmic membrane through which nutrients and waste products pass.

INVAGINATIONS
INCREASE SURFACE AREA

Growth Rate

The rate at which unicellular organisms grow under optimal environmental conditions varies dramatically. As a general rule the *growth rate increases as cell size decreases.* This relationship is most readily explained by the surface-to-volume

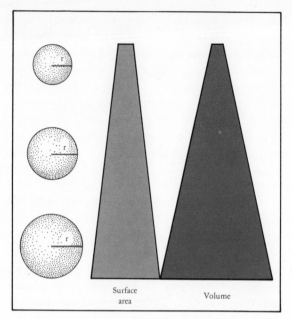

FIGURE 1-8
The relationship between surface area and volume of spheres. As
the radius of a sphere increases, there is a much greater increase
in the volume than in the surface area.

ratio of various cells. The larger the cell, the longer the time required for nutrients to pass into it and be metabolized for cell growth. However, within cell groups of approximately equal volume, there are some differences in growth rate. The reasons for these differences are not clear, but they may also be related to the fact that some organisms of the same size differ in their ability to take up nutrients from the environment.

SUMMARY

The combined work of a number of outstanding scientists established that the "animalcules" discovered by Anton van Leeuwenhoek about 300 years ago originated from preexisting organisms which were ubiquitous in nature. The relationship of microorganisms to other forms of life can be best described by the phrase "unity of biochemical principle and diversity of cellular structure and apparatus." The unity of biochemical principle is evident at every level of organization of living systems: molecules, cells, and organisms. Although the chemical elements and molecules, which comprise the basic unit of life, the cell, are uniform throughout the world, there is great diversity in the structures which carry out the basic functions of life. All cells can be divided into one of two distinct types: the simple cell type, called *procaryotic* and the more complex type, *eucaryotic*. The procaryotic cell lacks many of the structures of the eucaryotic. Procaryotic cells very often exist as single-celled organisms, whereas eucaryotic cells tend to be part of larger cellular aggregates, tissues, and organs, which in turn make up the organism.

1. Consider the major differences between eucaryotic and procaryotic cells.
2. What functions must cells be capable of performing in order to live?
3. What are the functions of DNA, RNA, and protein in the cell?
4. Discuss the reasons why different people obtained different results when they looked for life in infusions which had been boiled.
5. The following quote is attributed to van Leeuwenhoek (1719): "What does it matter whether we know these things or not." To what might he have been been referring?

FURTHER READING

BROCK, THOMAS (ed.), *Milestones in Microbiology.* Englewood Cliffs, N.J.: Prentice-Hall, Inc., 1961.

DOBELL, C., *Antony van Leeuwenhoek and His "Little Animals."* New York: Dover Press, 1960.

GREEN, D. E. and R. F. GOLDBERGER, *Molecular Insights into the Living Process,* New York: Academic Press, 1967.

LECHAVALIER, H. A. and M. SOLOTOROVSKY, *Three Centuries of Microbiology.* New York: McGraw-Hill, Inc., 1965.

MOROWITZ, H. and M. TOURTELLOTTE, "The Smallest Living Cells," *Scientific American* (March 1962). (Offprint #1005)

VALLERY-RADOT, RENE, *The Life of Pasteur.* Translated by R. L. Devonshire. New York: Garden City Books, 1926.

Chapter 2

Biochemistry of the Molecules of Life

The understanding of various aspects of the life of microorganisms at the molecular level has provided an insight into the chemical basis of life. In order to achieve some understanding of microbiology at the chemical level, it is necessary to have some knowledge of certain fundamental facts derived from physics and chemistry as well as microbiology.

The material in this chapter will serve as background as well as reference material for various aspects of microbiology covered throughout the text. For some, it may serve as a review of material already encountered. For others, it may be a first encounter with the chemistry of biological molecules. The discussion proceeds from the lowest level of organization—the atom—to the highly complex association among macromolecules.

FORMATION OF MOLECULES— CHEMICAL BONDS

ATOMS MAY BE HELD TOGETHER BY EITHER STRONG OR WEAK BONDS

Most atoms can associate with other atoms to form molecules. Some readily form chemical bonds with atoms of the same or other elements; a few do not combine easily and are relatively inert; they are relatively unimportant in biological processes. The chemical bonds which hold atoms together to form molecules are of various types. Each type is characterized by the arrangement of electrons as well as its strength or energy content.

Organic molecules are those which contain carbon except for carbon dioxide. All of these are held together by *covalent bonds*. The *sharing* of electrons by two atoms with each partner contributing electrons forms a covalent bond. The number of possible bonds that each element can form depends on the number of electrons that the atom requires to fill its outer shell. This information is given in Table 2-1 for the elements most important in the living world. Note that the single most important atom, C, forms four covalent bonds. Once the outer shell of electrons is filled, the molecule is highly stable and is not capable of forming additional covalent bonds. The dash or colon between two atoms represents two shared electrons (C—H or C:H).

The biologically important small organic molecules can be classified into major groups, based on the type of *functional group* or groups they possess (Table 2-2). This table also illustrates the fact that the most important covalent bonds are those between carbon and hydrogen (C—H), carbon and oxygen (C—O), carbon and nitrogen (C—N), oxygen and hydrogen (O—H), and nitrogen and hydrogen (N—H). Table 2-2 shows clearly that some molecules of biological importance contain several functional groups. For example, every amino acid contains both an amino as well as a carboxyl group, hence the appropriate name, amino acid. An acid which contains a keto group is often termed a keto acid; if the acid contains an hydroxyl group, it is termed an hydroxy acid. Unfortunately, many other compounds are commonly designated by trivial names which provide no clue as to what functional groups the molecules contain. Some examples of small molecules of biological importance that contain these functional groups are given. The biological significance of some of these molecules is discussed in later sections. Note that in some cases the electrons are

TABLE 2-1
Atomic Structure of Elements Commonly Found in the Living World

Element	Symbol	*Atomic Number* (Total Number Electrons)	*Diagram of* Outer Electron Shell	*No. of Possible Covalent Bonds*
Hydrogen	H	1	H	1
Carbon	C	6	C	4
Nitrogen	N	7	N	3
Oxygen	O	8	O	2
Phosphorus	P	15	P	3
Sulfur	S	16	S	2

The number of electrons required to fill the outer shell determines the number of possible covalent bonds. The number of electrons in a completed outer shell varies depending on the distance of the shell from the nucleus.

TABLE 2-2
Classification of Small Molecules by Functional Groups

Group	Name of Group	Example of Group in Biologically Important Molecule		Class of Molecule Found in
Methyl structure	Methyl	Methane structure	Methane	
—O—H	Hydroxyl	Ethanol structure	Ethanol (ethyl alcohol)	Alcohols
Carboxyl structure	Carboxyl	Acetic acid structure	Acetic acid	Acids
Amino structure	Amino	Glycine structure	Glycine	Amines and amino acids
Keto structure	Keto	Acetone structure	Acetone	Ketones
Aldehyde structure	Aldehyde	Acetaldehyde structure	Acetaldehyde	Aldehydes
—S—H	Sulfhydryl	Cysteine structure	Cysteine	A few amino acids
Ketone and Carboxyl structures	Ketone and Carboxyl	Pyruvic acid structure	Pyruvic acid	Keto acids
Hydroxyl and Carboxyl structures	Hydroxyl and Carboxyl	Lactic acid structure	Lactic acid	Hydroxy acid

shared in groups of four (for example, C=O). Some of these molecules may be referred to by more than one name. Specifically, inside the cell, salts of organic acids are present, rather than the acids themselves. Thus pyruvate rather than pyruvic acid and acetate rather than acetic acid are found. In practice, the name of the salt is used interchangeably with the name of the acid.

Although all of the bonds described in Table 2-2 are covalent, the bonds between various atoms can differ in two ways which have consequences for biological systems.

Bonds differ in their strength and in the distribution of the electrons shared between the atoms. The amount of energy required to disrupt the bond determines the strength of the bond. The stronger the bond, the more energy required to break it and the less likely the bond will be broken. The breakage of a bond joining atoms A and B is described by the following reaction:

DIFFERENCES BETWEEN
COVALENT BONDS

$$AB + energy \longrightarrow A + B.$$

Just as bonds can be broken by putting energy into the system, the reverse reaction can also occur:

$$A + B \longrightarrow AB + energy.$$

The formation of a covalent bond between A and B *releases* the same amount of energy as is required to break the bond. Thus the stronger the bond that is formed, the more energy released on its formation. In general, molecules always tend to form the strongest bonds possible, thus achieving the most stable situation. Most covalent bonds are extremely strong and thus do not break unless large amounts of *energy,* generally as heat, are supplied. Since biological systems function only within a narrow temperature range, they utilize protein *catalysts,* the *enzymes,* which can break these covalent bonds at physiological temperatures by a mechanism which is considered later. The energy in the most common covalent bonds is given in Table 2-3.

DIFFERENCES IN
STRENGTH OF BOND

Enzymes
Pages 142–146

TABLE 2-3
The Energies of Some Chemical Bonds

	Energy (kcal/mole)[a]
Covalent Bonds	
H—H	104
C—C	83
C—H	99
C—O	84
C—N	70
Hydrogen Bonds	
H—O····O	
H—N····O	2–10
H—N····N	
van der Waals Forces	1–2

[a] 1 kcal is the amount of heat required to raise the temperature of 1000 grams of water 1°C.

TABLE 2-4
Polar and Nonpolar Covalent Bonds

Type of Covalent Bond	Atoms Involved and Charge Distribution	
Polar	$\overset{-}{O}{-}\overset{+}{H}$	The O and N atoms have a
	$\overset{-}{N}{-}\overset{+}{H}$	stronger attraction for
		electrons than C and H, and
	$\overset{-}{O}{-}\overset{+}{C}$	so the O and N have a negative
	$\overset{-}{N}{-}\overset{+}{C}$	charge.
Nonpolar	C—C	C and H have equal attractions
	C—H	for electrons
	H—H	

DIFFERENCES IN
ATTRACTION OF
ELECTRONS—POLARITY

In addition to differences in strength of the covalent bond between different atoms, these bonds also differ in the *distribution of shared electrons* between the atoms—the phenomenon of *polarity*. Polarity results from the fact that different atoms have different affinities for electrons. In covalent bonds between similar atoms such as H—H, the electrons have no tendency to associate with one atom more than with the other atom. This is a *nonpolar* covalent bond and exists between atoms that have a similar attraction for electrons. A C—H bond is also nonpolar. In other cases, one of the atoms has a much greater affinity for electrons than does the other, and the electrons are not equally shared. The atom with the greater attraction for the electrons has a slight negative charge, leaving the other atom with a slight positive charge. Different atoms have different affinities for electrons and therefore different molecules will have different degrees of polarity. The atoms which have the greatest affinity in decreasing order are oxygen, nitrogen, carbon, and hydrogen. Thus in the O—H bond, the O has a much greater affinity for electrons than H, and will therefore have a negative charge, whereas the H atom will have a slight positive charge (Table 2-4). For this reason the water molecule is a highly polar molecule, having a slightly negative and two slightly positive charges at different sites on the water molecule (Fig. 2-1).

Ionic Bonds

An extreme case of polarity is represented by *ionic bonds* which are not covalent bonds. Some atoms attract electrons so much more strongly than other atoms that when both atoms are in close proximity the electrons completely leave one atom and become incorporated into the outer electron shell of the other. In this situation the electrons are not shared between the two atoms. The atom which gains the electron becomes negatively charged; the atom which gives up the electron becomes positively charged (Fig. 2-2). Ionic bonds are the most common type of bond in inorganic molecules.

Hydrogen Bonds

The phenomenon of polarity makes possible another type of bond, the *hydrogen bond*. This bond is formed when a hydrogen atom involved in a polar covalent

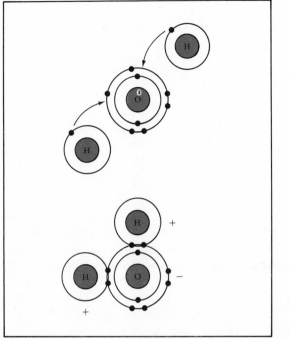

FIGURE 2-1
Formation of covalent bonds in water molecule. Oxygen has a greater attraction for the shared electron than do the hydrogen atoms. This results in the electrons being closer to the oxygen atom and conferring a negative charge on it. The hydrogen atoms in turn each have a positive charge.

bond, and therefore with positive charge, also interacts with the negative portion of another polar covalent bond. The negatively charged atom attracts the positively charged atom. Hydrogen bonds always involve an H atom bonded to either O or N (which gives the H a positive charge), and a negatively charged, covalently bound acceptor atom. The acceptor atoms are generally N or O bonded to H or C. One important example is the hydrogen bonds formed between water molecules (Fig. 2-3). The hydrogen bonds hold the water molecules together to form a shifting latticework. This bonding results in water being fluid with a relatively high boiling point. Hydrogen bonding is responsible for the properties of many other molecules that are considered, for example, DNA.

Several features distinguish hydrogen bonds from covalent bonds. The most important distinction is that *hydrogen bonds are much weaker than covalent*

HYDROGEN BONDS ARE WEAK BONDS

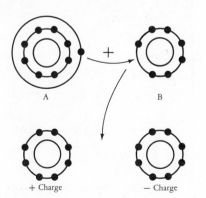

FIGURE 2-2
Ionic Bond. Atom A gives up an electron to Atom B. The result is that A acquires a positive charge and B a negative charge. Both atoms then have their outer electrons shells filled.

FIGURE 2-3
Water molecules held in a lattice work illustrating the hydrogen bonding between molecules.

bonds, having an energy of 2 to 10 kcal/mole (Table 2-3). Therefore, if atoms have the option of forming either a covalent or a hydrogen bond, they will always form the stronger covalent bond.

Van der Waals Interactions

All atoms will interact with one another no matter what their chemical nature, their charge properties, or their involvement in chemical bonds, *providing they can approach each other.* This type of bond, termed *van der Waals forces,* can occur only when the size and shape of molecules permit a close proximity of the atoms. If, however, the atoms come too close to each other, then the negative charges of the electrons on each atom will repel one another (Fig. 2-4). This means that pairs of atoms are attracted to a characteristic distance and no further. The strength of these van der Waals forces is small, even compared to hydrogen bonds (Table 2-3).

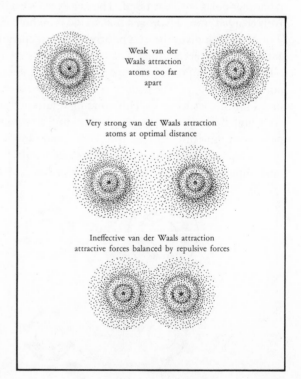

Weak van der
Waals attraction
atoms too far
apart

Very strong van der Waals attraction
atoms at optimal distance

Ineffective van der Waals attraction
attractive forces balanced by repulsive forces

FIGURE 2-4
Van der Waals Forces—the relationship between the distance and the attractive forces between atoms. Adapted from L. Pauling, *General Chemistry,* 2nd Ed., Freeman, San Francisco, 1958, p. 322, with permission.

Weak bonds have several important ramifications in biology. They are being constantly made and broken at room temperature, thereby indicating that there is enough energy in the movement of electrons even at room temperature to break weak bonds. Since the average lifetime of a *single* weak bond is only a fraction of a second at room temperature, cells do not need a special mechanism (enzymes) to speed up the rate at which weak bonds are made and broken. This is in contrast to the strong covalent bonds.

MANY WEAK BONDS CONFER STRENGTH

Because of their weakness, a single hydrogen or van der Waals bond is unable to bind molecules together. However, a large enough number of weak bonds can hold molecules together firmly. Thus only when there are a large number of weak bonds are they of biological significance.

One of the most important features of weak bonding forces is that they are effective only when the interacting surfaces are close to each other. Close proximity is only possible when the atoms have *complementary structures* such that a positive charge on one surface is matched by a negative charge on the other. The interacting molecules must have a lock-and-key relationship such that a protruding group of one molecule fits into a cavity of the other (Fig. 2-5). Because of this requirement for complementarity, great specificity exists as to which molecules will lie next to each other in the cell. Most molecules in the cell can make good weak bonds with only a small number of other molecules.

WEAK BONDS ALLOW FOR RECOGNITION BETWEEN MOLECULES

The crucial role that weak interactions play in biological phenomena are considered later in the text, so only an overview of the broad significance and diverse roles that these bonds play in microbiology is presented here.

Since water is a highly polar molecule, organic molecules that are also polar and therefore can form hydrogen bonds with water tend to be water soluble. Water pulls apart the polar compounds because of the strong attraction between

SIGNIFICANCE OF WATER AS A SOLVENT

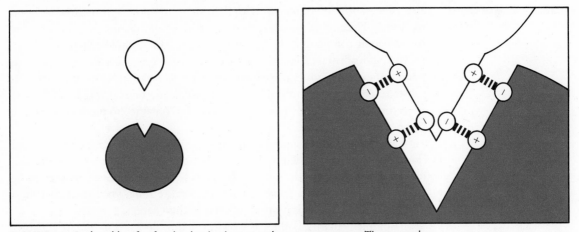

FIGURE 2-5   Lock and key fit of molecules that have complementary structures. The two molecules are complementary because the atoms on the "key" have the shape and charge to interact with atoms in the "lock." The interaction is manifest in the formation of many weak bonds which hold the complementary molecules loosely together.

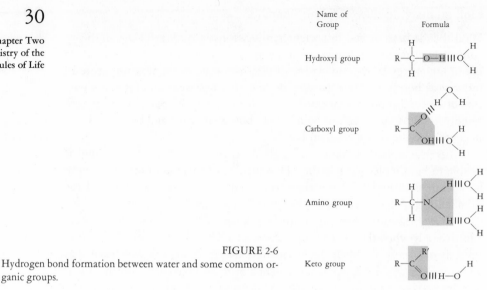

Name of Group	Formula
Hydroxyl group	
Carboxyl group	
Amino group	
Keto group	

FIGURE 2-6

Hydrogen bond formation between water and some common organic groups.

the water molecules and other polar or hydrophilic (water-loving) compounds, and in this way dissolves them (Fig. 2-6). Nonpolar compounds, such as benzene, are water insoluble because they cannot form hydrogen bonds with water. All molecules which enter bacteria and are metabolized must be polar since they must be soluble in water. Water has a strong tendency to exclude nonpolar or hydrophobic (water-fearing) groups. Nonpolar groups always attempt to arrange themselves so that they are not in contact with the polar water molecules. This association of nonpolar groups with each other to exclude water is termed *hydrophobic bonding,* although the weak bonds between the groups are primarily van der Waals forces.

Water and nonpolar compounds
Page 48—Fig. 2-26

The association of *enzymes* with their substrates occurs through weak bonding forces. The enzymes catalyze the conversion of one compound, the substrate, to a different compound, the product. The need for complementarity between enzyme and substrate explains why enzymes have great specificity in the compounds they will bind. It also explains why the association between enzyme and substrate can be made and broken so quickly, sometimes as often as 10^6 times per second. Thus 10^6 substrate molecules can be converted to product per second by one enzyme molecule.

Enzyme-substrate combination
Page 143—Fig. 7-5

Macromolecules, especially proteins and nucleic acids, have a three-dimensional structure that is determined to varying degrees by the weak bonds that form between atoms in the same molecule as well as weak bonds between molecules. This role of weak bonds is considered more fully when the structure of proteins and nucleic acids is discussed later in this chapter.

Antigens and antibodies
Page 369

Complementarity is the basis of interaction between *antibodies* and their *antigens.* Antibodies are protein molecules that are synthesized in the body of man and other animals in response to the presence of a foreign substance, the antigen. The antibody can combine with the antigen and thereby protect the body against any harmful effects the antigen might produce. In this association a protruding group of the antigen molecule fits into a cavity of an antibody molecule, thus allowing the formation of many weak chemical bonds. The fit is so precise and so many weak bonds are formed that the antigen-antibody complex is held together quite firmly. This illustrates the important fact that although each bond is weak, a sufficient number provides great strength. The reaction of antibodies and antigens is examined in detail in Chapter 20.

The synthesis of each of these macromolecules is a complex, multistep process which depends on complementarity and the formation of weak bonds between several molecules. These syntheses are discussed in Chapter 8.

Many structures in cells are highly complex, consisting of large numbers of different molecules arranged in a precise order. One reasonable explanation for the formation of such complicated structures is that molecules will associate only with a limited number of other molecules. The weak forces which hold these complex structures together suggest that complementarity plays an important role in determining which molecules can associate.

IMPORTANT MOLECULES OF LIFE

Small Molecules

All cells contain a variety of small molecules, both organic and inorganic; many of the small molecules occur in the form of charged ions. About 1 percent of the total weight of a bacterial cell is composed of inorganic ions, principally Na^+, K^+, Mg^{+2}, Ca^{+2}, Fe^{+2}, Cl^-, PO_4^{-3}, and SO_4^{-2}. Many of the positively charged ions are required in very minute amounts for the functioning of certain enzymes. The phosphate ion plays a vital role in energy metabolism and is discussed in Chapter 7. The organic molecules consist mainly of compounds that are being metabolized or have accumulated as a result of metabolism and those which serve as the building blocks of various macromolecules.

Alcohols, aldehydes, and organic acids are the most frequent products of metabolism. Table 2-5 lists some of the more common examples; the functional group characteristic of each compound is indicated by shading.

Macromolecules and Their Subunits

Three major classes of biologically important macromolecules are considered: polysaccharides, nucleic acids, and protein. The lipids, another class of molecules, not nearly as large as the macromolecules, also are briefly discussed.

As their name indicates, macromolecules are composed of a large number of atoms, generally more than 1000. Although they consist of a large number of atoms, all macromolecules have several structural features in common. The recognition and comprehension of these features make the structure of macromolecules relatively simple to understand. The application of these principles to each

GENERAL FEATURES
OF MACROMOLECULES

of the macromolecules is considered after the general features of macromolecules have been summarized.

All macromolecules are polymers—large molecules formed by the joining together of repeating small molecules, the subunits. Macromolecule synthesis occurs in two stages: first, the subunits are synthesized and then they are linked together in a systematic fashion.

TABLE 2-5
Some Common Products of Metabolism

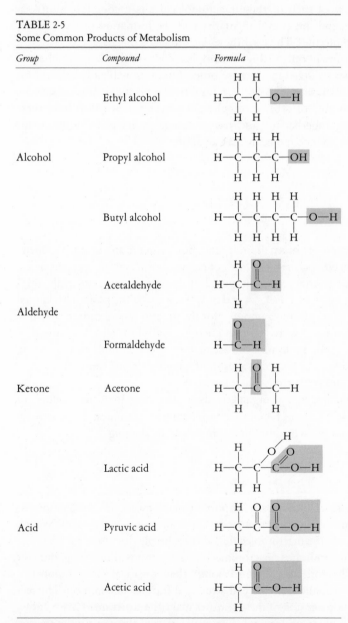

Group	Compound	Formula
Alcohol	Ethyl alcohol	
	Propyl alcohol	
	Butyl alcohol	
Aldehyde	Acetaldehyde	
	Formaldehyde	
Ketone	Acetone	
Acid	Lactic acid	
	Pyruvic acid	
	Acetic acid	

The shaded areas indicate the groups after which the compound is named.

The subunits are joined together by covalent bonds. However, the atoms that can connect the subunits are generally very restricted. The atoms which do take part in the joining form a part of the repeating series of identical chemical groups, the backbone of the macromolecule. Each class of macromolecule has a different backbone.

The individual subunits contain two more hydrogen atoms and one more oxygen atom than are found after the subunits are linked together. Thus the joining together of two subunits involves a chemical reaction in which H_2O is split out or removed. When the macromolecule is degraded into its subunits the reverse occurs, a water molecule is added. This type of reaction is common to the degradation of a large number of molecules. It is termed a *hydrolytic reaction* and only occurs through the action of specific enzymes. The more detailed aspects of macromolecule synthesis are covered later in this chapter and also in Chapter 8.

The subunits of any single macromolecule have common functional groups. For example, amino acids are the subunits of proteins. Although there are twenty different amino acids, each one is characterized by a carboxyl group and an amino group in the same position in the molecule.

Each of the macromolecules is now considered.

Proteins are polymers of amino acids. Twenty different amino acids are commonly found in proteins. Their formulas and names are given in Figure 2-7. All amino acids have several features in common, a carboxyl group (—COOH) and amino group (—NH$_2$) bonded to a terminal carbon atom. The same carbon atom also is bonded to a side chain that is characteristic for each amino acid. The side chain, which gives individuality to each amino acid, may be a very simple group such as —H or —CH$_3$, a longer chain of carbon atoms, or one of various kinds of ring structures, such as the benzene ring found in tyrosine and phenylalanine. Because of certain similarities in the side chains, the amino acids can be subdivided into several groups. (Fig. 2-7). For convenience, the side chain is given the designation "R" when no specific amino acid is being considered. Some of these side chains may also contain carboxyl or amino groups giving negative or positive charge properties to the amino acid. There is estimated to be between 2000 and 3000 different proteins in a bacterial cell, each different in its amino acid composition.

Whenever a carbon atom is bonded to four different atoms, the carbon atom with the four atoms can exist in two spatially different forms, one being the mirror image of the other. In all amino acids (except glycine) this situation occurs since the C atom which is bonded to the —COOH and —NH$_2$ group is also bonded to two other different groups. Thus this group of atoms can exist in either a "left-handed" or "right-handed" form known as the L-form or D-form, respectively (Fig. 2-8). *Only L-amino acids occur in proteins,* and therefore they are designated the *natural amino acids.* D-amino acids are rare in nature being found only in a few materials associated with bacteria—primarily cell walls and antibiotics.

The amino acids are bonded together by a *peptide bond,* a unique type of covalent linkage formed when the carboxyl group of one amino acid reacts with

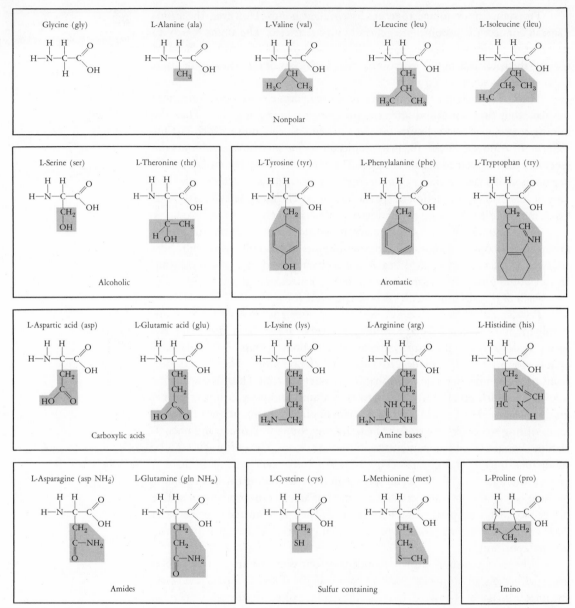

FIGURE 2-7   Amino acids. All amino acids have one feature in common: a carboxyl group and an amino group on the carbon atom next to the carboxyl group. Although the remainder of the molecule differs for each amino acid, they can be grouped because of certain features they have in common.

the amino group of the adjacent amino acid, with the splitting out of water. This reaction is diagrammed in Figure 2-9. The chain of amino acids which results when a large number of amino acids are joined by peptide bonds is often referred to as a *polypeptide chain*. (Fig. 2-10). One end of a polypeptide chain has a carboxyl group and the other end an amino group which are not joined in peptide linkage. These are designated the C terminal and N terminal ends respectively.

Proteins may vary in size from molecular weights of 6000 to about 7,000,000. The high molecular weight proteins are in fact *aggregates* of individual polypeptide chains which have associated with one another through weak bonding forces to form a highly complex, often very large molecule. A protein is defined as one or more polypeptide chains which are biologically functional. Some proteins consist of a single polypeptide chain; others are made up of a large number. (Fig. 2-11). Some proteins are arranged in a very interesting fashion, and the entire structure is large enough to be visible by a high-powered microscope. (Fig. 2-12). Often, different proteins associate with one another to form even larger structures, termed *multiprotein complexes* (Fig. 2-11). The structure of any

Mirror

FIGURE 2-8
Mirror image forms of an amino acid. The joining of a carbon atom to four different groups leads to asymmetry in the molecule. Thus, the molecule can exist in either the L or D form, one being the mirror image of the other.

C   —COOH   —H   R   —NH₂

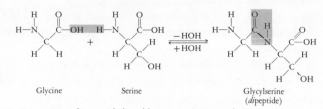

Glycine   Serine   Glycylserine
                                            (*di*peptide)

FIGURE 2-9   Formation of a peptide bond between glycine and serine.

protein is described in several terms. The sequence of amino acids determines the primary structure. This is the most important aspect of its structure, since it largely determines the function as well as the three dimensional structure of the molecule. Indeed the substitution of one amino acid for another is often enough to destroy completely the function of the protein. Proteins with the same primary structure are identical in all other respects.

A single polypeptide chain does not occur as an extended chain of amino acids. Rather it has a unique three-dimensional structure in which portions of the chain are folded. This folding and bending occur as a consequence of the molecule attempting to achieve its most stable three-dimensional structure. To achieve this stability, the molecule will fold in ways which allow hydrogen bond formation between the side chains of appropriate amino acids. In addition to hydrogen bonding, the side chains form additional weak linkages, including van der Waals and hydrophobic bonds and even some covalent bonds. The hydrophobic bonds are of great importance because they maintain the nonpolar hydrocarbon side chains tucked inside the molecule and not exposed to the aqueous environment. One covalent linkage of great importance is the combination of the sulfur atoms in two cysteine amino acid subunits to form a covalent S—S bond. Most proteins tend to have a *helical* three-dimensional structure which can be visualized as a chain wound like a helix around a rigid rod (Fig. 2-13). This regular structure is very stable primarily because the successive turns are held together by hydrogen bonds. Helical structures are encountered again in the structure of nucleic acids.

The weak bonds which are involved in the three-dimensional structure are readily broken if the protein is heated (energy put into the system). Once this happens the protein loses its three-dimensional structure and becomes *denatured* (Fig. 2-14). For example, boiling an egg results in a profound, readily observable effect on the protein (albumen) in the "egg white."

Nucleic acids are linear polymers in which the subunits are called *nucleotides*. All nucleotides are composed of three types of molecules: a nitrogen-containing ring compound covalently bonded to a 5-carbon sugar molecule, which in turn

Glycine   Serine   Valine   Alanine   Aspartic acid   Cysteine

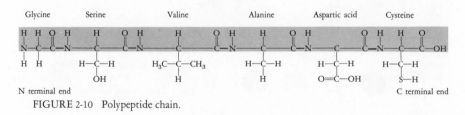

N terminal end   C terminal end

FIGURE 2-10   Polypeptide chain.

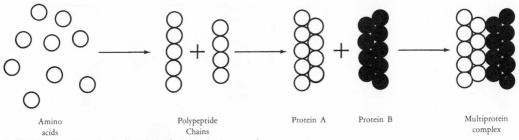

FIGURE 2-11 Steps in the buildup of multiprotein complexes.

is bonded to a phosphate molecule. There are five different nitrogen-containing bases which can be divided into two groups based on their structure, *purines* and *pyrimidines* (Fig. 2-15). There are two types of nucleic acids, ribonucleic acid (RNA) and deoxyribonucleic acid (DNA), which differ in the composition of their nucleotides. RNA contains the sugar ribose and DNA the sugar deoxyribose. An additional difference is that one pyrimidine in RNA is uracil and in DNA is thymine. Both contain the same two purines (adenine and guanine) and the pyrimidine cytosine. Thus each nucleic acid is composed of four particular kinds of nucleotide subunits (Fig. 2-16). The subunits are joined together by a covalent ester bond which joins the phosphate of one nucleotide to the sugar of the adjacent nucleotide. The phosphate bridge joins the number 3 carbon atom of one sugar to the number 5 carbon of the other so that a backbone composed

SUBUNITS OF NUCLEIC ACIDS ARE MONONUCLEOTIDES

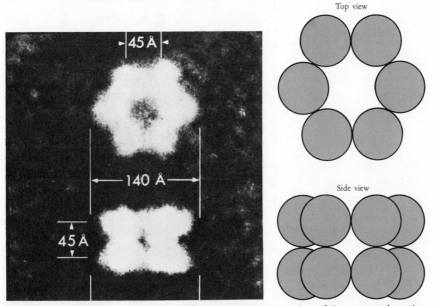

FIGURE 2-12 Electron micrograph and diagrammatic representation of the enzyme glutamine synthetase. The protein molecule is composed of 12 identical subunits forming two hexagons. The subunits bind together to form two closed circles which are attached to each other ($\times$ 2,085,000). (Photo courtesy of Dr. B. M. Shapiro, from R. C. Valentine, B. M. Shapiro, and E. R. Stadtman, *Biochemistry,* 7:2143 (1968), with permission.)

37

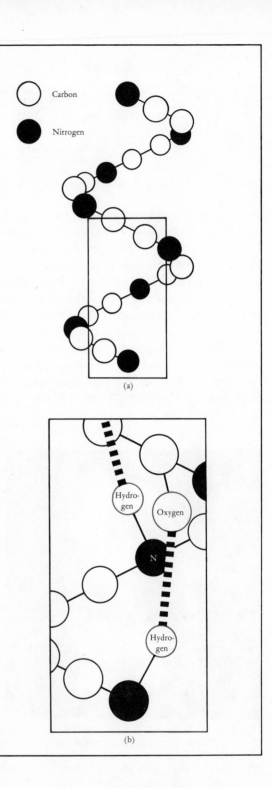

FIGURE 2-13

(a) Helical arrangement of polypeptide chains in protein. The turns in the helix are very regular and the regular arrangement is held together by hydrogen bonds between the O in the C—O group and the H in the N—H group. Although this helical arrangement, termed the alpha helix, is the most common, other configurations are found in some proteins. (b) Enlarged section of the polypeptide chain above showing the hydrogen bond between oxygen and hydrogen.

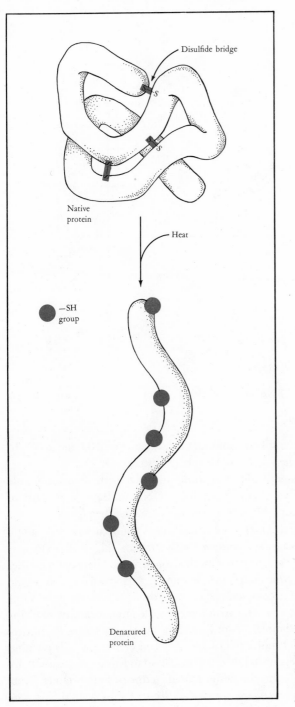

FIGURE 2-14

Heat denaturation of a protein molecule. The heating of proteins to high temperatures results in the disruption of the weak bonds as well as the S—S covalent bonds. The breaking of the S—S (disulfide) bonds results in the formation of S—H (sulfhydryl) groups.

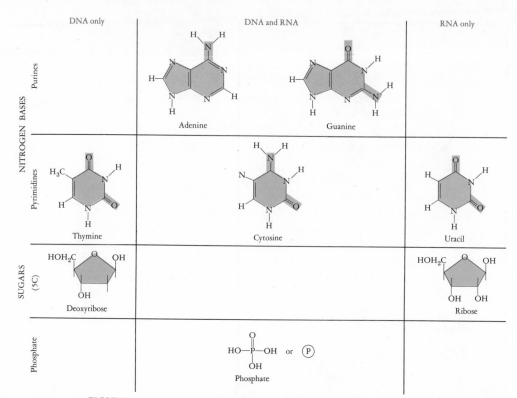

NITROGEN BASES

Purines

Adenine   Guanine

Pyrimidines

H_3C

Thymine   Cytosine   Uracil

SUGARS (5C)

HOH_2C OH

OH

Deoxyribose

HOH_2C OH

OH   OH

Ribose

Phosphate

$$HO-\overset{\displaystyle O}{\underset{\displaystyle OH}{P}}-OH \quad or \quad \textcircled{P}$$

Phosphate

FIGURE 2-15 Components of RNA and DNA. Carbon atoms are not represented in the ring structures.

of alternating sugar and phosphate molecules is formed (Fig. 2-17 and 2-18). In the reaction which joins the sugar and phosphate, water is split out.

The size of the polymer varies depending on whether the macromolecule is RNA or DNA. A typical bacterial cell has a DNA molecule composed of about 30 million nucleotides. RNA is considerably smaller, than DNA.

DNA IS COMPOSED OF TWO COMPLEMENTARY STRANDS ARRANGED IN A HELIX

DNA, and to a limited extent RNA, usually does not occur as a long, single strand. Rather it occurs as a *double-stranded helical structure* (Fig. 2-19). The two strands of this helix are held together by hydrogen bonds between complementary surfaces on the purine molecule (adenine) and the pyrimidine molecule (thymine). Since guanine is complementary to cytosine, it will therefore form hydrogen bonds with cytosine. This requirement for complementarity explains why, in all double-stranded molecules of DNA, the number of molecules of thymine equals the number of molecules of adenine, and cytosine equals guanine in the number of molecules. Since DNA is a double-stranded structure, its nucleotide subunit structure is generally designated in terms of both strands. Thus it is common to refer to *nucleotide pairs* or *base pairs*, which indicate the nucleotides which are opposite each other on both strands of the DNA polymer. This double-stranded structure again demonstrates how a large enough number of weak bonds can confer great stability on a molecule. The two strands will sepa-

HYDROGEN BONDS BETWEEN COMPLEMENTARY BASES

40

rate from one another only if they are heated to approximately 80°C, which is a temperature high enough to break most of the weak bonds. The complementarity of the purines and pyrimidines is important for another reason: the sequence of purines and pyrimidines on one strand dictates the sequence on the other strand. This has significance in the synthesis of new strands of DNA in the course of cell growth and division.

DNA carries the genetic information of the cell, coded in the sequence of purines and pyrimidines. This code is translated into a sequence of amino acids in protein molecules. RNA also functions in protein synthesis in several ways. The three different types of RNA are based on their function in the cell. These have different sizes and slightly different three-dimensional structures.

Carbohydrates are compounds containing principally carbon, hydrogen, and oxygen atoms in a ratio of approximately 1:2:1. The macromolecules are *polysaccharides*. These are linear polymers of *monosaccharides* which are single carbohydrate molecules. *Disaccharides* are molecules consisting of two monosaccharides joined together. The term *sugar* is often applied to monosaccharides and disaccharides. The chief distinguishing feature of carbohydrates is that they contain a large number of alcohol groups (H—C—OH) which accounts for their ratio of carbon, hydrogen, and oxygen atoms. They usually contain an aldehyde

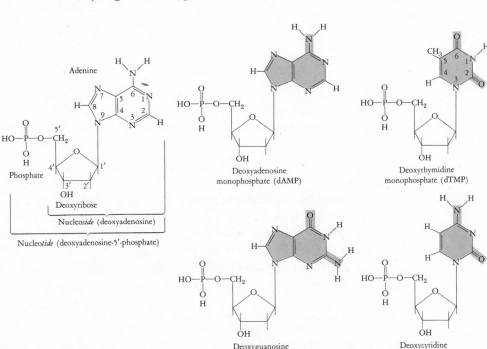

FIGURE 2-16 Subunits of DNA, the mononucleotides. Four mononucleotides are present in DNA; four in RNA. The nucleotides of RNA have the same structure except that ribose rather than deoxyribose and uracil in place of thymine are involved.

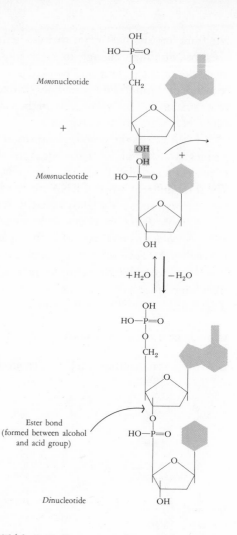

*Mono*nucleotide

*Mono*nucleotide

Ester bond
(formed between alcohol
and acid group)

*Di*nucleotide

FIGURE 2-17
Formation of ester bond in joining together two nucleotides.

and less commonly a ketone group (Table 2-2). Some contain an amino group, the amino sugars.

Monosaccharides are classified by the number of carbon atoms they contain. The most common in nature are the 5-carbon *pentoses* and the 6-carbon *hexoses*. The 5-carbon pentoses are represented by the sugars in nucleic acids, *ribose* and *deoxyribose* (Fig. 2-15). Common hexoses are *glucose, galactose,* and *mannose.* All of these hexoses contain the same number of atoms and the same kind of atomic groups. The difference in these sugars is due to different arrangements of the

H—C—OH in space (Fig. 2-20). Glucose and galactose have identical structures except for the space arrangement of the H and OH attached to carbon 4. Mannose and glucose differ with respect to carbon 2. These small changes result in three distinct sugars with different physical and chemical properties. The structural difference between fructose and other sugars is that fructose has a ketone group on carbon 2 (Fig. 2-20).

Sugars, like amino acids, have carbon atoms bonded to four different groups and so also can exist in two isomeric forms, D and L, which are mirror images of one another. By convention, if the —OH group on carbon 5 in hexoses (carbon 4 in pentoses) is written on the left, the sugar is L; if on the right, it is a member of the D-series. The majority of monosaccharides which occur in living organisms is of the D-configuration. This is the opposite of the situation with amino acids in which the L-form is the natural or biologically important form. Thus far the structures of the carbohydrates have been written as a linear molecule. This method of presenting the structure demonstrates graphically the several important characteristics that have been described. However, in aqueous solution, monosaccharides do not exist as a linear molecule, but rather occur as a closed ring as in Figure 2-20.

The most common *disaccharides* are the milk sugar, *lactose,* and the common table sugar, *sucrose.* Sucrose, which is derived commercially from sugar cane or sugar beets, is composed of one molecule of glucose and one of fructose. Lactose

Figures 2-15, 2-16, 2-17, 2-18, 2-19. Ring structures of sugars in nucleic acids

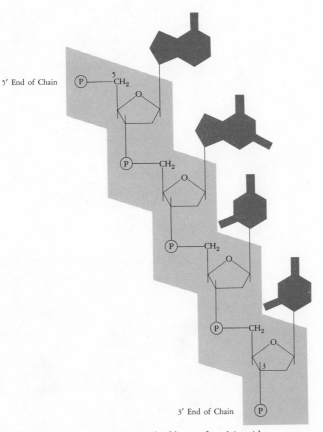

5' End of Chain

3' End of Chain

FIGURE 2-18 The repeating sugar-phosphate backbone of nucleic acids.

is composed of one molecule of glucose and one of galactose. The covalent bond joining the monosaccharides is termed a *glycosidic bond*. This linkage is formed in the reaction between a hydroxyl group of one sugar unit and the hydroxyl group of the other. In this reaction a molecule of water is split out (Fig. 2-21). The glycosidic bond in carbohydrates is analogous to the peptide bond in proteins and the ester bond in nucleic acids.

The general structural feature of a polysaccharide is the repeating mono-

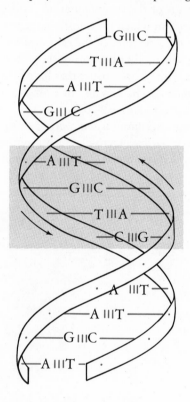

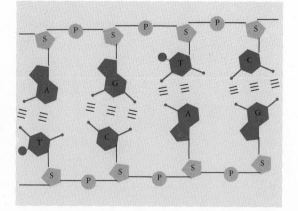

FIGURE 2-19
Helical structure of DNA and its appearance when not twisted. The helical DNA molecule can be looked on as a spiral staircase, in which the sugar-phosphate repeating sequences form the two railings and the purine and pyrimidine bases represent the stairs. The railings move in opposite directions, one going up, the other down. If the helix is untwisted, the molecule assumes the shape of a ladder. The fact that the two strands run in opposite directions is most readily seen in this untwisted form. The two strands are held together by hydrogen bonding.

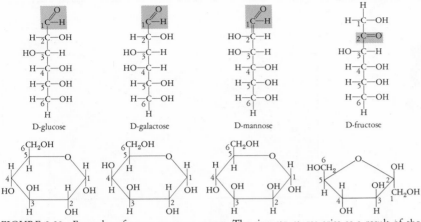

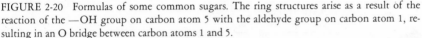

D-glucose   D-galactose   D-mannose   D-fructose

FIGURE 2-20 Formulas of some common sugars. The ring structures arise as a result of the reaction of the —OH group on carbon atom 5 with the aldehyde group on carbon atom 1, resulting in an O bridge between carbon atoms 1 and 5.

saccharide subunit, with D-glucose being the most frequent subunit. Polysaccharides are ubiquitous in nature and serve as storage forms for carbon and energy, as well as structural components of the cell. Cellulose, the principal structural component of plant cell walls, and which is also synthesized by some bacteria, is the most abundant organic compound in the world. It is a polymer of glucose subunits (Fig. 2-22). Glycogen, a storage product of animals and some bacteria, and dextran, which is also synthesized by bacteria, bear certain similarities in their structure to cellulose (Fig. 2-22). All of these polymers are made up of glucose residues. However, they are distinct molecules because they differ in the manner in which the glucose subunits are linked, the type of branching, and in the number of subunits. Polysaccharides which consist of the same subunits can differ in the following ways: (1) size of the polymer, (2) degree of branching of

MONOSACCHARIDES ARE
SUBUNITS OF
POLYSACCHARIDES

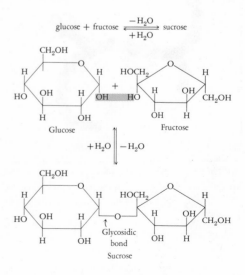

FIGURE 2-21
Formation of sucrose from glucose and fructose.

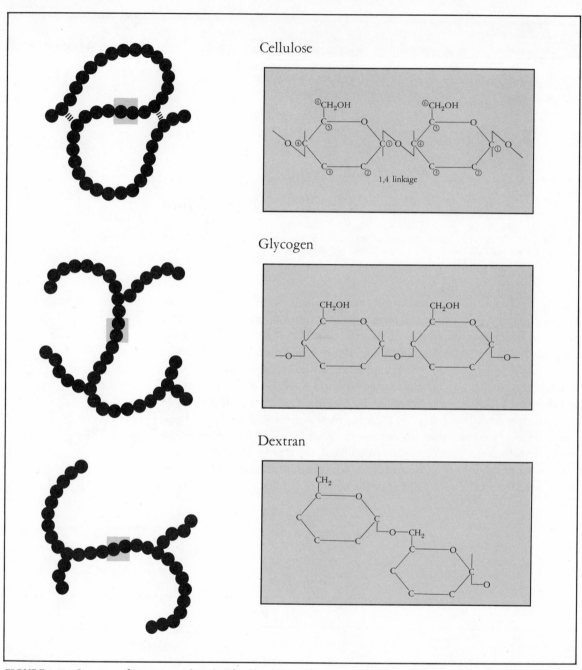

Cellulose

Glycogen

Dextran

FIGURE 2-22 Structure of important polysaccharides. The three molecules shown consist of the same subunit, glucose, but they are distinctly different molecules because of differences in the atoms which join the molecules together, the degree of branching, as well as the bonds involved in branching. Hydrogen bonds are responsible for some of the three-dimensional shapes of some of the polymers.

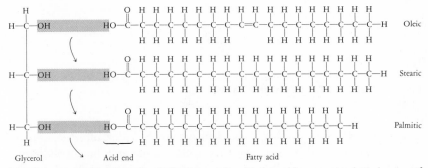

FIGURE 2-23 General formula of a fat. An ester bond is formed between the alcohol group of the glycerol and the acid group of the fatty acid.

the chain, and (3) the carbon atoms involved in the glycosidic linkage. Thus, a large variety of polysaccharides made up of the same subunits is possible.

The lipids represent a very heterogeneous group of biologically important molecules all of which have the property of being only slightly soluble in water but readily soluble in most organic solvents such as ether, benzene, and chloroform. The two general classes are the simple lipids and the compound. They are not macromolecules in the sense that the term has been used thus far, since they have molecular weights of no more than a few thousand. Furthermore, they have no well-defined subunits, but rather consist of a wide variety of substances of markedly different chemical structure.

LIPIDS ARE INSOLUBLE IN WATER

The most common of the simple lipids are the *fats*. Fats are esters of fatty acids and glycerol in which the bond forms between the carboxyl group of the acid and the hydroxyl group of the glycerol. The general formula for a fat is given in Figure 2-23. Note again the elimination of water in the formation of the ester bond. Glycerol has three hydroxyl groups which allow three fatty acid molecules to be bonded to it. These three fatty acids may be the same or different. There are many different fatty acids, but the most common in nature are palmitic (16 carbon atoms), stearic (18 carbon atoms), and oleic with 18 carbon atoms and some double bonds in the molecule. The fact that these molecules contain so many nonpolar groups, H—C—H, explains why the molecule is insoluble in water. *Waxes* are esters of fatty acids and alcohols *other* than glycerol.

FATS ARE ESTERS OF FATTY ACIDS AND GLYCEROL

Another very important group of lipids are the *steroids,* all of which have the four-membered ring structure shown in Figure 2-24. The name steroid is derived from *sterol,* compounds which contain hydroxyl (alcohol) groups. The most common sterol is *cholesterol* (Fig. 2-25). This sterol as well as others are important components of the cytoplasmic membrane of eucaryotic cells.

Compound lipids contain elements other than the carbon, hydrogen, and

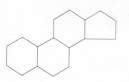

FIGURE 2-24
General structure of a steroid. The carbon atoms in the ring structures are not shown.

FIGURE 2-25
Cholesterol, a sterol. The carbon atoms in the
ring structures are not shown.

**MANY LIPIDS ARE
ASSOCIATED WITH
OTHER
MACROMOLECULES**

oxygen found in simple lipids. These elements may be sulfur, nitrogen, or phosphorus. Phospholipids contain the fatty acids and glycerol, but in addition a phosphate molecule bonded to a nitrogen-containing compound. Another group of compound lipids are the lipoproteins. Lipoproteins are loose complexes of proteins and lipids held together by weak bonding forces such as van der Waals, and not by covalent bonds. Lipopolysaccharides are molecules of lipids associated with polysaccharides through covalent bonds.

Lipids have several functions in living organisms. In microorganisms phospholipids are found in the cytoplasmic membrane and in many cell walls. In man and other animals fats act as the prime fuel reserve for metabolism as well as structural components of membranes. Most of the properties of cell membranes depend on the presence and structure of their lipid components. The lipids found in membranes are of the compound type which have unique biological properties with respect to their solubility properties in water. They are polar at one end (which contains the charged group) and nonpolar throughout the rest of the molecule which consists largely of C—H atoms. Therefore the latter portion of the molecule is insoluble in water, whereas the polar end is very soluble. The bimodal nature of many lipids results in their assuming a very ordered array when such lipids are placed in water. All of the charged groups are on the outside and the inside (Fig. 2-26). This bilayer arrangement represents one of the main features of membrane structure.

**LIPIDS HAVE TWO ENDS;
ONE CHARGED, THE
OTHER NOT CHARGED**

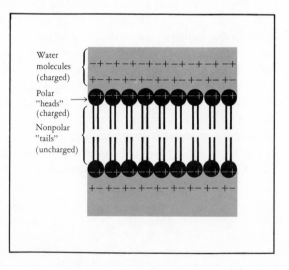

Water
molecules
(charged)

Polar
"heads"
(charged)

Nonpolar
"tails"
(uncharged)

FIGURE 2-26
Orientation of lipid molecules in water. The polar ends associate with the polar water molecules and the nonpolar hydrocarbon tails associate with each other. The exclusion of the water in the interior of the molecule is promoted by hydrophobic bonds between the nonpolar portions of the molecule.

This chapter presents some fundamental information required for understanding various facets of microbiology. Virtually all biological phenomena depend on the association of atoms and molecules. It is the precise, ordered arrangement of atoms and molecules which permits living organisms to function efficiently. If the arrangement is disrupted by chemicals or heat, death ensues. The forces or bonds involved in the association of molecules can be of several types. One classification depends on the strength of the bonds. The covalent bond, which involves the sharing of electrons between atoms, is a very strong bond. It cannot be broken unless energy is put into the system, or an enzyme is available which catalyzes the breakage of the bond. Since some atoms attract electrons more avidly than other atoms, the shared electrons in a covalent bond may be closer to one atom than the other atom. The partial positive and negative charges which result allow the formation of a weak type of bond, the hydrogen bond. Another weak bond, van der Waals forces, involves a nonspecific attraction of atoms, provided they are at the proper distance from one another. Although a single weak bond has no biological significance, a large number are capable of holding very large molecules firmly together.

Bond formation always results in the release of some of the energy of the atoms that are bonded. The more energy released, the stronger the bond. If several types of bonds are possible between atoms, the strongest bonds possible will form. The formation of the maximum number of bonds leads to maximum stability of the molecule.

A large number of organic molecules play important roles in life's processes. These molecules can be conveniently classified into the small molecules and the large molecules, the macromolecules, formed by the association of the small molecules. The small molecules can be grouped according to their functional group.

Macromolecules range in size from 5000 to 1 billion in molecular weight. Their basic structures have been elucidated largely because all macromolecules are composed of repeating subunits (Table 2-6). The subunits are joined together by covalent bonds in a reaction which involves the splitting out of a molecule of water.

Proteins are composed of amino acids bonded together in a peptide linkage. Protein molecules differ from each other in the sequence of the twenty kinds of amino acids which comprise the protein. The long protein chain assumes a three-dimensional configuration which permits the maximum number of bonds to be formed between amino acids in the chain. This structure is often in the shape of a helix.

The nucleic acids, RNA and DNA, are polymers of subunits, the nucleotides, which are composed of a molecule of a 5-carbon sugar, ribose or deoxyribose, a purine or pyrimidine, and a phosphate molecule. Ester bonds between the acid groups of the phosphate molecule and an alcohol group on the sugar join the nucleotides. DNA occurs as a double-stranded, helical molecule stabi-

TABLE 2-6
Structure and Function of Macromolecules

Name	Subunit	Bond Joining Subunits	Atoms in Bond	Some Functions of Macromolecule
Protein	Amino Acid	Peptide	$-\overset{\overset{\text{O}}{\|\|}}{\text{C}}-\overset{}{\underset{\underset{\text{H}}{\|}}{\text{N}}}-$	Catalysts; Structural portion of cell organelles
Ribonucleic Acid	Nucleotide	Ester	$-\text{O}-\overset{\overset{\text{O}}{\|\|}}{\underset{\underset{\text{O}-\text{H}}{\|}}{\text{P}}}-\text{O}-\overset{\overset{\text{H}}{\|}}{\underset{\underset{\text{H}}{\|}}{\text{C}}}-$	Various roles in protein synthesis
Deoxyribonucleic Acid	Deoxynucleotide	Ester	$-\text{O}-\overset{\overset{\text{O}}{\|\|}}{\underset{\underset{\text{O}-\text{H}}{\|}}{\text{P}}}-\text{O}-\overset{\overset{\text{H}}{\|}}{\underset{\underset{\text{H}}{\|}}{\text{C}}}-$	Carrier of genetic information
Polysaccharide	Monosaccharide	Glycosidic	$-\overset{\|}{\underset{\|}{\text{C}}}-\text{O}-\overset{\|}{\underset{\|}{\text{C}}}-$	Structural component of plant cell wall; Storage products

lized by hydrogen bonds between adenine and thymine and also guanine and cytosine.

Polysaccharides comprise a number of molecules of varying sizes. The monosaccharides, especially glucose, bond together by a glycosidic bond to form polysaccharides, such as starch and cellulose. Variations in the bonds involved in joining the glucose molecules together as well as the degree of branching account for differences in molecules composed of the same subunits.

The lipids are too small to be considered macromolecules. The two types are the simple and the compound. Simple lipids consist of glycerol bonded to three fatty acid molecules. The most common fatty acids in nature are palmitic, stearic, and oleic. They have unique biological properties since they are polar at one end (which contains the carboxyl group) and nonpolar through the rest of the molecule. The steroids are another group of lipids, and are characterized by a four membered ring structure. Cholesterol is a common steroid.

Compound lipids contain molecules which have either phosphorus or nitrogen in addition to fatty acids and glycerol. They are a very important component of all cell membranes.

QUESTIONS

1. Compare covalent and hydrogen bonds in terms of: (a) basis for bond formation; (b) biological significance; (c) strength; (d) compounds in which they are important.
2. Discuss the concept of complementarity in biological systems.

3. A bacterium ingests the following radioactive compounds. In what macro-molecules will the major elements in the compounds be found?

   (a) Nitrate (nitrogen)   (c) Phosphate (phosphorous)

   (b) Glucose (carbon)   (d) Sulfate (sulfur)

4. Compare DNA, RNA, and protein in terms of: (a) subunits, (b) bonding together of subunits, (c) components making up subunits.
5. What features do all macromolecules have in common? Would you consider lipids to be macromolecules based on these factors?

FURTHER READING

BENNETT, T. P. and E. FRIEDEN, *Modern Topics in Biochemistry—Structure and Function of Biological Molecules*. New York: The Macmillan Company, 1966. (Chap. 1, 2, 3, 7, 8, 9)

CRICK, F. H. C., "Nucleic Acids," *Scientific American* (September 1957). (Offprint #54)

DOTY, P., "Proteins," *Scientific American* (September 1957). (Offprint #7)

GREEN, D. E. and R. F. GOLDBERGER, *Molecular Insights into the Living Process*. New York: Academic Press, 1967. (Chap. 2 and 3)

KENDREW, J., "The Three-Dimensional Structure of a Protein Molecule," *Scientific American* (December 1961). (Offprint #121)

STEIN, W. and S. MOORE, "The Structure of Proteins," *Scientific American* (February 1961). (Offprint #80)

WHITE, E. H., *Chemical Background for the Biological Sciences,* 2d ed. Englewood Cliffs, N.J.: Prentice-Hall, Inc., 1970, (Chap. 1 and 2)

Chapter 3

The Protists—
An Overview

THE POSITION OF MICROORGANISMS
IN THE LIVING WORLD

MICROORGANISMS DO
NOT FIT INTO THE
ANIMAL OR PLANT
KINGDOM

For thousands of years man had observed two major groups of organisms around him. One group, classified as plants, was characterized by a type of nutrition dependent on sunlight for energy, had rigid cell walls, was rooted in the soil and therefore stationary. The other group, the animals, ingested its food and was motile. With the discovery of microorganisms major difficulties in classification arose. Some forms were motile, ingested their food, and were therefore classified as animals (protozoans). Others were nonmotile and photosynthetic and were naturally assigned to the plants (algae). However, a large number of unicellular forms remained which had properties of both groups—motile and photosynthetic or stationary and nonphotosynthetic. Perhaps most disturbing was the fact that in at least some single cell plantlike organisms the environmental conditions under which the organism was grown determined whether or not it was photosynthetic.

Thus it was difficult to fit single-celled organisms into either the plant or the animal kingdom. Although most of them are not photosynthetic, some are. Also many are motile and have rigid cell walls. The most reasonable solution to this dilemma seemed to be recognizing the insurmountable difficulties of the two-kingdom system and creating additional kingdoms.

Although most taxonomists do not disagree that more than two kingdoms exist, no single scheme of classification has met with universal approval. To some taxonomists the simplest and most satisfactory solution was advocated in 1866 by the German zoologist Haeckel, a student of Darwin. He proposed a third kingdom be established, the *Protista,* which would be on the same taxonomic level as the plant and animal kingdoms. This classification scheme will be followed in this text, because it has the merit of being relatively simple and yet draws on the fundamental differences in the structure and organization of cells. The only basis for an organism being classified as a member of the Protista is its simple biological organization characterized by a lack of extensive tissue formation. Included in this kingdom are *all bacteria, algae, fungi,* and *protozoa.* This group includes both microscopic, unicellular organisms and very large multicellular forms. It includes organisms that are procaryotic as well as those of eucaryotic cell type.

Animals and plants can be considered to represent a higher level of biological organization than the protista, because of their tissue differentiation. In actual fact there is not always a sharp dividing line between organisms that have tissues and those that have a tendency to form tissues. This is not surprising since the work of Darwin culminating with the publication of *Origin of Species* in 1859 suggested that there should be transition forms, rather than sharply delineated groups. The organisms that have properties of both plants and animals are also transition forms. Hence, although an organism is placed in the kingdom Protista, it may be more plantlike than animal-like in appearance (algae and fungi) or more animal-like than plantlike (protozoa).

The microbial world includes all members of the kingdom Protista. The taxonomic scheme outlined emphasizes the close relationship between microscopic and macroscopic forms of life which have no extensive tissue differentiation. This scheme further emphasizes that the basic biological dichotomy is not between the sizes of organisms but rather between the two basic cell types, eucaryotic and procaryotic.

A large part of this book is devoted to discussing the biology of various members of the Protista. Several chapters are devoted to discussing its major groups. However, at this time it is instructive to gain some appreciation of the broad range of the microbial world by discussing briefly the major groups of organisms that comprise this world.

MEMBERS OF THE PROTISTA— PROCARYOTIC FORMS

When we consider the members of the microbial world, the organisms that first come to mind are probably the bacteria and more specifically the kinds of bacteria that cause disease in man. The bacteria in fact do represent a very large group both in terms of variety and numbers of organisms. However, there are other organisms with unusual shapes and cell structure which, although not so well known to most people, are nevertheless of great interest and importance for the indispensable roles they play in nature.

The bacteria represent a very heterogeneous assemblage of organisms which have one property in common—they are procaryotic in cell structure. They are

PROTISTS HAVE NO
EXTENSIVE TISSUES

PROTISTS INCLUDE BOTH
EUCARYOTIC AND
PROCARYOTIC FORMS

distinguished from the other group of procaryotes, the blue-green algae, by the fact that the blue-green algae have one distinctive feature, they all gain energy by using the same type of photosynthesis associated with plants. Bacteria can be divided into a number of groups based on several criteria: (1) the means by which they move; (2) the degree of rigidity of the cell wall; (3) their degree of dependence on living cells for growth; (4) their means of reproduction; and (5) their nutritional requirements.

Eubacteria or True Bacteria

This group comprises a heterogeneous group of unicellular organisms that are generally associated with the term bacteria. The bacteria, however, cover a larger group than just the eubacteria. The heterogeneity of the true bacteria extends to their shape, size, structural components, metabolism, and a wide range of other physiological properties. The eubacteria share several features.

1. They all have rigid cell walls.
2. Their motility is due to the action of flagella. Some members, however, lack flagella and therefore are immotile.
3. They are all unicellular.
4. They multiply by binary fission. In this process one cell divides into two identical daughter cells.

EUBACTERIA HAVE
CHARACTERISTIC SHAPES

Eubacteria may be spherical, which in organisms are commonly referred to as *cocci* (singular, coccus); cylindrical, as in *bacilli* (singular, bacillus); helical, as in the *spirilla* (singular, spirillum) (Fig. 3-1). Note that the term bacillus is also the name of a genus of eubacteria. Although all cells of true bacteria divide by binary fission and are therefore functionally distinct, they may not always separate from each other. This adherence of cells results in a characteristic arrangement. Such arrangements depend on the planes along which the organisms divide (Fig. 3-2). Those dividing along only one plane form long chains; those dividing along several planes at random appear as clusters. If division occurs in sequence in two or three planes perpendicular to one another, cubical packets result. Since such arrangements are characteristic properties of certain genera, they are useful aids in the microscopic identification of eubacteria. Thus certain cocci tend to form long chains, characteristic of the genus *Streptococcus;* others, especially members of the genus *Staphylococcus,* tend to occur in clusters resembling bunches of grapes. The members of still other genera typically form cuboidal packets of eight or more cells.

Gram stain
Pages 72–73

The eubacteria are also classified into two groups on the basis of their staining properties when stained with a number of dyes. The organism is gram-positive if it retains a particular dye and gram-negative if it does not. The gram reaction, is correlated with many other properties of the cell.

Flagella arrangement
Page 57; Fig. 3-3

The arrangements of flagella are also characteristic of certain genera. Some of the true bacteria have a single flagellum at one end of the cell; some have a tuft. Other true bacteria have a flagellum at either end and still others have their

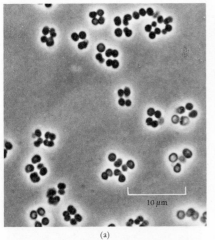

(a)

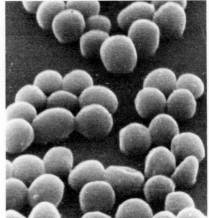

(b)

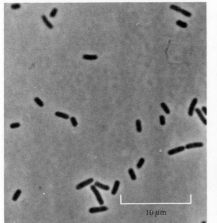

(c)

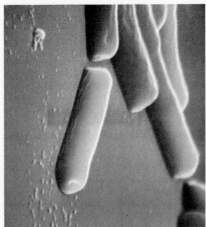

(d)

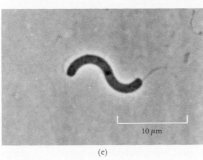

(e)

FIGURE 3-1 The three most common shapes of eubacteria as viewed through a phase contrast microscope (left) and a scanning electron microscope (right). (Top) Spherical (coccus); (center) cylindrical or rod-shaped (bacilli); (bottom) spiral-shaped (spirilla) (a, c, and e courtesy of Dr. J. T. Staley and J. P. Dalmasso). (b) (×30,000) (Courtesy of Dr. D. Greenwood; Greenwood, D. and F. O'Grady, *Science,* **163**:1076, 1963). (d) (×35,000) (Courtesy of Dr. N. Hodgkin)

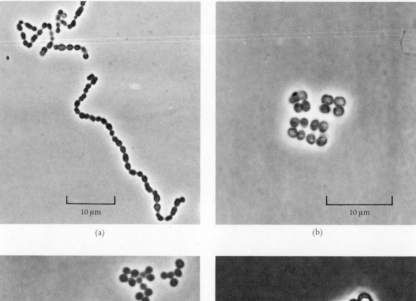

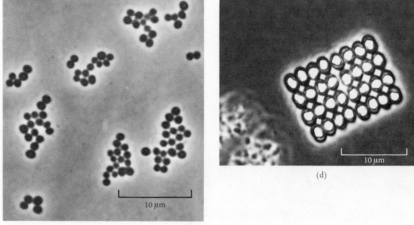

FIGURE 3-2   Arrangements of cocci into characteristic patterns. (a) Chains (streptococci); (b) cuboidal (sarcina); (c) clusters (staphylococci); (d) sheets of cells, one cell thick (*Thiopedia*). The light areas inside the cells (d) are gas vacuoles. (Courtesy of Dr. J. T. Staley and J. P. Dalmasso)

EUBACTERIA ARE
METABOLICALLY
VERSATILE

flagella inserted at many points along the side of the cell (Fig. 3-3). The eubacteria include a few members that are photosynthetic and others that obtain energy from the metabolism of inorganic compounds, but the vast majority gain energy from the degradation of organic compounds. Some require oxygen to grow, whereas others are killed by oxygen. The significance of these variations will become apparent when the nutrition and metabolism of bacteria are discussed. Since the eubacteria will be the main focus, we need only outline their relationship to the other members of the microbial world at this point. A more extensive classification scheme is given in Chapter 12. There are numerous groups of organisms which are obviously related to the eubacteria, but nevertheless possess distinguishing features that place them into other groups.

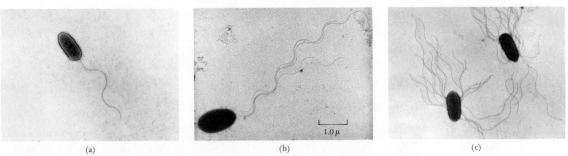

| (a) | (b) | (c) |

FIGURE 3-3 Arrangements of flagella. (Left) Single polar flagellum, *Pseudomonas aeruginosa* (polar flagellation) (×5250) (courtesy of Dr. V. Chambers); (center) tuft at one end (polar flagellation—lophotrichous) (courtesy of Dr. E. Boatman); (right) flagella inserted throughout *Proteus mirabilis* (peritrichous flagellation) (×3500) (courtesy of Dr. V. Chambers).

Other Groups of Bacteria

In addition to the eubacteria other groups of bacteria also exist. These groups possess distinguishing features of either cell wall structure, motility, or reproduction which distinguish them from the eubacteria and from one another. These features allow these bacteria to be conveniently divided into the groups given in Table 3-1. This table emphasizes that *the only feature common to all bacteria*

TABLE 3-1
Major Procaryotic Groups of the Microbial World

	Distinguishing Features
Bacteria	
Eubacteria or true bacteria	Rigid cell walls; unicellular; multiply by binary fission; if motile, by flagella
Prosthecate and budding bacteria	May have unusual shapes, appendages (prosthecae); most prosthecate bacteria divide by budding; most budding bacteria are prosthecate
Mycelial bacteria	Often moldlike in appearance
Filamentous (sheathed bacteria)	Individual cells enclosed in a common sheath
Mycoplasma or pleuropneumonia-like organisms (PPLO)	Very heterogenous group; properties distinct from bacteria: no cell wall and may not divide by binary fission, sterols in cytoplasmic membrane.
Gliding bacteria	Organisms have flexible cell wall and may have a developmental cycle
Spirochetes	Helical cells with flexible cell walls; movement by axial filament
Rickettsia	Obligate intracellular parasites; have "leaky" cytoplasmic membrane
Chlamydia	Obligate intracellular parasites which lack some enzymes required for energy production
Blue Greens	Flexible cell wall which allows them to glide; gain energy by photosynthesis; reproduction primarily by binary fission

is their procaryotic cell structure. Members of certain of these groups are important in pollution of water, causative agents of disease, and producers of antibiotics. A more detailed discussion of each of these groups is presented in Chapter 12 when a more complete overview of bacteria is given.

Blue-Green Algae or Blue Greens

The blue-green algae are procaryotic in cell structure and therefore are profoundly different from the eucaryotic algae. Therefore they shall be referred to as blue greens. The blue greens form a group of unicellular or filamentous, multicellular organisms (Fig. 3-4). The unicellular members are either rod-shaped or spherical and occur singly or as a group of cells embedded in a gelatinous matrix. The filamentous forms consist of long unbranched rows of cells held together by a common outer wall. Some features are common to most members: a characteristic pigmentation which allows them to derive energy from sunlight and a gliding type of motility, the mechanism of which is unknown. The process of photosynthesis results in the release of oxygen. They also have the ability to convert nitrogen gas to organic forms of nitrogen. The blue greens are the only group of organisms that can fix nitrogen and release oxygen in photosynthesis.

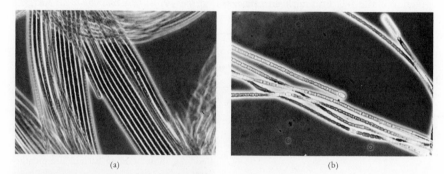

(a)   (b)

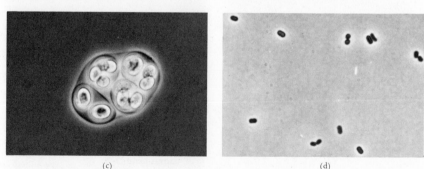

(c)   (d)

FIGURE 3-4 Blue greens. (a) *Oscillatoria*—a filamentous blue green associated with "blooms" in polluted water ($\times 250$); (b) *Anabaena*—there are constrictions between cells in the filament ($\times 300$); (c) *Gloeocapsa*—a unicellular blue green whose cells are grouped together by a multilayered sheath ($\times 600$); (d) *Anacystis*—A unicellular blue green; morphology identical to many bacteria ($\times 500$). (Courtesy of Dr. J. T. Staley and J. P. Dalmasso)

In some blue greens the pigment may be a type of chlorophyll plus additional pigments that impart characteristic colors. Although some of the blue greens are actually blue-green in color, others are black, brown, yellow, and red. The color of the Red Sea results from the presence of a red-pigmented blue-green group. Ponds or lakes which contain high concentrations of organic matter develop very high populations of these organisms, called *blooms*. Some species can grow at very high temperatures, up to 75°C, so that they are often found at hot springs such as those at Yellowstone National Park.

THE EUCARYOTIC MEMBERS OF THE PROTISTA

The eucaryotic protista include a vast array of highly divergent organisms, all characterized by a lack of cell differentiation into tissues. This portion of the protista can be subdivided into three major groups: algae, fungi, and protozoa. Since these are discussed individually in Chapters 13, 14, and 15, respectively, only their basic characteristics are summarized now. (Table 3-2)

Algae

The algae, as distinguished from the blue greens range in size from microscopic unicellular forms to the multicellular giant kelp, which can reach 150 feet in length (Fig. 3-5). The most significant feature of all algae is their ability to carry out photosynthesis; accordingly, all algae are pigmented. Their photosynthetic pigments are found in membrane-bounded structures in the cytoplasm called *chloroplasts* and they span the colors of green, yellow, brown, and red. The green scum often floating on lakes during the summer months is composed of algae.

Fungi

Fungi differ from the algae in that they do not carry out photosynthesis and lack photosynthetic pigments. Included in this group are mushrooms, yeasts, and molds. Members of the fungi assume a very large number of sizes and shapes. Although yeasts are unicellular, most fungi are "molds," some of which are multicellular organisms in which the cytoplasm of each cell contains a large number of nuclei. The multicellular molds are composed of a large number of

TABLE 3-2
Distinguishing Features of Eucaryotic Protists

	Distinguishing Features
Algae	Gain energy from sunlight (photosynthetic)
Fungi	Not photosynthetic, most multicellular, consisting of long filaments
Protozoa	Unicellular and nonphotosynthetic

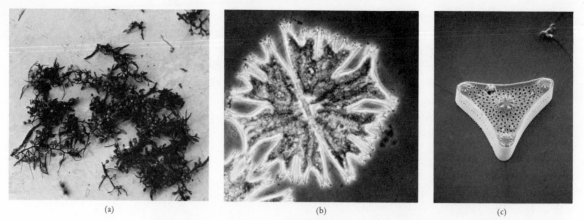

(a)　　　　　　　　　　　　　　　　　(b)　　　　　　　　　　　　　　　　(c)

FIGURE 3-5　Algae. (a) *Sargassum,* a macroscopic alga $\frac{1}{10}$ actual size (courtesy of Dr. J. T. Staley); (b) *Micrasterias,* a green alga composed of two symmetrical halves ($\times$350) (courtesy of Dr. J. T. Staley); (c) diatom organism which serves as food for aquatic animals, as viewed through a scanning electron microscope (courtesy of Dr. Norman Hodgkin).

long, fine filaments—mycelial growth which gives molds their gross appearance. In nature the mycelium is often embedded in soil or some other material, such as bread or moist leather, and may not be visible (Fig. 3-6). Many fungi form reproductive structures above the level of the medium in which they are growing. These structures generally identify the fungus; the mushroom, for example, is the portion of a fungus concerned with reproduction. Other reproductive structures are the black spherical bodies commonly found on "moldy bread." These structures release spores which then germinate to form another mushroom or mold.

There are probably as many as 200,000 species of fungi; about 50 are known to cause disease in man. Many others cause diseases in plants and animals. Because fungi are larger than bacteria, their role as causative agents of disease was recognized long before a similar role of bacteria was established. In 1836 Agostino Bassi demonstrated that a fungus caused a disease in silkworms, and from these studies he postulated in 1844 that diseases in man including syphilis and plague were caused by living parasites. However, it was not until 40 years later, in 1876, that Robert Koch unequivocally established that bacteria could cause a disease (anthrax in cattle).

Protozoa

The protozoa are colorless (nonphotosynthetic), unicellular, motile organisms (Fig. 3-7). Many members bear a strong resemblance to unicellular algae, and it is likely that they are descendants of algae which have lost their pigment. Although protozoa are unicellular, some of these single cells are particularly complex (see Chapter 15). These cells have probably achieved the ultimate complexity possible in a single cell. A higher level of organization can be attained only if the entire cell becomes differentiated, associates with other differentiated cells, and forms a tissue in a multicellular organism.

Protozoa occupy a wide variety of habitats in nature, including saltwater, freshwater, and soil, as well as within the bodies of other organisms. A number of these species cause disease in man and other organisms.

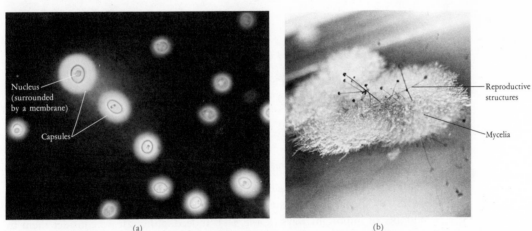

(a)

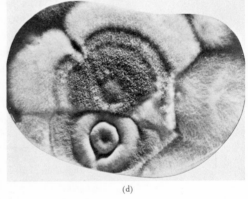

Reproductive structures

Mycelia

(b)

Gills

Stem

(c)

(d)

FIGURE 3-6   Fungi. (a) *Cryptococcus neoformans,* living cells of the yeast stained with India ink (negative stain) to reveal the large capsules which surround the cell ($\times 600$). (b) Aspergillus, a typical mold whose dark reproductive structures rise above the mycelium ($\times 7$). (Courtesy of J. P. Dalmasso) (c) Mushroom, a type of fungus whose spores are contained in the gills. The mushroom is the fruiting body; the reproductive structure and the mycelium from which it is produced spreads underground forming a diffuse mat perhaps 100 feet in diameter. (Courtesy of Dr. D. Stuntz) (d) Colonies of common air-borne molds.

(a)

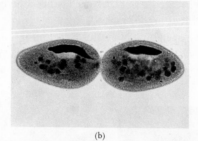
(b)

(c)

FIGURE 3-7   Protozoa. (a) *Amoeba,* whose cells continually change shape by extending pseudo-podia, their means of locomotion; (b) paramecium, a ciliated protozoan undergoing fission; (c) *Vorticella,* whose bell-shaped body is attached to a contractile stem and contains numerous cilia. (Courtesy of Carolina Biological Supply Company)

Chap. 16, 17   VIRUSES

Viruses are fundamentally different from any of the organisms discussed so far. Although they resemble the minute bacteria, the rickettsia and chlamydia superficially in also being obligate intracellular parasites, they differ from them in a number of important ways (Table 3-3). Indeed, although we consider them to be part of the microbial world, they are really not cells and therefore not organisms. A virus particle is actually a piece of genetic material, either DNA or RNA (never both), surrounded by a protective protein coat (Fig. 3-8). Viruses are probably best considered as a unique class of agents which possess a few of the properties associated with living forms. They appear to be the simplest agents that are capable of self-replication. However, this ability to replicate is dependent on the virus using the biosynthetic machinery of the host cell which it parasitizes since the virus has virtually no enzymes of its own. Thus viruses represent an extreme form of parasitism.

TABLE 3-3
Distinguishing Features between Eubacteria and Major Obligate Intracellular Parasites

Eubacteria	Rickettsia	Chlamydia	Viruses
Free-living	Obligate intracellular parasites	Obligate intracellular parasites	Obligate intracellular parasites
Contain both DNA and RNA	Contain both DNA and RNA	Contain both DNA and RNA	Contain either DNA *or* RNA
Multiply by binary fission	Multiply by binary fission	Multiply by binary fission	Nucleic acid and protein duplicate independently
Contain all enzymes required for independent existence	Probably contain all enzymes required for independent existence	Lack certain enzymes required for energy production	Contain very few if any enzymes

62

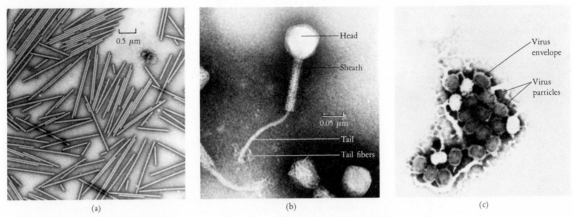

(a)   (b)   (c)

FIGURE 3-8 Viruses. (a) Tobacco mosaic virus that infects tobacco plants. A long hollow protein coat surrounds a molecule of RNA. (Courtesy of Dr. E. Boatman) (b) Bacteriophage, a virus that infects bacteria. This is a complex virus structurally and four main parts are readily observable. The nucleic acid (DNA) is contained within the hexagonal head. The virus attaches to the bacterial cell with its tail fibers. (Courtesy of Dr. E. Boatman) (c) Vaccinia, the virus used to vaccinate against smallpox ($\times$ 16,000). (Courtesy of Dr. M. Jevitz Patterson)

SUMMARY

The members of the microbial world can logically be classified as members of the kingdom Protista. Members of this kingdom may have either eucaryotic or procaryotic cells. The feature which distinguishes this kingdom from the animal and plant kingdom is that all members of the protista lack extensive tissue formation.

TABLE 3-4
Major Features of Members of the Microbial World[a]

Microbe	Mode of Movement	Reproduction	Rigidity of Cell Wall	Nutrition	Resting Stages
PROCARYOTES					
Eubacteria	Flagella, many nonmotile	Binary fission	Rigid	Most heterotrophic; some autotrophic	Endospores, cysts (some members)
Prosthecate and Budding Bacteria	Flagella	Mostly budding; some binary fission	Rigid	Most heterotrophic; some autotrophic (photosynthetic)	None known
Mycelial Bacteria	Nonmotile	Conidiospores, sporangiospores, fragmentation; binary fission	Rigid walls	Heterotrophic	None known
Filamentous (Sheathed bacteria)	Flagella	Binary fission	Rigid	Heterotrophic	None known

TABLE 3-4 Continued
Major Features of Members of the Microbial World[a]

Microbe	Mode of Movement	Reproduction	Rigidity of Cell Wall	Nutrition	Resting Stages
PROCARYOTES					
Mycoplasma (PPLO)	Nonmotile	At least one species divides by binary fission; mode of others unknown	Cell wall absent	Heterotrophic	None known
Gliding Bacteria	Gliding (mechanism unknown)	Short filaments of cell break off; binary fission	Flexible	Heterotrophic	Cysts
Spirochetes	By axial filaments	Binary fission	Flexible	Heterotrophic	None known
Rickettsia	Nonmotile	Binary fission	Rigid	Most are obligate, intracellular parasites; heterotrophic	None known
Chlamydia	Nonmotile	Binary fission	Rigid	Obligate, intracellular parasites; heterotrophic	None known
Blue Greens	Gliding (mechanism unknown)	Fission, fragmentation	Flexible if present	Autotrophic (photosynthetic)	Akinetes (some members)
EUCARYOTES					
Algae	Flagella, pseudopodia, many nonmotile	Fragmentation, spores, fission	Flexible if present	Autotrophic (photosynthetic	Cysts
Fungi	Immobile	Asexual spores; male and female gametes fuse to form diploid zygote	Rigid tubes enclose the cytoplasm	Heterotrophic	Spores
Protozoa	Flagella, pseudopodia, gliding, cilia	Fission, sexual reproduction between different mating types	May or may not have a thickened pellicle	Heterotrophic	Cysts
ACELLULAR					
Viruses	Nonmotile	Nucleic acid and protein replicate independently of one another	Not applicable	Not capable of independent metabolism; depend on host cell metabolism	None known

[a] All terms are defined in the Glossary

Although many members are multicellular and very large, all of the cells in the organism are essentially the same and have not differentiated into tissues. The procaryotes include the blue greens and bacteria. The bacteria can be further broken down into a number of groups based upon unique features in morphology, method of reproduction, and degree of parasitism with other living cells. The eucaryotes include the algae, fungi, and protozoa. The viruses are considered to be microorganisms for convenience, but in actual fact they are not cells and therefore are not really organisms.

The major features of members of the microbial world are summarized in Table 3-4.

QUESTIONS

1. Discuss the characteristics of an organism which places it in the kingdom Protista.
2. How can you justify including macroscopic forms in the microbial world?
3. What properties of eubacteria set them apart from all other bacteria?
4. Why do you think that rickettsia and chlamydia were once considered to be intermediate forms between bacteria and viruses? On what bases have they been reclassified?
5. You are given a single celled organism to identify. How would you determine whether it is a bacterium or a blue-green?

Chapter 4

Functional Anatomy of Procaryotic and Eucaryotic Cells

Van Leeuwenhoek used only a single lens to view his "animalcules" and the magnification with this simple microscope was limited to approximately 300-fold. The details of cell structure cannot be observed at this magnification and therefore more detailed studies of the internal structure of cells had to await the development of a microscope which could magnify objects more than a thousand times. However, even at this magnification the finer structure of procaryotic cells cannot be elucidated, and it was not until the development of the electron microscope in the 1930s that definitive studies on procaryotic cell structure could be done. The analysis of cell structure through microscopic techniques has been accompanied by a chemical analysis on the individual cell components that can be isolated in pure form. The combined chemical and microscopic approach has provided considerable insight into the function of the various structures of the cell and their role in the life of the organism.

MICROSCOPIC TECHNIQUES—INSTRUMENTS

Compound Microscope

The most commonly used instrument for observing any cell is the *light microscope,* so-named because the object is illuminated by visible light. All other light microscopes are variations on this instrument. The microscope used by van Leeuwenhoek, consisting of only a single magnifying lens, was termed a simple microscope. The light microscopes most commonly used today have two types of lenses, objective and ocular, and are called *compound microscopes*. The magnification achieved in such a microscope is the product of the magnification of each of the individual lenses. Most compound microscopes have a number of objective lenses, thereby making a number of different magnifications possible with the same instrument. A modern compound microscope, with its major components labeled, is shown in Figure 4-1. A good compound microscope can magnify objects up to 2000 times. The usefulness of a microscope depends not so much on the degree of magnification, but rather on the ability of the microscope to clearly separate or *resolve* two objects which are very close together. Indeed the renowned English microscopist, Robert Hooke, was not as successful as van Leeuwenhoek in making out the details of microorganisms, even though Hooke invented and used a compound microscope with a greater magnification than van Leeuwenhoek's. However, the lenses which Hooke employed suffered from serious optical defects and he could see only blurred images.

The resolving power of a microscope is defined as the minimum distance that can exist between two points so that the points are observed as separate entities. The resolving power determines how much detail can actually be seen. It depends on the quality of the lens, its magnification, and the preparation of the specimen under observation. The maximum resolving power of the best light microscope functioning under optimal conditions is 0.2 μm because the resolv-

RESOLVING POWER IS
MOST IMPORTANT
FEATURE OF A
MICROSCOPE

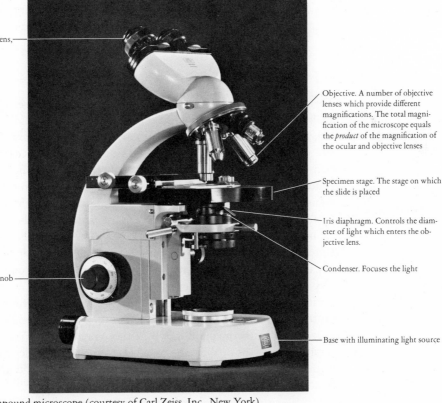

Eyepiece. A magnifying lens, usually about 10X

Objective. A number of objective lenses which provide different magnifications. The total magnification of the microscope equals the *product* of the magnification of the ocular and objective lenses

Specimen stage. The stage on which the slide is placed

Iris diaphragm. Controls the diameter of light which enters the objective lens.

Condenser. Focuses the light

Fine adjustment knob

Base with illuminating light source

FIGURE 4-1   Compound microscope (courtesy of Carl Zeiss, Inc., New York).

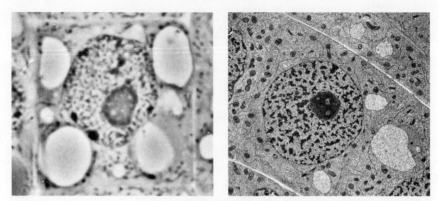

FIGURE 4-2   Comparison of the resolving power of a light microscope (left) and an electron microscope (right). The same preparation (onion root tip) was magnified 450 times. Note the difference in the degree of detail that can be seen at the same magnification. The magnification is the upper limit for the light microscope (phase contrast), but is a lower limit for the electron microscope. (Courtesy of Dr. W. A. Jensen)

ing power is limited by the wavelength of the light employed for illumination. Higher resolution can be achieved if a light other than visible light is employed. The shorter the wavelength, the greater the resolution. Thus microscopes have been developed which employ illumination which has a shorter wavelength than visible light, and this technique has improved their resolving power (Fig. 4-2). Two such instruments are the ultraviolet and the electron microscopes.

Phase Contrast Microscope

PHASE CONTRAST MICROSCOPY CAN REVEAL SOME STRUCTURES IN LIVING CELLS

This microscope is a variation of the compound microscope. Although bacteria are large enough to be seen with a microscope when suspended in a drop of liquid, they are very difficult to see because they are transparent, usually colorless, and so tiny that they may be even difficult to find. To overcome the problem of transparency in all types of cells, a special type of light microscope, the phase contrast microscope, has been devised. This microscope, probably the one most commonly used in research laboratories for observing living microbes, has special optical devices which increase the contrast between the bacteria and the surrounding medium. Advantage is taken of the fact that cells are denser than the surrounding medium and therefore illuminating light is slowed down more as it passes through the specimen as compared to the medium surrounding the specimen. The result is that even though the organisms are not magnified to any greater extent, they nevertheless stand out from the background and are clearly visible. Thus it is possible to observe living organisms clearly and study their movements in the medium in which they are growing. Figure 4-3a and b are photomicrographs of the same organism, one viewed under the ordinary light microscope and the other a phase contrast microscope. Note that even internal structures are discernable by phase contrast microscopy.

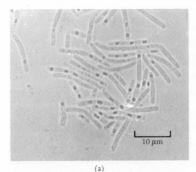

(a)

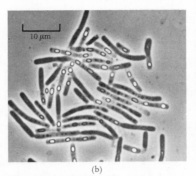

(b)

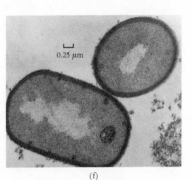

(c)

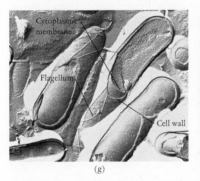

(d)

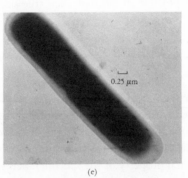

(e)

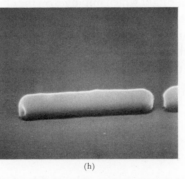

(f)

(g)

(h)

(i)

FIGURE 4-3 All of the photomicrographs shown here (except e, f, and g) are of *Bacillus megaterium.* They illustrate the details that are possible with each type of preparation and microscope. (a) Bright field (light) microscopy. The intracellular bodies are endospores. (Courtesy of Dr. J. T. Staley) (b) Phase contrast microscopy. (Courtesy of Dr. J. T. Staley) (c) Dark field microscopy (d) Fluorescence microscopy. The fluorescent material is bound to the cytoplasmic membrane. (Courtesy of Dr. B. A. Newton; Newton, B. A., *J. Gen. Microbiol.,* 12:226, 1955.) (e) Electron photomicrograph of *Bacillus fastidiosus.* (Courtesy of Dr. E. Leadbetter; Leadbetter, E. and S. Holt, *J. Gen. Microbiol.,* 52:299, 1968.) (f) Thin section of *B. fastidiosus;* electron photomicrograph. (Courtesy of Dr. E. Leadbetter; Leadbetter, E. and S. Holt, *J. Gen. Microbiol.,* 52:299, 1968.) (g) Freeze-etched preparation of *Bacillus subtilis.* (Courtesy of Dr. S. Holt) (h) Scanning electron photomicrograph. (Courtesy of Dr. N. Hodgkin) (i) Simple stained preparation, light microscopy. The light areas in some cells are endospores. (Courtesy of J. P. Dalmasso) (j) Negative stain. (Courtesy of J. P. Dalmasso)

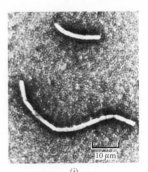

(j)

Another commonly employed method for achieving a marked contrast between living organisms and the background is darkfield microscopy. In this technique light is directed toward the viewer in such a way that the background is completely dark except for objects being viewed, which are brilliantly illuminated (Fig. 4-3c) Only light that is scattered by the specimen enters the ocular lens and is visualized. This technique makes objects and cells visible that are invisible by usual light microscopy. It is useful for observing very thin cells, in particular the organism causing syphilis, *Treponema pallidum,* an organism which is invisible using ordinary light microscopy. The advantage of observing living bacteria is that it is possible to estimate their true size, shape, and mobility undistorted by staining procedures.

Fluorescence Microscopy

Another type of microscope that has attained great importance in laboratories concerned with identifying microorganisms is the fluorescence microscope. This microscope is used to visualize objects which fluoresce, that is, emit light when a light of a different wavelength impinges on the object. The fluorescence may be a natural property of the specimen being viewed or it may result from a fluorescent compound being attached to a normally nonfluorescing material (Fig. 4-3d). The light which impinges on the object being viewed comes from a special lamp which emits light of only certain wavelengths. By using appropriate filters the light which finally reaches the object is of the proper wavelength so that it causes the fluorescent object to give off a light of a wavelength which often reaches the eye as a yellow-green glow. This microscope is commonly used to visualize fluorescent antibodies which attach to specific portions of specific cells. This technique has numerous applications in medical microbiology.

Electron Microscope

The light microscope is only capable of defining the gross anatomy of the bacterial cell—its size, shape, and a few of its largest components. The light microscope was developed almost to the zenith of its resolving power over a century ago. In order to increase the resolving power significantly a new type of microscope, the *electron microscope,* was constructed by Knoll and Hruska in Berlin in 1931. In this microscope a beam of electrons which has a wavelength only 1/10,000 that of visible light is focused by means of magnetic fields which function as lenses. Some of the electrons pass through the specimen, others are scattered, and still others cannot pass. The electrons impinge on an electron-sensitive screen, creating an image which is determined by the ability of the electrons to pass through various parts of the object being viewed. Since the resolving power of a microscope is proportional to the wavelength of radiation used, a resolving power of 0.0003 μm (3 Å) can be achieved (about 600 times better than the light microscope) (Fig. 4-3e). In order to increase the details of cell structure investigators often slice the specimen to be viewed into very thin sections, commonly with a diamond knife, and observe the "thin sections" (approximately 0.02 μm in thickness) (Fig. 4-3f). In the past specimens had to be fixed so that they were "bone dry" and extremely thin before they could be observed with the electron microscope. The fixation process, however, may introduce a large number of artifacts into the preparation; thus one of the main jobs of the electron microscopist is to interpret whether what he sees is actually present in the living cell or merely an artifact that results from vigorous treatment in the fixation process.

Recently, a technique (freeze-etching) has been devised which circumvents the need for chemical fixation and thereby presumably overcomes the problem of artifacts. The material is frozen and thin sections chipped off. The surface of the section is then coated with a thin layer of carbon, which is thin enough to be viewed with the electron microscope. The pictures of freeze-etched preparations can be quite dramatic (Fig. 4-3g).

A significant advance in microscopic techniques has been made recently with the development of a scanning electron microscope. The material to be observed is coated with a thin film of metal, and the electron beams scan back and forth over the surface. The most significant aspect is that relatively large specimens can be viewed and a three-dimensional figure is observed (Fig. 4-3h).

MICROSCOPIC TECHNIQUES—STAINING

To overcome the difficulty of observing living, transparent, often motile organisms, cells are frequently killed and then treated with one or more dyes which have a special affinity for one or more cellular components. Such treatment results in the entire organism, or part of it, achieving a marked contrast to the unstained background (Fig. 4-3i). In the staining procedure a drop of liquid containing the organisms is placed on a glass slide and allowed to dry. The dried film is then "fixed" onto the slide, either with a chemical fixative or more commonly by passing it over a flame. Fixation coagulates the cell protein. Stains are then applied to the "fixed" organisms, preferentially coloring the organisms. A wide variety of stains and staining procedures are currently in use, each one having its own particular use. Some dyes will only stain a particular cell component. Other staining procedures will stain one group but not another group of organisms, thereby serving to divide organisms into groups based on their staining characteristics. Stains can be divided into two major groups based on their affinity for cell components. *Positive stains* have a strong affinity for one or more cell components and color these components when added to fixed cell preparations. *Negative stains* cannot penetrate the cell envelope and thus make the cell highly visible by providing a contrasting dark background. Negative stains are generally employed on living cells to demonstrate surface structures which are not stained well by positive stains (Fig. 4-3j).

In this technique only a single dye is applied to the cells. This technique may be used to stain the entire cell or specific structures in the cell. Methylene blue, a basic dye, is frequently applied to a fixed suspension of bacteria in order to stain entire cells a blue color without staining the background material. This basic dye binds primarily to the RNA and DNA of the cell, and little evidence of internal structures is revealed. Acidic dyes, which include safranin, acid fuchsin, and congo red, stain basic compounds in the cell, primarily proteins carrying a basic charge. Sudan black is a dye that is very soluble in fat and is frequently used to identify the presence and location of fat droplets in bacteria.

Differential Staining Techniques

It is possible to separate bacteria into groups, depending on their ability to take up and retain certain dyes. The most widely used staining procedure for dividing virtually all bacteria into one of two groups is the *gram stain.* A Danish physician, Dr. Christian Gram, working in a morgue in Berlin, developed this staining method in 1884 in order to distinguish bacteria from mammalian cell nuclei in infected tissues. Gram was not satisfied with his method because not all bacteria retained the stain. What he considered to be the major defect in his staining method actually forms the basis of the most widely used characteristic test for bacteria. Who first had the idea of using the Gram stain as a diagnostic tool is unknown. The procedure involves the application of a basic purple dye, usually gentian or crystal violet (Fig. 4-4). All bacteria able to take up the dye are stained purple. A dilute solution of iodine is then added, which serves to decrease the solubility of the purple stain within the cell. An organic solvent, such as ethanol, is next added which readily removes the purple dye from some species of bacteria but not others. A red stain is then applied. Those bacteria which were decolorized by the ethanol appear red or gram negative. Those that retained the purple dye still appear purple (gram positive). The reason why some bacteria retain the basic purple dye and others do not is related to the chemical structure of the cell wall structure. The structure of the cell wall in gram-positive organisms is markedly different from that of gram-negative organisms. Since the structure of the cell wall correlates with many other distinctive properties of the cell, the gram reaction (that is, whether an organism is gram positive or gram negative) is correlated with other properties of the cell. The technique is generally applicable only to the eubacteria.

ACID FAST STAIN

Another differential staining procedure, the *acid fast stain,* is used to characterize a small group of organisms which are resistant to decolorization with an acidic solution of alcohol after being stained with a basic dye. Only a few groups of bacteria retain the dye, that is, are acid fast, but since one group causes tuberculosis, the acid fast stain has proven extremely valuable in diagnostic laboratories concerned with detecting this organism.

Differential stains are also employed to stain specific cell structures such as the cell wall, flagella, endospores, and nuclear bodies. The staining procedure for each structure is different and takes advantage of the chemical composition and properties of the structure.

Since staining generally requires that the cells be killed, their morphology can be easily distorted, producing artifacts. The fixation process tends to reduce

STEPS IN STAINING   APPEARANCE OF CELLS

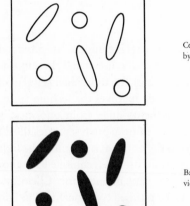

Cells fixed
by heating

Shape of cells
becomes distorted
and cells shrink
in size

Basic dye, crystal
violet applied

All cells
stain purple

Addition of
iodine

All cells
remain purple

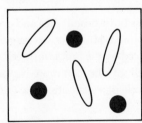

Addition of
alcohol

Gram positive cells
remain purple;
gram negative cells
become colorless

Addition of
counterstain,
often safranin
(red dye)

Gram positive cells
remain purple;
gram negative cells
appear red

FIGURE 4-4   The steps in the gram stain.

the size of the cells, whereas the addition of dyes tends to increase their size. Indeed, flagella (organelles of locomotion) are so thin that they can be observed with the light microscope only if their size is increased by a specific staining procedure.

BIOCHEMICAL ANALYSIS OF CELLS AND CELL STRUCTURES

The biochemical analysis of whole cells by chemical techniques has provided information on the exact chemical composition of the intact cell. In addition, it is now possible to separate physically (fractionate) the individual cellular structures from one another. The separation generally involves the technique of centrifugation at very high speeds. Most components differ from one another in density, and centrifugation for different lengths of time often separates the different components. The heaviest particles go to the bottom of the tube, the lightest remain near the top. A great number of variations on this technique have allowed each of the individual structures to be isolated, many in pure form. Chemical and physical analyses can then be performed on the purified components. The combination of gross microscopic observation of living cells with the phase contrast microscope, analysis of the intimate details of cell structure by electron microscopy, correlated with the chemical analysis of the individual cellular components has provided fundamental information on each of the component structures of both procaryotic and eucaryotic cells.

STRUCTURES OF PROCARYOTIC AND EUCARYOTIC CELLS

Since members of the microbial world encompass both procaryotes and eucaryotes, the major features of both cell types are explored. However, because of the special concern with bacteria here, emphasis is placed on the procaryotic cell type. An electron micrograph and diagram of a cell of the common bacterium *B. Megaterium* are given in Figure 4-5, and a cell of eucaryotic alga *Chlamydomonas* is diagrammed in Figure 4-6.

Comparison of Procaryotic
and Eucaryotic Cells
Page 15, Table 1-3

It should be apparent from these figures, as well as from the discussion in Chapter 1, that the eucaryotic cell is structurally far more complex than the procaryotic cell. However, because both cell types must perform essentially the same functions in order to live, apparently the simpler structures of the procaryotic cell can obviously carry out many of the same functions as are carried out by the complex structures in the eucaryotic cell.

The discussion in the remainder of this chapter will focus on the functional anatomy of procaryotic and eucaryotic cells—the functions that all cells must perform and the structures in the two cell types which perform these functions.

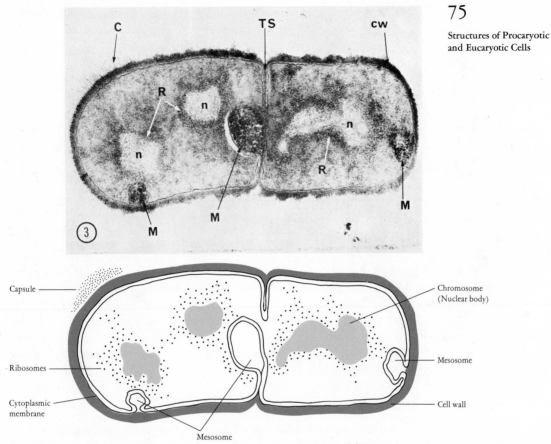

FIGURE 4-5 *Bacillus megaterium,* a representative procaryote, undergoing cell division.

The functional architecture of the cell must provide the means:

1. To enclose the internal contents of the cell and separate it from the external medium.
2. To store and replicate genetic information.
3. To synthesize cellular components.
4. To generate and store energy.
5. To allow for cell movement (optional).

ENCLOSURE OF CYTOPLASM

Most bacteria have two structures which surround the cytoplasm, the *cell wall* and the *cytoplasmic membrane,* and some may have a third, the *capsule.* Many eucaryotic cells have only a cytoplasmic membrane surrounding the cytoplasm. The discussion will proceed from the outermost (capsule) to the innermost layer (cytoplasmic membrane).

CELL ENVELOPES—
CAPSULE, WALL AND
CYTOPLASMIC
MEMBRANE

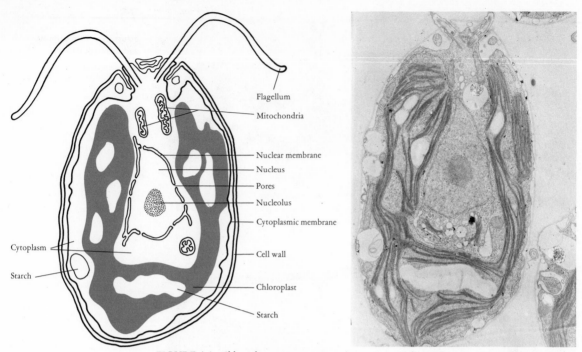

Flagellum

Mitochondria

Nuclear membrane

Nucleus

Pores

Nucleolus

Cytoplasmic membrane

Cytoplasm

Starch

Cell wall

Chloroplast

Starch

FIGURE 4-6 *Chlamydomonas,* a representative eucaryotic alga. (Photo courtesy of Dr. U. Goodenough; Johnson, U. and K. Porter, *J. Cell Biol.,* **38**:403, 1968.)

Capsule or Slime Layer

Colony
Page 101

This structure is a loose fitting, gelatinous structure which surrounds some bacteria. It is most readily demonstrated by negative staining. The cell is outlined as a light area against a darkened background (Fig. 4-7a). Colonies of bacteria which synthesize capsules are often relatively moist, glistening, and slimy in appearance when growing on solid medium (Fig. 4-7b). Capsules are produced by only certain species of bacteria, and often only when they are grown under certain nutritional conditions.

Amino Acids—Configuration
Page 33

Capsules vary in their chemical composition. Some are composed of polysaccharide, others of polypeptides consisting of only one or two amino acids. These amino acids are generally of the unnatural or D-configuration as opposed to the natural or L-configuration always found in proteins.

The advantages conferred by the presence of a capsule are not evident for most bacteria, although it definitely is protective under certain situations for some species. Probably the most well-studied capsule belongs to the organism causing bacterial pneumonia, *Streptococcus pneumoniae*. Only if this organism is encapsulated is it capable of causing disease. If the capsule is not produced, the organism is quickly destroyed by the defenses of the infected animal. Apparently the cells of the body which have the ability to engulf and destroy bacteria (phagocytic cells) cannot engulf the encapsulated organism, whereas the nonencapsulated organisms are quickly engulfed and destroyed.

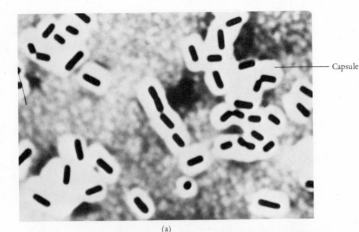

Capsule

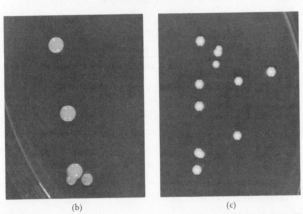

(a)

(b)   (c)

FIGURE 4-7 (a) Capsules as demonstrated by negative staining with India ink. The bacterial cytoplasm has been stained with a positive stain to increase the contrast (courtesy of Dr. V. Chambers). (Below) Colonial morphology of *Streptococcus salivarius* growing on medium with sucrose (b) and without sucrose (c). This organism forms a capsule if sucrose is present in the medium and this capsule imparts a larger, gum drop appearance to the colonies. (b and c courtesy of J. P. Dalmasso)

Bacterial Cell Wall

The bacterial cell wall deserves special attention for these reasons: (1) it is composed of subunits found no other place in nature; (2) the cell wall of certain organisms can produce symptoms of disease; and (3) the site of action of some of the most effective antibiotic medications is the cell wall.

The cell wall determines the shape of the organism; cylindrical-shaped cells have a cylindrical cell wall, spherical organisms have a spherical cell wall. If the cell is broken in two, the cell wall does not collapse but maintains its rigid shape (Fig. 4-8).

The chemical structure of the wall is responsible for its rigid nature. The backbone of the cell wall is the *mucocomplex,* consisting of two major subunits, N-acetyl muramic acid and N-acetyl glucosamine and a number of amino acids.

UNIQUE CHEMICAL STRUCTURE OF THE WALL

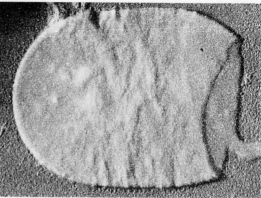

FIGURE 4-8
Rigidity of cell wall (courtesy of Dr. P. Gerhardt; Silvernale, J. H., H. L. Joswick, T. R. Corner, and P. Gerhardt, *J. Bact.*, **108**:482, 1971).

Formula Glucose
Page 45

ALTERNATING SUBUNITS
MAKE UP MUCOCOMPLEX

AMINO ACIDS CONNECT
LINEAR POLYMER

The N-acetyl muramic acid and N-acetyl glucosamine are chemically related to glucose (Fig. 4-9). These two subunits alternate to form a high molecular weight polymer. Although this polymer is found only in procaryotic cell walls, it bears a chemical relationship to the cellulose of plant cell walls and the chitin found in insects, crustaceans, and fungi. A chain of several amino acids is attached to each of the acetyl muramic acid molecules. The structural strength of the wall results from the connection of adjacent layers to each other by a bridge composed of amino acids. One end of the bridge is anchored in one layer of the polysaccharide chain; the other end of the bridge is covalently bonded to the adjacent polymeric layer. Only a few of the usual twenty amino acids found in proteins occur in the mucocomplex. One unusual amino acid is diaminopimelic acid which is related to the amino acid lysine (Fig. 4-10). It is found in no other place in nature other than bacteria and the blue-greens. Many of the amino acids in the cell wall are of the D-configuration.

The cell wall of gram-positive bacteria consists largely of layer upon layer of mucocomplex. The three-dimensional network, which is actually one large interconnected macromolecule, provides an unusually rigid, strong structure (Fig. 4-11b, 4-12b, and 4-13b). The wall of gram-negative organisms also contains a mucocomplex layer, but there are not as many layers. Consequently the wall of gram-negative organisms is more easily broken than is the wall in gram-positive cells. In addition to the mucocomplex, the wall of gram-negative orga-

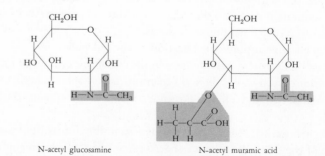

N-acetyl glucosamine   N-acetyl muramic acid

FIGURE 4-9   Chemical structure of N-acetyl glucosamine and N-acetyl muramic acid.

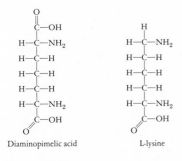

Diaminopimelic acid   L-lysine

FIGURE 4-10   Chemical structure of an unusual amino acid, diaminopimelic acid, and its common relative lysine.

nisms contains two additional layers: a lipopolysaccharide layer surrounds the mucocomplex and a layer of lipoprotein which represents the outermost layer of the cell wall (Fig. 4-11a, 4-12a, and 4-13a).

The lipopolysaccharide layer consists of two portions, a polysaccharide layer covalently bonded to the lipid fraction. The polysaccharide layer of different strains of the same species contain different sugars, some so unique that they are found in no other place in nature. From the large number of different sugars present, it is not surprising that over 1000 strains of *Salmonella* can be distinguished based on differences in the composition of the lipopolysaccharide layer.

The lipopolysaccharide layer is of great interest because of its toxic properties. Many gram-negative bacteria produce materials which are toxic for the host, thereby causing disease. The injection of the purified lipopolysaccharide isolated from some pathogenic gram-negative bacteria causes symptoms characteristic of the infection by the whole organism.

Lipopolysaccharide toxin
Page 365

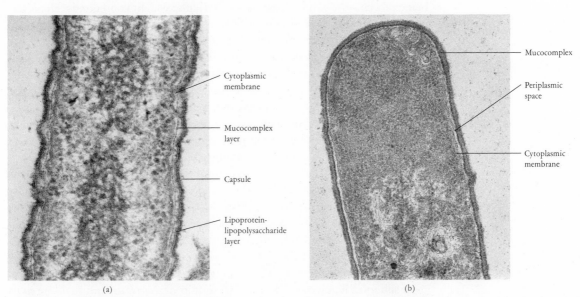

FIGURE 4-11   Electron photomicrographs of (a) gram-negative cell wall of the organism *Microscilla* (courtesy of Dr. V. Chambers) and (b) gram-positive cell wall of the organism *Bacillus subtilis* (courtesy of Dr. S. Holt). There are additional components to the wall other than the mucocomplex in gram positive organisms.

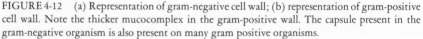

(a)   (b)

FIGURE 4-12   (a) Representation of gram-negative cell wall; (b) representation of gram-positive cell wall. Note the thicker mucocomplex in the gram-positive wall. The capsule present in the gram-negative organism is also present on many gram positive organisms.

The lipoprotein layer which forms the outermost portion of the cell wall in gram-negative bacteria contains the twenty amino acids commonly found in proteins, in contrast to the limited number of amino acids found in the cell wall of gram-positive organisms.

FUNCTION OF CELL WALL—TO PREVENT BURSTING OF CELL

The cell wall functions like a rigid girder framework on a dirigible to hold the cell together. Without the cell wall the cell would burst. Bacteria normally grow in dilute solutions, both in nature and in the laboratory. However, the bacterial cytoplasm is a very concentrated solution of inorganic salts, sugars, amino acids, and various other small molecules. There is a tendency for the *concentration* (the number of molecules per unit volume) of *each* of the low molecular weight compounds to equalize on the inside and outside of the cell. Theoretically this equalization can occur by (1) the movement of compounds out of the cell into the medium or (2) the movement of water from the medium into the cell. The first mechanism cannot operate freely because the cytoplasmic membrane prevents the exit of many small molecular weight compounds. However, it does allow water to penetrate freely, thus acting as a semipermeable membrane, allowing some but not other kinds of molecules to pass. Therefore water flows from the medium into the cell in an attempt to equalize the concentration of small molecules. The inflow of water exerts tremendous pressure, *osmotic pressure,* on all structures which enclose the cytoplasm. As a general rule, unless the cell has a rigid cell wall which cannot expand, it simply balloons in size until it bursts. The osmotic pressure in a gram-positive organism may be as much as 25 times the pressure of our atmosphere at sea level (1 atmosphere). In a gram-negative organism, the pressure may reach 5 atmospheres.

CELLS CAN EXIST WITHOUT A WALL

The cell wall becomes less important if the bacterium is in an environment which has a high concentration of low molecular weight compounds. In such an environment the tendency of the water to enter the cell is matched by its tendency to leave the cell. The result is that there is no net movement into the cell. Under these conditions the cell can exist without a cell wall. It is possible to destroy the cell wall by treating the cell with lysozyme, an enzyme which cleaves the chemical bonds between the N-acetyl muramic acid and N-acetyl glucosamine subunits. Figure 4-14 illustrates the results of this treatment. In dilute solutions water enters until the cell bursts. In high salt or sugar solutions the cell without its wall becomes spherical, its most stable shape. The wall-less cell, a *protoplast,* can metabolize but cannot multiply if it has lost all of its cell wall.

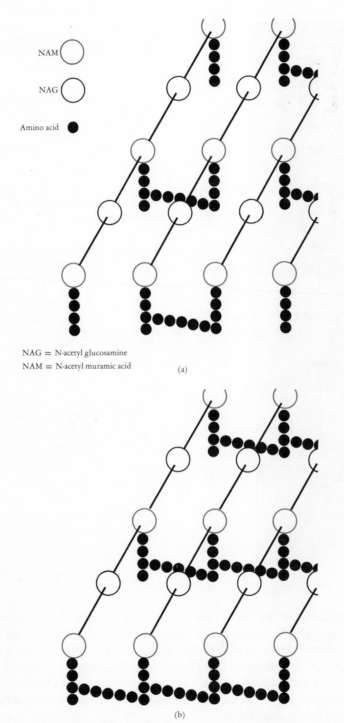

NAG = N-acetyl glucosamine
NAM = N-acetyl muramic acid

(a)

(b)

FIGURE 4-13   (a) Representation of molecular structure of mucocomplex portion of gram-negative cell wall. Note that the amino acid cross linking is incomplete. (b) Representation of molecular structure of mucocomplex portion of gram-positive cell wall.

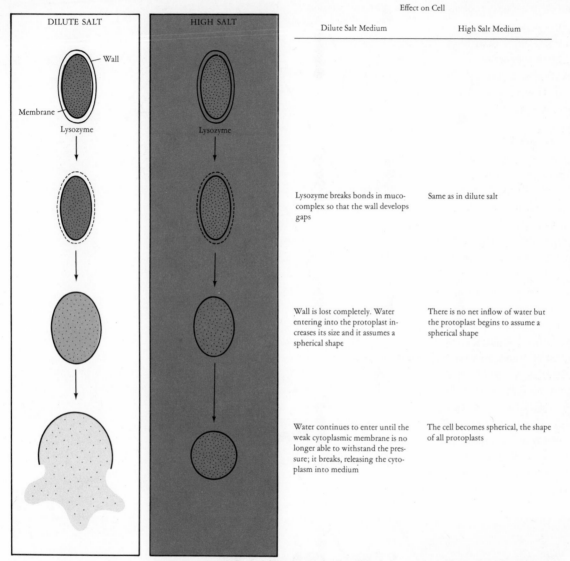

FIGURE 4-14 Effect of disruption of cell wall by lysozyme in a dilute salt and a high salt containing medium.

Gram-positive cells generally lose all of their cell wall by lysozyme treatment. Gram-negative cells require special treatment in order for the lysozyme to reach its site of action. Gram-negative cells which contain layers other than the muco-complex still have fragments of wall remaining, although not enough to prevent the cell from becoming spherical. These wall-deficient organisms are capable of limited multiplication. However, if the high salt solution is diluted, water enters the cell and it generally bursts.

Differences in cell wall composition between gram-positive and gram-negative bacteria probably account for their different staining characteristics. The gram-positive organism retains the purple dye after the cell is treated by iodine and washed with alcohol, whereas the alcohol washes out the dye-iodine complex in gram-negative cells. Apparently the gram-positive cell wall is relatively impermeable to the solvent (alcohol), whereas the gram-negative is not. This explanation may account for the observation that gram-positive cells frequently become gram-negative in old cultures, since it is known that enzymes which damage the cell wall are present in cells in old cultures. This damage alters its permeability to alcohol so that the dye-iodine complex is dissolved and then washed out. Further, if the solvent is applied for too long a time, the gram-positive organisms will appear gram-negative.

Penicillin kills bacteria by affecting the synthesis of the cell wall probably at several different stages (Fig. 4-15). This antibiotic interferes with one of the final enzymatic reactions in the synthesis of the mucocomplex, the joining of adjacent layers through the amino acids. As a result the three-dimensional network is not intact, the wall is greatly weakened, and the cells lyse. In a high salt medium, spherical protoplasts will form. Although penicillin was discovered in the 1920s, it was not until 30 years later that its mode of action began to become understood. With some clarification of penicillin's mechanism of action, a large number of hitherto unexplained experimental observations on the activity of penicillin suddenly fell into place: (1) why only growing cells are affected, (2) why the organisms lyse, (3) why gram-positive organisms are generally more susceptible than gram-negative, and (4) why mammalian cells are not affected. Only growing cells synthesizing cell wall are affected since penicillin results in the *synthesis* of a weakened cell wall. In many gram-negative organisms, penicillin has a more difficult time reaching its site of action because of the presence of additional wall layers covering the mucocomplex layer. This helps explain why some gram-negative cells are relatively resistant. All eucaryotic cells are completely resistant to penicillin because only procaryotic cells have mucocomplex in their cell wall.

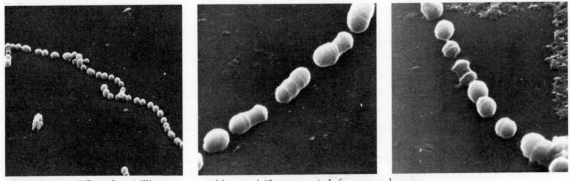

FIGURE 4-15   Effect of penicillin on gram-positive cocci (*Streptococcus*). Left, untreated; center and right, scanning electron photomicrographs showing appearance after two hours of treatment with 100 ug/ml of penicillin. (Courtesy of Dr. D. Greenwood; Greenwood, D., and F. O'Grady, Science **163**:1076, 1969.)

Tables 3-1 and 3-4
Pages 57 and 64

WALL-DEFICIENT FORMS
IN NATURE

WALL-DEFICIENT
ORGANISMS HAVE
UNIQUE PROPERTIES

Although the discussion thus far has emphasized the importance of the cell wall to bacteria, microorganisms which lack or are deficient in cell walls do occur in nature, the most notable being members of the genus *Mycoplasma*. These organisms grow very slowly and require a specially enriched medium, generally containing 20 percent serum. Some of the mycoplasma also have a requirement for sterols which are incorporated into the cytoplasmic membrane. The lack of the cell wall makes them extremely plastic. Many, however, are capable of existing in dilute media. It appears that the sterols in the cytoplasmic membrane confer enough strength on this structure to prevent lysis.

Another group of wall-deficient organisms, *L-forms* (named after the Lister Institute in London where they were first described in 1935), superficially resemble the mycoplasma. These organisms are simply bacteria, either gram-positive or gram-negative, which have lost part or all of their normal cell wall. This loss can occur in the body of a host or with some forms spontaneously outside the host. Wall-deficient forms can be artificially produced if cells are treated with penicillin or lysozyme. L-forms differ from one another in the remnants of cell wall that may remain, as well as in their ability to synthesize cell wall material. In all organisms, it is the mucocomplex layer which is most seriously affected. In some, this layer is completely eliminated, in others a small remnant remains. Some L-forms derived from gram-negative organisms, cannot synthesize any mucocomplex but can synthesize the lipopolysaccharide and lipoprotein layers. Others can synthesize the mucocomplex, but it does not have the regular structure shown in Figure 4-13a and b. If *all* of the wall material has been eliminated, then new wall cannot be synthesized. These latter wall-deficient forms can multiply in medium containing a high concentration of salt or sucrose. If the penicillin or lysozyme is removed, L-forms will either revert to the cell shape from which they were derived (revertible L-forms) or they may continue to grow as wall-less forms and never revert (stable L-forms). Some L-forms require a very concentrated medium to prevent the cell from bursting. Others, however, are able to grow in dilute medium perhaps because they can synthesize enough of their cell wall to retain their cytoplasmic contents. Although stable L-forms are virtually grossly indistinguishable from the mycoplasma, no one has yet succeeded in showing that any species of *Mycoplasma* is closely related to a species of eubacteria. From the study of ordinary eubacteria, L-forms, and *Mycoplasma* it is clear that there is a gradation in the ability of organisms to survive in ordinary dilute nutrient medium without a normal cell wall. Most bacteria will lyse. Some (mycoplasma) will be able to multiply very slowly if the medium is fortified with additional materials such as blood serum and perhaps sterols. Some L-forms are capable of multiplying in dilute medium. All wall-deficient organisms have several properties in common. They have no rigid cell shape, are extremely fragile, multiply relatively slowly, and are resistant to antibiotics which affect cell wall synthesis. Mycoplasma cause a number of diseases in domestic animals and some investigators feel that L-forms play a more important role in human disease than they are given credit.

Cell Walls of Eucaryotic Cells

The chemical structure of the cell walls of eucaryotes varies, but, in general, they are far simpler than bacterial cell walls. Most algae and fungi have rigid cell walls. In contrast, protozoa do not have a cell wall even though they do have a definite, specific shape. What confers this structural rigidity is not clear. Most algae and many fungi contain cellulose. Other fungi have walls also composed of repeating subunits of glucose, but since they are bonded together differently from that in cellulose, they are given another name, glucan. Fungal walls also often contain chitin, a polymer consisting of repeating units of N-acetyl glucosamine.

The cells of many higher eucaryotes, such as those of mammals, do not have a rigid cell wall. How can such cells exist in dilute solutions? The answer is not completely known, but apparently the cytoplasmic membrane is much more stable in eucaryotes lacking a cell wall than they are in procaryotes. Their membranes are similar in composition to members of *Mycoplasma* that also can survive in a dilute medium. Sterols, which are not found in the membranes of procaryotes except for the mycoplasma, probably also contribute to the resistance of mycoplasma and cell-wall free eucaryotes to osmotic lysis.

Cytoplasmic Membrane of Procaryotic Cells

This membrane forms the outer membrane in protoplasts. In normal procaryotes the osmotic pressure on the inside of the cell forces this thin, delicate, elastic membrane against the cell wall in dilute medium, although there is generally a space between these two layers, the *periplasmic space*. However, if the cells are placed in a medium of high salt concentration, the membrane will *shrink away* from the cell wall as water leaves the cell, and the space between the membrane and wall will increase dramatically (Fig. 4-16).

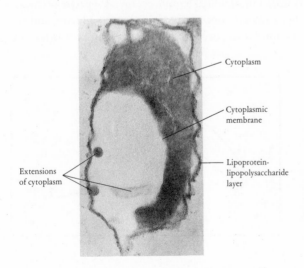

Cytoplasm

Cytoplasmic membrane

Lipoprotein-lipopolysaccharide layer

Extensions of cytoplasm

FIGURE 4-16
Escherichia coli plasmolyzed in 0.5 M sucrose. The high concentration of sucrose results in water leaving the cell and the membrane shrinking away from the rigid cell wall. The membrane seems to adhere firmly to certain areas of the wall. ($\times 67,500$) (Courtesy of Dr. D. Birdsell; Birdsell, D. C. and E. H. Cota-Robles, *J. Bact.,* **93**:427, 1967.)

Membranes isolated from procaryotic cells contain approximately 60 percent protein and 40 percent lipid, much of which are phospholipids.

The appearance of all membranes whether isolated from procaryotic or eucaryotic cells is virtually identical when viewed in the electron microscope—two dark bands separated by a light band—the total width being about 0.0075 μm (75 Å) (Fig. 4-11a & b). This similarity has resulted in the term *unit membrane* being applied to all membranes, implying that all membranes are identical in their molecular structure. However, the fact that membranes play so many different physiological roles suggests that although there may be a basic plan, differences must exist in membranes to allow them to perform different functions.

The cytoplasmic membrane of bacteria performs several functions vital to the life of the cell.

A large number of enzymes concerned with the degradation of foodstuffs and the production of energy are associated with the membrane.

FUNCTIONS OF
CYTOPLASMIC
MEMBRANE—TRANSPORT

The membrane is semipermeable. Generally, only low molecular weight materials (no greater than several hundred in molecular weight) can penetrate to the inside of the cell. The compounds enter and exit by one of two distinct processes—*passive diffusion* or *active transport*. In passive diffusion the molecules flow freely into and out of the cell without the cell expending energy. Diffusion will occur until the concentration of the particular molecule is the same inside as outside the cell (note that the concentration of *each* specific molecule is the same, not the sum total of *all* molecules). In active transport the cell expends energy to transport molecules into and out of the cell. In general, the cell transports more molecules into the cell than it transports out, the net result being an accumulation of molecules inside the cell (Fig. 4-17). The concentration of a particular nutrient may be more than a thousand times greater inside the cell than it is in the environment. The transport system, called a *permease,* is located in the membrane. A permease probably consists of a number of enzymes which catalyze a sequential series of reactions, some of which require energy. A separate permease exists for each nutrient. For example, one permease transports glucose and another lactose. Although the molecular mechanism of active transport is not un-

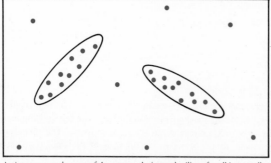

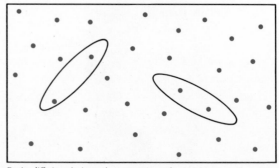

Active transport: because of the process, the internal milieu of a cell is generally much different than the external milieu of the cell's surroundings

Passive diffusion: the internal concentration and external concentration are identical

FIGURE 4-17   Active transport and passive diffusion.

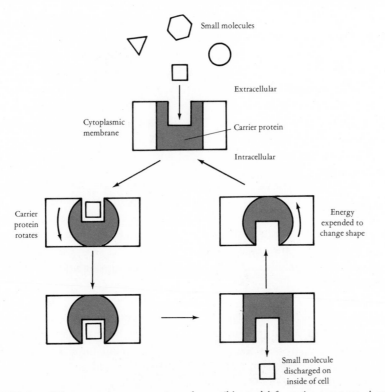

Small molecules

Extracellular

Cytoplasmic membrane

Carrier protein

Intracellular

Carrier protein rotates

Energy expended to change shape

Small molecule discharged on inside of cell

FIGURE 4-18 Diagrammatic representation of a possible model for active transport. A small molecule associates with the carrier protein, a permease, and changes its shape so that it can rotate, bringing the small molecule to the inside of the cell, where it is discharged. The discharge of the small molecule results in the carrier protein changing shape and in order to rotate, it must again change its shape. This requires the expenditure of energy. Different permeases exist for each of the differently shaped molecules.

derstood, one model that is consistent with experimental data is diagrammed in Figure 4-18.

The ability to concentrate nutrients allows the bacterial cell to maintain a relatively constant intracellular environment, even though the external environment changes drastically. Since bacteria live in environments in which many of the nutrients are in extremely low concentrations, free diffusion would not be sufficient to provide the concentration of nutrients inside the cell required for rapid growth. Thus bacteria have evolved a mechanism by which they can actively transport and accumulate the majority of nutrients.

CONCENTRATION OF NUTRIENTS INTRACELLULARLY

Bacteria also have the means for utilizing the high molecular weight compounds which cannot penetrate the cell wall or cytoplasmic membrane. The subunits of these polymers are potential nutrients. Most bacteria contain enzymes which degrade the large molecules into their subunits, which can then be transported into the cells, generally through the action of specific permeases. Thus proteins are broken down to amino acids by proteolytic enzymes or proteases. Other enzymes break down polysaccharides into their sugar subunits. Many en-

ENZYMES DEGRADE POLYMERS TO UTILIZABLE SUBUNITS

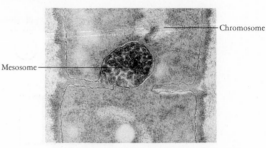

Chromosome

Mesosome

FIGURE 4-19

Mesosome, *Bacillus megaterium.* ($\times$ 36,400). An association between the DNA (chromosome) and the mesosome can be seen. (Courtesy of Dr. D. Lundgren; Ellar, D. J., D. G. Lundgren, and R. A. Slepecky, *J. Bact.,* 94:1189, 1967.)

zymes concerned with breakdown of large molecules are excreted by the cell into the medium, where they act. Since such enzymes are found outside the cell, they are called *extracellular enzymes* or *exoenzymes.* In the case of some bacteria, many degradative enzymes are located in the *periplasmic space.* Thus any large molecules which pass through the cell wall will come in contact with these enzymes, be broken down into smaller molecules, and transported through the membrane via the permeases.

In gram-positive bacteria the cytoplasmic membrane may form folds in the cytoplasm, termed *mesosomes* (Fig. 4-19). These membranous structures are not easily detected in some gram-negative bacteria for reasons that are not clear.

Mesosomes greatly increase the surface area of the membrane and therefore may increase the ability of cells to concentrate nutrients. In addition they seem to be located near the center of the cell, at the site at which the cross wall forms when the cell divides. Since the DNA of the nuclear body appears to be attached to the mesosomes, the mesosome may play a role in cross-wall formation and in the partitioning of the DNA of the parent into each of the two daughter cells.

The antibiotic *polymyxin,* produced by *Bacillus polymyxa,* combines with phospholipids in the membrane and disrupts its semipermeable character. As a result the cell contents leak out and the cell dies. Some other antibiotics function by this same mechanism.

Cytoplasmic Membrane in Eucaryotic Cells

The cytoplasmic membrane or plasma membrane of eucaryotic cells is similar in gross chemical composition and structure to the cytoplasmic membrane in procaryotic cells. The most striking difference is that the membrane is much more extensively arrayed in eucaryotic cells (Fig. 4-20). This series of internal membranes is termed the *endoplasmic reticulum.* These internal membranes, which are continuous with the cytoplasmic membrane, enclose a wide variety of structures found in eucaryotic but not in procaryotic cells. In addition the ribosomes are attached to the endoplasmic reticulum in eucaryotic cells, forming the rough endoplasmic reticulum (Fig. 4-20).

The cytoplasmic membrane in eucaryotic cells is also semipermeable. The fact that the external cytoplasmic membrane, which is the outer covering in

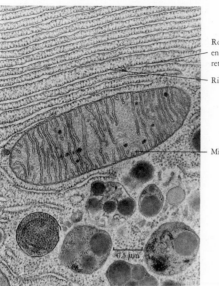

Rough endoplasmic reticulum

Ribosomes

Mitochondrion

FIGURE 4-20
Rough endoplasmic reticulum and mitochondrion in bat pancreas. The ribosomes on the membrane contribute to the rough appearance. (Courtesy of Dr. K. Porter)

many eucaryotic cells, is continuous with the internal membranes suggests that a series of channels exists between the surface of the cell and the interior cell structures.

A group of antibiotics, termed the *polyene antibiotics* because they contain a large number of double bonds, attaches to sterols in cell membranes and interferes with their semipermeable character. Since bacteria do not have sterols in their membranes, these antibiotics act primarily against eucaryotic cells such as fungi. They react almost as readily with mammalian cells, and thus are highly toxic.

SOME ANTIBIOTICS DISRUPT EUCARYOTIC CYTOPLASMIC MEMBRANE

STORAGE OF GENETIC INFORMATION

Procaryotic Cells

The major structure in which the genetic information of procaryotic cells is stored is the chromosome or nuclear body (Fig. 4-21). This structure is not sur-

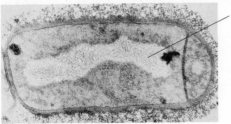

Chromosome

FIGURE 4-21
Chromosome of *Bacillus* megaterium. The granular appearance of the DNA is readily observable. ($\times$40,000) (Courtesy of Dr. D. Lundgren; Ellar, D. J., D. G. Lundgren, and R. A. Slepecky, *J. Bact.,* **94**:1189, 1967.)

FIGURE 4-22
The single chromosome of *Micrococcus lysodeikticus* as it appears in
the electron microscope after the cell has been gently lysed. Inside
the cell DNA may not have any free ends. (×12,000) (Courtesy
of Dr. A. K. Kleinschmidt)

rounded by a nuclear membrane. Each chromosome is composed of a single long
molecule of DNA, which when extended to its full length is about 1 mm long,
approximately 1000 times longer than the entire bacterium. The length of the
DNA can be most dramatically demonstrated by observing a cell which has
been gently lysed (Fig. 4-22). This chromosome is tightly packed into a volume about 10 percent the volume of the cell. The DNA molecule exists as a
closed circle in all bacteria in which it has been carefully studied.

Some genetic information in some bacteria is carried in the form of *extrachromosomal* DNA, which is not a part of the chromosome. These fragments
of DNA, only about 1 percent the size of the chromosome, often play an important role in chromosome transfer between bacteria. The genetic information that
determines whether a bacterial cell is resistant or sensitive to certain antibiotics
often resides in extrachromosomal DNA.

DNA transfer
Page 197

Resistance
Page 201

Eucaryotic Cells

The eucaryotic nucleus differs from the procaryotic chromosome in several respects. The eucaryotic nucleus is bounded by a nuclear membrane, which is actually a part of the cytoplasmic membrane (Fig. 4-6). There appear to be pores in
the membrane so that areas of the nucleus are actually in contact with the cytoplasm (Fig. 4-6). The DNA is packaged into a number of chromosomes, rather
than into the single chromosome that is characteristic of procaryotes. The quantity of DNA is much greater in the eucaryotic cell than in a typical bacterial cell,
ranging from approximately three times more in a yeast cell to 700 times more
in a cell of man.

The *nucleolus,* a structure found within the nucleus (Fig. 4-6) plays an important role in the synthesis of the RNA which comprise ribosomes.

Procaryotic Cells

Protein constitutes over 50 percent of the dry weight of the cell. Almost 90 percent of all of the energy that the cell utilizes in synthesizing cell components is used for the synthesis of proteins. The most conspicuous structures in the bacterial cell devoted to the synthesis of proteins are the ribosomes (Fig. 4-5). These structures are the workbenches on which amino acids are joined together to form proteins. Up to 15,000 of these small granules are found in the cytoplasm; the exact number depends on how rapidly the cell is synthesizing protein. The greater the rate of protein synthesis, the greater the number of ribosomes. Each granule is composed of two parts, and each part in turn is composed of two different macromolecules, protein and RNA. The latter is termed ribosomal RNA to distinguish this RNA from other types which also play a role in the synthesis of proteins (Fig. 4-23).

RIBOSOMES FUNCTION AS WORK BENCHES FOR PROTEIN SYNTHESIS

Many different kinds of protein, the ribosomal proteins, form a part of the ribosome structure. Ribosomes are characterized by their sedimentation properties when centrifuged at very high speeds in an ultracentrifuge. The faster they sediment, the heavier they are. Bacterial ribosomes are referred to as 70S ribosomes, the S being a unit of sedimentation.

Eucaryotic Cells

The ribosomes in eucaryotic cells are grossly similar to those in procaryotic cells except that they are larger and heavier than their procaryotic counterparts. They are designated 80S. The ribosomes are attached to the rough endoplasmic reticulum.

EUCARYOTIC RIBOSOMES LARGER THAN PROCARYOTIC

GENERATION OF ENERGY

Procaryotic Cells

Most cells obtain their energy by degrading foodstuffs. In procaryotic cells a large number of the enzymes involved in the generation of energy are attached to the cytoplasmic membrane (membrane-bound enzymes). Most other enzymes are located in the cytoplasm (soluble enzymes).

ORGANELLES OF ENERGY GENERATION

The photosynthetic bacteria have pigments as well as enzymes which are concerned with converting the energy of light into chemical energy. These pigments and enzymes are localized on the cytoplasmic membrane which occur in the form of vesicles or lamellae (singular, lamella) in the cytoplasm (Fig. 4-24). The name given to these membranous structures is *chromatophore* (Fig. 4-24).

CHROMATOPHORES

Eucaryotic Cells

Eucaryotic cells have specialized organelles, the *mitochondria* (singular, mitochondrion) which are concerned with the generation of energy. Mitochondria

MITOCHONDRIA

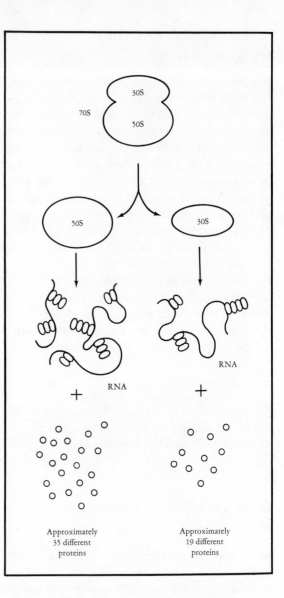

FIGURE 4-23

The procaryotic ribosome and its constituent parts. Under appropriate conditions of incubation, ribosomes dissociate into their
component parts. First, the 70 S ribosome dissociates into two sub
units, 30 S and the larger 50 S. These sub units in turn will break
down into their component parts, ribosomal RNA and structural
proteins. It is these structural proteins which bind antibiotics
which interfere with protein synthesis (Chap. 8). The ribosomes
from eucaryotic cells are larger (80 S) and contain different structural proteins. The dissociation of ribosomes into their two 50 S
and 30 S sub units is important in the process of protein synthesis.
(Chap. 8).

are highly complex structures about the size of a bacterial cell which consist of
a series of outer and inner membranes (Fig. 4-25). The enzymes concerned with
the generation of energy are localized in the inner membrane.

It has recently been shown that mitochondria also contain some DNA,
which carries the genetic information for at least some of the properties of the
mitochondrion. Other properties of the mitochondrion are determined by DNA
in the nucleus. Mitochondria also contain at least some of the machinery required
for the synthesis of protein, such as ribosomes. Surprisingly the ribosomes in the
mitochondria are 70S, which is characteristic of procaryotic cells rather than the
80S characteristic of eucaryotic cells. Evidence is accumulating that mitochondria

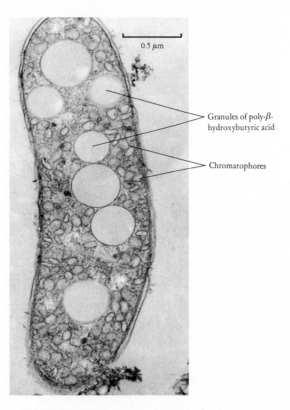

Granules of poly-β-
hydroxybutyric acid

Chromatophores

FIGURE 4-24
A thin section of a photosynthetic bacterium, *Rhodospirillum rubrum*. The cytoplasm is filled with chromatophores and inclusion granules. (Courtesy of Dr. E. Boatman)

arise from preexisting mitochondria, perhaps by the process of binary fission. Thus one mitochondrion can divide to form two, each of which can divide into two more mitochondria.

Like the photosynthetic procaryotes photosynthetic eucaryotes also contain pigments as well as enzymes which are involved in converting light energy into chemical energy. In eucaryotic cells the pigments and enzymes are enclosed in a

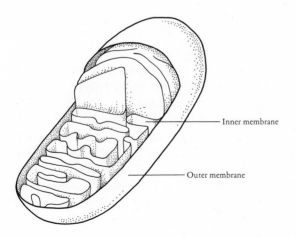

Inner membrane

Outer membrane

FIGURE 4-25
Diagrammatic representation of a mitochondrion (see Fig. 4-20). The enzymes are located in two positions. Some are in the fluid region inside and others are attached to the inner membrane, which because of its convolutions has an extraordinarily large surface area.

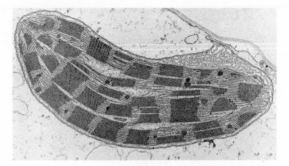

FIGURE 4-26

Cross section of chloroplast. The continuous membrane system contains chlorophyll and is the site of the reactions associated with light in photosynthesis. The material in which the membranes are embedded contains enzymes of photosynthesis, ribosomes, and DNA of the chloroplast. ($\times$10,000) (Courtesy of Dr. L. K. Shumway; Shumway, L. K. and T. E. Weir, *Amer. J. Bot.,* **54**:773, 1967.)

structure bounded by the cytoplasmic membrane, the *chloroplast* (Fig. 4-26). This structure bears a resemblance to the mitochondrion. It is enclosed by an outer membrane and is composed of a large number of closely packed internal membranes. Interestingly procaryotic-type ribosomes, as well as DNA and enzymes of protein synthesis, are also present in the chloroplasts. Chloroplasts also appear to arise from preexisting chloroplasts.

CHLOROPLASTS AND MITOCHONDRIA MAY BE REMNANTS OF BACTERIA

The fact that eucaryotic cells have organelles, such as mitochondria and chloroplasts which contain ribosomes identical to those of procaryotic cells as well as DNA and some machinery for protein synthesis, has given rise to the popular notion that mitochondria and chloroplasts were once independent organisms. This theory envisions that these organelles are descendants of free-living procaryotes that invaded another procaryotic cell and then lost some of the genetic information required for a free-living existence. Only certain remnants of their former self remain, the ribosomes, some DNA, as well as some enzymes required for protein and nucleic acid synthesis. Presumably the procaryotic cell evolved into the eucaryotic cell partly as a result of this association with intracellular procaryotic cells. Such a theory is strengthened by the observation that some eucaryotic cells living today have procaryotic cells living inside them.

CELL MOVEMENT

Procaryotic Cells

FLAGELLA PROPEL BACTERIA

The flagella (singular, flagellum), long protein filaments, helical in microscopic appearance, are responsible for the motility of most bacteria. The bacterial flagellum is composed of three parts: the *filament,* the *hook,* and the *basal body* (Fig. 4-27). The filament is a long helical structure external to the cell, which is composed of several chains of protein that are twisted together into a helix with a hollow core. The hook is the structure that is attached to the end of the filament. The basal body is attached to the hook and anchors the flagellum to the cell wall and membrane.

Exactly how flagella propel the cell is not known, but it is certain that movement is mediated by an expenditure of cellular energy. A popular idea is that an enzyme releases energy from the energy-rich compound ATP, and this energy is somehow converted into a physical change in the flagellum which propels the cell. However, all attempts to find such an enzyme in any of the parts of the flagellum have failed.

Flagella can be removed from bacteria by treatment in a blender. These nonflagellated cells remain viable and in time regenerate normal flagella. One puzzling unsolved question is how do flagella know when to stop growing?

Eucaryotic Cells

Many motile eucaryotes are propelled by either flagella or cilia. Although all flagella perform the same function, the flagella of eucaryotes are far more complex than that of procaryotes. The flagellum of eucaryotes also arises from a basal body within the cytoplasm. Nine pairs of hollow fibers which are arranged around a central pair constitute the flagellum (Fig. 4-28). The cytoplasmic membrane surrounds the portion which protrudes from the cell. The structure of cilia including those of some cells of the human body is identical to the structure of eucaryotic flagella and can be considered to be short flagella.

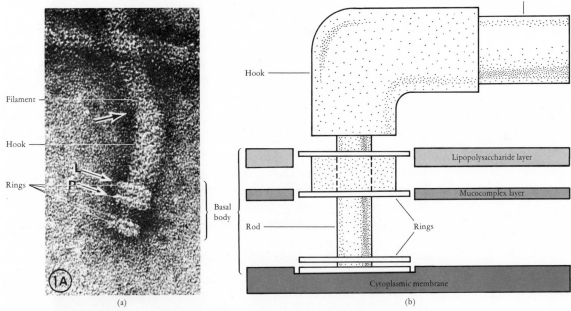

FIGURE 4-27   (a) A bacterial flagellum emphasizing the basal end ($\times$268,600). (Courtesy of Dr. J. Adler; DePamphilis, M. L. and J. Adler, *J. Bact.*, **105**:384, 1971.) (b) Model of a bacterial flagellum of *Escherichia coli*. The integration of the basal body of the flagellum into the cell wall and cytoplasmic membrane is shown. This basal body is composed of a number of rings which surround a rod. The filament extends from the cell. (Modified from DePamphilis, M. L. and J. Adler, *J. Bact.*, **105**:395, 1971.)

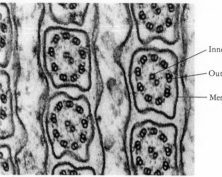

Inner tubules

Outer tubules

Membrane

FIGURE 4-28
Cross section of eucaryotic flagella from the protozoa, *Tricho-nympha*. Each flagellum consists of nine pairs of outer tubules, two inner tubules (9 + 2 arrangement) and an enclosing membrane. (✕65,000) (Courtesy of Dr. A. W. Grimstone)

ADDITIONAL STRUCTURES ASSOCIATED WITH PROCARYOTIC CELLS

Pili (Fimbriae)

Many gram-negative bacteria, (but no gram-positive apparently) possess hundreds of hairlike appendages which are considerably shorter and thinner than flagella (Fig. 4-29). Like flagella, they consist of protein subunits wound around one another, which generates a hollow core. They can also be removed without any loss of viability if the cells are agitated in a blender.

Pili
Page 197

The pili can be divided into a number of types based on their function. One group, termed the F or sex pili, are concerned with attaching one cell to another prior to the transfer of DNA from one cell to the other. Other pili may serve to keep bacteria near the surface of liquid or where the oxygen is most available. Some pili may serve to attach cells to other objects.

Inclusions of Bacteria

STORAGE PRODUCTS—
GLYCOGEN

Many bacteria form and store granules in the cytoplasm in the form of high molecular weight polymers. These granules, which serve as reserve food supplies, can be seen with special stains and are generally large enough to be readily detected by light microscopy. One common inclusion is glycogen, which serves as a storage form of both carbon and energy. Another storage form of both carbon and

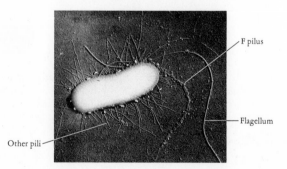

F pilus

Flagellum

Other pili

FIGURE 4-29
Pili of *Escherichia coli*. The variety of pili can be judged from the differences in their lengths. The F pilus can be readily identified since certain bacterial viruses specifically adsorb to it. Note that the flagellum is considerably longer than any of the pili. (✕11,980) (Courtesy of Dr. C. Brinton, Jr.)

energy is the polymer of poly-B-hydroxybutyric acid. Phosphate can be stored in the form of granular inclusions of PO_3^-, which are called *volutin* or *metachromatic granules*. The cell avoids the problem of increasing the osmotic pressure inside the cell by converting the numerous small molecules into a few large molecules.

The synthesis of inclusion bodies depends on the environment and the organism. If the environment contains an excess of a nutrient, then the cell will often convert the nutrient into a macromolecular form and store it until the time it is needed. As a general rule only one kind of reserve material is stored by a particular species.

Endospores

A very distinctive structure, the *endospore,* is often seen in certain species of gram-positive bacilli (Fig. 4-30). This structure is a *unique cell type* that is formed within the vegetative cell by a complex, highly ordered sequence of morphological changes termed *sporogenesis.* Under certain conditions the endospore can go through a sequence of events in which it is transformed into a vegetative cell—

$$\text{Vegetative cell} \underset{\text{germination}}{\overset{\text{sporogenesis}}{\rightleftarrows}} \text{Endospore}$$

the process of germination. The most distinctive feature of the endospore is that it has no metabolic activity. It is usually resistant to a variety of agents which

ENDOSPORE—A
RESISTANT,
NONMETABOLIZING
CELL

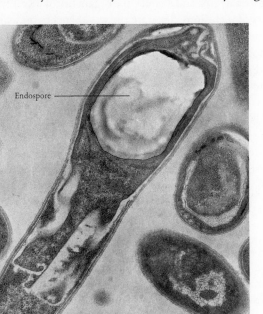

Endospore ————————

FIGURE 4-30 Endospore inside a vegetative cell of *Clostridium.* (Courtesy of Dr. J. Hoeniger; Hoeniger, J. F. M. and C. L. Headley, *J. Bact.,* **96:**1835, 1968.)

kill the vegetative cell, such as heat, drying, freezing, chemicals, and radiation. It is easily visible under the light microscope because it is highly refractile and does not take up stains as does the rest of the cell. Endospore formation represents an example of cellular differentiation in procaryotes, and the process of sporogenesis and germination are considered as a form of regulation of cellular activities.

SUMMARY

The elucidation of the detailed anatomy of microorganisms has been accomplished primarily through observations with the electron microscope, an instrument which can resolve objects only 0.0003 μm apart.

The structures of the bacterial cell can be divided into three major groups: (1) cell envelope, (2) appendages, and (3) cytoplasmic inclusions. Some of these structures are found in all bacteria; others are found only in certain bacteria. The cell envelopes include the capsule, cell wall, and cytoplasmic membrane. The cell wall primarily provides the rigid structure to the cell. The cytoplasmic membrane is responsible for the semipermeability characteristics of the cell envelopes. In addition it contains many enzymes concerned with energy metabolism. The appendages include the pili, which are organelles of attachment found only in gram-negative bacteria, and flagella, the organelles of motility. The ribosomes are another class of inclusions. They are the workbenches on which proteins are synthesized. The chromosome consists of one very long molecule of circular DNA in which most of the genetic information of the cell is coded. Some cells contain reserve food materials which cells synthesize in times of plenty and utilize when nutrients are in short supply.

Eucaryotic cells have several structures that are completely absent from procaryotic cells and other structures that are present in the procaryotic cells, but they are considerably simpler in the procaryotes. The chloroplasts and mitochondria are self-replicating organelles concerned with the generation of energy which are only present in eucaryotic cells. The nucleus, flagella, and ribosomes although present in both cell types are considerably more complex in the eucaryotic cells.

QUESTIONS

1. You have grown a large batch of bacterial cells and suspect that it has become contaminated with another organism. What could you do in a few minutes to determine whether your suspicions were well founded?
2. What envelopes enclose bacteria? What function does each serve and what is the chemical composition?
3. Penicillin is known to be most effective against gram-positive bacteria that are actively multiplying. Why is this so? Why is penicillin so selective in its action?

4. Consider the steps and mechanisms by which protein added to a nutrient medium can be utilized by bacteria.
5. Compare the structures in eucaryotic and procaryotic cells responsible for: (a) generation of energy, (b) storage of genetic information.

FURTHER READING

BARER, R., *Lecture Notes on the Use of the Microscope,* 2d ed. Oxford: Blackwell Scientific Publications, 1956.

BRACHET, J., "The Living Cell," *Scientific American* (September 1961). (Offprint #90)

CREWE, A. V., "A High-Resolution Scanning Electron Microscope," *Scientific American* (April 1971).

EVERHART, T. and T. HAYES, "The Scanning Electron Microscope," *Scientific American* (January 1972).

FOX, C. F., "The Structure of Cell Membranes," *Scientific American* (February 1972).

GOODENOUGH, U. and R. LEVINE, "The Genetic Activity of Mitochondria and Chloroplasts," *Scientific American* (November 1970). (Offprint #1203)

GRIMSTONE, A. V., *The Electron Microscope in Biology.* New York: St. Martin's Press, 1968.

MARGULIS, L., "Symbiosis and Evolution," *Scientific American* (August 1971). (Offprint #1230)

RACKER, E., "The Membrane of the Mitochondrion," *Scientific American* (February 1968). (Offprint #1101)

SATIR, P., "Cilia," *Scientific American* (February 1961). (Offprint #79)

SHARON, N., "The Bacterial Cell Wall," *Scientific American* (May 1969). (Offprint #1142)

Chapter 5

Dynamics of Microbial Growth under Laboratory Conditions

PURE CULTURE METHODS

The variety of organisms of different sizes and shapes found in most environments in nature represents a *mixed population* or *mixed culture*. Today we recognize that a mixed culture of bacteria represents different genera and species which have become adapted to a particular environment. In the 1860s, however, a different interpretation was favored by many bacteriologists—that bacteria have the capacity to change their shape and size, and perhaps even function. That bacteria having different shapes represent different species was demonstrated by separating all the progeny of a single cell from all other cells to obtain a *pure culture*.

KOCH PIONEERS PURE CULTURE METHODS

    The person who contributed the most to pure culture techniques was Robert Koch, a German physician, who succeeded in combining a medical practice with a remarkably successful and productive research career for which he received a Nobel Prize in 1905. Koch was primarily interested in isolating and identifying the bacteria that cause certain diseases. However, early in his career in the late 1870s, he recognized it was necessary to have simple methods for obtaining pure cultures of bacteria. Koch recognized that the isolation of pure cultures required a solid medium on which an isolated single cell could multiply in a limited area. *The population of cells which arise from a single cell in one spot is called a colony.* About 1 million cells are required for a colony to be easily visible to the naked eye.

    Koch initially experimented with growing bacteria on the cut surfaces of potatoes, but found that a lack of nutrients prevented growth of some bacteria. To overcome this difficulty Koch realized that it would be advantageous to be able to solidify any liquid nutrient medium. He conceived the idea of adding gelatin as a hardening agent. Since the gelatin-containing nutrient medium is liquid above 28°C, it was therefore only necessary to heat the medium above this temperature, pour it into a sterile container, and allow it to harden. Once hardened, a loopful of bacteria could be drawn lightly over the surface so as to deposit single bacterial cells at intervals. Each of the single cells divides, and after a day or two enough cells are present to form a visible colony (Fig. 5-1). A gelatinized nutrient offers certain problems, however. The medium must be incubated at temperatures below 28°C if it is to remain solid. Furthermore, gelatin itself can be degraded by many microorganisms, thus severely restricting its use. The perfect hardening agent came out of a household kitchen. Frau Hesse, the New Jersey-born wife of one of Koch's associates, suggested agar, a solidifying agent that her mother had used to harden jelly. Agar is a material extracted from certain marine algae. In contrast to gelatin, a 1.5 percent agar gel must be heated to above 80°C before it liquefies and is therefore solid over the entire range of temperatures at which bacteria grow. Once melted it can be cooled to about 40°C before it solidifies. This allows the incorporation of nutrients which might be destroyed at higher temperatures. It is a complex carbohydrate which is degraded by very few bacteria. In all these intervening years, no better hardening agent has ever been found.

STREAKING

Source of agar
Page 279

Another method for obtaining isolated single colonies which takes advantage of the properties of agar is the *pour plate method*. One merely mixes an appropriate number of bacteria with melted agar, at a temperature of about 42°C, and pours the agar containing bacteria into a sterile dish. The agar solidifies and traps the organisms in a defined space. The plate can then be incubated at an appropriate temperature and the single cells will multiply to form colonies. This technique is possible because bacteria are not killed at the low temperatures to which agar can be cooled and still remain liquid (42°C). If the original sample contains a very large number of organisms so that isolated, well-separated colonies cannot be obtained, then it is necessary to dilute the sample into water or other suitable diluent before mixing it with the agar. Figure 5-2 illustrates the pour plate method.

POUR PLATE METHOD

Another major technical advance made in Koch's laboratory was the development of a two-part glass dish which could be readily sterilized and maintained in a sterile condition. This dish served as a convenient container for solid medium during the incubation of the bacteria streaked on its surface. The dish remains

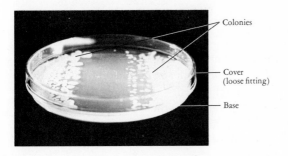

Colonies

Cover
(loose fitting)

Base

FIGURE 5-1
Petri dish containing a nutrient medium which supports the growth of bacterial colonies. Two different bacterial cultures were streaked over the solid surface giving rise to two different types of colonies. Colonies can vary in their appearance, in size, height, color, roughness of edges, and degree of dryness. (Courtesy of J. P. Dalmasso)

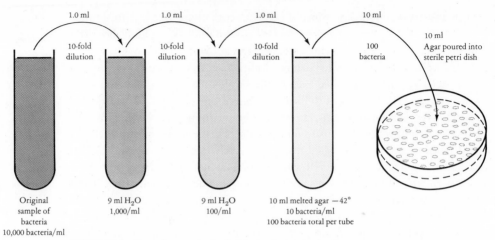

FIGURE 5-2   Pour plate method of achieving isolated colonies.

The significance of the pure culture technique to the development of microbiology cannot be overestimated. This method was essential for the isolation of pathogenic (disease-causing) as well as other bacteria, and it heralded a "Golden Age of Bacteriology." Within 20 years after the development of pure culture methods, the causative agents of most of the major bacterial diseases of mankind were isolated and characterized. In addition the concept that bacteria did indeed breed true was incontrovertibly established. Koch recognized that bacteria of different shapes and sizes could be considered different species. The fact that there were so many distinguishable forms with so many different morphological and metabolic properties had a tremendous impact on the development of the embryonic science of bacteriology.

CONCEPT OF "BREEDING TRUE" ESTABLISHED

Scientists working with microorganisms in the laboratory are very careful to insure that they are working with pure cultures. Otherwise it would not be possible to obtain consistent, interpretable results. However, it is also true that the understanding of interactions between different organisms with their natural environment will require using mixed populations.

Interactions between Organisms
Chap. 6

MEASUREMENT OF CELL GROWTH

Definition of Growth

Growth is the orderly increase in the quantity of all components of the bacterial cell. After a bacterial cell has almost doubled in size and in amount of each of its components, it divides into two daughter cells—the process of binary fission. The time required for one cell to divide into two is the *doubling* or *generation time*. Consequently, the growth of bacteria is measured in terms of an increase in the number of cells, rather than a significant increase in the size of a single organism.

Thus with bacteria the growth of a population is measured whereas growth of a multicellular organism is reflected primarily by an increase in the size of a single organism.

Measurements of Increase in Cell Mass

It is essential to be able to measure accurately the growth of a bacterial population, since only in this way can the effects of agents which inhibit or stimulate growth be assayed. Two parameters of growth can be measured, namely the increase in the mass of the population and the increase in cell number. Each technique has certain advantages and disadvantages, and different procedures may provide different information. It is important to be fully aware of the information each measurement provides as well as its limitations. Although measurements of cell mass and cell number generally yield comparable figures, this is not always the case. For example, although the doubling of the mass of a culture is generally accompanied by a doubling in cell number, under certain conditions an organism may accumulate reserve material or may synthesize capsular material without dividing. Furthermore, the size of cells, and therefore their mass, varies according to their rate of multiplication: the faster they multiply, the greater the mass of each cell.

There are several ways to measure cell mass. One method determines the weight of the population, or one component of the cell. A second method assays the mass by determining the amount of light scattered by a suspension of cells.

In one technique a measured volume of the culture is dried in an oven and then weighed. Another method involves the determination of one component of the cells, most commonly protein. The major problem with these methods for the study of unicellular organisms is their lack of sensitivity. Since each bacterial cell weighs about 50×10^{-13} grams it is difficult to weigh accurately a billion bacteria. Therefore this technique requires the processing of large volumes of media, ranging from 100 ml upward.

The most convenient and simplest method of determining the mass of cells is based on the fact that small particles, such as bacteria, scatter light which is passed through the cell suspension. The amount of light scattering is proportional to the mass of cells present, and since the bacterial cells are relatively uniform in size (mass), the number of cells can be measured from the amount of light that reaches a sensing device after passing through the suspension (Fig. 5-3). This method is far more sensitive than weighing the cells. However, it also has its limitations. A media containing a million bacterial cells (10^6) per milliliter is perfectly clear, and about 10 million cells (10^7) per milliliter gives only a barely discernible turbidity to the solution. Thus although a turbid solution suggests the presence of bacteria, a clear solution does not guarantee their absence. The presence of large numbers of bacteria in crystal clear media may have serious consequences in certain situations.

This technique is also subject to certain errors. Since the light scattering is proportional to cell mass, larger cells will scatter light more than smaller cells.

Pages 134–135

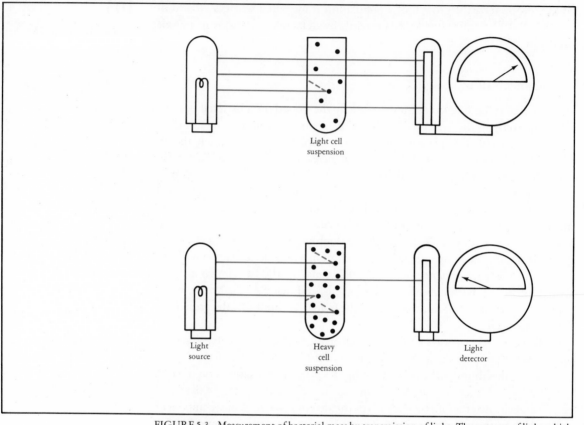

Light cell
suspension

Light
source

Heavy
cell
suspension

Light
detector

FIGURE 5-3 Measurement of bacterial mass by transmission of light. The amount of light which impinges on the light detector is proportional to the number of bacteria in the suspension.

Furthermore, since the mass of a cell depends on the conditions under which it is growing, the same number of cells growing under different conditions will scatter light to different extents. It is very important to relate cell number to cell mass under defined conditions of cell growth and species of bacteria.

Measure of Cell Numbers

The two common laboratory techniques are the plate count and the direct microscopic count.

COLONY COUNTS

The plate count method is based on the fact that a single living bacterial cell deposited on a solid nutrient medium will multiply and form a visible colony. The number of colonies which appear after a suitable period of incubation represents the number of living or *viable* cells in the original solution. Since it is difficult and inaccurate to count more than about 300 colonies per petri dish, it is generally necessary to dilute the original culture before pipetting a known volume on the solid medium. If the cells do not separate from each other following cell division, a common situation (for example, with the chain-forming

104

streptococci), then the viable cell count measures the colony-forming units and not individual cells. The viable cell method measures only living cells, but it is extremely sensitive since the presence of only a few cells per milliliter can be detected readily. However, it is slow because at least a one-day incubation is generally required before a colony is readily visible.

A much more rapid method of determining cell number involves counting the number of bacteria in an accurately known volume with a microscope. There are special slides called counting chambers that hold a known volume of liquid. One disadvantage of this method is that about 10 million bacteria per milliliter are required in order to gain an accurate count, and it does not distinguish living from dead bacteria.

COUNT OF INDIVIDUAL CELLS

PHYSICAL FACTORS INFLUENCING MICROBIAL GROWTH

A variety of environmental factors influences the growth of microorganisms. These can be divided into two broad categories: physical environment and nutritional factors. The physical environment includes temperature, pH, oxygen, and osmotic pressure. The nutritional environment comprises the sources of energy and constituents for synthesis of cell components.

As a group microorganisms will grow under a tremendous range of environmental conditions. The more adaptable organisms are widely distributed in nature. Other organisms can multiply only under a limited range of environmental conditions and are found only in restricted natural habitats. No single species is able to multiply over the entire range of conditions found in nature. As expected the majority of species live under conditions most commonly found on the earth's surface. Organisms, however, can be isolated from extreme environments. Such organisms have become adapted to their particular environment, and generally do not grow well if cultivated under ordinary conditions.

Temperature

Most bacteria grow over a temperature range of 30°C, with species having a well-defined upper and lower limit. As a general rule the optimum temperature for growth is close to the upper limit of its range. Growth slows as the temperature approaches the lower limit. Above the upper limit there is a very sharp drop in the speed with which cells grow (Fig. 5-4). This reflects the fact that temperature affects primarily the enzymes of the cell. The increase in temperature increases enzyme activity, and the cells grow faster. If the temperature rises too high, enzymes critical to the life of the cell are denatured and the cells grow more slowly or die.

Temperature and enzyme activity
Page 145

Bacteria are customarily divided into three groups based on their *optimal* growth temperatures (Fig. 5-5). *Psychrophiles* grow best between −5 and 30°C. Most actually grow slowly at the lower range and have their optimum near their

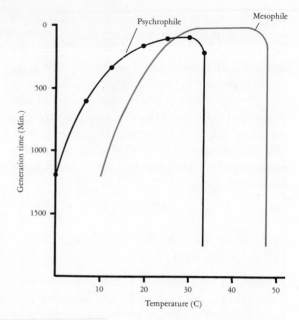

FIGURE 5-4
Effect of temperature on the generation time of a typical meso-
phile and a psychrophile. The generation time decreases as the
temperature is raised until a temperature is reached at which enzymes start to denature. Cell growth slows down precipitously at
this temperature.

TEMPERATURE AND
GROWTH

THREE GROUPS OF
ORGANISMS

upper limit of growth. The fact that organisms can multiply at low temperatures
often causes problems in food and blood storage in the refrigerator (4°C). *Mesophiles* grow best within the range of 15 to about 50°C, and constitute the majority
of bacteria. Most disease-causing bacteria adapted to growth in the human body
have their optimum between 35 and 40°C. Those found in the soil, a colder environment, have generally lower optimum, close to 30°C. The third group, the
thermophiles, have their optimum temperature between 50 and 60°C. It has recently been shown that some thermophilic bacteria isolated from hot springs can
actually grow at temperatures above 90°C. Apparently protein molecules from
thermophiles differ from the mesophiles since they are not denatured at the high
temperatures. Just what confers this heat resistance on proteins is not known.

Although it is convenient to classify organisms into these three groups, in
actual practice there is no sharp dividing line which separates the temperatures
at which organisms in each group will grow (Fig. 5-5). Furthermore not every
organism can grow over the entire range indicated for that group. The upper
limit of growth appears to be related to the acidity of the environment. Thus in
hot springs the upper temperature limit at which bacteria are found decreases as
the springs become more acid. In highly acid water the upper limit is 80°C; in
neutral or alkaline springs bacteria are found at temperatures over 90°C.

pH Measure

This is a measure of the acidity of a solution (Fig. 5-6). Most bacteria grow best
at a neutral pH 7, although they can tolerate ranges from pH 5 (acidic) to 8
(basic). Yeasts and molds grow best in an acid medium. Some bacteria can tolerate very high acid concentrations; some bacteria that form vinegar (acetic acid)

can actually live in 1N acid. The genus *Thiobacillus,* which accumulates sulfuric acid from its metabolic processes, can grow at pH values of 1 and even below. The pH inside the cell must be considerably higher than this, and the ability to grow depends on the cell's ability to keep the H^+ ions out of the cell. Many bacteria produce enough acid as a by-product in the breakdown of foodstuffs to inhibit their growth. Buffers are added to media to maintain certain pH levels. Buffers are generally supplied to the cell as a mixture of two salts of phosphoric

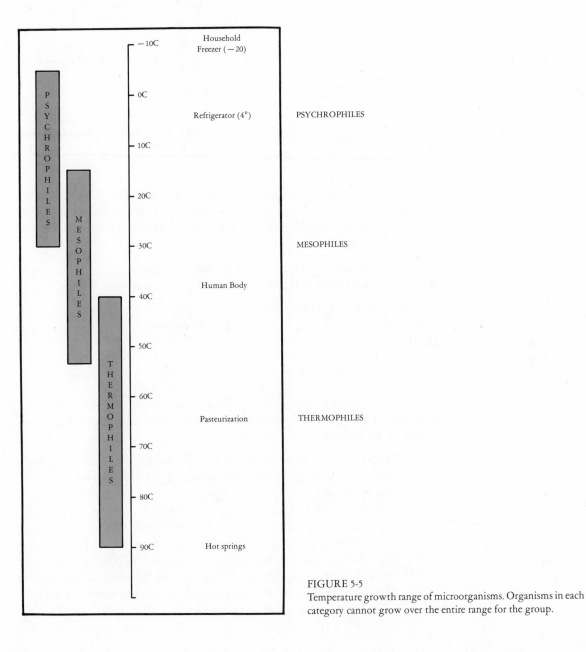

FIGURE 5-5
Temperature growth range of microorganisms. Organisms in each category cannot grow over the entire range for the group.

H$^+$ *Ion Concentration* (*moles per liter*)		pH		*Common Materials with* pH *Indicated*
1.0	10^0	0		
0.1	10^{-1}	1	Increasing Acidity	
0.01	10^{-2}	2		Human gastric contents
0.001	10^{-3}	3		Ginger ale and wines
0.0001	10^{-4}	4		Tomatoes, orange juice
0.00001	10^{-5}	5		Beans
0.000001	10^{-6}	6		Milk begins to taste sour
0.0000001	10^{-7}	7	Neutral	Pure water, human saliva
	10^{-8}	8		Sea water
	10^{-9}	9		
	10^{-10}	10		Soap solutions
	10^{-11}	11	Increasing Alkalinity	Household ammonia
	10^{-12}	12		
	10^{-13}	13		
	10^{-14}	14		

FIGURE 5-6 The pH scale is defined as the negative logarithm of the H$^+$ ion concentration. The use of logarithms is convenient since it expresses in a single number what would originally require many 0's to express. Note that there is a ten fold variation in the H ion concentration between the two successive figures in the pH scale.

MECHANISM OF BUFFERING ACTION

acid, commonly sodium phosphate (Na_2HPO_4 and NaH_2PO_4). These salts resist pH changes because they have the ability to combine chemically with the H$^+$ ions of strong acids and the OH$^-$ ions in alkaline solutions according to the following equations:

(1) H^+ (strong acid) + HPO_4^{2-} $\rightleftharpoons$ $H_2PO_4^-$ (weak acid)
(2) OH^- (strong base) + $H_2PO_4^-$ $\rightleftharpoons$ HPO_4^{2-} (weak acid) + HOH

Buffers maintain neutral solutions since in reaction (1) a strong acid is converted into a weak acid, and in reaction (2) the OH$^-$ pulls off a proton (H$^+$) to form HOH.

Oxygen

OXYGEN REQUIREMENTS

Microorganisms are classified into three major groups with respect to their requirement for oxygen.

1. *Obligate aerobes.* These organisms have an absolute requirement for oxygen. They grow best if they are grown in flasks which are continuously shaken in order to keep oxygen always available.

2. *Obligate anaerobes.* These organisms can only grow in the complete absence of oxygen and the slightest trace of oxygen may kill some organisms. The basis for the extreme toxicity of oxygen probably results from the fact that strict anaerobes lack enzymes which break down toxic derivatives of oxygen produced in metabolism. Some organisms require some oxygen, but not as much as found in the atmosphere. These organisms are commonly termed *microaerophilic.*

3. *Facultative anaerobes.* These organisms will grow either *with or without air,* although their growth is generally more rapid with air.

The classification of organisms into these categories is convenient, but in actual practice not easy to do. It is now recognized that conditions which formerly were thought to be anaerobic actually were not. Therefore organisms classified under these conditions are not correctly identified. Also, large groups of organisms grow best under conditions other than those listed. For example, certain medically important members of the streptococci grow best in a low concentration of air, but they will also grow under aerobic and strictly anaerobic conditions. Some people refer to these organisms as microaerophilic.

The requirement for oxygen reflects the metabolic pathways that the organism employs to gain energy. Some species break down foodstuffs (principally sugars) by a sequence of enzyme reactions which require free oxygen. These organisms are obligate aerobes. Organisms that metabolize sugars by a scheme that does not involve oxygen obviously have no need for it. Faculative anaerobes can use either pathway, depending on whether or not oxygen is available.

Oxygen and metabolism
Page 152

Osmotic Pressure

The osmotic pressure of a medium depends on the concentration of dissolved substances in solution. A dissolved substance commonly encountered in nature is sodium chloride (common salt). Most bacteria can grow over a broad range of salinity, because the cell is capable of maintaining a relatively constant internal salt concentration by specific permeases. However, if the high salt concentrations on the outside of a cell become too high, water is lost from the cell and growth is inhibited. This is the basis of food preservation by salting. Some organisms, called *halophiles,* actually require high salt concentrations for survival, and some may only grow in environments in which the concentration approaches saturation (above 30 percent). Halophiles isolated from the Dead Sea (salt concentration of 29 percent) contain a very high internal salt concentration which is required for the functioning of many of their enzymes as well as their ribosomes. The high external salt concentration required eliminates their need for a rigid outer cell wall, since there is no tendency for water to enter and burst the cell. This is probably why some halophiles do not contain any mucocomplex in their wall. Marine organisms require the salt concentration found in their natural environment (about 3.5 percent) in order to multiply. If the concentration varies much from this level, they die.

Permeases
Pages 86–88

Food preservation by salting
Page 653

NUTRITIONAL FACTORS INFLUENCING MICROBIAL GROWTH

In order for an organism to grow, not only must the physical environment be compatible but foodstuffs must be available which can serve as raw material for the synthesis of cell components and also as a source of energy.

Organisms can be divided into two large groups based on the foodstuff providing the carbon that is converted into cell constituents. *Heterotrophic* organisms utilize organic compounds as a source of carbon for synthesis as well as for a source of energy. A variety of organic compounds can serve as the energy source. In fact many of the compounds that serve as sources of carbon for biosynthetic processes can also serve as sources of energy. *Autotrophic* organisms utilize *carbon dioxide* as their *major* source of carbon for the synthesis of cell components. Generally this group gains its energy from either the sun (through photosynthesis) or by metabolizing *inorganic compounds*. A number of different inorganic compounds can serve as sources of energy, and this fact points up the nutritional diversity of the microbial world. Such simple substances as hydrogen gas (H_2), ammonia (NH_3), reduced iron (Fe^{2+}), nitrite (NO_2^-), and hydrogen sulfide (H_2S) can serve as sources of energy for some microorganisms.

Inorganic compounds in
metabolism
Pages 155–156; Table 7-2

Heterotrophs can utilize a variety of organic compounds as their source of carbon for biosynthetic needs, but are most commonly cultured in the laboratory on medium containing glucose. Members of the genus *Pseudomonas* are able to degrade over 90 different organic compounds. Most organic molecules, regardless of complexity, can be degraded by microorganisms. This versatility does not hold with any single species of microorganism, but rather with the microbial world as a whole. Any one species is capable of degrading only a limited number of organic compounds, but somewhere a microorganism exists which is capable of degrading virtually any particular organic molecule. In some cases the molecule may be broken down into small molecules, which can then be converted to the subunits of macromolecules. In other cases the compound is degraded and provides energy, but is not a source of carbon. However, it is becoming increasingly apparent that, although most biologically generated compounds in the environment can be broken down by microorganisms, a large number of man-made products cannot be degraded by any known organism.

Mineral Requirement

In addition to a source of carbon, all organisms require sources of each of the other elements found in cell constituents—primarily nitrogen, sulfur, and phosphorus. Most microorganisms, both heterotrophs and autotrophs, can utilize inorganic salts as a source of each of these elements.

Formulae—amino acids
Page 34; Fig. 2-7
Purines and pyrimidines
Page 40; Fig. 2-15

Microorganisms obtain nitrogen in a variety of forms for subsequent incorporation into amino acids, purines, and pyrimidines. Most organisms cannot utilize atmospheric nitrogen and require nitrogen provided in the form of an ammonium or nitrate salt, such as ammonium chloride (NH_4Cl), sodium nitrate ($NaNO_3$), or even more complex molecules.

The element sulfur is often provided as a sulfate salt (such as ammonium sulfate), which is in turn converted into the sulfur part of sulfur-containing amino acids, coenzymes, and other cellular constituents.

Phosphorus is present in nucleic acids and phospholipids, as well as in the key molecule of energy metabolism, ATP. It is generally supplied to the cell as a salt of phosphoric acid such as potassium phosphate (K_2HPO_4 and KH_2PO_4).

Growth Factors

Despite their similarity in chemical composition, microorganisms display a wide spectrum in their growth requirements. The growth requirements reflect the enzymatic capabilities of the cell. At one extreme are the obligate parasites such as the members of the genera *Rickettsia* and *Chlamydia* which lack some of the enzymes or structures necessary for an independent existence. At the other extreme are the autotrophs such as the photosynthetic bacteria which have enzymes that allow them to multiply with very simple nutrients—sunlight and inorganic salts. Heterotrophs do not have all of the enzymes present in autotrophs and therefore require organic compounds. However, all three groups have many enzymes in common. Comparative biochemical studies on a wide variety of organisms have made it apparent that there is a core of enzymatic reactions that most organisms can carry out. Many heterotrophs (and a few autotrophs) do not have all of the enzymes necessary for synthesizing some of the small molecules required for growth. These small organic molecules, called growth factors (such as amino acids, vitamins, purines, and pyrimidines, to name a few), must therefore be provided in the medium. The more enzymes concerned with the biosynthesis of small molecules which an organism lacks, the more growth factors required in the medium to allow growth. The need for a large number of growth factors is illustrated by the group called lactic acid bacteria, in which as much as 95 percent of the subunits which comprise cell material must be supplied.

The microorganisms' requirements for growth factors have been successfully exploited to isolate several vitamins long before anything was known about their chemical structure and metabolic function. A number of vitamins was recognized as unknown growth factors for microorganisms; their presence in extremely small amounts (10^{-12} gram) in rich, complex media could be detected by their ability to stimulate the growth of certain microorganisms. Wildiers, a Belgian scientist, was probably the first person to recognize the importance of growth factors. In 1901 he reported that a yeast would not grow unless the medium contained a mixture of organic compounds he labeled "bios." This was the same year that beriberi in animals was shown to be due to a lack of certain substances of importance in the metabolism of the central nervous system.

In 1912 a Polish born American biochemist, Casimir Funk, proposed the vitamin theory to explain certain nutritional diseases. It was recognized that man and microorganisms alike required essential growth factors in order to function properly. Much later, "bios" was identified as a mixture of several materials, including the vitamins biotin and pantothenic acid.

More recently vitamin B_{12} was isolated by purifying a substance that served as a growth factor for a species of *Lactobacillus*. The technique whereby microorganisms are used to assay the presence of any material is termed *microbiological*

Obligate parasites
Page 63; Table 3-4

RELATIONSHIP BETWEEN
GROWTH
REQUIREMENTS AND
ENZYME CONSTITUTION

ISOLATION AND
IDENTIFICATION OF
VITAMINS AS BACTERIAL
GROWTH FACTORS

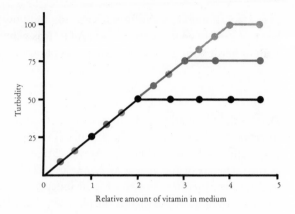

FIGURE 5-7

Microbiological assay. The vitamin in the medium limits the growth of the cells and the extent of growth achieved is directly proportional to the amount of the vitamin in the medium.

MICROBIOLOGICAL
ASSAY

assay. The exact quantity of any growth factor can be assayed by this technique by using a strain of microorganism which requires the growth factor for growth. The lactic acid bacteria as a group require a large number of amino acids, purines, pyrimidines, and vitamins for growth. If any one of these growth factors is completely left out of the medium and then an unknown amount added, the final level of cell growth is directly related to the amount of the growth factor present in the medium (Fig. 5-7). It is important to note that the *rate* of growth of the organism is identical in a medium with a limiting amount of a required nutrient as it is in a medium with an excess amount. The concentration inside the cell is essentially the same in both cases because of the ability of the permeases to concentrate low levels of the nutrient until all of the nutrient has become exhausted in the medium.

Permeases
Page 87

CULTIVATION OF BACTERIA IN THE LABORATORY
Basic Media

TYPES OF MEDIA

Microbiologists use two basic types of media for cultivating bacteria. One contains known amounts of pure chemical compounds; it is therefore a chemically defined or *synthetic medium.* The other type is a "rich" complex nutrient medium containing poorly defined materials such as digests of ground meat, plants, and fish.

SYNTHETIC MEDIUM

A synthetic medium commonly used to grow the bacterium *Escherichia coli* is given in Table 5-1. This species can obviously synthesize all of its cellular constituents from glucose and a few inorganic salts. Glucose also serves as the energy source. Although other organisms that require one or more growth factors will not grow, this glucose-salts medium can serve as the basic medium and various growth factors can be added to support growth of more fastidious organisms. For example, some members of the lactic acid bacteria will grow if the supplements noted in Table 5-2 are added to a glucose-salts medium. Other bacteria may have different nutritional requirements and will not grow even in this fortified medium. It often becomes necessary to add undefined nutrients, such as yeast and

TABLE 5-1
Synthetic Medium for Growth of *E. coli*—Glucose-Inorganic salts

Ingredient	Grams per Liter of Medium
Glucose	5
Dipotassium phosphate (K_2HPO_4)	7
Monopotassium phosphate (KH_2PO_4)	2
Magnesium sulfate ($MgSO_4$)	0.08
Ammonium sulfate ($NH_4)_2SO_4$	1.0
Water	1,000 ml.

1. The glucose is generally sterilized separately from the rest of the salts, since heating the glucose with the salts results in the conversion of glucose to materials which may be toxic to cell growth. When heated in the absence of the salts, this conversion does not occur.

2. For the preparation of solid medium, 1.5 percent agar (final concentration) is generally added.

meat extracts, to permit growth of some organisms. As a group pathogenic organisms are particularly exacting in their nutritional requirements. Their fastidious nature is probably the result of the fact that they have become adapted to a particular environment—the animal body—which contains a variety of rich nutrients. They have therefore been able to dispense with many of the enzymes concerned with the synthesis of these nutrients.

It is generally much more convenient to incubate cultures in a "rich" medium than it is to determine painstakingly the exact nutrient requirements step

UNDEFINED MEDIUM

TABLE 5-2
Synthetic Medium for Growth of a Lactic Acid Bacterium

Component	Grams per Liter	Component	Grams per Liter
Glucose	50	DL tryptophan	0.08
Sodium acetate	40	L-tyrosine	0.2
Ammonium chloride	6	DL-valine	0.5
Monopotassium phosphate	1.2	L-lysine	0.2
Dipotassium phosphate	1.2	DL-methionine	0.2
Magnesium sulfate	0.4	DL-isoleucine	0.5
Ferrous sulfate	0.02	Adenine sulfate	0.02
Manganese sulfate	0.04	Guanine hydrochloride	0.02
Sodium chloride	0.02	Uracil	0.02
DL-alanine	0.4	Xanthine	0.02
L-arginine hydrochloride	0.48	Thiamine hydrochloride	0.001
Asparagine	0.8	Pyridoxal hydrochloride	0.0006
L-Cystine	0.1	Pyridoxine hydrochloride	0.002
L-Glutamic acid	0.6	Pyridoxamine hydrochloride	0.0006
Glycine	0.2	Calcium panthothenate	0.001
L-Histidine hydrochloride	0.124	Riboflavin	0.0001
DL-Phenylalanine	0.2	Nicotinic acid	0.0002
L-proline	0.2	p-aminobenzoic acid	0.0002
DL-serine	0.1	Biotin	0.000002
DL threonine	0.4	Folic acid	0.00002

Adapted from Steel, Sauberlich, Reynolds, and Baumann, *J. Biol. Chem.* 177, 533 (1949).

TABLE 5-3
Rich Undefined Medium Nutrient Broth

Ingredient	Amount per Liter
Peptone (0.5 percent)	5.0
Meat extract	3.0
Distilled water	1000 ml

by step. This is especially true in laboratories in which a variety of different organisms are grown. For this reason a clinical diagnostic bacteriology laboratory relies almost solely on undefined media.

A typical rich medium, nutrient broth, is shown in Table 5-3. The peptones are breakdown products of plant, fish, and meat protein and consist primarily of peptides of various sizes. Meat extract is a water extract of lean meat and provides vitamins, minerals, and other nutrients. To make the medium richer, yeast extract may be added. Yeast extract, the broth of autolyzed yeast cells, serves as an excellent source of vitamins.

PROBLEMS IN DEVISING
A "UNIVERSAL MEDIUM"

It is theoretically possible to devise a single medium with the proper nutrients to support the growth of any bacterium, no matter how unique its growth requirements. In practice, no such single medium exists. One must not only contend with providing growth factors, but also many materials in rich medium are toxic to various bacteria. Pathogenic microorganisms seem to be especially sensitive. Other material must then be added to neutralize the toxins. Some organisms, notably mycoplasma, grow only in media containing a high concentration of serum (20 percent). Its major function is to supply serum albumin which binds small molecules such as fatty acids, metal ions, and detergents, all of which are toxic to many cells.

Mycoplasma
Page 57; Table 3-1
Page 64; Table 3-4

Members of the genus *Haemophilus* and certain other fastidious bacteria are generally grown on *chocolate agar,* a medium enriched with blood and then heated sufficiently to release the *hemin* from the denatured hemoglobin (giving a chocolate brown color to the medium). The heating also denatures an enzyme present in blood which destroys a growth factor required by many bacteria. Several industries are concerned primarily with the manufacture of hundreds of different types of media. Each one has been specially formulated to permit luxuriant growth of one or several groups of organisms.

Even with the availability of these rich media, however, there are still some microorganisms that have never been cultivated in the laboratory in the absence of living cells. These include most of the rickettsiae and the spirochete causing syphilis, *Treponema pallidum.* One large group of organisms which present special problems for cultivation are the anaerobes.

Cultivation Methods for Anaerobes

Since their growth is inhibited to varying degrees by free oxygen, special techniques are required for the cultivation of anaerobes. Two general methods are available. One involves incubating the cultures in jars from which the oxygen has been removed. In one common type of anaerobic jar the air is pumped out and replaced by hydrogen gas. An electrically heated gauze cage in the cover of the jar containing asbestos covered with platinum catalyzes the combination of most of the residual oxygen with hydrogen to form water. A more modern version of this jar employs a disposable hydrogen generator and a catalyst which functions at room temperature. The jar is simply closed and the hydrogen combines with the oxygen.

Various chemicals can be added to media which reduce the level of oxygen by chemically combining with it. These compounds include sodium thioglycollate, cysteine, and ascorbic acid. The medium is often boiled first to drive out the air before inoculation and then the surface can be covered with sterile vaseline and paraffin to keep out the oxygen.

The cultivation of anaerobic organisms presents a great challenge to the microbiologist. It is often difficult to provide suitable conditions quickly enough to strict anaerobes to keep them alive once they have been removed from their natural source. It seems certain that as techniques for culturing anaerobes are improved, many more such organisms will be isolated from habitats in which they had previously appeared to be absent.

Enrichment Cultures

An enrichment culture enhances the growth of one particular organism in a mixed population and thereby promotes its dominance. In practice this method allows one to isolate an organism from natural sources in pure culture. The technique is based on the well-documented concept that in any environment the organism best adapted will come to the fore. Thus it is often possible to manipulate the environment so as to favor the growth of a particular organism. The first consideration in setting up an enrichment culture is selecting the sample from which the organism is to be isolated. It should be selected from an area in nature in which the organism most likely occurs. The material is next inoculated into a medium that contains nutrients that favor the growth of the organism as much as possible, but not the other cells in the culture. The medium is incubated under conditions which favor the growth of the desired organism. For example, if a soil sample is inoculated into a medium in which nitrogen gas is the only nitrogen source, light is the energy source, and carbon dioxide is the carbon source, bluegreens will grow preferentially. If a culture is incubated at 60°C, only thermophilic organisms will grow. If the original sample is boiled before inoculation, the only organisms that will arise are spore-forming species of bacteria. By judicious culture methods, a skilled microbiologist can sometimes favor the growth of certain species so dramatically that a culture will consist of only a single species. It is relatively simple then to plate the organism out on agar and select a single colony as a pure culture.

The concept of selective enrichment has proven very helpful to bacteriologists, who are often faced with recognizing a minority member of a popula-

PROMOTION OF GROWTH OF SELECTED ORGANISMS

INHIBITION OF
GROWTH OF UNDESIRED
ORGANISMS

tion composed of many other species. For example, screening for an organism that has the ability to utilize a certain sugar as a carbon source can be done by simply including this material as the only source of carbon. One common, difficult problem is the selection of a very fastidious organism when the majority of the organisms is less demanding. One solution involves using *selective inhibitors* which can be added to medium in order to discourage the growth of some organisms in the population (providing the organism being sought is resistant to the inhibitor). For example, one diagnostic procedure for identifying typhoid fever involves isolating the organism, *Salmonella typhi,* from the stool, a habitat containing a large number of organisms related to *Salmonella.* The stool specimens are first incubated in selenite or tetrathionate broth, both of which inhibit the growth of the flora normally found in stools without having any effect on the multiplication of *Salmonella.* Many other examples involving other pathogens could be cited.

IDENTIFICATION OF
ORGANISMS BY THEIR
EFFECT ON MEDIUM

Once the organism becomes the dominant species, it may be tentatively identified by its pattern of growth on *indicator or differential agar medium.* These media contain a component which is changed in a recognizable and somewhat unique way. In cases of suspected typhoid fever clinical microbiologists in some laboratories plate the organisms on bismuth sulfite agar. *Salmonella typhi* forms hydrogen sulfide, which results in the formation of black colonies. Some bacteria produce enzymes, *hemolysins,* which break down blood. Colonies of the organism causing "strep" throat are readily detected on blood agar plates by their ability to destroy red blood cells and produce a clear zone around the colony (beta hemolysis) (Fig. 5-8). Certain dyes added to media will change color if the medium becomes acidic. A wide variety of other materials can serve as diagnostic aids when incorporated into media.

Streptococci
Pages 256–257

CELL DIVISION

Division of one cell into two occurs after a cell has duplicated its components. The process of cell division can be separated into several stages. First, the cyto-

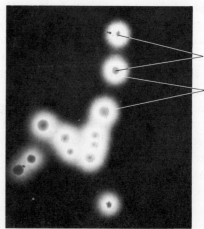

Colony

Zone of hemolysis

FIGURE 5-8
The colony excretes a substance which lyses the red blood cells (sheep) in the medium, resulting in complete clearing. (Courtesy of J. P. Dalmasso)

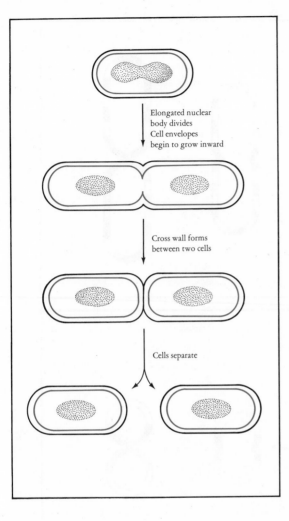

Elongated nuclear
body divides
Cell envelopes
begin to grow inward

Cross wall forms
between two cells

Cells separate

FIGURE 5-9
Diagrammatic representation of cell division.

plasmic membrane grows into the center of the cell, forming a transverse septum. Then the cell wall grows inward toward the center between the membrane, forming a cross wall thicker than the ordinary bacterial cell wall. Once the cell wall forms a completed septum and divides the cell in two, cell separation is initiated by cleavage of this thickened cell wall, each half forming part of the wall of each daughter cell (Fig. 5-9).

Although the overall process of cell division can be described in these simple terms, the regulation of numerous cellular processes that must be coordinated in the process of cell division is one of the most challenging and least understood areas in biology.

Cell Wall Synthesis

Before division the cell must duplicate its cellular components. In some species of bacteria, DNA may be synthesized throughout the life of the cell before it

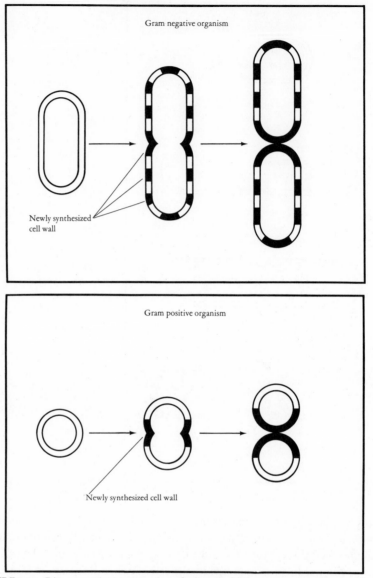

FIGURE 5-10 Diagrammatic representation of cell wall addition to growing gram positive and gram negative cells.

divides. Other species appear to synthesize DNA only during a portion of the lifespan of the cell. The cell also enlarges in size before division, which entails the synthesis of new cell wall material. One question which has been studied is where does cell wall synthesis take place—at the ends, the center, or throughout the old cell wall? The answer apparently depends on the organism. In *Streptococcus* and a species of *Bacillus,* both gram-positive organisms, the new wall was synthesized only at the sites of septum formation, that is, near the center. In several gram-negative organisms newly synthesized material was intercalated between layers of old cell wall (Fig. 5-10). Thus, in this case, cell wall synthesis occurred at various sites in contrast to the limited area of synthesis in the two gram-positive organisms. The reasons for this difference are not understood.

DYNAMICS OF POPULATION GROWTH

The growth of a bacterial population can be measured in terms of the increase in number of viable cells. The doubling or *generation time* conveniently expresses the rate of growth of the population. The increase in cell number in a growing bacterial population is logarithmic or exponential, since each cell gives rise to two cells, which in turn give rise to four cells, and so on. When the number of cells in a population is graphed against time of incubation, a straight line is generated if the ordinate is a logarithmic scale and the abscissa representing time of incubation is an arithmetic scale. The straight line indicates that there is an identical percentage of increase in the number of cells during any constant time interval (Fig. 5-11). Thus if the number of cells increases from 1 to 4 in one hour (a generation time of 30 minutes), there will be another four-fold increase in the population after an additional hour and another four-fold increase each succeed-

EXPONENTIAL INCREASE IN CELL NUMBER CHARACTERIZES BACTERIAL MULTIPLICATION

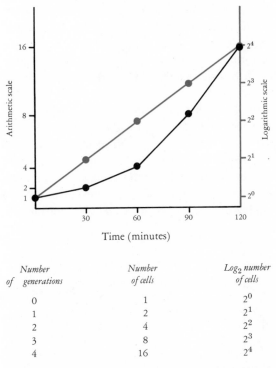

Number of generations	Number of cells	Log_2 number of cells
0	1	2^0
1	2	2^1
2	4	2^2
3	8	2^3
4	16	2^4

The number of cells at any specific time equals the original number of cells times 2^x, where x equals the number of generations. Each division raises the exponent of log_2 by one.

FIGURE 5-11
Relationship of cell number to number of generations plotted on arithmetic and log scales.

ing hour. Note, however, that the increase in the *number* of cells varies in the three intervals. The number of cells depends on the number present initially and the number of generations. Since many bacteria can divide every 30 minutes (and some every 10 minutes), the number of these cells in a population reaches high levels in a relatively short time. Therefore it is very convenient and sometimes absolutely necessary to plot the number of cells in the population on a logarithmic scale.

GROWTH RATE OF
FILAMENTOUS
ORGANISMS IS VERY
COMPLEX

Population growth in organisms that form filaments is quite different from organisms that divide by binary fission. In filamentous fungi, for example, growth usually occurs only at the tips of the filaments and is not exponential but arithmetic; that is, the increase in length of a filament is the same per unit time and is independent of the length of the filament. If branching occurs, however, then new ends are generated at which growth can occur. Thus population growth in fungi is difficult to describe in simple mathematical terms. Our discussion of population growth will focus on bacteria whose growth is not subject to these variations.

Phases of Bacterial Growth

If bacteria are transferred from one medium into another, the population undergoes a characteristic and predictable sequence in its rate of increase in cell numbers (Fig. 5-12). There are four readily distinguishable phases: (1) the *lag phase,* FOUR STAGES IN THE
GROWTH OF A
BACTERIAL CULTURE
in which there is no increase in the number of viable cells; (2) the *log phase,* in which the cell population increases logarithmically or exponentially with time; (3) the *stationary phase,* in which the total number of viable cells remains constant; and (4) the *death phase,* which is characterized by an exponential decrease in the number of viable cells.

When a culture of bacteria is diluted and placed in fresh medium, the number of viable cells does not increase immediately. However, cell growth takes place as indicated by an increase in cell mass. The cells enlarge in size and there is extensive macromolecule synthesis. This period is a "tooling up" stage: growth "TOOLING UP"
STAGE—LAG
without cell division during which time the cell is synthesizing and accumulating molecules it will require for cell division. Especially important are molecules of ATP which represent the supply of energy; ribosomes, required for protein

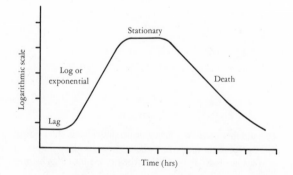

FIGURE 5-12
Bacterial growth curve. The duration of each phase depends on the environment and the particular organism.

synthesis; and any enzymes which the cells do not have that may be required for growth.

During the log phase the cells divide at their maximum rate. A convenient way of expressing the growth rate of a culture is in terms of the number of cell divisions which occur per hour. A culture with a generation time of 30 minutes will have a growth rate of two; a generation time of 60 minutes gives a growth rate of one. The larger the number, the faster the cells are multiplying. The increase in cell number does not occur in discrete jumps, but rather gradually with time, thus indicating that individual cells are at different stages in their division cycle; some have just divided, others have enlarged to the size at which they are about ready to divide, others are at intermediate stages in the division cycle. Thus cells in the population divide nonsynchronously.

Two parameters characterize the exponential growth phase of any culture: the rate of growth and the duration of the period of exponential growth. The faster cells multiply, the faster the growth rate, and in terms of the graph, the steeper the line or greater the slope of the line representing exponential growth.

The growth rate is influenced by all of the environmental conditions previously mentioned in this chapter. Obviously the optimal conditions for growth will vary for each particular organism. One important generalization that can be made is that the *rate of growth of a bacterium is proportional to its rate of energy metabolism.* All of the environmental and nutritional factors which influence the growth rate do so primarily by influencing the rate of ATP production by the cell.

The size a population reaches before it stops growing is controlled primarily by the environment but depends also on the genetic constitution of the organism. The generation time of many bacteria can be as low as 20 minutes. With this generation time the population will reach 10^9 cells in 10 hours starting with a single cell. The volume of an average bacterial cell is 1 μm^3. If the culture continued growing exponentially for 45 hours, its total cell volume would be greater than the volume of the earth. Fortunately bacterial cultures stop growing once they have reached a population density of approximately 10^9 cells per milliliter.

The reason populations cease increasing in number has traditionally been attributed to one of three possible mechanisms: physical overcrowding, limitation of a required nutrient, and accumulation of toxic products during growth. Physical crowding is not a reasonable explanation, since cells in colonies can grow on solid medium in which crowding is much more severe than in liquid medium. It has generally been possible to explain the stoppage of cell growth of all cases investigated by the exhaustion of a nutrient or accumulation of toxic products.

If cells growing in a synthetic medium exhaust one of the ingredients, growth ceases. This fact has been exploited as a tool for assaying the amount of any growth factor. Bacteria can exhaust a required nutrient even when growing in a rich nutrient medium. The material most commonly exhausted by aerobic organisms is oxygen. This gas is not very soluble in water, and once the popula-

Microbiological assay
Pages 111–112; Fig. 5-7

Acid formation
Page 151; Fig. 7-10

tion reaches a high cell density there are enough cells at the air–liquid interface to utilize all of it immediately. The cells beneath the surface of the liquid cannot grow.

Cells growing under anaerobic conditions generally run out of a usable energy supply when the population reaches high cell concentrations. They also produce a variety of waste products in their metabolism, many of which are toxic.

As a general rule the production of toxic materials by cells growing under anaerobic conditions limits the final level that the population can attain under otherwise optimal growth conditions. This explains why a facultative anaerobe capable of living under aerobic or anaerobic conditions will invariably grow to a higher cell density if it is grown under aerobic conditions. Not as much toxic material is likely to accumulate under aerobic conditions.

STATIONARY PHASE—
TOTAL NUMBER OF
CELLS REMAINS
CONSTANT

The stationary phase is reached when the total number of viable cells in the population no longer increases. This constancy may result from one of two different events, depending on the particular organism and nutritional conditions. Every cell in the population may stop multiplying, and none die. Alternatively, there may be a balance between cell division and cell death such that for every cell that divides another cell dies. In both cases, the total number of *viable* cells in the population remains constant. In the second case, however, the *total* number of cells, both living and dead, would increase.

The duration of the stationary phase varies, depending on the species of organism and environmental conditions. Some organisms remain in this phase for only a few hours; others for several days. Generally, if toxic products have accumulated in the exponential phase, the cells begin to die rapidly. Some cells lyse very quickly once they stop multiplying because of the action of a lytic enzyme present in the cell wall. This same enzyme may be concerned with the extension of cell walls during cell division.

DEATH PHASE—
EXPONENTIAL DECREASE
IN THE NUMBER OF
VIABLE CELLS

A decrease in the total number of viable cells in the population signals the onset of the death phase. Death of a cell is indicated if it is no longer capable of multiplying. A plot of viable cells on a log plot versus time on an arithmetic plot forms a straight line (Fig. 5-12). The same mathematical treatment given to the kinetics of growth in the exponential phase holds true here. There is an exponential decrease in the number of viable cells rather than an exponential increase. Thus if there is a tenfold decrease in the number of viable cells during the first hour of the death phase, an additional tenfold decrease is observed by the end of the next hour.

The vast majority of organisms in a culture are similar enough that death in a population may be treated as a statistical phenomenon. Since all organisms are equally likely to die in theory, which cells actually do die in any time period depends entirely on chance. The likelihood of a viable cell dying remains constant during most of the death phase and so the cells continue to die at a constant rate. However, as soon as the vast majority of the cells have died, the death curve flattens out (Fig. 5-12). This indicates that some cells in the population are more resistant than others to dying. Why a few cells in the population are more resistant than all other cells is not understood. This resistance is not transferred to daughter cells, however, since survivors that are diluted into fresh medium undergo the same stages of growth and die at the same rate as the original population.

An inherently more resistant form of the cell, the endospore, may be present in cultures of some bacteria during the death phase. These spores, which develop from vegetative cells during the stationary phase, would certainly be more resistant to dying in the death phase than vegetative cells.

If a population of cells in the death phase is diluted and inoculated into fresh medium, the cells do not continue to die but enter a lag phase and then the log phase. Since cells in the death phase are not primed for rapid multiplication, they will spend an extended period of time in "tooling up" before they can multiply. In contrast, cells transferred from the log phase culture into fresh medium will have only a very short lag phase, especially if the two media are similar in composition. If the media differ from each other, a longer time is required for adjustment.

Age of Culture Versus Age of Individual Cells

A bacterial culture is said to be "old" when it is no longer in the log phase of growth. However, any cells which have just finished dividing in the old culture would be young. Cells which divide by binary fission do not have a life cycle that includes youth, senescence, and death. Accordingly, the concept of age has an entirely different meaning as far as bacteria are concerned.

Practical Implications of Death Phase.

The steepness of the line which describes the exponential death phase is termed the *death rate*. The steeper the line, the greater the rate of cell death. No matter how steep the line, the killing always follows an exponential curve, even though it may be impossible to measure the number of viable cells at very close time intervals and experimentally verify that the decline is indeed exponential. A more detailed discussion of the principles of microbial killing and the effect of various physical and chemical agents on the death rate are covered in Chapter 33.

Growth of a Bacterial Colony

The growth and development of a bacterial colony involve many of the features of bacterial growth discussed in this chapter. A period of time elapses before a single cell on a solid medium starts to divide. The initial multiplication of cells may be exponential. However, after a very few divisions, the close proximity of cells to one another results in very crowded conditions. Further development of the entire colony is a very complex process. The crowding results in competition for available nutrients, which must diffuse from the medium into the cells. The cells on the periphery of the colony are favorably located to obtain oxygen from the air as well as nutrients diffusing into the colony. A gradient of nutrients is thus set up from the outside to the center of the colony. The center tends to be

DIFFERENT PARTS OF THE SAME COLONY ARE IN DIFFERENT GROWTH PHASES anaerobic, with only a limited supply of nutrients available. Thus the colony will enlarge through multiplication of cells at the periphery. The toxic end products of growth also influence the development of the colony. Toxic end products, such as acids, must also diffuse into the medium or they may kill cells in the immediate vicinity. Thus, within a single colony, there are cells which may be in the log, stationary, or death phase, depending upon their location. Cells at the edge may be growing exponentially, whereas those in the center may be in the death phase.

SUMMARY

This chapter has focused on the principles of microbial growth in the laboratory. Bacteria can be grown in a variety of media, each of which must provide sources of energy as well as the elements carbon, nitrogen, phosphorus, and sulfur in forms that can be used by the organism for the synthesis of cell material. An unusually broad spectrum of nutrients can serve these purposes for different microorganisms. At one end of the spectrum are the autotrophs—organisms which derive their energy from sunlight or from the metabolism of inorganic compounds and their source of carbon from carbon dioxide. At the other end are obligate parasites—bacteria which cannot multiply unless they are cultivated in living cells. Intermediate forms, the heterotrophs, represent the vast majority and gain their carbon for the biosynthesis of cellular components from organic sources. Heterotrophs gain their energy from metabolizing organic materials. Many organisms can use inorganic sources of nutrients such as nitrates, phosphates, and sulfates for the synthesis of proteins and nucleic acids, although some cells require certain preformed subunits, called growth factors. The complexity of nutrient requirements is inversely related to the organism's biosynthetic capabilities. Several environmental factors are also important to the growth of bacteria, including the temperature, presence or absence of oxygen, pH of the medium, and the osmotic pressure. Any single species can grow only within a limited range of conditions. However, as a group, microorganisms can grow over a very broad range. The growth of bacteria in the laboratory follows a predictable pattern, with four readily identifiable phases: lag phase, log phase, stationary, and death. During these phases, the culture undergoes a period of active metabolism without cell multiplication (lag), logarithmic increase in cell number (log), no further increase in the number of viable cells (stationary), and an exponential decrease in the number of viable cells (death).

QUESTIONS

1. Compare the various methods of assaying bacterial mass. Which methods would you use if you wanted to determine the percentage of viable cells in a culture? The total number of cells?

2. In looking at bacterial cells through a microscope, you observe that some cells appear larger than others. How would you proceed to determine whether or not the culture was pure?

3. You are attempting to isolate a halophilic aerobic spore former which can utilize N_2 gas and CO_2. How would you proceed to isolate such an organism?

4. Discuss the levels of synthesis of DNA, RNA, and protein during each stage of the growth cycle.

5. The bacteria you are culturing in a nutrient medium stop growing when they have reached a cell density of 10^8 cells per ml. What possible reasons might you give to explain the stoppage of growth? What experiments would you do to determine which of your explanations is correct?

FURTHER READING

STANIER, R., M. DOUDOROFF, and E. ADELBERG, *The Microbial World,* 3d ed. Englewood Cliffs, N.J.: Prentice-Hall, Inc., 1970.

Difco Manual of Dehydrated Culture Media and Reagents for Microbiological and Clinical Laboratory Procedures, 9th ed. Detroit, Michigan. Difco Laboratories, 1953.

Chapter 6

Dynamics of Microbial Growth in Nature

Chapter 5 was concerned with some principles of microbial growth as studied in the laboratory. The growth of microorganisms in nature is much more complex. Instead of growing as *pure cultures* under defined cultural conditions, organisms in nature exist as parts of *communities* under poorly defined environmental conditions that are often changing. In the laboratory the investigator provides the nutrients necessary to ensure the optimal growth of the culture. In nature food supplies are generally limiting, and intense competition between populations develops. Organisms may be highly dependent on the presence of other specific organisms. In the laboratory organisms are often grown in liquid culture, whereas in nature most microorganisms grow in association with solid surfaces—on soil particles, sediment particles in marine and fresh water, and on plants and animals. In this chapter some *principles* which govern the growth and development of populations in nature as well as the relationships between organisms and their environment are emphasized. Specific examples of these interrelationships will be considered throughout the text, which in large measure is concerned with the growth of organisms in their natural environment.

PRINCIPLES OF MICROBIAL ECOLOGY

All the living organisms of a given area, the community, and the nonliving environment of that area, interact to form an ecological system or *ecosystem,* the unit of biological organization above the community. Many major ecosystems exist in the world, each with its particular physical environment and spectrum of living organisms. These ecosystems include the seas, streams and rivers, lakes and ponds, marshes, deserts, tundras, grasslands, and various types of forests. Each ecosystem is populated with microorganisms which may be unique to the particular ecosystem. *Ecology* is the study of the relationship of organisms to each other and to their environment. The *function* that an organism performs in its particular *ecosystem* is termed its *ecological niche.*

Cycles in Nature

In order for a major ecosystem to sustain the life characteristic of that particular area, all of the chemical elements that make up living cells must be continuously recycled. A fixed and limited amount of these elements exists on the earth and in the atmosphere. However, these elements cannot be utilized for the synthesis of cell components unless they are in a specific form. Only if these elements are continuously recycled from the nonutilizable to the utilizable forms can there be enough to support the growth of new organisms. Thus there are recycling processes for nitrogen, phosphorus, sulfur, oxygen, and carbon. As an example of how these cycles function, the carbon cycle is diagrammed in Figure 6-1. Plants are eaten by animals which die and decompose through the action of numerous microorganisms. A number of microorganisms are capable of degrading the complex cell constituents to inorganic substances, such as nitrates, sulfates, and phosphates.

CARBON CYCLE

The carbon cycle revolves around CO_2, its *fixation* into organic compounds by green plants, and its regeneration into CO_2 primarily by microorganisms. The cycle begins with the fixation of atmospheric CO_2 by green plants by the process of photosynthesis. Some of the carbohydrate synthesized is utilized by the plant to obtain energy, with the release of CO_2 through the leaves or roots by a process termed *respiration,* a process which in many ways is the opposite of photosynthesis. The plants provide the animal world with organic material. The animals also respire and release CO_2. When animals and plants die, they are ultimately decomposed by heterotrophic microorganisms in the soil, with the production of more microbial cells and the decomposition products. These chemical transformations of carbon require the activity of organisms in three ecological niches, *producers, consumers,* and *decomposers.* The producers convert CO_2 into organic material. The consumers eat organic matter; the decomposers digest and convert cell components into small molecules utilizable by the consumers as well as the producers.

ORGANISMS PARTICIPATING IN CARBON CYCLE

In order for recycling to operate members of each functional group must be present in every major ecosystem for each element present in cell structures. The loss of any group would break the cycle. Members of the microbial world are indispensable for the operation of the carbon cycle as well as cycles of all other elements of biological significance. In the carbon cycle the producers are blue greens, algae and those bacteria which are capable of utilizing the carbon of CO_2 as the sole carbon source. Protozoa are typical consumers, pursuing and eating algae, bacteria, or other smaller protozoa. Most members of the microbial world are able to decompose organic material.

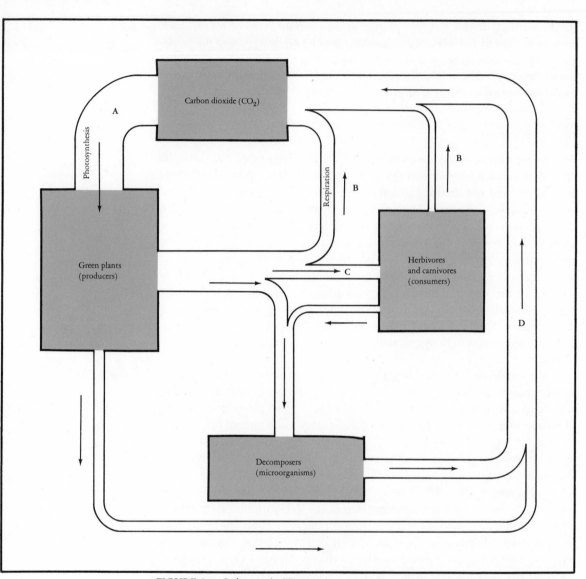

FIGURE 6-1 Carbon cycle. The 4 major steps in the pathway are lettered: **A**—The conversion of inorganic carbon (CO_2) into organic forms by green plants; **B**—The conversion of organic carbon into CO_2 through metabolic processes of plants and animals; **C**—The eating of green plants by animals; **D**—The conversion of organic forms of carbon into CO_2 through decomposition of dead plants and animals by microorganisms. The width of the tubing in the pathway gives an indication of the relative amount of material channeled through that pathway.

ELEMENTS ARE
RECYCLED; ENERGY IS
NOT

In addition to a supply of elements for the synthesis of cell components, all ecosystems must have access to sources of utilizable energy. In contrast to the recycling of chemical elements, energy cannot be reutilized. Once it is used it is no longer available for use. Thus there is a continual loss of energy from biological systems. To compensate for this outflow there must be a continual inflow. The energy which is being continually added to all ecosystems is solar or light energy. This form must be converted into chemical bond energy before it is useful. This conversion, mediated by chlorophyll-containing plants and microorganisms, makes the energy available to the consumers and decomposers. The requirement for solar energy explains why life does not exist everywhere in and on the earth. It is confined primarily to a thin shell in the upper levels of the soil and water and to the lower levels of the atmosphere. It does not thrive in deep soil or rocks or in the depths of the oceans. Any organisms present in these zones must depend for food on the organisms that have access to light.

Populations in Ecosystems

The regions of the earth inhabited by any form of life are termed the *biosphere,* and its major subunits are the *ecosystems.* Within the total biosphere a wide diversity exists both in the number of different types or species of organisms inhabiting particular ecosystems and in the total quantity of cell material present in any given area. What determines whether or not a particular organism will be found in a particular ecosystem? The fundamental requisite for all organisms in any given environment is that the organism be capable of multiplying. Otherwise it will soon disappear. In an environment conducive to life, such as the top few inches of fertile soil, there will likely be a large variety of different species as well as a large quantity of protoplasmic mass. Table 6-1 shows the relative abundance of microorganisms in fertile surface soil. The bacteria are represented by many different species, and it has been estimated that the top 6 inches of fertile soil may contain more than 2 tons of fungi and bacteria per acre. In harsh environments which are exposed to extremes in temperature, moisture, or other environmental conditions, relatively few species are capable of multiplying, and therefore existing.

The external environment has a more profound effect on conditions inside the microbial cell than it does on cells in higher organisms. Multicellular organisms contain sensitive controls in specialized organs which maintain reasonably constant internal conditions despite wide fluctuations in the external environment. In microorganisms the external and internal environment may be similar, although even bacteria have the ability to concentrate preferentially essential foodstuffs and keep out some potentially harmful chemicals.

ABILITY TO MULTIPLY CRUCIAL TO LIFE

Permeases
Pages 86–87

TABLE 6-1
Number of Organisms at Different Depths in the Soil—(organism per gram of soil)

Depth (cm)	Bacteria except Actinomycetes[a]	Actinomycetes	Algae	Fungi
3–8	9,750,000	2,080,000	25,000	119,000
35–40	570,000	49,000	500	14,000
135–145	1,400	0	0	3,000

[a] Actinomycetes are branching filamentous bacteria that are very plentiful in the soil.
Data taken from M. Alexander, *Introduction to Soil Microbiology,* John Wiley, New York, 1961.

Actinomycetes
Page 63; Table 3-4

The true environment of a microorganism is often difficult to identify and measure. The environment which immediately surrounds the microbial cell, the *microenvironment,* is that most relevant to the cell. The *macroenvironment,* which is the more readily measured gross environment, has probably far less effect on the cell. For example, to the bacterial cell living within a crumb of soil the environment inside the crumb is more important than the environment surrounding the crumb. The two environments may differ radically from each other. Indeed a variety of different environments may actually be found within any single crumb.

Competition for Survival

Although many different species may be capable of living in the same environment, it has been observed repeatedly that few species—sometimes only one— actually occupy a specific microhabitat. To return to the previous example, a large number of different species are found within the crumb of soil. If the crumb is carefully dissected into squares, about 70 μm on a side, only a single species is found living as a microcolony within any one of these squares. Thus the heterogeneity in the variety of species in the total crumb disappears when we consider each individual microenvironment. The species best adapted to live in each particular microenvironment is apparently the one that occupies it to the exclusion of all others. The organism best fitted to the particular environment is the one that persists.

The fitness of an organism as measured by its ability to compete successfully for its ecological niche or for its microhabitat is generally related to the *rate* at which the organism multiplies. If cells of two different species are introduced into a microhabitat which is suitable for both organisms, the organism which multiplies faster will soon become the dominant organism and the other species may disappear completely. The complete takeover by one of the species is especially likely when the two species have similar requirements for cell growth and when one of these requirements is in limited supply. Thus with a limited supply of food, the organism which multiplies faster will yield the greater cell population, thus utilizing most of the limited food supply. The more slowly growing species is deprived of more and more food, and it may soon be eliminated.

FASTER GROWING
ORGANISMS TEND TO
CROWD OUT SLOWER
GROWING CELLS

Perhaps nowhere in the living world is the competition more fierce and the results of competition more quickly evident than among microorganisms. This is especially true of organisms growing together in liquid culture. Since cells are often able to divide at least once every several hours, any small difference in the generation time will result in a very large difference in the total number of cells of each species which exist after a relatively short time. For example, if 10 cells each of organisms A and B are growing together in the same natural environment, under conditions in which the generation time of A is 100 minutes and B is 99 minutes, and if the cells reach about 10^9 cells, there will be 1.35 more cells of B than A. If the cells are diluted occasionally so that they can continue

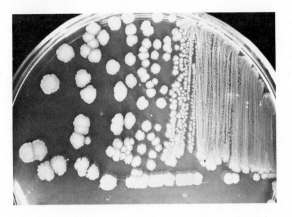

FIGURE 6-2
Competition for nutrients among bacteria. The closer the cells are
to one another, the smaller the colonies. (Courtesy of J. P. Dalmasso)

to multiply, the ratio of B to A will increase to the point where there are very
few A cells left.

In some cases, especially evident in colonies growing in fixed positions, the
cells only hamper each other's growth by competing for the same nutrients. Strict
competition can be observed when bacterial colonies grow very close to each
other on agar plates. The cells compete with neighboring cells, with the result
that colonies close to each other are much smaller than colonies which are well
separated from one another (Fig. 6-2).

Interdependence Between Populations

The existence of many populations depends on the presence of other populations.
For example, anaerobic organisms may exist in association with facultative anaerobes which use any available oxygen, thus creating anaerobic conditions. The
interrelationships between organisms in a medical context are covered more fully
in Chapter 19.

Changes in Microbial Populations

Environments do not remain constant, and changes in the environment bring
about changes in the population. The "fittest" organism several inches beneath
the surface of an untilled field probably fails to remain the fittest if the field is
plowed up, fertilized, and irrigated. If the organism is to survive under these new
conditions, it must in some way *adapt* to the changed environment. Cells can
adapt in a number of ways. The same cells may be able to synthesize new enzymes
which help them cope with the new environment. As a general rule, cells will
only synthesize the enzymes that they absolutely require for growth. To synthesize others is a waste of amino acids and the energy required for their synthesis.
Under changed environmental conditions, however, additional or different enzymes may become important. The cell now synthesizes these enzymes, and may
stop synthesizing other enzymes which have lost their usefulness under the

POPULATION CHANGES
MAY RESULT FROM
ADAPTATION OF ENTIRE
POPULATION

changed conditions. Frequently, the *preexisting* population changes. Thus a pre-existing *mutant* within a population may be especially well adapted to the changed environment. Whereas the mutant was in the minority in the old environment, it may become the dominant organism in the new environment.

The emergence of antibiotic-resistant organisms in hospital environments illustrates this type of adaptation. The antibiotic-resistant mutants always occur as a very small proportion of the total population. In the absence of antibiotics they generally multiply more slowly than the antibiotic-sensitive cells and therefore are *selected against*. However, in the presence of antibiotics, sensitive cells are killed and only antibiotic-resistant cells are capable of multiplying. Accordingly, an antibiotic resistant population replaces the sensitive population. The dramatic increase in the antibiotic-resistant cells in hospitals since the use of antibiotics is in part the result of this phenomenon.

Another possible alteration in the population in response to a changed environment is the replacement of the dominant species by another species originally present as only a minor species in the ecosystem. Whereas the latter species could not compete well with other organisms in the old environment, the new environment is much more favorable to its multiplication. Therefore it now becomes the dominant organism. One example of such a situation can be demonstrated on the skin. Gram-positive species of bacteria are the dominant organisms in the armpit (axilla), although small numbers of gram-negative organisms are also present. The gram-positive organisms excrete enzymes that degrade the greasy secretions of glands; the fatty acids produced inhibit the growth of the gram-negative cells and give rise to "body odor." If a deodorant containing an agent which kills gram-positive organisms is applied, the population gradually changes principally to a gram-negative one of about the same cell density as the gram-positive flora. The gram-positive flora returns as the use of the deodorant ceases.

In addition to changes in the environment which result from external sources, the growth and metabolism of organisms themselves may change the environment dramatically. Foodstuffs may become depleted and a variety of waste products, many of which are toxic, may accumulate. In some environments the changing conditions bring about a highly ordered and predictable *succession* of organisms: first one species of organism, then another, then a third becomes dominant. An example of such a succession is the one occurring in unpasteurized

milk. Such milk contains a variety of different organisms including various species of bacteria, yeasts, and molds which are derived mainly from the immediate environment surrounding the cow. Initially, the dominant organism is a bacterium (*Streptococcus lactis*) which breaks down the milk sugar lactose to lactic acid. The acid inhibits the growth of most other organisms in the milk and enough acid is eventually produced to even inhibit the further growth of the *Streptococcus*. Acid sours the milk and *denatures* (curdles) the proteins in it. However, other bacterial species (*Lactobacillus casei* and *L. bulgaricus*) are capable of multiplying in this highly acidic environment. These species break down the rest of the lactose and form more acid until their growth is also inhibited. Yeasts and molds, which grow very well in this highly acidic environment, now become the dominant group and convert the lactic acid into nonacidic products. Since all of the milk sugar has already been broken down by the bacteria, the major genera of bacteria, streptococci and lactobacilli, cannot resume multiplication because of the lack of food. Food, however, is still available in the form of protein (the casein) which can be utilized for energy by members of the genus *Bacillus* which excrete protolytic enzymes to digest the protein. The breakdown of protein, *putrefaction,* results in a product which is completely clear and very odorous. Thus the milk goes through a succession of changes with time, first souring and then finally putrefying. These changes are brought about by organisms which themselves have undergone a succession in their dominant population.

If milk is *pasteurized* after being drawn from the cow, a markedly different succession may occur. The pasteurization kills most of the nonendospore-forming organisms (streptococci, lactobacilli, yeasts, and molds) and the heat-resistant spores of the bacilli may predominate. These organisms break down the proteins, resulting in putrefaction. The milk does not sour, it rots (putrefies).

SUCCESSION IN
PASTEURIZED MILK

Quantitative Aspects of Growth of Populations in Nature

The number and types of microorganisms in many ecosystems remain relatively constant over a reasonably long period of time as long as the environment remains relatively constant. The types of organisms that are present are those that have adapted best to the ecosystem. Population size is generally determined by the outcome of the competition for limiting nutrients, since in natural environments the food supply is generally the limiting factor. Although the number of cells of each species remains relatively constant, the cells are not inert. Generally, there is a large turnover of the population, some cells may be killed and decompose, or are devoured while others multiply to take their place, keeping the total number constant. The population is actually in the log phase of growth, but the limitation in the amount of food available limits the size that the population can reach. In addition, the environmental conditions are seldom optimal for the growth of the microorganism, so its generation time is much longer than if it were growing under optimal conditions.

SIZE OF POPULATIONS
MAY REMAIN CONSTANT
THROUGH BALANCE OF
CELL MULTIPLICATION
AND DEATH

The two phenomena, cell turnover and exponential growth at a slow rate, can be duplicated and controlled in the laboratory by growing the cells in a device known as a *chemostat.* This device, diagrammed in Figure 6-3, consists of a growth chamber connected to a reservoir of nutrient medium and also to an outlet. The volume of liquid in the growth chamber remains constant since every drop that enters from the reservoir displaces an equal volume of culture that overflows. Thus there is a continual turnover of the population in the growth chamber. In nature this turnover would result from the action of predators, dying, and removal by environmental conditions. In the chemostat the rate of turnover can be easily controlled by the investigator by regulating the rate at which the drops flow into the system. The nutrient medium which enters the growth chamber

DIVISION RATE
CONTROLLED BY
LIMITATION OF
NUTRIENTS

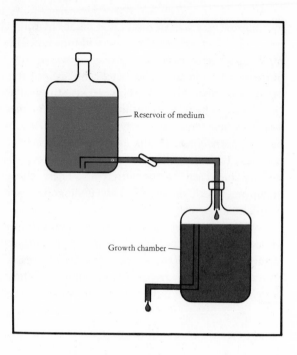

FIGURE 6-3
Chemostat. The cell concentration in the growth chamber is maintained at a constant level since the bacteria are limited in their growth by the limiting nutrient which flows from the reservoir. The time required for one volume of medium in the growth chamber to be replaced by one volume of medium from the reservoir is the generation time of the cells.

contains a limiting amount of a required nutrient. In nature this might well be the energy source. The organisms rapidly utilize all of this limiting nutrient and multiply. Without this nutrient the cells cannot grow and therefore the cell density (number of cells per milliliter) is proportional to the concentration of the limiting nutrient available. As long as the concentration of limiting nutrient remains constant, the cell density will not change.

Growth of Bacteria in a Hospital Environment

A natural habitat which illustrates a number of interesting and unusual aspects of bacterial growth is the reservoir of distilled water in mist therapy units commonly used in hospitals to combat lung diseases. In one published case approximately 10^7 viable cells per milliliter of viable *Pseudomonas aeruginosa* were found in the distilled water in these units. When these cells were diluted into sterile distilled water at a cell density of approximately 10^3 per milliliter and the culture incubated, after 48 hours there were approximately 10^7 cells per milliliter. The cell density remained at this level for at least 42 days.

These experimental results demonstrate in a practical sense some points of the growth of bacteria in nature. It is clear that the bacteria are able to grow in a medium which contains a very low level of nutrients. It is not clear what these nutrients are, but they are likely organic compounds that were absorbed either from the air or from inadequately cleaned units. The fact that bacteria can grow in such dilute medium indicates that they are able to effectively concentrate the nutrients inside themselves.

Species of *Pseudomonas* are commonly found in apparently inhospitable environments since members of this genus can utilize a wide variety of organic materials as sources of carbon and energy. Apparently the limitation in nutrients prevents the cell density from increasing much above 10^7 cells per milliliter. The cell concentration is not high enough to result in a noticeably turbid solution, but it is high enough to represent a definite menace if the contaminated water is sprayed into the lungs of a weakened patient. The generation time of the cells growing in the mist therapy units is about 4.5 hours, which is not the fastest the organisms are capable of growing, but nevertheless fast enough to reach high levels in a relatively short time. Their growth is slower than the optimum probably because of the lower than optimal growth temperature.

In this situation a dangerous organism can multiply without detection (no noticeably visible turbidity) in an inhospitable environment, distilled water. The organisms can then be transferred to other locations in the hospital via the airborne route, fomites, or personnel. It is not surprising that *Ps. aeruginosa* is an important causative agent of hospital-acquired infections.

SUMMARY

Microorganisms are ubiquitous in nature and are found in areas which will support no other forms of life. The organisms living in a particular environment have become adapted to the physical and nutritional aspects of the environment. Their interactions with other organisms may include competition, generally for limiting sources of food and interdependence on one another. Organisms fill ecological niches in any ecosystem, and may be classified as producers, consumers, or decomposers. The combined operation of these groups results in the recycling of the elements found in biological materials. This recycling is essential to the maintenance of life.

The organismal content of an ecosystem remains relatively constant as long as the environment does not change. However, although the numbers of each group of microorganisms do not change significantly, there is a continual turnover of organisms, some dying, others multiplying. The growth of microorganisms in nature is generally considerably slower than it is in the laboratory although the organisms are in the log phase of growth. Some of the quantitative features of bacterial growth in nature can be duplicated by a chemostat in the laboratory.

QUESTIONS

1. Why should we "Thank a Green Tree Today"?
2. By what experimental approach could you decide between adaptation of the whole population and mutation to account for changes in microbial populations in natural environments?
3. What qualities must members of the genus *Pseudomonas* have that allow them to grow in presumably distilled water?

4. What metabolic capabilities are exhibited by: (a) producers, (b) consumers, (c) decomposers?

5. In operating a chemostat, how would you increase the cell density? Decrease the doubling time?

FURTHER READING

ALEXANDER, M. J., *Microbial Ecology.* New York: John Wiley and Sons, Inc., 1971.

BOLIN, B., "The Carbon Cycle," *Scientific American* (September 1970). (Offprint #1193)

BROCK, T., *Principles of Microbial Ecology.* Englewood Cliffs, N.J.: Prentice-Hall, Inc., 1966.

FAVERO, M., L. CARSON, W. BOND, and N. PETERSON, "*Pseudomonas aeruginosa:* Growth in Distilled Water from Hospitals," *Science,* 173 (1971), 836–838.

MARPLES, M., "Life on the Human Skin," *Scientific American* (January 1969). (Offprint #1132)

Chapter 7

Metabolism

The growth of all organisms requires the synthesis of each and every component of cytoplasm. This synthesis involves the breakdown of foodstuffs to provide compounds which are converted to cell material and to supply energy. This energy is used to drive the large number of biosynthetic reactions by which the cell converts relatively simple molecules to the supramolecular structures that make up cytoplasm. One fantastic aspect of this entire operation is the unbelievable speed with which cell components are synthesized. For example, one *Escherichia coli* cell dividing every hour synthesizes 4000 molecules of lipid, almost 1000 protein molecules (each containing about 300 amino acids), and 4 molecules of RNA, *per second.*

All of this is made possible through the action of hundreds of different enzymes—the biological catalysts. They allow the cell to perform reactions at relatively low temperatures. These same reactions cannot be duplicated in a chemical laboratory except by employing considerably higher temperatures. Before we discuss the biology of energy generation and biosynthesis it will be helpful to consider the properties and mechanisms of enzymes.

CHEMICAL KINETICS

Enzymes are organic catalysts which speed up a chemical reaction without being used up or changed by the reaction. Thus if the conversion of a substrate to a product is measured at varying intervals, the data presented in Figure 7-1 are observed. In the absence of the enzyme, the *substrate* (the material on which the enzyme acts) remains essentially unchanged. In the presence of an enzyme, however, the substrate is converted to a product.

137

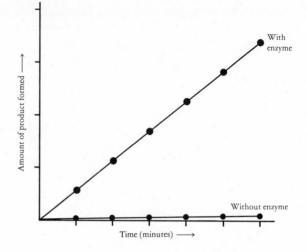

Covalent chemical bonds
Pages 23–25

FIGURE 7-1
Effect of enzyme addition on product formation.

How does an enzyme act? An understanding of enzyme action requires some simple, yet basic knowledge of the kinetics of chemical reactions. Much of this discussion relates to information given in Chapter 2, the formation of chemical bonds. *All reactions represent a rearrangement of molecules* whether or not they are catalyzed by an enzyme. Any rearrangement requires the breaking of some old bonds and the formation of new ones. Much of this discussion relates to a specific example, the formation of water (H—O—H) from oxygen (O—O) and hydrogen (H—H). When both gases are brought together, they will not react. However, if some form of energy is added to the system, such as a spark, a violent explosion erupts with the release of energy in the form of heat and light. Water (HOH) is formed. The reaction can be summarized:

$$2H—H + O—O \longrightarrow 2H—O—H + energy$$

Note that new bonds are formed, O—H in place of H—H and O—O. Two aspects of this reaction are considered. Why must energy be put *into* the system before a reaction occurs and why is energy released in the reaction? Understanding these features aids in understanding enzymatic reactions.

Activation Energy

When oxygen and hydrogen gas are mixed, the gaseous molecules collide with one another and with the walls of the container. Each molecule has a certain level of energy, that is *kinetic energy, the energy of motion.* This energy, however, is not sufficient at room temperature to allow the hydrogen atoms to break the strong covalent O—O bonds or for the oxygen atoms to push apart the H—H bonds. No new bonds are formed because atoms can only recombine with new partners after they have been partially separated. As the temperature increases, however, the hydrogen and oxygen molecules move more rapidly, and some collisions have sufficient energy to break apart the H—H and O—O bonds and form O—H bonds. An electric spark initiates the reaction, but once started, the heat energy given off by the reaction of a few molecules is sufficient to raise the kinetic energy of other molecules. A chain reaction is thus set in motion, eventually resulting in an explosion.

The reason energy must be first put into the system can be explained by the analogy described in Figure 7-2. The hydrogen and oxygen molecules are represented as balls located near the top of a hill. Potentially the molecules have considerable kinetic energy. However, this energy can only be expressed after they have rolled down the hill. But they are prevented from rolling because they are trapped in an "energy depression." If enough energy is supplied so that the molecules can surmount this hurdle, they will spontaneously roll down and generate kinetic energy. The amount of kinetic energy generated is much greater than the amount needed to bring the molecules out of their "energy depression." The amount of energy required to lift the molecules out of the depression is called the *activation energy. This is the site of action of enzymes.* Enzymes *lower* the activation energy to values which can be reached by the kinetic energy of the molecules at room temperature. Thus the enzyme-catalyzed reaction proceeds without any additional heat. The activation energy barrier is enormously important to life. Without it there would be a continuous rearrangement of all atoms and molecules and life could not exist. It would indeed be difficult to live in a world in which wood and coal could burst into flames spontaneously.

Just exactly how enzymes lower the activation energy is still not well un-

ENZYMES LOWER
ACTIVATION ENERGY

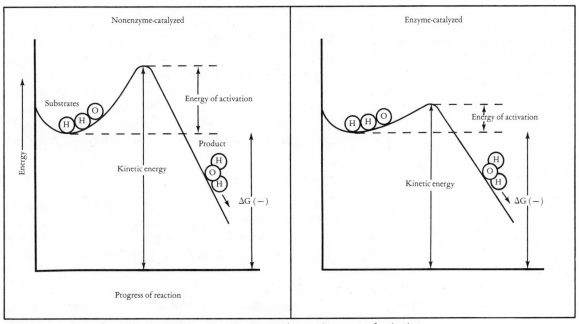

FIGURE 7-2 Energy barrier in chemical reactions. The enzyme lowers the energy of activation enough so that the energy in the molecules of H and O is sufficient to raise them to the top of the hill.

derstood. Part of their action can be explained by the fact that when enzymes combine with their substrate through weak bonding forces, they place a stress on those chemical bonds which will be broken. This stress weakens bonds in the substrate enough so that they are able to break and reform new bonds.

Free Energy

In the reaction of hydrogen with oxygen (returning to the analogy introduced in Fig. 7-2), once the molecules are elevated to the top of the hill, they will spontaneously roll down, releasing energy as they do so. In the terminology of chemical kinetics, there is a *decrease in free energy* as the molecules roll down the hill. The symbol for free energy is G; *a decrease* in free energy is noted as $-\Delta G$, an *increase* by $+\Delta G$. The letter G honors the great nineteenth century American theoretical physicist J. Willard Gibbs, of Yale University. This decrease in free energy is equal to the kinetic energy which is released as the molecules reach the bottom of the hill minus the amount of energy required to start them rolling (the activation energy). This relationship of the decrease in free energy to the activation energy and the kinetic energy is diagrammed in Figure 7-2.

The reason energy is released in the reaction

$$2H\text{---}H + O\text{---}O \longrightarrow 2H\text{---}O\text{---}H + \text{energy}$$

Bond strength
Page 25

involves the principle stated in Chapter 2. There are differences in the strength of bonds. The O—H covalent bond is stronger than either the H—H or O—O bond, and therefore energy will be released in its formation.

ENZYMES CATALYZE
ONLY REACTIONS THAT
WILL PROCEED SLOWLY
IN THEIR ABSENCE

All reactions are characterized by a decrease in free energy; that is, energy *must* be released. A reaction will proceed only if $A + B \rightarrow C + D$ yields a drop in free energy indicated as $-\Delta G$. In principle all reactions are reversible, although the reaction may proceed so strongly in one direction if there is a large drop in free energy that for all practical purposes it is irreversible. Thus in the reaction of H—H and O—O to form H—O—H, the reaction will proceed strongly in the direction of H—O—H formation. However, the reaction can be reversed if somehow the reaction gains a $-\Delta G$. This can be achieved if energy is put into the system:

$$\text{Energy} + 2H\text{---}O\text{---}H \longrightarrow 2H\text{---}H + O\text{---}O.$$

Many reactions of biological importance will proceed only if energy is supplied. In these reactions, which generally involve the biosynthesis of metabolites, an energy-yielding reaction $(-\Delta G)$ is *coupled* to the energy-requiring reaction $(+\Delta G)$, resulting in net loss of free energy $(-\Delta G)$. The reaction will then proceed. The coupling between an energy-requiring reaction and an energy-yielding reaction is mediated by a compound which is a component of the energy-yielding as well as the energy-requiring reactions. This compound is frequently adenosine triphosphate (ATP), a compound which contains chemical energy in the form of high-energy bonds. When high-energy bonds are broken, large amounts of energy are released, which can then be used by the energy-requiring reactions.

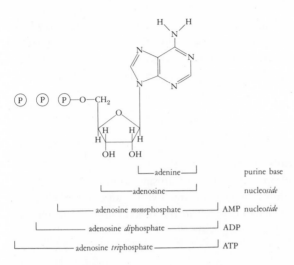

Subunits of RNA & DNA
Page 40; Fig 2-15

FIGURE 7-3
Formula of adenosine triphosphate (ATP).

Thus the cell breaks a high-energy bond which releases more energy than is actually required in the energy-requiring reaction. The sum of the energies released $(-\Delta G)$ and required $(+\Delta G)$ will yield a $-\Delta G$, and so the reaction will proceed.

High-Energy Bonds

All cells, plant, animal, and bacteria use the same chemical as the major storage form of chemical energy. This is the molecule ATP. This molecule contains three phosphate molecules which are linked to adenosine (Fig. 7-3). The most important feature of this molecule is that the breakage of either of the two terminal phosphate groups results in the release of large amounts of energy (Fig. 7-4). These high-energy bonds are shown by a wavy line instead of the usual straight line.

ATP IS MAIN ENERGY CURRENCY IN CELLS

Free Energy of Food Molecules

Since all covalent bonds do not have the same bond energy, the amount of free energy possessed by specific molecules varies greatly. The foods which are metab-

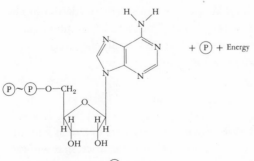

FIGURE 7-4
Energy is liberated when the high-energy phosphate bond of ATP is broken to yield ADP and inorganic phosphate.

olized by man and microorganisms contain a great deal of energy. There is a loss of free energy when the foods, such as glucose and other sugars, are converted to CO_2 and H_2O. The free energy released in this conversion is captured in part in the high-energy bonds of ATP and used for energy as the cell needs it.

ENZYMES

The enzymes represent probably the most thoroughly studied class of biologically important molecules. They deserve special attention since almost every chemical reaction that takes place within living organisms involves an enzyme. How an organism looks and how it behaves depend on the functioning of enzymes. If key enzymes no longer function, the organism dies.

Nomenclature

Unfortunately enzymes have not always been named according to a systematic rule. However, most enzymes end with the suffix *ase* and are commonly named after the substrate on which they act or the type of reaction that they catalyze. For example, an enzyme that degrades cellulose is termed a *cellulase;* the enzyme which degrades DNA is a *deoxyribonuclease.* The enzyme that polymerizes mononucleotides into an RNA polymer is called RNA *polymerase.*

Function

Enzymes lower the barriers which must be surmounted before chemical reactions can proceed. By lowering the activation energy, the speed of a reaction is greatly increased (Fig. 7-1). However, if the reaction has a $+\Delta G$, it will not proceed whether or not an enzyme is present. Conversely, if a reaction has a $-\Delta G$, it will proceed even in the absence of the enzyme. However, it may take a very long time before the products of the reaction can be detected by the methods presently available. In order to function the substrate binds to a specific portion of the enzyme, the *active* or *catalytic site,* to yield an enzyme-substrate complex. Next, the products are released leaving the enzyme unchanged. The substrates which can bind to any one enzyme are limited since a complementary relationship between the substrate and enzyme is required. This "lock-and-key" arrangement is diagrammed in Figure 7-5. This specificity requires essentially that every reaction in the cell be catalyzed by a different enzyme.

Structure of Enzymes

All enzymes are proteins. Each enzyme has a unique sequence of amino acids that determines its three-dimensional structure and therefore the reaction it will catalyze.

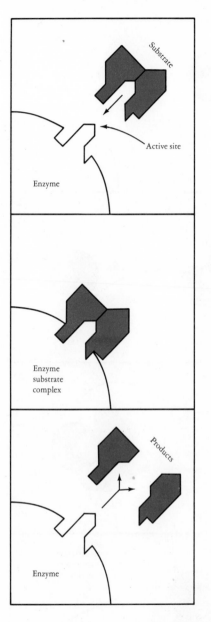

FIGURE 7-5
Enzyme catalysis. The requirement for a lock and key relationship
between enzyme and substrate.

Some enzymes contain a nonprotein portion that is essential for its functioning. If the nonprotein portion is an organic molecule that is readily separated
from the enzyme, it is termed a *coenzyme*. Coenzymes generally function in the
transfer of small molecules from one protein to another. Thus a coenzyme associates with an enzyme as it binds a substrate molecule, picks up a small molecule
or atom (such as a hydrogen atom) from the substrate, dissociates from that enzyme, and transfers the small molecule to another protein (Fig. 7-6). The coenzymes are not specific, but can associate with a number of different enzymes.

COENZYMES SERVE TO
CARRY MOLECULES
FROM ONE PROTEIN TO
ANOTHER

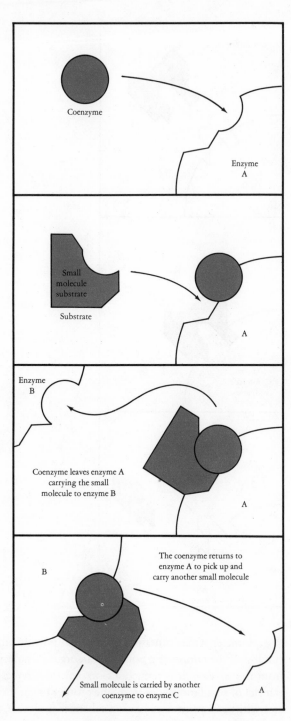

FIGURE 7-6

Coenzyme-enzyme interaction. The coenzyme carries molecules from enzyme to enzyme; thus one coenzyme interacts with a large number of different enzymes.

TABLE 7-1
Coenzymes

Name of Coenzyme	Vitamin from Which It Is Derived	Entity Transferred
Nicotinamide adenine dinucleotide (NAD)	Niacin	Hydrogen atoms
Flavin adenine dinucleotide (FAD)	Riboflavin	Hydrogen atoms
Coenzyme A	Pantothenic acid	Acetyl group
Thiamine pyrophosphate	Thiamine	Aldehydes
Pyridoxal phosphate	Pyridoxine	Amino group

However, each coenzyme is able to transfer only one type of small molecule or atom.

Coenzymes are comparatively small molecules, compared to enzymes, and are synthesized from vitamins by a series of enzymatic steps (Table 7-1). Since coenzymes, like enzymes, are recycled and therefore are required in minute quantities, the required dietary level of the vitamins, whether for man or microorganism, is very small. The lack of a vitamin which is converted to a coenzyme results in the nonfunctioning of all of the different enzymes that require the coenzyme to function.

Factors Influencing Enzyme Activity

The growth of any organism depends on the functioning of most of the enzymes that it contains. The environment exerts its influence on cell growth primarily by influencing the activity of enzymes. The optimum temperature, pH, and salt concentration for growth of any cell represents the best average for the functioning of all of the cellular enzymes.

EFFECT OF ENVIRONMENT ON BACTERIAL GROWTH REFLECTS EFFECT ON CELLULAR ENZYMES

In general, increasing the temperature promotes more rapid bacterial growth. This is true because an increase in temperature increases the energy of motion of molecules and the rates (speed) of reactions. However, if the temperatures become too high, the enzymes are denatured and the organism dies.

Effect of Inhibitors on Enzyme Activity

The most potent poisons known exert their effect by inhibiting enzyme function. Thus cyanides, arsenic, mercury, and the nerve gases all combine with and prevent the functioning of enzymes necessary for survival. The inhibition of enzyme function also plays an important role in the treatment of bacterial infections. The classic example is provided by the sulfa drugs. Sulfanilamide has a structure very similar to para-aminobenzoic acid (PABA) (Fig. 7-7). This latter compound is enzymatically converted to a required vitamin, folic acid. The structures of PABA and sulfanilamide are so similar that an enzyme involved in converting PABA to folic acid often combines with the drug rather than with PABA. Thus the synthesis of folic acid (and consequently bacterial growth) is inhibited. This

MODE OF ACTION OF SULFA DRUGS

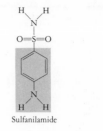

FIGURE 7-7

Structure of sulfanilamide and para-aminobenzoic acid (PABA).

Sulfanilamide   PABA

Sulfa drugs
Pages 564–565

is an example of competitive inhibition, since the drug competes with PABA for the active site of the enzyme (Fig. 7-8). The higher the ratio of sulfa molecules to PABA molecules, the greater the inhibition. Sulfa drugs are selective against bacteria and therefore of value in the treatment of certain diseases because man does not synthesize folic acid from PABA, whereas many bacteria do. Sulfa drugs are discussed further in Chapter 32.

ENERGY METABOLISM

This chapter has thus far considered the physicochemical principles which govern all chemical reactions as well as the functioning of enzymes in cellular reactions. With this background information, we will now consider the metabolic pathways that cells use to degrade foodstuffs and synthesize small molecules.

When a few yeast cells of the species *Saccharomyces cerevisiae* are placed in a glucose-salts medium and incubated for several days, a cursory glance at the flask and a smell of its contents are enough to convince anyone that profound changes have taken place. Whereas the contents of the flask was perfectly clear at the beginning of the incubation, it is now turbid, with bubbles of gas at the surface. In addition, the odor of ethyl alcohol emanates from the previously odorless liquid. The increase in turbidity and the formation of ethyl alcohol and gas reflect the entire spectrum of metabolic activity for the yeast cells. The cells transformed some of the glucose and inorganic salts into more yeast cells, which increases the turbidity of the medium. In addition, they have converted some of the glucose into ethyl alcohol and CO_2. The cell extracts energy from this degradation of glucose. The following equation summarizes the metabolism of glucose:

METABOLISM RESULTS
IN PRODUCTION OF
BOTH ENERGY AND
CYTOPLASM

$$C_6H_{12}O_6 \longrightarrow 2CH_3CH_2OH + 2CO_2 + \text{yeast cells} - \Delta G$$

glucose   ethanol   carbon
                              dioxide

Other organisms degrade glucose to products other than ethyl alcohol and CO_2. But in all instances the degradation results in the release of energy which the cell traps as chemical energy in the form of ATP. Thus the reactions involved in the breakdown of glucose have a large $-\Delta G$. The degradation is also accompanied by the synthesis of cell material. The glucose is metabolized to small molecules that are converted first to subunits and then to macromolecules.

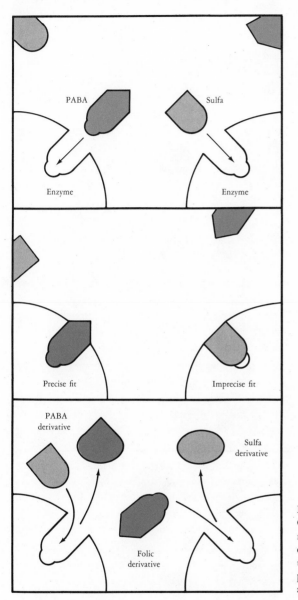

FIGURE 7-8

Competitive inhibition of folic acid synthesis by sulfa. The sulfa molecule occupies the same site on the enzyme as the PABA molecule and therefore competes with PABA for this site. Whether the enzyme will be occupied by a PABA or sulfa molecule depends on the proportion of these two molecules. Both the PABA and sulfa are converted to different products by the enzyme.

Glycolysis

There is a tremendous variety of compounds which cells can degrade. Such compounds include both organic and inorganic materials. Of all organic compounds the one most commonly used in the laboratory as a source of energy is glucose. This sugar can be degraded along a number of different pathways, the most common being termed *glycolysis*. The glycolytic pathway, is diagrammed in abbreviated form in Figure 7-9. In a series of reactions, one glucose molecule is degraded

GLYCOLYSIS IS THE MOST COMMON PATHWAY OF SUGAR METABOLISM

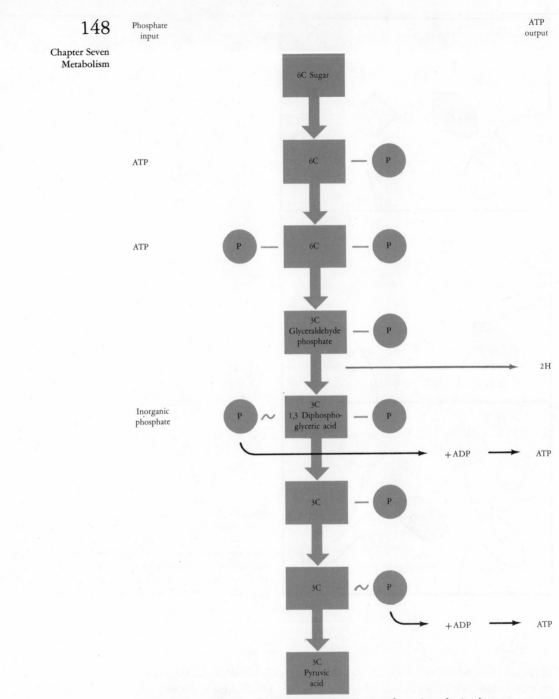

FIGURE 7-9   Glycolysis. This pathway involves the transformation of a 6-carbon sugar to two 3-carbon molecules. In the course of this transformation energy is extracted and transferred in the form of high-energy phosphate molecules to form ATP. Also 2H are removed. These atoms can react with pyruvic acid, also generated, to form a variety of products (shown in Figure 7-10).

to two molecules of pyruvic acid. These reactions involve a gradual drop in the free energy content of the molecule. It is very important that the drop in free energy be gradual—the consequences would be serious if the energy of glucose were released all at once. Consider the heat energy released when fuel burns in an ordinary furnace. The gradual drop requires the participation of a number of enzymes, each decreasing the free energy of its substrate a small amount. The rearrangement of the atoms in the molecules in the course of the breakdown concentrates the energy previously dispersed throughout the glucose molecule. The free energy difference between glucose and pyruvic acid is −52,000 cal/mole; the cell can trap a total of four ATP molecules from each molecule of glucose metabolized. However, since two high-energy bonds are consumed at the beginning of the pathway, there is a net gain of only two high-energy bonds.

Oxidation-Reduction Reactions

In addition to the two molecules of pyruvic acid and the two molecules (net) of ATP produced from each molecule of glucose which is degraded, one other product of glycolysis is produced—four pairs of *hydrogen atoms.* These hydrogen atoms are removed from compounds as they are being metabolized through the glycolytic pathway. Each hydrogen atom consists of one *electron* and a *hydrogen ion* (H^+). The removal of either an electron or H^+ from a compound results in the compound becoming oxidized and the removal is termed an *oxidation.* The biological importance of oxidation reactions is that the decrease in free energy of a molecule is associated primarily with oxidation reactions. The electrons and hydrogen ions do not remain free in solution but combine immediately with another compound. The compound that *accepts* the electrons and hydrogen ions becomes *reduced.* The compound that gives them up becomes *oxidized.* These two reactions (which occur virtually simultaneously) are termed *oxidation-reduction reactions.* Compounds that contain a high proportion of hydrogen atoms such as glucose are highly reduced; those that contain very few, if any, hydrogen atoms such as CO_2 are highly oxidized. Reduced compounds contain much more energy than oxidized compounds, and a potential foodstuff can be recognized as a compound that contains many hydrogen atoms. The degradation of foodstuffs involves a series of reactions in which highly reduced compounds become more and more oxidized. The decrease in free energy of a compound as it is oxidized is largely captured in energy of ATP.

HYDROGEN ATOMS ARE TRANSFERRED FROM ONE TO ANOTHER COMPOUND.

The hydrogen atoms (electrons and hydrogen ions) are removed by enzymes which have an associated coenzyme. This coenzyme, which is a derivative of the vitamin niacin, is *nicotinamide-adenine-dinucleotide* (NAD). When it picks up and carries the hydrogen atoms, it becomes reduced and is designated NAD_{red}. Since there is only a limited supply of the coenzyme in the cell in comparison to the large amounts of glucose that are broken down, NAD_{red} must transfer its hydrogen atoms to another compound becoming NAD_{ox} (which is synonymous with NAD) so that it can pick up more hydrogen atoms from other molecules. The acceptor for the hydrogen atoms varies, depending on the partic-

NAD TRANSFERS HYDROGEN ATOMS TO AN ACCEPTOR ATOM.

ular organism and the environment under which the glucose is being metabolized. However, the choice of acceptor for the hydrogen atoms determines the materials that will be synthesized from pyruvic acid.

Metabolism of Pyruvic Acid

PYRUVIC ACID CAN BE CONVERTED TO A VARIETY OF DIFFERENT COMPOUNDS

The metabolism of pyruvic acid can take several different directions, each leading to different end products. Since each reaction requires the participation of a specific enzyme, which not all organisms have, the end products depend to some degree on the species of microorganism. In all of these reactions the hydrogen atoms are taken from one organic compound in the glycolytic pathway and given to another organic compound. If the final hydrogen acceptor is an organic compound, the entire reaction sequence is termed a *fermentation*. These reactions can proceed under anaerobic conditions.

LACTIC ACID

Lactic acid fermentation in two molecules of NAD_{red} can reduce the two molecules of pyruvic acid to two molecules of lactic acid. NAD_{ox} is regenerated.

$$CH_3-\overset{\overset{O}{\|}}{C}-\overset{\overset{O}{\|}}{C}-OH + 2NAD_{red} \rightleftharpoons 2CH_3\overset{\overset{OH}{|}}{\underset{|}{C}}-\overset{\overset{O}{\diagup\!\!\|}}{C}-OH + 2NAD_{ox}$$

pyruvic acid   lactic acid

This reaction is the reaction involved in the souring of milk.

Another common reaction is the conversion of pyruvic acid to acetaldehyde and CO_2:

$$2CH_3-\overset{\overset{O}{\|}}{C}-\overset{\overset{O}{\|}}{C}-OH \rightleftharpoons 2CH_3-\overset{\overset{O}{\|}}{C}-H + 2CO_2$$

pyruvic acid   acetaldehyde

The acetaldehyde then becomes reduced to ethyl alcohol by another enzyme

ETHANOL + CO_2

$$CH_3-\overset{\overset{O}{\|}}{C}-H + 2NAD_{red} \rightleftharpoons CH_3-\overset{\overset{OH}{|}}{\underset{|}{C}}-H + 2NAD_{ox}$$

acetaldehyde   ethanol

This is the *alcoholic fermentation* carried out by yeast (*Saccharomyces*).

Pyruvic acid can be converted to a wide variety of other products of fermentation. Some of the final products of these reactions are given in Figure 7-10. In all cases the metabolism involves the same principles. The NAD_{red} must give up its hydrogen atoms to an organic molecule so that it can participate further in glucose metabolism. Pyruvic acid can be converted to a variety of products which are further reduced after accepting the hydrogen from NAD_{red}.

The metabolism of pyruvic acid has important consequences for many areas of microbiology. To some degree the products may help identify particular orga-

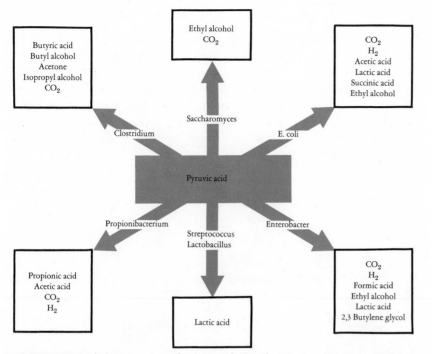

FIGURE 7-10   Metabolic conversions of pyruvic acid. This "key" intermediate in metabolism can be converted to a variety of end products, depending on the organism and the electron acceptors available.

nisms. Therefore one very important tool for identifying disease-causing organisms involves determining their products of metabolism.

Many of the metabolic products have great commercial value. Needless to say the ability of yeast to degrade a wide variety of grains with the production of ethyl alcohol has profound effects on man in all parts of the world. The formation of other reaction products has played important roles in world history. The state of Israel probably owes its existence in large measure to a fermentation process—the acetone-butanol fermentation carried out by certain Clostridia. A biochemist-microbiologist, Dr. Chaim Weizmann while at the University of Manchester in England, isolated an organism *Clostridium acetobutylicum* that converted sugars to acetone in large yields. This process was used by the British in World War I to produce the tremendous quantities of acetone required for the war effort. Dr. Weizmann came in contact with many British political figures as science and politics mixed, and it was his influence that was largely responsible for early events in creating of the state of Israel. Dr. Weizmann served as its first president.

Another very important feature of all fermentation reactions is that most of the energy of the glucose molecule remains untapped. Of the 686,000 calories of bond energy in each mole of glucose, approximately 634,000 calories still remain in the products of fermentation (Fig. 7-9).

ENERGY OF GLUCOSE
MOLECULE REMAINS
LARGELY INTACT
FOLLOWING GLYCOLYSIS

Obviously fermentation extracts only a very small fraction of the total energy available in the sugar. This remaining energy can be tapped if the pyruvic acid is further oxidized to CO_2 and H_2O. All of the energy in the glucose molecule is liberated. This pathway of oxidation of acetic acid is termed either the *Krebs cycle,* after the German biochemist working in England who pioneered in its elucidation, or the *citric acid cycle,* because the first compound formed in the cycle is citric acid (Fig. 7-11). Pyruvic acid is first converted to acetic acid, which is then oxidized to CO_2 and H_2O.

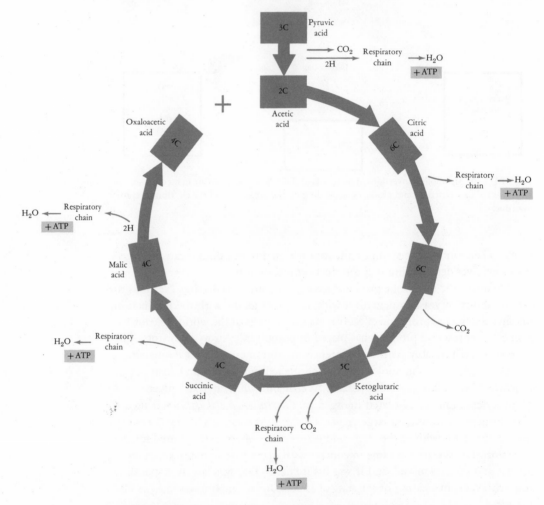

Summary-3C (Pyruvic acid) + O_2 $\longrightarrow$ 2C (Acetic acid) + 3ATP + CO_2

2C (Acetic acid) + $2O_2$ $\longrightarrow$ $2CO_2$ + 12ATP

FIGURE 7-11   Abbreviated form of Krebs cycle. In every turn of this cycle the 2C compound (acetic acid) is converted to two molecules of CO_2. The four carbon compound which combines with the 2C compound to initiate the cycle is always regenerated.

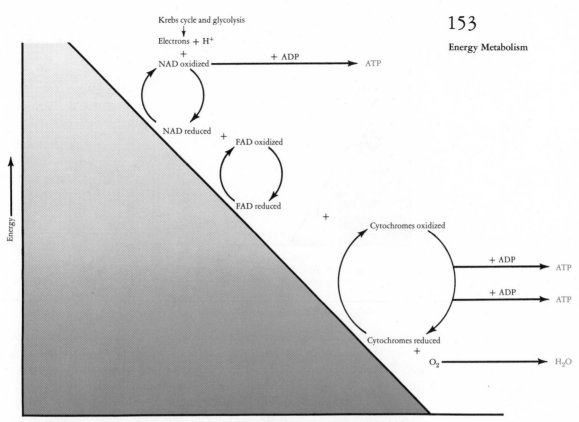

FIGURE 7-12 Diagrammatic representation of the formation of ATP molecules as the electrons and H atoms roll down hill from a high to a low energy level.

Respiration is the name given to the sequence of reactions in which an inorganic compound is the final electron and hydrogen acceptor. The large amounts of energy gained in respiration result from the transfer of electrons through a series of coenzymes termed the *respiratory chain* (Fig. 7-12). Each coenzyme, beginning with NAD, becomes reduced and then oxidized as it picks up and then transfers the electrons to the next member of the chain. The hydrogen ions (H^+) are simultaneously passed along the respiratory chain. The final compound which accepts the electrons and H^+ is usually oxygen, resulting in the formation of water.

HYDROGEN IONS AND ELECTRONS ARE PASSED FROM ONE MEMBER OF RESPIRATORY CHAIN TO ANOTHER

Exactly how the bond energy of ATP is generated as a result of the passage of electrons and hydrogen ions along the respiratory chain is not known. The electrons must have a high level of energy when they are first removed from the compounds in the Krebs cycle. As the electrons are passed along the chain they lose energy. The energy that they lose is converted into the high-energy phosphate bond energy of ATP by some unknown mechanism. The generation of ATP by this process of electron transport is termed *oxidative phosphorylation.*

ATP GENERATED IN COURSE OF TRANSFER

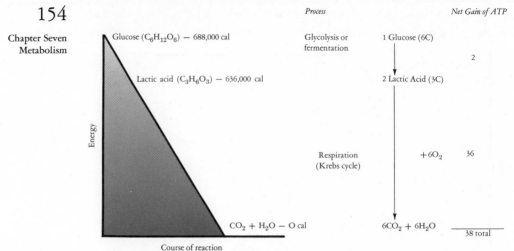

FIGURE 7-13   Energy gain in breakdown of glucose.

Energy Balance Sheet for Metabolism of Glucose

The energy balance sheet for the metabolism of glucose can now be summarized. Figure 7-13 illustrates the tremendous bonus in energy a cell obtains by oxidizing pyruvic acid to CO_2 and H_2O when compared to the energy that it extracts if pyruvic acid or another organic compound is the hydrogen acceptor.

Not all of the energy of the glucose molecule is captured as chemical energy in ATP, since no process is 100 percent efficient. Some of the chemical energy is wasted as heat energy. The cell, however, is remarkably efficient in capturing energy in high-energy bonds of ATP. A net total of 38 moles of ATP are produced in the oxidation of 1 mole of glucose to CO_2 and H_2O. Since one high-energy bond of ATP yields approximately 7000 calories, a total of 7000 $\times$ 38 or 266,000 calories are captured in the form of chemical energy. This represents 42 percent efficiency in energy conservation.

Types of Work

Cells perform basically three types of work: (1) chemical work, which involves the work required to synthesize chemical bonds, (2) work of transport and concentration of molecules from the outside to the inside of the cell, and (3) mechanical work, such as the movement of flagella. When any kind of work is done, the high-energy phosphate of ATP is hydrolyzed, yielding a large amount of energy. The energy released in this reaction drives those reactions which require energy.

Economies of Fermentation versus Respiration

A cell growing under anaerobic conditions must degrade about 20 times more glucose than a cell growing under aerobic conditions to obtain the same amount of energy from the glucose. Since the amount of work which a cell can perform depends on its supply of ATP, a cell growing under aerobic conditions can synthesize far more cell material per unit time (and therefore multiply more rapidly) than the same cell growing under anaerobic conditions. This is especially true if the cell suffers from a shortage of its energy source. In the laboratory the investigator generally supplies the organisms with an excess supply of foodstuffs, but in nature the energy supply generally limits the number of cells in any particular location. Anaerobic conditions also may limit growth because organic acids, commonly the end products of anaerobic metabolism, are often toxic to cells.

Limiting food supply
Page 133

Breakdown of Compounds Other Than Glucose

Many compounds, in addition to glucose, can be degraded to provide energy. Indeed anything that has chemical bond energy can be degraded and its energy converted into the phosphate bond energy of ATP if enzymes are available to catalyze the degradation. Most naturally occurring compounds are degradable by some microorganism, although the pathways by which these compounds are degraded are not known in most cases. It is certainly true, however, that degradation of organic compounds will often generate ATP—most likely through oxidation-reduction reactions.

PATHWAYS OTHER THAN GLYCOLYSIS FUNCTION IN CERTAIN ORGANISMS TO DEGRADE SUGARS AND OTHER METABOLITES

Most organic compounds, including some amino acids, lipids, and carbohydrates, can also be converted into the intermediates of either the glycolytic pathway or the Krebs cycle and thus enter the energy-yielding sequence of reactions.

Anaerobic Respiration

Respiration has been defined as an oxidation in which the terminal electron (or hydrogen) acceptor is an inorganic compound. The most common acceptor is oxygen, and therefore respiration usually occurs only under aerobic conditions. Some organisms, however, can utilize other inorganic compounds as electron acceptors if oxygen is not available. The most common acceptors are nitrate (NO_3^-), sulfate (SO_4^{2-}), and carbon dioxide (CO_2). This type of metabolism is termed *anaerobic respiration,* since it occurs in the absence of oxygen but does involve an inorganic terminal electron acceptor.

Utilization of Inorganic Compounds as Energy Sources

Autotrophic microorganisms can oxidize a variety of inorganic compounds and gain energy. Table 7-2 presents the relevant information about the major groups of autotrophic organisms. Many of these organisms and their metabolic processes are considered elsewhere in the text (Chapters 12, 34, and 36).

Many of these organisms can also gain their energy by oxidizing organic compounds and are termed *facultative autotrophs.*

TABLE 7-2

Energy Metabolism of Some Autotrophic Bacteria

Common Name	Source of Energy	Oxidation Reaction (Energy Yielding)	Important Features of Group	Common Genera In Group
		AEROBIC		
Hydrogen bacteria	H_2 gas	$H_2 + \frac{1}{2}O_2 \rightarrow H_2O$	Can also use simple organic compounds for energy.	Hydrogenomonas
Sulfur bacteria (non-photo-synthetic)	H_2S S	$H_2S + \frac{1}{2}O_2 \rightarrow H_2O + S$ $S + 1.5\,O_2 + H_2O \rightarrow H_2SO_4$	Some organisms of this group can live at pH of 0. Beggiatoa and Thiothrix can only use H_2S as energy source.	Thiobacillus Beggiatoa Thiothrix
Iron bacteria	Reduced iron (Fe^{2+})	$2\,Fe^{2+} + \frac{1}{2}O_2 + H_2O \rightarrow 2\,Fe^{2+} + 2\,OH^-$	Question as to whether Sphaerotilus will oxidize Fe^{2+}, although iron oxide is present in sheaths.	Sphaerotilus Gallionella
Nitrifying bacteria	NH_3 HNO_2	$NH_3 + 1.5\,O_2 \rightarrow HNO_2 + H_2O$ $HNO_2 + \frac{1}{2}O_2 \rightarrow HNO_3$	Important in nitrogen cycle. Important in nitrogen cycle.	Nitrosomonas Nitrobacter
		ANAEROBIC		
Denitrifiers	NO_3^-	$nH_2 + NO_3^- \rightarrow N_2$	Wide variety of heterotrophs can utilize NO_3 as an alternative electron acceptor to oxygen.	A few members of genus Pseudomonas Escherichia
Methane bacteria	H_2	$4H_2 + CO_2 \rightarrow CH_4 + 2H_2O$	Strict anaerobes which use carbonate as electron acceptor.	Methanobacterium
Desulfovibrio (genus)	H_2	$nH_2 + SO_4^{2-} \rightarrow S + H_2S$	Ability to reduce SO_4^{2-} is rare; all are obligate anaerobes. Desulfovibrio is a facultative autotroph responsible for the odor (H_2S) of polluted streams and the black color of mud flats (FeS).	Desulfovibrio

Enzyme Differences Between Aerobes and Anaerobes

ANAEROBES LACK
CERTAIN ENZYMES OF
RESPIRATORY CHAIN

Since anaerobes cannot utilize oxygen as a terminal electron acceptor, it is not surprising that they lack many enzymes concerned with oxygen utilization. Aerobes and many facultative anaerobes contain the intact respiratory chain through which electrons are passed to oxygen. Anaerobes lack a large portion of this chain (the cytochromes). If a strict anaerobe is exposed to air, the oxygen does combine with hydrogen through the action of the early enzymes of the respiratory chain. The combination of hydrogen with oxygen by the early enzymes of the respiratory chain does not result in the formation of water but hydrogen peroxide (H_2O_2), which is highly toxic to all cells. Most aerobic organisms contain the enzyme *catalase,* which degrades the H_2O_2 to H_2O and O_2. However, anaerobic organisms lack this enzyme. This is one reason even traces of oxygen may kill strict anaerobes. Anaerobes also lack many enzymes of the Krebs cycle, since these enzymes are superfluous to their metabolism.

156

The discussion thus far has focused on the mechanisms by which bacteria gain their energy by converting the chemical energy present in the bonds of either inorganic or organic molecules into ATP. These bacteria are the most common. Some organisms, however, have the ability to convert radiant or light energy into useful chemical energy in ATP molecules. This is the process of *photosynthesis*. This chemical energy is then available to *heterotrophic* organisms which can utilize the organic material in photosynthetic organisms as sources of energy.

PHOTOSYNTHESIS REQUIRES ENERGY AND SOURCE OF REDUCING POWER

The overall reaction of photosynthesis carried out by green plants, blue greens, and eucaryotic algae is summarized in the reaction

$$6CO_2 + 6H_2O \xrightarrow{\text{light}} C_6H_{12}O_6 + 6O_2$$

These organisms utilize CO_2 and H_2O, and in the presence of light synthesize a carbohydrate and release oxygen. The carbohydrate ($C_6H_{12}O_6$) represents not only the starch that plants synthesize but also all plant structures that the cells synthesize as they grow. This reaction therefore summarizes all of the biosynthetic reactions of the cell. It is the *reverse* of the reactions by which glucose is oxidized completely via glycolysis and the Krebs cycle:

PHOTOSYNTHESIS IS REVERSE OF RESPIRATION

$$C_6H_{12}O_6 + 6 O_2 \longrightarrow 6 CO_2 + 6 H_2O + \text{energy}$$

Since energy is gained when glucose is broken down ($-\Delta G$), energy must be provided when it is synthesized from its constituents ($+\Delta G$). Since hydrogen atoms are removed in the oxidation of carbohydrates, hydrogen atoms must be supplied for their synthesis.

Recent experiments have employed radioactive *isotopes* as tracers to follow the fate of each of the atoms in CO_2 and H_2O in photosynthesis. These experiments have conclusively demonstrated that the carbohydrate is synthesized from CO_2 and hydrogen of the water, while the O_2 originates from the oxygen of the water. The process of photosynthesis is more accurately shown by this equation:

$$CO_2 + H_2O \xrightarrow{\text{light}} C_6 H_{12} O_6 + O_2$$

The source of the hydrogen atoms for reducing CO_2 to $C_6H_{12}O_6$ is therefore the water. The formula of photosynthesis in bacteria and plants is the same, but the reactants are different. Whereas plants utilize H_2O as the source of hydrogen atoms for the reduction of CO_2, photosynthetic bacteria utilize other reduced compounds, either inorganic or organic, depending on the particular organism. Some groups utilize a reduced sulfur compound in place of water (H_2S in place of H_2O). In this case the hydrogen in the $C_6H_{12}O_6$ comes from H_2S, and sulfur, rather than oxygen, is released. Other bacteria use reduced organic compounds

Photosynthetic bacteria
Pages 238–239

as hydrogen donors. The general equation for photosynthesis which covers both bacteria and green plants may be restated as is indicated in:

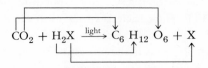

$$CO_2 + H_2X \xrightarrow{\text{light}} C_6H_{12}O_6 + X$$

LIGHT REACTIONS
INVOLVE CHLOROPHYLL;
DARK REACTIONS ONLY
ENZYMES PRESENT IN
MOST CELLS

The overall process of photosynthesis can be divided into two series of interrelated reactions: the "light" reactions which occur only in the presence of light and the "dark" reactions which do not require light. The light reactions are unique to photosynthetic organisms. In these reactions bacterial *chlorophyll* absorbs light energy. Electrons in the chlorophyll increase their energy by gaining the energy of the light. These electrons are transferred along the respiratory chain, and high-energy phosphate bonds of ATP are generated. Some of these electrons and hydrogen ions also serve to reduce compounds in the synthesis of $C_6H_{12}O_6$ from CO_2 and H_2O. Thus *the absorption of light by chlorophyll generates both chemical energy and reducing power.*

The formation of carbohydrate in photosynthesis can occur in the dark once light has generated ATP and reducing power. The dark reactions involve the incorporation of CO_2 into an organic molecule and then by additional enzymatic steps to glucose. These reactions are not unique to photosynthetic organisms but occur in many heterotrophic organisms as well as autotrophic organisms which gain energy by oxidizing inorganic compounds.

BIOSYNTHETIC METABOLISM

The discussion thus far has focused on the reactions by which cells obtain energy. Since glucose or some other energy source is often the only source of carbon available to cells, it must also serve as the source of carbon for the synthesis of all organic molecules within the cell. Many intermediate compounds of the glycolytic pathway and Krebs cycle serve as the starting compounds in the synthesis of amino acids, purines and pyrimidines, and lipids. Figure 7-14 illustrates in abbreviated form the various intermediates of glucose degradation which serve as the starting material for the synthesis of some subunits of macromolecules. If the intermediate is channeled along one pathway, it will be converted into a component of the cell's structure. If it is metabolized along the other pathway, it will serve to generate ATP. The synthesis of any amino acid requires the participation of a number of enzymes each catalyzing one reaction in the synthesis of the particular amino acid.

Formulae of some subunits
Page 40; Fig. 2-15
Page 34; Fig. 2-7

The same enzymes may be involved in the synthesis of several amino acids that are members of the same family. This is illustrated in the pathway leading to the synthesis of the three amino acids, tyrosine, phenylalanine, and tryptophan (Fig. 7-15), all of which have a benzene ring as part of their structure. This same

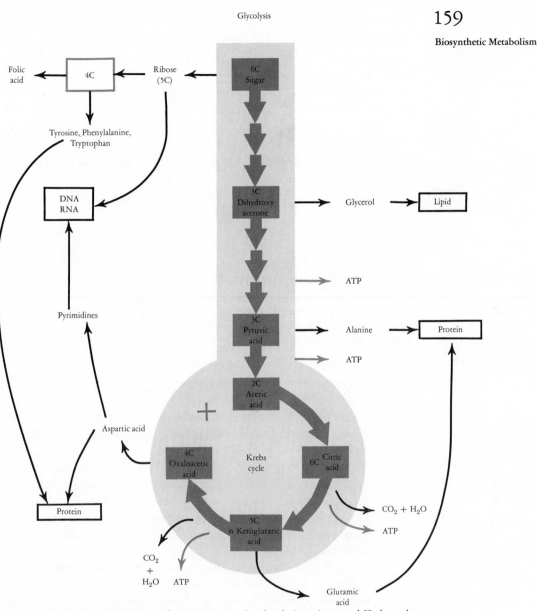

FIGURE 7-14 Summary of compounds originating in the glycolytic pathway and Krebs cycle leading to biosynthetic products. This diagram illustrates that most reactions involved in the degradation of a compound also serve to synthesize the beginning materials for the synthesis of macromolecules.

pathway is involved in the synthesis of several vitamins. This sequence of reactions illustrates features common to a large number of biosynthetic pathways.

1. A number of enzymes are common to the synthesis of all three amino acids and several vitamins. A defect in any one of these enzymes results

Formulae of amino acids
Page 34; Fig. 2-7

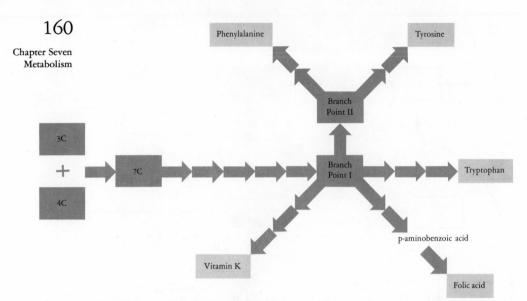

FIGURE 7-15 Biosynthetic pathway of aromatic acid biosynthesis. A complex biosynthetic pathway in which three different amino acids and several vitamins are synthesized. (The number of arrows do not reflect the number of steps actually involved.) The compound at Branch Point I is chorismic acid and at Branch Point II is prephenic acid.

in a requirement that all three amino acids and two vitamins be provided in the medium in order for growth to take place.

2. Energy must be supplied at several steps in the pathway. The high-energy compounds that take part are phosphoenolpyruvic acid (PEP) and ATP.

3. There are several branch points in the pathway. Chorismic acid can branch in two directions. Likewise, another compound in the pathway, prephenic acid, is metabolized to both tyrosine and phenylalanine. Chorismic acid also serves as the starting material for the synthesis of p-aminobenzoic acid, which in turn is converted to the vitamin folic acid and vitamin K.

4. The cell must regulate the flow of metabolites into the biosynthetic pathway commensurate with the availability of the energy supply. It makes no sense for the cell to channel metabolites into biosynthetic pathways if energy is not available to drive the many biosynthetic reactions that require energy. For example, in the aromatic acid biosynthetic pathway, the cell will utilize the phosphoenolpyruvic acid as a source of energy rather than use it in this biosynthetic pathway if energy sources of the cell are limited. The regulatory mechanisms which determine whether a metabolite will be converted into a subunit for macromolecule synthesis or degraded further for energy are also considered in Chapter 10.

One essential feature of amino acid biosynthesis is the introduction of nitrogen to form the amino ($-NH_2$) group. A number of enzymatic reactions catalyze the incorporation of ammonia (or the ammonium ion, NH_4^+) into an organic molecule. A key reaction links energy metabolism to biosynthesis. In this reaction α-ketoglutaric acid, a 5C intermediate compound in the Krebs cycle (Fig. 7-11), will react with NH_4^+ and be converted to the amino acid, glutamic acid:

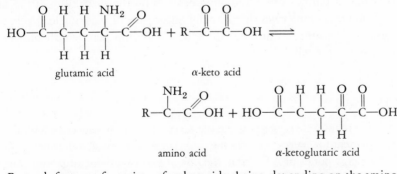

Since this reaction is reversible, glutamic acid can be readily converted to α-ketoglutaric acid. This latter compound enters the Krebs cycle and is transformed to CO_2 and H_2O with the production of energy.

Transamination Reactions

Glutamic acid plays a very important role in the synthesis of most other amino acids. It donates its amino group to a precursor of the amino acid being synthesized, an α-keto acid. This converts the α-keto acid to the amino acid, and simultaneously the glutamic acid is converted to α-ketoglutaric acid. The synthesis of most amino acids involves such a *transamination* reaction. The transamination reaction can be written as follows:

R stands for any of a variety of carbon side chains, depending on the amino

acid to be synthesized. For example, pyruvic acid ($CH_3-\overset{\overset{\displaystyle O}{\|}}{C}-\overset{\overset{\displaystyle O}{\|}}{C}-OH$) can be

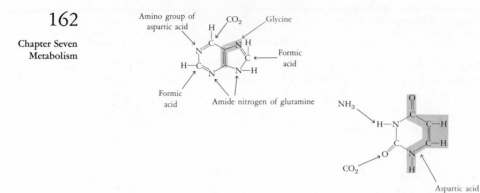

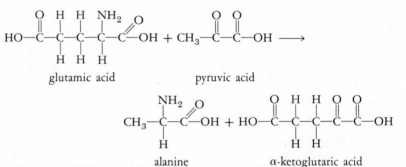

FIGURE 7-16 Origin of atoms for the biosynthesis of a purine molecule (top) and pyrimidine molecule (bottom).

converted to the amino acid alanine by transamination:

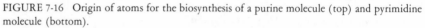

glutamic acid   pyruvic acid

alanine   α-ketoglutaric acid

Synthesis of Purines and Pyrimidines

The synthesis of these components of nucleic acids illustrates how compounds can be built up by the stepwise addition of a number of simpler molecules. The carbon and nitrogen atoms originate from a variety of sources (Fig. 7-16). The synthesis of purines and pyrimidines depends on the availability of the amino acids and other molecules that are component parts of these nucleic acid precursors. Their synthesis illustrates the interrelationship between all biosynthetic processes in the cell.

SUMMARY

All cells require energy to perform the myriad functions associated with life. The requirement for energy involves the conversion of chemical bond energy into the high-energy bonds of ATP, or in the case of photosynthetic organisms, the conversion of light energy into chemical bond energy in ATP. ATP serves as the energy currency in all cells. Heterotrophs obtain energy by oxidizing reduced organic compounds through a series of enzymatic reactions. The drop in free energy between the starting material and end products is largely captured in the form of high-energy phosphate bonds. One of the most common pathways for the breakdown of carbohydrates in all forms of life is the glycolytic pathway, which functions under both aerobic and anaerobic conditions. For every molecule of glucose metabolized, the products of glycolysis are two molecules of pyruvic acid, two molecules of ATP (net), and two molecules of NAD_{red}. Pyruvic acid can be metabolized in a variety of ways which yield a variety of products, including ethanol, CO_2, lactic acid, acetic acid, and several other acids and alcohols. Under aerobic conditions, the acetic acid may be metabolized through the Krebs cycle provided the cell has the necessary enzymes. If glucose is oxidized only to organic compounds, most of the energy available in the glucose molecule is untapped. However, if it is oxidized completely to CO_2 and H_2O, then about 40 percent of the energy originally present in the glucose is trapped as ATP. The ATP is generated in the course of transfer of electrons along the respiratory transport chain.

The pathways of degradation of glucose provide the starting compounds from which many small molecules are synthesized. Many steps in these biosynthetic reactions require energy, and the ATP generated in the breakdown of sugars drives these reactions.

QUESTIONS

1. The first step in glycolysis, the conversion of glucose to glucose-6-phosphate, requires energy. Explain how this reaction can proceed.
2. Under what conditions of oxygenation would a yeast cell (a facultative anaerobe): (a) grow to the highest cell yield, (b) produce the most alcohol, (c) have the shortest generation time. Explain your reasoning for each answer.
3. Consider why diseases of vitamin deficiency have widespread and far ranging effects on the body.
4. An organism may have some but not all of the enzymes of the Krebs cycle. Why would this be useful to the organism?
5. What feature do all inorganic metabolites have which can serve as sources of energy?

FURTHER READING

LEHNINGER, A., *Bioenergetics,* 2d ed. Menlo Park, Calif.: W. A. Benjamin, 1971.

———, "Energy Transformation in the Cell, " *Scientific American* (May 1960). (Offprint #69)

LEVINE, R. P., "The Mechanism of Photosynthesis," *Scientific American* (December 1969). (Offprint #1163)

PHILLIPS, D., "The Three-Dimensional Structure of an Enzyme Molecule," *Scientific American* (November 1966). (Offprint #1055)

STUMPF, P., "ATP," *Scientific American* (April 1953). (Offprint #41)

Chapter 8

Informational Macromolecules—Function and Synthesis

In Chapter 7 the mechanisms by which cells obtain energy and synthesize small molecules were explored. In this chapter the biosynthesis of several macromolecules, RNA, DNA, and protein, is considered. These macromolecules have several features in common. First, each embodies information that is contained in the sequence of the subunits that constitute the macromolecule. The biological function of these molecules depends on the exactness of this sequence. Second, the synthesis of each of these macromolecules requires *templates,* direct or indirect, macromolecules which serve as patterns for the synthesis of other macromolecules. Third, recognition of purine and pyrimidine base sequences through *complementarity* plays a vital role in the synthesis of all three classes of macromolecules. Fourth, the reactions require energy in order to proceed, and a high-energy bond is broken as each subunit bonds to its neighbor to form the macromolecule.

It may be helpful to refer to Chapter 2 for information on the structure of these macromolecules.

CONCEPT OF INFORMATION STORAGE AND TRANSFER

Biological macromolecules can contain information specified by the arrangement of subunits making up the *polymer.* For example, DNA contains four different subunits or bases. These may be thought of as a four-letter alphabet in which the sequential arrangement spells out a message. In all normal cells, DNA is the ultimate genetic language in which all the genetic information of a cell is written. On the other hand, proteins contain twenty different kinds of amino acid sub-

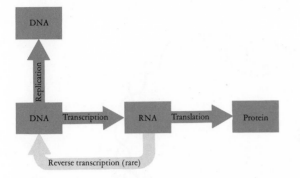

FIGURE 8-1
Flow of information in biological systems. Reverse transcription
occurs in a few tumor virus systems.

units, the arrangement of which constitutes another language written in a completely different alphabet.

Two basically different processes are involved in information storage in
cells. In one process the information is duplicated in an identical molecule and
passed on to the daughter cell. An example of this process is the replication of
DNA. The second process involves the transfer of information to another similar, but not identical molecule and the further transfer of this information to
other kinds of macromolecules containing different subunits. The second process
generally involves two stages, *transcription,* the transfer of information from one
language into a similar language, and *translation,* the transfer into the second
language. An example of this process is the transfer of information from DNA
to RNA (transcription) and then to protein (translation). This transfer of information is often referred to as the *central dogma* of molecular biology (Fig. 8-1).
It was believed to be unidirectional. Recently it has been shown, however, that
certain viruses that can cause cancer in animals can transfer information from
RNA to DNA using an enzyme with the descriptive name *reverse transcriptase.*
This finding has evoked great excitement since it raises the possibility that the
process may be crucial for the cancer-producing ability of these viruses.

CENTRAL DOGMA
OF MOLECULAR
BIOLOGY

Reverse transcriptase
Page 352

REPLICATION OF DNA

The bacterial chromosome is divided into a large number of genes, each gene
being the sequence of purine and pyrimidine bases which specifies one polypeptide chain.

In the process of DNA replication of a circular chromosome, the two
strands separate at the site at which the new strands are being synthesized, the
growing point. The separation of the two DNA strands at the growing point
forms a Y-shaped structure, referred to as the *replication fork* (Fig. 8-2). This fork
progresses around the circular DNA molecule, and as it moves two new strands
of DNA are synthesized, each one having a base sequence complementary to one
of the original strands of DNA (Fig. 8-3). When the fork has moved along the
entire DNA molecule, two complete double-stranded molecules are formed.
Since each double-stranded molecule contains one of the original strands and one

Base sequence complementarity
Page 44—Fig. 2-19

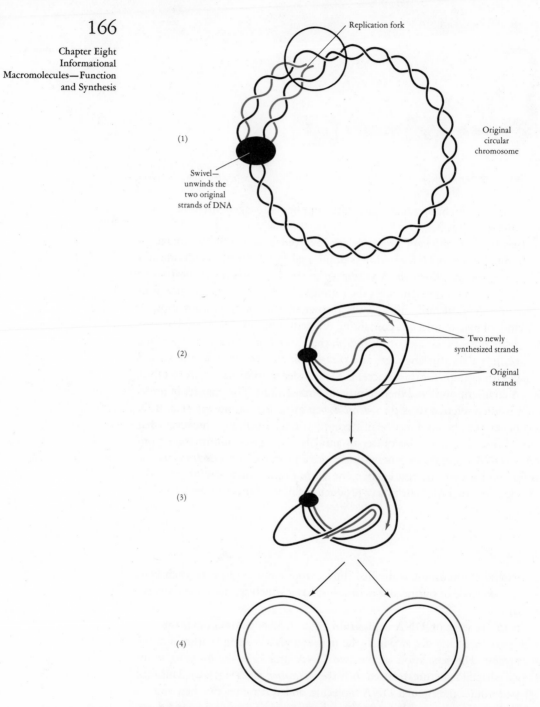

Replication fork

Original
circular
chromosome

(1)

Swivel—
unwinds the
two original
strands of DNA

(2)

Two newly
synthesized strands

Original
strands

(3)

(4)

Two DNA molecules identical to the original

FIGURE 8-2 The replication of a circular chromosome in bacteria. The DNA in (2), (3), and (4) is in the form of a double helix, as shown in (1).

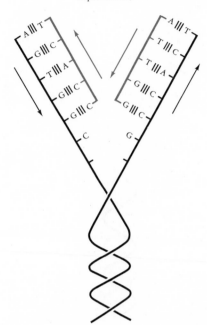

Replication fork

FIGURE 8-3

Replication of DNA. The two strands of DNA separate and two new strands are synthesized, each complementary to one of the original strands. Since the DNA molecule has one strand going down and the other up (indicated by arrows), the new strands which are synthesized must go in opposite directions also.

Direction of DNA
Page 44; Fig. 2-19

new strand, this process of replication is termed *semiconservative.* This type of replication results in only two of the cells in a bacterial population having DNA strands present in the original cell no matter how many cells have descended from this cell (Fig. 8-4). A number of components required for DNA synthesis must be present at the replication fork. These include the *enzyme,* DNA polymerase, the *substrates* for the reaction which are *deoxynucleoside triphosphates* of each of the purine and pyrmidine bases, and a *template* of DNA. The enzyme directs the incorporation of the purines and pyrimidines into a position such that the complementary relationship between purines and pyrimidines is maintained. The synthesis of any polymer which is synthesized on a template follows the same general pattern. This sequence of steps is diagrammatically represented in Figure 8-5.

DNA SYNTHESIS IS SEMI-CONSERVATIVE

Note that the structure of the DNA which is synthesized depends on the structure of the DNA which serves as the template and not on the enzyme carrying out the synthesis. The breaking of the high-energy bonds of the four different substrate molecules provides the energy required for bond formation between the mononucleotide subunits. Although the enzyme DNA polymerase I has been intensively studied for the last dozen years, it is still questionable whether this enzyme is really the one responsible for DNA replication inside the cell. Indeed, it now appears that more than one enzyme is involved.

DNA TEMPLATE DICTATES THE NATURE OF DNA SYNTHESIZED

Many other questions concerning DNA replication remain unanswered. What mechanism pulls the two strands of DNA apart? It must occur very rapidly since approximately 10 million nucleotides are incorporated into newly synthesized DNA every hour. What determines the site at which DNA synthesis be-

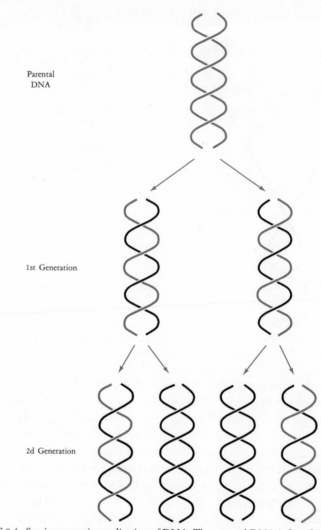

Parental
DNA

1st Generation

2d Generation

FIGURE 8-4 Semiconservative replication of DNA. The parental DNA is found in two progeny cells only, no matter how many cell generations occur.

gins? What determines the direction of DNA synthesis? The study of DNA replication still represents one of the most challenging areas of research in molecular biology.

THE EXPRESSION OF GENES

Transcription

All of the genetic information of a cell resides in the base sequence of DNA. The transcription process is the first step in gene expression. The transcription of the message encoded in the specific sequence of purine and pyrimidine bases in the

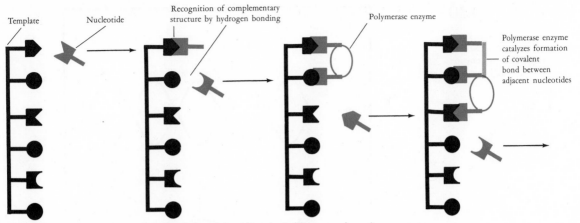

Template   Nucleotide   Recognition of complementary structure by hydrogen bonding   Polymerase enzyme

Polymerase enzyme catalyzes formation of covalent bond between adjacent nucleotides

FIGURE 8-5 Sequence in template-directed nucleic acid synthesis. This general reaction sequence diagrams the synthesis of DNA and RNA.

DNA proceeds by an enzymatic process which is analogous to the process of DNA replication. The enzyme *RNA polymerase* synthesizes a single-stranded RNA molecule which is complementary to *one* of the strands of DNA. The RNA molecule synthesized is termed *messenger RNA* (mRNA) and corresponds to one, or in some cases several, contiguous genes in length (Fig. 8-6). Genes vary in size, but a reasonable estimate is 1000 nucleotides. Thus the mRNA transcribed from such a gene is also 1000 nucleotides.

Translation

The sequence of purines and pyrimidines in mRNA is then translated into a sequence of amino acids in a protein (polypeptide) molecule coded by the transcribed gene. This process is the most elaborate biosynthetic mechanism known by far. It involves an ever-increasing number of enzymes, protein "factors" which are still being sought and identified, and two additional types of RNA, *transfer RNA* (tRNA) and *ribosomal RNA* (rRNA). About 90 percent of the total energy expended by a cell is dedicated to protein synthesis.

PROTEIN SYNTHESIS

Step 1 The twenty different amino acid molecules are "activated" by the formation of a high-energy bond between the amino acid and a specific amino acid carrier molecule, tRNA at the expense of energy in ATP. The tRNA carrying the amino acid is "charged"; when it is not carrying an amino acid, it is "uncharged." The activation of each different amino acid requires a different "activating enzyme." Transfer RNA molecules are relatively small, containing only about 70 nucleotides. A unique class of tRNA molecules exists for each amino acid, and in the case of some amino acids there may be as many as five different tRNA molecules which will bind the same amino acid. The significance of the multiplicity in tRNA molecules for the same amino acid is not clear. Each

EACH AMINO ACID BINDS TO A SPECIFIC tRNA MOLECULE

169

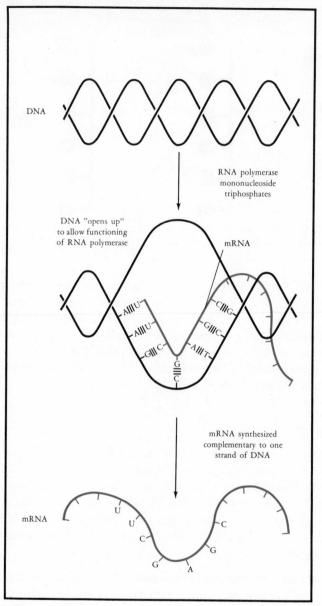

FIGURE 8-6 Gene transcription.

tRNA molecule functions as an *adaptor* between the mRNA and the amino acids. In order to perform its function of serving as an adaptor between nucleotide bases of RNA and amino acids, each tRNA molecule contains two unique sets of nucleotide sequences. One sequence determines the amino acid which bonds to the tRNA molecule. Another site on the molecule contains a sequence which is *complementary* to a sequence on the mRNA. This latter recognition site consists of three nucleotides and is termed the anticodon since it is complementary to the

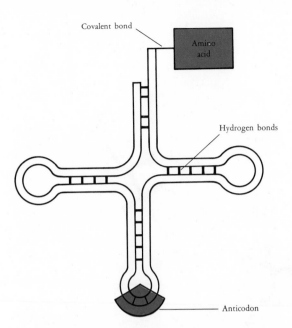

Covalent bond

Amino
acid

Hydrogen bonds

Anticodon

FIGURE 8-7
Molecule of tRNA. All tRNA molecules have an anticodon site
at one end of the molecule and a site which binds a specific amino
acid at another end. There are at least 20 different kinds of tRNA
molecules, each with a different anticodon and the ability to bind
a different amino acid.

codon, the set of three bases in the mRNA which codes for one amino
acid. A tRNA molecule is diagrammed in Fig. 8-7. All species of tRNA
thus far examined appear to exist in the shape of a cloverleaf.

Step 2 A ribosome binds to one end of the mRNA molecule, at a specific site.
This initial position of the ribosome on the mRNA is important since
it starts the translation of the mRNA in the correct frame. (The binding
of the mRNA to the ribosome occurs in such a fashion that three bases
of the mRNA are always positioned at a specific site.)

Ribosomes
Page 91; Fig. 4-23

Step 3 The charged tRNA molecules associate with the ribosome. The tRNA
molecule whose anticodon is complementary to the codon positioned on
the ribosome "locks" into place on the codon by the complementary fit
of the codon and anticodon. When the amino acid is thus fixed in position, an enzyme catalyzes the formation of a peptide bond between the
carboxyl and amino groups of adjacent amino acids. The amino acid is
then released from the tRNA, which is free to diffuse away from the
ribosome. The uncharged tRNA then binds another amino acid and returns to its proper codon site on the ribosome. The association of the
charged tRNA with the mRNA attached to a ribosome is diagrammed
in Figure 8-8.

tRNA MOLECULE BINDS
TO COMPLEMENTARY
SITE ON mRNA—PEPTIDE
BONDS FORM

Step 4 A "ratchet"-type mechanism moves the ribosome along the mRNA
through a distance of one codon or one reading frame so that the next
codon (or reading frame) is now in position to bind to a tRNA molecule. As the ribosomes move along the mRNA molecule, more ribosomes attach at the end. Five or six ribosomes can be attached to one
mRNA molecule to form a *polyribosome* (Fig. 8-9a and b).

RIBOSOME MOVES
ALONG mRNA

Step 5 As the peptide chain grows by the addition of amino acids in a typical
polypeptide chain, the steps 1 to 4 must be repeated 300 times for the

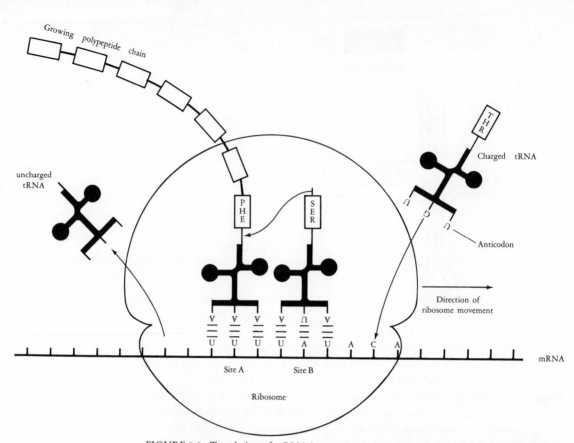

FIGURE 8-8 Translation of mRNA in protein synthesis. The ribosome binds to one end of the mRNA and moves along the mRNA three nucleotides at a time. A charged tRNA molecule binds to the appropriate codon through recognition by the anticodon. As the ribosome moves, the tRNA molecule charged with serine at site B will move to site A. Simultaneously, the serine will join in peptide linkage to the phenylalaline (phe), attaching the peptide chain to the serine tRNA. The uncharged phe tRNA will leave the ribosome to once again become charged and return. When the ribosome reaches the end of the mRNA, the polypeptide chain will be complete. Many ribosomes are attached to the same mRNA molecule.

POLYPEPTIDE
COMPLETED AS
RIBOSOME COMES TO
END OF mRNA
MOLECULE

synthesis of a complete chain. A complete polypeptide chain is synthesized on each ribosome, beginning when the ribosome attaches to the beginning (one end) of the mRNA molecule and being completed when the ribosome reaches the end. After the ribosome falls off the end of the mRNA it returns to the other end to begin the translation of the mRNA over again.

THE GENETIC CODE

The term *genetic code* describes the relationship between the sequence of purines and pyrimidine bases in the DNA and mRNA and the sequence of amino acids in the corresponding protein. Since there are only four different bases in the DNA, it is apparent that more than one consecutive base is required to specify

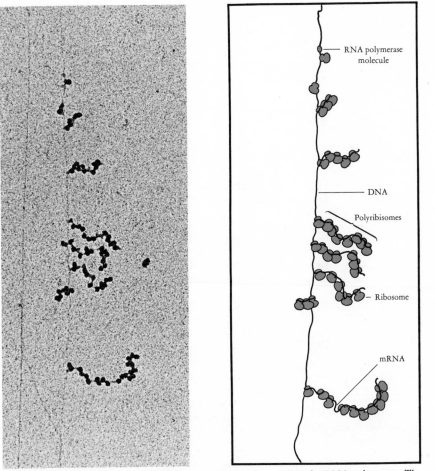

FIGURE 8-9 Genes in action. DNA is being transcribed into mRNA by RNA polymerase. The mRNA is bound to the ribosomes, thereby creating the polyribosomes. This figure illustrates how closely integrated the processes of transcription and translation are. (Photo courtesy of Dr. O. L. Miller, Jr. and B. A. Hamkalo; Miller, O. L. Jr., B. A. Hamkalo, and C. A. Thomas, Jr., *Science,* **169**:392, 1970)

the twenty different amino acids. From mathematical considerations alone the minimum number of bases in a code word is three, since the four bases used in groups of four can yield only $4 \times 4 = 16$ combinations. In groups of three, $4 \times 4 \times 4 = 64$ different amino acids could be specified. It happens that this apparently simple solution to the problem is the one adopted in nature. Thus a codon consists of three purine or pyrimidine mononucleotides.

Since there are sixty-four triplet combinations and only twenty amino acids, several nucleotide combinations may code for the same amino acid. However, no codon specifies more than one amino acid. In one of the most significant achievements of modern biology, the catalog of code words for each amino acid was worked out (Fig. 8-10). The person mainly responsible for cracking the genetic code was Dr. Marshall Nirenberg, an American biochemist working at the National Institutes of Health. In 1968 he shared the Nobel Prize for this work.

TRIPLET CODE—3 BASES CODE FOR 1 AMINO ACID

		Middle Letter						
	U		C		A		G	
First Letter U	UUU	Phe	UCU	Ser	UAU	Tyr	UGU	Cys
	UUC	Phe	UCC	Ser	UAC	Tyr	UGC	Cys
	UUA	Leu	UCA	Ser	UAA	Ochre	UGA	Umber
	UUG	Leu	UCG	Ser	UAG	Amber	UGG	Trp
C	CUU	Leu	CCU	Pro	CAU	His	CGU	Arg
	CUC	Leu	CCC	Pro	CAC	His	CGC	Arg
	CUA	Leu	CCA	Pro	CAA	Gln	CGA	Arg
	CUG	Leu	CCG	Pro	CAG	Gln	CGG	Arg
A	AUU	Ile	ACU	Thr	AAU	Asn	AGU	Ser
	AUC	Ile	ACC	Thr	AAC	Asn	AGC	Ser
	AUA	Ile	ACA	Thr	AAA	Lys	AGA	Arg
	AUG	Met	ACG	Thr	AAG	Lys	AGG	Arg
G	GUU	Val	GCU	Ala	GAU	Asp	GGU	Gly
	GUC	Val	GCC	Ala	GAC	Asp	GGC	Gly
	GUA	Val	GCA	Ala	GAA	Glu	GGA	Gly
	GUG	Val	GCG	Ala	GAG	Glu	GGG	Gly

FIGURE 8-10 Genetic Code. The dictionary of the genetic code words. The codons are read from left to right ($5' \rightarrow 3'$). The three nonsense codons (designated as three different colors) serve as punctuation marks between genes, since there is no tRNA which recognizes them. Note that there are many codons which specify the same amino acid. The significance of this feature of the code is not known.

EACH CODON SPECIFIES A SINGLE AMINO ACID, BUT 3 CODONS SERVE TO PUNCTUATE

Three of the sixty-four triplets do not code for any of the twenty amino acids, and these codons are said to be *nonsense.* These codons are very important because they serve to signal the end of a protein chain. Returning to the analogy between human languages and the genetic code, such termination signals correspond to punctuation. The requirement for efficient punctuation is a vital one due to the nature of the genetic code. Since all the bases are parts of codons, the code therefore may be said to be *commaless.* However, if there were no punctuation at all in the genetic code all of the proteins coded by a single mRNA molecule would be connected to one another. It is also crucial that the initiation of transcription begins at the correct codon and at the correct nucleotide *within* the codon. A sequence of bases may be read in three distinct ways depending on the point in the reading frame at which reading is initiated (Fig. 8-11). Depending on the site of initiation in the first codon, a completely different set of amino acids will be specified. This emphasizes the importance of precise initiation of transcription of DNA and translation of mRNA. Once the initial site transcription and translation are set, the reading frame is maintained by the sequential reading of three nucleotides at a time (Fig. 8-11).

The nature of the code also demands that the sequence of codons in a gene correspond directly to the order of amino acids in the protein coded for by that gene. This expectation has been experimentally verified in several systems. This finding also indicates that both transcription and translation proceed in an orderly linear fashion from one end of the gene to the other.

RNA POLYMERASE

This enzyme is structurally very complex and serves several functions. In recent years more people have studied it than any other single enzyme and entire scien-

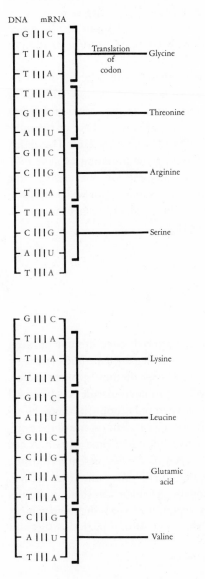

FIGURE 8-11

Importance of initiating translation of mRNA in the correct frame. A completely different amino acid sequence is generated depending on the frame in which translation is initiated.

Formulae of mononucleotides
Page 41; Fig. 2-16

RNA POLYMERASE HAS A
NUMBER OF FUNCTIONS

tific conferences have been devoted to its discussion. RNA polymerase is concerned not only with joining mononucleotide subunits to form the polymer RNA but also with selecting the site on the DNA at which synthesis is initiated and the site at which synthesis terminates. It is not surprising that an enzyme which fulfills all these functions is structurally complex. But for all of the information known about this enzyme, it is not clear how a sequence of amino acids in RNA polymerase recognizes a sequence of nucleotides in DNA which says to the enzyme "start transcribing here." The major portion of the enzyme, termed the *core enzyme,* catalyzes the formation of bonds between mononucleotides of RNA. A subunit of the enzymes, the *sigma factor,* which is loosely attached to the core enzyme, determines the site at which synthesis begins. The core enzyme can function without the sigma subunit, but has no specificity as to where it initiates RNA synthesis. The discovery and function of the sigma factor created much excitement in biology, because it has many applications for regulation of enzyme synthesis and cellular differentiation.

In all cells, at all times, certain functions are necessary, whereas others are superfluous. In terms of enzymatic functions certain enzymes are vital in some circumstances but unnecessary in others. Control over the regulation of different enzyme functions present in a cell is exerted in several ways. One of the most important is control over the rates at which enzymes are synthesized. This differential synthesis of enzymes operates by controlling the initiation of gene transcription. Some control over the initiation of transcription is exerted by the RNA polymerase enzyme itself, particularly by the sigma factor which controls the site on the DNA at which RNA synthesis is initiated. Other mechanisms for selective transcription are covered in Chapter 10.

QUANTITATIVE ASPECTS OF MICROBIAL PROTEIN SYNTHESIS

In general, mRNA in microorganisms has only a very limited lifetime between the time it is synthesized on the DNA until it is degraded to nucleotides by specific enzymes present in the cell. In *E. coli* the average lifetime is approximately 2 minutes. Since in the same organism an average protein molecule is synthesized in about 20 seconds, a given messenger is translated only about six times. However, the efficiency of translation is much greater since *each* ribosome of the polyribosome structure can translate the message to produce the same protein. A cell synthesizes proteins corresponding to the mRNA molecules present within the cell. However, the molecules of mRNA do not remain intact once they are synthesized. Rather the mRNA is being continually degraded into small molecules and resynthesized, so that the cell is able to respond to a changing environment by selectively transcribing different genes, thereby resulting in the synthesis of different proteins.

ANTIBIOTIC INHIBITORS OF PROTEIN SYNTHESIS

Several *antibiotic* drugs exert their effect by interfering with some step of protein synthesis. Unfortunately many antibiotics which interfere with protein synthesis are not selective enough to be clinically useful. This is not surprising since protein synthesis is a feature of all cells. Fortunately eucaryotic ribosomes, larger than bacterial ribosomes, are different enough in their structure that some selectivity in the action of some antibiotics is possible. However, bacterial type ribosomes are present in many of the cellular organelles of eucaryotic cells. This may account in part for the observation that side effects are often observed in patients undergoing treatment with antibiotics which affect procaryotic ribosomes. A more extensive discussion of antibiotics and their role in medical microbiology is covered in Chapter 32.

Chloramphenicol

Chloramphenicol interferes with peptide bond formation in bacterial or other procaryotic cells, but not on eucaryotic ribosomes. The drug is known to bind to a specific site on the ribosome, but the details of how this antibiotic acts are obscure.

Tetracyclines

This group of antibiotics also affects protein synthesis at the level of peptide bond formation. However, the precise step at which they exert their action appears to differ from that of chloramphenicol. The interference appears to be at the level of binding of the tRNA carrying its amino acid to the ribosome rather than the subsequent formation of peptide bonds.

Streptomycin, Kanamycin, Neomycin

These antibiotics, all of which have a similar chemical structure, bind tightly to a specific site on the ribosome. In so doing they alter its shape which inhibits protein synthesis and causes the genetic code to be incorrectly read. The effects are apparently accomplished by interfering with the recognition of codons in mRNA by the anticodons in tRNA.

Lincomycin

Lincomycin is an antibiotic which binds to the ribosome and inhibits the binding of tRNA to the messenger-ribosome complex. Its action is thus similar to the tetracyclines.

Eucaryotic ribosomes versus procaryotic ribosomes
Page 91

Chloramphenicol
Pages 559–560

Tetracyclines
Pages 558–559

Aminoglycosides
Pages 560–561

Lincomycin
Pages 561–562

Erythromycin

Erythromycin also binds to the ribosome and immediately stops protein synthesis. Its mechanism of action appears to be very similar to lincomycin.

ANTIBIOTIC INHIBITORS OF NUCLEIC ACID SYNTHESIS

Rifamycin

This antibiotic prevents bacterial RNA synthesis by combining with and thereby blocking the function of bacterial RNA polymerase, but not the polymerase from eucaryotic cells. Some derivatives of rifamycin inhibit the enzyme reverse transcriptase and there is some evidence that it may inhibit infection of certain animal tumor viruses as well as other viral infections.

Reverse transcriptase
Page 352

SUMMARY

The informational macromolecules, DNA, RNA, and protein, contain information in the sequence of the subunits composing the molecule. The information in DNA, the ultimate information, is passed onto succeeding generations through the process of DNA replication. This is a semiconservative replication

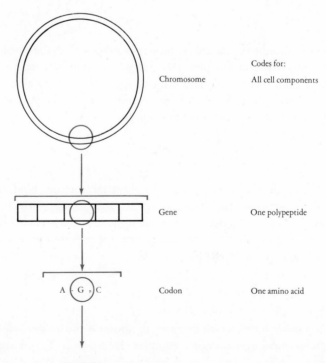

FIGURE 8-12 Levels of coding of DNA in the cell.

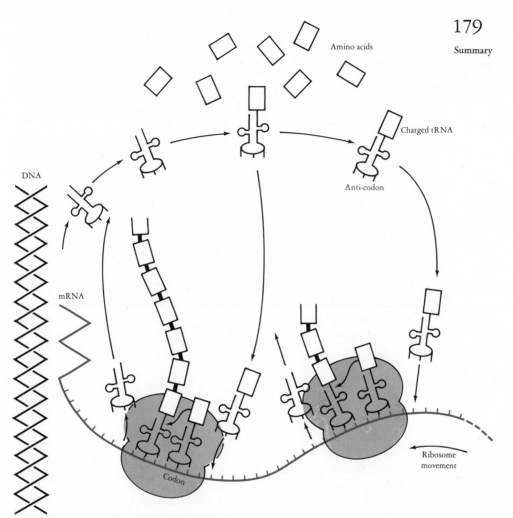

Amino acids

Charged tRNA

DNA

Anti-codon

mRNA

Ribosome
movement

Codon

FIGURE 8-13   Overall aspects of protein synthesis.

mechanism in which new strands are synthesized which are complementary to each of the original strands. The base sequence of DNA also codes for the sequence of amino acids in the corresponding protein. The relationship between the various functional components of DNA involved in information transfer is diagrammed in Figure 8-12. The mechanism of protein synthesis involves the transcription of the genetic message into mRNA, which binds to several ribosomes. Transfer RNA, carrying an amino acid, recognizes a codon of the mRNA and binds by the complementarity of the codon in mRNA and the anticodon of the tRNA. Peptide bonds form between adjacent amino acids, and the completed protein peels off the ribosome. This entire sequence is represented diagrammatically in Figure 8-13.

The most complex and inherently most interesting enzyme in this entire process is RNA polymerase. It consists of a core enzyme and several subunits.

Together the enzyme not only polymerizes the substrates of RNA into a macromolecule, but the subunits also determine the site at which RNA synthesis is initiated and the site at which a mRNA molecule terminates.

A large number of clinically useful antibiotics inhibit protein synthesis in bacteria. Many of them act at various steps in the translation process, by modifying the function of the procaryotic ribosome. Some antibiotics also act by inhibiting nucleic acid synthesis.

QUESTIONS

1. Consider the stages in protein synthesis in which complementarity plays a role.
2. Why does it make sense to the bacterial cell for mRNA to function for only a few minutes before it is degraded?
3. What factor is limiting in protein synthesis?
4. The faster bacterial cells multiply, the more chromosomes they contain. Why does this make sense to the cell?
5. Many antibiotics interfere with protein synthesis. What is the basis for this selectivity against bacteria? Why could eucaryotic cells also be affected?

FURTHER READING

BEADLE, G., "The Genes of Men and Molds," *Scientific American* (September 1948). (Offprint #168)

CAIRNS, J., "The Bacterial Chromosome," *Scientific American* (January 1966). (Offprint #1030)

CLARK, B. F. C. and K. MARCKER, "How Proteins Start," *Scientific American* (January 1968). (Offprint #1092)

CRICK, F. H. C., "The Genetic Code: III," *Scientific American* (October 1966). (Offprint #1052)

GORINI, L., "Antibiotics and the Genetic Code," *Scientific American* (April 1966). (Offprint #1041)

HOLLEY, R., "The Nucleotide Sequence of a Nucleic Acid," *Scientific American* (February 1966). (Offprint #1033)

KORNBERG, A., "The Synthesis of DNA," *Scientific American* (October 1968). (Offprint #1124)

NIRENBERG, M., "The Genetic Code: II," *Scientific American* (March 1963). (Offprint #153)

NOMURA, M., "Ribosomes," *Scientific American* (October 1969). (Offprint #1157)

WATSON, J. D., *The Double Helix.* New York: New American Library (Signet Book), 1969.

YANOFSKY, C., "Gene Structure and Protein Structure," *Scientific American* (May 1967). (Offprint #1074)

The Molecular Basis of Life. San Francisco: W. H. Freeman and Co., 1968. A collection of offprints of articles originally appearing in *Scientific American* including most of those listed here.

Chapter 9

Microbial Genetics

Bacteria have served as excellent experimental material for genetic studies. The organisms are small and therefore require very little laboratory space. Bacteria grow and divide quickly, allowing many experiments to be performed within a short time span. Under appropriate conditions billions of cells can be screened for a single characteristic by simple techniques that require little time and expense. Furthermore, the progeny of a single cell is essentially identical for all practical purposes. For these reasons the study of microbial genetics has made fundamental contributions to understanding microbiological phenomena as well as to many problems central to all of biology.

DNA-CELLULAR ORGANIZATION AND FUNCTION

There is tremendous diversity in the living world. This variation stems from two major factors: One factor is the unique genetic information each organism possesses. For example, bacteria give rise to bacteria and nothing else. Bacteria do not have the genetic information required to become a rabbit or a tree. The other factor is the different environments with which organisms come in contact. Two organisms with the same information in the DNA may appear remarkably different if they are exposed to two different environments. For instance, a facultative anaerobe growing in a glucose-salts medium will form different end products depending on whether it is grown in the presence or absence of air. The environment is controlling the functioning of enzymes. In addition, the environment can regulate gene functions. The sum of the genetic constitution of an organism is its *genotype*. The characteristics of an organism which are expressed within a

ORGANISMS GIVE RISE
TO SIMILAR ORGANISMS

181

given environment make up the *phenotype* of the organism. The genotype represents the potential of the organism; the phenotype describes what the organism actually is. Some of the mechanisms by which the environment acts on the genotype of the bacterial cell are discussed in Chapter 10.

The genotype of a cell is determined solely by the genetic information contained in the cell's chromosome or *genome*. The genome is divided into genes, each gene coding for the synthesis of one polypeptide chain.

THE NUCLEOTIDE
SEQUENCE IN DNA
DETERMINES AMINO
ACID SEQUENCE IN
POLYPEPTIDE CHAIN

Since an average polypeptide chain contains approximately 300 amino acids, the average gene contains over 900 nucleotide *base pairs* (since both strands of the double helix are included). The molecular weight of a gene is approximately 600,000 daltons (900 $\times$ 700, the molecular weight of a nucleotide pair). A *dalton* is the atomic weight of the hydrogen atom, 1.0. The molecular weight of the entire bacterial chromosome is approximately 2×10^9 daltons. Therefore there is enough DNA to code for approximately 3000 different proteins, or $(2 \times 10^9)/(6 \times 10^5)$. Since a bacterial cell contains only about 1000 different enzymes, it is apparent that much of the DNA has functions other than coding for enzymes.

MUTATIONS RESULT IN A
CHANGED SEQUENCE OF
BASES IN DNA

All the properties of a cell are dependent on the proteins that the cell possesses. The substitution of even one amino acid for another from the several hundred in the molecule may cause the protein to be nonfunctional. Thus any change in the sequence of purines and pyrimidines in the DNA results in a change in sequence of amino acids coded by that gene. A change in nucleotide sequence of the DNA is termed a *mutation*.

The scientists most responsible for providing the basic insight into how genes determine the properties of a cell are two Americans, George Beadle and Edward Tatum. In the early 1940s, working as a team, they demonstrated that some mutants of the bread mold *Neurospora* were unable to grow in a glucose-salts medium because of their inability to synthesize a specific amino acid. By the appropriate genetic crosses they were able to demonstrate that this requirement was passed on to progeny organisms as if there were a difference in one gene between the parent (which could grow on the glucose-salts medium) and the mutant. Beadle and Tatum postulated that the requirement for the amino acid for growth resulted from a mutation in a gene. However, it was not clear to Beadle and Tatum whether the genes themselves control biochemical reactions by acting as enzymes or whether the genes function by determining the specificities of enzymes. They recognized that it should be possible to isolate a variety of mutants by damaging the genes involved in their synthesis. They proceeded to isolate and identify a large number of these *biochemical mutants,* which have proved so useful for elucidating biochemical pathways.

GENE MUTATIONS

Although gene mutations occur with a low frequency, they can arise in a number of ways. The most common mechanism involves the process of DNA synthesis.

At the time of duplication the strands of the helix unwind and two new strands are synthesized, each one complementary to one of the original strands. However, in rare instances, an incorrect purine or pyrimidine is incorporated into the DNA during DNA replication (Fig. 9-1). It has recently been shown that mutations in a gene which codes for an enzyme concerned with inserting the correct purines and pyrimidines into position (DNA polymerase) can increase the frequency of mutation in all other genes from tenfold to a thousandfold. Apparently the defective enzyme is unable to insert the correct bases with the precision and accuracy of the normal enzyme.

Mutations can also result from the *deletion* of either a purine or pyrimidine nucleotide or an entire section of a gene. Nucleotides can also be added within a gene, resulting in a change in the sequence and thus a mutation. The addition or deletion of a purine or pyrimidine nucleotide results in a shift in the reading "frame" of the DNA when it is transcribed into mRNA (Fig. 9-2). Note that the addition or deletion of the nucleotide determines which amino acids are incorporated at all sites beyond the original site at which the purine or pyrimidine base was inserted or deleted. Thus these mutations are more severe than those which result from the substitution of one purine or pyrimidine base for another.

Since the frequency with which naturally occurring or *spontaneous mutants* arise in the population is extremely low, geneticists trying to isolate mutants often treat cultures with agents which increase their frequency. Any agent which increases the frequency of mutations is termed a *mutagenic agent* or *mutagen*.

Mutagenic Agents

Since mutations occur spontaneously when one purine or pyrimidine base does not pair with its complementary base in the duplication of DNA, any chemical treatment that can change the hydrogen-bonding properties of a base in the DNA will increase the frequency of mutation. Since the hydrogen-bonding properties are a function of the amino and hydroxyl groups of the purines and pyrimidines, any modification of these groups would be expected to increase the likelihood of mutations. A large number of simple, common chemicals are capable of modifying DNA, and depending on the particular agent and conditions of treatment, they can increase the spontaneous mutation frequency from tenfold to a thousandfold or more. For example, nitrous acid (HNO_2) removes some amino groups from certain purines and pyrimidines, and by this mechanism converts cytosine to uracil and adenine to hypoxanthine (Fig. 9-3). Hypoxanthine has the hydrogen-bonding properties of guanine, and uracil the bonding properties of thymine. Therefore, when the modified DNA replicates, the base incorporated opposite hypoxanthine will be cytosine and that opposite uracil will be adenine. As replication continues, two DNA strands are synthesized which have base sequences quite different from the original strands (Fig. 9-4).

Other mutagenic agents are the base analogs, compounds which structurally resemble the naturally occurring purine or pyrimidine closely enough to be incorporated into DNA in place of the natural base (Fig. 9-5). The analog,

DNA replication
Pages 165-168

MOST MUTATIONS
OCCUR DURING
REPLICATION OF DNA

Base sequence complementarity
Page 40

MODIFICATION OF
HYDROGEN BONDING
PROPERTIES OF BASES
RESULTS IN MUTATIONS.

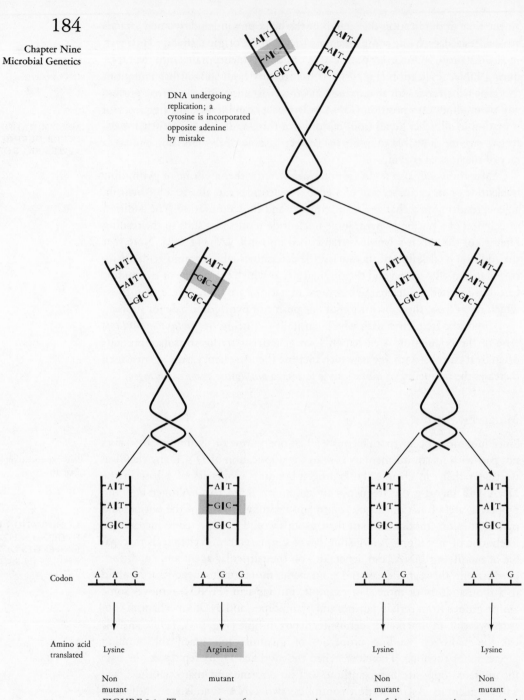

DNA undergoing
replication; a
cytosine is incorporated
opposite adenine
by mistake

FIGURE 9-1 The generation of a mutant organism as a result of the incorporation of a pyrimidine base (cytosine) in place of thymine in DNA replication.

however, does not have exactly the same hydrogen-bonding properties as the natural base.

Base analogs have been used in the treatment of tumors, because once they are incorporated into DNA, the macromolecule may not duplicate normally and the cells die. Indeed, all mutagenic agents are powerful killers. For every cell that is mutated by a mutagenic agent, several hundred are probably killed.

Certain types of radiation, *X-rays and ultraviolet light* (UV) in particular, are also powerful mutagens. Although the mutagenic effect of UV radiation was

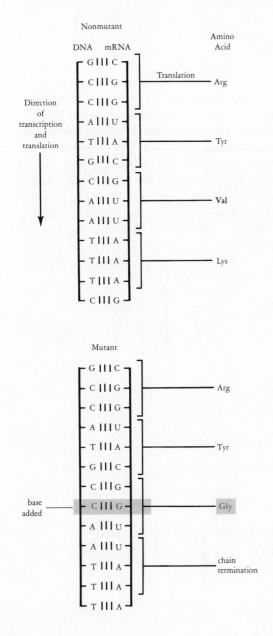

FIGURE 9-2
Production of mutation as a result of base addition. The addition of a nucleotide in the DNA results in a "frame shift" in the transcription of the DNA and a new triplet code word which is translated as a new amino acid. The deletion of a nucleotide would have essentially the same effect. The protein chain terminates when a triplet is produced which does not code for any one of the amino acids (Ch. 8).

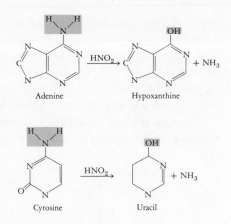

FIGURE 9-3

Removal of amino groups from adenine and cytosine by nitrous acid (HNO_2). The hydrogen atoms bonded to the C and N atoms in the ring are not shown. Can you see why hypoxanthine has the hydrogen bonding properties of guanine and uracil the hydrogen bonding properties of thymine?

MUTAGENIC EFFECT OF UV LIGHT

first recognized over 40 years ago, only within the last decade has its mode of action become understood. This radiation results in the formation of a covalent bond between adjacent thymine molecules on the same strand of DNA (intrastrand bonding) to form a thymine-thymine *dimer* (Fig. 9-6). This distorts the DNA strand backbone sufficiently so that the hydrogen-bonding properties of the purines and pyrimidines near the dimer are altered. The result is that incorrect purines and pyrimidines may be inserted in the new strand. Since UV light is a component of sunlight, cells are often exposed to this mutagenic and killing agent. To combat the effects of UV light many species of bacteria have enzymes that can repair the damage it causes. Some bacteria have an enzyme that in the presence of visible light is able to break the covalent bond joining the thymine

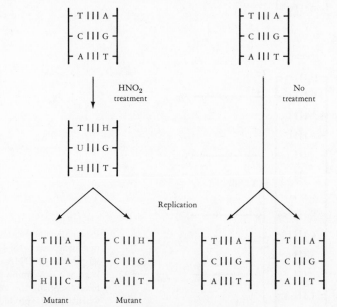

FIGURE 9-4 Mutation caused by nitrous acid treatment of DNA.

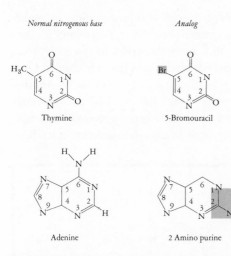

Normal nitrogenous base   Analog

Thymine   5-Bromouracil

Adenine   2 Amino purine

FIGURE 9-5
Common base analogs and the normal bases they replace in DNA.
The hydrogen atoms bonded to the carbon and nitrogen atoms
in the ring are not shown.

bases, the phenomenon termed *light repair* (Fig. 9-7). To prevent this repair, cultures treated with UV light to induce mutants are kept in the dark. Some bacteria also have enzymes that cut out or *excise* the damaged single strand of DNA and other enzymes that then repair the break by synthesizing a strand complementary to the undamaged strand. Since visible light is not required for the action of these enzymes, the term *dark repair* describes this process (Fig. 9-8).

REPAIR OF UV DAMAGE

Although mutagenic agents generally affect a specific purine or pyrimidine in a specific way, they do affect all the genes. No mutagen has yet been found which selectively mutates a specific gene. Since any agent that is capable of disrupting hydrogen bonds or removing amino groups is a potential mutagen, it is likely that many compounds in the environment are mutagenic.

Just as a gene can undergo mutation so that it no longer codes for a functional protein, in rare cases the same gene can undergo a *reverse mutation* or reversion so that the gene now codes for a functional protein.

Rate of Mutation

The *rate of mutation* is defined as the probability that a particular gene will mutate each time the cell divides. It is generally expressed as the negative exponent per cell division. Thus if there is one chance in a million that a gene will mutate when the cell divides, the mutation rate is 10^{-6} per cell division. Of course, the mutant will be recognized only if it is readily distinguished from the parent cells. Since the mutation rate of any single gene varies between 10^{-3} to 10^{-9} per cell division, each *gene* will be mutant in one in a million cells on the average. Thus the concept that all cells arising from a single cell are identical is not strictly true.

MUTATIONS ARE RARE AND OCCUR INDEPENDENTLY OF ONE ANOTHER

The advantage of working with organisms that can multiply to yield several billion cells per milliliter of medium now becomes apparent. Mutants in every single gene should be represented in the population. The problem is finding them. Fortunately some unusually simple but clever techniques have been devised that simplify the task of locating the proverbial needle in the haystack. One technique termed *direct selection* involves plating bacteria on a medium on which the mutant, but not the parent cell, can grow. Thus if one is searching for mutants that are resistant to an antibiotic, such as streptomycin, he can easily select this mutant directly by plating cells on a medium containing the antibiotic. Only the few cells in the population which are resistant will form a colony. Another type of approach is the *indirect selection technique,* illustrated by the ingenious *replica plating technique.* This was devised by the Lederbergs, a research team who made fundamental contributions in many areas of microbial genetics. It has been

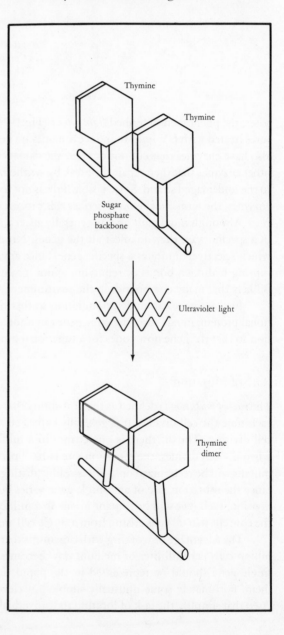

FIGURE 9-6
Formation of thymine dimers as a result of ultraviolet irradiation.

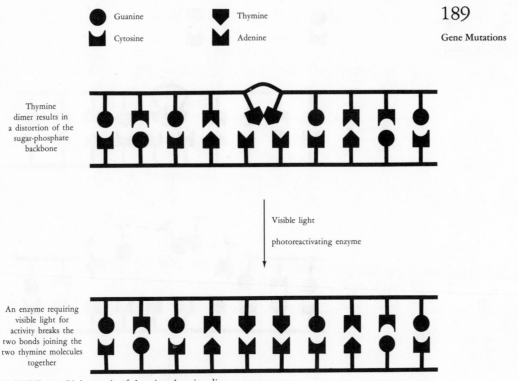

Guanine

Cytosine

Thymine

Adenine

Thymine dimer results in a distortion of the sugar-phosphate backbone

Visible light

photoreactivating enzyme

An enzyme requiring visible light for activity breaks the two bonds joining the two thymine molecules together

FIGURE 9-7   Light repair of thymine-thymine dimer.

invaluable in isolating mutants which require one or more nutrients. Since many bacteria, *Escherichia coli* in particular, can grow on a glucose-salts medium, it is possible to isolate a wide range of nutritional mutants. Mutants which require particular growth factors are termed *auxotrophs;* the parent cell which requires no organic metabolites other than glucose is termed a *prototroph.* The major problem with isolating the rare auxotroph is that a medium that supports growth of the auxotroph is also sufficient for growth of the prototroph. Thus finding the rare auxotroph in the midst of a very large number of prototrophs was always very tedious until the Lederbergs devised the technique of replica plating.

The technique of replica plating involves the transfer of a part of every colony from the "master" petri plate containing a rich medium (on which all cells will grow) to unsupplemented glucose-salts plates on which auxotrophic cells will *not* grow (Fig. 9-9). It becomes a simple matter to identify the colony on the master plate which does not appear on the unsupplemented medium. The nutritional requirement can be identified by replica plating the master plate to a medium containing each of the individual growth factors. A piece of sterile velveteen which contains thousands of tiny bristles serves to transfer bacteria from one plate to the other. Before the invention of replica plating, individual bacterial colonies had to be inoculated into the unsupplemented medium to determine whether or not each would grow.

REPLICA PLATING
TECHNIQUE

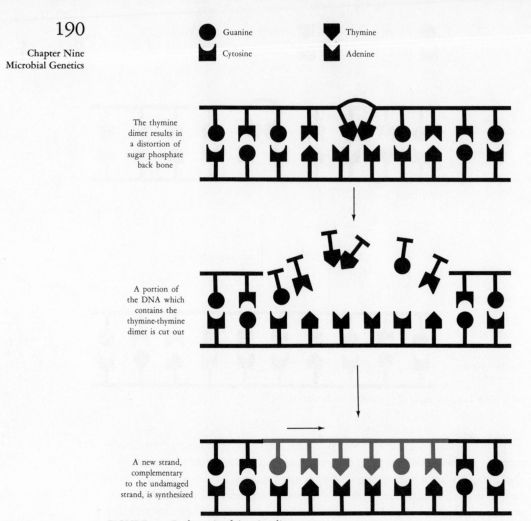

Guanine Thymine

Cytosine Adenine

The thymine
dimer results in
a distortion of
sugar phosphate
back bone

A portion of
the DNA which
contains the
thymine-thymine
dimer is cut out

A new strand,
complementary
to the undamaged
strand, is synthesized

FIGURE 9-8   Dark repair of thymine dimer.

**RATES OF MUTATION IN
SEVERAL INDEPENDENT
GENES**

Genes mutate independently of one another. Therefore the chance that two mutations will occur within the same cell is the *product* of the single rates of mutation (the sum of the exponents). For example, if the mutation rate to streptomycin resistance is 10^{-6} per cell division and that to penicillin resistance is 10^{-8} per cell division, then the chance that both mutations will occur within the same cell is $10^{-6} \times 10^{-8}$, or 10^{-14}—an infinitesimally small number. For this reason it is common practice in the treatment of certain diseases (tuberculosis in particular) to employ combined therapy, the simultaneous administration of two or more drugs. Any cells resistant to one antibiotic will be killed by the other.

Procaryotic cells contain one or several *identical* copies of each gene per cell and are thus *haploid*. In eucaryotic cells, each chromosome occurs in pairs and each member of the pair differs from its partner in its genetic information. These cells are *diploid*. In procaryotic cells any mutation in the cell's DNA will be ex-

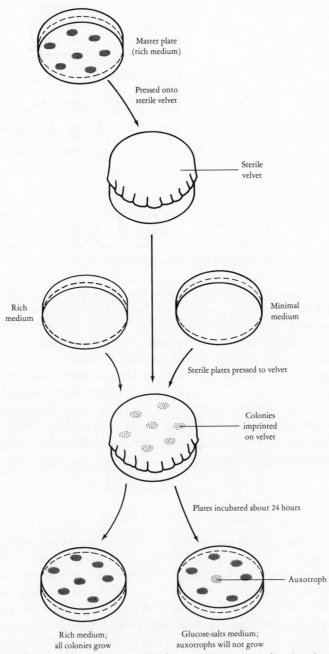

Master plate
(rich medium)

Pressed onto
sterile velvet

Sterile
velvet

Rich
medium

Minimal
medium

Sterile plates pressed to velvet

Colonies
imprinted
on velvet

Plates incubated about 24 hours

Auxotroph

Rich medium;
all colonies grow

Glucose-salts medium;
auxotrophs will not grow

FIGURE 9-9 Replica plating technique. A bit of each colony is imprinted on the velveteen, which is then pressed to both rich and glucose-salts media. Any colony originally growing on the rich medium which has a requirement for any growth factor will not grow on the glucose-salts medium.

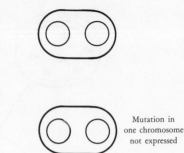

Mutation in
one chromosome
not expressed

FIGURE 9-10
Time lag in expression of mutations.

Mutant   Nonmutant

EXPRESSION OF
MUTATIONS IN HAPLOID
AND DIPLOID CELLS

pressed. Thus, a mutation which results in the synthesis of a nonfunctional enzyme required for growth causes the cell to require a growth factor. If there is
more than one chromosome per procaryotic cell, then the mutation is not expressed immediately since the nonmutant gene can synthesize enough of the
functional enzyme to allow growth of the cell in the absence of the growth factor. Expression of the mutant requires that cell division occur so that two mutant
genes reside in the same cell. Therefore, a period of time may be required between
the occurrence of a mutation and its expression (Fig. 9-10). In a eucaryotic cell,
a mutant gene will always be paired with a nonmutant partner in its homologous
chromosome and will remain disguised. Thus, the genetic analysis of mutations
is simplified in procaryotic cells.

MECHANISM OF GENETIC EXCHANGE

The fact that bacterial mutants arise and breed true was the first indication that
bacteria must have something analogous to the genes of higher organisms. The
isolation of mutants made it possible to determine whether or not *sexual* or *genetic recombination* occurred in bacteria. Sexual recombination results in the formation of a cell that has properties of both parents. Therefore it is possible to
look for sexual recombination only if strains of bacteria are available which differ
from one another in stable, easily recognized characteristics. The procedure used
by Joshua Lederberg, then a 20-year-old medical student, and Edward Tatum, in
looking for recombination in *E. coli,* illustrates the general approach of how genetic recombination can be recognized. A strain of *E. coli* which required threo-
nine, leucine, and thiamine (abbreviated thr⁻, leu⁻, thi⁻) but could synthesize
all other requirements for growth was mixed with a strain requiring phenylalanine, cystine, and biotin (phe⁻, cys⁻, and bio⁻). After incubation to allow
time for the bacteria to recombine, the mixture of cells was plated on a glucose-

salts medium without nutrients. On this medium only cells able to synthesize the six growth requirements can grow. Such recombinants must contain genes from each of the parental strains.

In this experiment about 100 colonies developed when approximately 10^8 cells of each type were mixed together and plated. When the cells of each parent were plated on the glucose-salts medium separately, no colonies appeared. This was not surprising since the three mutations would have to revert to prototrophy in the same cell in order for the cell to grow on the glucose-salts medium. The chance of this happening is about 1 in 10^{18} ($10^6 \times 10^6 \times 10^6$)! Therefore Lederberg and Tatum concluded that sexual recombination had indeed occurred. It is now known that a large number of genera of bacteria can undergo sexual recombination. The mechanism by which genes are transferred between cells can be divided into three categories. All mechanisms have certain features of the recombination process in common, but differ from one another in certain fundamental features. Some species of bacteria can undergo two of the processes, but thus far there is no compelling evidence that any single species can undergo all three. The three processes are the following:

THREE MECHANISMS DESCRIBED FOR GENE TRANSFER IN BACTERIA

1. *DNA mediated transformation.* Genes are transferred as "naked" DNA.
2. *Conjugation.* Genes are transferred across a protein "bridge" (pilus) from one cell to another.
3. *Transduction.* Bacterial genes are transferred by a bacterial virus.

Before discussing the specifics of each process, it is useful to discuss their basic similarities. In all three mechanisms the DNA is transferred from one cell, the *donor cell,* to another cell, the *recipient cell. Only* DNA is transferred and in most instances only a small fraction of the total DNA of the donor cell. Once inside the recipient cell the donor DNA is positioned along the recipient DNA so that their homologous genes are adjacent. One or more enzymes act on the recipient DNA, causing breaks and subsequent excision of a piece of DNA. The donor DNA is *integrated* into the recipient chromosome in place of the excised DNA. The details of this process are not yet understood, but it is likely that the same enzymes or enzymes similar in function to those involved in the repair of UV damage in DNA are also involved in the recombination process. The end result is that the chromosome of the recombinant cell contains DNA derived from the donor and the recipient cells. The piece or pieces of DNA which are excised from the recipient chromosome are probably broken down by DNA degrading enzymes. Recombination is precise; there is no increase or decrease in the number of nucleotides in the recombinant, only a precise substitution of some donor for recipient nucleotides.

SUMMARY OF THE VARIOUS STEPS INVOLVED IN GENETIC RECOMBINATION

ONLY A PORTION OF DONOR CHROMOSOME TRANSFERRED BETWEEN BACTERIA

In order to recognize that genetic exchange has occurred, it is necessary to distinguish the recombinant from the parent cells. Each of the three processes of exchange is a rare event generally affecting no greater than 1 percent of the cell population. It is therefore advantageous to be able to select *directly* for the recombinants by plating on a medium on which only the recombinants will grow. Since several billion cells can be plated on a single petri dish, it is possible to detect rare recombinants.

HISTORY OF BACTERIAL
TRANSFORMATION

The isolation and characterization of DNA as the transforming principle provided the first definitive evidence that DNA comprises the genetic material. The study of transformation dates back to the late 1920s when an English physician, Griffith, first observed the process in an organism that causes bacterial pneumonia, *Streptococcus pneumoniae*. This organism, as usually isolated in nature, is encapsulated by a polysaccharide. If a few encapsulated cells are injected into a mouse, the animal dies of infection. Mutants devoid of a capsule can be isolated in the laboratory. These organisms are readily killed by the host animal and therefore do not cause disease. Griffith inoculated one group of mice with nonencapsulated cells, a second group with heat-killed encapsulated cells, and a third with both living, nonencapsulated cells and heat-killed, encapsulated cells (Fig. 9-11). As expected the mice in the first two groups were unaffected. Much to Griffith's surprise, however, mice in the last group developed infection and died. Furthermore, he was able to isolate living encapsulated cells from the dead mice. It was shown later that when encapsulated cells were ground up and added to nonencapsulated cells in a test tube, living encapsulated cells would arise. But it was not until 1943 that a group of investigators at the Rockefeller Institute (Drs. O. Avery, C. MacLeod, and M. McCarthy) identified the active material which brought about this transformation of nonencapsulated to capsulated cells as DNA. It was this identification that proved that DNA was the genetic material and, together with the results of Beadle and Tatum published a few years earlier, suggested that DNA must function by determining the specificity of enzymes.

In hindsight the observations made by Griffith in 1928 can now be explained in terms of mutation, transformation, and selection. The heat-killed encapsulated cells carried the genetic information for synthesizing the capsule; the mutant, nonencapsulated strain had a defective gene so that it could not synthesize the capsule. Even though the encapsulated cells were killed, their DNA remained intact and could be transferred to the nonencapsulated recipient. Those cells which integrated the genes concerned with capsule formation from the donor were now capable of synthesizing the capsule. These cells, although initially few in number, were resistant to the body defense cells of the mouse and therefore multiplied. The nonencapsulated cells were rapidly selectively killed by the host animal, so that the bacterial population shifted in favor of the encapsulated cells. When these cells reached a high enough concentration, the mouse died of infection.

MECHANISM OF DNA
TRANSFORMATION

Since the original experiments with the pneumococcus, transformation has been observed in other organisms including *Bacillus subtilis, Neisseria meningitidis, Haemophilus influenzae,* and certain strains of streptococci and staphylococci. Investigations of transformation in these organisms have revealed certain common features (Fig. 9-12). When the donor cells are broken, the long, single molecule of DNA present in the cell is fragmented into about 100 pieces; each piece has a molecular weight of about 20×10^6 daltons and consists of at least 20 genes. One of the most amazing features of the transformation process is that this very large piece of DNA passes through the bacterial cell wall and membrane.

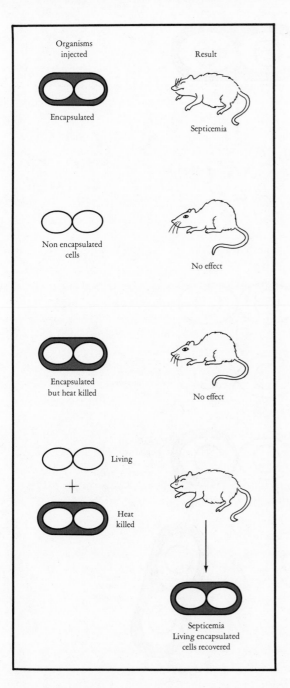

Organisms injected

Result

Encapsulated

Septicemia

Non encapsulated cells

No effect

Encapsulated but heat killed

No effect

Living

+

Heat killed

Septicemia
Living encapsulated cells recovered

FIGURE 9-11   DNA mediated transformation in pneumococci. The effects of injecting various preparations of pneumococci into mice.

Actually the recipient cells must be grown under rigidly controlled conditions in order to be able to take up this large fragment of DNA. These cells are said to be *competent*. What makes a cell competent is not known, but it seems likely that the cell wall may have to be modified so that the DNA can pass through. A cell will be transformed only for those genes which are on the chromosome

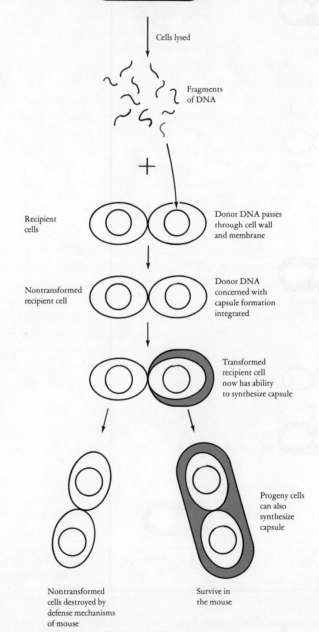

FIGURE 9-12   General features of DNA transformation.

fragment which enters it. Thus some cells in the population will be transformed for some characters; others will be transformed for others. Since the donor DNA comprises about 20 genes, the cell can be transformed for all of these characters at the same time. The frequency of transformation of any *single* gene is only about 1 percent of the number of recipient cells. However, if it were possible to somehow select for transformation of *all* recipient genes in the same experiment, it would be possible to show that most of the recipient cells were competent and had incorporated DNA from the donor.

DNA transformation provides a useful system for studying the effect of a variety of physical and chemical treatments on the biological functions of DNA. Thus it is possible to isolate and purify DNA from donor cells and then determine whether treatment with a certain chemical has any effect on the ability of the DNA to transform recipient cells. Mutagenic agents such as nitrous acid quickly destroy the transforming ability of a portion of the DNA. Mutations also arise in the recipient cells which have integrated the mutated genes.

Conjugation

This process differs from transformation in that cell-to-cell contact is required. This need can be shown experimentally by the following technique. If two different auxotrophic mutants are placed on either side of a filter through which bacteria cannot pass in a U tube, recombination does not occur. However, if the filter is removed, allowing cell-to-cell contact, the cells do recombine (Fig. 9-13).

There are two types of cells in any population of *E. coli* in which conjugation occurs. One is termed the F$^+$ (or male) cell, the other the F$^-$ (or female) cell. The F$^+$ cell has a few pieces of DNA in the cytoplasm which are not a part of the chromosome (*extrachromosomal DNA*). These F *particles* are absent in the female. Each particle, approximately 1 percent the size of the chromosome, multiplies independently of the chromosome. Since this extrachromosomal DNA carries genetic information, the F$^+$ cell synthesizes proteins that the F$^-$ cell does not. The most distinguishing proteins are the *sex pili* which serve to attach the F$^+$ to the F$^-$ cell (Fig. 9-14).

When a population of F$^+$ and F$^-$ cells are mixed together, the F$^+$ cells attach to the F$^-$ cells by means of the sex pili. Within minutes one of the F particles from the male cell enters the recipient female, perhaps passing through the hollow pilus. Since all F$^-$ cells in the population receive the F particle, the entire culture quickly becomes F$^+$. In this process the chromosome is *not transferred* and recombinants do not arise (Fig. 9-15).

The chromosome, however, *is transferred* to the F$^-$ cell by rare cells in the F$^+$ population. These cells are not F$^+$ but arise from F$^+$ cells as a result of the integration of the F particle into the chromosome. This cell type is termed Hfr, which is an abbreviation for *high frequency of recombination,* and only cells that have integrated the F particle can transfer their chromosome. Once integrated, the F particle remains integrated and replicates as part of the chromosomes. Consequently the progeny of an Hfr cell is also Hfr.

Extrachromasomal DNA
Page 89–90

Pili
Page 96

ONLY F PARTICLE IS TRANSFERRED IF DONOR CELL IS F$^+$

The F particle can be integrated at any location in the circular chromosome of *E. coli*. The chromosome, however, is not transferred as a circle but as a linear sequence of genes. The integration of the F particle apparently *mobilizes* the chromosome for transfer into the recipient cell. As the Hfr and F⁻ cells come in contact, the donor chromosome apparently breaks at the site of integration of the F particle. The gene that was *next* to the Hfr particle enters first, the *origin,* and the Hfr particle always enters last (Fig. 9-16). Since the F particle can become integrated at any site on the chromosome, there are a large number of different Hfr strains, each one identified by the genes which are transferred first (Fig. 9-13). The order of gene transfer is constant for each Hfr strain and depends on the site at which the F particle becomes integrated into the chromosome.

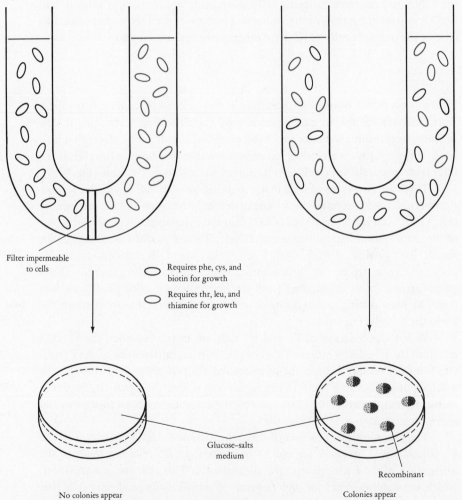

Filter impermeable to cells

Requires phe, cys, and biotin for growth

Requires thr, leu, and thiamine for growth

Glucose–salts medium

Recombinant

No colonies appear

Colonies appear

FIGURE 9-13 Demonstration of requirement for cell to cell contact for recombination in conjugation. In order to grow on the glucose-salts medium, the bacteria must be able to synthesize all amino acids and vitamins.

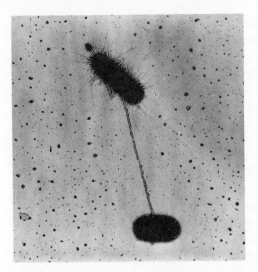

FIGURE 9-14
Sex or F pilus holding together a donor and a recipient cell of
Escherichia coli during DNA transfer. The "dots" on the pilus are
bacterial viruses that have adsorbed to the pilus. Some investigators believe that the DNA is transferred through the pilus, so
that the pilus should be considered an organelle of transport.
(Courtesy of Dr. C. Brinton, Jr. and J. Carnahan)

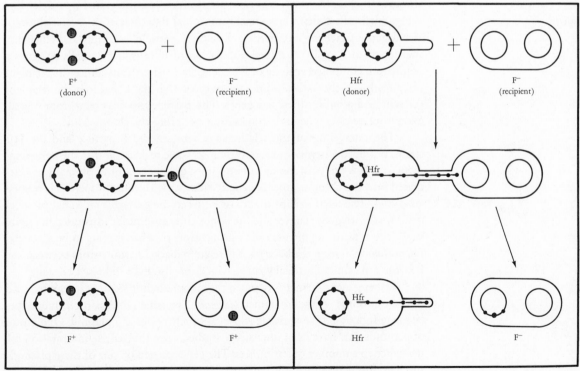

FIGURE 9-15
Transfer of the F particle results in recipient cell becoming F$^+$
and the donor remaining F$^+$.

FIGURE 9-16
Transfer of a portion of the chromosome of the donor cell.
Since the Hfr fragment remains in the donor cell, the recipient
cell remains F$^-$.

In contrast to the rapid transfer of the F particle into F⁻ strains, the transfer of the entire chromosome of Hfr strains into F⁻ cells requires about 90 minutes. The rate of transfer is constant during the entire process, that is, 30 minutes after conjugation exactly one-third of the donor chromosome will be transferred into the recipient cell; after 45 minutes, one-half. However, the pilus holding the Hfr and F⁻ cells together is not strong, and in most cases it breaks long before the entire chromosome is transferred. Thus only the genes close to the origin are generally transferred into the F⁻ cell. Therefore the Hfr gene is not transferred and consequently the vast majority of the recipient cells remain F⁻.

Thus two different types of DNA can be transferred in conjugation: the extrachromosomal F particle if the donor is F⁺ and a portion (or, more rarely, all) of the chromosome if the donor is an Hfr strain. Another type of extrachromosomal particle which some strains are capable of transferring is a combination of the F particle and a segment of the chromosome. This extrachromosomal particle is termed F′ (F prime), and the strains which transfer these particles are termed F′ strains. The F′ strains arise in the following manner. Just as the F particle can become integrated into the chromosome, in rare instances it can detach from the chromosome and thus again become extrachromosomal. The detachment of the F particle often involves not only the F⁺ DNA but also some of the chromosomal DNA which is in close proximity to the integrated F particle (Fig. 9-17). Such cells contain an extrachromosomal F particle with several chromosomal genes attached. This particle, like the F particle, is rapidly and efficiently transferred to F⁻ cells. A mixed population of F′ cells and F⁻ cells becomes a population composed entirely of F′ cells within a short time. Since the recipient cell already has the chromosomal genes carried on the F′ particle, the recipient cell will be diploid for these few genes. This particle remains extrachromosomal, except in rare cases when it becomes integrated into the chromosome.

The unraveling of the relationship between the F particle and the Hfr strains provided a new concept to biologists—the realization that the same fragment of DNA can exist in one of two forms, either extrachromosomally or integrated into the chromosome. A genetic element with these properties is termed an *episome*. The relationships that exist between the extrachromosomal particles concerned with gene transfer and the donor chromosome are illustrated in Figure 9-17. The F factor represents only one example of what appears to be a general phenomenon among the bacteria. There are remarkable similarities between the F factor and certain bacterial viruses, which can also exist in these two states.

Other genetic elements occur in the cytoplasm but are *never* integrated into the chromosome. These are termed *plasmids*. The same DNA particle in one species may function as an episome and in another species as a plasmid. One group of plasmids that have been intensively studied carry the genetic information for resistance to a number of antibiotics. The genes carried on one of these plasmids specifies production of the enzyme penicillinase in *Staphylococcus aureus*. This enzyme destroys certain forms of penicillin, and therefore any strains which carry this plasmid are penicillin resistant. These plasmids can also be transferred from one strain of staphylococci to another by the process of transduction and transformation but not by conjugation.

EPISOMES

F factor
Pages 197–198
Viruses
Pages 320–323

Transduction
Pages 323–325

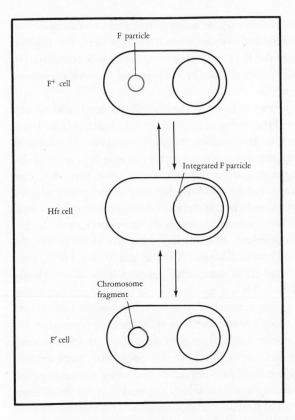

FIGURE 9-17
Relationship of F⁺, Hfr, and F′ cells.

Genetic elements which superficially have many features of the F′ particles
are the *R factors*. These are extrachromosomal genetic elements which are composed of two parts, a *resistance transfer factor* (RTF) analogous to the F factor,
and multiple genes concerned with drug resistance (*R genes*) (Fig. 9-18). For example, a strain of bacteria carrying the R factor is commonly resistant to the four
widely used antibacterial medicines: sulfanilamide, streptomycin, chlorampheni-

R FACTORS
SUPERFICIALLY SIMILAR
TO F′ FACTORS

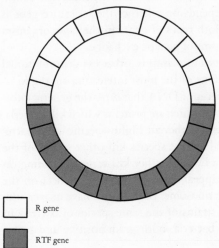

FIGURE 9-18
Diagrammatic representation of R factors. There are many individual genes concerned with resistance (R genes). The R genes
are transferred only if they are attached to the RTF genes.

R FACTORS CAN BE
TRANSFERRED BETWEEN
DIFFERENT GENERA

col, and tetracycline. Other strains may be resistant to certain combinations of these drugs, and others are resistant to more drugs than these four. The unusual and distinguishing feature of the R factor is that it can be rapidly transferred to antibiotic-sensitive cells in the population by conjugation, thereby conferring resistance on these recipient cells.

The R factor was discovered in Japan in the late 1950s when it was noticed that a very high percentage of the strains of *Shigella*, which cause bacillary dysentary, was resistant to at least four drugs, rather than only one or two. Furthermore the patterns of resistance suggested that resistance did not arise by a series of discrete steps in which the organisms became resistant to first one, then two, then three, and so on, of the drugs. Further work by Japanese investigators revealed that resistance to several drugs could be transferred to sensitive bacteria *as a unit*. This transfer can occur not only between strains in the same species but also between other closely related organisms which include members of the genera *Shigella, Salmonella, Escherichia, Yersinia, Klebsiella, Serratia,* and *Proteus.* The R factor can also be transferred to members of some other genera which are not closely related such as the genus *Vibrio.* Thus if cells in any one of these genera contain the R factor, it can be very rapidly transferred to other cells in any of these genera by cell-to-cell contact. In today's world the widespread use of antibiotics has strongly selected for mutants that are antibiotic-resistant. This selection process, however, does not explain how and where R factors originated. Some investigators feel that R factors are merely F' particles which carry drug-resistance genes. However, there is no convincing evidence which suggests that the drug-resistance loci were ever once a part of the bacterial chromosome. It is abundantly clear that R factors as well as penicillinase plasmids represent a serious threat to antibiotic therapy. Thus an ever-increasing number of drug resistant strains are continuously emerging in the population both from mutation and selection of antibiotic resistant cells as well as from the transfer of drug-resistance genes.

CHROMOSOMAL
RESISTANCE GENES

Although the discussion thus far has emphasized plasmids and episomes as the site of genes specifying resistance to antibiotics, many chromosomal genes also confer resistance to the same antibiotics. In many cases the exact mechanism of resistance to the same antibiotic depends on whether the resistance gene is chromosomal or extrachromosomal. Both types of genes in the same organism can be transferred by the same mechanisms of genetic exchange.

In addition to the F and R factors and plasmids, other extrachromosomal elements are present in some bacteria. One of the most interesting are the *bacteriocin factors.* These are small circular pieces of DNA that carry the genetic information for the synthesis of the *bacteriocins,* which are proteins which kill bacteria. There are a large number of bacteriocins, and they are highly specific in the bacteria they kill. The bacteriocins produced by one species kill other strains of the same or related species. The mechanism by which they kill varies depending on the particular bacteriocin. One group appears to have its primary effect on the cytoplasmic membrane; another on the ribosome. Bacteriocins have proven useful for distinguishing between certain strains of the same species of bacteria in clinical diagnostic laboratories. Many bacteria, both gram-positive and gram-

Clinical use of bacteriocins
Page 416

negative, synthesize bacteriocins, but the most intensively studied are the *colicins* produced by *E. coli*. Although an organism may possess the colicinogenic factor, its genes are not always expressed. However, if the culture is exposed to a variety of agents which affect DNA synthesis, for reasons that are not known, colicins are produced.

Some of the colicinogenic factors are episomes; others are plasmids. Like the F particle these factors can promote chromosome transfer if they are integrated; in the extrachromosomal state they are rapidly transferred to recipient cells.

Transduction

In this mechanism of genetic exchange, bacterial DNA is carried within a bacterial virus, termed a *bacteriophage,* from the donor to the recipient cell. Since this process involves the life cycle of the bacteriophage, the discussion of transduction is deferred to Chapter 16.

Significance of Gene Transfer to Bacteria

DNA-mediated transformation and conjugation have been described in relatively few species of bacteria (Table 9-1). Thus conjugation has been shown to occur in only about a dozen species of gram-negative bacteria. Since sex pili so far have only been observed in gram-negative organisms, presumably this process will not operate in gram-positive organisms. Transformation has been demonstrated in the laboratory in about a dozen species, both gram-positive and gram-negative. Transduction is probably the most widespread mechanism of genetic exchange, having been demonstrated in virtually all species in which it has been actively sought. The fact that conjugation and transformation have been found in relatively few species does not necessarily mean that these processes do not occur in other species. It may only mean that other species have not yet been investigated or that if they have, the proper conditions for demonstrating the exchange have not yet been found.

The fact that these exchange mechanisms have been demonstrated in the laboratory under carefully controlled conditions does not necessarily mean that

TABLE 9-1
Some of the Species in Which Gene Transfer Has Been Observed

Transformation	Transduction	Conjugation
Bacillus subtilis	Escherichia coli	E. coli
Streptococcus pneumoniae	Salmonella typhimurium	S. typhimurium
Neisseria meningitidis	Bacillus subtilis	P. aeruginosa
Haemophilus influenzae	Pseudomonas aeruginosa	
Rhizobium meliloti	Staphylococcus aureus	
Micrococcus radiodurans		
Staphylococcus aureus		

they play a significant role in nature. Nevertheless, it is easy to imagine how gene transfer and recombination can serve a very useful function in microorganisms. The major source of variation within a bacterial species results from mutation. Since bacteria are haploid, the mutation becomes evident very quickly. Because of the rapid multiplication of bacteria by binary fission, selection within the population operates very rapidly. For example, in a hospital environment in which a large variety of antibiotics are constantly present, a mutation to streptomycin resistance will confer a selective advantage on the mutant organism. The antibiotic-sensitive organisms will be killed by the antibiotic and the population will change very quickly from a population in which the resistant cells are in the minority to one in which the resistant organisms predominate. However, only a *single* gene will undergo mutation at any one time. In contrast, since a considerable number of genes can be transferred at the same time, *groups* of genes can be tried out in new environments and in different combinations with genes of the recipient cells. A greater number of genetic backgrounds are available on which the selection process can operate. Probably the best example of the importance of gene transfer to bacterial survival is the transfer of the R genes. This transfer must be important in nature since strains carrying these factors are extremely widespread in any country in which antibiotics are commonly used.

Significance of Gene Transfer to the Microbial Geneticist

LOCATING POSITION OF
GENES ON THE
CHROMOSOME

An analysis of gene transfer provides the means whereby genes can be located or mapped on the bacterial chromosome. The basic idea behind all mapping theory is this: the closer two genes are located on the chromosome, the more likely they are to be transferred together from the donor to the recipient cell; the farther apart they are, the less chance they will be transferred together. In the process of transformation the donor bacterial chromosome fragments and only those genes which are close enough to each other to be on the same fragment of DNA, will be transferred into the same recipient cell. In conjugation the frequency with which genes are transferred into the recipient cell is directly related to their distance from the gene which enters first. It is actually possible to time the entrance of genes into the recipient cell. The distance between genes is inversely proportional to the difference in minutes that separates the entrance of the two genes of the donor into the recipient cell. Two genes very close to one another enter the recipient cell almost simultaneously; those widely separated from one another enter at widely divergent times.

By measuring the frequency with which genes are transferred either together or separately, it is possible to map a large number of genes on the chromosomes of a variety of organisms which undergo genetic exchange. The most intensively mapped chromosome is that of *E. coli;* the locations of over 100 genes have been determined. Unfortunately since genetic maps can only be constructed in those organisms which undergo genetic recombination, we have no information on the location of genes in the vast majority of microorganisms.

From the data which have been collected thus far, several interesting facts on the map location of bacterial genes have emerged. Organisms which are closely related have very similar genetic maps; the genes which specify the same functions are located at approximately the same location on the chromosome. In unrelated organisms, such as *E. coli* and *B. subtilis,* the genetic maps are quite different. In all bacteria, genes which code for enzymes in the same pathway are often clustered, that is, occur next to one another. This phenomenon is so prevalent in bacteria that it cannot have occurred by chance but must have been strongly selected for in the course of evolution. The selective advantage that such gene clustering confers on the cell can only be speculated on at this time. One popular notion is that the functioning of many genes can be efficiently controlled if they are located close to one another. The means of this control is considered in Chapter 10.

SEXUAL RECOMBINATION IN EUCARYOTES

The processes of transformation, transduction, and conjugation display the fundamental features of genetic recombination; namely the union of the genetic material of two individuals to form the genetic material of another individual. One feature of each of these processes that makes them unique in the biological world is that the only part of the donor cell which enters the recipient cell is DNA, and in most cases only a small portion of the entire chromosome at that. Since the entering DNA is very quickly incorporated into the recipient cell, replacing a segment of the recipient DNA, the recombinant cell remains haploid. In eucaryotic organisms, an actual *fusion* of the nuclei of the two parental cells takes place rather than a replacement of the recipient DNA by donor DNA. The two cells that fuse are termed *gametes,* and the product of their fusion is a *zygote.* The diploid zygote has two copies of each chromosome (2n); the haploid gametes only one (1n). Each species of organism has a characteristic number of chromosomes. In man, there are 23 pairs (total, 46). Before the organism can reproduce sexually (that is, when the gametes undergo fusion to form a zygote), the number of chromosomes must be reduced from 2n to 1n. Thus, in human beings, both the egg and sperm have 23 chromosomes. The process by which this reduction in chromosome number occurs is termed *meiosis* (Fig. 9-19). This process involves a number of steps, the first being a *duplication* of the DNA by the process of *mitosis.* In this process, each nucleus is duplicated and gives rise to two identical nuclei (Fig. 9-19). The process of mitosis precedes the division of one eucaryotic cell into two when these cells divide asexually by fission. In meiosis, after the DNA has been duplicated a series of steps occur in which the cells undergo two divisions, resulting in each cell having a 1n number of chromosomes. The processes of mitosis and meiosis share many features in common but also differ in a number of respects. In particular, in meiosis homologous chromosomes line up along their entire length (*synapse*) whereas this does not occur in mitosis.

    The stages in the life cycle of eucaryotic protists at which meiosis occurs

IN PROCARYOTES, ONLY PORTION OF DNA IS TRANSFERRED.

IN EUCARYOTES, NUCLEI FUSE

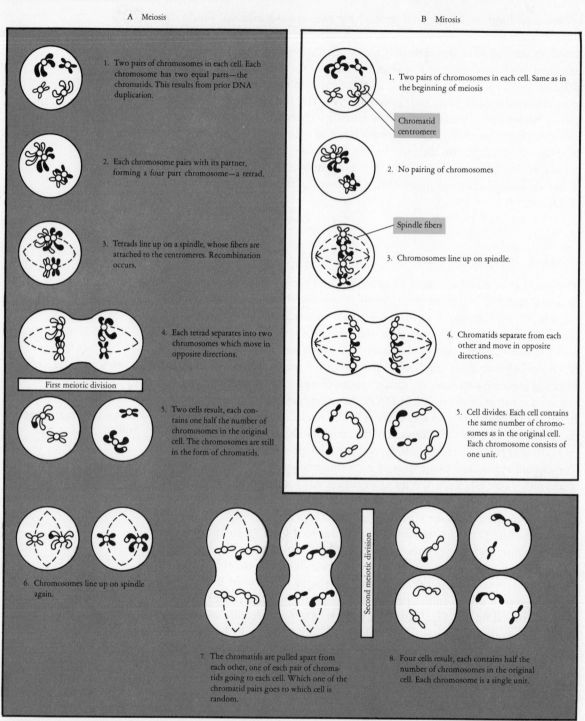

A   Meiosis

B   Mitosis

1. Two pairs of chromosomes in each cell. Each chromosome has two equal parts—the chromatids. This results from prior DNA duplication.

1. Two pairs of chromosomes in each cell. Same as in the beginning of meiosis

Chromatid
centromere

2. Each chromosome pairs with its partner, forming a four part chromosome—a tetrad.

2. No pairing of chromosomes

Spindle fibers

3. Tetrads line up on a spindle, whose fibers are attached to the centromeres. Recombination occurs.

3. Chromosomes line up on spindle.

4. Each tetrad separates into two chromosomes which move in opposite directions.

4. Chromatids separate from each other and move in opposite directions.

First meiotic division

5. Two cells result, each contains one half the number of chromosomes in the original cell. The chromosomes are still in the form of chromatids.

5. Cell divides. Each cell contains the same number of chromosomes as in the original cell. Each chromosome consists of one unit.

6. Chromosomes line up on spindle again.

Second meiotic division

7. The chromatids are pulled apart from each other, one of each pair of chromatids going to each cell. Which one of the chromatid pairs goes to which cell is random.

8. Four cells result, each contains half the number of chromosomes in the original cell. Each chromosome is a single unit.

FIGURE 9-19   Meiosis and Mitosis. In meiosis, one diploid cell gives rise to 4 haploid cells. In mitosis, one cell divides into two cells each having the same DNA content as the original cell.

differ, depending on the organism. In some organisms, the life cycle of the organism is characterized primarily by the diploid phase. In other organisms, the zygote undergoes meiosis very quickly after it is formed so that the organism is haploid during most of its life cycle. The processes of meiosis and mitosis are fundamentally the same in all organisms and both sexes.

One eucaryotic alga that has proven very useful in genetic studies is the single-celled *Chlamydomonas*. This organism reproduces both sexually and asexually (Fig. 9-20). In the sexual phase, two different haploid mating types (+ and −) fuse to form the diploid zygote, which is encysted in a thick cell wall. The zygote undergoes a period of maturation (6 days duration) and then undergoes meiosis resulting in the formation of four haploid vegetative cells, two + and two − . Each of these cells can divide by mitosis to form clones of identical haploid cells (asexual reproduction). Under certain conditions however, the vegetative cells undergo a process of differentiation and turn into gametes, these being of the same mating type as the vegetative cells from which they were derived. The gametes fuse and the life cycle is repeated. In the case of this alga, the haploid phase is the dominant one in the life of the organism.

In *Chlamydomonas* chemical differences exist between the gametes, so that a + gamete will only mate with a − gamete. In other organisms, all gametes are identical and any gamete will fuse with any other gamete.

LIFE CYCLE OF A
EUCARYOTE,
CHLAMYDOMONAS

SUMMARY

Variations can arise in a single microbial cell as a result of two phenomena: changes in the nucleotide sequence of the DNA (mutations) and changes in the environment that determine which genes and enzymes will function. Changes in the nucleotide sequence of a DNA molecule generally arise when the improper base is inserted into the growing DNA chain as the DNA is being replicated. These variations are very rare, but can be increased by mutagenic agents which may increase their frequency over a thousandfold.

By mixing mutants with different genotypes, three mechanisms of genetic exchange have been demonstrated in bacteria: transformation, in which the donor DNA is "naked;" transduction, in which the donor DNA is carried from the donor to the recipient cell inside a bacteriophage; and conjugation, in which the donor DNA passes from the donor to recipient cell following cell-to-cell contact. The three mechanisms of gene transfer have basic features in common. In all cases the donor DNA enters the recipient cell, pairs with the homologous region of the recipient chromosome, and then is integrated. A portion of the recipient chromosome is replaced by the donor chromosome. In DNA transformation highly purified DNA can pass through the cell wall and membrane of the competent cell. In conjugation there is unidirectional transfer of DNA from the Hfr to the F⁻ cell. If the donor cell is F⁺, then only the F particle is transferred. If the donor cell is an Hfr strain, the chromosome, or more commonly a portion of it, is transferred into the F⁻ cell.

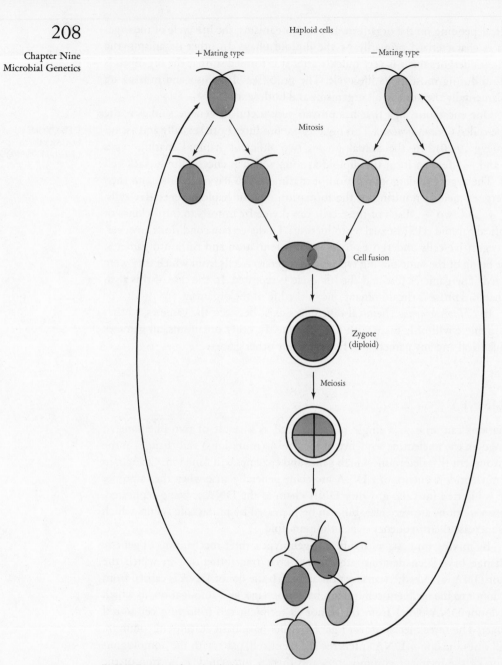

FIGURE 9-20   Life cycle of *Chlamydomonas*. This single-celled eucaryotic protist reproduces sexually by cell fusion and asexually by fission.

All processes of genetic exchange occur at a relatively low frequency, and it is therefore necessary to plate the recipient cultures on selective medium in order to detect recombinants.

In eucaryotes, sexual recombination results from the fusion of haploid nuclei to form a diploid zygote. Many organisms multiply asexually as diploids, with their nuclei dividing by the process of mitosis. Other eucaryotes reduce their chromosome complement immediately after formation of the zygote to 1n and multiply asexually.

QUESTIONS

1. Mutants which involve macromolecule synthesis are generally lethal to the cell. How might such mutants be isolated? (HINT: Consider the fact that proteins are prone to denature at a high temperature.)

2. You are interested in obtaining a mutant in which one amino acid is substituted for another in the mutant protein. Would you employ a mutagen which is known to cause base additions, deletions, or substitutions to produce such a mutation? Why?

3. Many mutations are silent, in that changes in the DNA do not result in changes in the protein coded by the corresponding gene. Why is this true (see Fig. 8-10)?

4. When two auxotrophic strains are mixed together, recombinants arise. What are the possible mechanisms by which this occurs (4) and how would you experimentally determine which process is responsible?

5. How can you argue that genetic recombination in bacteria has any significance when it is such a rare event?

FURTHER READING

BENZER, S., "The Fine Structure of the Gene," *Scientific American* (January 1962).

DEERING, R., "Ultraviolet Radiation and Nucleic Acid," *Scientific American* (December 1962). (Offprint #143)

HARTMAN, P. and S. SUSKIND, *Gene Action,* 2d ed. Englewood Cliffs, N.J.: Prentice-Hall, 1969.

HANAWALT, P. and R. HAYNES, "The Repair of DNA," *Scientific American* (February 1967). (Offprint #1061)

LEVINE, R. P., *Genetics,* 2d ed. New York: Holt, Rinehart and Winston, Inc., 1968.

TAYLOR, J. H., "The Duplication of Chromosomes," *Scientific American* (June 1958). (Offprint #60)

TOMASZ, A., "Cellular Factors in Genetic Transformation," *Scientific American* (January 1969). (Offprint #1130)

WATANABE, T., "Infectious Drug Resistance," *Scientific American* (December 1967). (Offprint #1087)

WOLLMAN, E. and F. JACOB, "Sexuality in Bacteria," *Scientific American* (July 1956). (Offprint #50)

Chapter 10

Enzyme Regulation

ENERGY CONSIDERATIONS IN CONTROL SCHEMES

ENERGY SUPPLY OFTEN
IN LIMITED SUPPLY IN
NATURE

Biosynthetic reactions often require energy. Since the supply of available energy is generally the limiting factor in bacterial growth in nature, it is imperative that a cell synthesize the maximum amount of cell material from a limited supply of energy if it is to survive. The discussion of control mechanisms in much of this chapter focuses on the control of one amino acid pathway leading to synthesis of the amino acid histidine. The general principles rather than the details of the control systems will be emphasized. The principles for the regulation of this biosynthetic pathway apply to all amino acids, purines, pyrimidines, vitamins, and other metabolites.

The biosynthesis of one molecule of histidine requires the expenditure of a great deal of energy—41 high-energy phosphate bonds. This figure represents the sum of two values: the number of molecules of ATP which are actually utilized in the pathway of histidine synthesis and the potential number of molecules of ATP that could have been synthesized if the carbon precursors of histidine were metabolized to CO_2 and H_2O to provide energy rather than the amino acid. In addition, the cell must expend energy as well as carbon and other elements to synthesize the enzyme molecules which catalyze the ten enzymatic steps in the synthesis of this amino acid (Fig. 10-1).

How can the cell conserve the energy required for this synthesis? By utilizing any histidine that might be available in the environment rather than synthesize its own. This requires the expenditure of only a few molecules of ATP per molecule of histidine that enters the cell: one at least involved in the transport process itself and those required for the synthesis of the permease system. Once

Permease
Pages 86–87

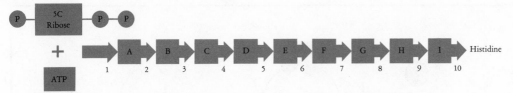
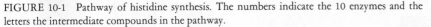

FIGURE 10-1   Pathway of histidine synthesis. The numbers indicate the 10 enzymes and the letters the intermediate compounds in the pathway.

synthesized this system can be utilized again and again. It is little wonder that cells which are subjected to such intense competitive pressures as are bacteria have evolved complex but highly efficient systems for the uptake of metabolites.

MECHANISMS OF CONTROL OF ENZYMATIC ACTIVITY

Cells control the synthesis of metabolites by two mechanisms. In both cases the end product of a pathway inhibits its own synthesis. Thus any histidine in the environment will enter the cell and shut off the synthesis of histidine. This inhibition occurs in two entirely independent ways by two different mechanisms. (1) The end product of a pathway *inhibits the activity* of a preformed enzyme of the pathway involved in its synthesis. This mechanism is termed *feedback* or *end product inhibition.* (2) The end product of a pathway *inhibits the synthesis of the enzymes* of the pathway involved in its synthesis. This is termed *end product repression.* Thus histidine inhibits the activity of an enzyme concerned with histidine biosynthesis and also inhibits the synthesis of the enzymes in the histidine biosynthetic pathway.

TWO DISTINCT CONTROL MECHANISMS— CONTROL OF ENZYME ACTIVITY AND CONTROL OF ENZYME SYNTHESIS

Feedback Inhibition

In addition to the *active* or *catalytic* site which all enzymes have, the first enzymes of most biosynthetic pathways have a second site which can combine with the end product of the pathway. This second site is termed the *allosteric site* and the end product is termed a *feedback inhibitor.* The combination of the allosteric site of the enzyme with the end product of the pathway through weak bonding forces changes the conformation or shape of the enzyme molecule so that the enzyme is no longer capable of catalyzing its usual reaction (Fig. 10-2). By this mechanism the end product prevents the formation of the product of the first enzyme step of the pathway, thereby effectively shutting down the entire pathway. Consider again the biosynthesis of histidine which involves ten enzymatic steps (Fig. 10-1). Any histidine in the medium combines with the first enzyme and changes the enzyme's shape. The product of the first enzymatic reaction is not synthesized. Therefore the substrate of the second enzyme is not synthesized and the pathway is shut down instantaneously.

Enzymes
Pages 142–146

END PRODUCT COMBINES *REVERSIBLY* WITH FIRST ENZYME AND INACTIVATES IT

211

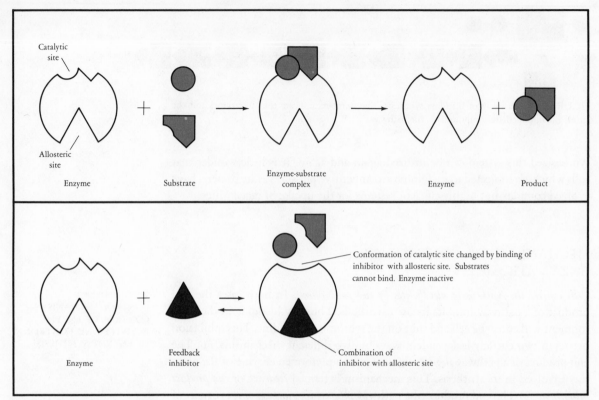

FIGURE 10-2 Effect of feedback inhibitor on catalytic activity of enzyme. The combination is reversible and if the level of inhibitor drops, the reaction goes in the reverse direction, so the enzyme becomes functional again.

FIGURE 10-3 The photograph shows two colonies which excrete histidine and a "halo" of histidine-requiring cells growing around the excreting colonies. Note that the closer the cells are to the colonies excreting histidine, the larger the histidine-requiring colonies. (Courtesy of J. P. Dalmasso)

212

The importance of feedback inhibition in the regulation of end product synthesis is illustrated by observing what happens when feedback control does not function. Mutants can be isolated in which the histidine does not change the shape of the first enzyme of histidine biosynthesis although the catalytic site is normal. Such mutants are termed *feedback-resistant.* These mutants invariably overproduce and actually excrete large amounts of histidine into the medium. This phenomenon is readily detected by growing a few colonies of the feedback-resistant mutant on histidine free medium seeded with a very large number of cells of a histidine-requiring mutant. The excretion of the histidine by the feed-back-resistant mutant allows the histidine-requiring cells to grow around the colony, giving rise to a halo of background cells (Fig. 10-3).

A wide variety of feedback-resistant mutants which overproduce and excrete metabolites have been isolated. Such mutants are useful to industrial concerns interested in isolating the metabolite on a commercial scale.

MECHANISMS OF CONTROL OF ENZYME SYNTHESIS

End Product Repression

Although feedback inhibition effectively shuts off the synthesis of the end product, it still allows some waste of energy and carbon since feedback inhibition has no effect on the synthesis of enzymes of the pathway. Therefore another mechanism, *end product repression,* comes into play. The end product represses or prevents the synthesis of the enzymes concerned with the synthesis of that particular end product. Returning to the example of histidine synthesis, the presence of histidine in the medium inhibits the synthesis of all ten enzymes of histidine biosynthesis. This inhibition results in a saving of the many molecules of ATP which must be expended for the synthesis of these enzyme proteins. However, this regulatory mechanism is slow to act in regulation since all of the enzyme molecules in the cell are still functional. Only when the number of enzyme molecules per cell decreases as a result of cell division does this mechanism operate.

The mechanism by which an end product represses the synthesis of enzymes is not completely understood at the molecular level. Indeed it appears that the details vary in different systems, although the general aspects are reasonably clear. To return to the regulation of histidine biosynthesis, the ten genes which code for the ten enzymes are linked to one another on the chromosome in *Salmonella typhimurium* (Fig. 10-4). In addition, another gene termed a *regulatory gene* codes for a protein that is concerned with regulating the function of these ten genes. In this system, there actually appear to be several genes concerned with regulation. The protein product of the regulatory gene is a *regulatory* or *repressor protein.* This protein has the ability to recognize and bind to a specific nucleotide sequence at the extreme end of the sequence of histidine genes. The nucleotide sequence to which the regulatory protein binds is the *operator region* and is adjacent to the region at which RNA polymerase initiates gene transcription of the

END PRODUCTS
CONTROL THE *RATE*
OF ENZYME SYNTHESIS

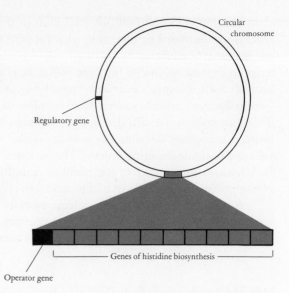

FIGURE 10-4
Genes of histidine biosynthesis. All of the genes which code for enzymes of histidine biosynthesis are closely linked on the chromosome. The regulatory gene is not located close to the histidine genes. In actual fact there are several genes concerned with the regulation of histidine biosynthesis located at different sites on the chromosome.

REPRESSION PREVENTS
INITIATION OF
TRANSCRIPTION

ten genes of histidine synthesis (Fig. 10-5). The binding of the repressor protein to the operator region prevents the RNA polymerase enzyme molecule from transcribing the ten genes of histidine synthesis. Therefore none of the enzymes concerned with histidine biosynthesis can be synthesized. What role does histidine play in repression? Although the regulatory gene continually synthesizes repressor protein, this protein is unable to combine with the operator region unless it is "activated." Histidine combines with and activates the repressor protein. How the protein is activated is not clear. It seems likely, however, that the repressor protein is an allosteric protein and so can exist in two different shapes. In one shape it does not bind to the DNA. However, if the end product binds to the repressor protein and changes its shape, it now is able to bind.

GENE CLUSTERING AND
REGULATION

Since all of the genes of histidine synthesis are closely linked (Fig. 10-4), the entire cluster is transcribed as a unit into one large mRNA molecule (Fig. 10-5). Thus a single operator region at one end can control the synthesis of this single mRNA molecule and therefore the synthesis of all ten enzymes. Gene clustering thus provides a very efficient mechanism for regulating the synthesis of all of the enzymes of a pathway as a unit. The term *operon* defines a cluster of genes which are controlled as a unit. Genes, however, which code for biosynthetic pathways are not always clustered on the chromosome. When they are not, there must be an operator region next to each gene that codes for a biosynthetic enzyme of the pathway. However, there is still only one repressor protein that binds to all of the operator regions to shut off enzyme synthesis. Operons are very common in bacteria, but are rare in eucaryotic cells. The reason for this difference is not clear.

A summary of the mechanisms that the end product employs to control its own synthesis is diagrammed in Figure 10-6.

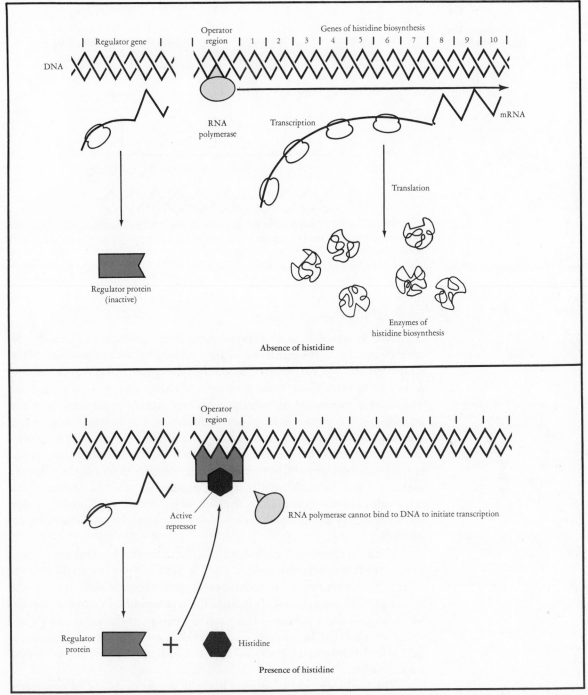

FIGURE 10-5 Diagrammatic representation of the regulation of synthesis of enzymes of histidine biosynthesis. The nature of the regulator protein is not clear, but there are apparently several different genes concerned with synthesizing proteins involved with regulation of histidine enzyme synthesis. Transfer RNA also plays a role, as yet unclear, in the regulation.

215

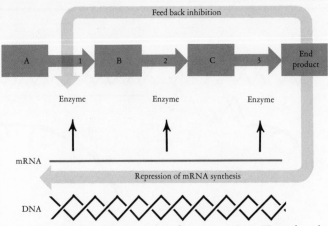

FIGURE 10-6 Effect of end product on control of its own synthesis. The end product feed back inhibits the activity of the *first* enzyme of the pathway and prevents the synthesis of *all* enzymes in the pathway.

Enzyme Induction

Genes can be turned on or off, depending on the environment in which the cells are growing. So far the regulation of gene activity by the end products of biosynthetic pathways has been explored. Inactive genes, however, can also be turned on by metabolites in the environment so that they now are transcribed. The enzymes synthesized as a result of genes being turned on are termed *inducible enzymes,* and the process is called *enzyme induction.* The low-molecular weight compound which activates gene transcription is called the *inducer.* Inducers are usually compounds that the cells are potentially capable of utilizing as a source of energy, carbon, or both, but which are not normally present in the cells environment. Until the compound is actually present, the cell will synthesize only very small amounts of the enzymes required for its breakdown. Thus the cell does not waste energy and carbon by synthesizing enzymes whose substrates are not available.

CELLS SYNTHESIZE ONLY THE ENZYMES THEY NEED

The most thoroughly studied case of induced enzyme synthesis involves the enzymes of lactose degradation in *E. coli.* This system serves as a model for most other inducible systems. Lactose, the most abundant sugar in milk, is a disaccharide of glucose and galactose. It induces the synthesis of two enzymes required for its breakdown: a permease which actively transports the sugar into the cell and an enzyme (β-galactosidase) which breaks down the sugar into glucose and galactose. A third enzyme is also induced, but its significance in lactose metabolism is unclear.

Lactose
Pages 43–44

The mechanism of induction is basically similar to that of enzyme repression (Fig. 10-7). The same genetic elements are involved: a regulatory gene which synthesizes a repressor protein, an operator region which binds the repressor protein, and the two genes which code for the permease and β-galactosidase.

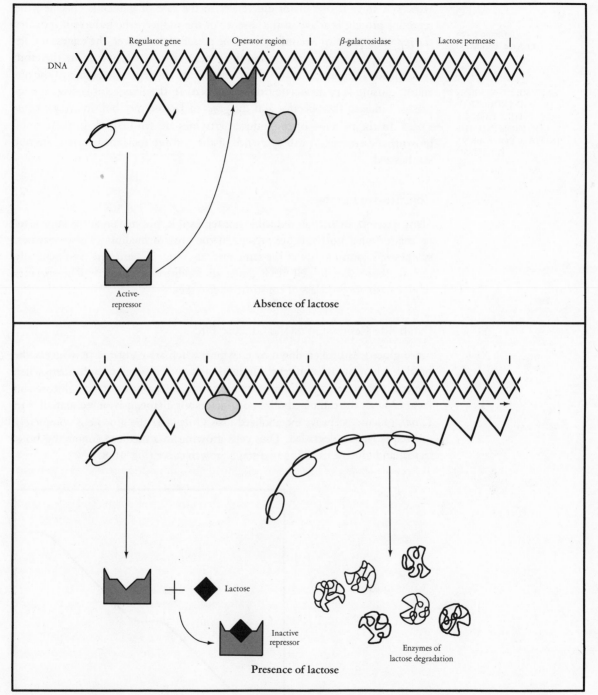

FIGURE 10-7   Induction of a degradative enzyme system. Note that the same genetic elements take part as in the regulation of a biosynthetic pathway.

ENZYME REPRESSION
AND INDUCTION
FUNCTION BY
CONTROLLING THE
INITIATION OF mRNA
SYNTHESIS

However, there is one major difference. In the case of inducible enzymes, the repressor protein is active in the absence of the inducer; the inducer inactivates the repressor protein. Accordingly, in the presence of inducer, the repressor does not bind to the operator region and the genes of the lactose operon are therefore transcribed. The inducer is also able to inactivate any repressor bound to the operator, causing it to dissociate from the DNA. In the absence of lactose, the repressor binds to the operator and the *genes* of lactose metabolism *are not transcribed.* In the presence of the inducer, enzymes are synthesized. In the case of biosynthetic enzymes, in the presence of the end product, the enzymes are not synthesized.

Constitutive Enzymes

Many enzymes are neither inducible nor repressible, but remain at the same level no matter what nutrients are present in the growth medium. These enzymes, synthesized continuously at the same rate, are termed *constitutive,* and generally include those that the cell needs under all conditions of growth. The enzymes of glycolysis are one class of constitutive enzymes.

Catabolite Repression or the Glucose Effect

Since glucose is broken down by enzymes which are constitutive, whereas the breakdown of lactose requires the induction of two enzymes, what happens when the cell is grown in a medium containing limiting amounts of *both* glucose and lactose? Does the cell metabolize both sugars simultaneously or sequentially? In *E. coli,* glucose is always metabolized first. Only after the glucose is completely utilized is lactose degraded. Thus cells growing in a medium containing both lactose and glucose exhibit a two-stage growth curve (Fig. 10-8).

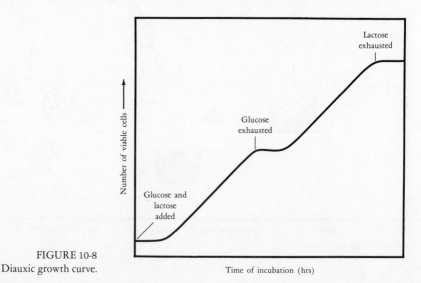

FIGURE 10-8
Diauxic growth curve.

The first period of growth is at the expense of glucose, and the enzymes of lactose metabolism are not induced. After all the glucose has been degraded, a short period of time lapses during which the enzymes of lactose synthesis are induced. The second period of growth involves the degradation of lactose. This effect is only one manifestation of a far-reaching effect that glucose has on induced-enzyme synthesis. Glucose represses the synthesis of a very large number of inducible enzymes. This general repression termed *catabolite repression* or the *glucose effect* is not completely understood, but some interesting features have been uncovered. The addition of glucose to a growing culture results in an almost immediate release of an important nucleotide from the cell, 3', 5'-cyclic AMP (Fig. 10-9). The significance of this nucleotide was not previously appreciated in *E. coli,* although it had been known for a number of years that in mammalian systems many hormones functioned by regulating the synthesis of cyclic AMP. This latter compound in turn is directly responsible for the effects of many hormones. This effect is likely related to the fact that certain important enzymes seem to function only in the presence of cyclic AMP. Therefore it is interesting that cyclic AMP is also important in regulatory processes in bacteria and other protists. In bacteria this nucleotide is required for the initiation of transcription of a large number of inducible enzyme systems most of which are involved in energy metabolism.

Glucose inhibits enzyme induction by causing the loss of cyclic AMP from the cell by some unknown mechanism. Mutants which lack the enzyme concerned with the synthesis of cyclic AMP cannot synthesize most inducible enzymes unless cyclic AMP is added to the culture. However, the story of induced enzyme synthesis is even more complicated. This nucleotide does not function alone, but requires a protein to which it must bind to carry out its function of promoting selective transcription. The protein which binds the cyclic AMP also apparently binds to the DNA to facilitate transcription. Thus the diagram of the regulation of inducible enzyme synthesis is incomplete (Fig. 10-7). Not only must lactose be present to inactivate the repressor protein but cyclic AMP must be available to bind to another protein which binds to the DNA to promote gene transcription. The American biochemist, Dr. Earl Sutherland, who isolated, identified, and recognized the general significance of cyclic AMP in both animals and bacteria was awarded the Nobel Prize in Medicine in 1971.

Catabolite repression serves a very useful function in bacteria. It requires the cell to utilize the best available source of energy. Glucose is generally the most readily utilizable source and therefore it inhibits indirectly the synthesis of en-

CYCLIC AMP REQUIRED
FOR INITIATION OF
TRANSCRIPTION OF
GENES CONCERNED
WITH DEGRADATION

A PROTEIN REQUIRED
FOR FUNCTIONING OF
CYCLIC AMP

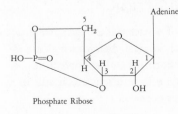

FIGURE 10-9
Cyclic AMP. The phosphate molecule is bound to carbon atoms 3 and 5 thereby giving the molecule its name.

zymes which metabolize poorer sources, such as lactose. However, if cultures are grown under conditions in which glucose is *not* being metabolized at a rapid rate, then it does not function as a catabolite repressor. Furthermore, glucose is not a catabolite repressor in species which cannot metabolize this sugar readily.

Cyclic AMP and cholera
Page 480

Cyclic AMP has been implicated as playing a role in a number of other bacterial systems. The synthesis of the enzyme required for bacterial bioluminescence is under the control of cyclic AMP and there is now strong evidence that the level of cyclic AMP plays an important role in the symptomatology of the disease cholera. Further, bacteria require cyclic AMP in order to synthesize flagella and mutants unable to synthesize it cannot synthesize flagella. It is quite possible that these seemingly unrelated effects have a commonality in their relationship to energy metabolism.

Regulation of Cell Growth

Media
Pages 112–114

Bacteria are often exposed to changing nutrient conditions and will grow at different rates depending on the nutrients in their environment. In a rich nutrient medium they will use the preformed nutrients and channel their energy into the synthesis of macromolecules. In a glucose-salts medium considerable energy must be spent in the synthesis of subunits of the macromolecules. Therefore growth will be slower in a glucose-salts medium than in a rich nutrient medium. In all cases, however, the rates of synthesis of RNA, DNA, and protein parallel each other. The cells are in a state of *balanced growth*. Before a cell can divide into two cells, DNA, RNA, protein, and all other cellular components must double in amount. Therefore it is not surprising that the synthesis of one macromolecule is geared to the synthesis of the others. Thus if a cell is starved for an amino acid, the synthesis of RNA and DNA soon ceases.

The interrelationship among RNA, DNA, and protein synthesis can be observed when cells growing in a glucose-salts medium are diluted into a rich-nutrient medium. First, there is an immediate increase in the synthesis of ribosomal protein and ribosomal RNA, but no immediate significant increase in the overall rate of protein synthesis. This suggests that the synthesis of protein is limited by the number of ribosomes available. Until more ribosomes are synthesized, the rate of protein synthesis cannot increase. The rate of DNA synthesis also slowly increases, resembling the synthesis of protein. Eventually, the synthesis of the three macromolecules becomes balanced at an increased rate.

REGULATION OF BACTERIAL SPORULATION

Endospore
Pages 97–98

In Chapter 4 the composition and properties of a bacterial *endospore* were described. The progressive physiological and morphological changes which are involved in the conversion of a *vegetative cell* to the dormant endospore requires the participation of all the regulatory mechanisms we have discussed, acting in a specific sequence to turn some genes on and other genes off.

The development normally begins several hours after a population of vegetative cells enters the stationary phase of growth. The sequence of morphological changes as viewed with the electron microscope is given in Figure 10-10 with a diagrammatic representation of each stage. The nuclear bodies or chromosomes in the vegetative cell fuse to form a long rod-shaped body of DNA. The DNA divides into two parts, one part entering the region of the vegetative cell destined to become the endospore and the rest remaining in the *sporangium* (the portion of the vegetative cell not destined to become the endospore). The cytoplasmic membrane grows across the cell near one pole, forming a septum that separates the DNA into two parts. The septum continues to grow, surrounding, and thereby separating the portion of the cell destined to become the endospore from the sporangium. This precursor of the endospore is termed a *forespore* and has a characteristic transparent appearance under the light microscope. The forespore undergoes further development, becoming surrounded by additional outer layers. First a *cortex* which consists primarily of *mucocomplex* material is laid down. Several layers of *spore coat* are synthesized which surround the cortex. In addition, some spores are surrounded with a loose-fitting coat.

In the formation of the mature endospore the cytological transformations are accompanied by profound changes in the chemical and enzymatic machinery of the vegetative cell. Thus many enzymes which are not detected in the vegetative cell appear during certain stages of sporulation. In contrast many enzyme systems which are characteristic of the vegetative cell are present at very low levels or even completely absent from the spore. In addition, a variety of new structural components appears in the spore that is not present in the vegetative cell. All of these facts indicate that a considerable amount of genetic information concerned specifically with the process of sporulation is not expressed during vegetative growth. Perhaps fifty genes concerned specifically with sporulation are located at widely separated sites on the chromosome.

What controls which genes are transcribed? The answer relates to the mechanisms by which cells undergo differentiation, one of the central questions in biology. The answer is not entirely known, but recent investigations have provided some clues. First, it appears that each step in the process requires the participation of the preceding step. For example, if an early spore enzyme is defective, the development of the spore will cease at that stage and none of the enzymes or spore structures of later stages of development will be synthesized. Second, it has been shown that the RNA polymerase of the vegetative cell is modified in the course of sporulation. Thus it appears that in a sporulating cell a modification of the polymerase occurs that renders it unable to transcribe genes associated with vegetative growth but able to transcribe genes associated with sporulation. This could account for the fact that certain genes associated with vegetative growth are turned off and genes associated with sporulation are turned on.

Catabolite repression by glucose may also be responsible for repressing the synthesis of a variety of enzymes of sporulation. In order to sporulate, bacteria must be growing at either a very slow rate or have stopped growing altogether. Since glucose starvation results in sporulation, it seems possible that glucose may

SEQUENCE OF EVENTS IN
ENDOSPORE
FORMATION

Mucocomplex
Pages 77–78

RNA polymerase
Pages 175–176

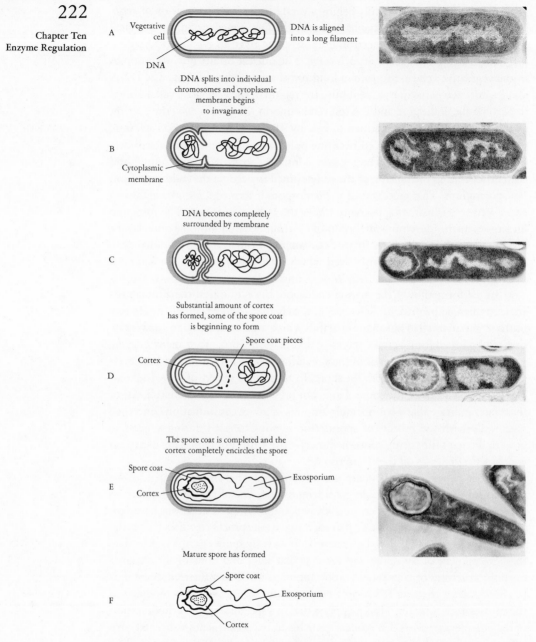

A Vegetative cell

DNA

DNA is aligned into a long filament

DNA splits into individual chromosomes and cytoplasmic membrane begins to invaginate

B Cytoplasmic membrane

DNA becomes completely surrounded by membrane

C

Substantial amount of cortex has formed, some of the spore coat is beginning to form

Spore coat pieces

Cortex

D

The spore coat is completed and the cortex completely encircles the spore

Spore coat

Cortex

Exosporium

E

Mature spore has formed

Spore coat

Exosporium

Cortex

F

Free endospore

FIGURE 10-10   Major steps in sporulation as observed through the electron microscope. The time from the first to the last picture was 16 hr. (Photos courtesy of Dr. L. Santo; Santo, L., H. Hohl, and H. Frank, *J. Bact.*, **99**:824 (1969))

function as a catabolic repressor for at least some spore enzymes. Starvation for nitrogen also results in the conversion of vegetative to sporulating cells. Rapid growth and sporulation seem to be mutually exclusive aspects of cell development. Perhaps nature has employed this elaborate control system involving starvation for required nutrients to ensure that sporulation will occur when conditions are no longer favorable for vegetative growth.

Spore Germination

A bacterial spore is able to remain dormant for many years. However, it can break the period of dormancy very quickly under the proper environmental conditions and develop into a vegetative cell. The first stage of this process is termed *germination* (Fig. 10-11). The factors required to trigger germination vary from species to species. In some, heating the spores to 60 to 80°C (*heat shock*) will initiate germination. In other species, specific chemicals such as alanine or adenosine will trigger the process. The mechanism by which these chemicals and heat shock act is not clear nor is it clear why these conditions should induce a cell to germinate. The first evidence of germination generally can be observed within minutes after heat or chemical treatment. The most characteristic properties of the spore are quickly lost, including resistance to staining and heat. These changes are accom-

ENVIRONMENTAL CONDITIONS DETERMINE WHETHER SPORES WILL BREAK DORMANCY

Heat shock
Page 8

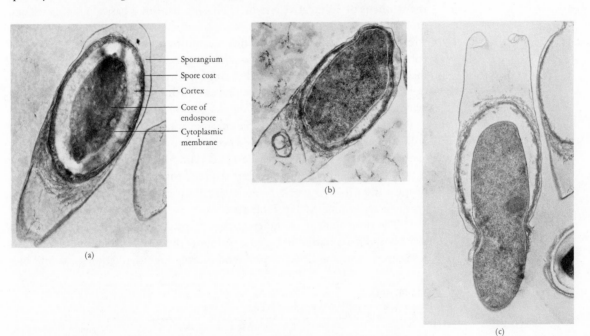

Sporangium
Spore coat
Cortex
Core of endospore
Cytoplasmic membrane

(a)

(b)

(c)

FIGURE 10-11 Sequence of major events in germination and outgrowth of an endospore.
    (a) A resting spore prior to heat activation, encased in the sporangium.
    (b) Deformed vegetative cells developing within the spore coat. The cortex has almost completely dissolved; materials lost from the spore. Germination
    (c) The elongated rod has ruptured both the spore coat and the sporangium as it emerges. Outgrowth
(Courtesy of Dr. J. Hoeniger; Hoeniger, J. F. M. and C. L. Headley, *J. Bact.*, **96**:1835 (1968))

panied by a disappearance of the cortex and loss of soluble organic materials from the spore. If the medium in which germination takes place is adequate to support vegetative growth, the germinated spore will continue to develop into a vegetative cell by a process called *outgrowth*. During outgrowth the proteins and structures of the vegetative cell are synthesized and the products of the genes of sporulation are not.

SUMMARY

In this chapter the two basic mechanisms by which bacteria control the flow of metabolites through metabolic pathways were considered: end product inhibition and end product repression. In the first control system the end product *inhibits the activity* of a preformed enzyme (generally the first) of a biosynthetic pathway. This results in an *immediate* shutdown of the pathway. The second control mechanism involves the control of enzyme synthesis by controlling gene transcription. The end product activates a repressor protein which binds to an operator gene, thereby preventing transcription of all the genes involved in the synthesis of the particular end product. In addition to these regulatory mechanisms for the control of biosynthetic pathways, cells have a mechanism to control the synthesis of degradative enzymes. This mechanism is basically similar to end product repression, except that in enzyme induction the substrate *inactivates* an already active repressor protein. Thus the substrate prevents the repressor from combining with the operator region, and the genes are transcribed. These control mechanisms are specific in that only a single pathway is controlled by an inducer or repressor. Glucose, however, can prevent the induction of most inducible enzymes which are concerned with the generation of energy. Indeed, if cells are growing on glucose, most inducible enzymes concerned with energy generation will not be synthesized in the presence of their substrate. Glucose apparently exerts its effect by controlling the level of cyclic AMP in the cell. This nucleotide as well as a protein which binds it are required for the initiation of transcription of most inducible enzymes. These salient features of the regulation of enzyme synthesis are summarized in Table 10-1.

The development of the endospore from the vegetative cell represents a process of differentiation in bacteria. A highly ordered sequence of enzyme reactions must occur, resulting in profound morphological, physiological, and bio-

TABLE 10-1
Features of Inducible and Repressible Enzyme Systems

	Inducible Enzyme Systems	Repressible Enzyme Systems
Name of controlling metabolite	Inducer	Repressor
Effect of small metabolite	Inactivates inducer	Activates repressor
Effect of cyclic AMP	In many, stimulates transcription by RNA polymerase	No known effect
Pathways affected	Degradative	Biosynthetic

chemical changes to the cell. In the process of sporulation some genes are turned off and others are turned on. The mechanism by which this occurs is not known although modifications of RNA polymerase have been implicated.

QUESTIONS

1. Tryptophan is added to a culture of *E. coli* growing in a glucose-salts medium. Explain the effect which tryptophan exerts on the synthesis of tryptophan by the cells. Give the detailed mechanisms by which tryptophan exerts its effect.
2. Why must the effects that end products exert on their own synthesis be reversible?
3. If you isolated a mutant which synthesized a defective repressor protein of histidine biosynthesis, how would you recognize it? What would be the phenotype of a strain which had a mutation in the operator gene of histidine biosynthesis?
4. Explain the economy a cell obtains by having repressible and inducible enzymes.
5. What would be the phenotype of a bacterium that was unable to synthesize cyclic AMP?

FURTHER READING

BECKWITH, J., "Regulation of the Lac Operon," *Science,* **156** (May 5, 1967), 597.

CHANGEAUX, P., "The Control of Biochemical Reactions," *Scientific American* (April 1965). (Offprint #1008)

PASTAN, I. and R. PERLMAN, "Cyclic Adenosine Monophosphate in Bacteria," *Science,* **169** (July 24, 1970), 339.

PTASHNE, M. and W. GILBERT, "Genetic Repressors," *Scientific American* (June 1970).

SUSSMAN, A. S. and H. O. HALVORSON, *Microbial Dormancy.* New York: Harper and Row, 1966.

II

The Microbial World

Beggiatoa species (courtesy of J. T. Staley and J. P. Dalmasso)

Chapter **11**

Classification
of Microorganisms

PROBLEMS IN CLASSIFICATION
OF MICROORGANISMS

The taxonomy of microorganisms is difficult and complex for a number of reasons. First, the number of different organisms is immense and the various groups of microbes are very different from one another. In fact, some, like Protozoa, fall within the province of zoology as others such as algae are dealt with by botanists. In these cases the microorganisms are classified according to conventions developed for animals or higher plants which are not necessarily appropriate for unicellular organisms.

The purpose of all classification schemes is to group organisms with similar properties and to distinguish those which are different. There are several basic differences between bacterial taxonomy and the taxonomy of higher plants and animals. In both taxonomic schemes the species is the basic taxonomic unit.

CONCEPT OF SPECIES   However, the definition of a species differs between higher organisms and bacteria. For higher organisms a species may be fairly precisely defined as an *inbreeding group having a limited geographic distribution.* This results in individuals having distinct morphological characters which distinguish them from individuals of other species. These same criteria cannot be applied to bacteria, because morphological characters are limited and sexual reproduction is seldom obligatory and is limited to a relatively few genera. Therefore in bacteria a species can be defined only in terms of a population of cells or *clone* all derived from a single cell. The members of the clone are virtually indistinguishable from one another, but do differ from a clone of another species, generally in several features. The major problem in the taxonomy of bacteria becomes one of trying to decide how different two clones must be for their members to be classified as different species. As more organisms are studied in detail the divisions between species tend to become blurred. The "lumpers" represent a group of taxonomists who prefer to see as few species as possible. In contrast the "splitters" are those who would have any clone recognized as a distinct species if it differed from all others in only a few characteristics. Just as for higher organisms, several species of bacteria are grouped into a genus and several genera are included within a family. Families in turn are grouped into orders; many orders constitute a class. However, it is extremely difficult to deal with these broad relationships.

This scheme of classification was patterned after that of higher plants and animals in which relationships can be assessed through studies of fossil records, embryological development, and morphological criteria. Such information provides the vertebrate and plant taxonomist with a large amount of information necessary to establish a broad classification which implies evolutionary relationships. Bacteria, which have only the faintest fossil record, no embryological development, and only a limited range of morphologies, present formidable problems to being classified according to a scheme which reflects their evolutionary descent. Therefore the classification of bacteria is completely arbitrary, based on organisms sharing certain properties. Thus one scheme of classification is based on the observation that most bacteria can be grouped into one of three morphological groups. Another scheme emphasizes the final metabolic products which are produced from the degradation of glucose. However, which criterion should be used in grouping organisms together is open to question. Is a rod-shaped organism which produces only lactic acid from glucose more closely related to a coccus with the same metabolic patterns or to another rod-shaped organism which does not produce lactic acid? There is no correct answer with our present state of knowledge.

BACTERIA CLASSIFIED ACCORDING TO THE PROPERTIES THEY SHARE, NOT ACCORDING TO EVOLUTIONARY RELATIONSHIPS

Therefore the classification scheme turns out to be one of convenience, consisting of descriptions of organisms, in terms of morphology, staining characteristics, nutritional and metabolic properties, among others. For bacteria these descriptions are contained in a reference text, *Bergey's Manual of Determinative Bacteriology,* the Bible of bacterial taxonomy. This book is now in its seventh edition. Since this manual does group organisms according to shared properties, it is not surprising that each edition is revised extensively as ideas change on the relative importance of different shared properties and as new methods of examining bacteria are developed. Since Bergey's Manual is a dictionary of descriptions of generally accepted species of bacteria which have been discovered, it is very useful in bringing a uniformity in the nomenclature of bacteria. However, its taxonomic scheme does not represent a scheme based on evolutionary relationships.

Until techniques are devised which can show relationships between organisms, classification schemes will be merely arbitrary, subject to the changing practices of taxonomists. Fortunately, several modern approaches to taxonomy provide hope that evolutionary relationships between organisms can be established more precisely than has thus far been accomplished. Some of these techniques are now considered; the more conventional techniques as exemplified by the approaches used in clinical microbiology laboratories are considered in various chapters throughout the text.

NUMERICAL TAXONOMY

It is immediately obvious that a taxonomist who performs a large number of convenient and rapid tests with a group of organisms is soon faced with a large body of data. Thus, if fifty tests are run on ten organisms, 500 pieces of data are generated, each describing whether a certain character is present or absent. Two alternative approaches are possible for handling these vast amounts of data. The traditional taxonomist attempts to weight the various characteristics assuming some to be of much greater importance than others. Although it seems justified to consider the shape of a bacterium as a more basic feature than whether or not the organism grows on potato starch agar, it is also true, however, that subjective judgments must be made in order to assign priorities to different characters.

Strain Number	1	2	3	4	5	6	7
1	100						
2	5	100					
3	10	95	100				
4	0	90	95	100			
5	80	15	35	15	100		
6	70	25	40	10	80	100	
7	95	10	20	10	90	75	100

(a)

	Strain Number	A				B		
		1	7	5	6	3	2	4
A	1	100						
	7	95	100					
	5	80	90	100				
	6	70	75	80	100			
B	3	10	20	35	40	100		
	2	5	10	15	25	95	100	
	4	0	10	15	10	95	90	100

(b)

FIGURE 11-1 Similarity matrices. Seven strains of organisms were tested in 100 different ways, each test resulting in a positive or negative result. Each strain is then compared to the other strains by determining the similarity coefficient, the percentage of the total characters tested which are held in common in each of the strains. In (a) the strains are arranged randomly; in (b) they are arranged so that the organisms with similar matrices are grouped together. Note that the seven organisms fall into two unrelated groups. (Modified from Mandelstam and McQuillen, *Biochemistry of Bacterial Growth,* Oxford: Blackwell Scientific Publications, Ltd, 1968. By permission of the publisher.

The proponents of numerical taxonomy hold that although some characteristics such as shape and form are more basic descriptions of an organism than an arbitrarily chosen set of biochemical characteristics, if *enough* characters are examined, the need for weighting different characteristics will be minimized. These principles imply that no intuition is required to construct a meaningful taxonomy and that the problem is purely logistic in the sense that data must be collected on as many features as possible and the vast amount of results must then be analyzed. Thus numerical taxonomy yields results which are unbiased by subjective judgments.

The final result of a classification by numerical taxonomy is expressed in terms of affinities or similarities coefficient, defined as the percentage of the total characters tested which are held in common between two strains (Fig. 11-1). In calculating the similarity coefficient, it is important not to include any characters which are negative for both organisms, for the lack of a character should not imply relationship. For example, the absence of flagella in two organisms need not imply similarity, although presence and absence in two strains which are otherwise related are useful facts. It is also important not to score characters which have a common basis. When a large number of characters are examined in a large number of strains, a computer program is used to calculate similarity coefficients for each pair of organisms and to construct a similarity matrix. On the basis of this matrix it is possible to arrange the strains in a hierarchy so that those which have more than a 90 percent similarity are termed a single species, and other more distinct groups are classified into different species and perhaps different genera.

Flagella
Pages 94–95

Numerical taxonomy has been used with some success with several groups of bacteria. For some groups of organisms the agreement with more classical methods is quite impressive, whereas for others the agreement is poor. In other cases the results of numerical taxonomy have been compared with results based on genetic and molecular methodology. Again agreement is sometimes impressive and sometimes not.

GENETIC AND MOLECULAR APPROACHES TO TAXONOMY

Recent developments in methods of classification are based on attempts to more directly compare the information resident in the DNA of a group of organisms. Since the genetic information of a cell is encoded in its DNA, the relatedness of two organisms is directly related to the gross composition as well as to the sequence of purines and pyrimidines (base sequence) in the DNA. As the base sequence of the DNA diverged organisms diverged in evolution. Different proteins were then synthesized which resulted in organisms which no longer could mate with each other. Thus modern approaches to an evolutionary taxonomy have often involved a comparison of base sequences in DNA, the ability of two organisms to mate, or a comparison of the amino acid composition of a specific protein or group of proteins.

The structure of double stranded DNA is such that the proportion of adenine (A) equals the proportion of thymine (T) and that of guanine (G) equals that of cytosine (C). The relative proportions of the four bases are usually expressed as the percent GC. This is calculated from the equation:

Structure of DNA
Page 44; Fig. 2-19

$$\frac{\text{moles G + moles C}}{\text{moles G + moles C + moles A + moles T}}$$

The relative proportion of AT and GC base pairs varies widely among different organisms. In fact, these variations in the base composition are of considerable value for classification. *Escherichia coli* DNA has 50 percent GC and both human and *Bacillus subtilis* DNA have 40 percent GC. These numbers mean that the three DNAs contain 50 percent or 60 percent AT base pairs.

COMPARISON OF GC
CONTENT

　　Knowledge of the base composition of DNA has revealed some very interesting facts. The base composition of DNA is extremely variable, ranging from about 22 percent GC to about 78 percent GC. However, *organisms known to be related by other criteria have DNA base compositions which are similar or identical.* Thus if the base ratios of two organisms are widely dissimilar, that is, a difference of more than 10 percent GC, they cannot be closely related. However, it is important to realize that a similarity of base composition does not necessarily imply a relationship. This is because many base sequences are possible even though the proportion of bases is identical. Thus in the example given above one should not conclude that human beings are closely related to *B. subtilis* (both have 40 percent GC DNA), whereas *E. coli* is unrelated to either one. Obviously, the correspondence between the two 40 percent GC DNAs is adventitious and provides no real clue for classification.

　　The application of these principles to bacterial taxonomy is illustrated by a summary of base composition measurements on many bacteria (Fig. 11-2a). Although the range of base composition is highly variable, the range of composition for a known group of related bacteria is much more limited. Thus all

Pseudomonad group
Page 250

the strains of the pseudomonad group studied had base compositions between 60 and 70 percent GC (Fig. 11-2b). In fact, since base compositions can be measured to within 1 percent GC, very careful determinations allow the taxonomist to divide the group into several subgroups. When this was done, the division correlates well with other taxonomic criteria such as metabolic capabilities.

　　The study of the composition of DNA has been especially valuable for various groups of bacteria since it has revealed that certain strains classified in a group are in fact completely misplaced. For example, several types of bacilli which were formerly classified in the genus *Bacillus* have been removed from the genus and reclassified. Unfortunately, the method is limited in its resolution and is less useful for more highly evolved organisms. This is primarily a result of the fact that the variation is more limited for eucaryotic microorganisms such as fungi or algae, and is practically nonexistent among higher plant or vertebrate DNAs, all of which have virtually identical percentages of each purine and pyrimidine base. For this reason the use of DNA measurements has been extended by the development of methods sensitive to DNA base sequence as well as base composition.

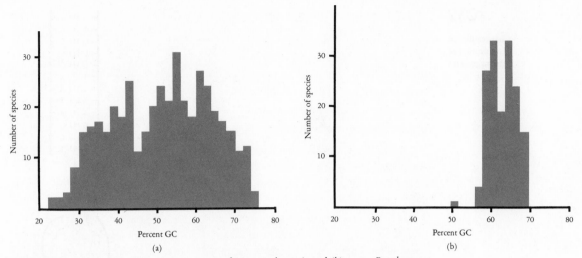

FIGURE 11-2 Base composition measurements of (a) many bacteria and (b) genus *Pseudomonas*. The range of GC content ranges from 22 to 75 percent when a large number of different genera are studied (a). When only members of the genus *Pseudomonas* are considered, the range narrows considerably (b).

Hybridization of DNA

When double-stranded DNA is heated, the two strands of complementary sequence dissociate or denature. If this mixture of single strands is incubated for extended periods at the appropriate temperature, they can reassociate to reform a double-stranded molecule whose physical properties are very similar to those of the original native double-stranded DNA (Fig. 11-3). The efficiency with which this reassociation or renaturation takes place depends on many factors, such as the temperature and the ionic strength of the solution. When such conditions are optimal, the reassociation can be complete. Since the reassociation reaction depends intimately upon complementarity of base sequence, the same procedures can be employed to assess the extent of similarity of base sequence between the DNAs of two related organisms. In this case a double strand is formed from two single strands, each originating from a different organism. Alternatively one of the two strands may be RNA rather than DNA since the base sequence of an RNA molecule is complementary to that of one of the two strands and identical to that of the other. This approach, which is referred to as *nucleic acid hybridization,* is a direct and powerful method for assessing genetic relatedness between organisms.

MEASUREMENT OF SIMILARITY IN BASE SEQUENCE OF DNA

Nucleic acid hybridization reactions are usually studied by using radioactive DNA from one organism and nonradioactive DNA from other organisms. For example, a culture of one bacterial strain may be grown in a medium containing a radioactive nucleic acid precursor such as C^{14} adenine. Then the radioactive DNA is prepared from this culture and unlabeled DNA from other strains of interest. Often the unlabeled DNA is imbedded in a semisolid matrix after it has been denatured. This prevents the single strands from reassociating with each

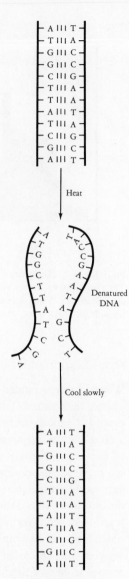

FIGURE 11-3
Dissociation and reassociation of double stranded DNA.

other. The radioactive DNA is denatured, sheared into small fragments and then added to the unlabeled DNA. In general, some of the labeled DNA fragments form double strands, while others fail to combine (Fig. 11-4). This distinction reflects different extents of genetic relatedness at various genetic loci. If two organisms are closely related, the base sequences in their DNA will be similar and the single strands will associate to yield a double-stranded molecule. The relative extent of reaction can easily be computed and used as a measure of relatedness (Table 11-1). For example, in a test of the cross reaction of DNA from several related species, *Salmonella typhimurium*, and *Shigella dysenteriae* appeared much more closely related to *Escherichia coli* than did *Proteus vulgaris* or *Serratia marcescens*. DNA from a bacterium of the genus *Pseudomonas* which is not in the same group showed little or no relatedness by DNA hybridization.

Recent developments of this methodology permit a quantitative evaluation of the number of base changes which have accumulated by mutation in the evolution of two organisms since their descent from a common ancestor. Such data may be obtained from studies of the physical properties of the double strands formed by two different DNAs. Minor differences in the two DNAs attributable

Organisms
Pages 252–253

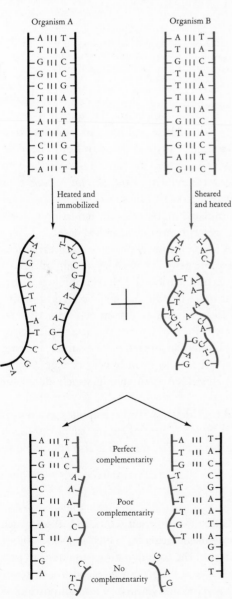

FIGURE 11-4
Measuring relatedness between organisms by the similarity in their DNA base sequences. Whether the DNA of organism A will hybridize with organism B depends on the DNA base sequences being complementary to one another. The single stranded can be readily separated from double stranded DNA.

TABLE 11-1

DNA Hybridization between *E. coli* DNA and DNA of Other
Related Bacteria.

Source of Unlabeled DNA	Relative Binding of DNA (%)
Escherichia coli	100
Salmonella typhimurium	71
Shigella dysenteriae	71
Exterobacter aerogenes	51
Proteus vulgaris	14
Serratia marcescens	7
Pseudomonas aeruginosa (unrelated to the genera above)	1

The reference DNA labeled with C^{14} was prepared from *E. coli*. The amount of
binding with homologous DNA was arbitrarily assigned the value of 100 percent
and binding with other DNA expressed relative to this. Data from B. J. McCarthy
and E. T. Bolton, *Proc. Natl. Acad. Sci. U. S.* **30**, 156 (1963).

to accumulated mutations will place two bases other than the normal A-T and
G-C pairs opposite one another (Fig. 11-4). These will not pair, and failure to
pair deforms the helix and reduces its resistance to dissociation by heat. The more
perfectly the complementary strands fit together, the more resistant the two
strands are to dissociation by heating. Measurement of the *thermal stability* thus
provides a means of measuring the extent to which these mismatched bases are
present, and thus the number of accumulated mutations. The reduction in thermal stability is a direct measure of the percentage of mismatched bases.

The several variants of the DNA hybridization approach show great potential. The techniques can be used to explore genetic homology among groups of
organisms within which no genetic recombination and thus no biological hybridization is possible. The method is quite flexible and may be used to distinguish closely related organisms as well as to explore distant evolutionary relationships. This flexibility derives partly from the fact that the base sequences of
different genes diverge in evolution at highly variable rates. For example, the base

Ribososomal RNA
Page 91

sequence of ribosomal RNA is very resistant to evolutionary change so that
similarities of base sequence exist in this RNA even among widely divergent
groups of organisms.

At the present time nucleic acid hybridization has been applied as a means
of classifying many groups of bacteria. Some groups of eucaryotic protists such
as fungi have also been examined in this way. The method is also applicable to
higher plants and animals.

Proteins as Tools for Molecular Taxonomy

The structure of a gene can be described in two different terms, its base sequence
or the amino acid sequence of the protein molecule for which it codes. Thus a
comparison of the amino acid sequences of the corresponding protein from two
organisms elucidates the relationship between the genes which code for this protein type. Although this approach is precise, it is limited to a small portion of

the genetic information of the cell in contrast to the overall estimation of DNA homology made by DNA hybridization. Unfortunately, however, although the complete amino acid sequences have been determined for many proteins, few of these are of microbial origin. The comparison of amino acid sequences of proteins such as hemoglobin has been used with some success as a molecular parameter for the taxonomy of mammals.

However, even when the complete amino acid sequence of a protein is not available, the proteins can be compared by indirect means. For example, if the proteins have enzyme activity, various parameters of catalytic function, such as pH optimum, heat stability, and sensitivity to inhibitors, may be used. When two organisms are closely related, a given enzyme is usually similar in structure and function. Alternatively the physical properties of two enzymes may be compared in several ways. Molecular weights may be measured by sedimentation in an ultracentrifuge or by other methods and charge properties determined by the relative migration of the molecules through an electric field (electrophoresis). Unrelated organisms may produce enzymes which perform the same catalytic function but differ in the length of the polypeptide chain and the mixture of amino acids which comprise it. This will be revealed as molecular weight or charge differences. On the other hand, closely related organisms will produce similar proteins.

GENETIC APPROACHES TO TAXONOMY

In certain groups of bacteria where genetic exchange occurs, relationships may be explored by what amounts to biological hybridization. This is the approach which involves the formation of hybrids between two species, used in the taxonomy of higher plants. In various groups of bacteria, transduction, transformation, or conjugation, whichever is appropriate, are used as taxonomic indices. In principle each of these types of genetic exchange, although mediated by different processes, depends on the integration of a piece of DNA originating from a donor into the chromosome of the recipient. Since this process depends on the DNA of the donor and recipient being complementary to each other, the *efficiency* of the process is a measure of genetic relatedness.

Extensive studies of genetic relatedness as indicated by transformation have been made in several groups of bacteria. For example, DNA is obtained from various strains of *Bacillus* and used in attempts to transform a particular reference strain of *B. subtilis*. Several different genetic markers may be scored so that genetic homology is measured in various parts of the genetic map. Thus the relative efficiency with which streptomycin resistance is transmitted to a recipient measures the genetic homology in the region of the genetic map surrounding that particular gene. Such experiments demonstrate that particular areas of the genetic map, such as those around the gene that determines the resistance or sensitivity of the strain to streptomycin, are very resistant to evolutionary change compared to other genes.

Enzymes
Pages 142–146

Transformation
Pages 194–197

236

Chapter Eleven
Classification of
Microorganisms

Organisms
Pages 252–253

Genetic transfer among different strains by conjugation has been studied between members of the same group. Hybrids may be obtained within the closely related organisms *Escherichia, Shigella,* and *Salmonella.* These three organisms also appear closely related by DNA hybridization. Indeed, it is fair to question whether these three genera which are so closely related by DNA hybridization techniques should be consolidated into one single genus rather than three. However, no biological hybridization is obtainable between *Escherichia* or *Salmonella* and *Enterobacter, Serratia* or *Proteus* even though these are detectably related by DNA hybridization. This example points out the fact that *considerably higher degrees of base sequence homology are necessary for recombination to form biological hybrids than are required for DNA hybridization in the test tube.*

SUMMARY

One major difficulty in microbial taxonomy is the problem of defining a microbial species, the basic unit of all taxonomic schemes. For convenience, a microbial species is usually defined as a group of strains, clones, or cultures which have many properties in common and yet are easily distinguished from other species. The traditional schemes of bacterial classification are convenient. Organisms which share common properties are grouped together. However, since the members of each group share properties with members of other groups, the decision as to which properties take precedence in classification is quite arbitrary. Thus traditional classification schemes are useful for providing a description of each species; such taxonomic schemes make no claim to represent a scheme based on evolutionary relationships.

Probably the most accurate method for assessing relatedness is based on the nature of the DNA in organisms. Closely related organisms have similar base compositions. A more quantitative estimate of their relatedness can be gained by determining the degree of base sequence homology in their DNA or RNA. This can be accomplished *in vitro* by studying the degree of reassociation of single-stranded DNA of the two organisms or by determining the efficiency of the organisms to undergo genetic recombination with one another.

QUESTIONS

1. One spherical organism is capable of undergoing sporulation. What techniques would you use to decide whether this organism is more closely related to cocci or to a misshapen rod? Which results would you tend to believe the most?

2. Do you think that the ability of an organism to grow in air is more important than its ability to grow on lactose as a carbon source? Why?

3. Using nucleic acid hybridization techniques, an investigator found that two organisms appeared to be closely related when the hybridization was done at

a low temperature but distantly related when done at a higher temperature. What is the basis of this observation?

4. Why may episomes be transferred between species even though the chromosomes of these organisms will not undergo genetic recombination?

5. Discuss the reasons why the most widely used book on the taxonomy of microorganisms undergoes continual revisions. What data will be required before these revisions will no longer be necessary?

FURTHER READING

BREED, R. S. and others (ed.), *Bergey's Manual of Determinative Bacteriology,* 7th ed. Baltimore: The Williams and Wilkins Co., 1957.

DAYHOFF, M. O., "Computer Analysis of Protein Evolution," *Scientific American* (July 1969). (Offprint #1148)

DICKERSON, R., "The Structure and History of an Ancient Protein," *Scientific American* (April 1972).

MANDELSTAM, J. and K. McQUILLEN(ed.), *Biochemistry of Bacterial Growth.* New York: John Wiley and Sons, Inc., 1968. (Appendix A)

SOKAL, R., "Numerical Taxonomy," *Scientific American* (December 1966). (Offprint #1059)

SPIEGELMAN, S., "Hybrid Nucleic Acids," *Scientific American* (May 1964). (Offprint #183)

Chapter 12

Bacteria

Procaryotic forms
Pages 53–58
Bergey's manual
Page 227

As indicated in Chapter 3 bacteria represent a very large subdivision of the protists, and one diverse in habitat and biochemical capability. *Bergey's Manual of Determinative Bacteriology,* 7th edition, lists 12 orders, 51 families, and 207 genera of bacteria. It is far beyond the scope of this text to cover even a fraction of these organisms, and there is no reason to burden a student with memorizing the details of a classification scheme which is somewhat arbitrary and subject to change. It does seem important to us, however, to present an overall view of the bacterial members of the protists so that the student may appreciate the tremendous diversity of bacteria. This chapter should also serve as a reference for grouping bacterial genera and species that are considered elsewhere.

The classification scheme presented here is based primarily on nutrition, morphology, staining, and motility. Table 12-1 summarizes the major groups with their salient characteristics. This classification scheme has the merit that it is based on easily discernible characteristics of the organisms and avoids Latinized names, except for genus and species. A large number of organisms are covered elsewhere and in most instances we only make reference to these organisms in this chapter.

THE MAJOR GROUPS OF BACTERIA

Photosynthetic Bacteria

Photosynthesis
Page 157
Fig. 4–24, Page 93

The photosynthetic bacteria gain their energy from solar energy through the process of photosynthesis. The details of bacterial photosynthesis and how it differs from that of plants and algae are given in Chapter 7. Phototrophic bacteria are found in wet soil and aquatic and marine habitats which lack oxygen but in which light penetrates. As indicated elsewhere, purple sulfur bacteria and green sulfur bacteria utilize reduced inorganic sulfur compounds such as hydrogen sulfide for the electrons necessary to reduce carbon dioxide to cell material. The purple nonsulfur bacteria employ organic materials such as alcohols or fatty acids as hydrogen (electron) donors instead of reduced sulfur compounds, and are therefore heterotrophic. Some purple nonsulfur bacteria are able to use aerobic metabolic processes in the dark, and thus are not obligately photosynthetic. Bacterial photosynthesis can occur only under anaerobic conditions. To achieve the location that provides optimum conditions for growth, some photosynthetic bacteria can change their position in a fluid environment by using their polar flagella or by altering the relative number of gas flotation vacuoles in their cytoplasm.

Like blue-greens some purple sulfur and green sulfur bacteria can convert nitrogen gas into organic compounds (nitrogen fixation). In addition to an appropriate hydrogen donor, the requirements for growth of these autotrophs are simply molecular nitrogen, carbon dioxide, light, water, and some minerals. It seems probable that some types of photosynthetic bacteria evolved at a time before the earth had an oxygen-containing atmosphere. By converting solar energy to chemical energy for the production of organic compounds from nitrogen and carbon dioxide, they may have helped pave the way for emergence of heterotrophic forms. At the present time their ecological role as the primary producers of organic from inorganic material is probably insignificant in comparison with algae.

Algae
Pages 59, 266

All the major morphological types are represented among the bacterial photosynthesizers: rods, cocci, and helical forms. Thus it seems likely that the phototrophic bacteria are of diverse origin.

Gliding Bacteria

The "gliding bacteria" represent a group of small rod-shaped organisms that have a thin flexible cell wall that allows the cells to slide along solid surfaces or show a flexing, twisting motion when in suspensions. They lack flagella, and the mechanism by which they move is unclear.

One of the best known genera of this group is *Beggiatoa* (Fig. 12-1a), found in sulfur springs, black mud, and polluted water. The morphological similarity to certain blue-greens is striking even though their metabolic apparatus differs sharply in that the blue-greens are photosynthetic. The evolutionary relationship of these algal and bacterial groups is obscure.

Non-photosynthetic sulfur
bacteria Table 7-2;
Page 156

Members of other genera (such as *Thiothrix*) are quite similar to those of *Beggiatoa,* except that they have long filaments composed of chains of cells enclosed in a common wall. The filaments attach to solid surfaces, often in clusters (Fig. 12-1b). The filaments themselves are nonmotile, but short segments possessing gliding motility separate from the tips and disperse. Sometimes several of the motile segments will congregate in a rosette on a solid surface, secrete a material

TABLE 12-1
The Major Groups of Bacteria

Bacterial Group	Mode of Motility	Morphology	Nutrition	Staining	Other Distinctive Features
Photosynthetic					
Purple sulfur	Flagella, if motile	Rods, cocci, and helices	Photosynthetic; autotrophic	Gram-negative	Sulfur granules deposited intra-cellularly[a]
Green sulfur	Flagella, if motile	Rods[a]	Photosynthetic; autotrophic	Gram-negative	Sulfur granules deposited extra-cellularly
Purple non-sulfur	Flagella, if motile	Rods, helices[a]	Photosynthetic; heterotrophic	Gram-negative	No sulfur granules
Gliders					
Filamentous sulfur	Glide on solid substratum	Multicelled filaments	Oxidize reduced sulfur compounds	Gram-negative	Sulfur granules deposited intra-cellularly
Nonfruiting myxobacteria	Glide on solid substratum	Long rods	Heterotrophic	Gram-negative	
Fruiting myxo-bacteria	Glide on solid substratum	Short rods that form microcysts	Heterotrophic	Gram-negative	Can form elaborate fruiting structures
Sheathed	Flagella	Multicelled filament enclosed in sheath	Heterotrophic	Gram-negative	Sheath may become encrusted with iron
Appendaged or Budding	Flagella, if motile	Unicellular rods, vibrios, cocci; some have appendages and some divide by budding	Heterotrophic[a]	Gram-negative	
Spirochetes	Axial filaments	Helical; flexible wall	Heterotrophic	Most too thin to stain well	Cell is quite flexible and bends easily
Spiral or Curved	Flagella, if motile	Bent rods or helical; rigid wall	Heterotrophic	Gram-negative	Cell is rigid and does not bend
Strictly aerobic gram-negative rods	Polar flagella if motile[a]	Rods	Heterotrophic, nonfermentative	Gram-negative	One large group are called "pseudomonads"
Facultatively anaerobic gram-negative rods	Polar or peri-trichous flagella if motile	Rods, some very short	Heterotrophic	Gram-negative	Many pathogens; peri-trichously flagellated called "enterobacteria"
Strictly anaerobic gram-negative rods	Polar flagella if motile	Curved and straight rods, some spindle shaped	Heterotrophic	Gram-negative	
Nonphotosynthetic autotrophs	Flagella, if motile	Rods, spheres, or helices	Use CO_2 as major carbon source	Gram-negative	Reduced inorganic compounds supply energy
Gram-negative cocci	Nonmotile	Spherical, fre-quently in pairs	Heterotrophs; many parasites and pathogens	Gram-negative	Includes causative agents of gonorrhea and bacterial menin-gitis
Gram-positive cocci	Nonmotile[a]	Spherical, some-times in chains, packets, or clusters	Heterotrophic	Gram-positive	Includes important pathogens
Endospore formers	Peritrichous flagella if motile	Rods[a]	Aerobic, facul-tative or anaer-obic heterotrophs	Gram-positive[a]	Spore is very re-sistant to heat
Nonspore forming gram-positive rods	Peritrichous flagella if motile	Rods	Heterotrophic	Gram-positive	Diverse group

TABLE 12-1 *(continued)*

Bacterial Group	Mode of Motility	Morphology	Nutrition	Staining	Other Distinctive Features
Branching	Flagella if motile	Branching rods and nonseptated filaments; some produce spores on aerial hyphae	Heterotrophic	Some acid fast	Includes important pathogens; often referred to as "actinomycetes," or "mycelial bacteria"
Mycoplasmas	Nonmotile	Irregular shape due to absence of cell wall	Heterotrophic; most require sterols	Gram-negative	Include the only procaryotes with sterol-containing cytoplasmic membranes
Obligate intracellular	Nonmotile	Small, short rods	Heterotrophic; must grow intracellularly[a]	Gram-negative	Some have a leaky cytoplasmic membrane

[a] There are minor exceptions.

that attaches them to the surface, and then to develop into long filaments. Some of these organisms are also important in the sulfur cycle.

Multicellular gliding bacteria are also found in the mouths of human beings and other animals. Members of this group are represented in the genus *Simonsiella*, which occur as short chains of flattened cells in a ribbonlike arrangement (Fig. 12-1c).

Myxobacteria are also gliding bacteria, but in contrast to those discussed previously they occur primarily as single cells. Most are found in soil or decomposing organic matter, although marine forms are also known. Myxobacteria are subdivided into two groups depending on whether or not they produce *fruiting bodies,* which are globular or complex structures consisting of masses of cells supported above the solid substrate on stalks (Fig. 12-2). Fruiting myxobacteria are most commonly found on bark and decaying wood and in soil or manure. Some grow at the expense of other species of bacteria which are lysed and digested, whereas others utilize cellulose as a nutrient. They have one of the most complex

MYXOBACTERIA

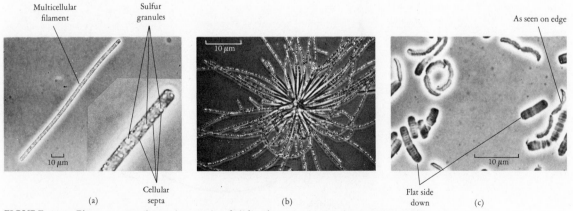

FIGURE 12-1 Phase contrast photomicrographs of gliding bacteria. (a) *Beggiatoa* species (courtesy of J. T. Staley and J. P. Dalmasso). (b) *Thiothrix* species. Note rosette arrangement. (Courtesy of F. Palmer and E. Ordal) (c) *Simonsiella* species (courtesy of J. T. Staley and J. P. Dalmasso).

241

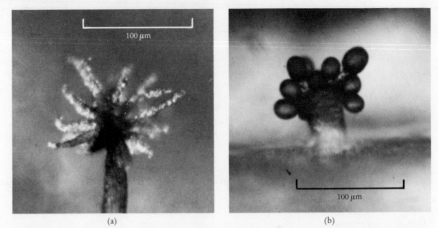

(a)   (b)

FIGURE 12-2 Two species of fruiting myxobacteria. (a) *Chondromyces apiculatus*. (b) *Stigmatella aurantiaca*. (Courtesy of M. Dworkin and H. Reichenbach)

developmental patterns of any procaryotic cells (Fig. 12-3). On solid media flat colonies spread over the surface as a result of gliding motion of the cells. Formation of fruiting bodies is thought to occur when some nutrient has been exhausted. Cells then aggregate at different points and pile up. A stalk is produced, surmounted by the fruiting body containing most of the cells. Cells within the fruiting body differentiate into resting forms called *microcysts*. Microcysts are re-

MICROCYSTS

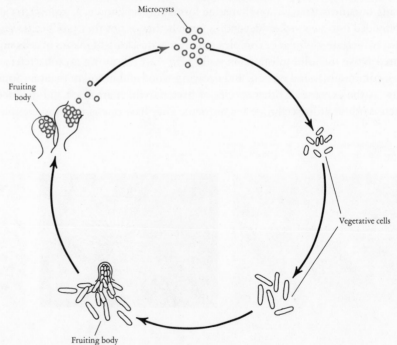

FIGURE 12-3 Life cycle of a fruiting myxobacterium.

sistant to heat, drying, and radiation, relative to the vegetative cells, but are not as resistant as bacterial endospores. The microcysts germinate when conditions become favorable for growth.

Nonfruiting myxobacteria are represented by the genus *Cytophaga*, which consists largely of aerobic cellulose decomposers. The common soil cytophagae employ cellulose as their sole source of carbon and energy. Some myxobacterial species attack other polysaccharides such as chitin and agar. Some nonfruiting myxobacteria form microcysts. Another important nonfruiting myxobacterium is the aquatic organism commonly known as *Chondrococcus columnaris,* an important fish pathogen which causes extensive loss of salmon and related species.

Although the gliding bacteria are grouped together on the basis of their unique type of motility, they are a diverse group in terms of evolutionary relationships. For example, organisms of the genus *Cytophaga* contain DNA that is 30 to 55 percent guanine + cytosine, while in the various fruiting myxobacteria the guanine + cytosine ratio is between 67 and 70 percent.

Percent GC
Page 230

Sheathed Bacteria

Sheathed bacteria represent another group of multicelled filamentous organisms. They differ from similar organisms discussed in the section above in that the chain of bacterial cells is enclosed in a sheath composed of a lipoprotein-polysaccharide complex. The sheath is chemically distinct from the bacterial cell walls. In fresh-water habitats containing iron or manganese, deposition of ferric hydroxide or manganese oxide may occur on the sheaths. *Sphaerotilus* (Fig. 12-4) is a very widespread genus in this group. The filaments of members of this genus often produce masses of brownish scum that may be seen beneath the surface of polluted streams; they interfere with sewage treatment processes by plugging pipes. Reproduction occurs by binary fission, polarly flagellated bacilli being released from the ends of the sheath. These bacilli then reproduce, secreting a new sheath.

Compare *Thiothrix* and *Beggiatoa,* Page 239 and Fig. 12-1; Page 241

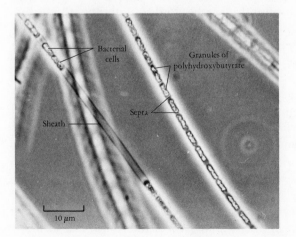

FIGURE 12-4
Sphaerotilus species, a sheathed bacterium. (Phase contrast photomicrograph courtesy of J. T. Staley and J. P. Dalmasso.)

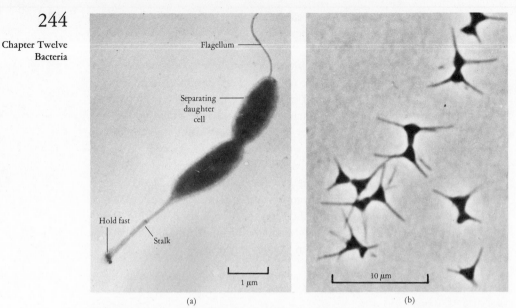

FIGURE 12-5   Prosthecate bacteria. (a) *Caulobacter* species and (b) *Ancalomicrobium* species (Courtesy of J. T. Staley and J. P. Dalmasso)

Prosthecate and Budding Bacteria

Prosthecate and budding bacteria are characterized by appendages which project from the bacterial cells (Fig. 12-5). For example, members of the genus *Caulobacter* and related genera form stalks (prosthecae) which attach to solid material by an adhesive area (a holdfast) at the tip of the appendage. These bacteria reproduce by elongation and division; the daughter cell lacks a prostheca and swims away by means of a flagellum. As the daughter bacterium grows it develops a prostheca at the site of flagellation, and then attaches by the holdfast, completing the growth cycle (Fig. 12-6). In some organisms of this group the holdfast is on the main portion of the cell rather than on the prostheca. Other prosthecate bacteria lack a holdfast. In prosthecate bacteria the cytoplasm actually extends into the prosthecae, thus differentiating them from many other bacteria which have appendages that do not contain cytoplasm.

Budding bacteria may or may not be able to attach to surfaces. Their distinguishing feature is *budding,* an unusual mode of reproduction among bacterial species (Fig. 12-7). A photosynthetic budding species, *Rhodomicrobium vannielii* (Fig. 12-8), is classified with the photosynthetic group. Budding bacteria are widely distributed in soil and water.

Spirochetes

The spirochetes compose a group of helical and wave shaped bacteria which have
AXIAL FILAMENTS   flexible cell walls and move by means of unique structures called axial filaments

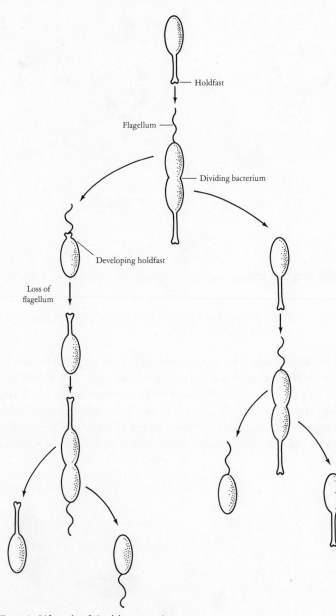

FIGURE 12-6 Life cycle of *Caulobacter* species.

(Fig. 12-9). Electron microscopic study of some of these organisms has revealed that each axial filament is actually composed of two fibrils identical in structure to flagella, one originating from each pole of the organism. The fibrils extend toward each other between two layers composing the cell wall and apparently overlap in the midregion of the cell. The layer of cell wall material covering the axial filament tends to separate easily in laboratory studies, and is often referred to as a sheath. As with flagella the molecular basis for movement due to axial

Flagella
Page 95

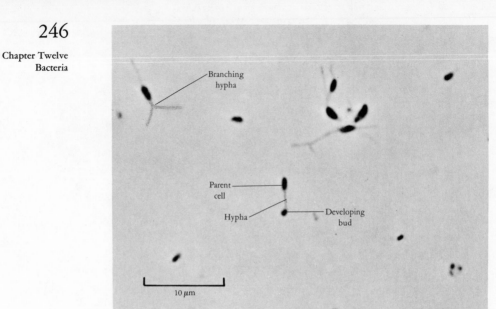

FIGURE 12-7 *Hyphomicrobium* species, a budding bacterium (courtesy of J. T. Staley and J. P. Dalmasso).

filaments is not yet clear. The smaller spirochetes are very slender and therefore difficult to see by the usual microscopic methods (Fig. 12-10). Many are also difficult or impossible to cultivate and their classification is based largely on morphology and ability to cause disease. Difficulty in cultivation probably relates to the stringent anaerobic growth requirements of many spirochetes. Some species are widespread in aquatic habitats and others parasitize warm-blooded animals.

Anaerobic culture methods
Page 114

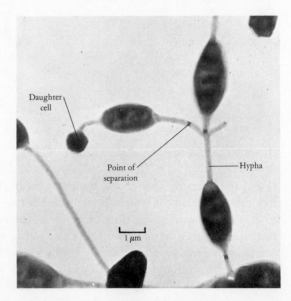

FIGURE 12-8
Rhodomicrobium vannielii. Since this budding bacterium is photosynthetic it is classified with the photosynthetic rather than the budding bacteria. (Courtesy of H. Douglas; Duchow and Douglas, *J. Bact.,* **58**:409, 1949)

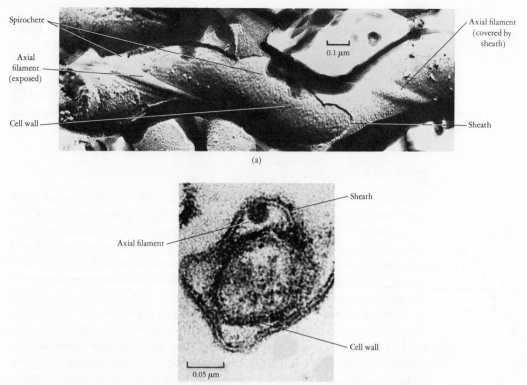

(a)

(b)

FIGURE 12-9 Electron micrographs showing the axial filament of a spirochete of the genus *Leptospira;* (a) longitudinal view, (b) cross section. (R. K. Nauman, S. C. Holt and C. D. Cox, *J. Bact.,* **98**:264, 1969).

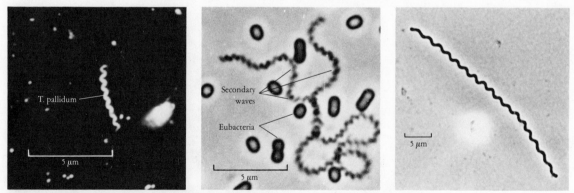

FIGURE 12-10 Spirochetes. (a) Darkfield photomicrograph of *Treponema pallidum,* the cause of syphilis (courtesy of F. Schoenknecht and P. Perine) (b) Phase contrast photomicrograph of a very long free-living spirochete showing secondary waves. Note the size relative to the eubacteria also present in the photograph. (Courtesy of J. T. Staley and J. P. Dalmasso). (c) Although helical, this organism is not a spirochete; it is a member of a genus of multicellular gliding bacteria and lacks an axial filament. (Phase contrast photomicrograph courtesy of J. T. Staley and J. P. Dalmasso).

247

Syphilis
Page 495

One group of spirochetes includes only nonpathogenic species. Organisms in this family are often extraordinarily long, up to 500 μm in length. Examples are members of the genus *Spirochaeta* which were probably observed by van Leeuwenhoek in 1683 in spittle from his mouth.

Another group of spirochetes comprises both pathogenic and nonpathogenic species. Included are the medically important genera *Treponema, Borrelia,* and *Leptospira,* which are the causes of syphilis, relapsing fever, and leptospirosis, respectively.

Spiral and Curved Bacteria

Gram stain
Page 72

Cholera
Page 479

Spiral and curved bacteria are quite different from spirochetes, although some confusion might arise because of their name. For the most part members of this group are larger in diameter than spirochetes, easily stainable (gram-negative), and motile by means of polar flagella. They have the thicker, more rigid cell wall characteristic of eubacteria. Most species are nonpathogenic and are common saprophytic inhabitants of fresh- or saltwater, living on decaying organic matter. One important pathogenic member, *Vibrio cholerae,* causes cholera (Fig. 12-11).

One of the most interesting genera in this group is *Bdellovibrio.* The bdellovibrios are obligate parasites of other bacteria. The generic name *Bdellovibrio* describes the organism's behavior and shape. They attach to other bacterial cells (bdello, from the Greek word for "leech") and have a comma shape like bacteria of the genus *Vibrio. Bdellovibrio bacteriovorus,* the most intensively studied species of this group, is approximately 0.25 μm wide and 1 μm long. It has a single thick flagellum at one end of the cell (Fig. 12-12). *Bdellovibrio* attacks the host bacteria by striking them at high velocity so that their nonflagellated end attaches to the host cell. A hole is then made in the cell wall through which *Bdellovibrio* pene-

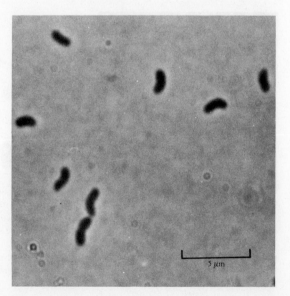

FIGURE 12-11
Vibrio cholerae, a curved bacterium, which causes cholera. (Phase contrast photomicrograph courtesy of J. T. Staley and J. P. Dalmasso).

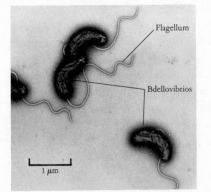

Flagellum

Bdellovibrios

1 μm

FIGURE 12-12
Bdellovibrio bacteriovorus. Note the small size and single thick flagellum. (S. C. Rittenberg, *J. Bact.,* **109**:432, 1972)

trates. The organism multiplies between the cell wall and the cytoplasmic membrane, with digestion of the contents of the host cell occurring as the bdellovibrio multiplies. The progeny then lyse the cell wall and leave the empty host
cell to attach to other bacteria and reinitiate the cycle (Fig. 12-13). Rare variants
of *B. bacteriovorus* can be isolated which can grow on a nutrient medium in the
absence of bacteria.

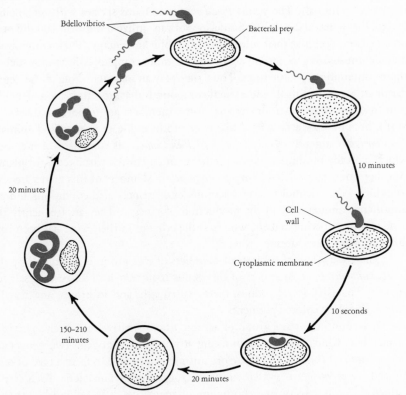

Bdellovibrios

Bacterial prey

10 minutes

20 minutes

Cell
wall

Cytoplasmic membrane

10 seconds

150–210
minutes

20 minutes

FIGURE 12-13 Life cycle of *Bdellovibrio bacteriovorus.* Note that the bdellovibrio multiplies between the cell wall and cytoplasmic membrane of the bacterial cell upon which it preys. (Adapted
from an illustration courtesy of S. C. Rittenberg)

One surprising fact concerning *Bdellovibrio* is that it was discovered as recently as 1962, even though this type of organism can now be easily isolated from soil or sewage. Viruses, which are not even visible with the light microscope, were discovered in 1892 and rickettsiae in 1909. How did *Bdellovibrio* escape the detection of so many scientists using sophisticated microscopic techniques for all those years? The major reason for this oversight is probably that there was not an *a priori* reason to believe that such a creature existed. Although many people had undoubtedly observed this organism while examining the microbial flora in soil, there was no inkling that it was anything more than a typical, albeit small, bacterial cell.

Strictly Aerobic Gram-Negative Rods

The strictly aerobic gram-negative rods include a number of important groups. One of the best known, the pseudomonads, is composed of members which are motile by polar flagella and often producing nonphotosynthetic pigments. Although classified as strict aerobes, a few can grow anaerobically in the presence of nitrate that acts as a final electron acceptor (anaerobic respiration). Many species of pseudomonads are free-living and harmless, and others can produce disease in plants or animals. The genus *Pseudomonas* contains species with remarkably diverse biochemical capability. Some species can utilize more than 100 different organic compounds as their sole source of carbon and energy. These compounds include unusual sugars, amino acids, and far more complex compounds such as those containing aromatic rings. Thus, they play an important role in the degradation of many man-made and natural compounds that are refractory to degradation by most other microorganisms. Some members are important threats to health because they attack food plants or produce disease in man and animals. *Xanthomonas,* a genus closely related to *Pseudomonas,* is composed of members which typically produce a yellow pigment, and are often pathogenic for plants. Another important related genus is *Acetobacter.* Members of this genus are able to oxidize ethyl alcohol to acetic acid under extremely acid conditions and are important commercially in the production of vinegar. They are frequently the bane of amateur wine-makers, who find the ethanol in their wine has been oxidized to acetic acid during storage.

Anaerobic respiration Page 155;
Pseudomonas Page 134

The wide range of habitats of members of this group is illustrated by the genus *Halobacterium.* Members of this genus require at least 12 percent salt for growth. They live in salt ponds, or on salted fish, and in brines, and usually produce brightly colored pigments.

A second important group of aerobic gram-negative rods is characterized by members which are capable of fixing atmospheric nitrogen. Two genera are *Azotobacter* and *Beijerinckia.* These organisms are also able to form a type of resting cell, a *cyst,* which is similar to the microcysts of myxobacteria. Each cyst is formed through division and shortening of vegetative cells, followed by the secretion of a thick protective wall which surrounds the cell (Fig. 12-14). Thus these bacteria represent one of the few groups of unicellular bacteria able to produce a resting cell. Cysts are resistant to drying and ultraviolet irradiation, but are not highly resistant to heating. These cysts differ from another form of resting cell, the endospore, in both their formation and their degree of resistance to deleterious agents. Whereas the formation of endospores involves a sequential series of steps involving the regulation of the expression of a large number of genes, the formation of microcysts involves primarily changes in the cell wall. The enzyme content of the cyst is very similar to the cell from which it arose. Members of the two genera, *Azotobacter* and *Beijerinckia,* are important in supplying fixed nitrogen for plant growth in certain situations.

Another important group among the aerobic gram-negative rods includes the genera *Brucella, Bordetella,* and *Francisella,* members of which are parasites of animals and produce serious diseases in human beings and animals. These organisms are characterized by their minute size, typically less than 1 μm in length, and all except one species are nonmotile.

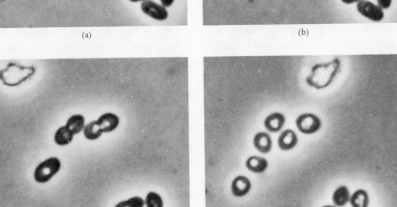

(a)   (b)

(c)   (d)

5 μm

FIGURE 12-14 Development of cysts of *Azotobacter* (a) thirty minutes after inoculation; (b) two and one half hours after inoculation; (c) three and one half hours after inoculation; (d) eight hours after inoculation. (H. L. Sadoff, E. Berke and B. Loperfido, *J. Bact.,* **105**:184, 1971).

The group of facultatively anaerobic gram-negative rods is composed largely of bacteria which are able to ferment carbohydrates. The most important group, commonly termed the enterobacteria, includes a number of medically important species as well as plant pathogens and saprophytes. These bacteria are flagellated, or nonmotile fermentative microbes which are oxidase negative, meaning that they lack the respiratory enzyme, cytochrome C oxidase. The oxidase test is easily performed in the laboratory and is an important test in distinguishing among groups of organisms that grow in air. *Escherichia coli* is probably the most extensively studied member of this group. Most strains of *E. coli* ferment the disaccharide lactose, and ability to ferment lactose or lack of it is an important feature in identifying *E. coli* and other enterobacteria. Two closely related genera, *Salmonella* (which includes causes of typhoid fever and intestinal infections) and *Shigella* (causes of dysentery) generally do not ferment lactose. Other enterobacteria are exemplified by the closely related genera *Enterobacter* (formerly called *Aerobacter*) and *Klebsiella*. Species of these genera usually ferment lactose promptly, and together with lactose-fermenting escherichiae are often referred to as *coliforms*. *Serratia,* a related genus but does not ferment lactose, is unique in that many of its members produce a red pigment, prodigiosin. Members of *Serratia* grow readily on bread and it is very likely that they were responsible for "miracles" in which bread was purported to become bloody. One such episode occurred in 332 B.C. among the armies of Alexander the Great laying seige to the city of Tyre. The red spots inside the bread were interpreted to mean death of the inhabitants of the city. Alexander's forces were thus encouraged to complete the conquest.

Proteus, another genus of the enterobacteria, contains two species which characteristically swarm over the surface of solid media. *Proteus* species, like *Escherichia,* are principally intestinal organisms of man and other animals. Another group of enterobacteria is typified by the genus *Yersinia* which contains the cause of plague (the "black death"). A final group of importance is exemplified by members that cause plant diseases. This is a very heterogeneous group, but recent studies indicate that many of the species belong in the genus *Enterobacter.*

The different enterobacteria are distinguished partly on the basis of biochemical tests such as the ability to utilize citrate as the sole carbon source, to produce acetoin, to deaminate phenylalanine, or produce hydrogen sulfide (Table 12-2). Since members of all the genera mentioned can cause diseases in man, the precise identification of biochemical characteristics is often of great importance.

Identification of structural components of the cell may also be important in distinguishing between members of the enterobacteria. The general antigenic structure of an enterobacterial cell is shown in stylized form in Figure 12-15. Note that the typical organism has three main components which may be termed antigens since they stimulate production of specific antibodies when the organisms are injected into an animal (Chapter 20). The three types of antigens are chemically distinct. The *O* (cell wall antigen) is a lipoprotein-lipopolysaccharide

TABLE 12-2
Representative Genera of Four Subgroups of Enterobacteria

Genera	Characteristic Biochemical Reactions
Escherichia Shigella	Fail to utilize citrate as sole carbon source; do not produce acetoin during glucose fermentation (negative Voges-Proskauer test).
Enterobacter Klebsiella Serratia	Utilize citrate as sole carbon source; produce acetoin during glucose fermentation (positive Voges Proskauer test).
Proteus	Deaminate the amino acid phenylalanine.
Salmonella	Produce hydrogen sulfide; do not deaminate phenylalanine.

complex, the *H* (flagellar antigen) is only protein; and the *K* (capsular antigen) is polysaccharide. Each of these substances varies in chemical composition from species to species and even within species. Identifying the specific kinds of O, H, and K antigens of an organism can therefore be very helpful in identifying its species or subspecies.

In addition to the enterobacteria, other groups of organisms are also facultatively anaerobic gram-negative rods. The genus *Flavobacterium* represents one such group. Flavobacteria produce yellow colonies, often inhabit water faucets, and are commonly encountered as laboratory contaminants. One species may cause a fatal meningitis (infection of the surface of the brain and spinal cord) in babies. Another group, represented by the genus *Haemophilus,* requires complex growth factors and includes members which can cause respiratory and nervous system diseases. Another genus of importance is *Pasteurella. Pasteurella multocida* causes a plaguelike disease of rabbits and frequently causes infection of human beings following animal bites. Finally, the genus *Aeromonas* contains a number of species pathogenic for aquatic animals and occasionally man, and members of the genus *Photobacterium,* found on decaying marine fish, are luminescent (Fig. 12-16).

Cyclic AMP and
bioluminescence
Page 220

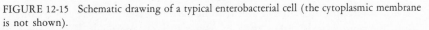

FIGURE 12-15 Schematic drawing of a typical enterobacterial cell (the cytoplasmic membrane is not shown).

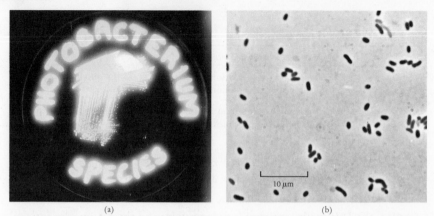

(a)   (b)

FIGURE 12-16 *Photobacterium* species. (a) This photograph was taken using light produced by the bacteria themselves. (b) Phase photomicrograph of cells. (Courtesy of J. T. Staley and J. P. Dalmasso)

Strictly Anaerobic Gram-Negative Rods

The upper respiratory tract, gastrointestinal tract, and lower urogenital system of man and other animals normally harbor tremendous numbers of strictly anaerobic gram-negative rods. In the large intestine of man anaerobes may exceed aerobes by more than 10 to 1. Some of these anaerobic species are difficult to cultivate and may be killed by only a few minutes exposure to air. For this reason accurate information on their relative numbers in different locations is scant. In medical bacteriology about 90 percent of the gram-negative anaerobes recovered in cultures are members of the genus *Bacteroides,* but several other anaerobic gram-negative genera are recognized. For example, members of the genus *Fusobacterium,* rods with pointed ends, are present in enormous numbers in trenchmouth (Vincent's angina).

Anaerobic gram-negative rods are important both in the external environment and as parasites of man. The sulfur cycle depends in part on reactions carried out by anaerobic gram-negative rods represented by the genus *Desulfovibrio.* This genus includes a group of bacteria which respire anaerobically utilizing sulfate ion as the final electron acceptor (anaerobic respiration). They thus may be the chief sources of hydrogen sulfide in areas where sources of reduced sulfur are relatively scant. Hydrogen sulfide, it will be recalled, is utilized by some of the photosynthetic bacteria as an electron donor in the fixation of carbon dioxide. Although *Desulfovibrio* is a curved rod, it is grouped here rather than with the spiral and curved bacteria because of its requirement for anaerobic growth conditions. The photosynthetic bacteria, many of which are also obligately anaerobic gram-negative rods, are placed in a separate group, discussed previously.

Sulfur cycle
Page 612

Nonphotosynthetic Autotrophic Bacteria

All known nonphotosynthetic autotrophic bacteria generate energy by oxidizing inorganic compounds rather than by using organic materials. Two genera of nonphotosynthetic autotrophs, *Beggiatoa* and *Thiothrix,* were included with the gliding bacteria because of their peculiar motility. Most other nonphotosynthetic autotrophic bacteria have the unique ability to grow in the dark on a medium containing only inorganic substances; however, some require vitamins or other growth factors. Autotrophic bacteria play essential roles in the cycling of nitrogen and sulfur compounds.

One group, the nitrifying bacteria, includes members of the genus *Nitrosomonas* (small gram-negative rods) which oxidize ammonia to nitrite, and members of the genus *Nitrobacter* (gram-negative budding bacteria) which oxidize nitrite to nitrate.

Nitrogen cycle
Page 602

The second group, which oxidizes sulfur or sulfur compounds as a source of energy, is represented by the genus *Thiobacillus* (Fig. 12-17), members of which play an important ecological role in the sulfur cycle and are found widely distributed in soil and water. Some species oxidize hydrogen sulfide, thereby tending to prevent the buildup of toxic levels of this compound in soils. The sulfuric acid produced when these organisms oxidize reduced sulfur substrates tends to lower the pH and counteract any excessive alkalinity in the natural environments in which they occur.

Acid production by *Thiobacillus*
Page 107

Acidic freshwaters may contain dissolved iron in a reduced state (Fe^{++}). This ferrous ion can be oxidized by some bacteria and the energy gained used for fixing carbon dioxide into cellular components. Some members of the genus *Thiobacillus* can do this as an alternative to utilizing sulfur compounds. Other genera, such as *Siderocapsa* and *Siderococcus,* may also utilize reduced iron. Like the

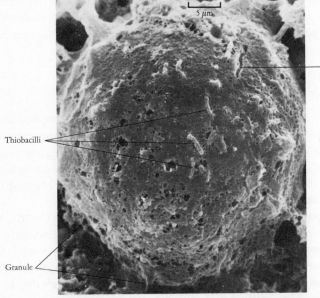

5 μm

Furrow left by thiobacillus

Thiobacilli

Granule

FIGURE 12–17 *Thiobacillus denitrificans* grown on a granule of sulfur. (Scanning electron micrograph courtesy of J. LeGall).

Carbon cycle
Pages 127, 598

heterotrophic organism *Sphaerotilus,* iron salts are deposited about the organisms.

Another group of nonphotosynthetic autotrophs, the anaerobic hydrogen-utilizing bacteria, grow in areas where there is an abundance of decomposing organic material under anaerobic conditions. These organisms are therefore found in the digestive systems of cattle and similar animals as well as in marshes and ponds. Their metabolic processes result in the reduction of carbon dioxide to methane ("marsh gas"), which accounts for the name "methane bacteria" often applied to this group. Both rodlike and coccoid forms occur, and three genera, *Methanobacterium, Methanococcus,* and *Methanosarcina,* are recognized.

Gram-Negative Cocci

Gram-negative cocci are commonly seen among the normal microbial flora of animals and less commonly in the external environment. A few gram-negative cocci have been included among the bacterial groups discussed above because of some other dominating characteristic. Some bacterial species which are rod shaped frequently produce coccoid variants in cultures, but are also excluded from this group in this discussion. The remaining gram-negative cocci include both aerobes and anaerobes. The genus *Neisseria* consists of aerobic gram-negative diplococci (with opposing sides of the paired cells flattened) which give a positive oxidase test. Pathogenic members of the genus are discussed in the sections on gonorrhea and meningitis.

Gonorrhea
Page 492
Meningitis
Page 504

Anaerobic gram-negative cocci are represented by the genus *Veillonella,* which also colonize mucous membranes of human beings. They are often difficult to cultivate from these sources, probably because they are strict anaerobes. They have little virulence and very rarely cause infections. Their importance in the ecosystem represented by the mucous membrane is largely undefined, although some can utilize acid by-products of other bacteria and thus may participate in maintaining a pH near neutrality.

Gram-Positive Cocci

The gram-positive cocci include species of major importance in medical and veterinary microbiology as well as in industrial and food-producing processes. These cocci are subdivided into three groups: one, microorganisms that grow in air and are arranged in clusters or packets; another in which the bacteria are similar but occur in pairs or chains, and a third, composed of the anaerobic counterparts of organisms in the first two groups.

Staphylococcal diseases
Pages 430, 517

Two important genera of aerobic organisms which grow in clusters, *Staphylococcus* and *Micrococcus,* are very commonly represented among normal microbial flora of human beings. Further discussion of members of this group is found in the section on skin diseases. Members of the genus *Streptococcus* grow in chains. They are interesting because many grow readily in air even though they lack the enzyme catalase. They are able to do this because they possess another enzyme, peroxidase, which also degrades toxic hydrogen peroxide. Streptococci lack cytochromes, and they transfer electrons to molecular oxygen through flavoproteins, enzymes of the respiratory chain. Since little or no energy is trapped in this process, streptococci generally do not show an increased growth in air, in contrast to the facultative anaerobes which possess cytochromes. Streptococci convert the sugars they metabolize principally to lactic acid. They are commonly included in a group termed the "lactic acid bacteria." Streptococci may be characterized by the changes produced when their colonies grow on blood agar. Some streptococci, called alpha streptococci, show a green discoloration ("alpha hemolysis") around the colonies. Others, called beta streptococci, cause complete destruction (beta hemolysis) of surrounding red blood cells, while still others, gamma streptococci, show no changes about their colonies. These effects on blood agar medium vary depending on the atmosphere of incubation and the species of animal supplying the blood, but nevertheless the changes are useful in distinguishing certain groups of streptococci. For example, the *viridans streptococci* compose a large group of alpha streptococci normally residing in the mouth and throat, while most of the streptococci causing human infections are beta. Medically important streptococci are discussed in detail elsewhere.

LACTIC ACID BACTERIA

Streptococcal diseases
Pages 432, 448

Endospore Forming Bacteria

Species capable of forming an endospore are gram-positive and with one exception all are rod shaped. The endospore formers can be divided into those that grow in air and those requiring anaerobic conditions.

Endospores
Page 220

Species that grow in air are represented by the genus *Bacillus*. Members of this genus are widespread in soil and are frequent dust-borne contaminants of laboratory media. A few strains produce antibiotics, and one species is pathogenic for man and animals, causing the disease *anthrax*.

For most of the anaerobic spore formers, even low concentrations of oxygen inhibit growth, and with some, faint traces of oxygen are lethal. Since the anaerobic spore formers lack cytochromes, hydrogen peroxide is formed in air as the result of the transfer of electrons to molecular oxygen by flavoprotein enzymes. Since the organisms lack both catalase and peroxidase, no means exist for preventing the buildup of toxic levels of hydrogen peroxide. The spores are not harmed by oxygen. Members of the genus *Clostridium* are well-known anaerobic spore formers, commonly represented among nitrogen fixers of the soil and among bacterial inhabitants of the intestines of man or other animals. Medically important anaerobic spore formers are discussed elsewhere.

Clostridial diseases
Pages 483, 518

Nonspore-forming Gram-Positive Rods

Nonspore-forming gram-positive rod-shaped bacteria are very widespread and include several diverse groups. The first group is represented by the genus *Lactobacillus*. The members of this genus share a number of properties with streptococci, forming chains, producing lactic acid from glucose, not producing catalase, and generally requiring rich media for growth. These properties are characteristic of the lactic acid bacteria of which they are a member. Colonial morphology is often also similar to streptococci. Lactobacilli are almost universally present among the microbial flora of the mucous membranes of human beings, and are especially abundant in the vagina during the childbearing years. They have virtually no disease-producing ability, although they may contribute to dental caries under certain conditions. These bacteria tend to prefer acidic and relatively anaerobic conditions. In addition to their animal habitats they are commonly present in decomposing plant material, and in milk and other dairy products.

Another group of nonspore-forming rods are distinguished by the fact that they divide in such a way that the bacterial cells arrange themselves at angles to one another rather than in chains (Fig. 12-18). The cells of this group, termed the coryneform bacteria, are also usually club shaped or irregular in outline, rather than strictly cylindrical. One genus is *Corynebacterium,* a name derived from *coryne,* a Greek word for "club." Members of this genus are often beaded in appearance when stained due to the presence of intracellular granules of volutin, a phosphate compound. A few are motile and almost all are catalase positive, which distinguishes them in yet another way from the lactobacilli. Some medically important members of this genus are discussed elsewhere. The anaerobic counterpart of members of the coryneform bacteria is the group termed propionic acid bacteria, so-named because of their ability to produce relatively large amounts of propionic acid from glucose. One species, *Propionibacterium shermanii,* first isolated from Swiss cheese by Professor J. Sherman, produces carbon dioxide during fermentation which is responsible for the holes, and propionic acid which contributes to the flavor. Another species, *P. acnes,* is almost universally present as an inhabitant of certain skin glands.

The nonspore-forming gram positive rods of the genus *Arthrobacter* represent an important group of soil bacteria. Strains of some species of *Arthrobacter* are responsible for the degradation of insecticides and herbicides in the soil. Members of this genus may undergo transitions during growth in which the shape of the cells is markedly altered. In the stationary phase the cells are essentially spherical, but when such a culture is diluted into fresh medium, the spherical cells elongate and form the irregular rods characteristic of coryneform bacteria. The rods revert to spherical cells when the culture reaches the stationary phase (Fig. 12-19). The molecular basis for these transitions probably lies with varia-

Diphtheria, a corynebacterial
disease
Page 449

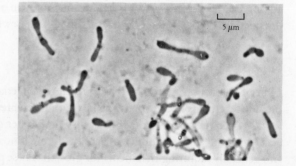

5 μm

FIGURE 12-18

Phase contrast photomicrograph of *Corynebacterium diphtheriae* after 18 hours incubation on Loefflers medium. *C. diphtheriae* causes diphtheria. (Courtesy of J. T. Staley and J. P. Dalmasso).

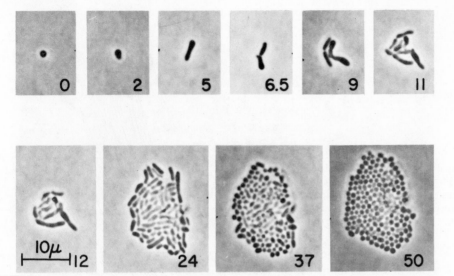

FIGURE 12–19   Phase contrast photomicrographs of *Arthrobacter crystallopoietes* colonial development from a single cell. Note the change in shape from coccus to rod to coccus. The numbers indicate time in hours $10\mu = 10\mu$m. (J. C. Ensign and R. S. Wolfe, *J. Bact.,* **87**:924, 1964).

tions in the structure of the cell wall, perhaps in the degree of cross-linking of mucopeptide layers. Such changes in the cell wall may also account for the unusual property of cells of *Arthrobacter* to stain gram negatively in young cultures, whereas the spherical cells of old cultures are gram-positive. This is the opposite reaction that most cultures undergo on aging.

Mucocomplex
Page 77

Branching Bacteria

Some bacteria may form branching multicellular filaments somewhat reminiscent of molds. Several diverse families are grouped together on this basis. They are often referred to as actinomycetes or mycelial bacteria.

The simplest morphological group with regard to the degree of branching is represented by the genus *Mycobacterium.* The members generally branch very little if at all, are acid fast in their staining reaction, and are aerobic (Fig. 12-20). Although several species of *Mycobacterium* infect man or other animals, the vast majority are saprophytes, able to grow on non-living organic material. A few are difficult or impossible to cultivate *in vitro,* but most of them have simple nutritional requirements and are readily grown in the laboratory. Some pathogenic species have generation times exceeding 12 hours. Their principal ecological role is probably the degradation of lipids and waxes in dead plants and animals. Because of the waxy nature of their cell walls, mycobacteria take up little stain in the gram-staining process; although they usually appear faintly blue, it is not correct to refer to them as gram-positive because of their resistance to staining. Mycobacteria are one of the very few groups which show the acid fast staining property. Among the pathogenic species of mycobacteria are the causative agents of tuberculosis and leprosy.

Tuberculosis
Page 461
Leprosy
Page 507

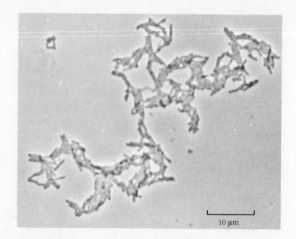

FIGURE 12-20
Mycobacterium smegmatis. This acid fast bacterium is a normal resident of human beings, especially in the genital area. Note the tendency to parallel arrangement and the absence of distinct branching characteristic of most mycobacteria. (Phase contrast photomicrograph courtesy of J. T. Staley and J. P. Dalmasso)

The second major subgroup of the branching bacteria is represented by the genus of soil organisms called *Nocardia* (Fig. 12-21). Nocardiae are also aerobes and some are acid fast. However, in contrast to most mycobacteria, they are definitely gram-positive, and show extensive branching. As a culture ages, these filaments fragment into rod-shaped organisms with appearance similar to the coryneform bacteria. Indeed there is other evidence that nocardiae and corynebacteria are closely related. Several species of nocardiae can infect human beings, usually entering the body by inhalation (*Nocardia asteroides*) or through a break in the skin. In healthy people these infections often resemble tuberculosis in being chronic and gradually advancing. In persons with an underlying disease such as cancer, infection may spread rapidly to the bloodstream and all parts of the body.

A third major subgroup of branching bacteria is represented by members which closely resemble nocardiae except they are generally nonacid fast and anaerobic, although some are able to grow in low concentrations of oxygen. One genus of this group, *Actinomyces,* has as its normal habitat the mucous membranes of the upper respiratory and gastrointestinal tracts of human beings and other

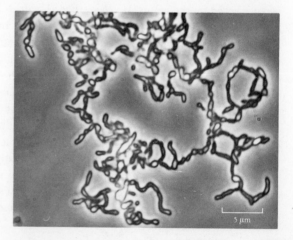

FIGURE 12-21
Phase contrast photomicrograph of *Nocardia asteroides.* Note the presence of branching. (Courtesy of J. T. Staley and J. P. Dalmasso)

animals. A pathogenic species of *Actinomyces* is discussed in Chapter 29. *Bifido-bacterium,* another genus of this group is similar, but in contrast to *Actinomyces,* little branching occurs. It commonly occurs as the principal fecal organism of breast fed babies. It tends to disappear when the infant changes to other diets.

The genus *Streptomyces* characterizes a fourth group of branching bacteria. Like nocardiae these are aerobic soil organisms. They differ from members of the genus *Nocardia* in being nonacid fast and in failing to show fragmentation of mycelium into coryneform rods. Instead the streptomycetes reproduce by formation of specialized oval or spherical cells at the ends of the branches (Fig. 12-22). These reproductive cells are called *conidia* and are readily detached and scattered.

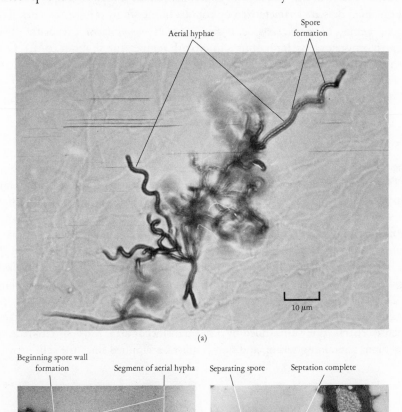

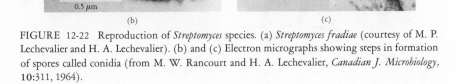

FIGURE 12-22 Reproduction of *Streptomyces* species. (a) *Streptomyces fradiae* (courtesy of M. P. Lechevalier and H. A. Lechevalier). (b) and (c) Electron micrographs showing steps in formation of spores called conidia (from M. W. Rancourt and H. A. Lechevalier, *Canadian J. Microbiology,* **10:**311, 1964).

Antibiotics from actinomycetes
Table 32-2; Page 555

Upon reaching a suitable medium the conidia germinate and produce a new mycelium. Many of the eucaryotic fungi also reproduce by the production of conidia. The importance of some strains as producers of antibiotic medications is documented in Chapter 32. An aquatic family of mycelial bacteria is represented by the genus *Actinoplanes,* similar to *Streptomyces* except that the reproductive cells are enclosed in a saclike structure at the end of a specialized branch. The sac ruptures to release motile coccoid cells which swim away by means of flagella.

Another group of branched bacteria contains two genera of medical importance. The first, *Micromonospora,* appear somewhat similar to *Streptomyces,* except its members are characterized by conidia borne singly on the branches. The second genus, *Thermoactinomyces,* is similar to *Micromonospora,* except the thermoactinomycetes have high optimum growth temperatures (about 65°C). One member of the *Micromonospora* produces the important antibiotic gentamicin, whereas several thermoactinomycetes grow in moldy hay and produce an allergic type of lung disease in persons who inhale the hay dust (Farmers' lung).

Mycoplasmas

The mycoplasmas are sometimes called pleuropneumonia-like organisms or PPLO. Because they *lack cell walls* they are plastic, easily deformed organisms which can pass through filters retaining most other bacteria. Nevertheless different species have characteristic shapes (Fig. 12-23). Most species of mycoplasmas are unique among procaryotic cells in that their cytoplasmic membranes contain sterols, which stabilize the membranes and protect against osmotic lysis. Sterols are normally found in membranes of eucaryotic cells, many of which also lack rigid cell walls. Growth of mycoplasmas on solid media is quite distinctive (Fig. 12-24). The colonies are often small, requiring a lens or dissecting microscope for good visualization. The organisms grow down into the agar, often producing a dense central zone surrounded by a flat translucent area. Most species require a medium containing serum, and sterols must be supplied since the cells usually

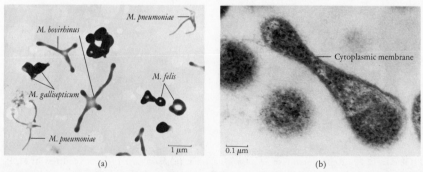

(a)   (b)

FIGURE 12–23 (a) Electron micrograph showing the morphology of different species of mycoplasma and (b) a thin section of *M. bovirhinus* showing absence of a cell wall. (Courtesy of Edwin S. Boatman).

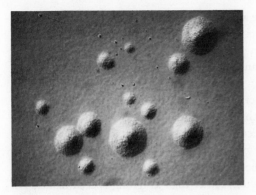

FIGURE 12-24
Colonies of *Mycoplasma pneumoniae*. Note the dense central portion characteristic of most mycoplasma colonies. (Courtesy of G. Kenny and F. Schoenknecht).

cannot synthesize them. Most strains also require vitamins, amino acids, purines, and pyrimidines. Carbohydrates can generally be utilized as sources of energy. Both aerobic and anaerobic species are recognized, and mycoplasmas appear to be a very diverse group in terms of their hereditary relationships. Some scientists feel that they are sufficiently distinctive to represent a group separate from the bacteria. They are responsible for a number of animal and plant diseases, and one species, *Mycoplasma pneumoniae,* can cause pneumonia in man.

Mycoplasmal pneumonia
Page 458

Obligate Intracellular Bacteria

The obligate intracellular parasitic bacteria fall principally into two groups: rickettsiae and chlamydiae.

Rickettsiae are tiny rodlike bacteria averaging about 0.5 × 1.1 μm in size (Fig. 12-25). They are unicellular and reproduce by binary fission. Some rickettsiae have leaky cytoplasmic membranes, which may explain the failure of most of them to multiply on cell-free media, since vital materials would be lost from the bacterium. This leakiness may however also serve to facilitate the transfer of nutrient and other molecules from the cytoplasm of the host cell. Most rickettsiae survive poorly outside the host cell. However some, notably *Coxsiella burnetii,* the cause of Q fever, can remain viable extracellularly for extended periods of time. The cell wall of *C. burnetii* may be different since it has been reported to be gram-

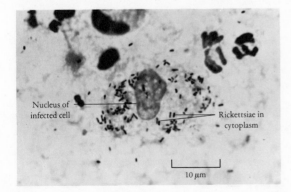

Nucleus of infected cell

Rickettsiae in cytoplasm

10 μm

FIGURE 12-25
Rickettsia rickettsii growing within a cell of an infected rodent. (Courtesy of Dr. Willy Burgdorfer)

positive, whereas other rickettsiae are gram-negative. Rickettsiae typically infect arthropods, such as ticks or insects, and may in some instances cause little harm to them, being transmitted from generation to generation through the eggs. Infected arthropods may bite mammalian hosts and thereby infect them with rickettsiae.

Chlamydiae represent another group of obligate intracellular bacteria. They are spherical and slightly smaller than rickettsiae (Fig. 12-26). Unlike rickettsiae, chlamydiae are not known to infect invertebrates, but they do infect a wide variety of birds and mammals. Only one genus, *Chlamydia* (formerly *Bedsonia*), is presently recognized, and it has been customary to name the different types according to the diseases they produce, rather than assigning special names. For example, the TRIC agents are responsible for the diseases *trachoma* and *inclusion conjunctivitis.* Trachoma, affecting 400 million people, is the world's leading cause of blindness. The agent of inclusion conjunctivitis is responsible for a mild eye disease contracted by newborn infants during passage through the birth canal or by persons swimming in contaminated pools. The organisms normally inhabit the lower urinary tract of human beings. Another of the chlamydiae causes the venereal disease, *lymphogranuloma venereum.* It is closely related to the ornithosis agents, which are transmitted to human beings from various species of birds and fowl and produce pneumonia. Since psittacine (parrot-like) birds were sources of the first recognized epidemics, the term psittacosis is often used instead of the preferred term, *ornithosis.* Chlamydiae, like rickettsiae, appear to have lost the capacity for extracellular growth by evolutionary adaptation to intracellular parasitism. However, unlike rickettsiae, they have even lost the means for generating energy and must obtain ATP from their host cells.

Intracellular parasites
Table 3-3; Page 62

Apparent close relatives of some obligate intracellular bacteria grow on cell-free media. For example, *Rickettsia quintana,* the cause of trench fever, a louse-borne infection, can be grown on blood agar. A group of bacterial parasites of red blood cells, represented by *Bartonella bacilliformis,* has often been included in the same large group as rickettsiae because some of them must infect living cells for growth and they are transmitted by arthropods. However, *B. bacilliformis* can readily be grown on cell-free media, and is a strictly aerobic, polarly flagellated gram-negative rod.

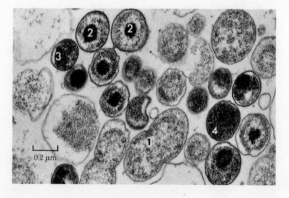

FIGURE 12-26

Chlamydiae growing in a tissue cell culture. The numbers indicate development from dividing form (1) to mature infectious bacterium (4). (Courtesy of R. Friis)

Bacteria comprise a large, diverse, and well-studied group of microorganisms. Classification into genetically related groups is not broadly feasible at the present time. This chapter divides bacteria into seventeen groups on the basis of morphology, nutrition, staining, and motility. Appendix I lists the groups and some medically important genera. For comparison, Appendix II outlines the widely known classification of bacteria given in the 7th edition of Bergey's Manual.

QUESTIONS

1. How do bacteria differ from other microorganisms?
2. Give the different mechanisms employed by bacteria to change position in response to external stimuli.
3. Are any obligate intracellular bacteria known to have close relatives that grow on cell-free media? Would you expect that such organisms could be found? Explain.
4. Defend the idea that mycoplasmas should be put in a group separate from bacteria and other procaryotes.
5. What are the principal characteristics of the lactic acid bacteria? The enterobacteria?

FURTHER READING

BREED, R. S. and others (ed.), *Bergey's Manual of Determinative Bacteriology,* 7th ed. Baltimore: The Williams and Wilkins Co., 1957.

COWAN, S. T. and K. J. STEEL, *Manual for the Identification of Medical Bacteria.* London: Cambridge University Press, 1966.

EDWARDS, P. R. and W. H. EWING, *Identification of the Enterobacteriaceae,* 3d ed. Minneapolis: Burgess Publishing Co., 1972.

MCELROY, W. D. and H. H. SELIGER, "Biological Luminescence," *Scientific American* (December 1962), 76-89.

POINDEXTER, J. S., *Microbiology: An Introduction to Protists.* New York: The Macmillan Co., 1971.

ROUECHÉ, B., "A Man Named Hoffman" in *Annals of Epidemiology.* Boston: Little, Brown and Co., 1967. (The story of anthrax)

STANIER, R. Y., M. DOUDOROFF, and E. A. ADELBERG, *The Microbial World,* 3d ed. Englewood Cliffs, N.J.: Prentice-Hall, Inc., 1970

Chapter 13

Algae

The algae, like bacteria, are ubiquitous. Although the blue-green algae (blue-greens) are procaryotic, all other algae are eucaryotic protists. In general, algae are characterized by the presence of chlorophyll-*a* and other pigments, and they carry out plantlike photosynthesis.

OCCURRENCE

Algae are abundant in both fresh and salt water and many inhabit soil, plants, or other terrestrial locations. Unicellular algae make up most of the ocean's phytoplankton, sometimes called the pastures of the sea. In contrast to pastures on dry land where multicellular plants are found, marine pastures of the open ocean have not favored the emergence of a predominance of multicellular forms for a number of reasons. For one thing nutrients are not as concentrated in the sea as they are in soil, and single cells which can move about may have an advantage over static multicellular organisms. Also, single cells that can move with the currents are well adapted to survival in strong ocean currents that continuously circulate the water.

On land, algae grow in the soil, on tree trunks or other plants, and in surface films on rocks. They reproduce rapidly when conditions are favorable, and huge numbers of algae in a "bloom" can color the water of ponds or areas of several square miles in the sea.

Because of their adaptability and their photosynthetic capacities, algae often can grow where no other life can exist. Species of blue-greens have been found growing about half a mile from the site of a 20 kiloton atomic explosion within three months after the blast. Algae were also among the first living forms to repopulate the lifeless remnant of Krakatoa after that Indonesian island was almost obliterated by volcanic eruptions. There they provided organic materials necessary for the establishment of higher forms of life, just as their predecessors must have pioneered in the first emergence of terrestrial life.

Some algae can be cultured in the laboratory in much the same way as are bacteria. Media are prepared to simulate the environmental nutrients in which the algae normally live; for example, marine species are grown in an artificial seawater medium. Light of the proper wavelength is supplied, and temperatures, gases, and other factors of the environment are adjusted to reproduce the natural habitat as nearly as possible. Mixed cultures, in which algae of more than one kind live together, are easily prepared, usually in the presence of bacteria or many other microbial species. Of course, for metabolic studies or other experiments when pure cultures are essential, single cells or spores must be isolated and allowed to grow alone. The preparation of pure cultures is considerably more difficult than preparing mixed cultures.

MORPHOLOGY

Algae occur in a variety of forms, some of which are diagrammed in Figure 13-1. The single cells of algae may be nonmotile, or motile either by means of ameba-like pseudopodia (rhizopodial algae, Fig. 13-1a) or flagella (flagellate algae). Often the cells are associated in colonies, filaments, or other structures. Chains of algae may form by continuous division in one plane, producing filaments which are usually enclosed by a mucilaginous sheath (Fig. 13-1b). Often the filaments branch by lateral divisions of the cells. Hollow-tube forms (Fig. 13-1c) do not produce cross walls, but instead become coenocytic, having many nuclei within a single, nonseptate organism.

The cell walls of algae vary considerably according to phylum, as indicated in Table 13-1. Different algal species secrete various kinds of mucilages, all of them polysaccharide in composition. The mucilage has several functions. It may aid in motility, and it can be a matrix providing for one type of colony formation by nonmotile forms (Fig. 13-1). The mucilage can adhere to rocks or plants and hold the colony in positions advantageous for growth.

PHYSIOLOGY AND METABOLISM

Algae are aerobic photosynthetic organisms, classified into seven major phyla. Table 13-1 lists the principal pigments of each phylum of algae. The photosynthetic pigments absorb light energy, and both their color and the amount of energy they take up depend on the wavelength of light they can utilize. Three major kinds of pigments are found in algae: chlorophylls, carotenoids (carotenes and xanthophylls), and phycobilins (Fig. 13-2). Slight differences in the molecu-

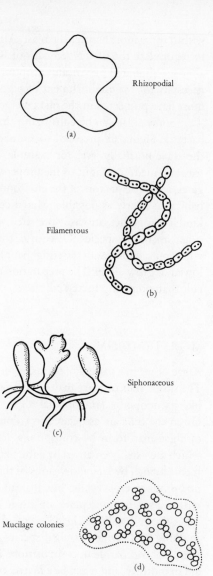

Rhizopodial

(a)

Filamentous

(b)

Siphonaceous

(c)

Mucilage colonies

(d)

FIGURE 13-1
The gross morphology of various algae.

lar structure of the chlorophylls (*a, b,* etc.) determine the wavelengths of light they can employ, and consequently the amount of energy absorbed. Chlorophyll-*a,* present in all algae, absorbs red and blue light and transmits a green color, which may be masked or altered by other pigments. For example, some algae appear brown because they contain a relative abundance of xanthophylls and carotenes that transmit brown light and mask the green color. Others appear reddish or purplish because of their phycobilins. Photosynthetic pigments of the procaryotic blue-greens are arranged on membranes within the cytoplasm, but the photosynthetic pigments of more complex eucaryotic algae are found in membrane-enclosed chloroplasts.

Chloroplasts
Page 94

The nature of the substances stored by various phyla of algae may be related

TABLE 13-1
Characteristics of Seven Major Phyla of Algae

Phylum	Usual Habitat	Principal Pigments[a]	Storage Products	Cell Walls	Mode of Motility (If Present)	Modes of Reproduction
Cyanophyta	Fresh- and saltwater; soil; hot springs; lichens	Phycocyanin; phycoerythrin	Starchlike glycogen; glycoproteins	Pectin; cellulose. May contain diaminopimelic acid	Nonmotile, although some may slide or oscillate	Asexual only, by binary fission, hormogonia, or spore formation
Chlorophyta	Fresh- and saltwater; soil; tree bark; lichens	Chlorophyll-b; carotenes; xanthophylls	Starch (α-1, 4-glucan)	Cellulose and pectin	Mostly nonmotile (except one order), but some reproductive elements may be flagellated	Asexual, by multiple fission; spores; or sexual
Euglenophyta	Freshwater	Chlorophyll-b; carotenes; xanthophylls	Fats; starchlike carbohydrates	Lacking, but have elastic pellicle	One to three anterior flagella	Asexual only, by binary fission
Chrysophyta	Fresh- and saltwater; soil; higher plants	Carotenes	Starchlike carbohydrates (β-1, 3-glucan); oils	Pectin, often impregnated with silica or calcium	Unique diatom motility; one, two, or more unequal flagella	Asexual or sexual
Pyrrhophyta	Mostly saltwater but common in freshwater	Carotenes; xanthophylls	Starch; oils	Cellulose and pectin	Two unequal lateral flagella in different planes	Asexual; rarely sexual
Phaeophyta	Saltwater	Xanthophylls, especially fucoxanthin	Starchlike carbohydrates; mannitol; fats	Cellulose and pectin; alginic acid	Two unequal lateral flagella	Asexual, motile zoospores; sexual, motile gametes
Rhodophyta	Mostly saltwater but several genera in freshwater	Phycoerythrin and other phycobilins; carotenes; xanthophylls	Starchlike carbohydrates	Cellulose and pectin; agar; carrageenan	Nonmotile	Sexual, gametes; asexual spores

[a] In addition to chlorophyll-a, which is present in algae of all phyla.

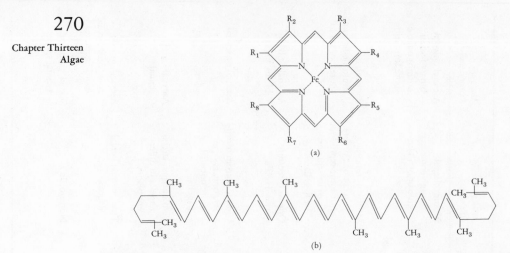

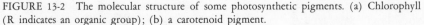

FIGURE 13-2 The molecular structure of some photosynthetic pigments. (a) Chlorophyll (R indicates an organic group); (b) a carotenoid pigment.

to their evolutionary development toward higher plants. For example, green algae, which have a number of properties in common with green plants, store a plantlike starch. Algae of other phyla polymerize the same simple sugars in different ways to produce other kinds of starches. Some algae store oils or fats, a property which is probably directly related to the survival of these species; those which make up much of the marine phytoplankton often store fat globules that make them buoyant and help keep them floating in the sea at levels where light for photosynthesis is available.

REPRODUCTION

Most algae can undergo vegetative reproduction by cell division or fragmentation without the production of any specialized structures. Asexual reproduction, also common among the algae, involves the formation of specialized cells called spores which germinate without uniting with other cells. Asexual spores may be either motile cells without cell walls (zoospores) or nonmotile cells enclosed by a wall.

The modes of sexual reproduction in algae vary so greatly, even within groups, that it is difficult to make generalizations. A few specific examples are included in later sections.

PROPERTIES OF THE PHYLA OF ALGAE

Some of the general characteristics of the major phyla of algae are summarized in Table 13-1. This classification is based on a number of properties, including the principal pigments of each group, their storage products, method of motility, and modes of reproduction.

The phylum Cyanophyta comprises the blue-greens. Members of this procaryotic group contain (in addition to chlorophyll-*a*) varying proportions of phycocyanin, phycoerythrin, and other pigments that are responsible for their colors. Although most species are various shades of blue-green, some are yellow, red, violet, or other colors. A starchlike material is their principal storage product.

The distinction between photosynthetic bacteria and blue-greens is based largely on differences in the pigments and in the photosynthetic process between the two groups. The *algae carry on typical plant-type photosynthesis. Some of the blue-greens can fix nitrogen from the air, a property that other algae lack.*

Photosynthesis, Pages 157, 158

The gliding motion observed in some filamentous blue-greens appears to depend on the extrusion of mucilage against a solid substrate, with consequent physical changes in the mucilage that propel the organism.

Cell walls, if present, contain cellulose or pectins, and (in some species) diaminopimelic acid. Both colonial and filamentous forms of blue-greens are common (Fig. 13-1).

Vegetative reproduction is common, and sexual reproduction has not been observed in blue-greens. Within the phylum Cyanophyta single cells divide by fission; in some genera cells in groups called hormogones act as reproductive elements.

Microcystis aeruginosa is an example of a blue-green which divides by simple fission. Single cells grow within a loose colony, enveloped in secreted mucilage (Fig. 13-1d). The cells divide randomly in various planes. As cells divide, portions of the colony separate and some newly formed cells are liberated to form new colonies. This species, one of many that can cause "water bloom" in fresh water, is notorious for its toxins.

The algae which reproduce by means of hormogones are represented by a species of *Rivularia* (Fig. 13-3). Its filaments are rows of cells with a common outer gelatinous sheath. In order to reproduce, short lengths (hormogones) break off from the ends of old filaments and glide away to form new ones. This alga grows in rounded, jellylike masses of mucilage on rocks and grasses of salt

(a)

(b)

FIGURE 13-3
Rivularia grows in colonial form on rocks and grasses (a). Each colony is composed of radiating filaments of single cells (b).

marshes (Fig. 13-3a). If a mass is examined under the microscope, it is found to consist of successive layers of sheathed filaments, within the mucilage, radiating from the attachment site (Fig. 13-3b). Some algae of the same order form resting spores that are enclosed by thick cell walls and can remain dormant for long periods of time, germinating when the environment becomes favorable.

The Eucaryotic Algae

Six major phyla comprise the eucaryotic algae. Both fresh and saltwater, soil, plants, and tree bark serve as natural habitats for green algae of the phylum Chlorophyta. All members of this phylum contain chlorophyll-*a* and -*b*, carotenes, and xanthophylls, and form starch as a storage product. Most of the vegetative forms of green algae are haploid. The cells can divide vegetatively, or they may divide asexually to form flagellate zoospores.

A common green alga often found in ponds is the single-celled, motile *Chamydomonas*. The single cells are oval in shape (Fig. 4-6) and typically contain one chloroplast. Two identical flagella protrude from the anterior portion of the cellulose cell wall. They have two excretory contractile vacuoles, a red eyespot, and a pigment-containing light receptor that permits the organism to move toward light. During asexual reproduction, the haploid cells lose motility and divide to give 4 (or, in some species, 8 or 16) zoospores which escape from the wall of the original cell to become independent and motile.

Another genus of the phylum Chlorophyta which has been of great value in the study of molecular biology is the huge, single-celled alga *Acetabularia* (Fig. 13-4), which may be as tall as 10 cm. The single nucleus near the base of its stemlike stalk can be removed by cutting the stalk. Sections of the alga that contain a nucleus can regenerate a new cap (or leaflike portion) readily, but sections that lack a nucleus have limited powers of regeneration. *Acetabularia* cells of different species, each with a distinctive cap, can be grafted together, as indicated in Figure 13-4. When two nucleated stalks of different species are grafted, the resulting hybrid will form a cap with some characteristics of each species. However, if a stalk without a nucleus is grafted to the nucleated stalk of a different species, the hybrid will form a cap characteristic of the nucleated species. In other words, the nucleus clearly determines the ability to regenerate a cap and the nature of the cap that is made. Experiments such as these have been valuable in studying the relative functions of nuclear and cytoplasmic constituents of the cell.

A small phylum of unicellular flagellates, the Euglenophyta, includes a few marine forms, but its members are predominantly freshwater forms. Members of the prototype genus *Euglena* have a single obvious anterior flagellum, but close examination may reveal a smaller vestigial second flagellum. A rigid cell wall is lacking, and an elastic pellicle covers the cell. Chlorophyll and carotenes are contained in chloroplasts, and the cells store fats and a starchlike carbohydrate. The cells divide by binary fission and sexual reproduction very rarely, if ever, occurs.

Many biologists regard members of the phylum Euglenophyta as protozoans for reasons discussed in Chapter 15.

CHLOROPHYTA
(GREEN-ALGAE)

Chlamydomonas
Pages 207, 208

EUGLENOPHYTA

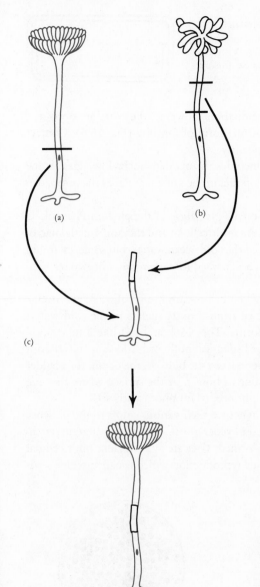

FIGURE 13-4
Cells of different species of *Acetabularia* (a,b) can be grafted together (c). The cap formed by the grafted alga is characteristic of
the species from which the nucleus was derived (d). Experiments
such as these have shown that the nucleus determines both the
ability to regenerate cellular constituents and the nature of the
regenerated cell components.

The phylum Chrysophyta includes several classes. In one class some genera
have calcium-containing hard walls; another class (Bacillariophyceae), members
of which contain silica in their walls, is of particular interest because it includes
the diatoms. The abundant diatoms are found in plankton or in films on rocks,
plants, and wood within both salt and fresh water. They have a unique morphology, being constructed like a petri dish (centric diatoms) or a covered box
(pennate diatoms), as illustrated in Figure 13-5. Their pectin-containing cell walls
are impregnated with siliceous compounds to make an extremely hard shell-like
covering called a frustule. Pores in the frustule are covered by a membrane in a

CHRYSOPHYTA

FIGURE 13-5

The construction of a diatom frustule. The hard coverings of diatoms (frustules) resemble covered boxes or dishes. When the cell divides each daughter cell retains half of the parent frustule and subsequently forms another corresponding half.

number of species; in many pennate diatoms, however, the pores are open in a slit, or raphe, associated with an unusual form of motility (Fig. 13-6). Contact with a solid surface is required for motility.

A third class of Chrysophyta includes rhizopodial ameba-like algae. Some of these can engulf food with pseudopodia in the same manner as the protozoan amebae.

PHAEOPHYTA (BROWN ALGAE)

Xanthophyll and carotene pigments give algae of the phylum Phaeophyta their brown color. The brown algae are multicellular and they are usually macroscopic. They are the sessile seaweeds of the cold ocean waters, but some are found in warmer areas such as the Sargasso Sea, named for *Sargassum,* a brown seaweed floating in it.

Members of the genus *Nereocystis* are common brown algae of the Pacific Coast which are attached at depths of from 6 to 24 meters in the cold waters between Alaska and southern California. They look and feel like long, tough, brown rubber tubes, with an enlarged spherical bulb, or bladder, at the anterior end (Fig. 13-7). Leaflike ribbons extend from the bulb. In the ocean, the bladder is filled with gas and holds the leaflike ribbons near the surface where they are exposed advantageously to the sunlight needed for photosynthesis.

RHODOPHYTA (RED ALGAE)

Algae of the phylum Rhodophyta owe their various colors to the presence of several pigments, especially the red phycoerythrin. They store a unique starch-like carbohydrate. Flagella are never formed. They are unicellular or multicellular and have complicated modes of sexual reproduction which usually involve sepa-

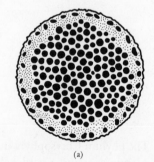

(a)

FIGURE 13-6

Diatoms. (a) Centric; (b) pennate, with a central slit (raphe) through which mucilage is extruded. Diatoms with a raphe exhibit a characteristic gliding movement.

(b)

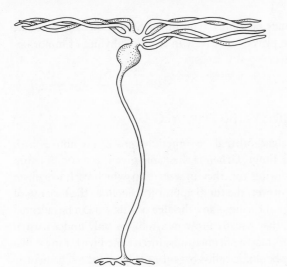

FIGURE 13-7
The kelp *Nereocystis* has a bulb-like float that keeps the alga near
the surface of the water, where the leaflike projections are exposed
to light and can carry out photosynthesis.

rate male and female organisms. Nonmotile cells from the male are carried by
the water to the egg-bearing female, and a zygote is formed.

Members of Rhodophyta are seaweeds which grow abundantly, attached in
coastal areas of warm saltwaters as well as very deep in the ocean, sometimes as
deep as 260 meters. One class (Corallinaceae) has hard, calcareous cell walls. Some
of the earliest algal fossils are of this coralline class, and its members help to make
up the present-day coral reefs of tropical waters. Photosynthesis is essential for
life of the reef because the coralline algae, along with some species of green algae,
are the primary producers of nutrients for the corals; therefore reefs develop only
in shallow waters where sunlight can penetrate. The algae also help to precipitate
calcium carbonate in the reef by removing carbon dioxide, thereby shifting the
equilibrium of the following equation to the right:

$$Ca(HCO_3)_2 \rightleftharpoons CaCO_3\downarrow + H_2O + CO_2\uparrow$$

calcium bicarbonate   calcium carbonate + water + carbon dioxide.

The dinoflagellates of the phylum Pyrrhophyta are primarily marine organisms and most of them are unicellular. They contain carotenes and xanthophylls
and store oils and starch. Asexual reproduction is the general rule in this phylum.

One genus of dinoflagellates, *Gonyaulax,* may be lethal to man. Some species of this genus produce a neurotoxin which causes a paralysis known as "paralytic shellfish poisoning," associated with "red tides." This problem may occur
wherever conditions allow abundant proliferation of *Gonyaulax* or a number of
other toxin-producing algae.

Gonyaulax and a few other species of marine algae are luminescent by means
of a biochemical reaction similar to the one responsible for light emission by
fireflies.

Not all of the eucaryotic algae can be classified into the major phyla. A
small group of organisms which have not been positively classified are considered

PYRRHOPHYTA
(DINOFLAGELLATES)

ALGAE OF UNCERTAIN
CLASSIFICATION

by some to constitute a separate phylum, Cryptophyta. Some other algae that are difficult to classify have been placed tentatively in another phylum of minor importance, Chloromonadophyta.

LICHENS AND OTHER SYMBIOTIC ALGAE

In addition to the coralline algae, other algae sometimes grow in symbiosis with certain organisms, including fungi. Lichen is the name given to a combination of an alga and a fungus growing together in a state in which each organism benefits. In a few lichens, however, the fungi appear to benefit at the expense of their algal partners. Although the fungus and the alga of a lichen can be cultured separately in the laboratory, they usually grow in symbiosis only under natural conditions. The algal partners may be filamentous, intertwined with fungal filaments (hyphae), or they may be single-celled organisms supported by the hyphae of the fungus. Either blue-greens such as *Nostoc* species or green algae such as *Trebouxia* species are found in lichens.

The supporting fungal hyphae provide a safe haven for the alga, offering protection from the environment as well as nutrients extracted from the rocks, soils, or other surfaces on which they grow (Fig. 13-8). The contributions of the algae are equally important. A blue-green may be able to fix nitrogen from the air to provide organic materials for the whole lichen derived from air, water, and a few minerals supplied by activities of the fungus. A green alga does not fix nitrogen but will be the primary producer, able to utilize carbon dioxide and sunlight to produce carbohydrates for the pair. The fungus is able to absorb nu-

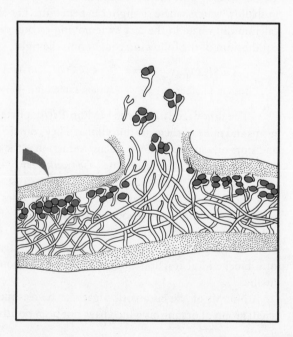

FIGURE 13-8
This lichen consists of algal cells (in color) entwined within mycelial filaments of the fungal partner.

FIGURE 13-9
A species of *Cladonia,* the lichen known as reindeer moss. (Courtesy of CCM: General Biological, Inc., Chicago.)

trients from its partner by means of tiny rootlike fungal projections (haustoria) which appear to penetrate the algal cell.

Although they have a widespread distribution, lichens are not often found near large cities because they are readily killed by the smokes and gases produced by civilization. However, they frequently grow in areas where many other kinds of organisms or vegetation cannot survive. Consequently, lichens are important as a source of food, especially in extreme northern climates. The reindeer moss which supports herds of reindeer are lichens of the genus *Cladonia* (Fig. 13-9). Their stalked cuplike structures grow as high as 15 cm, and are commonly found on ground which may be frozen during much of the year. *Cladonia* species and other lichens are used as food for man in various parts of the world.

Some of the lichens also make a significant contribution to soil formation by gradually eroding rocks or other strata on which they live. Unique acids made by the lichens aid in this breakdown of materials and also have other functions, such as protecting the lichens from destruction by insects and other small animals.

In addition to coral reefs and lichens there are other examples of algal symbiosis. One species of green algae has been reported which lacks photosynthetic pigments but contains blue-greens within its cytoplasm. In this symbiotic arrangement the blue-greens supply the photosynthetic apparatus for both organisms.

ROLE OF ALGAE IN THE FOOD CHAIN

In the aquatic environment, algae are the primary producers in the food chain, and thus serve as the principal food source for smaller animals that participate in the chain. It is possible to show a direct correlation between the numbers of algae and fish present in a given area, because fish thrive on the rich food supplies derived from algae (Fig. 13-10).

In fresh waters, and on land, a similar chain of events takes place—the algae synthesizing carbohydrates, proteins, and other foodstuffs, while at the same time producing oxygen which supports animal respiration.

FIGURE 13-10 World-wide fishing areas (in color) reflect the abundance of algae which serve as primary producers in the aquatic food chain. Algae are found in largest numbers above the continental shelves or in currents flowing from these areas, because more nutrients are available there than in other parts of the ocean.

In addition to their vital activities in the food chain and in the production of oxygen, there are a number of other ways in which algae are of use to man.

Diatomaceous earth (diatomite, kieselguhr) consists of the frustules of dead diatoms. This silica-containing material is a mild abrasive, and is the polishing agent of many metal polishes; its resistance to temperatures of 600°C or higher makes it a valuable heat-insulating material; and it is valuable in making a kind of filter.

The red algae yield two polysaccharide products of major economic importance, agar and carrageenan, useful as jelling or emulsifying agents. Alginates are other useful polysaccharides, derived from some of the brown algae such as the genus *Macrocystis*. Huge beds of *Macrocystis* off the coast of California, estimated to cover about 100 square miles, are "farmed" for the production of alginates.

Experimentally it has been proven that it is possible to cultivate green algae of the genus *Chlorella* as a food; by adjusting cultural conditions nutrients can be produced that are either high in fat or are more than half protein (dry weight). It is calculated that a dry weight yield of 17.5 tons per acre per year can be achieved. At present, however, the costs outweigh the benefits, since it costs about eight times as much to produce the algal food as it does to grow soya beans. The possible benefits of *Chlorella* cultivation to space travel are being investigated. During long space voyages, food and oxygen would have to be replaced and wastes consumed. It is feasible to use *Chlorella* or other algae to harness solar energy and produce oxygen from products of the metabolism of other microorganisms which can degrade human wastes and recycle essential elements.

Vitamins A and D produced by diatoms are of importance, both medically and commercially. Fish eat diatoms and concentrate the ingested vitamins in their livers. Man then extracts the vitamins from fish livers (for example, cod-liver oil) for human consumption.

Algal toxins, referred to previously, probably represent the most significant danger afforded by the algae. Although birds and mammals are occasionally affected by algal toxins, fish are by far the most frequent victims. In Israel, for example, extensive use of fertilizers which drain from fields into fish ponds has enriched the ponds to the point where algal blooms are common. As a result thousands of fish have been killed each year. Consequently extensive studies on the nature of the toxins have been undertaken and several of the toxins have been well characterized. Perhaps some day man will be able to utilize these toxins for medicinal purposes as snake venoms and other toxins have been utilized already. If that day comes, the benefits of algae to man will be even greater than they are today.

SUMMARY

Algae are ubiquitous photosynthetic organisms. Many are aquatic, but others occur in soil, and on vegetation, trees, and rocks. They may be multicellular or unicellular and are often arranged in colonies, filaments, or other forms.

The characteristics of algae vary considerably among phyla. The phylum Cyanophyta comprises the blue-greens, which are procaryotic and may have cell walls resembling those of bacteria. The blue-greens differ from bacteria principally in that they carry out typical plant-type photosynthesis, quite distinct from bacterial photosynthesis. Vegetative reproduction is common among the blue-greens, which do not reproduce sexually.

The eucaryotic algae are classified into six major phyla. These include Chlorophyta (the green algae), Euglenophyta (unicellular flagellates), Chrysophyta (several classes, including the diatoms), Phaeophyta (the brown algae), Rhodophyta (the red algae), and Pyrrhophyta (the dinoflagellates).

Algae may exist in a number of symbiotic relationships, for example, with fungi in lichens or with corals in coral reefs. The food chains of the ocean depend on algae as the primary producers. In addition algae or their products are put to many uses and algae are thus economically important. Some algae produce toxins that are lethal for man or other animals.

QUESTIONS

1. What major characteristics distinguish the blue-greens from other algae?
2. Compare and contrast photosynthesis as carried out by algae and by bacteria.
3. Can algae fix nitrogen from the air and if so, what kinds of algae have this capability?
4. How are the seven major phyla of algae distinguished?
5. What might the world be like if all algae were killed?

FURTHER READING

CARSON, R. L, *The Sea Around Us.* New York: New American Library, 1950, Pp. 169.

ECHLIN, P., "The Blue-Green Algae," *Scientific American,* **214** (June 1966), 75–81.

GIBOR, A., "Acetabularia: A Useful Giant Cell," *Scientific American,* **215** (November 1966), 118–124.

ISAACS, J. D., "The Nature of Oceanic Life," *Scientific American,* **221** (Mar. 1969), 146–162.

PRESCOTT, G. W., *The Algae: A Review.* Boston: Houghton Mifflin Co., 1968.

ROUND, F. E., *The Biology of the Algae.* London: Edward Arnold & Co., 1965. Pp. 269.

SHILO, M., "Formation and Mode of Action of Algal Toxins," *Bact. Rev.,* **31** (1967). 180–193.

Chapter 14

Fungi

Yeasts, molds, and mushrooms are part of the major group of eucaryotic protists called the fungi. These simple organisms lack chlorophyll and are nonphotosynthetic. Most of them live as saprophytes on dead organic materials, often in soil, but some are commonly parasitic on living plants or animals. The fungi include many important plant pathogens but only a few species (less than 100) that are pathogenic for animals and man.

OCCURRENCE

The fungi are found in all parts of the world, wherever organic materials are available as nutrients. Some species are very specialized; for example, they may occur naturally only on a particular strain of one genus of plants; others are extremely versatile metabolically and are able to subsist on many kinds of unlikely media such as leather, cork, hair, wax, ink, or even synthetic plastics, especially polyvinyls. Other species are able to grow in high concentrations of salts, sugars, or acids that would kill most bacteria; these fungi are responsible for spoiling pickles, preserves, and certain other foods.

A few kinds of fungi can grow at temperatures as high as 50°C. Fungi are usually able to become dormant and to survive low temperatures much more readily than high temperatures. Some actually grow below the freezing point and can rot bulbs and destroy grass in frozen ground.

Fungal reproductive cells, called spores, are found virtually all over the earth and are also abundant in the air. These minute forms occur in tremendous numbers in air near the earth's surface, and have also been recovered from air currents at altitudes of more than seven miles.

281

TABLE 14-1
Some Characteristics of the Major Classes of Fungi

Group	Usual Habitat (Representative Groups)	Mycelia	Asexual Spores	Sexual Spores
Phycomycetes	Water (watermolds); soil, bread, etc. (black bread mold)	Nonseptate	Sporangio-spores, occasionally conidia	Various forms; may have flagel-late gametes; zygospores; oöspores
Basidiomycetes	Soil (mushrooms); grains (rusts, smuts)	Septate	Commonly not formed	Basidiospores (borne on clublike basidium)
Ascomycetes	Fruits, other organic media (true yeasts, Neurospora)	Septate	Conidia	Ascospores (borne in saclike ascus)
Deuteromycetes (Fungi Imperfecti)	Soil (most of the pathogens for man)	Septate	Conidia Arthrospores	Absent

CLASSIFICATION

Organisms are classified as fungi on the basis of their morphology, physiology, and modes of reproduction. The principal characteristics of fungi are mycelial growth (described below), spore formation, and the lack of chlorophyll. The fungi are divided into four major classes (Table 14-1) largely on the basis of their method of sexual reproduction, or lack thereof.

MORPHOLOGY

Fungal cells and filaments are enclosed by rigid cell walls that usually contain the polysaccharide chitin. A typical reproductive spore is a single cell about 3 to 30 μm in diameter, depending on the species, although some are as small as 1 μm and others may be 300 μm long. When a mold spore reaches a favorable environment, it germinates and sends out a projection called a germ tube (Fig. 14-1). This tube continues to grow into a long filament called a *hypha* which branches and intertwines. The complex tangled mass of hyphae which results is known as *mycelium*. The cottony growths of molds on bread and fruits are familiar examples of mycelia. Mycelial formation is a principal distinguishing characteristic of many, but not all, fungi.

The mycelium is well suited for its function of collecting and utilizing

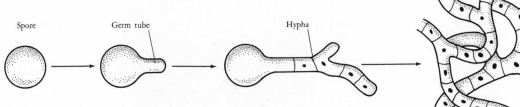

FIGURE 14-1 Spores of fungi germinate by forming a projection from the side of the cell (a germ tube) which elongates to form hyphae. As the hyphae continue to grow, they form a tangled mass called mycelium.

foodstuffs. Fungi must absorb nutrients through their rigid cell walls. Hyphae (about 2 to 10 μm in diameter) are able to grow and advance in the medium, and rapidly form mycelia with a large surface area through which materials can be absorbed. Sometimes tiny filaments called haustoria grow out from the hyphae. Although the haustoria appear to penetrate plant cells that are parasitized by fungi, they do not actually enter the cytoplasm. Instead they invaginate the membrane of the host cell, allowing intimate contact, and facilitate the uptake of nutrients by the fungus.

Three of the classes of fungi (the Ascomycetes, Basidiomycetes, and Deuteromycetes) form mycelia that are divided into cell-like compartments with one or two nuclei each (septate mycelia). However, the dividing cross wall, or septum, is not completely closed and cytoplasm can circulate from one compartment to another. The other class, the Phycomycetes, has mycelia without cross walls (nonseptate), so that the entire mycelium may actually consist of a single multinucleated cell.

Yeast is the name given to single-celled ascomycetes or deuteromycetes which lack a mycelium, and mold refers to fungi with a mycelium. The true yeasts are never mycelial, but other species known as *dimorphic fungi* are capable of growing in either the yeast or the mold form, depending on environmental conditions. A rich medium, 37°C temperature, and increased carbon dioxide in the atmosphere favors growth of the yeast form; mold forms of dimorphic fungi tend to grow under less favorable conditions. The morphology of a typical yeast cell is shown in Figure 14-2.

PHYSIOLOGY AND METABOLISM

In order to absorb nutrients through their cell walls, fungi secrete a wide variety of enzymes which degrade organic materials, especially complex carbohydrates, into small molecules that are readily absorbed.

Most fungi are aerobic, but some of the yeasts are facultatively anaerobic

283

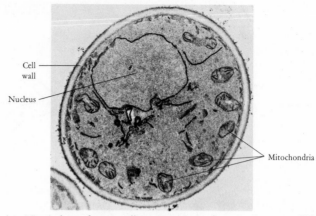

FIGURE 14-2 Morphology of a yeast cell as seen with the electron microscope.(**X**48000) (Courtesy of Dr. E. Boatman and Dr. H. C. Douglas)

and carry out important fermentation reactions. There are no obligately anaerobic fungi. Light is usually not necessary for growth; in fact, sunlight will often kill the vegetative cells. However, spore forms are generally resistant to ultraviolet rays of sunlight. Although the water molds are aquatic, the majority of fungal species prefer a slightly moist environment in equilibrium with a relative humidity of about 70 percent or more. Most fungi thrive at temperatures ranging

TABLE 14-2
Some Characteristics of Fungal Spores

Type of Spore	Characteristics	Examples
Asexual Spores		
Conidiospores	Single-celled (microconidia) or multicellular (macroconidia): formed by extrusion of single cells from specialized areas of hyphae	*Penicillium* *Alternaria*
Arthrospores	Single-celled; formed by disjointing of hyphal cells	*Coccidioides*
Sporangiospores	Single-celled; formed within sacs at end of hyphae	*Rhizopus*
Blastospores	Buds; formed on yeast cells	*Candida*
Chlamydospores	Thick-walled single cells; highly resistant to adverse conditions	*Candida*
Zoospores	Single-celled; motile with flagella	*Saprolegnia*
Sexual Spores		
Ascospores	Single cells in ascus (often eight per ascus)	*Neurospora* (class Ascomycetes)
Basidiospores	Single cells borne on basidium (often four per basidium)	*Agaricus* (class Basidiomycetes)
Zygospores	Large spore encased in a thick wall	*Rhizopus* (class Phycomycetes)
Oöspores	Develop within oögonium (one to 20 or more per oögonium)	*Saprolegnia* (class Phycomycetes)

from 20°C, or approximately room temperature, to 35°C, just below body temperature.

The pH at which molds can grow varies widely, ranging from as low as 2.2 to as high as 9.6, but they usually grow best at an acid pH of 5.0 to 6.0. Yeasts have an even more acid pH optimum of 4.5 to 5.0, with a range from about 3.0 to 7.5.

REPRODUCTION

Both asexual and sexual reproductive spores of fungi occur in a number of forms, as indicated in Table 14-2, and diagrammed in Figure 14-3. Most sexual spores of fungi are borne on specialized structures called *fruiting bodies*. Sexual repro-

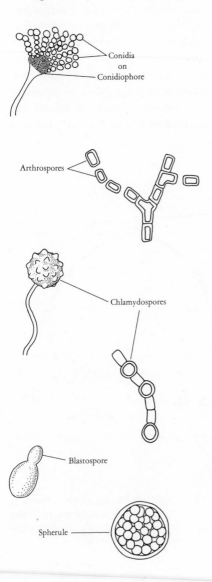

Conidia
on
Conidiophore

Arthrospores

Chlamydospores

Blastospore

Spherule

FIGURE 14-3
The morphology of various kinds of fungal spores.

duction, common among the fungi, often occurs after many generations of asexual division. The Basidiomycetes, however, usually reproduce sexually and do not often form asexual spores. Examples of modes of sexual reproduction are given below in discussions of the various classes of fungi.

SOME CHARACTERISTICS OF THE CLASSES OF FUNGI

Phycomycetes

Members of this class are often called the lower fungi. Some of them, the watermolds, are aquatic and range from simple to more complex species; other groups are terrestrial. The simpler watermolds are unicellular and do not produce mycelium; however, other species develop mycelial growth. Watermolds reproduce asexually by means of motile zoospores, which have one or two flagella and resemble protozoa. More highly developed genera, such as *Allomyces,* reproduce both sexually and asexually. The life cycle of *Allomyces* is described in Figure 14-4.

A familiar example of a terrestrial phycomycete is the genus *Rhizopus* that includes the common black bread mold. It reproduces by means of asexual sporangiospores as shown in Figure 14-5a. Sexual reproduction is also common in *Rhizopus* (Fig. 14-5b).

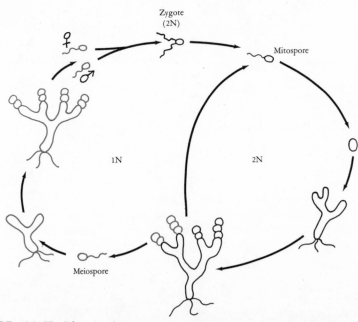

FIGURE 14-4 The life cycle of an aquatic phycomycete. Male and female gametes (upper left) fuse to form a diploid spore (mitospore) which germinates to form diploid mycelium (lower right). Following meiosis, haploid spores (meiospores) are formed in one type of sprorangia. Diploid spores arise from another type of sporangia and repeat the diploid cycle.

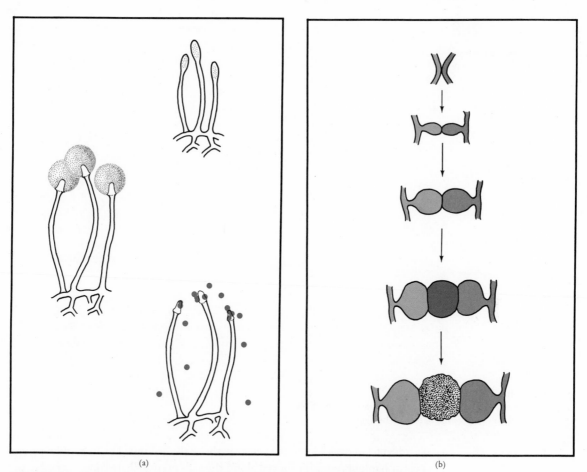

FIGURE 14-5 (a) Asexual sporangiospores of *Rhizopus*. The spores develop within a sporangium, which ruptures to release the mature spores. (b) Sexual reproduction in *Rhizopus*. Two mycelia of opposite sex (+ and −) come together and each forms a side branch. Following division of the two branches, the cells in contact fuse to form a zygospore, which becomes surrounded by a thick, black wall.

Basidiomycetes

Members of this class produce sexual spores on a specialized structure called a *basidium* (Fig. 14-6). Well-known Basidiomycetes include the mushrooms, puffballs, bracket fungi of trees, and the rusts and smuts that destroy grains. The familiar mushrooms are fruiting bodies of large subsurface mycelia and they bear huge numbers of sexual spores. Figure 14-6 illustrates the gilled structure of these fruiting bodies in which spores are borne on the sides of the gills.

The life stages of a mushroom, *Agaricus campestris,* are shown in Figure 14-7. Spores that land in favorable soil germinate to form haploid mycelium which spreads through the soil. Two haploid mycelia fuse to give rise to diploid mycelia. The mycelium secretes enzymes that break down complex organic compounds into simple nutrients it can absorb. After a period of growth, the fruiting bodies

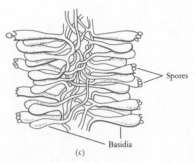

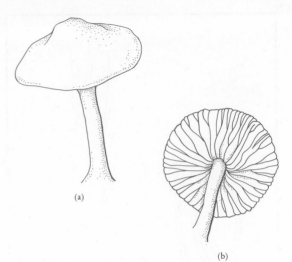

FIGURE 14-6

(a) A typical Basidiomycete fruiting body; (b) the underside of the cap is composed of radiating gills; (c) a magnified view of the surface of a gill, showing a mass of basidia, bearing spores.

(formed from diploid mycelia) emerge above the ground. The tremendous numbers of haploid spores produced along the gills eventually become air-borne and each can give rise to a new organism.

Ascomycetes

This very large class includes most of the yeasts, and certain other well-known fungi such as morels and truffles. The sexual spores of Ascomycetes, formed

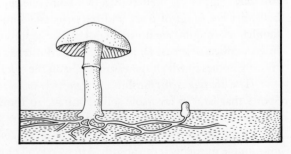

FIGURE 14-7

The meadow mushroom, *Agaricus campestris,* has an extensive underground mycelium. The fruiting body emerges as a small button mushroom (right), which rapidly grows into a typical mushroom with a gilled cap (left).

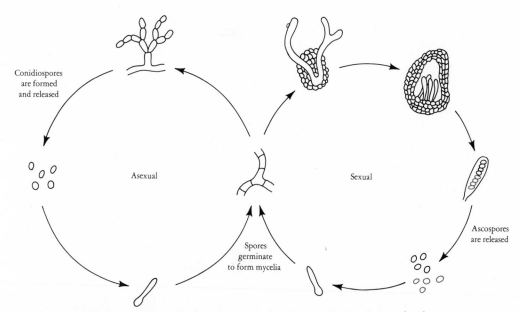

Industrial uses of yeasts, Pages 660—663; 669—671

Conidiospores
are formed
and released

Asexual

Sexual

Spores
germinate
to form mycelia

Ascospores
are released

FIGURE 14-8 The life cycle of an ascomycete (a species of *Neurospora*). During the sexual cycle, cells of opposite mating types (indicated by colors) fuse and give rise to ascospores enclosed within asci (singular: ascus).

within a structure called an ascus, often resemble peas in a pod (Fig. 14-8). A simplified version of the life cycle of a species of *Neurospora* is shown in the figure.

The life cycle of the yeast most used for making bread, wine, and beer, *Saccharomyces cerevisiae,* is outlined in Figure 14-9. Actually, the cycle is more complex than indicated. This yeast divides asexually by budding, but it is also capable of conjugation and sexual reproduction.

Deuteromycetes (Fungi Imperfecti)

This group has been called the taxonomic dump-heap of the fungi, because it comprises those species which lack an observed perfect, or sexual, stage and which therefore cannot be placed in any other class. These fungi apparently have become so adept at maintaining their species without utilizing sexual processes that they have either lost the capacity for sexual reproduction or else use it so seldom that it has not been observed in the laboratory. From time to time sexual sporulation is observed in fungi that have previously been considered Fungi Imperfecti. In these cases, the fungus usually is found to be of the class Ascomycetes.

Molds of the genera *Penicillium* and *Aspergillus* are deuteromycetes of widespread distribution and commercial importance (Fig. 14-3). In addition, most of the fungi that are pathogenic for man are deuteromycetes. They form asexual spores, frequently of several varieties within the same species, and these spores are often helpful in identifying pathogenic fungi.

289

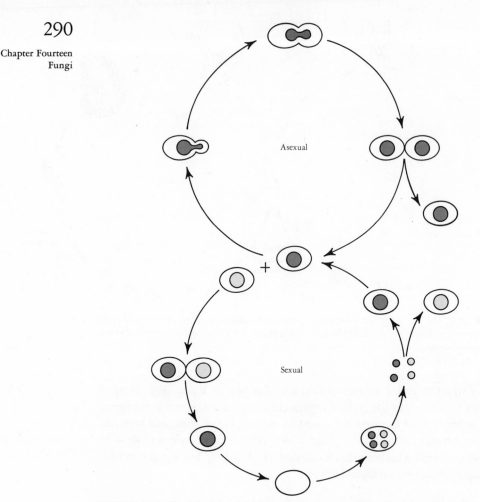

FIGURE 14-9 The life cycle of the yeast *Saccharomyces cerevisiae*. During asexual reproduction, the yeast cells divide by budding. During sexual reproduction, two cells of opposite mating types (indicated by colors) fuse, and an ascus is formed. By the process of meiosis, four haploid ascospores are formed, two of each mating type.

An overall view of the fungi of major medical importance, all of them members of the group Deuteromycetes, is presented in Table 14-3, and some of them are illustrated in other chapters.

Medically important fungi
Pages 464–467; 499–500; 523

SYMBIOTIC RELATIONSHIPS OF FUNGI

Lichens
Pages 276, 277

In addition to their association with algae in lichens, fungi have developed many other interesting symbiotic relationships. Mycorrhizas, fungal symbioses of particular importance, are literally fungus roots formed by the intimate association between fungi and the roots of certain plants. The fungi involved commonly occur only in association with the roots, and they often allow their partners to

TABLE 14-3
Some Deuteromycetes (Fungi Imperfecti) of Major Medical Importance

Genus or Genus and Species	Disease Produced	Important Characteristics
Epidermophyton Microsporum Trichophyton	Superficial fungal infections	Infect superficial skin, hair, or nails. Cause tinea (infections such as athlete's foot, ringworm, and others)
Histoplasma capsulatum	Histoplasmosis	Dimorphic yeast. Infections described in Chapter 25
Coccidioides immitis	Coccidioidomycosis	Large spherules (15–75 μm) filled with endospores; infections described in Chapter 25
Blastomyces	Blastomycosis	Blastospores
Sporotrichum	Sporotrichosis	Rosette-arrangement of conidia (Chapter 29)
Cryptococcus neoformans	Cryptococcosis (frequently meningitis)	Usually nonpathogenic but can cause severe or fatal infections
Candida albicans	Candidiasis (thrush, moniliasis)	Usually nonpathogenic but can invade (Chapters 24 and 27)

grow where the plants could not otherwise survive. Many trees have mycorrhizal roots, such as the conifers that are able to live in sandy soil by means of extensive mycelial mycorrhizas. Truffles form mycorrhizas with either oak or beech trees. Indian pipe and some other mycorrhizal plants have lost the ability to form chlorophyll and thus have become completely dependent on their fungal partners.

Certain insects also depend on symbiotic relationships with fungi. For example, some ants and termites grow fungi for food, nourishing the fungi with leaves which they carry into their nests. Also, wood-boring beetles inoculate fungal spores into wood where the fungi can grow and can subsequently serve as food for the beetles.

UTILIZATION AND PRACTICAL IMPORTANCE

Many of the fungi are important commercially. They participate in the production of many alcoholic beverages, foods and chemicals and of some antibiotics and drugs. Fungal plant pathogens are of great economic importance. In addition, fungi are essential as degraders of organic wastes.

Fungi in soil
Pages 594–595
As plant pathogens
Table 34-5, Page 614

THE SLIME MOLDS (MYXOMYCETES)

The unattractive name of slime molds has been given to a group of unique eucaryotic protists which bear some resemblance to both fungi and protozoa. Four diverse orders comprise this group: two of them are amebae during their vegetative phase and the other two are acellular. Members of all four orders often produce funguslike fruiting bodies and they have in common the property of forming slimy vegetative bodies which lack differentiation into tissues—thus the name, slime molds.

The *vegetative plasmodium* of acellular slime molds *is a multinucleate mass of protoplasm* which oozes like slime over the surface of the substrate, ingesting microorganisms, fungal spores, and small particulate matter. Although it looks like slime, the plasmodium actually has a structure very similar to fungal mycelium. It is in fact one very large multinucleate cell (up to 7 cm in diameter) enclosed by a cytoplasmic membrane within which the cytoplasm can flow. However, in contrast to fungal mycelium, the cytoplasm of slime molds is not surrounded by a rigid cell wall and so is free to ooze over the substrate. As long as environmental conditions remain favorable for vegetative development, the plasmodium continues to enlarge. However, under conditions of starvation and in the presence of light, the plasmodium develops into dozens of small, intricately constructed, and often brilliantly colored fruiting bodies, each one consisting of many sporangia situated on top of a stalk (Fig. 14-10). Each sporangium contains many uninucleate spores, formed by the walling off of sections of the plasmodium. These resting spores are generally blown by wind, and in a moist environment

FIGURE 14-10 Fruiting bodies of slime molds. (a) Species of *Stemonitis* growing on a log (courtesy of J. P. Dalmasso); (b) *Dictydium cancellatum;* (c) *Metatrichia vesparium;* (d) *Lepidoderma tigrinum.* (b, c, and d courtesy of Dr. E. F. Haskins)

they germinate. Each spore gives rise to one or more cells which can divide vegetatively for a time. Under appropriate conditions, two sexual cells fuse to form a zygote, which in turn develops into a new plasmodium.

The cellular slime molds resemble acellular forms in that they also have a life cycle which includes a vegetative as well as a fruiting body stage. They differ from the acellular in that the vegetative stage of the cellular slime molds consists of uninucleate amebae which are almost indistinguishable from true amebae. When the food supply becomes exhausted, some of the amebae begin to excrete a hormonelike substance, now known to be cyclic AMP, which attracts other amebae, causing them to stream toward a common point. When a large number of them have come together, a "pseudoplasmodium," or slug, is formed. This community of amebae, each of which retains its separateness, acts as a single unit. The slug lifts itself, its posterior end flattened on the surface, and its anterior end forming a nipplelike structure. The amebae at the top begin to migrate downward through the center of the slug, secreting cellulose as they migrate, thereby forming a cylinder from the top to the base of the slug. When the cellulose cylinder reaches the base, the amebae at the base begin to migrate up the stalk and finally form a mass at the tip. Each of these amebae becomes enveloped in a thin cellulose wall to form the mature spores, which are held together by a droplet of viscous liquid. When released, the spores germinate and develop once again into amebae. This life cycle, in certain respects, resembles that of the myxobacteria.

Myxobacteria
Pages 241–243

The various slime molds are important links in the food chain within soil. They ingest bacteria, algae, and other organisms as nutrients, and in turn serve as food for larger predators. Slime molds have been valuable also as unique models for studying cellular differentiation during their aggregation and especially during the formation of fruiting bodies.

SUMMARY

Yeasts, molds, and mushrooms are included among the fungi, which are nonphotosynthetic, eucaryotic protists. A majority of them are saprophytes, living on dead organic materials, but some are parasitic on living plants or animals. Fungi are classified largely on the basis of their method of sexual reproduction, or lack thereof, into four classes: Phycomycetes, Basidiomycetes, Ascomycetes, and Deuteromycetes.

Mycelial growth is characteristic of many fungi. Mycelial filaments and other fungal forms are enclosed by a rigid cell wall which usually contains chitin. Fungi reproduce by means of asexual spores, sexual spores, or both.

Phycomycetes, often called the lower fungi, may be aquatic or terrestrial. The aquatic watermolds reproduce asexually by means of motile zoospores. More highly developed genera, such as *Allomyces,* reproduce asexually and sexually. The terrestrial phycomycetes are represented by the common black bread mold *Rhizopus,* which reproduces by means of asexual sporangiospores, or by formation of sexual zygospores.

Basidiomycetes produce sexual spores on a specialized structure called a basidium. Mushrooms are familiar examples of Basidiomycetes. The mushroom is actually the fruiting body formed from large subsurface mycelia.

The class Ascomycetes includes most of the yeasts, as well as certain other fungi. Members of this class form sexual spores within a saclike structure called an ascus.

The Deuteromycetes, or Fungi Imperfecti, are so named because they lack an observed perfect, or sexual, stage. They comprise genera of widespread distribution and commercial importance, such as *Penicillium* and *Aspergillus,* and also most of the fungi of medical importance.

Fungi participate in a number of important symbiotic relationships; for example, with algae they form lichens and with the roots of certain plants they form mycorrhizas.

Fungi are of great significance in certain commercial processes, such as the manufacture of alcoholic beverages, some foods, antibiotics, and other products. Fungi also may be important plant pathogens and are essential as degraders of organic wastes.

QUESTIONS

1. Discuss the pathogenicity of fungi for animals and plants and compare this with the disease-producing capabilities of bacteria.
2. What are fungal spores and how do they function?
3. How are the four major classes of fungi distinguished?
4. Consider the differences in physiology and metabolism between bacteria and fungi.
5. Discuss several symbiotic relationships of fungi with other organisms.

FURTHER READING

AINSWORTH, G. C. and A. S. SUSSMAN, *The Fungi. An Advanced Treatise.* Vols. I and II. New York: Academic Press, 1965.

BONNER, J. T., "Hormones in Social Amoebae and Mammals," *Scientific American* (June 1969).

CHRISTENSEN, C. M., *The Molds and Man. An Introduction to the Fungi.* Minneapolis, Minnesota: University of Minnesota Press, 1951.

FULLER, J. G., *The Day of St. Anthony's Fire.* New York: Macmillan Co., 1968.

Human Mycoses. Scope Monograph by The Upjohn Company, Kalamazoo, Michigan, 1968.

KAVALER, LUCY, *Mushrooms, Molds, and Miracles.* New York: Signet Books, John Day, Inc., 1965.

McKENNY, M. and D. E. STUNTZ, *The Savory Wild Mushroom.* Seattle: University of Washington Press, 1971.

WILSON, J. W., and O. A. PLUNKETT, *The Fungous Diseases of Man.* Berkeley and Los Angeles, California: University of California Press, 1965.

Chapter 15

Protozoa

In addition to algae and fungi the eucaryotic protists include another major category, the protozoa. There are not always clear dividing lines among algae, fungi, and protozoa and many species have the characteristics of more than one group. The slime molds, for example, have some characteristics of all three groups. Another example is the unicellular flagellates Euglenophyta, which are often considered to be protozoa. Most euglenae have chlorophyll and carry out photosynthesis, clearly properties of the algae; on the other hand, some species lack photosynthetic pigments and all of the species in this group are capable of existing on complex nutrients when in the dark, both characteristics of the protozoa. Mutant strains of the algal genus *Chlamydomonas,* which have lost chlorophyll, might be classified in the protozoan genus *Polytoma,* but often protozoologists consider all *Chlamydomonas* to be protozoa.

Euglenophyta
Page 272

In general, however, protozoa differ from algae and fungi in various ways: by their lack of photosynthetic capabilities, their motility, means of reproduction, and other characteristics. Most of the protozoa are microscopic unicellular organisms, with some as small as 2 μm in diameter, but other species are easily visible.

The life cycles of protozoa are sometimes complex, involving more than one habitat or host. It is usual to find morphologically distinct forms of the same species at different stages of the life cycle; they are said to be polymorphic (Fig. 15-1). This polymorphism can be compared to differentiation of various cell types in higher organisms.

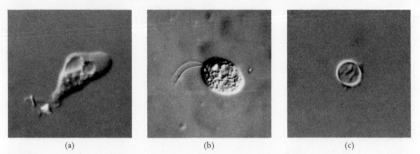

(a)   (b)   (c)

FIGURE 15-1 Polymorphism in a protozoan. This species of *Naegleria* may infect man. In human tissues the organism exists in the form of (a) an ameba (10 to 11 μm in its widest diameter). After a few minutes in water the flagellate form (b) appears. Under adverse conditions a cyst (c) is formed. (Courtesy of M. Roth and Dr. F. Schoenknecht)

OCCURRENCE

All protozoa require large amounts of moisture, no matter what their habitat may be. Various species are found in marine, freshwater, or terrestrial environments, and a number are parasitic, that is, living on or in other host organisms. The hosts for protozoan parasites range from simple organisms, such as algae, to complex vertebrates including man.

Marine protozoa make up part of the zooplankton, where they feed on algae of the phytoplankton. Other marine protozoa live attached to objects at the bottom of the water or along the shores. Freshwater species occupy similar habitats in lakes, streams, ponds, and even puddles of stagnant water.

On land, protozoa are abundant in the soil as well as in and on plants and animals. Some specialized protozoan habitats include the gut of termites, roaches, and ruminants such as cattle.

MORPHOLOGY

Although polymorphic life cycles are common in the protozoa, the mature stages of each species are characteristic of the class to which it belongs. The four classes are listed in Table 15-1.

Organelles of locomotion in protozoa are either flagella, cilia, or pseudopods, depending on the species. The cilia differ from flagella in that they are shorter and their basal bodies are connected by fibers which some believe may coordinate the ciliary motion. In ciliates the multiple cilia move in sequence, creating a wavelike beating. In addition to being organelles of locomotion cilia also help to propel food into the cell. Pseudopod formation in amebae is effected by streaming of the cytoplasm within the cell. Pseudopods, like cilia, are also used for food collection and ingestion.

Many protozoa form fibrillar or skeletal structures to give shape or rigidity to the cell. Slender fibrils in the cytoplasm of some species, such as *Amoeba proteus,* are associated with pseudopod-mediated motion. The ciliates commonly possess

TABLE 15-1
Some Properties of the Major Classes of Protozoa

Class	Mode of Motility	Mode of Reproduction
Mastigophora	Flagella (one or more)	Longitudinal fission
Sarcodina	Pseudopods (some have flagella also)	Binary fission
Sporozoa	Often nonmotile; may have flagella at some stages; creeping motion in some	Multiple fission; asexual reproduction in one host; some species have sexual reproduction in a second host
Ciliata	Cilia (multiple)	Transverse fission, asexual; also sexual reproduction involving micronucleus

a thickened pellicle consisting of two outer layers of membranes with a less dense middle layer (Fig. 15-2).

Foraminifera of the class Sarcodina often have shells (sometimes called *tests*) impregnated with calcium or silicon compounds. Hard coverings known as agglutinated shells are made of an organic material in which granules of hard, inorganic materials are embedded. Frequently they are composed of polysaccharide material, similar to the chitin of fungal cell walls, encrusted with granules of inorganic silicon-containing quartz or calcium-containing limestone, as seen in Figure 15-3. One species of protozoa ingests sponge spicules and later deposits them in its agglutinated shell.

Periods of shell growth, alternating with rest periods when no shell is formed, give rise to many-chambered forms of shells enclosing some of the foraminifera (Fig. 15-4). Other exquisitely beautiful geometric forms are seen in the radiolarians and heliozoans (Fig. 15-4). These species produce radial skeletons that are within the cell and extend to the outside. They are usually composed of silica, but some species have skeletons made of strontium compounds. The spicules of the skeletons cause a low specific gravity of the whole organisms, allowing

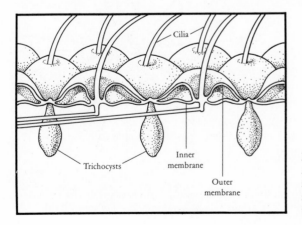

FIGURE 15-2
A cross section of the pellicle of *Paramecium* as seen by electron microscopy. The many cilia emerge through the inner and outer membranes of the pellicle. Undischarged trichocysts extend below the pellicle.

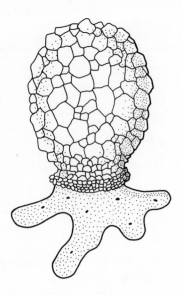

FIGURE 15-3

Difflugia, an ameba with an agglutinated shell (test) composed principally of small sand grains.

them to float in plankton. Flotation is also aided by the formation of globules of oil or fat storage products.

Certain of the protozoa have specialized intracellular organelles that function in defensive and offensive roles. For example, in some of the ciliates miniature dartlike structures (Fig. 15-5) are ejected from the cell as a defense mechanism. Other species contain similar organelles which can paralyze or poison their targets.

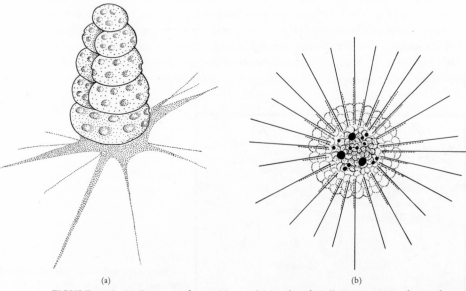

(a)   (b)

FIGURE 15-4 (a) Protozoan form with a multichambered shell, characteristic of some foraminifera. (b) A freshwater heliozoan. The marine radiolarians have forms similar to the heliozoan.

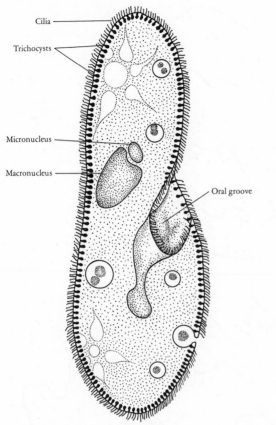

Cilia

Trichocysts

Micronucleus

Macronucleus

Oral groove

FIGURE 15-5
The morphology of *Paramecium*. The trichocysts are dartlike
structures that can be ejected as a defense mechanism.

NUTRITION

Protozoa are generally nonphotosynthetic and must assimilate preformed nutrients. Members of the classes Mastigophora and Sporozoa usually take in organic
materials through their cytoplasmic membranes. Other protozoa engulf and
ingest (phagocytize) particulate matter (Fig. 15-6). Ameboid organisms of the
class Sarcodina phagocytize solid food and enclose it in membrane-bound vesicles
containing a variety of enzymes. As a result, both food and the enzymes capable
of digesting it become enclosed in the same vacuole, where digestion takes place.
Products of digestion can then be used by the cell in various metabolic processes.
It is of interest to note that virtually the same sequence of events occurs in some
human cells that can phagocytize. Pinocytosis is a process analogous to phagocytosis, except that soluble materials, rather than particles, are engulfed in tiny
vacuoles (Fig. 15-6). Both pinocytosis and phagocytosis are used by many protozoa as a means of ingesting nutrients.

Vacuoles containing indigestible particles or undigested residue are ejected
through the cytoplasmic membrane in a reverse phagocytic process. In some species they are egested through a fixed area of the cell. Freshwater forms of protozoa

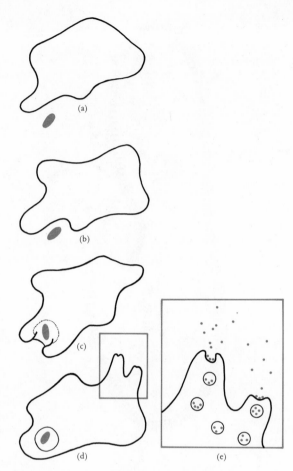

FIGURE 15-6

Phagocytosis and pinocytosis by an ameba. Phagocytosis (a through d) entails the engulfment and ingestion of particulate material. Pinocytosis (e), an enlargement of part of the cell shown in (d), is a similar ingestion of soluble molecules.

often contain contractile vacuoles which appear to function in regulating osmosis by discarding excess water or wastes.

Many of the protozoa are strict aerobes or facultative anaerobes. Only a few strictly anaerobic species have been identified. Thus aerobic metabolic pathways are frequently followed in utilizing absorbed or ingested nutrients. Excess nutrients are converted into food reserves of various kinds, such as carbohydrates, polyphosphates, and lipids.

ENCYSTMENT

Many protozoa have the capacity to form *protective cysts*. When conditions are unfavorable for continued growth, the *vegetative forms (trophozoites)* become dehydrated and enclosed within a thickened, warty covering. Some cysts have remained viable in dried materials for 40 or 50 years. This form of protozoa is important in dissemination, especially in a number of parasitic species, because it allows the organisms to survive until they reach a new, suitable host.

Protozoa may divide by means of binary fission, budding, or a process of *multiple fission* called *sporulation* (which is quite different from fungal spore formation). Sexual reproduction is also common, and a few specific examples are discussed in following sections. Some protozoa are diploid during most of their existence, in contrast to the algae which commonly have haploid vegetative cells.

REGENERATION

Many protozoa have remarkable abilities to regenerate when they are injured or mutilated. Some of the ciliates can regenerate a whole cell capable of reproduction from less than 10 percent of the original cell volume, provided the nucleus is intact. Besides its obvious advantage to the injured cell, this property has proven of value to biologists who study cell function.

BEHAVIOR

Certain components of protozoan cells can function as receptors for stimuli, causing the cells to express the property of irritability. For example, some protozoa (like certain algae) have pigment-containing eyespots that respond to light. In addition, cilia and flagella not only propel but can also act as tactile organelles, causing protozoa to retract from or advance toward various objects that are encountered.

SOME CHARACTERISTICS OF THE CLASSES OF PROTOZOA

Mastigophora

These are morphologically the simplest of the protozoa. They occupy diverse habitats, either free-living or as parasites. Typically the cells are ovoid or elongated, and are characterized by the presence of one or more flagella.

A well-known example of this class is the species *Trypanosoma gambiense,* which along with a closely related species is the causative agent of African sleeping sickness. These parasites infect tens of thousands of persons in the world each year. Their morphology is illustrated in Figure 15-7, as they appear in one stage of development in the human bloodstream. As do many other protozoa, *T. gambiense* has a complex life cycle (Fig. 15-8), and requires two hosts.

Sarcodina

Members of this class possess one or more pseudopods and are characteristically ameboid. Some have skeletons or shells. They reproduce by binary fission. The amebae, members of this class, do not reproduce sexually; however, others of the Sarcodina do.

Amoeba proteus, a freshwater species of Sarcodina (Fig. 15-9), is a relatively large protozoan, over 300 μm in length when extended (as compared with *T. gambiense* which is about 20 μm long). Extensive studies of phagocytosis and pinocytosis have been carried out using *A. proteus.* This has also been a useful organism for studying the relationship between nucleus and cytoplasm and the effects of each on cell function because *A. proteus* has a very large nucleus which can be removed and replaced with a nucleus from another ameba. Whereas *A. proteus* does not cause human disease, a few species of smaller amebae are pathogenic for man.

Also included in the class Sarcodina are the foraminifera with their multichambered shells (Fig. 15-4a); species with aggregation shells (Fig. 15-3); and the radiolarians and heliozoans (Fig. 15-4b), described previously.

Sporozoa

Some of the Sporozoa form pseudopods. Flagella are found only in gametes, if they occur at all. Many species can develop cysts which protect them during transfer from host to host. All members of this class are parasitic, some for a single host and others for more than one host during their life cycle. An example of those which live in a single host is *Monocystis* which parasitizes earthworms (Fig. 15-10). Examples of the group of Sporozoa requiring more than one host are the malarial parasites of the genus *Plasmodium.*

Malaria
Pages 538–541

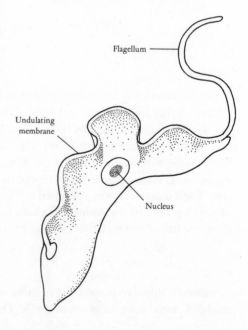

Flagellum

Undulating membrane

Nucleus

FIGURE 15-7
A trypanosome as it appears in the human bloodstream.

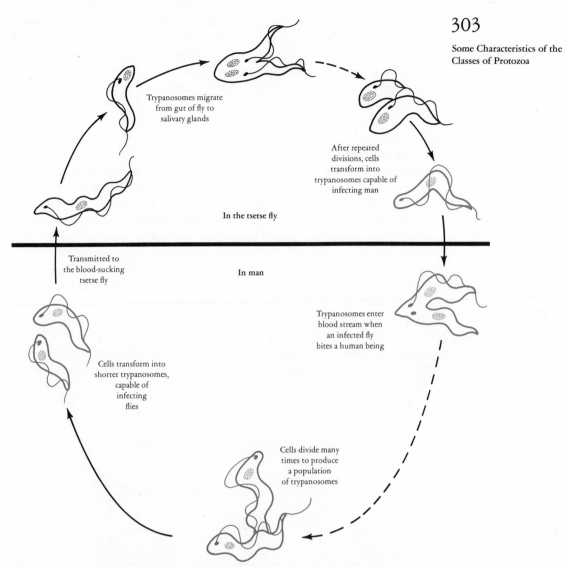

Trypanosomes migrate
from gut of fly to
salivary glands

After repeated
divisions, cells
transform into
trypanosomes capable of
infecting man

In the tsetse fly

Transmitted to
the blood-sucking
tsetse fly

In man

Trypanosomes enter
blood stream when
an infected fly
bites a human being

Cells transform into
shorter trypanosomes,
capable of
infecting
flies

Cells divide many
times to produce
a population
of trypanosomes

FIGURE 15-8 Diagram of the life cycle of a species of *Trypanosoma* that causes African sleeping sickness. Development within the blood-sucking tsetse fly is required in order to produce forms that can infect man; therefore the disease is transmitted only by the bite of an infected fly.

Ciliata

Ciliates represent the most complex, highly developed forms of unicellular organisms. About 6000 species are known. The class can be divided into two groups: the members of one group have cilia generally over only part of the cell, and members of the other group have cilia evenly distributed over the entire cell. *Paramecium* is a familiar example of the latter type.

The various species of *Paramecium* are generally found in fresh water. They are large protozoa, about 150 μm in length (Fig. 15-5). Parallel rows of cilia cover the exterior of the cell, and their rhythmic beating is responsible for the motility of the organism. When the *Paramecium* is not moving, cilia around the oral groove create a whirlpool effect which helps bring food into the mouth.

One important characteristic of the class Ciliata, readily seen in paramecia, is the presence of a large macronucleus and one or more small micronuclei. The macronucleus is also multiple in some ciliates. Both kinds of nuclei are necessary because the macronucleus is concerned with cell growth and most cellular functions, whereas the micronucleus is involved in cell heredity and sexual reproduction. The macronucleus probably has many copies of all of the genetic information for, in some species, it contains up to 500 times as much DNA as the micronucleus. Because they contain many copies of each chromosome, macronuclei cannot carry out the reduction process to produce haploid cells; therefore the micronucleus is necessary for the replication of genetic information and its distribution to daughter cells. The complex method of reproduction in *Paramecium* is shown in Figure 15-11. It can be seen that during asexual reproduction the cell divides by transverse fission and that conjugation of two cells occurs during sexual stages.

Interesting cytoplasmic inclusions are found in some paramecia. For example, *P. bursaria* is host to the green alga *Chlorella*. In this symbiotic relationship the alga has a safe haven and the paramecium benefits from algal photosynthesis. Other inclusions found in the cytoplasm of some paramecia are self-replicating particles of nucleic acid, probably remnants of bacteria. It is thought that origi-

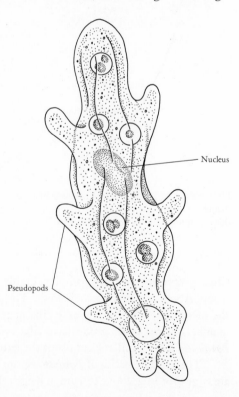

FIGURE 15-9

Diagram of *Amoeba proteus,* a large free-living ameba found in fresh water.

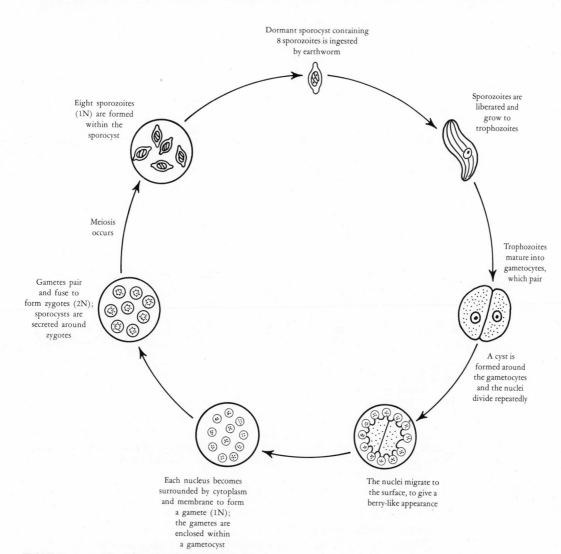

Dormant sporocyst containing
8 sporozoites is ingested
by earthworm

Eight sporozoites
(1N) are formed
within the
sporocyst

Sporozoites are
liberated and
grow to
trophozoites

Meiosis
occurs

Trophozoites
mature into
gametocytes,
which pair

Gametes pair
and fuse to
form zygotes (2N);
sporocysts are
secreted around
zygotes

A cyst is
formed around
the gametocytes
and the nuclei
divide repeatedly

Each nucleus becomes
surrounded by cytoplasm
and membrane to form
a gamete (1N);
the gametes are
enclosed within
a gametocyst

The nuclei migrate to
the surface, to give a
berry-like appearance

FIGURE 15-10   The life cycle of a species of *Monocystis,* a gregarine protozoan that infects earthworms. All of the stages are haploid except for the zygote.

nally certain bacteria lived in symbiosis within the paramecium and that during evolution these bacteria gradually lost portions of their cells until only part of the nucleic acids remains now, replicating along with the host cell. One example of these organelles is the kappa particle of *P. aurelia.* Paramecia containing kappa particles are called killers, because they produce a toxin that kills sensitive paramecia lacking the particles. Another example of these symbiotic particles, the lambda particle, has been grown *in vitro* and shown to be a gram-negative rod-shaped bacterium.

Another very fascinating ciliate is *Stentor.* These large (about 1 mm) trumpet-shaped protozoa are barely visible to the unaided eye, but a view through a low-power microscope reveals a beautiful symmetry and gracefulness

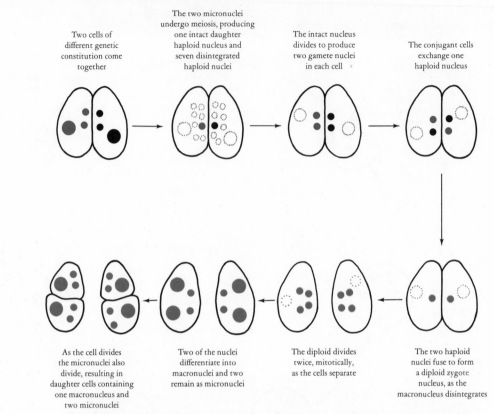

Two cells of different genetic constitution come together

The two micronuclei undergo meiosis, producing one intact daughter haploid nucleus and seven disintegrated haploid nuclei

The intact nucleus divides to produce two gamete nuclei in each cell

The conjugant cells exchange one haploid nucleus

As the cell divides the micronuclei also divide, resulting in daughter cells containing one macronucleus and two micronuclei

Two of the nuclei differentiate into macronuclei and two remain as micronuclei

The diploid divides twice, mitotically, as the cells separate

The two haploid nuclei fuse to form a diploid zygote nucleus, as the macronucleus disintegrates

FIGURE 15-11 Reproduction in *Paramecium* species can occur by the process of conjugation. Cells of different genetic constitution are indicated by the color of the micronuclei. The daughter cells produced after conjugation contain a mixture of genes from each of the original conjugating cells.

of movement. They are large enough to be worked with easily and consequently have been studied extensively.

The only ciliate that is pathogenic for man is *Balantidium coli* (Fig. 15-12). It lives in the gastrointestinal tract, and both vegetative forms (trophozoites) and cysts are passed in the feces. The cysts can survive in soil or water, often long enough to be transmitted to another human host.

PRACTICAL IMPORTANCE

The protozoa are important to man in many ways. Many participate directly in the food chain, largely by feeding on bacteria and algae and in turn serving as food for larger species. They play a major role in maintaining an ecological balance in the soil by devouring tremendous numbers of bacteria and other microorganisms. For example, a single paramecium can easily eat as many as 5 million bacteria in one day. This function of protozoa is important in sewage disposal

306

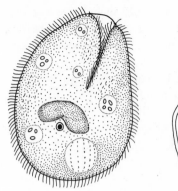 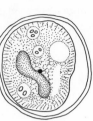

FIGURE 15-12
Balantidium coli, the only ciliate known to infect man. Tropho-zoite (left) and cyst (right) forms.

also, because most of the food consumed is used for energy; the solids are largely converted to carbon dioxide and water, resulting in a large decrease in total solids.

The major threat posed by protozoa derives from their ability to parasitize, and often to kill, a wide variety of animal hosts. In the United States, human infections with the sporozoan *Toxoplasma gondii* and the flagellate *Trichomonas vaginalis* are common. In many parts of the world, malaria and trypanosome infections are prevalent. Malaria alone has been one of the greatest killers of man throughout the ages. At present, at least 150 million people in the world contract malaria each year, and 1½ million die of it. These are conservative estimates; one authority states that today over 300 million people have some kind of malarial infection.

SUMMARY

The protozoa are eucaryotic protists which lack chlorophyll and cell walls, although some of them have a thickened pellicle or a shelllike covering. They are widespread in aquatic and terrestrial habitats and many species are parasitic, their various hosts ranging from simple algae to man.

Most of the protozoa are aerobic or facultatively anaerobic and only a few strictly anaerobic species have been identified. Nutrients are frequently ingested by phagocytosis or pinocytosis and wastes excreted by the reverse process. Many protozoa have the capacity to form protective cysts which enable them to survive for long periods in an unfavorable environment. Reproduction may be either asexual, sexual, or both, depending on the species and environmental conditions.

Some protozoa have cellular components which permit them to respond to stimuli therefore expressing the property of irritability. Cilia or flagella propel cells which possess them and also act as tactile organelles.

The four classes of protozoa are Mastigophora, Sarcodina, Sporozoa, and Ciliata. Organisms of the class Mastigophora are characteristically ovoid or elongated and possess flagella. An example is *Trypanosoma gambiense* which causes

African sleeping sickness. Members of the class Sarcodina include the amebae, foraminifera, radiolarians, and heliozoans. All of the class Sporozoa are parasitic. The malarial parasite is an example. Protozoa of the class Ciliata possess cilia, and are exemplified by the familiar *Paramecium*.

Protozoa are of practical importance to man in many ways. They participate in the food chain and play a major role in maintaining an ecological balance in the soil. Protozoa have been valuable as subjects for biological research. Their major threat to mankind comes from the ability of some of them to parasitize human and animal hosts, causing disease or death.

QUESTIONS

1. How do protozoa differ from algae and fungi?
2. Discuss several modes of locomotion that occur among the protozoa.
3. Describe several protective mechanisms that various protozoa have developed.
4. Compare and contrast a paramecium and a typical bacterium.
5. Consider the kinds of information that can be gained from experiments using *Amoeba proteus*.

FURTHER READING

CHEN, T. T. (ed.), *Research in Protozoology,* Vols. I–IV. New York: Pergamon Press, 1967–1971.

CORLISS, J. O., "Systematics of the Phylum Protozoa," in Florkin, M. and B. T. Scheer (ed.), *Chemical Zoology,* Vol. I. New York: Academic Press, 1967, pp. 1–20.

CURTIS, H., *The Marvelous Animals—An Introduction to the Protozoa.* New York: Natural History Press, 1968.

DOGIEL, V. A. *General Protozoology,* Rev. ed., trans. by G. I. Poljansky and E. M. Chejsin. London: Oxford University Press, 1965.

GARNHAM, P. C. C., *Malaria Parasites and Other Haemosporidia.* Oxford: Blackwell Scientific Publishers, 1966.

HALL, R. P., *Protozoa—The Simplest of All Animals.* New York: Holt, Rinehart and Winston, Inc., 1964.

JAHN, T. L. and F. F. Jahn, *How to Know the Protozoa.* Dubuque, Iowa: William C. Brown Company, Publishers, 1949.

KIDDER, G. W. (ed), *Protozoa,* Vol. I of Florkin, M. and B. T. Scheer (ed), *Chemical Zoology.* New York: Academic Press, 1967.

KUDO, R. R., *Protozoology,* 5th ed. Springfield, Ill.: Charles C Thomas, Publishers, 1966.

PITELKA, D. R., *Electron-Microscopic Structure of Protozoa.* New York: Pergamon Press, 1963.

TARTAR, V., *The Biology of Stentor.* New York: Pergamon Press, 1961.

VICKERMAN, K. and F. E. G. Cox, *The Protozoa.* Boston: Houghton Mifflin Company, 1967.

Chapter 16

Viruses: Their Nature and Methods of Study

During the late nineteenth century many infectious agents were identified as being bacteria, fungi, or protozoa. Most of these were readily seen with the microscopes available at that time and often they could be cultured *in vitro*. It was observed, however, that some infectious agents (such as those responsible for pox diseases of man, foot and mouth disease of cattle, and mosaic disease of tobacco plants) could not be grown outside of host cells, that is, were obligate intracellular parasites. They were too small to be seen with the light microscope—in fact so small that they were able to pass through filters which retained almost all known bacteria. Therefore these agents were at first called filterable viruses, virus being a general term (meaning poison) that had been applied to any and all infectious agents. Soon the adjective was dropped and the word virus came to designate only these filterable, obligately intracellular, infectious agents. For years it was erroneously thought that viruses were simply very small bacteria, and many unsuccessful attempts were made to grow them in cell-free media.

As experimental data accumulated it became apparent that viruses have certain features that are characteristic of complex chemical substances rather than of cells. For example, it was found that tobacco mosaic virus could be precipitated from a suspension by ethyl alcohol and still remain infective; yet this sort of treatment destroyed the infectivity of bacteria and other cells. In 1935 tobacco mosaic virus was crystallized, emphasizing that the physicochemical properties of viruses and cells are quite different, because crystals can be formed only by purified chemical compounds and not by cells or by mixtures of dissimilar molecules. Nevertheless the crystallized tobacco mosaic virus retained the capacity to infect cells and cause the replication of more virus—clearly properties associated with life.

309

Efforts to learn more about the nature and characteristics of these curious agents, some of which infected animals and other plants, were greatly aided by the discovery of yet other viruses which infect only bacteria. Early in the twentieth century, Twort in England and d'Hérelle in France, working independently, demonstrated filterable agents which infect and destroy bacteria. These *bacterial viruses* became known as *bacteriophages,* or simply phages. Because of the relative ease of growing bacterial cells, phages can be cultivated much more readily than many other viruses that infect animals or plants. Therefore bacteriophages have been studied extensively and the knowledge gained has contributed enormously to an understanding of viruses in general and to a degree understanding of all organisms at the molecular level.

Finally, the contributions of the electron microscope to virology cannot be overemphasized. Since only a few of the largest viruses are visible in the light microscope, the development of techniques for staining and visualizing viruses by electron microscopy has been vital for an appreciation of virus morphology and its relation to the infective process.

THE NATURE OF VIRUSES

As suggested by the early work with tobacco mosaic virus, and proven by innumerable other experiments and observations, all viruses are chemical particles which differ from cells in many ways. Each virus particle, called a *virion,* consists of a single type of nucleic acid (DNA or RNA) surrounded by a protein coat (*capsid*) and in some cases also by an outer layer or envelope which contains carbohydrates and lipids (Fig. 16-1). The presence of *either DNA or RNA, but not both,* is an important property which distinguishes viruses from cells. Thus there are RNA viruses and DNA viruses. *The nucleic acid may occur as either double-stranded DNA, single-stranded DNA, single-stranded RNA or double-stranded RNA.* Other important features which distinguish viruses from cells include the lack of components necessary for energy generation and protein synthesis (for example, ribosomes). Certain viruses do contain enzymes involved in nucleic acid synthesis. Generally, however, the enzymatic capabilities of viruses are extremely limited and are confined to enzymes involved in their entry into cells and replication of their own nucleic acid. Thus viruses are unable to replicate independently, but they must invade a living cell and utilize the cellular ribosomes, energy sources, and certain other components in order to replicate new virions. This process frequently causes profound changes in the host cell, often death of the cell.

The replication of viruses is unique. Whereas a cell customarily divides to give rise to two cells identical to the original, a single virion can direct the synthesis of dozens or even up to many thousands of similar virions within a host cell. It is characteristic of all viruses that at one stage during replication they exist as nucleic acid only, devoid of their protective protein capsid and envelope. A comparison of the properties of viruses and cells is summarized in Table 16-1.

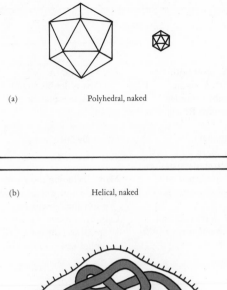

(a)   Polyhedral, naked

(b)   Helical, naked

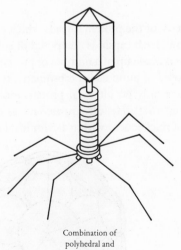

(c)   Enveloped

(d)   Combination of
polyhedral and
helical, naked

FIGURE 16-1
Virus morphology.

TABLE 16-1

Comparison of Viruses and Cells

	Viruses	Cells
Nucleic acid composition	Either DNA or RNA, never both (double-stranded DNA, single-stranded DNA, double-stranded RNA, or single-stranded RNA). No ribosomes	RNA *and* DNA; DNA is double-stranded; Ribosomes present
Reproduction takes place	Only in living cells: animal cells, plant cells, bacteria, algae, and fungi. Obligate intracellular parasites	Generally "free-living"
Enzyme content	Very few if any. Most concerned with penetration into and exit from host	Many, including those concerned with energy metabolism and biosynthesis
Replication	Nucleic acid and protein separate and are synthesized independently of each other	Cellular integrity is maintained and the entire cell replicates as a single unit

VIRUS ARCHITECTURE

Virus particles are generally either polyhedral or helical in structure, or sometimes rather complex combinations of these two shapes (Fig. 16-1). Polyhedral viruses often appear to be almost spherical. Closer examination, however, shows that the capsids are actually composed of identical subunits arranged in patterns of icosahedral symmetry (that is, 20-sided polyhedrons in which each side is an equilateral triangle). The American architect, Buckminster Fuller, has popularized the same icosahedral structure in his famous geodesic domes. He has found that this sort of arrangement permits a given volume to be enclosed by identical subunits having the smallest surface area.

The symmetry of viruses is a property of the protein capsids which enclose and protect the viral nucleic acid genome. Each capsid is composed of subunits called *capsomeres*. In turn each capsomere is made up of a number of protein molecules. Although a capsid may be composed of hundreds of capsomeres, the simplest possible icosahedral virion contains only 60 identical protein molecules arranged in five identical capsomeres (Fig. 16-2). Helical viruses, such as the tobacco mosaic virus (Fig. 16-3), consist of nucleic acid within a cylindrical capsid

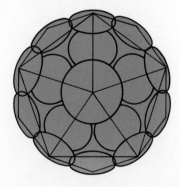

FIGURE 16-2

Diagram of an icosahedral virion, showing the symmetrical arrangement of the capsomeres.

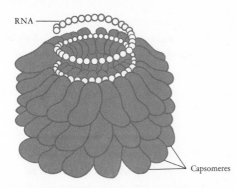

RNA

Capsomeres

FIGURE 16-3
Tobacco mosaic virus has a helical symmetry. The capsid, formed
of many capsomeres, surrounds the helical strand of viral RNA.

composed of many identical capsomeres in a spiral arrangement. Many virions
are of much more complicated morphology (Fig. 16-1c). The nucleic acid of
some viruses, the *enveloped viruses,* is contained within a helical or polyhedral protein capsid, which in turn is surrounded by a membranous outer structure. The
membranous structures may be complex and consist of several layers of lipid and
protein.

Some of the bacterial viruses are also structurally complex. For example,
T-even bacteriophage which infects certain strains of *Escherichia coli* is composed
of a polyhedral head attached to a helical hollow tail (Fig. 16-1d). The nucleic
acid of this phage is a single molecule of double-stranded DNA which is tightly
packed within the head. Both the efficiency of packing and the incredible amount
of DNA contained within the tiny virion are indicated in Figure 16-4, which
shows DNA released by gentle disruption of the phage head.

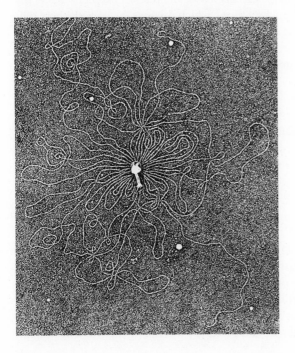

FIGURE 16-4
Osmotically shocked T-even bacteriophage. The single molecule
of DNA containing the genetic information of the phage
has been released from the phage coat (center). (Courtesy of
Dr. A. K. Kleinschmidt; from Kleinschmidt, A. K., D. Lang,
D. Jacherts, and R. K. Zahn, *Biochim. Biophys. Acta,* **61**:857,
1962)

Viruses are of many different size classes. The smallest ones are similar in size to large protein molecules or ribosomes, and their nucleic acid content is only a few genes. The more complex enveloped virions may be larger than some of the most minute bacteria.

VIRUS REPLICATION

The basic steps in replication are similar for all viruses that have been examined, whether they infect bacterial, plant, or animal cells. First the viral nucleic acid must enter the cell. In the case of bacterial and animal viruses the virion adsorbs specifically to receptors on the host cell. After adsorption, the viral nucleic acid penetrates into the cell. Either free nucleic acid (in the case of some bacteriophages) or whole virions (for all other viruses) enter the cell. Replication of viral nucleic acid and synthesis of other viral constituents follows. These constituents are made separately within the host cell and are then assembled into complete virions during the stage of virus maturation. The process is similar to assembly line production in a factory where automobiles, for example, are manufactured from parts made separately and then efficiently assembled. Finally, the newly formed mature virions are released from the host cell. The method of release varies depending on the virus in question: cells may be lysed, releasing many mature virions, or virus particles may be gradually extruded from the living host cell.

The details of replication in T-even phage have been studied extensively, both morphologically and on a molecular basis, and offer an excellent model for the general process of infection by a phage which destroys, by lysis, the host cell. The term T-even phage designates any of a group of phages which includes T2, T4, and T6. Data collected from studies of any one of these phages are generally applicable to all three. The steps described below are diagrammed in Figure 16-5.

Adsorption

When a suspension of T-even phage particles is mixed with a susceptible strain of *E. coli,* the phage collide by chance with bacteria. Fibers at the end of the tail are the adsorption sites of the phage which bind to specific receptors on the bacterial cell wall (Fig. 16-5).

Penetration

An enzyme (phage lysozyme) located in the phage tail degrades a small portion of the cell wall. The tail sheath then contracts and the tail core mechanically penetrates the cell wall. The tip of the phage tail must then open so that viral DNA in the head is free to pass through the channel of the phage tail. The DNA is literally injected through the cell wall and, by some unknown mechanism, penetrates the cytoplasmic membrane and enters the interior of the cell. The protein

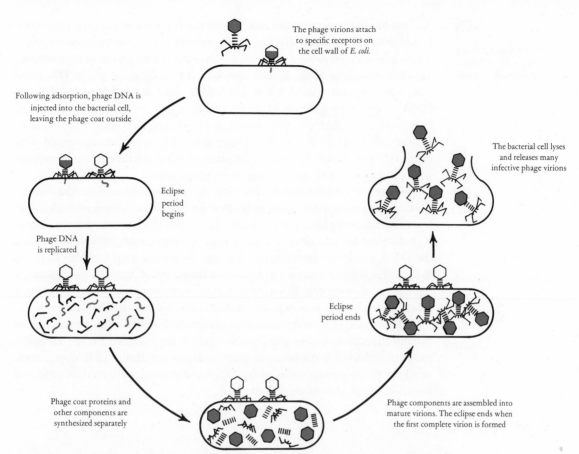

The phage virions attach to specific receptors on the cell wall of *E. coli.*

Following adsorption, phage DNA is injected into the bacterial cell, leaving the phage coat outside

Eclipse period begins

Phage DNA is replicated

Phage coat proteins and other components are synthesized separately

Phage components are assembled into mature virions. The eclipse ends when the first complete virion is formed

Eclipse period ends

The bacterial cell lyses and releases many infective phage virions

FIGURE 16-5   Steps in the replication of T-even phage during the infection of *Escherichia coli.*

coat of the phage remains outside the cell, having completed its functions. The head protein protects the DNA while it is outside a host cell, and the tail proteins serve to attach and penetrate the bacterial cell wall, allowing viral nucleic acid to enter the host cell.

Replication of Phage DNA

The introduction of the phage DNA has several immediate ramifications. Within minutes all transcription from the host chromosome ceases. In fact the host DNA is actually destroyed within minutes after infection. Subsequently all RNA synthesized is mRNA transcribed from the phage DNA. By this mechanism the phage subverts all metabolism of the bacterial cell to its own purpose—the synthesis of phage. Host enzymes continue to function, however. They supply energy for phage replication through the breakdown of glucose, as well as synthesize the subunits of protein and nucleic acid of the replicating phage; they even function in the synthesis of phage nucleic acid and protein coats. The production of phage proteins also requires the participation of bacterial ribosomes. In addition to enzymes of the host, many enzymes in the infected cell are *phage-specific, since they are encoded in the phage DNA and are not present in the uninfected cell.* Without these phage-specific enzymes, the synthesis of phage DNA and phage protein coats could not proceed. One such enzyme degrades the host DNA. Another phage-induced enzyme catalyzes the synthesis of a pyrimidine base (hydroxymethyl cytosine) which is incorporated in place of cytosine in T-even DNA. At least ten other phage-induced enzymes are concerned with phage DNA synthesis. In addition the phage DNA codes for enzymes concerned with the synthesis and the structure of the phage coat.

During the replication of any virus the protein and nucleic acid components separate completely from each other. In fact the separation of nucleic acid from its protective protein coat is an obligatory and characteristic feature of virus reproduction. In the case of T-even phages the components separate at the time the DNA is injected into the cell, leaving the protein coat outside. Thus the T-even phage DNA has two independent functions: it provides a template for replication of more viral DNA, and it serves as the template for mRNA which functions in the synthesis of phage-induced enzymes and capsid protein.

This mode of reproduction accounts for the fact that during one phase of the infection process, the *eclipse period,* there is no evidence for any complete phage particles inside the host cell. If bacterial cells are disrupted during the early period of phage reproduction, phage DNA and protein coats are detectable, but no whole phage virions are present.

Maturation

The maturation process involves the combination of phage protein and phage DNA to form the mature infective progeny virions. Maturation of T-even phage is a complex multistep process. The phage tail, built up in a specific stepwise fashion, attaches to the head, which has formed from the aggregation of protein subunits. Before the tail attaches, the head becomes packed with the long thin molecule of phage DNA (Fig. 16-6).

Release

During the latter stages of the infection period, another phage-induced enzyme (coded for by the phage DNA) makes its appearance. This is the phage *lysozyme* that digests the cell wall from within, resulting in cell lysis and release of many complete phage virions. It is significant that this enzyme is never synthesized early in the infection process, presumably because it would lyse the host cell before any mature phage could be formed. There are controls regulating the regions of the viral nucleic acid that will be transcribed into mRNA at any time in the infection process. There are "early" messages which are translated into "early" proteins (enzymes of DNA synthesis), and "late" messages for "late" proteins (coat proteins and lysozyme). This control of transcription is probably mediated

Intact *E. coli* cell at the time of infection. The light areas within the cell represent bacterial DNA.

By two minutes after infection changes in the bacterial DNA are apparent. Empty phage coats remain attached to the bacterial cell walls.

At eight minutes after infection, bacterial DNA has been destroyed and phage DNA is being made (light areas).

By 12 minutes, condensed phage DNA is apparent (dark areas).

At 30 minutes the bacterial cell is nearly full of condensed DNA and completed phage heads. After assembly of many mature virions, the bacterial cell lyses and releases the phage particles.

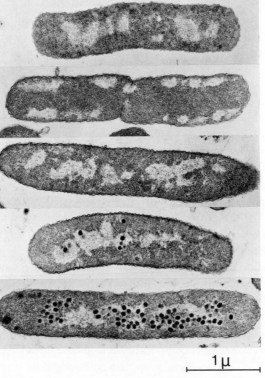

1 μ

FIGURE 16-6   Cells of *E. coli* sectioned for electron microscopy during infection with T-even phage. (Courtesy of Dr. E. Kellenberger)

by changes in the enzyme RNA polymerase which in some manner determine the site at which the polymerase will initiate transcription. Viruses which are then released infect available susceptible cells, and the process of virus replication is repeated.

RNA polymerase
Pages 175, 176

HOST RANGE OF VIRUSES

It is probable that any kind of cell can be infected by some viruses. To date, viruses have been shown to infect cells of at least some species of algae and fungi, and of most species of bacteria, plants, and animals. However, any given virus has a limited host range and generally will infect cells of only one or a few species, and often just a few strains of a species. Thus the T-even phage will infect only certain strains of *E. coli,* and polioviruses naturally infect only cells of man and a few related higher primates. This limited host range reflects the fact that animal and bacterial viruses must interact with specific receptor sites on the host cell surface in order to invade the cell. The receptor site varies in chemical properties and location depending on the particular virus. In the case of phages the receptor site is usually a chemical constituent of the bacterial cell wall. Some phages, however, attach to receptors on pili and others to sites on flagella. Animal viruses attach to specific receptor sites on the cytoplasmic membranes of appropriate host cells. Mutations in host cells that result in altering or deleting receptor sites render the cells resistant to attack by the particular virus that can no longer attach and infect the cell.

The attachment of virus to host cell receptors is a chemical interaction. Its specificity depends on a close complementary fit between the molecules of the virus and the others of the receptor; the two are held together by many weak chemical bonds (such as hydrogen bonds and van der Waals forces) which form when they associate. The complementary fit must be close in order to allow formation of enough of these weak bonds to hold virus and cell together.

Weak bonds
Pages 26–31

In the laboratory it is possible to bypass the initial steps of adsorption and penetration and to infect cells with viral nucleic acid alone. For example, the RNA of polioviruses will cause virus replication when introduced into its usual host cells and also in a number of additional cell types which the whole virion cannot penetrate; however, in certain other cells it will not cause virus production, even though it can be proven that the poliovirus RNA enters the cell. Therefore it is apparent that specific adsorption merely helps viruses to penetrate barriers such as cell walls and cytoplasmic membranes. Although specific adsorption is important in determining host range, other factors also play a part; for example, the ability of the viral nucleic acid to survive within the host cell and utilize cellular components during viral replication.

These general principles governing host range of viruses have been studied on the molecular level with the T-even bacteriophages. The chemical nature of host cell receptors and phage attachment sites has been elucidated. In addition the host range of these phages may also be determined by the chemical structure of their DNA. The T-even DNA contains glucose bonded to specific areas of the molecule (glucosylated DNA). Several enzymes are required for the incorporation of glucose into the DNA; some are bacterial and others are phage induced. Normally, host *E. coli* cells contain these necessary bacterial enzymes and glucosylated phage DNA is replicated. However, if the phage invades *E. coli* cells which lack these enzymes because of a mutation in the bacteria, the DNA of the progeny phage is not glucosylated. The progeny phage cannot replicate in new host cells because *E. coli* has an enzyme (a DNAse) which degrades any DNA which is not glucosylated. This enzyme rapidly destroys progeny phage DNA lacking attached glucose. The phage can replicate in certain other gram-negative rods. These observations indicate that *the bacterial host is capable of modifying the structure of phage DNA and consequently the host range of the progeny phage* (Fig. 16-7). Other examples of this phenomenon of *host-controlled modification* of viruses have been studied in bacteriophage systems. However, analogous examples in animal and plant virus systems have not yet been described. It is significant to note that the DNAse of *E. coli* mentioned above distinguishes between DNA molecules that are apparently identical except for the presence or absence of glucose at areas on the large molecule. This provides the cell with a mechanism for selectively destroying any foreign DNA which may enter.

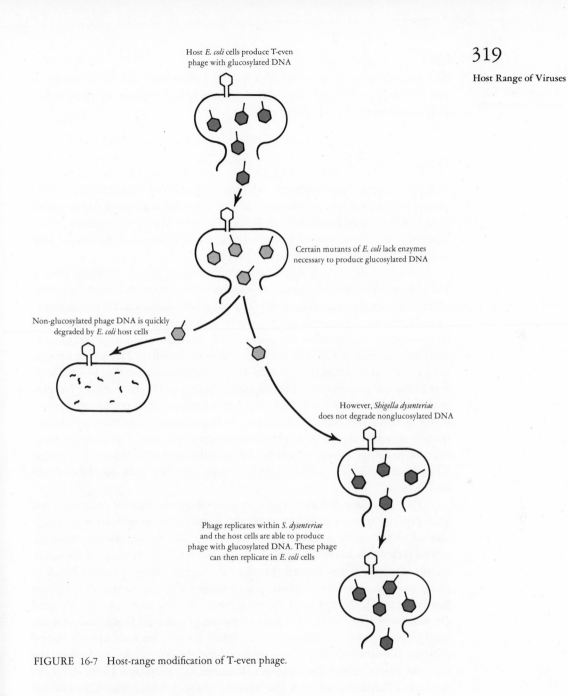

Host *E. coli* cells produce T-even
phage with glucosylated DNA

Certain mutants of *E. coli* lack enzymes
necessary to produce glucosylated DNA

Non-glucosylated phage DNA is quickly
degraded by *E. coli* host cells

However, *Shigella dysenteriae*
does not degrade nonglucosylated DNA

Phage replicates within *S. dysenteriae*
and the host cells are able to produce
phage with glucosylated DNA. These phage
can then replicate in *E. coli* cells

FIGURE 16-7 Host-range modification of T-even phage.

VIRUS-HOST RELATIONSHIPS

Relationships between viruses and host cells are poorly understood in the case
of many viruses; however, in bacteriophages three major types are recognized: the
lytic, the lysogenic, and a state where viruses are released without cell lysis. Other
less well-defined relationships undoubtedly exist also.

Lytic

The lytic relationship is exemplified by the T-even infection described above. Here the virus is referred to as a *virulent bacteriophage* because infection results in lysis and death of the infected cell.

Lysogenic

Whereas virulent bacteriophages can only replicate and cause cell lysis, other phages exist which are capable of *either replicating and causing cell lysis or having their DNA integrated with host cell DNA.* These are known as *temperate phages.* The phage DNA when integrated with host cell DNA is called *prophage,* and the bacterial cell carrying a prophage is a *lysogenic cell.*

The most thoroughly studied temperate bacteriophage is the phage lambda (λ) of *E. coli.* This phage is similar in appearance to T-even, and the initial stages of infection are the same (Fig. 16-8). The phage adsorbs and its DNA penetrates into the interior of the cell, leaving the protein coat on the outside. Then one of two events occurs: either the phage DNA replicates and phage specific proteins are synthesized with a lytic cycle ensuing or the phage DNA may become integrated into a specific region of the bacterial chromosome (Fig. 16-9). Therefore in any culture infected with temperate phage, some bacteria will be lysogenized and others will be lysed. As a prophage the phage DNA is replicated as part of the bacterial chromosome and gives little evidence for its existence; consequently it is difficult to know when a bacterium is lysogenic. Occasionally, however, the prophage leaves the host cell DNA and begins to replicate whole phage. Thus the release of infectious phage in a culture is the best indicator that the cells are lysogenic.

For some time this ability of temperate phages to become integrated and their capacity to exist as either integrated prophage or as ordinary lytic phage seemed very mysterious. During the last two decades the availability of sophisticated techniques has made it possible to gain some understanding of the mechanisms involved. It was learned that short nucleotide sequences in the DNA of phage and host are either identical or quite similar, allowing pairing of the phage and bacterial DNA, and then the integration of the phage into the bacterial chromosomes (Fig. 16-9). One gene of the integrated viral DNA codes for the synthesis of a protein repressor that is released into the bacterial cytoplasm and prevents the synthesis of proteins required for the lytic cycle. As long as the repressor is synthesized the phage is maintained as a prophage. However, if the synthesis of repressor stops or the repressor protein is inactivated, an enzyme coded for by the virus is synthesized which excises the viral DNA from the bacterial chromosome. This excised DNA replicates, phage proteins are synthesized, mature phage is formed, and the cell lyses. Under ordinary conditions of bacterial growth the release of the phage DNA is a rare event, occurring perhaps once in 10,000 divisions of a lysogenic bacterium. However, if a lysogenic culture is treated with any agent that interferes with DNA replication (such as ultraviolet

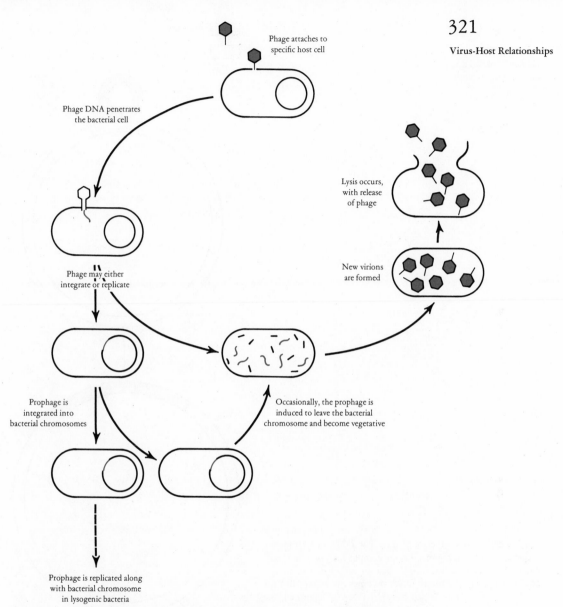

Phage attaches to
specific host cell

Phage DNA penetrates
the bacterial cell

Phage may either
integrate or replicate

Prophage is
integrated into
bacterial chromosomes

Prophage is replicated along
with bacterial chromosome
in lysogenic bacteria

Occasionally, the prophage is
induced to leave the bacterial
chromosome and become vegetative

New virions
are formed

Lysis occurs,
with release
of phage

FIGURE 16-8   Temperate phage cycle. The lysogenic cycle is diagrammed on the lower left and
the lytic cycle on the right.

light), all of the prophage will enter the lytic cycle. This process, termed *induction,* results in complete lysis of the culture.

The repressor produced by a prophage prevents the infection of a lysogenic cell by the same phage carried by the lysogenic cell. The cell is "immune" to infection by this phage but not other phages.

Except for this property of being immune to reinfection with the same phage, lysogenic and nonlysogenic cells usually appear identical. However, in a

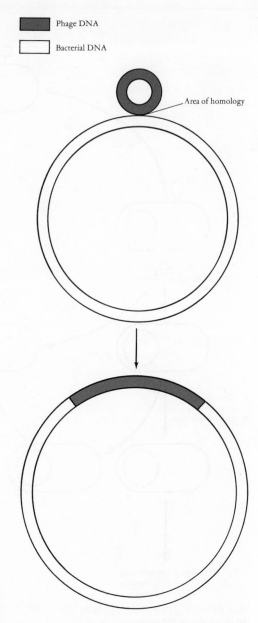

Phage DNA

Bacterial DNA

Area of homology

FIGURE 16-9
The integration of temperate phages. The DNA of a temperate
phage contains some regions that are complementary to host bacterial DNA (areas of homology). Base-pairing occurs in an area
of homology and the phage DNA becomes integrated into the
bacterial chromosome. When bacterial DNA is replicated, the
phage DNA is also replicated.

few well-documented cases, prophages have had profound effects on the properties of the host cell, a phenomenon termed *lysogenic conversion.* For example,
Corynebacterium diphtheriae, the causative agent for diphtheria, can only synthesize the toxin responsible for this disease if it is lysogenized with a particular
phage. Similarly, only certain strains of lysogenic Group A streptococci produce
the toxin that causes scarlet fever, and a recent report indicates botulinum toxin
may be synthesized only by lysogenic strains of *Clostridium botulinum.* In all of
these cases loss of the prophage results in loss of the toxin-producing ability.

Diphtheria
Pages 449, 450;
scarlet fever, Page 448

Temperate bacteriophages are episomes and their relations with host cells bear many similarities to the F⁺–Hfr situation. As the Hfr can be excised from the chromosome and become an F⁺ particle, the prophage can be excised from the bacterial DNA to generate the replicating phage termed *vegatative phage*. In a few cells a piece of the bacterial DNA remains attached to the piece of phage DNA which is excised. The bacterial genes replicate as part of the phage DNA, become incorporated into mature phage, and are released following lysis of the bacterial cell. When the phage infects another bacterial cell, both phage and bacterial DNA may be integrated into the new host chromosome. Thus a lysogenic cell is produced which now has bacterial genetic information from the previously lysogenized cell. This mechanism of genetic transfer is called *transduction* (Fig. 16-10). *Lambda* is a restricted or specialized transducing phage because it always becomes integrated at a particular region on the host DNA and can transfer only the limited group of genes that are physically linked to the region in which the prophage is integrated.

Episomes
Page 200; F⁺-Hfr,
Pages 197–200

Other phages, the generalized transducing phages, can transfer any region of the bacterial chromosome. In this latter case bacterial genes are accidentally packaged into the phage heads in place of phage DNA in the course of phage maturation. The heads often contain only bacterial DNA (Fig. 16-11). Thus, in generalized transduction the phage DNA is not necessarily integrated into the host cell chromosome as it is in specialized transduction.

Transducing phages are termed *defective phages* since the phage DNA is never complete. In specialized transducing phage, such as lambda, the bacterial genes which are integrated into the phage DNA replace a few phage genes. In generalized transducing phage a segment of the bacterial DNA is packaged in the phage head in place of part or all of the phage DNA. The resulting defective phage cannot proceed through its entire replicative cycle because it lacks some of the genetic information required for various parts of the cycle. For example, if the gene that specifies the synthesis of lysozyme is not present, the phage will not be released from the infected cell. In order for a defective virus to complete its infective cycle, it is necessary that the missing genes be supplied. This can be done by infecting the same cell in which the defective virus is replicating with another virus particle which can supply the missing functions. The second phage in this example can supply the functional gene for the synthesis of lysozyme, thereby lysing the cell and allowing defective phage to escape from the infected cell (Fig. 16-12). It is highly probable that infection of cells in nature by phages and other viruses may be far more common than it appears to be because special efforts are necessary to recognize defective viruses.

Usually only a few of the total number of phage particles carry bacterial genes; nevertheless the process of transduction is important as a means of genetic transfer. In contrast to mutation in which a single gene is changed at a time, during transduction a number of genes may be incorporated at one time resulting in several changed characteristics in the cell. Even though it is a rare event— perhaps occurring with about one in a million lambda phages—transduction can lead to a rapid change in the characteristics of a host bacterial population if the genetically changed bacteria are more capable of surviving and thus are selected.

The lysogenic relationship is of profound importance in other ways. Lysogenic conversion has proven to be of significance in the production of the few human diseases mentioned above (diphtheria, scarlet fever, and botulism) and it is probable that other similar situations remain undiscovered. Perhaps most important is the fact that a number of viruses which cause cancer in animals may

The DNA of a temperate phage penetrates into the bacterial host cell

The phage DNA may become integrated with host cell DNA as a prophage

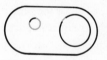

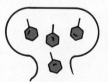

When the prophage is induced to leave the bacterial chromosome it may exchange a bit of DNA, thereby carrying with it a few bacterial genes

The phage that are replicated are defective because they lack viral genes that have been replaced by bacterial DNA

The defective phage DNA penetrates new host cells but cannot cause the production of new phage particles

Bacterial genes introduced into the new host cell are integrated into the DNA, become a part of the bacterial chromosome, and are replicated along with the rest of the bacterial DNA

FIGURE 16-10
Specialized transduction by phage.

When phage infects a host cell it may cause the degradation of host DNA into small fragments.

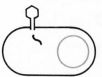

During maturation of the virions, a few phage heads may envelope fragments of bacterial DNA instead of phage DNA

When this bacterial DNA is introduced into a new host cell it can become integrated into the bacterial chromosome, thereby transferring several bacterial genes at one time

FIGURE 16-11
Generalized transduction by phage.

be defective viruses. For this and other reasons it is strongly suspected, although not proven, that some human cancers may be caused by viruses which can establish a state in man analogous to lysogeny in bacteria.

Virus Release Without Cell Lysis

Both T-even and lambda lyse, and therefore kill, the host cell when they are released. A group of bacterial viruses are known which are released without harming the host cell. These are represented by the filamentous phages. They have a long protein coat enclosing one molecule of single-stranded DNA. Their hosts are gram-negative bacteria which possess sex pili. These phages adsorb to the tips of the pili, but it is not clear how they reach the inside of the cell after they adsorb. However, once inside, the viral DNA (separated from the protein) replicates and viral protein is synthesized. Several aspects distinguish this infection

Defective viruses (those that lack a functional gene) cannot replicate all viral components. Consequently, no complete virions are formed in cells infected with a defective virus.

In cells infected with different defective viruses, complementation can occur. Each virus supplies the gene missing from the other virus and complete virions are formed

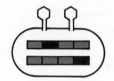

FIGURE 16-12

Complementation. Defective phages can cooperate to produce complete virions.

from infection by T-even phages. The virus does not completely take over host metabolism since the infected bacteria continue to multiply. Also, neither an intracellular pool of virus nor even the phage coat can ever be detected in the cytoplasm. This suggests that the phage coat once synthesized is stored in the membrane of the cell, and the virus is assembled as it is extruded from the cell. Virus is continuously extruded and is not released in a burst as is T-even phage when the cell lyses. Infected bacteria are *carrier cells;* they can be subcultured and stored in the same way as noninfected strains, and they will continuously release virions into the medium. Many animal viruses also release virions without killing the cell.

Nonlytic release of animal viruses
Pages 339, 341

SOME METHODS USED TO STUDY VIRUSES

Virus Purification

For many purposes it is necessary to have quantities of purified viruses. These include chemical and physical studies as well as the production of vaccines. With bacterial or plant viruses this is usually relatively easy since it is possible to obtain large quantities of infected cells. The purification process consists of separating the viral particles from the constituents of the host cell. One method that is often used involves breakage of the host cells and sedimentation in an ultracentrifuge. Centrifugation at low speed removes particles of cell debris, whereas higher speeds of centrifugation are necessary to sediment the smaller viral particles.

In animal viruses the primary difficulty is not so much in purifying the virions but in obtaining enough infected cells. Some viruses still defy efforts to cultivate them except in animal hosts, while others are more conveniently grown in cell cultures or embryonated chicken eggs.

Culture of Animal and Human Host Cells

Animal and human cells have been grown under laboratory conditions since the early part of the century. However, the methods have become routine only during the last 25 years since the advent of antibiotics (used to suppress growth of bacteria and fungi in cell cultures) and of more sophisticated aseptic techniques. Currently almost any type of animal cell that normally divides in the body can be grown or maintained in culture. Such cell cultures provide the most convenient means for studying the growth of animal viruses or producing them in large quantities.

Animal cells may be cultured *in vitro* in several different ways. Slices or fragments of organs can be kept alive for days or weeks before they begin to die. During this period such organ cultures may act as hosts for virus growth. More commonly, animal tissue is dissociated into suspensions of single cells by homogenization or brief treatment with the proteolytic enzyme, trypsin. These suspensions may be cultivated as cell cultures of two distinct types. In a *monolayer culture* the cells settle down on the surface of a glass or plastic container, adhere, and begin to divide. Under the best conditions a new generation of cells is produced by division at approximately daily intervals. When normal cells in culture contact other cells they are inhibited in their growth, a phenomenon called *contact inhibition.* As a result many normal cells tend to grow in a monolayer attached to the surface of the culture vessel. In other cases, depending on the cell type or the constituents of the growth medium, the cells do not adhere to the surface but remain in suspension. To prevent cells from settling out on the bottom of the vessel, these *suspension cultures* normally are stirred with a magnetic stirrer. The composition of media for cell culture depends on the type of cell in question and may be very complex. Often blood serum must be added.

Cell cultures differ considerably in longevity. Many cells are either unable to divide or are able to divide only a few times in culture before dying. At the other extreme some cells have been propagated continuously for thousands of cell generations over more than 25 years, and are seemingly immortal. These differences in lifetime provide a convenient classification of cell cultures into three broad categories. *Primary cell cultures are those initiated with cells taken directly from the body.* They have only a limited growth period *in vitro* and normally consist of a mixture of cell types. Monkey kidney primary cell cultures are com-

Properties of tumor cells
Pages 347–349

monly used as hosts for the growth of poliovirus and a number of other animal viruses. Some types of cells (such as fibroblasts from fetal tissues) may be established as cell cultures which survive for about 100 cell divisions. At the end of this period they inevitably die for reasons that are yet obscure, but possibly connected with senescence. Since the normal chromosome number is maintained throughout this period of growth, they are often referred to as *diploid cell cultures.* The term *continuous cell line* is given to a culture of cells that divides indefinitely. These cell lines may be established from cancers or from variants occurring in diploid cell strains as a result of mutation. In addition, certain types of animal virus can convert or transform diploid cells into continuous line cells. These continuous line cells have several properties in common with cancer cells and are distinguishable from normal cells. They will, in fact, often give rise to a tumor if injected into animals. Furthermore, their morphology and biochemical abilities are quite different from the normal cells from which they have derived. In most cases they have abnormal numbers of chromosomes and changes in chromosome morphology. Despite these abnormalities continuous cell lines are extraordinarily useful to the virologist, because they are easy to propagate and will support growth of many different animal viruses.

Before the development of methods for cell culture, the most useful host for the cultivation of animal viruses was the embryonated chicken egg. Embryonated eggs are infected through an opening made in the egg shell about a week or two after fertilization. Depending upon the virus in question, growth occurs in one or more of the embryonic membranes or the yolk sac or in the embryo itself.

Quantification of Viruses

For any detailed study of viruses it is necessary to develop methods for accurately determinating their number. The methods used vary greatly, ranging from the counting of virions to studying the consequences of their interaction with living cells. A summary of the most commonly used methods follows.

Electron Microscopic Counting If reasonably pure preparations of virions are obtainable, their concentration may be readily determined by counting particles in a specimen prepared for the electron microscope. A great inherent disadvantage in this method is that it cannot distinguish between infective and noninfective viral particles.

Plaque Method The detection of plaques is at once the most flexible, the most general, and the most informative method for quantification since it is a simple and reproducible assay for infectivity. The method was first perfected for bacterial viruses and consists of the following operations. A sample of bacteriophage is mixed with a concentrated suspension of host bacteria and a few milliliters of melted agar at about 44°C. The agar is then poured into a petri plate, where it solidifies into a thin sheet containing a random distribution of viruses and bacte-

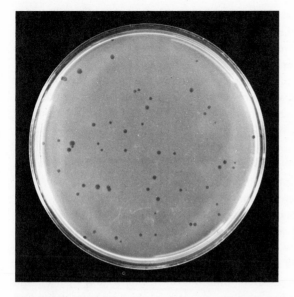

FIGURE 16-13
Phage plaques. The opaque background represents a lawn of bacteria. Each clear area is a plaque formed as a result of lysis of the bacteria by the progeny of a single phage particle.

ria. Each viral particle infects a bacterium, multiplies, and releases several hundred new virions. These infect other bacteria in the immediate vicinity, which again release viral particles. After a few multiplication cycles, all the bacteria in an area surrounding the initial viral particle are destroyed, leaving a clear or plaque area (Fig. 16-13). During the same period of time, uninfected bacteria in regions of the plate without viruses multiply rapidly giving dense, opaque areas. Thus the net result of the plaque assay consists of clear plaque areas contrasting with areas of growth. Since with dilute viral suspensions each plaque corresponds to a single viral particle in the initial sample, the number of plaques is a direct assay for virus concentration. The plaque is analogous to the bacterial colony.

Plaque assays are equally useful for quantitating animal viruses that are lytic (Fig. 16-14). A sample of a viral suspension is added to a monolayer culture of appropriate host cells. The viral particles adsorb within a few minutes, whereupon the medium is replaced with a solid agar gel to restrict the diffusion of progeny particles. Again each originally infective particle gives rise to a localized focus of infected cells that may be recognized within a few days by the naked eye.

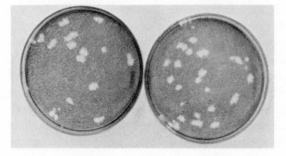

FIGURE 16-14
Plaques caused by infection of human embryonic tonsil cells with herpesvirus, type 1. (Courtesy of Dr. B. Wentworth, from Wentworth, B. B. and French, L. *Proc. Soc. Exp. Biol. Med.* 131:590, 1969)

Plaques produced by animal viruses are recognizable in various ways. Harmful changes in host cells are called *cytopathic effects*. In cases in which the virus kills the cells, plaques are readily revealed by staining with a dye that distinguishes live from dead cells. Plaques then appear as colored spots or as uncolored holes in an otherwise colored background. Alternatively, cells may not be killed but may show other cytopathic effects. Tumor-inducing viruses produce foci of cells that grow in an unrestrained fashion to produce colonies distinguishable from the majority of the cells.

The Pock Method   An older method, similar in principle to the plaque method, involves an assay of local infections (manifest as pocks) in the chorioallantoic membrane of chick embryos. Some 48 hours after infection, pocks appear (opaque areas in the membrane). Such pocks are a result of infection of a single cell with a virus. This method is semiquantitative at best.

Quantal Assays   Combined with dilution *quantal assays* can yield an approximation of viral concentration. Several dilutions of the preparation are made and administered to a number of animals, cell cultures, or chick embryos depending on the host specificity of the virus (Fig. 16-15). Sufficient time is allowed for the virus to destroy the whole cell culture or kill the animal, as the case may be, and the proportion of infected cell cultures or animals is determined. As in the plaque method several different criteria of infection may be used. At the highest dilutions no signs of infection are visible, whereas at the lower dilutions all tests are positive. The titer of virus or endpoint is taken to be that dilution at which 50 percent of the inoculated hosts are infected. This is referred to as the ID_{50} or infective dose or LD_{50}, lethal dose.

Hemagglutination   A large variety of viruses are able to agglutinate red blood cells by chemical reactions between virions and surface components of the cells (*hemagglutination*). A virion attaches to two red cells simultaneously and causes them to clump together. Sufficiently high concentrations of virus cause large aggregates to form. Hemagglutination assays can be measured using serial dilutions of the viral suspension mixed with a standard red blood cell suspension. The highest dilution showing complete agglutination is taken as the endpoint or titer.

One group of animal viruses able to agglutinate red blood cells is the myxoviruses, as exemplified by the influenza virus. The interaction between virus and cell receptor is always specific, and investigation of the influenza–red cell combination has led to increased understanding of this kind of interaction. It has been found that the cell receptors for influenza virus are mucoproteins (that is, proteins containing certain sugars). Hemadsorption involves the adsorption of virus to certain groups in the host–cell receptor sites. The influenza virus also contains an enzyme, neuraminidase, that specifically cleaves these groups from the mucoprotein on the surface of the host cell. Thus the virus particles first adsorb to red cells and subsequently are released by action of the viral neuramini-

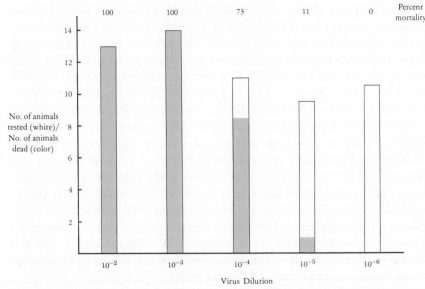

FIGURE 16-15 Calculation of the LD_{50} titer by the Reed-Muench method. From these experimental data the LD_{50} is calculated by determining the proportionate distance between the two dilutions nearest 50% mortality, as follows:

$$\frac{\text{Proportionate}}{\text{distance}} = \frac{(\text{\% mortality at dilution next above 50\%}) - 50\%}{\text{\% mortality at dilution next above 50\%} - \text{\% mortality at dilution next below 50\%}}$$

Thus in this example,

$$\frac{\text{Proportionate}}{\text{distance}} = \frac{73 - 50}{73 - 11} = \frac{23}{62} = 0.4$$

Next, the proportionate distance is multiplied by the dilution factor (the negative logarithm of the dilution steps used). In this example the dilutions are ten-fold and the dilution factor is therefore $-\log$ of 10, or -1.

$$\frac{\text{Proportionate distance factor}}{\text{corrected for dilution}} = (0.4)(-1) = -0.4$$

Negative log of LD_{50} titer $\quad$ = negative log of the dilution next above 50% plus the proportionate distance factor corrected for dilution

$$= (-4.0) + (-0.4) = -4.4$$

LD_{50} titer $= 10^{-4.4}$

dase. If appropriate proportions of virions and red cells are mixed, the virus particles can attach to more than one cell simultaneously, resulting in bridges and lattice formation linking the red cells together in clumps.

ARE VIRUSES LIVING?

This discussion clearly shows that viruses are not cellular organisms, since they have no cellular organization, but rather are obligate parasites at the genetic level. Whether or not they can be considered living depends on the definition of life. Since there is no universally accepted definition as to what constitutes life, it is not surprising that there is no universally accepted view as to whether or not viruses are alive. If life is considered to be a complex set of processes initiated by the transcription of nucleic acid, then a virus has both living and nonliving stages. It is alive when it is replicating inside an infected cell and dead as a virion outside a cell. By this consideration only, a transforming fragment of DNA must also be alive once it becomes integrated into the DNA of a recipient cell. There is, however, at least one important difference between the transferability of bacterial-transforming DNA and the viral nucleic acid. The transfer of the transforming DNA occurs at a low frequency and is not an essential part of the existence of the species, but the transfer of the viral nucleic acid is an essential part of viral existence and occurs with a very high frequency, at least with bacterial viruses. Viruses have evolved to entities highly specialized for transfer of their genetic material. Thus they are "more living" than fragments of DNA. The fact that they are entities highly specialized for transfer also makes them more like organisms than other cellular organelles, such as mitochondria and chloroplasts that also contain protein and nucleic acid. The question of whether or not viruses are living, then, is more philosophical than scientific.

SUMMARY

Viruses are obligate intracellular parasites that require a host cell in order to replicate. Each virus particle, or virion, consists of either RNA or DNA, surrounded by a protein coat. Some viruses have an additional exterior envelope. A virion contains only a few enzymes concerned with viral replication. Viruses lack enzymes concerned with energy production and they contain no ribosomes or other organelles essential for replication. Consequently they depend on host cells to supply these needs.

Virus particles are generally either helical or polyhedral in structure, or sometimes combinations of these two forms of symmetry.

Virus replication is unique. Most viruses attach to specific receptors on host cells—the process of adsorption. Penetration of viral nucleic acid into the cell follows. Replication of the various viral components is the next step, followed by assembly into mature, complete virions. The period during which no infective virions are present within the host cell while the viral nucleic acid directs the replication of viral constituents is called the eclipse phase. The eclipse ends when the first mature infective virion is formed. Eventually, mature virions are released from host cells. This may occur in a burst when the host cell lyses and releases many mature virions, or mature virions may be released gradually over a period of time, without lysis of the host cell.

Viruses are specific with regard to host cells. The host range of a virus depends in part on the presence of specific receptors on the host cell. However, other factors which determine host range include the ability of the viral nucleic acid to survive within the host cell and to subvert cellular metabolism.

Three kinds of virus-host relationships are recognized among bacterial viruses and probably occur in other kinds of viruses as well. These are the lytic relationship, the lysogenic relationship, and the case where viruses are released without cell lysis. The lysogenic relationship involves a temperate virus that is an episome. Depending on certain conditions, a temperate virus can either replicate within the cell, resulting in the formation of many new virions and lysis of the cell, or alternatively the temperate viral nucleic acid may become integrated within the host cell chromosome. During the latter state of integration the virus is known as a prophage, and the host bacterium is described as a lysogenic bacterium. When the prophage leaves the host chromosome, it can once again replicate and become lytic. The lysogenic relationship is responsible for several important biological phenomena, including transduction of genetic information from one host cell to another and lysogenic conversion.

Viruses can be studied by cultivating them in specific host cells, either bacterial, plant, or animal cells depending on the virus. Quantification of viruses may be achieved by various methods including electron microscopic counting, plaque assay, quantal assay, and hemagglutination.

The intriguing question of whether or not viruses are actually living is unanswerable.

QUESTIONS

1. Define the term virion and describe a typical virion.
2. Compare the mode of replication of a virus with the reproduction of a bacterium.
3. Man is susceptible to the influenza virus, cause of the "flu." Can fungi get "the flu," and if not, why not?
4. What are the differences between the lytic and lysogenic relationships of phages and bacteria?
5. Describe the plaque method for studying viruses and its uses.

FURTHER READING

HORNE, R. W., "The Structure of Viruses," *Scientific American,* 208 (January 1963), 48–56.

KELLENBERGER, E. "The Genetic Control of the Shape of a Virus," *Scientific American,* 215 (December 1966), 32–39.

LURIA, S. E., "The Recognition of DNA in Bacteria," *Scientific American,* 222 (January 1970), 88–102.

WATSON, J. D., *Molecular Biology of the Gene,* 2d ed. New York: W. A. Benjamin, Inc., 1970.

WOOD, W. B. and R. S. EDGAR, "Building a Bacterial Virus," *Scientific American,* 217 (July 1967), 61–66.

Chapter 17

Animal and Plant Viruses

Much of the basic biology of viruses elucidated in bacteriophage systems applies also to animal and plant viruses; however, there are also certain differences unique to each group. In this chapter a general approach to the classification of viruses is presented followed by a discussion of animal viruses, their variety, mode of replication, and effects on host cells. In addition, a brief introduction to plant viruses is included.

CLASSIFICATION OF VIRUSES

The taxonomy of viruses has been subject to change over the years. As more is learned about the properties of viruses, classifications change. For this reason only the principles of viral taxonomy are considered here. A current classification scheme is included in Table 17-1 for reference. A survey of presently recognized groups of animal viruses can be found in the appendix. Figures 17-1 through 17-3 show the morphology of several well-known viruses that infect man.

Survey of groups of animal viruses
Pages 676–681

The most widely used taxonomic criteria depend upon the structure of the virus itself. Four major criteria are used: (1) the nature of the nucleic acid—DNA or RNA, single-stranded or double-stranded; (2) particle structure—helical, icosahedral, or complex; (3) presence of a viral envelope (enveloped) or absence of an envelope (naked); (4) dimensions of the particle. Beyond these physical characteristics, other criteria (immunologic, cytopathologic, or epidemiologic) are used to subdivide the groups. Such a classification provides great convenience and utility, although it is not necessarily a natural or evolutionary based scheme.

334

TABLE 17-1
Classification of Viruses

Nucleic Acid	Capsid Shape	Envelope	Dimensions of capsid (nm)	Bacterial	Plant	Animal	Special Features
RNA	Polyhedral	Naked	20–25	Coliphage[a] f2	Bushy stunt virus	Picornaviruses	
			18–38				
			70–77		Wound tumor virus	Reoviruses	Double-stranded RNA
	Helical	Naked	17 × 300		Tobacco mosaic virus		
		Enveloped	30–75			Some of the Arboviruses	
			90–300			Myxoviruses	Multiple-piece genome
			37–220			Paramyxoviruses	
DNA	Polyhedral	Naked	22	Coliphage øX174			Single-stranded DNA
			18–26			Parvoviruses (Picodnaviruses)	Some have single-stranded DNA
			40–57			Papovaviruses	
			70–80			Adenoviruses	
			140			Tipula insect viruses	
		Enveloped	120–250			Herpes viruses	
	Helical	Naked	5 × 80	Coliphage fd			Single-stranded
	Complex (polyhedral and helical)	Naked	Head 60 × 90 Tail 17 × 120	Coliphages T2, T4, T5, T6			Polyhedral heads Helical tails
	Complex	Enveloped	200 × 350			Poxviruses	Brick-shaped, capsid shape unknown

[a] A coliphage is a bacteriophage that infects *E. coli.*

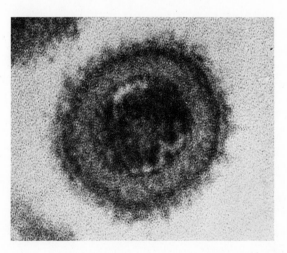

FIGURE 17-1
The morphology of herpesvirus. The nucleic acid and protein show icosahedral symmetry and are surrounded by an envelope. (Courtesy of Dr. B. Roizman, from *Hospital Practice,* April 1972)

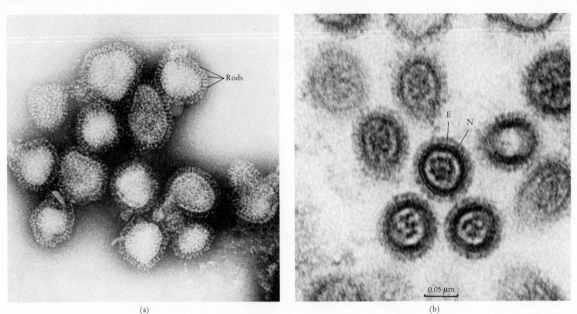

(a)

Rods

(b)

E N

0.05 μm

FIGURE 17-2   Morphology of influenza virus. (a) Electron micrograph of influenza A virions of the Hong Kong strain. Note the spikes or rods inserted into the envelope. (Courtesy of Dr. Frederick Murphy, from the cover of *Science, 163,* No. 3865, 1969) (b) Thin section of influenza virions. The exterior envelope (E) and internal nucleic acid and protein (N) can be distinguished. (Courtesy of Dr. E. Boatman)

INTERACTION OF VIRUSES WITH ANIMAL CELLS AND TISSUES

In many cases the ultimate result of infection of an animal cell by a virus is death of the cell. The various kinds of cell damage which may lead to death or abnormal appearance of cells in culture are referred to as cytopathic effects. Damage to the cell is caused not only by the formation of vast numbers of viral particles; much

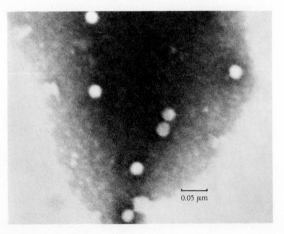

FIGURE 17-3
Morphology of rhinovirus. These are typical viruses of the picor-navirus group. The small naked virions consist of RNA enclosed in an icosahedral capsid. (Courtesy of Dr. E. Boatman)

0.05 μm

more important are the effects of virus-specified proteins on normal cellular processes. For example, it is quite common for such viral proteins to induce changes in the permeability of one or more of the cellular membranes. Again, some viral proteins can specifically inhibit host DNA, RNA, or protein synthesis. Cells may develop abnormalities in chromosome structure.

A most important response of some viral-infected cells is the synthesis of *interferon,* a protein which interferes with the replication of viruses. This protein is induced by viral infection, but it is coded for by the cellular DNA. Interferon is released from infected cells and acts to protect other neighboring uninfected cells. It acts intracellularly to inhibit viral replication, in contrast to antibody which acts extracellularly to inactivate virus. Normal cell functions are unaffected by interferon activity. Most viruses will initiate this response from cells and, in fact, the response initiated by one virus will act upon other simultaneous or subsequent viral infections. The response to produce interferon appears to be triggered by the presence of double-stranded RNA, viral replicative form (described below), or even added synthetic RNA. This natural defense mechanism may be artificially induced by administering synthetic RNA, and protection of laboratory animals from some viral diseases has been accomplished. This presents an exciting future possibility for antiviral chemotherapy.

Adsorption of Animal Viruses to Cells

Like bacterial viruses, animal viruses attach to host cells by means of a complementary association between attachment sites on the virion and receptor sites on the cell surface. When such receptors do not exist, as in cells of other types or other species, or when they are destroyed by an experimenter by means of enzymatic treatment, no attachment occurs. Animal viruses usually do not contain specific appendages corresponding to bacteriophage tail fibers; rather, it appears that attachment sites are distributed over the surface of the virion. The chemical composition of receptors varies according to the virus in question. For example, certain mucoproteins act as receptors for influenza virus, whereas specific lipoproteins are the receptors for poliovirus. Within a given species of animal receptors occur in some cell types but may be absent or few in number in some other cell types.

Uncoating of the Virus

Complete virions are engulfed into animal cells by phagocytosis and are subsequently uncoated. This is true for virions of all sizes, shapes, and taxonomic assignment. The particles exist temporarily within membrane-bound vacuoles and later may be released within the cytoplasm or nucleus. In the case of most viruses, the uncoating process, whereby the viral nucleic acid is released from the coat, is poorly understood and apparently quite variable. Uncoating appears to be the result of degradation of capsid proteins by enzymes of the host cell.

With viruses containing more than one coat or layer such as the enveloped poxviruses, uncoating is a more complex, stepwise process. For these viruses the outer coat is removed in the cytoplasmic vacuoles. Later, after the partially un-coated particles have penetrated the cytoplasm, a specific uncoating enzyme, specified by the viral DNA and present in the virion, appears and removes the inner core protein.

Replication of DNA Viruses

Subsequent to the uncoating step the details of the replication cycle of DNA and RNA viruses differ sufficiently to justify separate summaries of subsequent events. With some DNA-containing viruses, such as smallpox virus, infection leads to an inhibition of host DNA synthesis. Cellular DNA is not broken down and is therefore unavailable as a source of precursors for viral DNA synthesis. Other viruses such as the oncogenic (cancer-inducing) papovaviruses may cause an increase in the rate of cellular DNA synthesis, a feature which appears to be related to the mechanism of carcinogenesis itself. The mode of viral replication depends to some extent on the nature of the host cell. This is clearly seen in one group (adenoviruses) which are virulent for human cells but oncogenic for certain other animal cells.

DNA-containing viruses and cancer Pages 352–354

  The period of eclipse between infection and the appearance of progeny virus may be divided into several stages. The first involves the synthesis and translation of early mRNA. Just as for bacteriophage, parts of the genetic material of DNA-containing animal viruses are transcribed early and others later. The early messengers code for enzymes, particularly those involved in DNA synthesis such as thymidine kinase and DNA polymerase. Next, viral DNA is replicated by a process which depends on the prior synthesis of these early enzymes. In all cases except that of poxviruses viral DNA is synthesized in the cell nucleus. In the case of the poxviruses, the largest and most complex viruses known, the site of DNA replication is the cell cytoplasm. Some DNA viruses also possess genetic functions which permit the shutdown of host RNA and protein synthesis.

  At this stage late messengers are produced that code for viral capsid proteins and other proteins involved in regulation or assembly. Maturation occurs over a considerable period of time. Some viruses are released from the cell by egestion without cell death, but others are released when the cell dies and disintegrates. Again, for the larger viruses, the process is more complex. In particular, the maturation of poxviruses appears to involve changes after the components are assembled within a viral membrane. The DNA is enclosed by many layers of membrane which later sort themselves out into inner and outer membranes.

Replication of RNA Viruses

Viral RNA, replicating within cells, must face the problem of competition in a world dominated by DNA. This consideration forces RNA viruses to adopt one of several novel mechanisms for replication (Fig. 17-4). With single-stranded RNA viruses the predominant route of replication involves the production of a *replicative form* consisting of double-stranded RNA; a new strand complementary in base sequence to the infecting strand is produced (Fig. 17-4a). This intermediate form acts as a template for new viral RNA strands. Other RNA viruses have adopted an even more novel stratagem in which a DNA copy of the infecting RNA strand is synthesized. However, mechanisms for replicating RNA are lacking in noninfected cells, so that an infecting RNA molecule must specify the synthesis of the required enzyme before it can replicate, or alternatively the viral particle itself must carry such an enzyme. As examples of ways in which these problems are mastered, the replication of a single-stranded RNA virus, poliovirus, is first considered and later others in which the solution differs.

In poliovirus RNA the infecting strand can act as both mRNA and template for the replication of the viral RNA. The incoming strand rapidly attaches to host cell ribosomes, where it serves as a messenger for protein synthesis. At this time host RNA and protein synthesis are severely inhibited, apparently as a result of synthesis of a new viral-specified protein. Among the proteins produced at this time is RNA synthetase, an enzyme which catalyzes the replication of the viral RNA—a process which commences within half an hour after infection. Viral RNA synthesis proceeds exponentially for about 3 hours. One novel feature of the translation of poliovirus messenger is that the entire RNA molecule codes for one very large polypeptide chain which is later cleaved into several proteins, including the RNA synthetase, the capsid proteins, and the proteins responsible for the curtailment of host RNA and protein synthesis. When a large pool of mature virions has accumulated by association of the capsid proteins with the RNA, the cell eventually lyses and releases them.

The double-stranded RNA of certain viruses, such as reovirus, consists of seven to ten pieces. Each is apparently transcribed into a messenger for protein synthesis and replicated by an enzyme or enzymes carried within the virion itself. Thus infected cells contain several species of single-stranded polyribosome-associated RNA, corresponding to the double-stranded fragments of the viral RNA.

All RNA-containing animal viruses, except certain small ones (picornaviruses) and the double-stranded reoviruses, are enclosed by a lipoprotein envelope. This is acquired at the time of release from the cell membrane by a process termed *budding* (Fig. 17-4b). The nucleocapsid (nucleic acid enclosed by capsid) is expelled through the plasma membrane which has been altered by the incorporation of viral protein. Thus only extracellular RNA-containing viruses contain the complete envelope.

PLANT VIRUSES

A great number of plant diseases are now recognized as viral infections. Many of these are of considerable economic importance, particularly those occurring in crop plants. Virus infections are particularly prevalent among perennial crop plants or those propagated vegetatively such as potatoes. Other crops in which

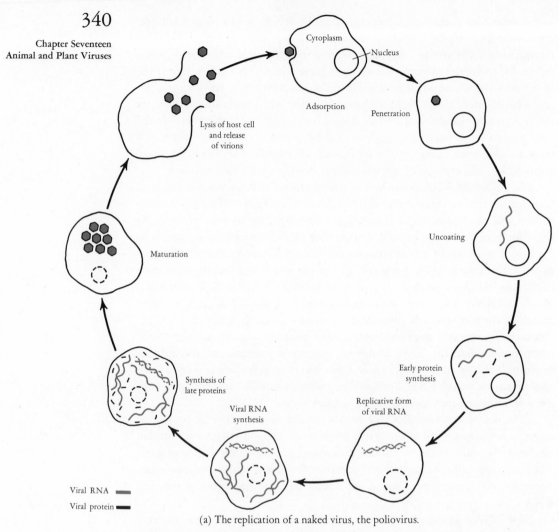

Cytoplasm

Nucleus

Adsorption

Penetration

Lysis of host cell
and release
of virions

Uncoating

Maturation

Early protein
synthesis

Synthesis of
late proteins

Replicative form
of viral RNA

Viral RNA
synthesis

Viral RNA
Viral protein

(a) The replication of a naked virus, the poliovirus.

FIGURE 17-4   The replication of single-stranded RNA-containing viruses.

viruses cause considerable loss of productivity include wheat, soy beans, and sugar beet. A serious virus infection may reduce productivity by more than 50 percent.

Infection of plants by viruses may be recognized through various signs (Fig. 17-5). Localized abnormalities may occur in which there is a loss of green pigments and complete leaves may turn yellow; in addition, rings or irregular lines often appear on the leaves and the fruit. Individual cells or specialized organs of the plant may become necrotic and calluses or tumorous growths may appear. Usually plants become stunted in growth, although in a few instances growth is stimulated leading to deformed structures. In the vast majority of cases plants do not recover from virus infections; unlike animals plants are not capable of developing specific immunity to rid themselves of invading viruses. On occasion, however, infected plants develop new growth in which visible signs of infection are absent even though the virus is still present. The reasons for this are not un-

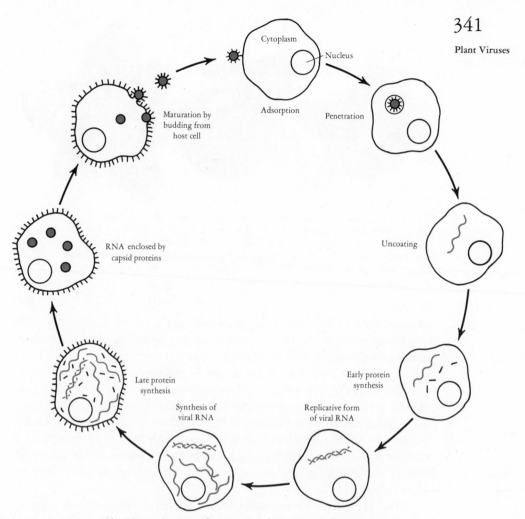

(b) The replication of an enveloped virus, the influenza virus.

derstood. In severely infected plants virus particles may accumulate in enormous quantities. For example, as much as 10 percent of the dry weight of an infected tobacco plant may consist of tobacco mosaic virus.

In a few specialized instances plants have been purposely maintained in a virus-infected state. The best known example is in tulips, where a virus transmitted through the bulbs can cause a desirable color variegation of the flowers. The virus was transmitted in this way for a long time before an infectious agent was even suspected.

Virus Multiplication

The production of such large quantities of plant viruses has facilitated some aspects of their study. In fact, purification of and chemical analysis of plant viruses

FIGURE 17-5 Lesions caused by plant viruses. *Nicotiana glutinosa* plants 11 days after inoculation with lettuce necrotic yellows virus. On the right is a healthy plant for comparison; left, a plant given a large dose; and center, a plant given a small dose of the virus. Visible effects of the viral infection include loss of pigment in some areas, necrotic lesions, and distortion and stunting of infected leaves. (Courtesy of Dr. J. W. Randles; Randles, J. W. and D. F. Coleman, *Virology,* 41:459–464, 1970)

advanced considerably more rapidly than with animal viruses, culminating in the crystallization of tobacco mosaic virus. However, other aspects of the study of plant viruses, particularly their multiplication and interaction with their hosts, suffer from severe experimental limitations. Methods of quantification are inefficient and irreproducible and plant cell cultures are rarely available. The initiation of infection is cumbersome and inefficient, usually involving mechanical damage to the leaves. In other cases it may not be possible to infect host plants directly, but only by grafting or through an insect vector. All these limitations make it virtually impossible to analyze the growth cycle of a plant virus in any detail. Most of our knowledge of plant virus multiplication therefore derives from indirect evidence.

Tobacco mosaic virus,
Figure 16-3; Page 313

The stages in plant virus multiplication are best summarized with reference to tobacco mosaic virus, the most widely studied plant virus. The virion is a rigid rod some 300 nm in length. Each particle has 2200 identical protein capsomeres surrounding a single-stranded RNA molecule. Usually the virus infects leaves damaged either under natural conditions or experimentally through rubbing the leaves with an abrasive such as carborundum. The efficiency of infection is very low; only about one in a million particles is effective. These infections lead to localized lesions in the leaves indicative of groups of dead cells in which virus multiplication has taken place.

The first step in multiplication involves separation of viral RNA from the capsid protein. RNA replication appears to occur in the cell nucleus through an intermediate double-stranded replicative form. In contrast, the subunit proteins are synthesized in the cytoplasm, implying that newly replicated viral RNA migrates to the cytoplasm to act as mRNA. Finally, RNA and protein assemble to form crystals within the plant cell or are carried throughout the plant slowly by means of intracellular connections or rapidly through the circulatory system.

Many plant viruses are extraordinarily stable; tobacco mosaic virus apparently retains its infectivity for up to 50 years. This stability is an important feature for maintaining the virus since processes of infection are, in general, very inefficient. The same feature is important in considering the difficulty of eradicating a viral infection. Viruses may spread from plant to plant by a variety of means. Some are transmitted through soil contaminated by prior growth of infected plants. For perhaps some 10 percent of the known plant viruses, transmission is possible through seeds or tubers or by pollination of flowers on healthy plants with pollen from diseased individuals. Seed transmission may occur in some hosts but not in others. Infections may also spread after grafting healthy plant tissue onto diseased plants. Another more exotic transmission mechanism is effected by plants such as the parasitic vine, dodder, which establishes connections with the vascular tissues of host plants and transmits virus from plant to plant.

Other important mechanisms involve vectors of various types. These include insects, worms, man, soil, and fungi. For example, tobacco mosaic is a serious disease of the tobacco crop. There are no known insect vectors; instead man himself is the major vector. In this instance virus is transmitted to healthy seedlings on the hands of workers, who have previously picked up the virus from infected plants or tobacco products. However, the most important plant virus vectors are probably insects, and insect control is a potent tool for controlling spread of viruses.

Specific Examples of Plant Viruses

The *tobacco necrosis* virus is an icosahedral virus representative of a large class of similar particles infecting plants of many species. A special feature of this particular virus is its association with a satellite virus which is much smaller—in fact one of the smallest known viruses with a diameter of only 18nm. The satellite virus is itself noninfectious, but in combination with tobacco necrosis virus both viruses multiply. Apparently the very small RNA molecule of the satellite is sufficient to code for the capsid protein and only one other protein. All other viral functions required for infection and replication must be supplied by the companion virus.

TOBACCO NECROSIS VIRUS

The *alfalfa mosaic* virus is representative of a class of plant viruses having bacilliform shape. It is also an example of a multiparticle virus. At least four different particles appear to be required to establish infection, apparently because the complete nucleic acid does not occur in any one particle. This system offers an advantage in terms of increased genetic flexibility, although it carries obvious disadvantages in terms of a requirement for a high enough infective dose to ensure inclusion of all four particle types.

ALFALFA MOSAIC VIRUS

The *rice dwarf* virus and *wound tumor* virus are both arthropod-borne and they have double-stranded RNA, whereas all other known plant viruses contain single-stranded RNA. The particles are icosahedral and resemble the double-

RICE DWARF VIRUS

stranded RNA viruses of animals (reoviruses). They are transmitted by insects carrying virus particles in their mouths as a result of feeding on infected plants. However, in a great many cases the relationship among plant, insect, and virus is considerably more complex and specific, involving much more than mechanical transfer.

Arthropod Transmission of Plant Viruses

Several distinct types of arthropod vector transmission are recognized. First, *external or temporary transmission* involves the association of virus with the external mouth parts. The ability to transmit virus lasts only a few days. Second, in *circulative transmission* the virus circulates in the body of the insect and may be infective for the lifetime of the insect. Third, the *transmission may involve actual multiplication of the virus within the insect.* In this case the viral agent is truly an insect virus as well as a plant virus. In many instances, such as leaf hoppers, viruses persist through a large number of insect generations and may be transmitted to plants at any time. The existence of insect transmitted plant viruses raises several interesting questions about viral evolution. In particular, it seems that plant and animal viruses may not be so different as appears at first sight. Several of these viruses infecting plants and insects show similarities with animal viruses. Potato yellow dwarf virus is very similar to animal myxoviruses, and wound tumor virus of plants is morphologically very similar to reoviruses of animals.

Control of Plant Virus Infections

Normally, no attempts are made to cure virus infections of plants, since treatments are both ineffective and uneconomic. Instead, prevention of the spread of virus infection is the preferred tactic. This is accomplished by burning or otherwise destroying infected plants and sterilizing the soil. An exception to this general rule exists for long-lived plants such as fruit trees, where heat treatment of young plants is sometimes used to eliminate viruses. For diseases transmitted by insect vectors destruction of the vector itself by insecticides may be effective. Alternatively new genetic strains of the plant in question may be developed which are resistant to the virus or unappealing to the insect vector.

INSECT VIRUSES

For a long time plant viruses carried by and multiplying within insect vectors were thought to be without serious effect on their insect hosts. However, several cases are now known in which the insect vectors are themselves damaged or even killed. In addition to these kinds of viral infections, insects themselves are subject to their own viral diseases. In nature, these are often important as a means of control of insects that are for one reason or another considered pests by man.

Virus diseases of insects are commonly revealed by inclusions either in the nucleus or in the cytoplasm of cells. Such inclusions contain protein, usually of high molecular weight and determined by the viral nucleic acid. At the termination of the infection when the insect dies these inclusion bodies make up a considerable part of the cell volume. Viral particles consisting of double-stranded DNA virions surrounded by two membranes occur with these inclusions. These viral particles are evidently infective and represent the disease-producing agents. Although viral infection normally leads to progressive deterioration of the insect host, many cases of viral latency are known. In the latent state viruses produce no obvious signs of illness, but virus production and disease may occur suddenly through external stimulation even after the infection has been passed to several generations of the host.

SUMMARY

Animal viruses are generally classified on the basis of their nucleic acid content, capsid structure, presence or absence of an envelope, and the dimensions of the virion.

Virus replication depends on properties of both the virus and the host cell. Infected cells may produce interferon, which interferes with subsequent viral replication. In general, animal viruses adsorb to specific receptors on host cells and are phagocytized by the cell. Within the cell the virus is uncoated and viral components are replicated separately. The components are then assembled into virions. Enveloped viruses gain their envelope from either the nuclear or the cytoplasmic membrane of the host cell. Mature virions are released by budding from the cell, by reverse phagocytosis, or by lysis of the host cell.

Plant viruses replicate in a manner similar to animal viruses. Viruses, however, usually enter plant cells as a result of injury or they are introduced by insect vectors.

Insects not only act as vectors for plant viruses which may multiply in the insect host but they also may be infected by specific insect viruses.

QUESTIONS

1. What criteria are used to classify viruses? How does such a classification compare with evolutionary based schemes for classifying animals?
2. Why is the artificial induction of interferon being studied intensively? How does interferon act?
3. What are some of the differences between replication of phages within bacterial host cells and animal viruses within animal cells?
4. Compare and contrast the replication of the naked virion of polio virus and the enveloped virion of influenza virus.
5. How are viruses transmitted among plants?

FURTHER READING

FENNER, F. J., *The Biology of Animal Viruses.* New York: Academic Press, 1967.

FENNER, F. J. and D. O. WHITE, *Medical Virology.* New York: Academic Press, 1970.

HILLEMAN, M. R. and A. A. TYTELL, "The Induction of Interferon," *Scientific American,* 225 (July 1971) 26–31.

LURIA, S. E. and J. E. DARNELL, *General Virology,* 2d ed. New York: John Wiley & Sons, 1967.

III

Microbes and Man

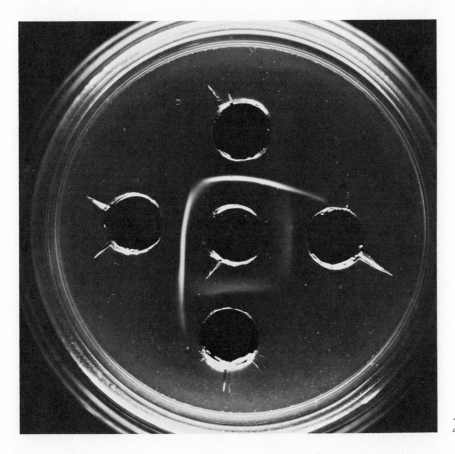

Antigen-antibody precipitation in agar gels (courtesy of Dr. M. Tam)

Viruses and Cancer

The most common result of the infection of an animal cell by a virus is a cytopathic effect which disturbs the metabolism of the cell and may eventually destroy it. In such cases virus multiplication occurs and the progeny virus may infect neighboring cells. It is this cytopathic effect which is responsible for the symptoms of viral disease.

Other types of virus can produce quite different changes in cells leading to the eventual appearance of a tumor. Tumors result from the rapid or unrestricted growth of malignant cells to produce various conditions, generally known as cancer, which can result in death of the animal. The term *malignancy* is usually defined in this context as the capacity of tumor cells to grow progressively and in an unrestrained fashion. The process by which normal cells are altered to become malignant cells as a result of virus infection is termed *transformation*. This designation is somewhat unfortunate, since it may be confused with the unrelated process of bacterial transformation (Chapter 9). These transformed cells often undergo a dramatic change in morphology and metabolic properties which precedes their rapid multiplication.

Bacterial transformation
Page 194

At the present time no viruses have been unequivocally shown to be the cause of human cancer. Nevertheless, the large number of well-documented examples of virus-induced tumors in other animal species have suggested that at least some forms of human cancer are likely to be of viral origin. In this chapter we consider the role of viruses in the induction of tumors and the mechanism by which this may occur.

Chapter Eighteen
Viruses and Cancer

A large variety of types of abnormal growth in animals are described by the general terms, *tumor* or *neoplasm*. They are often characterized by rapid growth and, more importantly, by the degree to which the tumor displays autonomy and independence from the normal control mechanisms of the rest of the body. Thus these growths occur in an unrestrained fashion and may intrude upon or envelop normal tissues of the body. Another important characteristic of tumors is that they fail to respond to the control mechanisms which determine organ and tissue morphology. In some cases the cells are easily distinguishable from normal cells, although in others they may not present any obvious differences.

Tumors vary considerably in the effect which they have on the organism. At the one extreme, *benign* tumors are those whose growth is localized and not invasive into neighboring tissues or organs. Moreover, they have no tendency to give rise to secondary tumors in other parts of the body. Benign tumors may interfere with the functioning of organisms by reasons of sheer size. However they may be completely removed by surgery. At the other extreme are malignant tumors, or cancers, which can invade normal tissues and organs and destroy them. This process may be accomplished by *metastasis,* a process which involves the separation of cells from the main body of the tumor and their passage to other parts of the body to initiate the growth of secondary tumors.

BENIGN VS MALIGNANT
TUMORS

The growth and properties of tumor cells may be studied in experimental animals in a variety of ways. One of the major tools of tumor research involves the use of transplantation. Thus a tumor which has grown in one animal may be transferred to another animal by grafting some of the tissue or injecting some of the cells. This transplantability is, of course, subject to the same genetic and immunological barriers which are encountered in the grafting of normal tissues. For this reason, tumor biology experiments require highly inbred stocks of laboratory animals which are produced by repeated brother-sister matings so as to produce genetic homogeneity.

Transplantation
Page 401

Another powerful experimental method involves the growth of tumor cells as a cell culture. In this case the cells are maintained in a complex medium in the laboratory, where their metabolic and morphological properties are easily studied. Furthermore, as is discussed later, the infection and transformation of cells by tumor viruses may be accomplished in culture and the resulting malignant cells may subsequently be transferred to animals, where they result in tumors.

Several quite different kinds of agents can play a role in the induction of tumors. A well-known example is provided by carcinogenic chemicals which occur in tobacco and other smoke. The evidence which connects them as causal agents of human lung cancer is now overwhelming. In fact, these carcinogenic agents may be used to produce many different kinds of cancer in laboratory animals. In addition, sources of radiation, such as ultraviolet light and X-rays, are effective inducers of tumors. It is well known that humans who are exposed to excessive doses of such irradiation, for example, X-ray technicians or sailors, are more prone to develop certain kinds of cancer. Both in the case of the chemical *carcinogens* and radiation, it is likely that the malignancy is brought about by direct effect of these agents upon the genetic material, the nucleic acids of the cell, leading to a heritable change.

Finally, there are the tumor viruses, the subject of this chapter. Partly because of the many cases of viral-induced tumors in animals and partly because of the experimental advantages, research on tumor viruses is currently among the most active of all those areas of research encompassed in tumor biology.

EARLY HISTORY OF TUMOR VIROLOGY

There are many highlights in the history of tumor virology. In fact, over the past 50 years, the field has displayed several waves of activity and popularity. The idea that viruses may be involved in cancer dates from the earliest discoveries of viruses themselves, although the proposition that human cancer may be of viral origin was distinctly unpopular until the last ten years or so. The first successful transfer of the capacity to induce a tumor was reported in 1908 by Ellerman and Bang. They showed that leukemias of chickens could be transferred by means of cell extracts. However, at that time there was no apparent connection between this blood disease and tumors. Soon after, however, Rous, in 1911, demonstrated a similar transfer of a chicken sarcoma (connective tissue cancer). Even then, however, this important finding was generally regarded as a peculiarity of the domestic fowl.

The viral induction of a mammalian tumor was first reported by Shope in 1932, who demonstrated the transfer of cell-free filtrates of a rabbit papilloma (wartlike tumor of epidermal tissue). These tumors were generally benign in the wild cottontail rabbit population in which they occurred naturally, although when the virus was transferred to the domestic rabbit, malignant skin tumors called *carcinomas* appeared. A few years later Bittner discovered a virus in mice which could induce mammary gland carcinomas and which was transmitted from mother to progeny through the milk. This discovery was very important for the development of the field because of the great advantages of the mouse as a laboratory animal. The discovery of many other mouse tumor viruses followed thereafter, most of which are responsible for leukemias as well as solid tumors, called *lymphomas* and more recently *sarcomas*.

Even before the viruses themselves were studied or any biochemical studies were initiated, several important characteristics of virus-associated tumor induction (or *oncogenesis*) became apparent. In many instances a considerable period may elapse between introduction of the virus and the appearance of tumors. For example, mice which are infected with the Bittner virus as sucklings may not develop a mammary tumor for some months. Furthermore, although the disease is contagious, the route of infection may be obscure as in this case. In addition, the efficiency of viral transformation is often extremely low, so that the vast majority of viruses infect cells without transforming them. Likewise, many animals

CHARACTERISTICS OF
VIRUS-ASSOCIATED
ONCOGENESIS

infected with tumor viruses do not develop tumors. The reasons for resistance are not at all well defined although it is clear that immunology plays a protective role.

As the methods became available these viruses were purified and their composition studied. Both the chicken and the mouse viruses were found to contain RNA. In the late 1950s, several new tumor viruses discovered in rodents and primates were found to contain DNA. One example of this is simian virus 40 (SV_{40}), which was discovered in the cultures of Rhesus monkey kidney cells used for growing large quantities of poliovirus. Perhaps most surprising was the finding that human adenoviruses, common in respiratory infections, could induce tumors in various rodents.

The following discussion presents a brief summary of the special features of some of the known tumor viruses.

ROUS SARCOMA VIRUS

Rous sarcoma virus (RSV) is one of a large group of viruses prevalent in domestic fowl. Some of these viruses are apparently nonpathogenic, while others cause leukemia, and the Rous virus produces sarcomas. In fact all stocks of RSV contain large numbers of nonpathogenic viruses. This is because the virus itself is apparently incapable of specifying the synthesis of some component of the viral particle. Thus complete RSV can only be produced under conditions of coinfection with another virus which supplies the viral protein. Thus the RSV resembles antigenically that virus which supplies the virus coat. This feature of RSV growth is referred to as *defectiveness* and the other virus is known as the *helper*. This explains why several different strains of RSV exist: each contains the RSV genome surrounded by a slightly different protein shell provided by a different helper.

DEFECTIVE AND HELPER VIRUSES

This group of avian leukosis viruses all possess the same structural features. They contain RNA, which represents about 2 percent of the particle weight. The viruses are relatively large (Table 18-1) and contain a large lipid envelope similar to the myxoviruses. All of the viruses may be grown in chick embryos or in tissue cultures of embryonic chick cells.

Myxoviruses
Page 335; Table 17-1

Cell Transformation

A very interesting important feature of the Rous sarcoma virus is the high efficiency with which it can promote cell transformation. The addition of virus to a culture of chicken fibroblasts results in the appearance of foci of transformed cells. These can easily be recognized and distinguished from normal cells.

The normal fibroblasts tend to grow in parallel arrays and cease multiplication when they have produced a confluent monolayer. This property is characteristic of all normal cells and is referred to as *contact inhibition,* a term which implies that cell-contact regulates cell proliferation. On the other hand, trans-

TABLE 18-1
Physical Properties of Some Tumor Viruses

	Nucleic Acid	Molecular Weight	Particle Diameter (nm)	Other Properties
Rous sarcoma	RNA (single-stranded)	10×10^6	80	Contains 35% lipid in envelope
Mouse leukemia	RNA (single-stranded)	13×10^6	100	Has envelope similar to myxoviruses
Polyoma	DNA (circular double-stranded)	3.5×10^6	45	Icosahedrons; no envelope
SV_{40}	DNA (circular double-stranded)	3.5×10^6	45	Icosahedrons; no envelope
Adenoviruses	DNA	20×10^6	70	Icosahedrons; contains proteins other than capsomere

formed cells do not obey these restraints and grow irregularly, forming colonies of cells containing more than one layer. These foci may easily be recognized by microscopic examination and provide a means of quantifying the number of RSV particles. In fact it has been shown that *all* cells which are infected with RSV are transformed whether or not RSV are produced in the transformed cells. Thus each focus of transformed cells results from the multiplication of a single transformed cell that may be produced by infection with a single RSV particle.

When cells are infected with a large number of RSV particles, they are transformed and they also produce infectious RSV particles. This is apparently because the contaminating helper virus multiplies and supplies the necessary viral coat components. Other cells, however, particularly those infected by small numbers of RSV particles, become transformed but produce no infectious virus. These nonproducer cells may, however, produce labile, incomplete viral particles and can release infectious RSV following later superinfection by a helper virus.

The Rous virus can even transform mammalian cells, either in the animal (producing tumors) or in tissue cultures. These cells are also nonproducers. However, they evidently contain the viral genome as evidenced by the fact that if hamster cells transformed by RSV are transferred to chicks, they produce the characteristic tumors. Moreover, these tumors will produce mature virus if infected with a helper.

TRANSFORMATION OF MAMMALIAN CELLS

Persistence of the Viral Genome

These various observations of the biology of the transformed cell show that the viral genome is present in all the cells and that it must multiply along with the cell as the colony of cells or the tumor grows. It has been suggested that the viral genetic material is integrated with that of the host cell in some way. Until recently this was rather difficult to visualize since the viral genome consists of RNA rather than DNA. However, it now appears that after infection, the RNA

Reverse transcription
Page 165

genome is copied into a DNA form and that this is physically integrated with host DNA.

This process is accomplished by an enzyme which can be isolated from intact virions of Rous sarcoma virus. The enzyme called RNA-dependent DNA polymerase exhibits considerable specificity for the RNA genome of the virus. The initial reaction involves the formation of an RNA/DNA hybrid according to the following scheme:

$$\text{Viral RNA} + \begin{matrix} \text{dTTP} \\ \text{dCTP} \\ \text{dATP} \\ \text{dGTP} \end{matrix} \xrightarrow[\text{enzyme}]{\text{Polymerase}} \text{RNA/DNA hybrid}$$

Subsequently the DNA strand can act as a template for the synthesis of a complementary DNA strand. It is presumed that at this stage the double-stranded DNA copy of the viral genetic information can recombine with the host chromosome and enter the integrated state. At this time the viral genes behave as host genes and may be expressed by transcription in the normal way.

It is clear that the viral genome persists and that it enters this cryptic state with a 100 percent efficiency. All infected cells are transformed and transmit the viral genome to their progeny. This efficiency contrasts with the very low rate of transformation of other tumor viruses. Whether this difference is directly correlated with the defectiveness of the RSV genome and inability to produce viral particles remains to be seen.

The discovery of this novel enzymatic mechanism has several important implications. A similar enzyme activity has been demonstrated in other RNA tumor viruses, such as those responsible for mouse leukemia. No such enzyme activity exists in nononcogenic viruses. An interesting exception to this general rule is the existence of the same enzymatic activity in the so-called slow RNA viruses responsible for some neurological disorders. Thus it seems that the existence of such an enzyme activity will be a useful diagnostic tool for detecting oncogenic virus infections. Furthermore, the very novelty of the RNA-dependent DNA polymerase reaction engenders some hope that specific inhibitors may be developed as therapeutic agents.

POLYOMA AND SV_{40} VIRUSES

Polyoma virus is so-called since it can produce a wide variety of different kinds of tumors. SV_{40} was discovered as a passenger virus in cultures of Rhesus monkey kidney cells used to grow poliovirus, and was later shown to produce sarcomas in baby hamsters. The two viruses are basically similar structurally and in biological properties and will therefore be considered together as examples of DNA tumor viruses.

The viral particles are simple icosahedrons of about 45 nm containing 72 capsomere protein subunits. In both cases the genome is circular double-stranded

DNA of about 3.5×10^6 daltons. The two viruses, however, do not appear to be closely related antigenically and their DNA molecules differ in overall base composition.

Infection of Cells

In both cases the viruses infect some types of cells productively with the appearance of mature infectious virus, whereas others are transformed. For example, SV_{40} infection of green monkey kidney cells leads to virus production, whereas hamster or human embryonic cells are transformed. There are, however, some cases where both productive infection and transformation occur simultaneously.

In productive infection large numbers of viral particles are produced in the cell nucleus (several millions per cell). These are eventually released as the cells disintegrate. An antigen (distinguishable from the capsid protein) appears in the cell nucleus very early in infection. This same antigen is present in transformed cells or tumors, and is therefore useful as a means of identifying the prior infection by the virus. This antigen, referred to as the T-antigen, is virus-but not cell-specific. Thus the T-antigens of the same cell infected by SV_{40} or polyoma are different, whereas different cells infected by the same virus contain the same T-antigen. Moreover, cells transformed successively by both SV_{40} and polyoma contain the two T-antigens. Thus the T-antigen is specified by the viral genome and may correspond to one of the early proteins synthesized in certain bacteriophage infections.

Antigens
Pages 30, 369

Phage specific enzymes
Page 316

Transformation

Many kinds of cell cultures may be transformed by both viruses. Foci of rapidly growing cells in random arrays are produced. However, although a single virus particle may transform a cell, the efficiency of this process is extremely low. The efficiency with which productive infection is initiated may be 10,000 times that with which cells are transformed. This suggests that only a small fraction of the infective particles are capable of transformation, a situation which contrasts markedly with that for Rous virus.

The morphological and biochemical characteristics of transformed cells are determined both by the cell type and the infecting virus. Since cells infected with both viruses display both sets of new properties, it is evident that the virus itself is responsible for the altered properties of transformed cells. Transformed cells do not contain detectable quantities of virus or of capsid protein, although they must contain at least part of the viral DNA.

Integration of the Viral Genome

Recent experiments have demonstrated directly that transformed cells contain viral DNA and that this viral DNA is transcribed into viral messenger RNA.

These experiments are of several types, each relying on recent advances in molecular and cell biology.

Molecular hybridization
Page 231

1. RNA molecules transcribed from the viral genome can be detected both in polyoma and SV_{40} transformed cells. By the use of molecular hybridization techniques, it can be shown that a small proportion of the RNA molecules produced in such cells have the base sequence characteristic of the virus genome in question. Moreover, in cells transformed successively with both SV_{40} and polyoma, RNA molecules are apparently transcribed from both viral genomes. These experiments demonstrate the presence of at least a part of the viral genome in the transformed cell.

2. Similar experiments using radioactive polyoma DNA show more directly that the virus DNA itself is present. Again, using molecular hybridization techniques, it can be demonstrated that polyoma DNA reacts with cellular DNA extracted from transformed mouse cells much more efficiently than it does with normal mouse DNA. This reaction therefore provides a means of quantifying the amount of polyoma DNA actually present in the genome of the transformed cell. Current estimates vary, but the amount is placed between 2 and 60 copies of the viral genome per cell, assuming that the complete viral DNA molecule is present.

3. The fact that the complete viral genome is present has been most convincingly demonstrated by the liberation of mature infectious polyoma virus after transformed cells have been fused with other cells. This procedure releases the virus genes from their cryptic state in the transformed cells and allows them to specify the biochemical steps necessary for virus multiplication.

Lysogenization
Page 320

The actual state of the virus DNA molecule in the host genome is not yet clearly understood. It is presumed that the DNA is integrated and covalently bound to host DNA. The integration of viral DNA therefore appears to be analogous to the process of lysogenization by bacterial viruses. Insertion presumably occurs after an initial pairing event and the viral DNA becomes covalently linked at each end to the host DNA. This mechanism would be consistent with the circularity of polyoma DNA, a feature which appears to be necessary for lysogeny to occur. In the same way as for lysogenic bacteriophages, release of the viral genome to initiate viral multiplication would be a comparatively rare event, and one which can be triggered by radiation or physiological change. However, the molecular events involved in transformation are undoubtedly more complex than outlined in these few sentences. For example, infection with tumor viruses induces host DNA synthesis which appears to be crucial for subsequent cell transformation.

The widespread occurrence of virus-induced tumors in animals is sufficient to predict the likelihood that at least some types of human cancer are of viral origin. We can confidently expect that proof of this will be forthcoming within the next few years and that this will permit the development of methods for the prevention and treatment of many types of human cancer.

One current candidate for the role of human cancer virus is Epstein-Barr or EB virus. This herpes virus may be isolated from cell cultures of an unusual lymphoma which is common in patients in parts of Africa. However, actual causation of the tumor by the virus has yet to be established. Perhaps of greater significance is the fact that patients suffering from infectious mononucleosis, glandular fever, possess antibodies directed against this same virus. Thus it is possible that the virus is a causative agent of this particular form of lymphoma, Burkitt's lymphoma or mononucleosis or both.

Herpes viruses,
Table 17-1; Page 335

At the present time it appears that viruses may be causally implicated in human breast cancer. Virus particles have been observed in samples of human milk virtually identical in structure to mouse mammary tumor (Bittner) virus. The occurrence of such particles is closely correlated with the incidence of breast cancer in the families of the women concerned. Even more significant is the fact that these particles contain the RNA-dependent DNA polymerase characteristic of animal tumor viruses. It is, however, impossible at the present time to estimate what fraction of human cancer conditions is attributable to virus infection, since there are many other obvious ways in which the genetic material of the cell may be permanently altered.

Some of the approaches to the elucidation of the role of viruses in human cancer will be obvious to the student from the preceding brief account of animal tumor viruses. The rationale for several of the most promising approaches is outlined below.

1. *Examination of tumors for the presence of viruses and attempts to cultivate these using cell cultures.* This approach suffers two obvious drawbacks. First, viruses are ubiquitous, and the majority of those observed will be merely passengers. Second, if we are to take account of the analogies from the Rous or polyoma virus situations, it is very likely that the oncogenic virus in question is absent from the tumor and represented only by its genome.

2. *A survey of known viruses for tumorigenic activity.* It has often been suggested that oncogenic activity may be manifested by viruses well known for other properties such as production of infectious diseases. Known viruses can be tested for their ability to transform human cells and to induce tumors in animals. The basic difficulty with this approach is, of course, the inability to test directly for tumor production in humans.

3. *The detection of the presence of the genome of a known virus in tumor cells.* This approach is based on the expectation that the viral genome may be inte-

EXPERIMENTAL
APPROACHES

grated into a cell in a functionally incomplete form. If so, its presence may be detectable by virtue of the production of a viral antigen or by direct assay for the base sequences characteristic of the DNA of the virus in question. For example, by methods analogous to those which have proved successful in demonstrating the presence of the polyoma genome in tumor cells, it should be possible to ask whether any human tumors contain the genome of any known human virus. This approach rests upon the sensitivity of nucleic acid hybridization methods.

4. *Animal tumor virus genomes as probes.* This approach is currently being pursued by Dr. Spiegelman and his colleagues at Columbia. Again, it rests upon the sensitivity of DNA/RNA hybridization and the supposition that the genomes of human tumor viruses may be very similar to those of known animal tumor viruses. For example, it has already been shown that human mammary tumors contain RNA molecules which are very similar to the RNA genome of the mouse mammary tumor (Bittner) virus. Similarly, RNA is detectable in some leukemias and sarcomas which correspond to a mouse leukemia virus genome. The findings suggest that, though the actual viruses responsible for these human neoplasms have yet to be isolated, they are closely related to the well-studied rodent tumor viruses.

5. *Release of cryptic viruses from tumor cells by various means.* If only the viral genome is present, it is possible that this may be released and revealed as viral particles by irradiation, treatment with drugs, superinfection by other viruses, or fusion of the tumor cell with another cell. The use of each of these methods is suggested by the experience gained with animal tumor viruses.

SUMMARY

Although viruses have not been unequivocally shown to be the cause of human cancer, it seems highly probable that at least some forms of human cancer will be shown to be of viral origin. This supposition rests on several observations. There is widespread occurrence of virus-induced tumors in animals. Virus particles have been observed in human milk which are virtually identical in structure to a virus that is responsible for mouse mammary tumor.

The study of tumor-causing viruses of animals has revealed some characteristics which will prove useful in the search for viruses causing tumors in man. Thus in the Rous and polyoma virus situation, only the viral genome is present in the tumor. Presumably the nucleic acid of the virus is integrated into the host DNA. In the case of RNA viruses a unique enzyme which synthesizes DNA from the viral RNA apparently is responsible for synthesizing the DNA which is then integrated. Also a virus may be oncogenic in one host and cause no harmful effects in a closely related host. This creates the possibility that oncogenesis may be manifested by viruses which are now recognized as causing infectious diseases. Indeed, the human adenoviruses which are common in respiratory infections can induce tumors in various rodents.

357

Although viruses undoubtedly are the cause of certain human tumors, it seems unlikely that they are the cause of all tumors. A wide variety of chemicals as well as X-ray and ultraviolet irradiations increase the frequency of cancer. It is likely that these agents bring about the malignancy by directly affecting the DNA of the cell, leading to a heritable change.

QUESTIONS

1. Discuss the reasons why a viral origin of cancer has gained popularity in the past 10 years.
2. Compare the major features of the known tumor viruses. Do they have any features in common?
3. Discuss the evidence that the viral genome is integrated into the host DNA.
4. How has the discovery of the enzyme RNA-dependent DNA polymerase helped explain the role of viruses in cancer?
5. Compare a lysogenic bacterial cell with a cell transformed by an oncogenic virus.
6. How have the techniques of nucleic acid hybridization been useful in implicating viruses in cancer production?

FURTHER READING

GALLO, R. C., "Reverse Transcriptase, the DNA Polymerase of Oncogenic RNA Viruses," *Nature*, 234 (1971), 194.

GREEN, M., "Oncogenic Viruses," *Ann. Rev. Biochem.*, 39 (1970), 707.

"New Evidence as the Basis for Increased Efforts in Cancer Research," *Proc. Nat. Acad. Sci.*, 69 (April 1972), 1009.

TEMIN, H. M., "Mechanism of Cell Transformation by RNA Tumor Viruses," *Ann. Rev. Microbiology*, 25 (1971), 609.

————, "RNA-Directed DNA Synthesis," *Scientific American* (January 1972), 25.

Chapter 19

Interactions Between Men and Microbes

THE NORMAL FLORA

Since microorganisms are so common, it is not surprising that some of them grow abundantly on surfaces of the human body, including the skin and gastrointestinal tract. Microorganisms that are well established on the external and internal surfaces of the body without producing overt disease are often referred to as *resident* or *normal flora,* and those that inhabit man only sporadically, *transient flora.* The number of different species living in close association with man is very large and includes microorganisms which have not yet been adequately studied or characterized. Examples of the principal resident species are *Staphylococcus epidermidis, Propionibacterium acnes,* and *Pityrosporum ovale,* found on the skin; the viridans group of streptococci, *Neisseria catarrhalis,* and *Corynebacterium hoffmani* in the throat; and *Bacteroides, Lactobacillus* species, and enterobacteria in the intestine.

Viridans streptococci Page 257; enterobacteria Page 252

The composition of this complex community at any time represents a dynamic balance of many opposing forces which are constantly changing in response to external influences and as a result of the activities of the human host. The microbial flora and the host comprise an ecosystem, each member of which may be influenced in some way by the others. For example, activity which increases perspiration and thus skin moisture and nutrients encourages proliferation of skin organisms, sometimes resulting in skin disease. Intestinal microbial populations may change with variations in diet, gastric acidity, and the degree of intestinal activity.

The terminology used to define associations among the large and dynamic community of microorganisms living with the human host is confusing and inconsistent from one biologist to another. In order to ensure that our meaning is clear, several definitions are given in the following paragraphs.

Symbiosis simply means living together, and *symbionts* are organisms which live together in a particular environment. It is the practice to subdivide symbiotic relationships into *commensalism,* indicating that the microorganism benefits from the association, whereas the host is neither benefited nor harmed; *mutualism,* a state in which both the host and the microorganism derive benefit; and *parasitism,* a state in which the microorganism benefits at the expense of the host. Many symbionts of man may be classified into more than one of these categories depending on a number of conditions. For example, *Staphylococcus aureus,* growing in the nostrils is a harmless commensal. When aspirated into the host's lung or introduced into his surgical wound, however, it may become parasitic. *Bacteroides fragilis,* growing at its normal site in the intestine, is commensal, but if it gains access to the bloodstream it is parasitic and may cause fatal illness.

Whether or not a given microorganism plays a parasitic role depends on the effectiveness of host defenses as well as on the ability of the microbe to circumvent or neutralize the host's defenses. Some microorganisms may even be truly mutualistic in certain circumstances, benefiting the host by producing vitamins or by discouraging colonization with parasitic forms. Studies, however, indicate that most animals, including man, do perfectly well without a resident flora provided they are protected from environmental microbes.

A great variety of host and microbial factors interact in determining the effect of organisms on the human body, whether these organisms are established members of the resident flora or new arrivals from the environment or other hosts. *Colonization* simply implies establishment of microbes on a body surface. If the microbe breaches the surface, enters body tissues, and multiplies, then *infection* is said to have occurred. An infection which causes noticeable impairment of body function is called an infectious *disease.*

A *pathogen* is any microorganism or virus capable of producing disease. *Pathogenic* means disease-causing, and *pathogenicity* is the ability to cause disease. An *opportunist* is a pathogen which is able to cause disease only in hosts with impaired defense mechanisms, such as might result, for example, from wounds or alcoholism. A *nonpathogen* is a microbe which never causes disease, or does so only under very unusual circumstances favoring the microorganism against the host. These terms should never be used in the absolute sense because *pathogenicity always depends on both host and microorganismal factors.*

Virulence refers to degrees of pathogenicity of different strains within a species. The different degrees of pathogenicity are the result of attributes of the organism such as possession of a capsule. Thus pneumococci that have a capsule are *virulent;* those lacking a capsule are *avirulent* (not virulent). The word virulent is also used in a quantitative sense indicating that an organism has more virulence than other strains of the same species. The implication is that such organisms are more likely to cause disease, or more likely to cause severe disease, than the other strains. In some instances the attribute responsible for virulence is known (such as the capsule of pneumococci), and in other instances it is not. However, the relative virulence of different strains can be compared by determining the minimum doses required to produce disease in laboratory animals of matched age and genetic background. Virulence measured in this way does not necessarily apply to human beings. For example, pneumococci with one type of capsule fail to produce disease in laboratory mice, but are nevertheless pathogenic for human beings.

PROTECTIVE EFFECT OF MICROBIAL COMPETITION

Yeast
Page 283

Infection appears to occur more readily when competing microbes are eliminated or suppressed. Enhanced penetration of microorganisms may be related to their relatively high concentration, resulting from the unrestricted growth which can occur in the absence of microbial competition. For example, the yeast *Candida albicans* is commonly found in relatively small numbers in the intestinal tract. However, if an antibiotic suppresses the competing normal bacterial flora in the intestine, the concentration of the candida may increase manyfold and the organism is then prone to invade the host. Another example of the ability of large numbers of organisms to overcome host defenses and invade is seen in certain eye infections. Eyes are frequently bombarded with saprophytic organisms of the *Bacillus* and *Pseudomonas* genera, with no evidence of any deleterious effects.

Bacillus, Page 257;
Pseudomonas, Page 250

However, if contaminated eye medication, in which one of these organisms has grown to large numbers, is introduced into the eyes, serious invasion and destruction of the eye may result.

Although synergistic growth of microorganisms (two or more together growing better than alone) may occur on or in the tissues of the human host, mutual inhibition among the resident flora is more generally the rule. It is likely that much inhibition by one microorganism of another results simply from competition for foodstuffs. However, other mechanisms are known; for example, the pathogen *Corynebacterium diphtheriae* is inhibited by hydrogen peroxide produced by the viridans group of streptococci. In addition, fatty acids produced by the breakdown of oily skin secretions by normal skin flora are inhibitory for many pathogens. Other microbes produce antibiotics or antimicrobial proteins (bacteriocins) of more restricted activity than antibiotics. In other instances acid byproducts of the metabolism of one organism may be inhibitory for others. Although most of the mechanisms can be easily demonstrated *in vitro,* their importance in the ecosystem is still a matter of debate. It is established, however, that the suppression of one component of normal flora may allow other components to produce disease, or may facilitate invasion by exogenous parasites.

Corynebacterium, Page 258;
bacteriocins, Page 202

HOST FACTORS INFLUENCING RESISTANCE TO INFECTION

A number of mechanical and physiological functions of the host tend to restrict the growth of microorganisms on body surfaces and also to isolate or destroy them should they enter the tissue. Complex mechanisms are also operative to minimize the entrance of microorganisms into body cavities such as the lung and bladder. Mechanisms more or less restricted to a single organ system are discussed in subsequent chapters dealing with infectious diseases. Host factors which operate throughout the body are discussed below.

Tissue Factors

Foreign substances (including microbes) entering host tissue initiate a complex series of responses referred to as *inflammation.* The principal events of the inflammatory process are shown in Table 19-1. Tissue injury resulting from the invasion causes release of enzymes which result in cleavage of peptide substances called *kinins* from specialized plasma proteins. The kinins result in the enlargement of blood vessels in the local area and increased permeability of their walls, allowing loss of fluid from the vascular system into the tissue spaces adjacent to the injury. At the same time clotting of the fluid is initiated. White blood cells (leukocytes) adhere to the dilated blood vessels near the area of tissue damage, migrate through the vascular walls, and engulf (phagocytize) the offending substance. These cells (phagocytes) may destroy the foreign substance by the action of digestive enzymes. Initially, cells with multilobed nuclei (polymorphonuclear leukocytes) appear. Many of these die following phagocytosis, releasing antimicrobial substances. Later, larger mononuclear cells (macrophages) arrive. These also have a phagocytic function and, in addition, participate in the specific immune response. Other cells produce specific proteins called *antibodies* which react with microbes or other foreign material and thereby promote phagocytosis and de-

INFLAMMATION

Antibodies
Pages 30, 369

TABLE 19-1
Principal Components of the Inflammatory Process

Event	Effects
Tissue injury	Release of kinins, prostaglandins, histamines, and other mediators to act on adjacent blood vessels.
Blood vessels dilate and show increased permeability to plasma which may clot	*Swelling* of the tissues results from the leakage of plasma. *Elevated temperature* of the region may occur as a result of increased blood flow through the dilated vessels. *Redness* may appear for the same reason. *Pain* may result from increased fluid in the tissues and from direct effect of mediators on sensory nerve endings.
Circulating white blood cells adhere to the walls of the altered blood vessels	The white blood cells migrate chemotactically through the vessel walls and to the area of injury. They are responsible for phagocytosis of foreign material and tissue debris, and for initiating antibody production.

struction of the offending agent. These cells that respond during inflammation are termed collectively *inflammatory cells*. Of interest is the fact that the inflammatory response is less effective against microbes when an inert foreign body is present. Thus numbers of *Staphylococcus aureus* which could otherwise be eliminated easily by tissue defenses persist and even multiply in the presence of a surgeon's suture. Although inflammation is the first line of defense against invading microbes, when it is intensive or widespread the process can itself cause marked impairment of body functions.

Besides kinins, other mediators of inflammation include histamine and prostaglandins. In fact, the anti-inflammatory medicines antihistamine and perhaps aspirin act by opposing the respective actions of these two substances.

If the invading microbe escapes the localizing and destructive action of the inflammatory response and enters the blood or lymph vessels, it may still be removed and destroyed by collections of macrophages distributed along the circulatory system, especially in the lymph nodes, spleen, liver, and bone marrow. These phagocytic cells are collectively known as the *reticuloendothelial* (mononuclear-phagocyte) *system*.

Nonspecific Factors in Body Fluids

In addition to the inflammatory response and the action of phagocytes of the reticuloendothelial system, other nonspecific mechanisms contribute to defending the host against infection. For example, tissues and body fluids such as saliva and blood plasma contain compounds able to kill many species of bacteria. Two of these substances, lysozyme and beta-lysin, are also found in high concentrations in cellular elements of the blood from which they are released by a variety of factors. These include inflammatory reactions and entrance of bacteria into the bloodstream. Lysozyme degrades the mucopeptide composing bacterial cell wall, whereas beta-lysin attacks the cytoplasmic membrane of the bacterial cell. These and other substances may work along with antibody to lead to the destruction of invading bacteria.

Mucocomplex
Page 77

Another important group of substances are the *interferons*. These proteins help limit viral infections and perhaps those due to other parasites as well. Their induction and mode of action were discussed previously.

Interferons
Page 337

IMPAIRMENT OF DEFENSE MECHANISMS

The factors above are important in maintaining the ecological balance in favor of the host. A significant impairment in any of these may result in the extension of microorganisms into host tissue and the establishment of disease. For example, some people are born without the capacity to produce functional phagocytes or normal antibodies. Others lose these capabilities later in life. Still others, such as those receiving heart or kidney transplants from another person, may be given medicines which impair the functioning of the immune response in order to prevent rejection of the graft. Some patients are given antiinflammatory medicines such as corticosteroids to control the symptoms of one disease, only to contract infection because of interference with the inflammatory response. In other cases the defects in the protective mechanism of the host are less clear, as with cancer, sugar diabetes, or malnutrition. In all these cases members of the resident flora assume an overtly parasitic role more frequently than they would with hosts having normal defensive mechanisms. Such people can sometimes be protected from microbial invasion for long periods by being placed in a sterile plastic-walled chamber, with their resident flora suppressed by antimicrobial medicines. Of course, all their food and even the air they breathe must be freed of microbes to prevent introducing pathogenic organisms resistant to the action of the antimicrobial agents.

MICROBIAL PROPERTIES WHICH INFLUENCE PATHOGENICITY

The pathogenic effect of microorganisms stems from their ability to alter the structure and function of the tissue cells of the host. Such alterations are accomplished through actual penetration of the tissue by the parasitic organisms or by the penetration of their toxic by-products. These harmful effects of microorganisms are related to a number of factors. Some organisms cause remarkably little damage to host tissue, but because of their ability to stimulate a marked inflammatory response, normal function of the host is seriously impaired. This is true in most cases of pneumonia due to *Streptococcus pneumoniae,* where the tremendous outpouring of fluid and inflammatory cells into the air sacs of the lung produces impairment of respiratory function. The ability of many organisms to grow within host organs or tissues is related to their ability to circumvent host defense mechanisms. For example, as indicated in Chapter 4, many pathogenic bacteria possess a large capsule which inhibits engulfment by phagocytes. Some capsules inhibit engulfment while others do not, but the reasons for the difference is not known. Some organisms produce substances which kill or injure the inflammatory cells, thus interfering with their phagocytic function. At least one organism, *Borrelia recurrentis* (a cause of relapsing fever), is for a time able to bypass the antibody defense mechanism of the host by readily producing mutants with a different surface chemical structure. These mutants account for the recurrence of disease after the host antibody has surpressed the original strain.

Capsules
Page 76

Toxins

Pathogenic microorganisms are often armed with poisonous substances (*toxins*) which produce harmful effects on the host (Table 19-2). The toxins produced by bacteria fall into two groups depending on their chemical properties. The first group of toxins are proteins and are produced by certain species of both gram-positive and gram-negative bacteria. In general, they have the following properties:

1. They are inactivated by heat (60°C to 100°C for 30 minutes);
2. They are potent in microgram quantities;
3. They are synthesized and accumulate intracellularly, and then are released into culture media in large concentrations;
4. Each has a distinctive effect;
5. They can be converted into *toxoids,* substances which lack toxic properties but can stimulate immunity to the toxin.

EXOTOXINS

Toxins with these characteristics are generally called *exotoxins.* Important exotoxin producers are: the gram-positive *Corynebacterium diphtheriae,* the cause of diphtheria; *Clostridium tetani,* the cause of tetanus (lockjaw); and *Clostridium botulinum,* the cause of a kind of paralysis (botulism). The causative bacteria of diphtheria and tetanus actually have little invasive tendency, but toxin absorbed from a localized area of infection is responsible for the symptoms; therefore the disease can be prevented by immunizing against the toxin. In botulism, the bacterium is grown in food and releases its toxin there; generally it is not infectious. One important exotoxin, the *enterotoxin* of some strains of *Staphylococcus aureus,* resists boiling temperatures for 30 minutes or more but the others are generally inactivated at 60°C to 100°C.

PROTEIN TOXINS OF
GRAM-NEGATIVE
BACTERIA

Highly potent protein toxins are also found in certain gram-negative pathogens and they are also referred to as exotoxins. In general, they differ somewhat from the examples above in that they are not so readily released into culture media and immunization with their toxoids usually does not protect against the disease. Some important gram-negative toxin producers are: *Vibrio cholerae,* the cause of cholera; *Bordetella pertussis,* the cause of whooping cough; *Shigella dysenteriae,* a cause of dysentery; and *Yersinia pestis,* the cause of plague. In some of these diseases the role of the toxin is in doubt; the toxin effect is only demonstrable in selected species of laboratory animals.

Besides the exotoxins mentioned above, species of bacteria may release other products that could possibly contribute to virulence. Some of these products have traditionally been called toxins, and others are only known by the kind of effects they produce *in vitro.* In contrast to exotoxins they are generally not

TABLE 19-2
Important Properties of Bacterial Toxins

	Exotoxins	*Endotoxins*
Bacterial source	Gram positive and negative species	Gram negative species
Location in bacterium	Synthesized in cytoplasm and released from cell	Component of the cell wall
Chemical nature	Protein	Lipopolysaccharide
Ability to form toxoid	Present	Absent

TABLE 19-3
Some Bacterial Extracellular Products that May Contribute to Virulence

Description	Function	Producing Bacterium (example)
Phospholipase	Breaks down lecithin, a lipid component of mammalian cell membranes	*Clostridium perfringens*
Hemolysin	Destroys red blood cells	*Clostridium perfringens*
Coagulase	Clots plasma	*Staphylococcus aureus*
Collagenase	Breaks down collagen, a tissue fiber	*Staphylococcus aureus*
Hyaluronidase	Breaks down hyaluronic acid, a tissue component	*Staphylococcus aureus*
Lipase	Breaks down fat	*Staphylococcus aureus*
Deoxyribonuclease	Breaks down DNA	*Staphylococcus aureus*
Leukocidin	Kills white blood cells (leukocytes)	*Staphylococcus aureus*

responsible for any major effects in diseases although it seems likely that they contribute in a minor way. A number of nonpathogens release similar products, emphasizing the auxilliary role the substances probably play in disease. Many are enzymes. Several examples are given in Table 19-3 and in the chapters on specific diseases.

Bacterial endotoxins (not to be confused with the enterotoxin of *S. aureus*) differ greatly from exotoxins in that:

1. They are only found in gram-negative bacteria in which they represent a component of the outer portion of the cell wall. Many gram-negative organisms, both pathogens and nonpathogens, possess them rather than only a few bacterial species as with exotoxins.
2. The toxic effect, principally fever and damage to the circulatory system, is common to all endotoxins.
3. They are lipopolysaccharides, not protein.
4. They cannot be converted to toxoids.
5. They are not released into culture media except in old cultures showing disruption of the bacterial cell walls.

Microbes other than bacteria may also produce toxins. For example, a substance similar in action to endotoxin has been demonstrated in some fungi, and viruses may also show toxic properties. To date relatively little is known of many of these substances and their role in infectious diseases.

INFECTIOUS DISEASES

The preceding paragraphs have dealt with many of the factors that define the role of microorganisms colonizing the body. When microbial numbers and virulence outweigh host defenses, infection occurs and disease is a likely consequence. In subsequent chapters numerous infectious diseases are discussed along with the organisms causing them. These organisms are the traditional "pathogens" which frequently are responsible for diseases in normal persons without previous exposure to them. It will be well to keep in mind, however, that many other microorganisms have the capacity to cause disease in an abnormal host or in a normal host under special conditions favoring the microbe.

LATENT INFECTIONS

Herpes viruses,
Page 335; Table 17-1

In some instances the parasitic tendencies of the microorganism are finely balanced with the host defenses, and a microorganism may live in host tissue without manifesting its presence. Agents may remain latent for many years, only revealing their pathogenic capacity when the balance shifts in their favor. Such is the usual case with the virus of herpes simplex ("cold sores") and may also occur with chicken pox virus and the intracellular bacterium, *Rickettsia prowazekii,* the cause of typhus. Another example is *Mycobacterium tuberculosis,* the cause of human tuberculosis. The mycobacteria are often confined within a small area of tissue by host defense mechanisms, but may eventually begin growing and destroying tissue when the host experiences overwork, malnutrition, antiinflammatory medicines, or other factors poorly understood.

SUMMARY

Large numbers of microorganisms of many varieties inhabit the body surfaces in symbiosis with human beings. Members of this resident or normal flora show commensalism, mutualism, or parasitism, their roles varying with external and host factors. Normally, these microorganisms keep each other in check by poorly understood mechanisms, and when this balance is disturbed, parasitism may be enhanced and result in disease of the host. This may occur following administration of an antibiotic active in killing or suppressing some components of the normal flora but not others, or by alterations in host physiology.

Invasion of the tissues of the host, infection, occurs when the aggressive capacities of microorganisms overcome host defenses. In addition to the specific defenses of each organ system of the body, the defensive actions of inflammation and antibody response work to stop the spread of microorganisms and eliminate them from tissue. Virulent microbes more readily overcome body defenses than others, often because they possess extracellular products which degrade components of tissue or interfere with phagocytosis. Some possess powerful toxins which can cause serious disease even though they have no apparent role in assisting microbial invasion.

Organisms which cause disease may be transmitted from other persons, animals, or the environment. Another major source of disease-producing organisms is the normal flora of the individual. Since disease results from an imbalance between body defenses and the disease-producing capabilities of the organism, under some circumstances even saprophytes can be pathogenic.

QUESTIONS

1. Defend the idea that the normal microbial flora is beneficial to man.
2. Defend the idea that the normal microbial flora is a threat to human health.
3. Discuss the ways by which the body defends itself against microbial invasion.
4. Discuss the microbial factors which foster pathogenesis.
5. Discuss the difference between infection and infectious disease.

FURTHER READING

AJL, S. J., S. KADIS, and T. C. MONTIE (eds.), *Microbial Toxins,* Vol. 1. New York: Academic Press, 1970.

ALLISON, ANTHONY, "Lysosomes and Disease," *Scientific American,* 217 (November 1967), 62.

BRAUDE, A. I., "Bacterial Endotoxins," *Scientific American,* 210 (March 1964), 2.

BROCK, T. D., *Principles of Microbial Ecology.* Englewood, N.J.: Prentice-Hall, 1966.

BURNET, MacF., *Natural History of Infectious Disease.* London: Cambridge University Press, 1962.

PIKE, J. E., "Prostaglandins," *Scientific American,* 225 (November 1971), 84.

ROSEBURY, T., *Microorganisms Indigenous to Man.* New York: McGraw-Hill, Inc., 1962.

WOOD, W. B., "White Blood Cells v. Bacteria," *Scientific American,* 184 (February 1951), 48–52.

Chapter 20

Immunology: Antigens and Antibodies

Many of the defense mechanisms discussed in Chapter 19 are examples of the natural or innate immunity inherent in the genetic makeup of the host. The study of these mechanisms, however, is only a small part of that discipline known as immunology.

THE SCOPE OF IMMUNOLOGY

Long before microorganisms were discovered it was known that one attack of certain illnesses, such as smallpox, left the survivor specifically immune to the same disease. Before 1800 Jenner capitalized on the common knowledge that people who had been exposed to cowpox were often not susceptible to the similar but much more serious human disease, smallpox. He introduced a method, similar to that in use today, of vaccinating with cowpox to induce immunity against smallpox. During the nineteenth century, Pasteur, Koch, and others were quick to investigate the possibilities of immunizing with microorganisms that cause anthrax, rabies, and other diseases. They found that immunization was possible if the organisms used either were modified so that they were no longer capable of causing disease (attenuated) or were killed. Thus it was that immunology developed within the discipline of microbiology.

Since the time of Koch and Pasteur, tremendous progress has been made in understanding the theoretical basis of immunological processes. Although certain nonspecific defenses are found in very primitive animals, the capacity for highly specific antibody-mediated immunological interactions has arisen during the evolution of complex vertebrates, and is lacking in animals below the level of the vertebrates in the phylogenetic scale. The specificity of these antibody-mediated reactions has opened exciting areas of investigation into the means by which molecules recognize one another. It has become apparent that the same kinds of specific mechanisms may be responsible for a wide variety of reactions that are not related to antimicrobial defense. Under certain circumstances immunological interactions can cause tissue damage (allergic reactions) in the host. They may also account for the failure of organ transplants. Specific immunological reactions can be used in typing blood, as an aid in diagnosing many diseases, in classifying bacteria, and even in identifying human beings. This last application has been used in criminal investigation.

The scope of immunology has broadened to include a number of subdisciplines, including immunogenetics, immunochemistry, and others that are not directly concerned with body protection. The following brief consideration of immunology indicates how the same basic mechanisms may account for the diversity of phenomena covered by these broad areas.

NATURAL (INNATE) IMMUNITY AND ACQUIRED IMMUNITY

Some of the characteristics of both innate and acquired immunity are presented in Table 20-1. Innate immunity is inborn, independent of previous experience, and often depends on the activities of phagocytes and of the nonspecific defense factors discussed in Chapter 19. Acquired immunity, on the other hand, is gained in response to a foreign agent. Although examples can be cited where *acquired immunity* is nonspecific, as a rule it *depends on the acquisition of antibodies, and is therefore highly specific* (the *specific immune response*). For instance, immunity is acquired following infection or immunization with one type of poliovirus; specific antibodies are formed which protect against reinfection with the same type of poliovirus but offer no protection against infection with mumps or measles virus, streptococci, or other microorganisms.

ANTIBODIES AND ANTIGENS

It is impossible to define antibodies adequately without, at the same time, defining antigens. *An antibody (Ab) is a protein that is produced in the body in response to the presence of an antigen and can combine specifically with that antigen. An antigen (Ag) is a substance that can incite the production of specific antibodies and can combine with those specific antibodies.* The simple diagram in Figure 20-1 illustrates some of the properties of antigen and antibody molecules. The antibody-reaction sites on antibody molecules represent the areas of the molecules that combine with antigen. The antigenic-determinant sites on the surface of antigen molecules are the specific chemical groups that combine with antibodies. The specificity of an antigen or antibody molecule depends on the size and shape of the antigenic

TABLE 20-1
Examples of Innate and Acquired Immunity

Innate	Inborn as a result of the genetic constitution of a species or an individual; independent of previous experience		For example, immunity of man to distemper viruses of cats and dogs; protection provided by phagocytes and protective substances such as lysozyme
Acquired, following exposure to a foreign agent	Naturally acquired	Active	Antibodies acquired following natural exposure to a foreign agent, for example, after a case of poliomyelitis; long-lasting, specific
			Interferon produced following exposure to viruses; temporary; nonspecific
		Passive	For example, placental transfer of immunity to poliomyelitis from mother to fetus; temporary; specific
	Artificially acquired	Active	Acquired following immunization, for example, with poliovirus vaccine, long-lasting; specific
		Passive	Acquired by administration of protective antibodies; for example, passive transfer of antibodies against poliovirus; temporary; specific

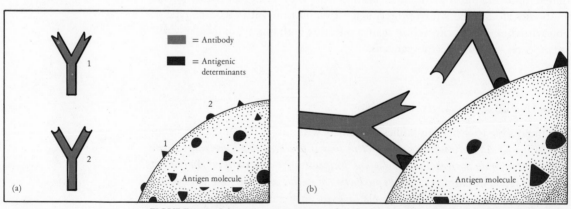

FIGURE 20-1 Schematic representation of antibody and antigen molecules. The antibody molecules represented each have two antibody reaction sites that are complementary to the antigenic determinant for which they are specific. Note that a single antibody molecule has reaction sites for only one specificity of antigenic determinant. The antigen molecule has more than one kind of antigenic determinant, so it is multispecific, and more than one of each kind of determinant, so it is also multivalent.

determinant or the corresponding antibody-reaction site, because there must be a *close complementary fit* between the two. The valency of a molecule refers to the number of specific antigenic determinants present on a molecule of antigen, or of specific antibody-reaction sites on a molecule of antibody.

The diagram in Figure 20-1 indicates that a molecule of antigen may have more than one kind of antigenic-determinant site; in other words, antigen may be multispecific. Each kind of antigenic-determinant site contributes a single specificity. The sites are drawn as projecting areas, because there is evidence to indicate that these chemical groupings project from the surface of the antigen molecule. As shown in the diagram, antigen molecules are often multivalent, that is, have more than one antigenic determinant of a particular kind.

Antibody molecules are diagrammed as monospecific, because all known *antibodies have the property of combining with only one kind of antigenic determinant.* Figure 20-1 shows antibody molecules that are bivalent, that is, capable of reacting with two antigenic determinants of the same kind. In man more than 80 percent of all antibodies are bivalent. In any event, a single antibody molecule is always monospecific regardless of the valency. The antibody-reaction site is drawn as an inverted region, in agreement with a great deal of physicochemical evidence indicating that this site is actually recessed.

Figure 20-1 shows that there is a close configurational "lock-and-key" fit between antigenic-determinant 1 and antibody 1, and between antigenic-determinant 2 and antibody 2. Thus antibody 1 would not be expected to combine with antigenic-determinant 2, which lacks the close complementary conformation.

The need for a close fit between the antigenic-determinant and the antibody-reaction site is understandable when the nature of the binding between the two molecules is examined. The forces that result in the antigen-antibody interaction are weak and short-range, such as the van der Waals forces and hydrogen bonds. A number of such bonds must participate to hold the two molecules together. As would be expected with weak noncovalent binding forces, antigen-antibody interaction is reversible under the proper circumstances; however, the combination of close fit and the simultaneous activity of a number of weak bonds usually produces a stable interaction under normal body conditions.

Some additional properties of antibodies and natural antigens are included in Table 20-2. Antibodies are immunoglobulins (Ig) usually found in the gamma globulin portion of serum when blood proteins are separated. The table shows that antibodies are large molecules. Five major classes of immunoglobulins have been distinguished, as outlined in Table 20-3.

The Structure of Antibodies

Antibodies, like other proteins, are made up of polypeptide chains, each kind of chain coded for by a particular gene. The five classes of immunoglobulins differ in the number and kinds of chains in the molecule.

Immunoglobulins of the most abundant class, IgG, have been studied ex-

TABLE 20-2
Important Properties of Antibodies and Natural Antigens

Property	Antibodies	Antigens
Specificity	Always monospecific	Usually multispecific
Valency	Bivalent (or multivalent, depending on class of immunoglobulin)	Multivalent
Chemical nature	Immunoglobulin (protein, usually gamma globulin)	Protein ⎫ Polysaccharide ⎭ Often Lipid ⎫ Nucleic acid ⎭ Occasionally
Molecular weight (approximate)	160,000 to 900,000	Greater than 10,000

tensively. A diagrammatic representation of the structure of an IgG protein molecule is given in Figure 20-2. Each IgG molecule comprises two identical halves, each half consisting of a heavy (H) and a light (L) polypeptide chain held together with both covalent and noncovalent bonds; in turn the two halves are similarly bonded together. Thus there are 2 H and 2 L chains per molecule of IgG. The Y shape in Fig. 20-2 indicates flexibility around a central "hinge" area, as has been observed with electron microscopy (Fig. 20-3).

The H and L polypeptide chains have both variable and constant sequences of amino acids, and the two identical antibody-reaction sites are formed by variable portions of both H and L chains (Fig. 20-4). Thus the amino acid sequence in one part of the molecule varies from antibody of one specificity to another, but the constant amino acid sequences are virtually the same in all human antibodies of the same class. The constant portion of the molecule is responsible for a number of biological properties of the molecules (Table 20-3).

TABLE 20-3
Properties of the Various Classes of Immunoglobulins

Class	IgG	IgM	IgA	IgE	IgD
Ability to combine with specific antigen	+	+	+	+	+
Functions:					Unknown
Protection within the body	+	+			
Protection of mucous membranes			+		
Responsible for atopic allergies				+	
Biological properties:					
Complement-fixation	+	+			
Placental transfer	+				
Secretion into saliva, mucus, and other external secretions			+		
Specific attachment to phagocytes	+	+			
Specific attachment to mast cells and basophils				+	

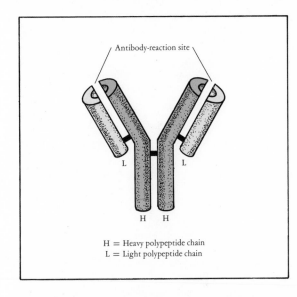

Antibody-reaction site

L   L

H   H

H = Heavy polypeptide chain
L = Light polypeptide chain

FIGURE 20-2
A molecule of human Immunoglobulin G is made up of two heavy polypeptide chains and two light polypeptide chains. There are two antibody-reaction sites on each molecule where combination with specific antigen can occur.

The structure of larger and more complex classes of immunoglobulin has not yet been fully determined. The class IgM, for example, is known to consist of five subunits (Fig. 20-5), each of which resembles IgG. The H chain is characteristic and different for each immunoglobulin class.

The study of immunoglobulin structure has contributed greatly to an understanding of the ways in which antibody molecules function. It has also given insight into the genetics and evolution of immunological responses.

The Structure of Antigens

Most antigens, like antibodies, are macromolecules. Substances with molecular weights of less than 10,000 are usually not antigenic. Antigens may be proteins,

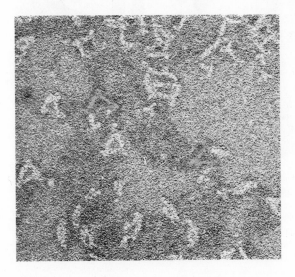

FIGURE 20-3
Electron micrograph showing rabbit antibodies (IgG) that have combined with a specific hapten. The Y shape of the antibodies is readily apparent. Immunoglobulin molecules are flexible as shown by similar photographs of antibodies combined with larger antigen molecules, in which the Y shape forms different angles around the central hinge area. (✕ 500,000) (Courtesy of Dr. N. M. Green)

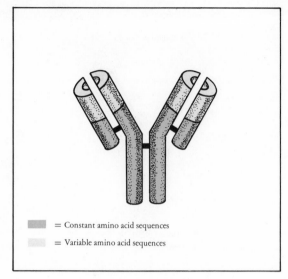

FIGURE 20-4
Variable amino acid sequences at the ends of both heavy and light polypeptide chains cooperate to make up the antibody-reaction sites. The variability accounts for the wide range of different antibody specificities.

= Constant amino acid sequences

= Variable amino acid sequences

polysaccharides, complex molecules combining various substances (for example, lipopolysaccharides), or occasionally lipids or nucleic acids; however, proteins and polysaccharides are usually much better antigens than lipids and nucleic acids. Another important characteristic of antigens is that they are usually foreign to the host that forms the specific antibodies. If this were not so, an individual would respond immunologically against his own body constituents, perhaps

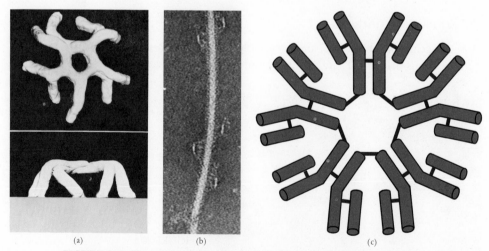

(a)　　　　　(b)　　　　　(c)

FIGURE 20-5   The structure of Immunoglobulin M. (a) Model of the proposed structure of IgM. When viewed from the top, it is apparent that the molecule consists of five Y-shaped units similar to IgG. The side view of IgM attached to an antigen gives the impression of a "staple." (b) Actual electron micrograph of IgM attached to specific antigen determinants on a flagellum of *Salmonella paratyphi*. Note the "staple" appearance. (c) A schematic representation of IgM. Note that each Y-shaped subunit consists of two heavy and two light polypeptide chains. (Photos courtesy of Dr. E. A. Munn and Dr. M. J. Hobart, (a) from A. Feinstein and E. A. Munn, *Nature,* 224:1307, 1969)

leading to tissue damage. Under unusual circumstances this can occur, resulting in autoimmune diseases. It is important to note that the body components of one host may be antigens in other hosts to whom they are foreign.

Low molecular weight substances by themselves are not often antigenic, but they can serve as *haptens*. *A hapten is a substance which can react with specific antibodies, but cannot incite the production of antibodies unless it is coupled to a large carrier molecule.* Thus a hapten may be simply an antigenic determinant, or a part of one. For example, penicillin is a small molecular weight substance that is not antigenic alone; in the body, however, it can couple with large protein molecules of the host to form an antigenic hapten-carrier complex. Antibodies are produced in response to this complex, and some of them can react with penicillin alone. Obviously, the ability of many small molecules to act as haptens greatly increases the possibility of immunological reactions against foreign substances.

Many factors, including genetic constitution, are involved in determining whether or not an individual host responds to an antigen. The failure to develop immunity to a potential antigen, known as immunological tolerance, can be induced in various ways. Recognition of one's own tissues (self) and consequent failure to respond immunologically to substances of self is a form of tolerance. Before birth the antibody-producing tissues are relatively inactive; substances that are encountered during this period and that persist may not stimulate an immune response. The individual may become tolerant to such substances and remains tolerant as long as the substances remain accessible to potential antibody-forming cells.

IMMUNOLOGICAL
TOLERANCE

ANTIBODY PRODUCTION

The mechanisms of antibody production have not been fully elucidated; it is known, however, that immunoglobulins are produced in lymphoid tissues by specialized cells called lymphocytes and plasma cells. Animals below the vertebrates in the phylogenetic scale lack these cells and apparently do not have the capacity to form antibodies.

In man lymphoid tissues are widespread throughout the body and include the lymph nodes, spleen, tonsils, and others. The principal cells of these tissues, lymphocytes, also circulate in the blood and lymph (Table 20-4 and Fig. 20-6). Many phagocytic cells called macrophages are also found in all lymphoid organs. Macrophages can engulf foreign material and remove it from the circulation. In addition, they cooperate with lymphocytes and plasma cells during the immune response, as described below. Lymphoid tissues are strategically located to protect the body against invasion by virtually any route. Obviously this distribution has evolved as a major contribution to survival of the species.

It is estimated that a single human being can make antibody molecules of about a million different specificities. Exactly how the body can make so many different kinds of specific antibodies is not fully known. It is well established that a single antibody-forming cell generally can synthesize antibodies of only one

TABLE 20-4
Cells Involved in Body Defense

Cell Type	Morphology	Location in Body	Functions
Polymorphonuclear neutrophils (PMN, poly)	Lobed nucleus; granules in cytoplasm; ameboid appearance	Account for part of the leukocytes in the circulation; few in tissues except during inflammation	Phagocytosis and digestion of engulfed materials
Monocytes; macrophages	Single nucleus; abundant cytoplasm	Macrophages present in all tissues and in lining of vessels; monocytes are less mature circulating forms.	Phagocytosis and digestion of engulfed materials; can participate in killing foreign cells that are not engulfed
Lymphocytes	Single nucleus; little cytoplasm	In lymphoid tissues (such as lymph nodes, spleen, thymus, appendix, tonsils); also in the circulation	Participate in immunological responses
Plasma cells	Single nucleus pushed to one side of ovoid cell; cytoplasm packed with ribosomes	In lymphoid tissues	Antibody synthesis
Basophils; mast cells	Lobed nucleus; large granules in cytoplasm contain histamine	Basophils in circulation; mast cells present in most tissues	Release histamine and other mediators of inflammation

specificity. A clone, or family of cells which results from the proliferation of one cell, makes antibody molecules of a single specificity.

When antigen enters the body, it soon reaches the circulation and is channeled through the various depots of lymphoid tissue, where it is eventually removed from the circulation. Many antigens, especially antigen-containing particles such as microorganisms, are ingested by macrophages at these sites. Inside the macrophage, large particles may be digested by enzymes into smaller molecules, some of which are antigenic and can reach the potential antibody-forming cells, the lymphocytes and plasma cells (Fig. 20-7). The process of digestion of ingested materials within the macrophage is analogous to that described in amebae (Chapter 15). Macrophages probably lack the capacity to synthesize antibodies, but often appear to be important in the antibody response because they prepare antigens and present them to other cells.

The available evidence suggests that only certain lymphocytes can interact with each antigen and respond to it. Reaction with the antigen causes the cells to proliferate and differentiate into antibody-producing cells. It is thought that

Phagocytosis and digestion by
amebae
Pages 299, 300

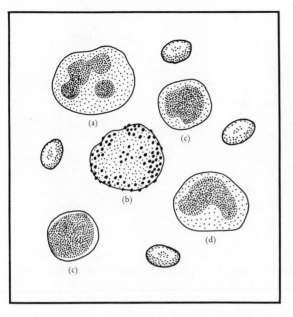

FIGURE 20-6
White blood cells that participate in body defense. (a) Polymorphonuclear neutrophil (also referred to as PMN, poly, or
granulocyte); (b) basophil; (c) lymphocytes; (d) monocyte.

antibody or antibodylike molecules of a given specificity are bound to the surfaces of potential antibody-forming cells, and that each of these cells has the
predetermined capacity to respond to a single specific antigenic determinant.
When antigen interacts with this antibodylike molecule, the cell is stimulated
to proliferate and differentiate. The mechanisms responsible for these changes are
not yet clear, but perhaps membrane changes following the interaction stimulate
cell division.

Of the many millions of lymphocytes in the body, only a limited number
recognize each specific antigenic determinant; however, a multitude of antigens
can be recognized as foreign by different members of the large population of
lymphocytes of a normal human being.

Postulated development of the
immune response,
Figure 21-2, Page 399

THE INTERACTION OF ANTIGENS AND HUMORAL ANTIBODIES

Humoral antibodies, that is, those which are free in the blood and other body
fluids, aid in body defense in many ways. The term *humoral immunity* refers to
the *in vivo protection provided by the activities of humoral antibodies*. These
activities can be measured *in vitro* by means of *serological reactions*.

An increase in amount, or a *rising titer,* of specific antibodies in the serum
is often useful in diagnosing disease. Early during infection the titer of specific
antibodies may be very low or nonexistent; however, by ten days or two weeks
the titer increases due to the antigenic stimulus of the infecting organisms.
Therefore in a case of suspected typhoid fever, for instance, serum samples may

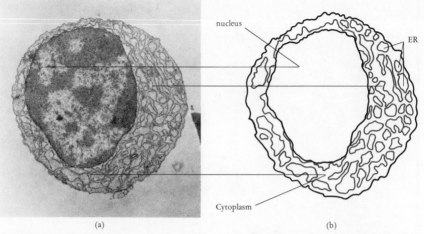

(a)   (b)

FIGURE 20-7   (a) Electron micrograph of a plasma cell; (b) diagram of the same cell, indicating the channels of rough endoplasmic reticulum (ER) in the cytoplasm, where antibody molecules are synthesized. Plasma cells are responsible for the synthesis of most antibody molecules. (Photo courtesy of Dr. U. Storb)

be taken as soon as possible during the illness and again approximately two weeks later. An appreciable increase in titer of antibodies specific for the typhoid organism indicates that the illness was indeed typhoid fever.

Although the antigen-antibody interactions described below can be measured *in vitro,* they are of primary significance *in vivo,* where they function in protecting the host against microorganisms and other foreign substances. The importance of humoral antibodies is clearly indicated by the observation that those rare individuals who lack the capacity to produce antibodies cannot survive unless they are continually supplied with humoral antibodies by means of repeated injections of pooled gamma globulin.

The Precipitin Reaction

Antibodies that combine with soluble antigen *in vitro* to form a visible precipitate are called *precipitins.* This is simply a functional term, describing an observable activity of these antibody molecules which may under different circumstances be capable of causing a variety of other kinds of activities. The precipitin reaction occurs in two stages. During the first stage, which takes place within seconds, antigen and antibody molecules combine to form small primary complexes. The second stage, which takes minutes to hours, involves the formation of precipitates by means of cross-linking in a lattice formation (Fig. 20-8). The first stage merely requires collision of antigen and antibody molecules, which then spontaneously combine; however, the second stage is dependent on the presence of certain ions. Whereas most chemical interactions involve the combination of fixed proportions of each reactant, it is remarkable that antigens and antibodies combine in *multiple proportions.* That is, an antigen molecule is usually

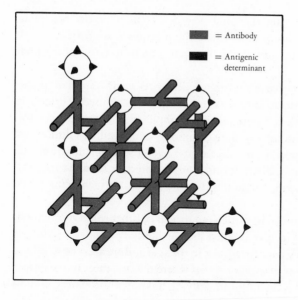

= Antibody

= Antigenic
determinant

FIGURE 20-8
Lattice formation by antigen and antibodies leads to precipitation.
Bivalent antibodies combine with multivalent antigens, linking
the two in an insoluble complex lattice formation that precipitates.

multivalent and has many antigenic-determinant sites available to combine with
bivalent antibodies; therefore one large antigen molecule may bind from one to
as many as 40 antibody molecules, depending on the antibody concentration.

The property of antigen-antibody combination in multiple proportions is
well illustrated by the quantitative precipitin test. In this test constant amounts
of antiserum are placed in each of a series of tubes and antigen is added in increasing amounts. Thus the first tube contains very little antigen, with an excess
of antibody; the second tube contains the same amount of antibody, with more
antigen; and so on until the last tubes contain an excess of antigen. At some
point between the zones of antibody excess and antigen excess, a *zone of optimal
proportion* is established, in which all the antigen and antibody molecules are
combined in lattice formation in the precipitate (Fig. 20-9). If all the tubes are
centrifuged and the precipitates removed, supernatant fluid from the first tubes
can be shown to contain free antibodies, and supernatant fluid from the last tubes
contains free antigen. Antibody molecules are known to have an average nitrogen
content; therefore the actual amount of antibody in an antiserum can be deter-

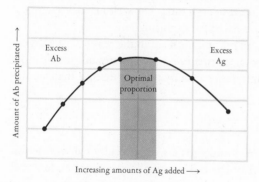

FIGURE 20-9
A precipitin curve. The maximum amount of precipitate forms
near the zone of optimal proportion where no free antigen (Ag)
or antibody (Ab) exist. Soluble Ag-Ab complexes form in the
region of Ag excess.

mined indirectly by measuring the total nitrogen content of the precipitate near the point of optimal proportion where the precipitate is greatest. Simple calculations are then used to convert the nitrogen content to terms of antibody protein.

Although the quantitative precipitin test illustrates many important points about antigen-antibody interactions, it is too complex to be used routinely in practical laboratory situations. Consequently a simplified qualitative precipitin reaction (the tube precipitin test) is more commonly employed. A solution suspected of containing antigen is carefully layered over a specific antiserum in a narrow tube. If a line of precipitate forms near the interface between the antiserum and the saline solution it indicates that the specific antigen being tested for was, indeed, present.

Precipitation tests can also be carried out in agar or other gels. One method, involving double diffusion in agar, is called the Ouchterlony technique. Antigen and antibody are placed in separate wells cut in the agar and are allowed to diffuse toward each other. A line of precipitate forms where the two meet in the proper proportions. This method is useful for identifying unknown substances, as indicated in Figure 20-10a.

Immunoelectrophoresis is a variation of the precipitation in gel technique which combines precipitation with electrophoresis, a method for separating mixtures of proteins (Fig. 20-10b).

The Agglutination Reaction

Agglutination reactions are similar in principle to precipitation reactions; however, *the antigen is particulate* rather than soluble, and as a result much larger aggregates of antigen and antibody are formed. Figure 20-11, which illustrates an agglutination reaction between red blood cells and specific antibodies (agglutinins), indicates how readily the large agglutinates can be seen. For comparison the figure also shows a preparation without any agglutination.

In testing for agglutination of bacteria or other particulate antigens, the suspension of antigen can be mixed with a small amount of serum on a slide, rocked for a minute or two, and observed for visible agglutination. Frequently, it is more informative to measure the titer of antibodies in the serum (Fig. 20-12). The usual procedure is to make serial dilutions of serum in test tubes. An equal volume of particulate antigen is added to each tube. The well-mixed contents of the tubes are incubated and then observed for agglutination.

In vivo, both agglutination and precipitation aid in the removal of antigens from the circulation, because the agglutinates or precipitates are much more quickly and efficiently removed and disposed of by phagocytic cells than are single molecules or particles.

Complement-Fixation Reactions

Bacteria or other cells may sometimes undergo *lysis* as the result of an interaction between the antigen that is a part of the cell, specific humoral antibodies, and

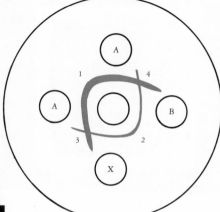

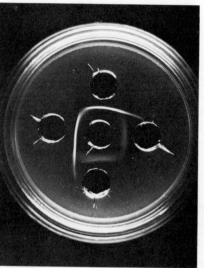

(a)   (b)

FIGURE 20-10   Antigen-antibody precipitation in agar gels. (a) Using the Ouchterlony double diffusion in agar method, antigen (Ag) is placed in one well cut in the agar and antibody (Ab) in another well. Each diffuses through the agar and a line of precipitate is formed where the two meet in proper proportions. This method can be used to determine whether an unknown substance is identical with or shares antigenic determinants with a known substance. If the known and unknown are identical in terms of antigenic determinants, the lines of precipitate fuse (a reaction of identity); if they are not the same, the lines do not fuse but cross each other (a reaction of nonidentity). In the center well are anti-A Ab and anti-B Ab; Ag A and Ag B are in outer wells as shown. X indicates an unknown Ag. Fusion of the lines at 1 and 2 shows reactions of identity (Ag A = Ag A; Ag B = Ag X). Reactions of nonidentity at 3 and 4 indicate that Ag A and Ag B are not the same and that Ag A and Ag X are not the same. (Photo courtesy of Dr. M. Tam) (b) The immunoelectrophoresis technique combines double diffusion in gel with electrophoresis. A mixture of many protein antigens, in this case human serum, is partially separated by electrophoresis before diffusion occurs, thereby permitting many more lines of precipitate to be distinguished. Human serum was subjected to an electric current; the various serum proteins present migrated to different areas, depending on their electric charges and other properties. Following electrophoresis, rabbit antiserum against human serum was placed in the trough below and the two diffused. More than 30 different proteins in serum, each represented by a distinct line of precipitate, can be distinguished by this method. (Photo courtesy of Dr. B. C. Gilliland)

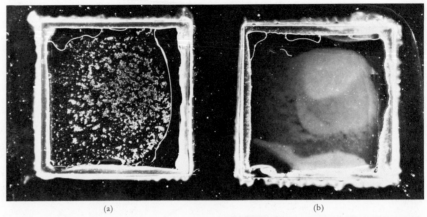

FIGURE 20-11 Agglutination of erythrocytes. (a) Agglutinated red blood cells and antibody. (b) Nonagglutinated red blood cells (control).

complement (*C*). Actually, C is not a single substance, but a very complex system of at least 11 serum proteins that participate nonspecifically in a temporally defined sequence in various immunological reactions. Cell lysis thus depends on specific antibody acting together with nonspecific C. As a rule C is fixed by antibody only after an antigen-antibody interaction has occurred. Complement can become fixed to antigen-antibody complexes in a number of ways to cause results

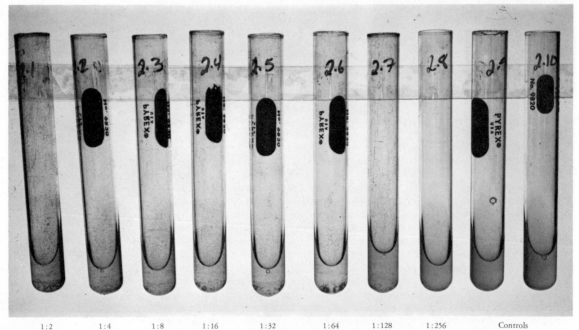

| 1:2 | 1:4 | 1:8 | 1:16 | 1:32 | 1:64 | 1:128 | 1:256 | Controls |

FIGURE 20-12 Tube agglutination test. Dilutions of serum are mixed with equal volumes of antigen-containing particles. The control tubes contain serum alone or particles alone. The test shown has visible agglutination through 1:128 and none in dilution 1:256; therefore the titer is 128. (Courtesy of Dr. B. Gilliland)

other than cell lysis. Many tests have been devised to utilize complement-fixation as an indirect measure of antibody concentration in a given material. The advantage of using complement-fixation is that some antibodies do not give a visible reaction after combining with antigen; however, if they fix C after reacting with antigens, these antibodies can be assayed by determining the amount of C that has been fixed. Figure 20-13 diagrams the test procedure, which is divided into two separate reactions. In the first, the antigen, antibody, and C are allowed to combine; however, the reaction is not visible. In the second reaction, an indicator

Positive complement-fixation reaction

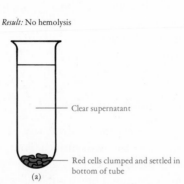

Test system:
Antigen +
Heated test serum containing specific antibodies +
Unheated serum as a source of complement
} Ag-Ab-C complex

Indicator system added:
Red blood cells +
Antibodies specific for the red cells
} Red cell-Ab complex but no C left to lyse red cells because C was fixed by test system

Result: No hemolysis

— Clear supernatant

— Red cells clumped and settled in bottom of tube

(a)

Negative complement-fixation reaction

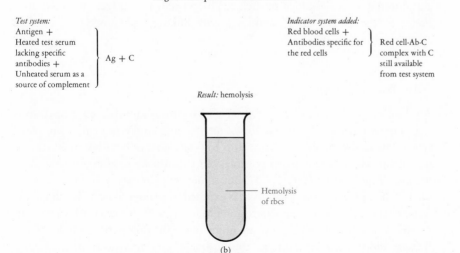

Test system:
Antigen +
Heated test serum lacking specific antibodies +
Unheated serum as a source of complement
} Ag + C

Indicator system added:
Red blood cells +
Antibodies specific for the red cells
} Red cell-Ab-C complex with C still available from test system

Result: hemolysis

— Hemolysis of rbcs

(b)

FIGURE 20-13   Complement-fixation test procedure.

system is added that will give a visible indication of the amount of C left which has not combined in the first reaction and hence is free to combine with the indicator system.

It is first necessary to heat the patient's (test) serum (56°C, 30 min) in order to inactivate the unknown amount of C present. Serial dilutions of the serum being tested are then mixed with equal volumes of the antigen preparation and an appropriate amount of C. After the specific antibodies have combined with antigen and have fixed C, the indicator system is added. Usually this consists of sheep red blood cells (SRBCs) plus specific antibodies against SRBCs. When SRBCs, antibodies against SRBCs, and C combine, the various components of C act sequentially, resulting in enzymatic degradation of small areas of SRBC membrane. Tiny holes are made in the membranes and the red cells are lysed (hemolysis), giving rise to a readily visible change in color. If the C has been fixed by the first (test) reaction, it is not available to cause hemolysis in the second (indicator) reaction. However, if no antigen-antibody-C interaction occurred in the test system, C remains available to cause hemolysis in the indicator reaction. The amount of hemolysis can be quantitated to give a measure of the amount of C fixed in the test reaction and, indirectly, the amount of specific antibody present in the test serum.

The Coomb's Antiglobulin Test

The Coombs' test is another indirect way of demonstrating antibodies that do not give a visible reaction with antigen. For this test, antiserum is made in goats, rabbits, or other foreign species against human serum globulin (Coombs' antiglobulin serum). Because human antibodies and complement are serum globulins, the Coombs' serum will react with them. Suppose a few human antibody molecules specific for certain red blood cells have combined with the red cells but have not been able to agglutinate them under the conditions employed. When antiglobulin serum is added to such red blood cells it will react with the antibodies on the red cell surface and cause cross-linking, and consequently, visible agglutination (Fig. 20-14).

Virus Neutralization

Virus plaque formation
Pages 328, 329

Antibodies against viruses can be detected by various methods. One of them is virus neutralization. The interaction of neutralizing antibodies with virus particles results in loss of ability of the virus to infect host cells, either in intact animals or in cell cultures. When possible, cell cultures are used for the test, as in the case of poliovirus (Fig. 20-15), but sometimes laboratory animals must be used. When a susceptible cell culture is exposed to poliovirus particles alone, the cells become infected and are destroyed, producing a visible cytopathic effect. If, however, neutralizing antibodies are mixed with the poliovirus particles before they are added to the cell cultures, the antibodies prevent virus from combining with host receptors and infecting the cells.

Other tests used for measuring antiviral antibodies include the hemagglutination-inhibition reaction. Some viruses contain substances that will cause the agglutination of normal red blood cells (hemagglutination). Specific antiviral antibodies can react with the substances and prevent hemagglutination reactions; thus they can be measured in the hemagglutination-inhibition test.

The Interaction of Antigens and Humoral Antibodies

Hemagglutination of viruses,
Pages 330, 331

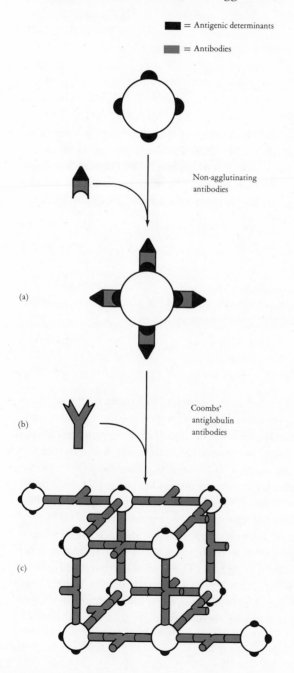

■ = Antigenic determinants

■ = Antibodies

(a)

Non-agglutinating antibodies

(b)

Coombs' antiglobulin antibodies

(c)

FIGURE 20-14
The Coombs' antiglobulin test. (a) Antibodies sometimes combine with antigen in such a way that agglutination or precipitation cannot occur to give a visible reaction; (b) specific antiglobulin antibodies can be used to link the antigen-antibody complexes together; (c) the result is lattice formation and a visible reaction.

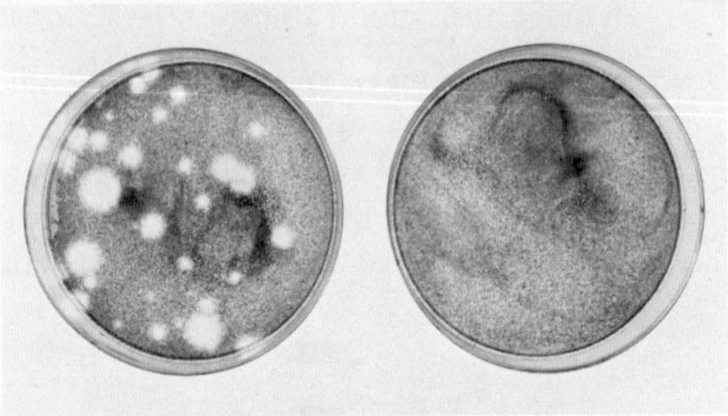

FIGURE 20-15
Virus neutralization test. (a) When a susceptible cell culture is exposed to poliovirus particles, cells become infected and are destroyed, producing visible clear areas (plaques) in a lawn of cells. (b) If specific neutralizing antibodies are mixed with polioviruses before exposure to the cells, the viruses cannot combine with host cell receptors. As a result, the cells are not infected and no plaques appear in the lawn of cells.

(a)   (b)

Toxin Neutralization

Antibodies specific for soluble microbial toxins are called antitoxins. They cause toxin neutralization by combining with toxins in such a way as to block the reaction of the toxic site on the molecule with its target cell (Fig. 20-16). Antitoxin molecules may also cause the formation of a harmless toxin-antitoxin complex or precipitate, which *in vivo* is readily removed from the circulation by phagocytic cells.

Opsonization

Humoral antibodies can often coat bacteria or other particles, causing them to be engulfed, or phagocytized, more efficiently. This phenomenon is called opsonization, that is, a process that promotes phagocytosis.

CELLULAR IMMUNITY

The immunity mediated, not by humoral antibodies, but rather by the activities of specifically immune cells is termed *cellular immunity*. The immune cells which participate may be either lymphocytes or macrophages carrying cell-bound antibodies on their surfaces. An example of cellular immunity is seen in tuberculosis, where the causative organism (*Mycobacterium tuberculosis*) lives within macrophages of the host. Although humoral antibodies against *M. tuberculosis* are often present, they do not control the infection. Resistance depends on effective cellular immunity, mediated by changes in the macrophages that harbor the organisms (Fig. 20-17). The macrophages are stimulated and increase their metabolic activity to produce large quantities of enzymes, such as lysozyme, which act against *M. tuberculosis* within the macrophage. This kind of antimicrobial cellular immunity is important in many other kinds of infection besides tuberculosis. Some fungi and the bacteria that cause brucellosis, leprosy, and a number of other diseases live within macrophages and are also controlled by changes in the macrophages.

386

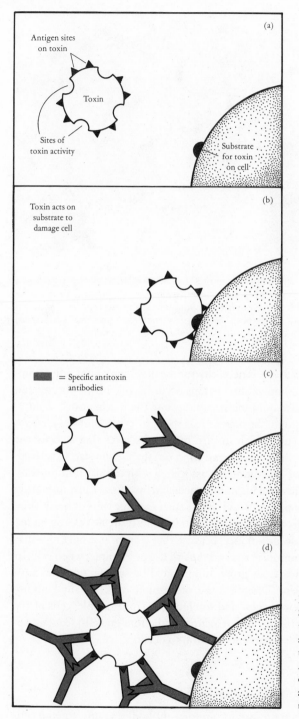

FIGURE 20-16
Steric hindrance of toxin activity by antitoxin action. (a) Sites of toxin activity, indicated by inverted areas on the toxin molecule, are specific for substrate on the cell surface. (b) Interaction of toxin with substrate leads to cell damage or death of the cell. (c) Antitoxin antibody molecules are specific for everted antigenic determinant sites on the toxin molecule. (d) Reaction of toxin and antitoxin leads to steric hindrance of the sites of toxin activity, thereby blocking the toxic action of the molecules.

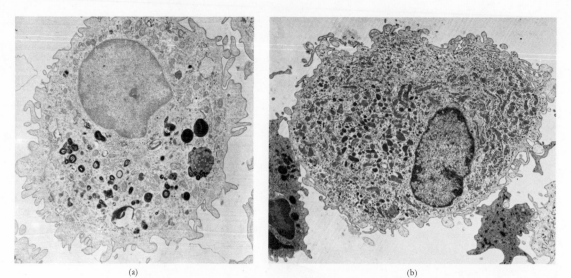

FIGURE 20-17 Normal and activated macrophages. (a) A macrophage from the lung of a normal nonimmunized rabbit. Few phagocytic vacuoles (dark areas) are present. (b) A macrophage from the lung of a rabbit five days after immunization with mycobacteria. The cell is highly activated, as evidenced by the abundant phagocytic vacuoles and lysosomes (dark areas) present in the cytoplasm. (Courtesy of Dr. Q. N. Myrvik and E. S. Leake)

Specific lymphocyte activity is necessary for the development of cellular immunity. It is known that complex interactions occur between lymphocytes and macrophages during cellular immune responses, but the mechanisms of interaction are not well understood. At present it is thought that a series of changes occur in specific lymphocytes, which in turn make substances that activate the macrophages. This interaction also occurs during certain allergic states (delayed-type hypersensitivities) that are commonly associated with cellular immunity.

Delayed-type hypersensitivities
Pages 398–401

Antitissue cellular immunity
Pages 401-403

In addition to antimicrobial immunity mediated by cells, cellular immunity may be directed against foreign cells and tissues such as transplants of heart, skin, or kidney. The mechanisms of antitissue cellular immunity are not clearly understood but are of the utmost importance, because they govern not only rejection of tissue transplants but immunity to cancers as well. For example, when cellular immunity fails, many cancers can grow, uninhibited by specific humoral antibodies which may be present. Direct contact of immune cells (usually lymphocytes but occasionally macrophages) and foreign cells (for example, cells of the transplant or of a tumor) results in destruction of the foreign cells. Contact appears to be essential in order for cell destruction to occur.

IMMUNOLOGICAL ENHANCEMENT

It seems paradoxical that certain specific humoral antibodies against foreign cells (called enhancing antibodies) may oppose an effective antitissue cellular immunity. These humoral antibodies have the capacity to react with foreign cells and

388

yet not harm them; however, in so doing they may interfere with the immune cell-target cell contact that is necessary for cell destruction. They may also act in other ways to inhibit the development of cellular immunity. The result is that survival of the foreign tissue is enhanced by the action of specific humoral antibodies—a phenomenon referred to as immunological enhancement.

THE ANAMNESTIC RESPONSE

The immune response, whether it is predominantly cellular or humoral, is much faster and more efficient after a secondary or subsequent exposure to the antigen than after the first exposure. A *primary response* follows the first contact of the host with the antigen, during which time lymphocytic cells must proliferate extensively before an effective response is achieved. The *secondary response* designates the events which occur when an antigen is contacted again, whether it is for the second or the hundredth time; it is accelerated and more intense, and is called an *anamnestic response* (memory response). One possible explanation for the greater extent of the secondary response is that during the primary response a number of specific cells proliferate, and some of these survive to give a larger number of cells called memory cells, that is, cells which "remember" the specific antigen and are ready to react and respond to it. Figure 20-18 illustrates the differences in extent and in timing of humoral antibody production during the primary and secondary responses.

The anamnestic response is the basis of immunization methods to prevent disease. The immunizing agents prepare the host to give a prompt and accelerated response when the antigen is again encountered, perhaps in the form of virulent organisms. A table of commercially prepared immunizing agents is presented in Appendix IV, pages 682 and 683.

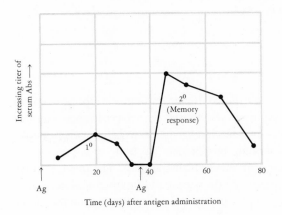

FIGURE 20-18
Primary and secondary immune responses. Following a secondary dose of antigen (Ag) an accelerated memory (anamnestic) response occurs, with much higher titers of antibodies (Abs) which persist for a longer period of time.

CONTROL OF ANTIBODY SYNTHESIS AND DEGRADATION

The anamnestic response described allows the rapid synthesis of antibodies in response to antigenic stimuli such as pathogenic microorganisms and thereby obviates the necessity of maintaining large quantities of each specific antibody at all times. Antibodies are produced only following specific antigenic stimulation of appropriate lymphoid cells, and persistence of antigen is required for continued proliferation of antibody-forming cells. When antigen is no longer present and available, antibody formation lasts for a limited time and then gradually decreases.

Almost all the antibodies possessed by the newborn (except in the case of infections of the fetus) are of maternal origin, hence are IgG molecules that have been produced by the mother and have crossed the placenta to the fetus (because only antibodies of the class IgG are able to cross the placenta). These are soon replaced by antibodies of the various classes, synthesized by the infant. The amounts of immunoglobulins increase during childhood to reach adult levels at about 12 years of age or earlier.

Immunoglobulins in the blood of adults normally remain within a remarkably constant range of approximately 1500 to 2000 mg per 100 ml of blood plasma. Following prolonged or intensive antigenic stimulation, the blood level of immunoglobulins may become elevated. More often, however, antibodies combine with and cause rapid removal of the antigen, thereby effectively preventing the synthesis of excess amounts of immunoglobulin.

Once formed, antibody molecules have a finite lifespan, as do all protein molecules. They may be lost from the body in secretions or excretions, and those which are not lost are eventually degraded by proteolytic enzymes. The *half-life* of antibody molecules varies considerably, depending on their class and their location (that is, whether they are free, cell-fixed, or transported into secretions). The average half-life of most antibody molecules (IgG) in human circulation is estimated to be 20 to 25 days. Serum concentrations of some immunoglobulins appear to exert a regulatory influence on the rate of their catabolism: when the total concentration is high, these immunoglobulins are degraded more rapidly than when the concentration is lower.

SUMMARY

Immunity may be either innate or acquired. Innate immunity is inborn and often depends on the activities of phagocytes and nonspecific defense factors. Acquired immunity is gained in response to a foreign agent and usually involves specific antibodies. An antibody is a protein, produced in the body in response to an antigen, which can combine specifically with that antigen. The interaction between determinant sites on antigen and the corresponding reaction sites on antibody molecules requires close complementary fit between the two and the formation of a number of noncovalent bonds. The interaction is highly specific and it is reversible.

Antibody molecules have portions that consist of variable sequences of amino acids, accounting for the tremendous variety of specificities found among antibodies. They also have constant sequences of amino acids that are responsible for certain biological properties of the molecules, such as the ability of some to pass the placenta and to fix complement. The five known classes of antibody molecules differ in structure and, to some degree, in function.

Antigens are commonly large molecules, foreign to the host who responds to them by forming antibodies. Haptens are low molecular-weight substances that can combine with specific antibodies, but cannot stimulate the production of antibodies unless the hapten is coupled to a high molecular-weight carrier molecule.

Antibodies are synthesized by lymphoid cells, principally plasma cells. It is postulated that antigen stimulates a few lymphoid cells which can specifically respond to the antigen by proliferating and differentiating into a clone of antibody-producing plasma cells. Antibodies synthesized by plasma cells and secreted into body fluids are known as humoral antibodies.

Humoral antibodies aid in body defense by interacting *in vivo* with foreign antigen in various ways, leading to the destruction or removal of foreign materials. The interaction of antigen and antibody may lead to precipitation, agglutination, opsonization, complement-fixation, virus neutralization, or toxin neutralization. These reactions can be observed *in vitro* also and serve as means of demonstrating the presence of antibodies in body fluids, such as blood, or in tissues.

The protection provided by humoral antibodies is called humoral immunity. The immunity mediated, not by humoral antibodies but rather by the activities of specifically immune lymphoid cells, is known as cellular immunity. The development of cellular immunity depends on the activities of specific lymphocytes. Both lymphocytes and macrophages cooperate in effecting cellular immunity.

The immune response occurs faster and is more efficient following a secondary stimulus than after a primary stimulus with any given antigen. The accelerated secondary response is called the memory, or anamnestic, response.

Antibody synthesis and degradation are subject to a variety of controls which are poorly understood at present.

QUESTIONS

1. Make a diagram that illustrates the important characteristics of antigens and antibodies.
2. Are the bonds responsible for antigen-antibody interactions covalent or noncovalent? How might the interactions differ if both covalent and noncovalent bonds participated? Would this be likely to benefit or harm the host?

3. The constant amino-acid sequences of antibody molecules are responsible for many biological functions. List some of these functions. What is the role of the variable amino-acid sequences of antibody molecules?

4. What is the cause of the precipitin reaction? In a quantitative precipitin test why might some tubes that contain small amounts of antigen and large amounts of antibody fail to show any precipitate? Draw an explanatory diagram.

5. Describe a complement-fixation test. What are its advantages over the simpler agglutination test?

FURTHER READING

CARPENTER, P. L., *Immunology and Serology,* 2d ed. Philadelphia: W. B. Saunders Co., 1965.

EDELMAN, G. M., "The Structure and Function of Antibodies," *Scientific American,* 223 (August 1970), 34–42.

NOSSAL, G. J. V., *Antibodies and Immunity.* New York: Basic Books, 1969.

PORTER, R. R., "The Structure of Antibodies," *Scientific American,* 217 (October 1967), 81–87.

WEISER, R. S., Q. N. MYRVIK, and N. N. PEARSALL, *Fundamentals of Immunology for Students of Medicine and Related Sciences.* Philadelphia: Lea and Febiger, 1969.

Chapter 21

Immunology: Hypersensitivities

Antibody production and other immunological mechanisms have developed during thousands of centuries of evolution; consequently most of these mechanisms have obvious survival value. Nevertheless we find that immunological responses are not always protective, but may sometimes harm or even kill the individual making the response. *The term hypersensitivity has been applied to immunological reactions that cause tissue damage in the host. A synonym for hypersensitivity is allergy.* Even though it is often difficult to imagine why man should possess the capacity to develop allergic reactions, recent evidence suggests that at least some allergies offer definite advantages for survival, along with their more readily apparent drawbacks.

IMMEDIATE-TYPE AND DELAYED-TYPE ALLERGIES

Hypersensitivities can be classified into two types: immediate and delayed. Immediate-type reactions occur rapidly after challenge of a sensitized individual with specific antigen, and are maximal within minutes to a few hours, depending on the type of reaction; delayed-type reactions, however, are not apparent until six to twelve hours after challenge and reach their peak at two to three days. Immediate-type hypersensitivity reactions depend on the activities of humoral antibodies, whereas those of delayed hypersensitivity result from the activities of specifically sensitized lymphoid cells, rather than free humoral antibodies. As a rule with immediate-type hypersensitivities, if blood serum is collected from a person or an experimental animal that is allergic to a specific antigen and is transferred to a recipient of the same species, the recipient becomes allergic temporarily to that antigen as a result of the transfer of specific antibodies in the serum. Delayed hypersensitivities, on the other hand, can be transferred with living lymphoid cells from an allergic individual, but not with serum; therefore some activities of these lymphoid cells (specifically sensitized lymphoid cells) must be responsible for this type of allergy.

Some of the characteristics of the major categories of hypersensitivities are outlined in Table 21-1. Immediate sensitivities are classified into three categories on the basis of the mechanisms responsible for the allergic reaction. An understanding of the mechanisms is important in order to know how to treat each kind of allergy. This chapter is concerned with the IgE-mediated and cell-mediated hypersensitivities because they account for a large majority of all allergic reactions. Immediate cytotoxic hypersensitivities in which cells are destroyed directly, are exemplified by blood transfusion (discussed later in this chapter).

The immunological response to any antigenic stimulus probably will be many-faceted, involving protective and hypersensitivity responses of various kinds. However, most hypersensitivity reactions are predominantly of one of the four types listed in Table 21-1, with varying smaller degrees of other responses occurring simultaneously. Thus "serum sickness" is a disease that usually results chiefly from an immediate-type immune-complex reaction, although some components of the other types of responses may also be demonstrable. This disease represents an allergic reaction to foreign blood serum, such as horse antitoxin

TABLE 21-1
Some Characteristics of the Major Types of Hypersensitivities

Type	Time of Reaction, Following Challenge of Sensitized Individual with Antigen	Mediated by Activities of	Examples
Immediate			
Atopic or IgE-mediated	Immediate, within seconds or minutes; fades by an hour or several hours	IgE antibodies, fixed to mast cells in tissues and to basophils	Hayfever, hives, asthma, anaphylaxis
Cytotoxic	Immediate, within seconds or minutes	Humoral antibodies and complement,[a] reacting with antigen that is a part of a cell	Blood transfusion reaction with mismatched blood
Immune complex	Immediate, within minutes to few hours	Humoral antibodies and complement,[a] reacting with soluble antigens that are not part of a cell	Serum sickness, Arthus-type reactions, farmer's lung (Chap. 22 and 25)
Delayed			
Cell-mediated	Delayed; first visible at about 6 hours, peaks at 24 to 48 hours; gradually declines over a period of days.	Specifically sensitized lymphoid cells	Positive tuberculin skin test; poison ivy rash

[a] Complement may not always be necessary

that is given to counteract the toxin of tetanus. Humoral antibodies of the class IgG are formed, which combine with the soluble horse serum antigens and complement; the soluble immune complex formed in this way interacts with polymorphonuclear white blood cells, leading to release of enzymes that are directly responsible for the tissue damage.

Immediate-Type Hypersensitivity (IgE-Mediated or Atopic)

Anyone who has ever seen a person with hayfever, hives, or allergic asthma is familiar with the manifestations of atopic allergy; these conditions are so common that almost everyone recognizes them.

Hayfever occurs when a suitably sensitized person inhales specific antigen; the itching, running eyes, sneezes, and dripping nose that result are evidences of extensive damage to cells in the upper respiratory tract. *Hives* is a skin disease characterized by redness, itching, and the formation of wheals; the skin lesions in some ways resemble mosquito bites but are much larger and more troublesome. Usually, but not always, hives occur after ingestion of the antigen; for example, a person allergic to lobster may experience intense itching and "break out with hives" soon after eating even small amounts of lobster. Occasionally hives can be a severe disease. The localized swelling may produce giant hives; when the swelling impinges on the larynx, it may even obstruct the air flow, but this occurs rarely. *Eczema* is another skin condition often caused by atopic allergies.

Asthma, with its symptoms (wheezing, coughing, and difficulty in breathing, particularly in exhaling) is perhaps not as familiar to many as are hayfever and hives; asthma, however, is a fairly common and sometimes serious affliction.

Anaphylaxis, a reaction of shock, is another important manifestation of atopic allergy. It may be fatal, but it is fortunately extremely rare. Anaphylaxis is similar to other atopic allergic reactions, such as hives and asthma, but instead of being localized it is manifested systemically throughout the body. Swelling occurs, not just in the skin but also in the bronchi (resulting in difficulty in breathing) and in many other internal tissues. Death often ensues from a combination of causes, including generalized loss of fluid from the vessels because of increased vascular permeability (shock). Within seconds or minutes after exposure to the antigen, the symptoms and signs of anaphylaxis become apparent, for example, itching and flushing of the skin, difficulty in breathing, and subsequent collapse as a result of shock.

In the United States today penicillin is the substance most often responsible for anaphylactic reactions. As mentioned previously this small molecular weight antibiotic by itself is nonantigenic; however, penicillin can act as a hapten. It can combine within the body with certain large molecular weight proteins; the resulting hapten-carrier complex is capable of inciting production of antibodies, some of which are specific for penicillin alone. Penicillin is often injected, rather than taken by mouth, so that a large quantity of the antibiotic gets into the bloodstream very quickly and is immediately circulated throughout the body. If

Haptens
Page 375

the recipient has previously become allergic to penicillin, an immediate reaction can occur systemically, and may be fatal within minutes. Although anaphylaxis is extremely rare in patients receiving this antibiotic, less severe allergic reactions to penicillin are not uncommon and are estimated to occur in as many as 10 percent of recipients. Because of the danger of anaphylaxis, any person who has had a reaction to penicillin, no matter how mild, should avoid subsequent doses.

For many years the mechanisms common to the various atopic allergic diseases were obscure. Many investigators, who observed that these allergic manifestations are often intensified by emotional upsets, were convinced that these diseases were caused by emotional and psychological factors. However, evidence accumulated over the years that antibodies are responsible for this kind of allergic reaction. It was not until 1965 that the class of immunoglobulins which is responsible for atopic allergies was discovered and named IgE. This class was extremely difficult to isolate and identify because it is present in the blood serum in almost infinitesimal quantities. In fact the amounts of IgE present in normal blood plasma represent only about one fifty-thousandth of the amounts of IgG present!

The property of IgE antibodies responsible for their unique activity is their capacity to attach to a certain type of cells, called mast cells and basophils, found in almost all parts of the body. These cells are particularly abundant in the nose, lungs, and skin, which helps to explain the frequent localization of this type of reaction to these areas.

Figure 20-6
Page 377

Mast cells and basophils (Fig. 20-6) contain many large granules packed with histamine and other substances. The attachment of IgE to mast cells is not an antigen-antibody interaction; rather it is a cytophilic association in which a specific site on the constant amino-acid portion of the IgE molecule attaches to a receptor on the cell as shown in Figure 21-1. Once attached, the IgE molecule can survive for many weeks and its antibody-reaction sites are readily available to interact with specific antigen. The person with an atopic hypersensitivity, such as an allergy to ragweed pollen, will have many IgE antibodies specific for this pollen fixed to mast cells throughout the body, and especially in the upper respiratory tract. When not exposed to the antigen, the antibodies are harmless, but when ragweed pollen is inhaled, it combines rapidly with the cell-fixed IgE. Within seconds the antigen-IgE reaction on cell surfaces leads to the release, from the granules, of histamine and other chemicals which mediate the tissue damage. The result is an attack of hayfever. If the ragweed antigen was accidentally injected into the bloodstream of a highly allergic individual, anaphylaxis and death might occur within minutes.

Histamine is the principal mediator in hayfever and hives; therefore antihistamine drugs are very effective in treating these diseases. Other mediators are more important in asthma; consequently antihistamines are of limited value in counteracting asthmatic attacks. It is easy to show that the symptoms of hives are caused by the action of histamine simply by injecting a small quantity of histamine into the skin; a reaction identical to hives occurs immediately. Other factors leading to histamine release may lead to the development of hives or hayfever; however, antigen-IgE interactions are a most frequent cause.

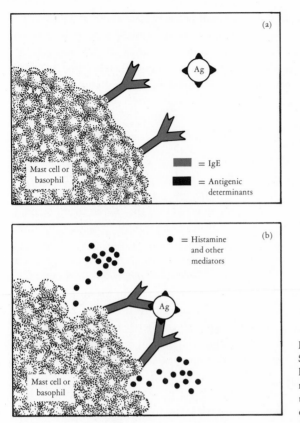

FIGURE 21-1
Schematic representation of an Immunoglobulin E-mediated
hypersensitivity state. (a) IgE molecules form a cytophilic attachment to mast cells or basophils. (b) A molecule of antigen bridges
two molecules of IgE, causing release of mediators from granules
of the cell.

Skin tests are often done to determine the substances to which a person is
sensitive, or allergic. Extremely small quantities of antigen are injected directly
into the skin. If the subject is sensitive, an immediate reaction resembling hives
occurs: a fluid-containing wheal forms, surrounded by a reddened area. The reaction reaches a peak within 20 to 30 minutes and fades within a few hours. Great
care must be taken to avoid inducing anaphylaxis while skin testing highly allergic persons.

In view of the danger of anaphylaxis following exposure to antigen, it
seems paradoxical that one of the common ways of treating severe allergies involves multiple injections of specific antigen. This procedure is often referred to
as desensitization. The aim is to make the patient less hypersensitive (or allergic)
to the offending antigen. Starting with extremely small doses, the amount of
injected antigen is very gradually increased over a period of months. Usually the
patient becomes less sensitive or nonreactive to the particular antigen. The rationale is that repeated doses of antigen cause the formation of IgG and other
classes of antibody. The IgG antibodies then combine with the antigen before
it can react with IgE, thereby preventing the atopic reaction. There is no solid
proof that this is true; it is known, however, that IgG antibodies are formed, and
the procedure often seems to be effective whatever the mechanism.

It has been observed that only a small percentage of those exposed to potential antigens, such as ragweed pollen, for example, become allergic to them. Furthermore there is a familial tendency to develop atopic allergic diseases. Although the reasons for these observations are not known, it is thought that these "allergic individuals" are genetically predisposed to produce abnormally large amounts of IgE antibodies.

Delayed-Type Hypersensitivity (Cell-Mediated)

It is likely that most people in the United States have been skin tested with tuberculin, so that the prototype of cell-mediated allergic reactions, the positive tuberculin skin test, may be familiar. In this test proteins from tubercle bacilli (the causative agents of tuberculosis) are injected into the skin in the same manner as described for the atopic skin tests. Instead of an immediate reaction, however, individuals who are allergic to this substance usually have a delayed-type reaction that first becomes apparent as a reddened area about 12 hours after the antigen is injected, and gradually becomes *thickened, or indurated,* reaching a *peak at 24 to 48 hours.* There is no formation of wheals, as seen in the atopic reaction. A positive tuberculin skin test does not indicate that the individual has tuberculosis; it simply means that he has been exposed to tubercle bacilli enough to become allergic to the bacterial protein. The test is useful in diagnosing doubtful cases of tuberculosis, in epidemiological studies, and several other ways. Similar tests are used in diagnosing other diseases. If these were the only applications of cell-mediated allergic reactions, we would not be particularly concerned here with understanding them. However, delayed-type hypersensitivity reactions are intimately involved in other major areas, such as transplantation rejection, cancer immunology, autoimmune diseases, and cellular immunity to many microorganisms. Delayed-type hypersensitivity usually accompanies cellular immunity, and some have equated the two. Immunity, however, leads to protection and hypersensitivity causes damage to the host. Although the mechanisms may be the same, the results differ.

Obviously if delayed-type reactions have such widespread importance, it is essential to understand how they occur, what mechanisms are at work, and how they differ from immediate-type sensitivities. Unfortunately large gaps exist in our understanding of delayed-type hypersensitivities. At the present time many investigators are studying this problem, for if cell-mediated allergic reactions could be manipulated and controlled at will, it would permit tremendous progress in the areas mentioned.

Lymphocytes are the cells that initiate delayed-type hypersensitivity reactions (Fig. 21-2). When lymphocytes from an allergic individual (*sensitive lymphocytes*) are cultured *in vitro* in the presence of specific antigen, they synthesize and secrete into the medium a variety of products with biological activities. These products are called collectively *lymphokines* (Table 21-2). The table indicates how each lymphokine functions *in vitro,* but proof of *in vivo* activity is lacking as yet. Nevertheless, it seems likely that lymphokines are the mediators of reactions of delayed-type hypersensitivity and cellular immunity.

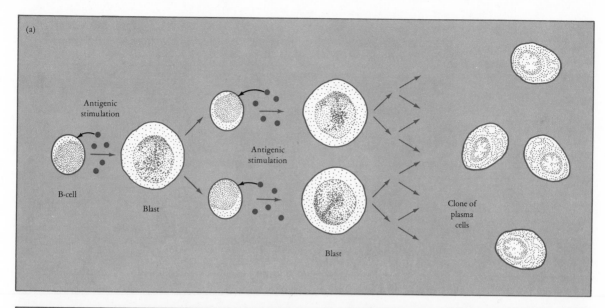

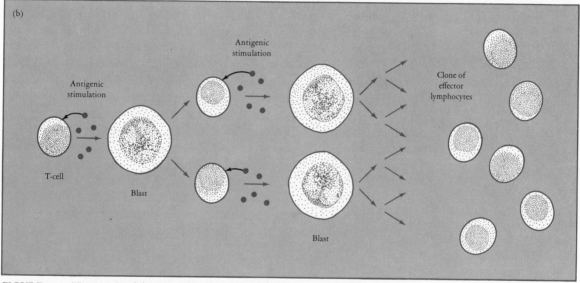

FIGURE 21-2 The postulated development of the immune response. (a) Under the influence of the lymphoid tissue of the gut, stem cells from the bone marrow become B-cells. Upon stimulation with specific antigen, B-cells enlarge to become blast cells which divide to form two small lymphocytes. The process is repeated if antigen is available. During proliferation the B-cells differentiate, finally yielding a clone of plasma cells synthesizing antibody molecules. (b) Under the influence of the thymus, stem cells become T-cells. Following antigen-stimulated proliferation, T-cells yield clones of effector lymphocytes that are active in cellular immunity.

TABLE 21-2

Lymphokines (Products of Sensitized Lymphocytes)[a]: Probable Mediators of Delayed-type Hypersensitivity and Cellular Immunity

Lymphokine	Activity in Vitro	Probable Action in Vivo
Chemotactic factor	Attracts macrophages	Causes influx of macrophages into area where antigen is present
Migration-inhibitory factor	Inhibits migration of macrophages	Keeps macrophages immobilized in area near antigen
Lymphotoxin	Kills many different kinds of cells nonspecifically	Kills foreign cells; also kills cells of host, resulting in tissue damage
Macrophage-activating factor	Causes macrophages to become metabolically active and to synthesize many degradative enzymes	Results in macrophage activation; activated macrophages are able to kill foreign cells and to degrade ingested materials efficiently

[a] Lymphokines are synthesized and released by sensitized lymphocytes following stimulation by specific antigen. Only a few of the lymphokines are listed in this table.

Lymphocytes and their products are not the only factors in delayed allergic reactions; macrophages and other cells also participate. When an antigen, such as tuberculin, is injected into the skin of a sensitized person, it is soon encountered by a few sensitive lymphocytes, for these cells circulate constantly. It is thought that upon recognizing and interacting with antigen, the sensitive lymphocyte is stimulated to transform and to divide, and during the process the lymphokines are secreted. These substances act on many other cells in the vicinity; for example, some of the lymphokines attract macrophages and immobilize them at the site. The resulting accumulation of cells is in large part responsible for the induration of the skin at 24 to 48 hours. If this hypothetical sequence of events is truly what happens during delayed reactions, as now seems likely, then the delay of 12 hours or so before the reaction becomes visible represents time required for the transformation of lymphocytes and the subsequent steps.

Antimicrobial cellular
immunity
Pages 386, 388
The question of how delayed hypersensitivity is related to cellular immunity has not been fully answered. It is known that a state of delayed hypersensitivity is usually present when there is effective cellular immunity and it seems likely that the products of stimulated lymphocytes may lead both to protection against microorganisms and to tissue damage in the host. In tuberculosis, effective cellular immunity depends on changes in macrophages (host cells that harbor the tubercle bacilli). These cells become activated, probably at least in part through the action of lymphokines. Because of their huge quantities of enzymes, activated macrophages are able to kill and digest many of the intracellular bacilli.

One other kind of cell-mediated hypersensitivity reaction that is of particular interest are reactions of *contact hypersensitivity*. The prototype is the familiar poison ivy rash. When a person is first exposed to poison ivy, no rash develops, but sensitivity to a particular hapten in the plant may be established. Upon reexposure to poison ivy, perhaps a year later, the allergic reaction is triggered. The allergic individual may remain sensitive for many years or may gradually lose sensitivity if the antigen is not encountered. The delayed reaction begins hours after encountering poison ivy, reaches a peak after two or three days, and gradually subsides if the hapten does not remain present. Contact allergies can be caused by many kinds of substances, some of them simple chemicals such as the chromium salts used in tanning leather or the nickel in watch bands.

Perhaps the distinction should be made here between substances that are toxic and those which are allergenic. Certain soaps are good examples; some contain toxic materials that damage the skin and cause a rash on virtually all who are exposed; others may be very mild and harm only the skin of a few individuals who are allergic to some of their constituents. In the latter case the rash may be extensive and severe, even though the soap is nontoxic for normal people.

MEDICAL PROBLEMS INVOLVING HYPERSENSITIVITY STATES

Having examined some of the basic mechanisms of immediate-type and delayed-type hypersensitivities, we can now discuss some problems in terms of the kind of allergy involved, and the balance between protective and allergic properties of the immunological response.

Transplantation

It is common knowledge that successful organ transplantation depends largely on overcoming or bypassing the body's very efficient mechanisms for rejecting or destroying foreign materials. Most tissue grafts in man are transplanted from a donor to a genetically different human recipient who has not been exposed to donor antigens before (primary grafts). There are differences in the way various kinds of grafts are rejected, but here we are concerned only with the common situation found in transplanting human tissues. As a rule primary graft rejection does not result from the activities of humoral antibodies, but from the action of sensitized lymphocytes acting in much the same sequence described in delayed hypersensitivity reactions.

Several methods are used to partially avoid or overcome the rejection reaction. Avoidance is preferable, and therefore tissue matching is done. Human tissue can be classified into major groups, called histocompatibility groups. These are analogous to the major blood groups, discussed later. In the laboratory the donor tissue is typed for major histocompatibility antigens and for major blood group antigens. If these match or are compatible with those of the recipient, the rejection response will be much weaker than if they are mismatched. Even so immunosuppression must be used to overcome the immune response to numerous weaker mismatched antigens.

Immunosuppression is a descriptive term meaning the suppression of immune responses in a nonspecific manner. It differs from immunological tolerance, which is lack of response to a specific antigen. Immunosuppression results in an overall immunological depression, so that the response to many or even all antigens is inhibited. In transplantation specific tolerance to all the antigens of the transplant is the goal; unfortunately, since no one knows how to achieve this, nonspecific immunosuppression must be used. It is like dropping a bomb to kill a colony of ants; the ants are eliminated, but the accompanying damage is overwhelming. With immunosuppression the protective responses to many antigens are reduced; consequently infections are a major cause of death in transplantation patients. In situations, such as kidney transplantation, where it is possible to choose donor organs that are fairly well matched, a minimum of immunosuppression may be sufficient to overcome the rejection process. Consequently a majority of kidney transplants are highly successful. Other organs, such as the heart, cannot always be carefully matched, and therefore more immunosuppression must be used than with well-matched grafts.

There are many ways to bring about immunosuppression, but most of them depend on preventing the proliferation of lymphoid cells. This may be done with certain kinds of irradiation, drugs, or by means of antilymphocyte globulin, prepared from antiserum made in horses or other animals against human lymphocytes. Usually transplant recipients are given a combination of immunosuppressive agents.

Mention has been made of enhancing antibodies which—paradoxically—seem to enhance the survival of foreign cells rather than oppose them. In the following section concerning immunity to tumors, this mechanism is discussed in more detail. It appears that enhancing antibodies may sometimes occur during primary organ transplantation, resulting in prolonged survival of the graft; however, at present this mechanism is not understood well enough to know how to encourage the formation of enhancing antibodies and at the same time overcome other immune responses that destroy the graft. When, and if, this is ever achieved the problem of transplantation rejection may essentially be solved.

Immunity to Tumors

The nature of tumors
Pages 347–349

Tumor cells arise from host tissue; therefore it might be expected that they would be recognized as self and would not incite an immune response. However, this is not the case, for it is well established that tumor cells have unique antigens that cause them to be recognized as foreign. A cellular immune response analogous to transplantation rejection takes place, with tumor cells being killed following direct contact with immune lymphoid cells. It is probable that this protective response eliminates many cells that would otherwise grow into detectable tumors. There are several possible explanations for the occurrence of tumors, in spite of an active immune response against them. First, tumor cells proliferate actively and they may simply outgrow and overwhelm the capacity of immune cells to destroy them. Furthermore, tumor cells commonly have a coating substance on their surfaces (somewhat analogous to the capsules of bacteria) that inhibits effective contact with immune cells. Perhaps most important, however, is the activity of enhancing antibodies specific for the tumor cells.

Recently it has been found that in all the tumor systems studied, both in animals and in man, the following results hold true. Subjects with tumors have immune lymphocytes (in their circulation and in lymphoid tissues) that are capable of killing or inhibiting tumor cells, and thus of inhibiting colony formation of specific tumor cells *in vitro*. For example, when blood lymphocytes are collected from an experimental animal with a tumor and mixed with its own tumor cells in the test tube, the tumor cells are killed or inhibited, as compared with control tumor cells mixed with lymphocytes from a normal animal. Figure 21-3a illustrates the experimental procedure. If, however, blood serum containing specific humoral antibodies against the tumor is added to the mixture, the ability of lymphocytes to kill specific tumor cells may be abrogated (Fig. 21-3d). When the tumor is regressing, indicating an effective immunological response, addition of the animal's serum does not decrease the killing of tumor cells by sensitized lymphocytes. On the other hand, when the tumor is progressing, addition of the animal's serum counteracts or blocks the killing of tumor cells by lymphocytes, leading to enhanced growth of the tumor cells. This has been shown to result from the presence of specific humoral factors, probably enhancing antibodies or antigen-antibody complexes. It appears that fatal progressive tumors (cancers) can evade cellular immune mechanisms because of the interference of enhancing antibodies. Here again, as was the case with transplantation, understanding how to control the immunological response, both cellular and humoral, would perhaps provide a rational approach toward controlling a major medical problem.

Blood Grouping and Blood Transfusion

Blood transfusion is a form of tissue transplantation, with foreign cells being transferred to a recipient of the same species. The reason it is feasible to transfuse blood so freely, first discovered by Landsteiner and his co-workers early in the twentieth century, is that the strongest and by far the most common transfusion incompatibilities are caused by the few antigenic types of a single, relatively simple blood group system known as the ABO system. It is important to remember that there are hundreds of different antigens in blood plasma or serum and on both erythrocytes and leukocytes; in fact there are dozens of minor blood group systems of red cell antigens in addition to the major ABO system. Fortunately most of these antigens do not incite a strong antibody response; otherwise blood transfusions would not be practical.

Because of the complexities of blood group antigens, this discussion is limited to simplified explanations of two of the most important systems, the ABO and the Rh (Rhesus) systems.

In the ABO system the presence or absence of only two antigens, called A and B, accounts for the four major blood groups, as shown in Table 21-3. Antigens A and B are polysaccharides found in abundance on the surfaces of erythro-

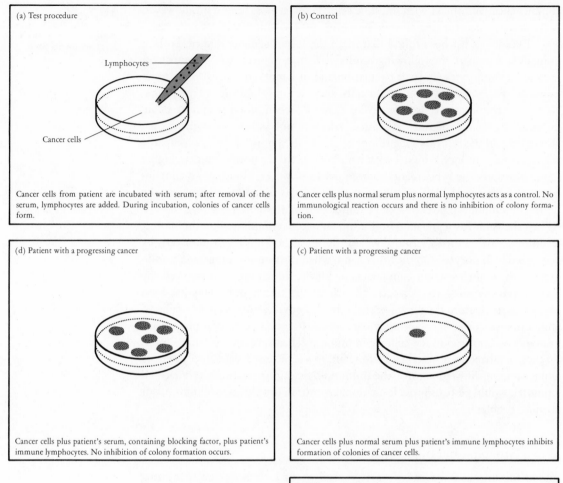

(a) Test procedure

Lymphocytes

Cancer cells

Cancer cells from patient are incubated with serum; after removal of the serum, lymphocytes are added. During incubation, colonies of cancer cells form.

(b) Control

Cancer cells plus normal serum plus normal lymphocytes acts as a control. No immunological reaction occurs and there is no inhibition of colony formation.

(d) Patient with a progressing cancer

Cancer cells plus patient's serum, containing blocking factor, plus patient's immune lymphocytes. No inhibition of colony formation occurs.

(c) Patient with a progressing cancer

Cancer cells plus normal serum plus patient's immune lymphocytes inhibits formation of colonies of cancer cells.

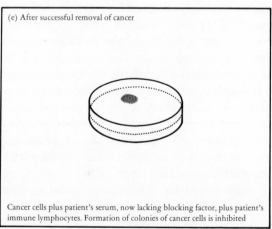

(e) After successful removal of cancer

Cancer cells plus patient's serum, now lacking blocking factor, plus patient's immune lymphocytes. Formation of colonies of cancer cells is inhibited

FIGURE 21-3

The colony inhibition test, used to study immunity to cancer. (a) The test procedure involves mixing cancer cells, serum, and lymphocytes and observing the formation of colonies of cancer cells. Inhibition of colony formation indicates an effective immunological reaction against the cancer. (b through e) The results of this test, using various combinations of either immune or normal (nonimmune) lymphocytes and immune or normal serum, indicate that lymphocytes from a patient with cancer have the ability to kill cultured cells from the same cancer. However, blocking factor may be present in the serum and able to interfere with killing by the lymphocytes. After removal of the cancer, blocking factor soon disappears from the serum and the lymphocytes are fully capable of killing specific cancer cells.

TABLE 21-3
Some Characteristics of the ABO Blood Group System

Blood Type (Phenotype)	Antigen(s) Present on the Erythrocytes	Antibodies Normally Present in the Plasma	Percent in a Mixed Caucasian Population[a]
O	Neither A nor B	Both anti-A and anti-B	45
A	A	Anti-B	41
B	B	Anti-A	10
AB	Both A and B	Neither anti-A nor anti-B	4

[a] The incidence of ABO blood groups may vary greatly between genetically different populations.

ABO BLOOD GROUP
SYSTEM

cytes of the appropriate blood groups. Type O red cells lack both A and B antigens; type A red cells have A (but not B) antigen; type B erythrocytes possess B (but not A) antigen; and type AB red cells have both A and B antigens on their surfaces. It is a general immunological rule, mentioned previously, that a normal individual will not produce antibodies against his own antigens; therefore, if either A or B antigen, or both are present, antibodies against that antigen will not be produced. A unique feature of this system is that (contrary to the usual situation) antibodies are normally present against A and B antigens that are lacking from the red cells. These are called natural antibodies, meaning antibodies that are formed without *deliberate* antigenic stimulation. This means that a type O person has in his plasma or serum both anti-A and anti-B antibodies, even when he has never been exposed to foreign blood. Similarly, a type A person will have anti-B antibodies, and a type B person has anti-A antibodies. Of course, type AB individuals lack both anti-A or anti-B antibodies in their plasma because both A and B antigens are on their red cells; producing either of these antibodies would endanger their own erythrocytes.

These basic facts about the ABO system are sometimes stated as *Landsteiner's principle:* whenever antigen A or antigen B, or both is missing from the red cells of a normal individual, the corresponding specific antibody for the missing antigen(s) is present in the plasma.

Natural antibodies account for a very small part of the total immunoglobulins. It is thought that natural anti-A and anti-B antibodies are probably formed as the result of constant accidental stimulation with substances in the environment that contain A and B antigenic determinants. These are known to be simple sugars, chemical groupings that are widespread in many foods, dusts, and other substances to which man is constantly exposed; unless they are recognized as being identical with components of self, antibodies are formed against them.

Table 21-3 includes a summary of the plasma antibodies normally present in human beings of each major blood group. There are some subgroups and exceptions to the simplified information included in this table, but for the most part this is a less complex blood-group system than many others.

In transfusing blood the antigens on donor red cells and the antibodies in recipient plasma are of greatest importance. This is because transfusion reactions are usually the result of destruction of donor erythrocytes by recipient antibodies. Certainly the reverse situation can occur, and antibodies in the donor plasma can react with erythrocytes of the recipient, but this is much less likely to happen, simply because of the quantities of each. When a pint of blood is transfused, the donor plasma is immediately diluted by the 10 to 12 pints of blood in the adult recipient; therefore the donor must have an uncommonly large amount of antibodies for enough to be transferred to cause a detectable reaction. Donor red cells, however, are constantly bathed in recipient plasma, so they have a high risk of being promptly destroyed by the recipient's antibodies. With these facts in mind it is possible to predict from Table 21-3 which blood group might, in theory, serve as a universal donor, capable of donating blood to recipients of any ABO group. It must be a group with red cells that lack both A and B antigens, and that hence could not be destroyed by interaction with plasma of any group; it is obvious that this could only be group O. Indeed type O individuals are sometimes referred to as universal donors. Conversely type AB persons have been called universal recipients, because their plasma contains neither anti-A nor anti-B antibodies to react with transfused cells. This concept has been referred to as "the myth of universal donors and recipients," because in practice it does not always work. The fault is not in the reasoning as given, but rather it stems from the fact that there are many other minor blood group systems. Thus before transfusions are given, the bloods of donor and recipient must be allowed to interact in laboratory tests to minimize the possibility of reactions *in vivo* that might result from other antigen-antibody interactions.

The genetics of the various blood group systems has been extensively studied by immunogeneticists and immunohematologists. Many practical applications have been helpful in law, anthropology, and other diverse areas. For example, knowledge of the genetics of the ABO system allows determination of ABO blood types to be accepted as evidence in court under certain conditions. The ABO blood group locus is found on one pair of chromosomes; the ABO genes on each chromosome of the pair are codominant, that is, the A and B genes are both expressed if both are present. Because one chromosome of each pair is inherited from the mother (maternal) and one from the father (paternal), an individual's blood type is determined by the ABO genes obtained from both parents (Fig. 21-4). Table 21-4 shows the possible genotypes of individuals of the major ABO blood groups and of their parents. The simplest case is seen with type O individuals. Their genotype can only be OO, and obviously they must have

FIGURE 21-4
Codominant inheritance of ABO blood groups. One pair of chromosomes in each diploid somatic (body) cell carries the genes determining the ABO group. Ova and sperm cells are haploid (IN), containing only one chromosome of each pair and therefore only one gene for the ABO group. At mating the ova and sperm combine to give a random distribution of all possible combinations. The ABO genes are codominant and both are expressed. Consequently, the overall frequency of blood groups in offspring from this combination will be 25 percent each of groups O, B, A, and AB.

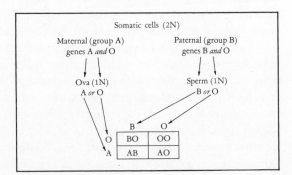

TABLE 21-4
Possible Parental Genotypes[a] for Individuals of the Major
Blood Groups

Blood Type (Phenotype)[b] of Individual	Possible Genotype(s) of Individual	Possible Parental Genotypes
O	OO	OO
		AO
		BO
AB	AB	AB
		AA
		BB
		AO
		BO
A	AA	AA
	AO	AO
		BO
		OO
		AB
B	BB	BB
	BO	BO
		AO
		OO
		AB

[a] Genotype—genetic constitution, determined by both of a pair of genes.
[b] Phenotype—expressed characteristics, determined by the dominant gene or genes of a pair.

received one O gene from each parent, so the only possible parental genotypes are those with at least one O gene. When both parents are type O, there is no possibility that the children will be anything except type O (Fig. 21-5a). The situation becomes slightly more complex with other blood types (Fig. 21-5b and c), but it is still a simple matter to calculate the probability of children of a given blood group being born of any ABO genetic mating.

The ABO blood groups are accepted as evidence in cases of disputed parentage only when they exclude an individual as being the parent. Thus if a type O woman bears a type O child, a man with AB blood can be excluded as being the father of the child, since he could have transmitted only an A gene or a B gene and the child has neither. If the male in question were type A, B, or O, he could not be excluded as a possible father on the basis of the phenotype alone.

The Rh (Rhesus) system is the most important of the minor blood group systems. Although many antigens are involved, the strongest one can be called the Rh antigen. If it is present on the red cells, a person is Rh-positive; if it is lacking, the person is Rh-negative.

The Rh-positive adult has no cause to worry about the most important Rh blood group problems; he has the Rh antigen and hence cannot make anti-Rh antibodies. He can receive either Rh-positive or Rh-negative blood transfusions with impunity, since the Rh-positive blood is compatible and Rh-negative red cells simply lack the Rh antigen. It is the Rh-negative person who may experi-

RHESUS BLOOD GROUP
SYSTEM

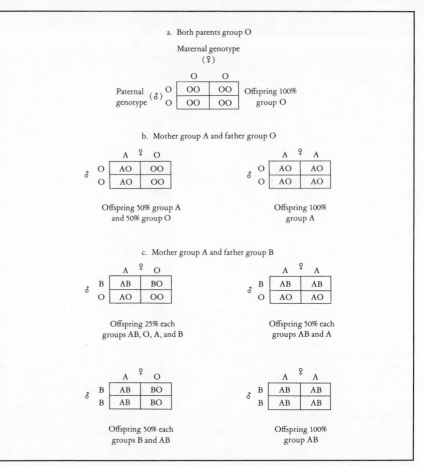

a. Both parents group O

Maternal genotype
(♀)

Paternal
genotype (♂)

	O	O
O	OO	OO
O	OO	OO

Offspring 100%
group O

b. Mother group A and father group O

	A ♀	O
O	AO	OO
O	AO	OO

Offspring 50% group A
and 50% group O

	A ♀	A
O	AO	AO
O	AO	AO

Offspring 100%
group A

c. Mother group A and father group B

	A ♀	O
B	AB	BO
O	AO	OO

Offspring 25% each
groups AB, O, A, and B

	A ♀	A
B	AB	AB
O	AO	AO

Offspring 50% each
groups AB and A

	A ♀	O
B	AB	BO
B	AB	BO

Offspring 50% each
groups B and AB

	A ♀	A
B	AB	AB
B	AB	AB

Offspring 100%
group AB

FIGURE 21-5   Probabilities of offspring of various ABO blood groups from different matings.

ence Rh blood group problems as a result of being immunized against the Rh antigen, either from transfusions, or, in the case of females, from pregnancies. The antibodies that are formed by a pregnant woman most often damage her offspring. The disease that results is called hemolytic disease of the newborn, or simply Rh disease (Fig. 21-6).

When an Rh-negative mother carries an Rh-positive fetus there is very little passage of Rh-positive cells across the placenta to the mother; a few fetal cells get into the mother, but usually not enough to stimulate a *primary* antibody response. At birth, however, the baby's blood cells often enter the mother's circulation in sufficient quantities to incite a vigorous immune response. Also, abortion procedures are very likely to introduce large numbers of blood cells from the fetus into the mother. The anti-Rh antibodies formed by the Rh-negative mother cause her no harm; they cannot damage her cells (which lack Rh antigens); but problems, however, may arise when this mother carries another Rh-positive fetus. This time, the few Rh-positive cells that get from fetus to mother are enough

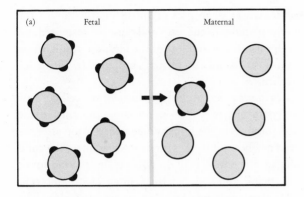

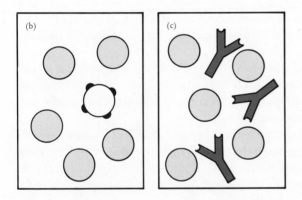

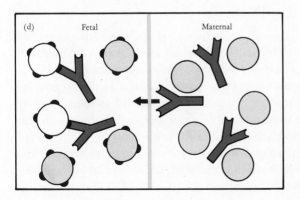

FIGURE 21-6
Events leading to Rh disease (hemolytic disease of the newborn).
(a) Fetal red blood cells bearing the Rh antigen enter the maternal
circulation; (b) the Rh antigenic determinants become available
to stimulate an immune response; (c) anti-Rh antibodies are
formed; (d) during subsequent pregnancy with an Rh-positive
fetus, the anti-Rh antibodies can cross the placenta, leading to
destruction of fetal red blood cells and Rh disease.

to initiate an *anamnestic response*. Large quantities of anti-Rh antibodies of the class IgG are produced, cross the placenta to the fetus and cause extensive red cell damage in the child. Although miscarriage and loss of the fetus can result, often the damage is not apparent until very soon after birth, as indicated by the name, hemolytic disease of the newborn.

The child usually survives *in utero* because the products of red cell destruction can be eliminated by certain enzymes, present in large amounts in the mother but only in very small amounts in the fetus and newborn. Soon after the exchange of materials between mother and fetus is interrupted at birth, the child becomes acutely ill. Not only are the products of erythrocyte destruction toxic, but also the baby is seriously anemic and may not have enough red blood cells to survive. In this critical situation, exchange transfusions are life-saving. The baby's blood is gradually withdrawn while fresh *Rh-negative blood is added*. The benefits are twofold: first, toxic products, maternal antibodies, and the infant's Rh-positive red cells are all removed, and second, the red cells are replaced by Rh-negative erythrocytes which cannot react with any remaining anti-Rh antibodies. The child does not produce large numbers of Rh-positive red cells until the transfused cells age and are removed from the circulation, a matter of weeks. By this time, the maternal anti-Rh antibodies have also decreased to a harmless level.

Anti-Rh antibodies are often given to susceptible Rh-negative mothers at or soon after the birth of an Rh-positive child. This procedure can prevent the mother from becoming immunized by Rh-positive cells at the time of birth and thereby prevent Rh disease in her subsequent children.

Although it might be expected on the basis of overall incidence of Rh-positive and Rh-negative people in the general population that nearly 15 percent of all children would develop Rh disease in fact less than 1 percent are affected for many reasons, including the following: First, many Rh-positive males are heterozygous and transmit an Rh-negative gene, so that children of an Rh-positive father and an Rh-negative mother are often Rh-negative and consequently not subject to Rh disease. Second, the mother may not become immunized, even when incompatibilities exist.

Autoimmune (Autoallergic) Diseases

It was stated earlier that occasionally an individual may, for various reasons, fail to recognize certain of his own components as self and may respond immunologically to them. The result is commonly known as autoimmune disease.

It is well established that the immune system is immature during fetal development. Most antigens that are encountered by potential antibody-forming cells during fetal existence induce a state of *specific immunological tolerance,* that is, a lack of ability to respond immunologically to those antigens. This state of tolerance exists as long as the antigens remain accessible to potential antibody-forming cells; however, if the antigen is removed, the individual gradually develops the ability to respond to that antigen.

The extent of protection or tissue damage that results from an immunological response depends on a combination of factors: the nature and dose of the antigen, its route of entry and fate in the body, the numbers and state of the lymphoid cells that respond, the classes of antibodies formed, the sites of interaction between the antibodies and antigen, and numerous others. In virtually no case is the response purely antibody production or only cell-mediated hypersensitivity or any other single state. As more is learned about immunological mechanisms, it may become apparent that hypersensitivity reactions, even though they damage tissues, may still offer essential protection in conjunction with conventional antibody-mediated immunity.

The subject of immunity against various microorganisms is discussed in subsequent chapters. It is well to remember that the basic concepts discussed in this chapter must be applied when choosing immunization procedures. Immunizing antigens must be administered so that they increase the protective benefits and decrease the chances of inciting hypersensitivity reactions.

SUMMARY

The term hypersensitivity has been applied to immunological reactions that cause tissue damage in the host. A synonym for hypersensitivity is allergy. Allergic reactions may be classified into two broad categories: immediate-type and delayed-type.

Immediate-type hypersensitivities include atopic allergies such as hayfever, hives, and allergic asthma. These allergies are mediated by a particular class of immunoglobulins, IgE, which are cytophilic for mast cells. Interaction of antigen with cell-fixed IgE causes the release of chemical mediators such as histamine from the mast cells, resulting in the observed reactions. Other kinds of immediate-type hypersensitivities result from the action of antigen with humoral antibodies of classes other than IgE.

Delayed-type hypersensitivity is mediated by specific sensitized lymphoid cells, principally lymphocytes. The mechanism of lymphoid activity is similar, or perhaps identical, to the action of lymphocytes during cellular immunity. The tuberculin skin test is the prototype of delayed-type allergic reactions. Contact hypersensitivity is another form of delayed-type hypersensitivity, exemplified by the familiar poison ivy rash.

Hypersensitivity reactions form the basis of transplantation rejection, immunity to tumors, blood transfusion reactions, and a variety of other medical problems.

The immunological response to any foreign antigen usually involves a combination of humoral and cellular responses, and of both protective and hypersensitivity reactions.

1. List some differences between immediate and delayed-type hypersensitivities. How do these relate to humoral and cellular immunity?
2. Describe atopic allergies. What role does IgE play in causing this kind of allergic reaction?
3. What kinds of cells participate in reactions of delayed-type hypersensitivity? What is their probable mode of action?
4. Consider some of the practical reasons for studying hypersensitivities. List at least five important applications of such studies.
5. What is the chance that a man with type O blood is the father of a type AB baby whose mother is type A? Could this blood type evidence be used in court to force him to pay support for the child? Explain.

FURTHER READING

CLARK, C. A., "The Prevention of Rhesus Babies," *Scientific American*, 219 (November 1968), 46–52.
GELL, P. G. H. and R. R. A. COOMBS (eds.), *Clinical Aspects of Immunology*, 2d ed. Philadelphia: F. A. Davis, 1968.
GOOD, R. A. and D. W. FISHER (eds.), *Immunobiology*. Stamford, Conn.: Sinauer Assoc., Inc., 1971.
NOSSAL, G. J. V., *Antibodies and Immunity*. New York: Basic Books, 1969.
WEISER, R. S., Q. N. MYRVIK, and N. N. PEARSALL, *Fundamentals of Immunology for Students of Medicine and Related Sciences*. Philadelphia: Lea and Febiger, 1969.

Chapter 22

Epidemiology of Infectious Diseases

The field of science which deals with factors determining the frequency and distribution of disease is called epidemiology. As the population of the world increases the importance of the epidemiology of microbial infections is underscored. This importance arises not only as a result of the increasing density of human populations which facilitates spread of infectious diseases but also because of mass production and distribution practices. A single food supplier, for example, may send his products not only to a community but to several states or nations. If the food is contaminated with a pathogenic microbe, the resulting illness may be widespread. Also efforts employed in infectious disease control, such as vaccination or chlorination of a water supply, often involve millions of people in a single program. Our species tends to rely increasingly on such measures, as well as highly organized surveillance of public health, to control infectious diseases. It is well to note that such reliance on advanced technology and social organization is vulnerable to war, insanity, and other catastrophies. A single irrational act may affect thousands of people.

EPIDEMICS: DETECTION AND GENESIS

In a given area the occurrence within a short period of time of an unusually large number of infections by a single agent is called an epidemic. The following sections elucidate a few factors involved in the origin, spread, and control of epidemic diseases of microbial origin.

One may suspect an epidemic if an unusually large number of cases of a given disease appear. However, many different agents produce the same disease and, conversely, a single agent may produce a wide variety of diseases. For example, more than 80 different infectious agents can produce the common cold, whereas a single agent, poliovirus, can produce illness ranging from the symptoms of a cold to fatal paralysis. The factors involved in any given epidemic can usually be elucidated only when a sizable number of cases due to the same agent have been segregated. Sometimes an infectious agent such as *Vibrio cholerae* produces a disease which is highly characteristic of the agent. Indeed, many years ago, a serious epidemic of cholera in England was solved by noting that almost all persons with the disease reported drinking at a single water source, while almost all those escaping the disease drank water from other sources. This occurred before the causative agent of cholera had ever been described. On the other hand, in most instances it is necessary to identify the causative agent with great precision, even individual strains within a species.

Cholera
Pages 479, 627

Precise Identification of Infectious Agents

In some instances careful identification of species may be essential to identify the existence of epidemic spread. For example, in one large hospital it was the practice to name a group of lactose nonfermenting enterobacteria "paracolons." Since a rather large number of species fall into this group and occur in cultures from sick people, it did not seem particularly remarkable that patients occasionally died with such organisms in their bloodstream. However, with the introduction of new techniques in the microbiology laboratory of the hospital, it was noted that the vast majority of lactose nonfermenting bacteria from blood cultures fell within a single species, *Serratia marcescens*. The observation that almost all patients with paracolon in their blood were infected with the same species prompted special investigations and led to the discovery that hospital equipment was contaminated with *Serratia* and infecting the patients.

Enterobacteria
Page 252

The value of identification of types within a species is shown by another episode. An unusually large number of infections due to *Streptococcus pyogenes* appeared over the course of several months, but since they were reported from different hospitals among patients attended by many different medical and other personnel, it was not clear whether an epidemic existed. When the surface antigens of the streptococci were identified by using antisera, it was found that about three-quarters of the patients were infected with the same type of *S. pyogenes*, whereas the others were infected by miscellaneous types. Moreover all of the patients with the epidemic type of streptococcus had been attended by the same physician, who proved to be a carrier. In addition to streptococcal epidemics, those due to *Staphylococcus aureus* or *Pseudomonas aeruginosa* are even more difficult to trace, because these species of organisms are so commonly present on normal people or in the environment. For staphylococci, individual strains can be

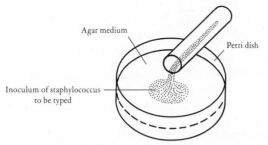

a. Spreading an inoculum of *S. aureus* over the surface of agar medium

b. Dropping bacteriophage suspensions on the inoculated surface. Twenty-three different suspensions are thus deposited in a fixed pattern. After incubation, different patterns of lysis are seen with different strains of *S. aureus*.

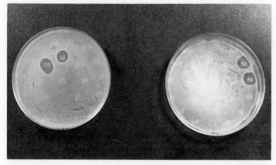

c. Photograph of two different strains of *S. aureus*, 53/54 and 80/81.

FIGURE 22-1   Bacteriophage typing of *Staphylococcus aureus*.

identified by bacteriophage typing (Fig. 22-1), and *Pseudomonas* strains can be distinguished using bacteriocin typing (Fig. 22-2). Greater precision can be obtained using both antibiotic susceptibility patterns and bacteriophage typing, as shown in Table 22-1 for *S. aureus*. The first eight patients shown very likely had the same (epidemic) strain of staphylococcus since all the isolates had the same antibiogram and the same phage type. Patients 9 and 10 and patients 11 and 12 very likely were not part of the epidemic since their isolates differed either in phage type or antibiogram.

Testing of antibiotic
susceptibility
Page 570

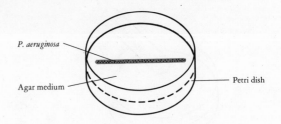

a. Strain of *P. aeruginosa* inoculated in a straight line and allowed to grow.

b. The growth of *P. aeruginosa* is being scraped off the medium. Any remaining bacteria are subsequently killed with chloroform vapor. Known strains of *P. aeruginosa* are then streaked at right angles to the area from which growth has been removed. The known strains are allowed to grow.

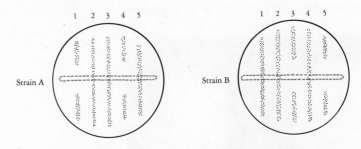

c. Two different strains of *P. aeruginosa* distinguished by their differing pattern of killing against the same set of five known strains. *P. aeruginosa* strain A kills known cultures 1 and 4; strain B, 3 and 5.

FIGURE 22-2   Bacteriocin typing of *Pseudomonas aeruginosa.*

Spread from Reservoirs

The reservoir of any infectious agent is the sum of all potential sources of the agent. For example, the reservoir for diphtheria, syphilis, and typhoid fever is other human beings; for botulism and coccidioidomycosis, the soil; for more than 100 other diseases, animals. The routes of spread from reservoirs may involve vectors (arthropods or other animals), vehicles (food or other nonliving materials), and human beings (through direct person-to-person contact). When agents spread from sources relatively distant from man (either in distance or ge-

Coccidioidomycosis
Page 464

TABLE 22-1
Use of Antibiogram and Bacteriophage Type in
Distinguishing Epidemic Strains of Staphylococci

| Patient | Staphylococcus | |
	Antibiogram[a]	Bacteriophage Type
1	RRSRR	80/81
2	RRSRR	80/81
3	RRSRR	80/81
4	RRSRR	80/81
5	RRSRR	80/81
6	RRSRR	80/81
7	RRSRR	80/81
8	RRSRR	80/81
9	RRSRR	54/83
10	RRSRR	54/83
11	RSSSR	80/81
12	RSSSS	80/81

[a] R—resistant, S—sensitive to the antibiotics penicillin, tetracycline, chloramphenicol, erythromycin, and streptomycin, respectively.

netic relatedness), it is sometimes possible to detect their spread before significant human involvement occurs. For example, encephalitis viruses of wild birds and mosquitoes may be detected by placing a "sentinel" (such as a chicken) in a cage which would expose it to the mosquito vectors. The sentinel animals can then be tested periodically to see whether the encephalitis arborvirus has infected them. The use of sentinels is especially helpful in this kind of disease since extensive spread among the human community can occur without detection due to the high ratio of infected cases to those with overt disease. The time it takes to detect the spread of an infectious agent is also influenced by the incubation period of the disease. With relatively long incubation periods extensive spread can occur before the first cases of disease appear. The *dose* of infecting agent received by susceptible people is important because smaller doses generally result in a longer incubation period and a higher percentage of asymptomatic infections.

Encephalitis
Page 541

The extent and speed of spread are influenced by the *mode of transmission*. A dramatic example of the importance of this factor was the spread in 1963 of typhoid fever in a ski resort in Switzerland. As many as 10,000 people were exposed to drinking water containing small numbers of typhoid bacilli. The resulting long incubation period allowed widespread dissemination of the agent by skiers flying home to various parts of the world. Altogether more than 430 cases of typhoid developed in at least six countries.

MODE OF TRANSMISSION

Air as a Vehicle of Infectious Agents

Microorganisms are discharged into the air by sneezing, coughing, talking, and singing. Many are enclosed within large droplets of saliva or mucus, most of

DROPLET NUCLEI

which drop quickly to the ground within a distance of 3 or 4 feet. Large-size droplets are important under crowded conditions such as schools and military barracks. Ideally, desks or beds should be spaced 8 or 10 feet apart to minimize transfer of infectious agents by this route. Smaller droplets quickly dry, leaving one or two organisms attached to a thin coat of dried material. These "droplet nuclei" remain suspended in the air by minor currents for indefinite periods, and because of their small size may escape the trapping mechanisms of the upper airways and enter the lungs. Under usual conditions, however, droplet nuclei are rapidly diluted by the movements of the air and only highly infectious agents such as the viruses of measles and smallpox are readily disseminated by this manner. That one cannot rely on the dilution factor under conditions of inadequate ventilation is shown by a recent hospital epidemic of tuberculosis. In this epidemic an unsuspected case of tuberculosis was cared for in a large ward for two and one-half days. During this time 13 out of 44 persons working in the ward became infected with the tubercle bacillus. As is often the case the air-conditioning unit merely recirculated most of the room air without filtering it.

MICROBES IN THE AIR
OF BUILDINGS

Large buildings present special problems of airborne transmission of disease agents. Not only are respiratory pathogens disseminated as a result of sneezing, coughing, and talking, but persons with skin infections such as boils may discharge pathogens as body movement shakes clothing and rubs off skin cells. The number of organisms in air can be estimated by using machines which propel measured volumes of air against the surface of medium in petri dishes. The number of colonies which result per cubic foot of air sampled rises proportionately to the number of people in a room. Partly for this reason most modern public buildings have elaborate ventilation systems to give a constant change of air. In some hospitals the air flow can be regulated so that the operating room is under a relative positive pressure to prevent contaminated air from flowing into the room, whereas microbiology laboratories may be placed under negative pressure so the escape of microbes to other parts of the building cannot occur. Special problems may arise because of the pumping action of elevators and the chimney-like effect of a laundry chute extending from the warm air (and dirty laundry) in the basement. Air-conditioning systems themselves may be the source of infectious agents, because the location of the air intake may be such that contaminated dust can be drawn in. In addition, molds, actinomycetes, and other organisms may grow in humidifying devices within the ductwork and be distributed throughout the building. Some people have expressed concern that during wartime, saboteurs could infect large numbers of people by this route. Militarists have suggested that in the event of threatened bacteriological warfare, vaccines could be sprayed into the air intake systems of public buildings.

SURVIVAL OF MICROBES
IN AIR

The survival of organisms in air varies greatly with the species and with conditions such as relative humidity, temperature, and exposure to light. Subdued light promotes survival of microbes, but the effects of temperature and humidity vary with the species. Organisms in still air and those in large particles eventually fall to the floors of buildings, but are readily resuspended in air with dust particles. Thus sweeping is not permitted in most modern hospitals and mopping with a dampened mop is used instead. Although the use of ultraviolet light or filtration can markedly reduce the number of viable organisms in air, these methods are generally expensive and unnecessary. Under usual conditions satisfactory control of airborne infection is achieved by good ventilation and dust control. Filtration and ultraviolet irradiation do, however, have important uses in laboratory work and some special types of hospital units where the risk of microbial contamination is great.

Person-to-Person Transmission

Hands are important agents in the transfer of disease-producing organisms, and in modern hospitals they generally exceed the importance of air. Hands readily become contaminated from contact with the nose, the area of skin infection, or fecal discharge. Dysentery bacilli readily contaminate hands through toilet tissue. Hands are thus of prime importance in transfer of organisms from one person to another, directly and through intermediaries, such as foods and fomites. Scrubbing and washing are of major importance in reducing the numbers of potentially pathogenic microbes on the hands and thus lowering the possibility of transferring an infectious dose of the organisms.

HANDS ARE IMPORTANT TRANSMITTERS

Carriers

The human reservoir is often much larger than the size of the sick population would suggest. In fact the most important transmitters of infectious agents are those who have mild or inapparent infections. In contrast to those who are ill, these people move freely about in their daily activities, the unsuspected transmitters of epidemic agents. In addition to persons with mild or atypical disease, the human reservoir includes persons still in the incubation period who have not yet become ill, persons who have recovered from the illness, transient carriers, and chronic carriers—all of whom may disseminate disease agents. With many diseases the infectious agent continues to be present in body secretions for the convalescent period, gradually decreasing in concentration over days to months. Transient carriers are those who become contaminated with an agent only for a brief time, whereas chronic carriers continue to excrete agents constantly or intermittently for months, years, or even a lifetime. After a disease is controlled in a population, the presence of chronic carriers poses a threat of recurrence which continues for many years. Under conditions of good sanitation where spread is difficult, deaths from the diseases of advancing age gradually eliminate chronic carriers.

Water and Food As Vehicles

Contamination of water and food has been responsible for innumerable epidemics, frequently involving thousands of people. Waterborne epidemics result in a large number of cases because of the number exposed, but the percentage of people who develop disease is generally low. Moreover, the incubation period is apt to be long. These features of waterborne epidemics result from the enormous dilution of the infecting agent which generally occurs. Attack rates (the number of cases developing per hundred people exposed) with food tend to be higher because of the lack of dilution, and food may be nutrient for the agent, allowing multiplication. Therefore a sizable increase in the numbers of the agent may occur before the food is eaten by the recipient. Feces, directly or indirectly, is the usual source of water and food contamination. Foods can be infected at their source (pork with *Trichinella*, eggs with *Salmonella*, milk with *Brucella*), or during preparation, as with the ameba, *Entamoeba histolytica*, or bacteria such as *Salmonella* and *Staphylococcus*.

Trichinella
Page 547
E. histolytica
Page 482

FACTORS INFLUENCING THE CHARACTER OF EPIDEMICS

Acquired Immunity

The attack rates of an epidemic disease agent are influenced by previous exposure of the population to that agent or agents antigenically related to it. Influenza, for example, is unlikely to spread in a population where 90 percent of the people had previously been infected by the same agent and acquired antibody against it. A strictly human disease of this type requires a continuous source of susceptible hosts. Resistance to epidemic disease may also be acquired by exposure to related microbes, as for example, exposure to cowpox protected the milkmaids of Jenner's time against smallpox. The age distribution of a population influences the attack rate partly because of this factor. A high proportion of old people in a population may mean a better chance of low disease rates with many infectious agents because older people with existing immunity may act as barriers to the spread. A high proportion of young children may give a better chance of epidemic spread, although some antibodies cross the placenta and give protection for the first three to six months of life. This protection of infants depends on the immune status of their mothers and the nature of the agent. Antibodies protective against *Bordetella pertussis* (the cause of whooping cough), for example, do not cross the placenta. The persistence of acquired immunity also varies greatly with the agent, perhaps lifelong for poliovirus or two years with rhinovirus.

RELATIONSHIP OF AGE
AND ATTACK RATE

Rhinovirus
Pages 446, 679

Genetic Background

Natural immunity in laboratory animals varies greatly with their genetic background and the same is probably true of human beings, although there is little scientific proof for this. Differences in the incidence of epidemic disease among different racial groups are often pronounced, but it is (usually) difficult to determine to what extent the differences are due to genetic factors or to differing cultural and environmental factors.

Other things being equal, infectious diseases tend to hit economically disadvantaged groups harder. Malnutrition and fatigue may contribute to greater susceptibility to some diseases since this is true with laboratory animals. Paradoxically, economically deprived people often fare better during some epidemic diseases (for example, poliomyelitis), because overcrowding, inadequate sanitary practices, and lack of medical care promote their constant spread to the very young. In the young, maternal antibody (through placental transfer) and other factors often favor the development of mild or unnoticed infections and lasting immunity.

Poliomyelitis
Page 509

A wide variety of cultural traits influence markedly the rates of disease. For example, people in some European countries drink little water, and thus largely escape waterborne typhoid fever. Various cultural or ethnic groups savor dishes prepared from raw meat or fish, and are thus subject to high attack rates of *Trichinella* and other parasitic worms.

Virulence of the Epidemic Strain

Marked variations in the incidence of disease appear among different epidemics with the same microbial species. The spread of *N. meningitidis* in a population will produce low disease rates one time, whereas at another high rates of meningitis and significant mortality will occur. Similarly, one bacteriophage type of *S. aureus* may spread among hospital personnel and patients with little disease resulting, whereas another bacteriophage type causes boils, pneumonia, and many infections of surgical wounds. It is likely that such differences are sometimes due to differences in virulence among strains within a species or subspecies. Unfortunately good laboratory tests of virulence for human beings are available for only a few microbial pathogens, so the importance of this factor cannot readily be quantified.

Virulence
Page 359

Intensity of Exposure

The probability of developing infection and disease may be low or absent if the concentration of microbial pathogens is low. On the other hand, there are probably no human infections for which immunity is absolute. Heavy exposure to a microbial pathogen may produce serious or even fatal disease in a person who has specific antibodies and immunity to ordinary doses of the agent. *Even immunized persons should therefore take precautions to minimize exposure to epidemic agents.* This principle is especially important to laboratory workers who may be exposed to unnaturally high inocula as a result of accidents or faulty technique, and to medical workers who attend patients with infectious diseases.

IMMUNITY IS RELATIVE

Increasing Host Resistance

Methods which promote good general health appear to result in increased resistance to infection by some agents (such as tuberculosis) and less development of severe disease. Host resistance can also be increased by inducing specific immunity against pathogenic agents by using vaccines and toxoids (Chapter 39). The chief problem is to maintain that immunity once the possibility of recurrent natural exposure becomes low. Fortunately for diseases such as tetanus, diphtheria, and smallpox, "booster" injections of toxoid or vaccine usually produce a prompt anamnestic antibody response, even in adults who were last immunized in childhood. Finally, host resistance may be increased—even though temporarily—by passive immunization (administration of gamma globulin to prevent infectious hepatitis, for example), or by administration of certain medications. As an example of the latter, persons exposed to smallpox and influenza may be protected by giving medicines which interfere with viral penetration of host cells.

Hepatitis
Page 475

Reduction of Reservoirs and Vectors

Control of infectious agents can sometimes be achieved most readily by an attack on the reservoir. In the case of *Plasmodium vivax* (a cause of malaria) or the encephalitis viruses, reduction of the mosquito host population reduces both the reservoir and the vector. Searching out and treating patients with malaria decrease the reservoir. In the case of *Schistosoma* (a parasitic worm), treating patients and poisoning the snail intermediate host with copper sulfate helps decrease the reservoir. Carriers of *Salmonella typhi* can be detected by screening populations for antibodies to the Vi capsular antigen and instituting treatment. Control of tuberculosis can similarly be achieved through identification of cases by the use of X-rays and skin tests followed by medical treatment. Modern processing of sewage eliminates much of the pathogenic potential of sewage. Purification of water sources by filtration and chlorination and destruction of foods or other materials containing epidemic agents may also be highly effective.

Malaria
Page 538
Schistosoma
Page 549
S. typhi carriers
Page 481

Segregation, Isolation, and Quarantine

Separating persons with infectious disease from susceptible people interferes with transmission and is helpful in preventing spread of certain diseases. For example, a singing teacher with tuberculous infection of his larynx would soon infect his entire class unless he were segregated from them. Complete isolation may be helpful in a highly infectious disease such as smallpox, and medical attendants need to cover themselves with gowns, gloves, and masks to prevent contact. Quarantine of susceptible unexposed groups is also helpful in some cases where the group can be reliably identified, perhaps a village near an area experiencing plague. For many epidemic situations, however, segregation, isolation, and quarantine are inadequate by themselves in gaining control of the contagion. The failure of these measures results from the fact that it is usually difficult to identify more than a fraction of the infected persons. The remainder has mild or atypical illnesses which go undetected, and they continue dissemination of the epidemic agent to susceptible people.

Control of Infectious Diseases in Hospitals

As indicated earlier, hospitals pose special problems in infection control. In fact a hospital can be regarded as a high population density community made up largely of uniquely susceptible people into which the most virulent microbial pathogens of the region are continually introduced. Extensive use of antimicrobial medications in hospitals fosters the selection of strains which resist therapy, and the hospital staff are likely to become carriers of these strains. Moreover modern medical practices often require transgressions of normal host defenses in the form of surgical wounds, interference with normal microbial flora, and medications which impair the inflammatory response and the immune mechanism. For these reasons many modern hospitals employ a specially trained nurse or doctor to investigate infections acquired in the hospital. The hospital epidemiologist also keeps abreast of regional infectious diseases (such as influenza) which might pose a special threat to hospitalized patients, supervises surveillance of hospital equipment and medications to ensure that they do not become contaminated with microbial pathogens, and helps prevent spread of resistant hospital pathogens (such as *Staphylococcus aureus*) to the families of patients.

THE HOSPITAL
EPIDEMIOLOGIST

The Public Health Network

Specific approaches to infectious disease control such as those mentioned above are pale in significance compared to the modern public efforts which put them into practice. Massive programs involving international cooperation are promoted by the World Health Organization, and national institutions such as the United States Public Health Service exercise constant surveillance. Doctors, nurses, laboratory workers, and teachers participate in reporting infectious diseases to public health authorities. Expert teams are ready on short notice to investigate outbreaks and to assist in instituting control measures. The public health network extends far beyond such organizations, however. Education and maintenance of community awareness involve teachers and communications media. Politicians, engineers, industrialists, and home economists all participate. The dramatic declines in incidence of the terrifying scourges of former years could not have occurred without the collaboration of many units comprising our social structure.

SUMMARY

Epidemiology is the science dealing with factors determining the frequency and distribution of disease. The rapidly increasing density of human populations, the popularity of mass production and distribution methods, and the magnitude of artificial (as opposed to natural) immunization techniques are some of the modern factors influencing disease occurrence. Epidemic spread of infectious agents is not always easy to detect, and precise identification of the causative organism is usually of great importance. The nature of reservoirs of infection and routes of spread from them are important to understand. Vectors, air, water, food, and people are the principal modes of transmission. Control involves reduction of the reservoir, interfering with transmission and protecting susceptible hosts by increasing their resistance or separating them from exposure. The extent and distribution of diseases caused by microbes are influenced by many characteristics of the population being exposed, including age distribution, previous exposure to the same or related agents, and cultural practices. Effective control of epidemics due to infectious agents requires participation of many sectors of our social structure.

QUESTIONS

1. Discuss the importance of precise identification of microbial strains in defining an epidemic and tracing its source.
2. Give the principal reservoirs of microbial pathogens.
3. How can the size of the infecting dose influence the early detection and scope of an epidemic?
4. Discuss the principal means by which epidemics can be controlled and prevented.
5. Are infected persons who lack symptoms important in epidemics? If so, how so?

FURTHER READING

COLBECK, J. C., *Control of Infections in Hospitals.* Hospital Monograph Series #12. Chicago: American Hospital Association, 1962.

CORRIGAN, M. J. and L. E. CORCORAN, (eds.), *Epidemiology in Nursing.* Washington: Catholic University Press, 1961.

GREENBERG, B., "Flies and Disease," *Scientific American,* 213 (January 1965), 92.

ROUECHE, BERTON, *Eleven Blue Men and Other Narratives of Medical Detection.* New York: Berkeley Medallion, 16th printing, 1970.

TAYLOR, I. and J. KNOWELDEN, *Principles of Epidemiology.* Boston: Little, Brown and Co., 1964.

THOMPSON, L. R., *Microbiology and Epidemiology.* Philadelphia: W. B. Saunders Co., 1962.

THOMPSON, MORTON, *The Cry and the Covenant.* Garden City, New York: Doubleday and Co., 1949.

WILLIAMS, R. E. O., R. BLOWERS, L. P. GARROD, and R. A. SHOOTER, *Hospital Infection: Causes and Prevention.* London: Lloyd-Luke Medical Books, Ltd., 1966.

Chapter 23

Skin Infections

The major part of the body's confrontation with the outside world occurs at the surface of the skin. This tough, flexible, and by no means inert outer covering is remarkably resistant to infection as long as it is physically and functionally intact. On the other hand, because of its exposed position, it is frequently subject to damage by mechanical, thermal, and chemical injury. As indicated in Chapter 21 external substances can stimulate hypersensitivity reactions (allergies) which frequently involve the skin. Likewise hormonal or other changes arising within the host may alter normal skin function. Infections of the skin and subcutaneous tissue most frequently arise because of such abnormalities, although a few organisms are capable of invading apparently unbroken skin directly from the environment. Microbes also frequently infect the skin from within, carried by the bloodstream after initial entrance to the body via the respiratory or gastrointestinal system.

Eczema
Page 395

ANATOMICAL AND PHYSIOLOGICAL CONSIDERATIONS

The surface layer (Fig. 23-1) of the skin is composed of flat scalelike material consisting of cells containing *keratin,* a tough protein which is also found in hair and nails. The more superficial cells are dead and continually peel off, to be replaced from a deeper layer consisting of cells more nearly polyhedral in shape. The latter become flattened and die as keratin is formed within them. These superficial layers are called the epidermis. Quite obviously, from the microbial point of view, the outer level of the skin is far from the smooth durable surface

425

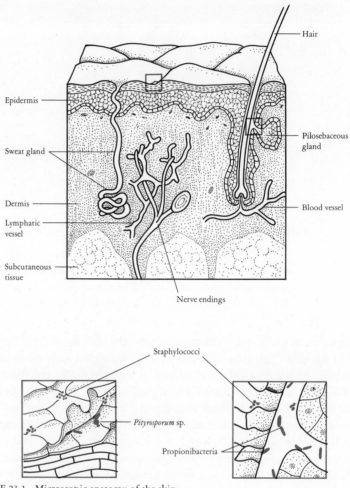

FIGURE 23-1 Microscopic anatomy of the skin.

it appears to us (Fig. 23-2). Supporting the epidermis is a second layer of skin cells composing a matrix through which a large number of tiny nerves, blood, and lymphatic vessels penetrate. This layer, called the dermis, blends in a very irregular fashion with the fat and other cells which make up the subcutaneous tissue.

Almost completely traversing the dermis and epidermis are the fine tubules of sweat glands and the hair follicles. These two kinds of structures are obvious potential passageways through which microbes might pass beneath the skin to reach the deeper body tissues. Feeding into the sides of hair follicles are tiny pilosebaceous glands that produce an oily secretion. This secretion normally flows up through the follicle and spreads out over the skin surface.

The secretions of the sweat glands and sebaceous glands are very important to the microbial residents of the skin since they supply water, amino acids, and lipids which serve as nutrients for microbial growth. Breakdown of the lipids by

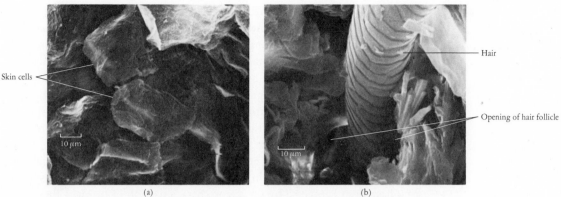

FIGURE 23-2 Scanning electron micrographs of normal skin. (a) Surface of the epidermis. (b) Hair follicle. (Papa, C. M. and B. Farber, *Arch. of Derm.*, **104**:262, 1971; C. M. Papa, M.D., Johnson and Johnson Co., Research)

the normal skin microbial residents results in fatty acid by-products, and these inhibit many potential disease-producers. In fact the normal skin surface is a rather unfriendly terrain for most pathogens, being too dry, too acid, and too toxic for survival.

Factors which derange the normal functioning of the skin tend to nullify these antibacterial properties. For example, an allergic condition which results in weeping of plasma onto the surface of the skin neutralizes the antimicrobial action of the fatty acids. Interestingly enough, for reasons not yet clear, several disorders of body hormones also seem to make the skin more prone to invasion.

THE MICROBIAL INHABITANTS OF THE NORMAL SKIN

The microorganisms living amongst the various components of the normal human skin (Table 23-1) are rather startlingly numerous. It is very difficult to remove and count the viable microbial inhabitants, since they come off in clumps or clusters, each of which may produce only one colony when grown on laboratory medium. In one study a sharp knife was used to scrape the superficial layer of the skin over the shoulder blade, and more than 200,000 viable organisms were

NUMBERS OF MICROBES ON THE SKIN

TABLE 23-1
Principal Members of the Normal Skin Flora

Name	Characteristics
Diphtheroids	Pleomorphic, non-motile, gram-positive rods of the genera *Corynebacterium* and *Propionibacterium*.
Micrococci and Staphylococci	Gram-positive cocci arranged in packets or clusters; micrococci are non-fermentative, staphylococci, fermentative.
Pityrosporum species	Small yeasts which require oily substances for growth.

427

found per square centimeter. Depending on the body location, amount of skin moisture, and other factors, gentle washing with a detergent solution has yielded counts of superficial bacteria ranging from only a few organisms per square centimeter on the back to more than 10 million on the scalp and axilla (armpit).

In recent years there has been renewed interest in studying the skin flora because of the possible derangements which might occur as a result of space flights. The United States National Aeronautics and Space Administration has supported extensive studies of the effects of space diet, high atmospheric oxygen, and confinement in space suits with minimal opportunity for bathing. Under these conditions shifts have been found to occur in the various components of the skin microbial flora and skin disease can result. Unfortunately studies of this kind have been hampered by the vague characterization of many of these organisms, so that potentially significant changes could be obscured by failing to distinguish among different types.

Diphtheroids

The vast majority of the organisms constantly found inhabiting the normal human skin fall into three main groups. The first group can be conveniently labeled "diphtheroids," or organisms resembling, if to a limited degree, the diphtheria bacillus *Corynebacterium diphtheriae*. Diphtheroids are all small grampositive pleomorphic ("many shaped") rods of very low virulence. They are nonmotile, do not form spores, and are distinguished from the diphtheria bacillus by different morphology, fermentation patterns of carbohydrates, and by their nonproduction of toxin.

Corynebacterium
Propionibacterium
Page 258

The diphtheroid species found on the skin in largest numbers is *Propionibacterium acnes* (formerly called *Corynebacterium acnes*), present on virtually all human beings studied. Surprisingly enough *P. acnes* is either anaerobic or microaerophilic. The primary site of growth in the skin, however, is not on the surface cells but within the hair follicle canals. Their growth is enhanced by the oily secretion of the sebaceous glands, and they are usually present in large numbers only in areas of the skin where these glands are especially well developed—the face, upper chest, and back. These are the areas of skin where acne develops, and the constant association of these organisms with acne undoubtedly inspired their name. The concentration of organisms in the acne lesions is often higher than normal, and may contribute to high production and release of fatty acids from the follicle into the tissue, with resulting inflammation and scarring. *Propionibacterium acnes* can be divided into several groups by their patterns of fermentation of glucose, mannose, fructose, lactose, sucrose, mannitol, glycerol, and other carbohydrates. Propionic, acetic, and other volatile acids are produced, which ally them closely with a vast group of propionibacteria in nature.

The more superficial skin areas also contain large numbers of aerobic diphtheroids, representing an even more heterogeneous group than the anaerobic ones. They are especially numerous in moist areas, such as the axillae and between the toes. These organisms show a wide variety of shapes and biochemical actions, and a few have been assigned species names. Some, in the species *Corynebacterium xerosis*, metabolize fatty acids and may contribute to maintaining the normal level of skin fatty acids.

Micrococci and Staphylococci

The second large group of microorganisms universally present on the normal skin are gram-positive cocci which occur in groups or clusters and resemble the pathogen *Staphylococcus aureus*. They are distinguished from that organism by their differing action on sugars and by their nonproduction of coagulases. Like diphtheroids they are nonpathogens. Also like diphtheroids they are a large heterogeneous group, the taxonomy of which is a matter of continuing debate. Current practice is to divide the aerobic members into several genera, the most important of which are *Staphylococcus* and *Micrococcus*. On the normal skin the most important member of the former genus is *Staphylococcus epidermidis*. Micrococci are separated from staphylococci by inability to ferment glucose. Both genera contain several subgroups based on metabolism of carbohydrates and production of the enzyme phosphatase. All produce colonies with white, gold, or yellow nondiffusible pigment and are catalase positive. Staphylococci and micrococci compose the largest group of skin bacteria able to grow in air. Strictly anaerobic organisms of similar morphology (peptococci) also occur on the skin, although much less frequently.

Staphylococci
Micrococci
Page 256

The chief importance of these micrococci and staphylococci is probably in preventing colonization of the skin by pathogens and in helping in the regulation of the other skin flora. In some instances at least they have been shown to produce diffusible antimicrobial substances highly active against *P. acnes* and other gram-positive bacteria, and this may be a factor in maintaining the ecological balance on the skin. Their relative resistance to salt and drying also gives them an advantage over potential pathogens in the competition for nutrients.

Pityrosporum

The third group of organisms inhabiting the normal skin are not bacteria but tiny yeasts. For a long time they were seen in scrapings from skin but could not be grown in culture. Their appearance was round, oval, or sometimes a short fat rod, constricted in the middle, giving rise to the name "bottle bacillus." It was then discovered that growth would occur if fats such as olive oil were incorporated into media, and their fungal nature was also established. They are now known to be asporogenous yeasts of the genus *Pityrosporum*. Two species are generally recognized and are almost universally present on human skin. The first, *P. ovale*, is an oval organism about 4 μm in length, which divides by the formation of a large broad-based bud. A wall, or septum, is then formed between the bud and the parent organism and separation ensues. The second species, *P. orbiculare*, is a spherical, thicker-walled organism about 3 μm in diameter which divides (in a more conventional fashion for yeasts) by pinching off polar buds. The two

Yeasts
Page 283

species also differ in the kinds of fats which can be used for growth. As with some of the diphtheroids large numbers of these organisms are found in certain skin conditions, and attempts have been underway for years to define what role they play in the causation of dandruff and *tinea versicolor*. The latter is a skin disease common in tropical countries and among poor people of temperate zones. There are no symptoms other than complaints about a patchy increase in pigment in light-skinned or decrease in dark-skinned people and a slight scaliness.

Pityrosporum orbiculare can be cultured in large numbers from the patches of involved skin. In contrast to normal skin, however, the organisms growing in masses on the skin surface put down tubelike projections (pseudomycelia) between the skin scales, and even into the layer of living cells. With some other yeasts, production of pseudomycelia is taken to mean parasitic invasion rather than saprophytic growth, and the same may be true with *P. orbiculare.* As for the reason for this behavior it is not yet clear whether it is the result of a peculiarity of the host, or less likely, whether virulent strains will subsequently be identified among the human *Pityrosporum* groups now recognized.

DISEASES OF THE SKIN CAUSED BY BACTERIA

Furuncles and Carbuncles

The familiar pimple is sometimes the result of an infection of the skin. Presumably infection occurs when virulent organisms are rubbed into the opening of a pilosebaceous gland, from which they penetrate the wall of the follicle. An inflammatory response results, followed by the formation of pus and death of tissue. Pressure increases as the result of breakdown of host materials, with resulting increase in osmotically active particles. Protrusion outward through the superficial skin layers follows, with eventual rupture, discharge of pus, and healing. With pimples the process is all relatively close to the surface; there is not much swelling or effect on blood vessels and nerves to give pain. If, however, the infection is deeper in the dermis, a boil (furuncle) results, with consequent more dramatic and disturbing results. The deeper infections may even spread to the area around adjacent follicles, with pus being pushed by the increasing pressure through the tissues to discharge at multiple points. This kind of infection is called a carbuncle, and is the most dangerous because of the possibility that the infection will be carried by the blood vessels to vital areas such as the bone, heart, or brain. At each location where infection is established, tissue destruction and pus formation occur as with a boil. This is why it is more than flirting with danger to squeeze pimples, boils, or carbuncles.

Although pimples may or may not be the result of infection, boils and carbuncles are almost always caused by the pyogenic (pus causing) bacterium *S. aureus* (Table 23-2). Stained smears from pus show gram-positive spherical bacteria averaging a little less than 1 μm in diameter. These are arranged in variable fashion as singles, pairs, clusters, or sometimes chains of three or four. Some of

TABLE 23-2
Characteristics of *Staphylococcus aureus*

Identification	Extracellular Products	Pathogenic Potential
Gram-positive cocci arranged in clusters, cream colored colonies, produce catalase and coagulase and most ferment mannitol; strain identification using bacteriophages.	Hemolysins, leukocidin, hyaluronidase, deoxyribonuclease, staphylokinase, proteinase, lipase, penicillinase, and others.	A prominent cause of boils, wound infections, abscesses, impetigo.

the white blood cells show masses of staphylococci within the cytoplasm of the cell. *S. aureus* grows well on laboratory media, producing colonies which are usually cream colored. Pure cultures of *S. aureus* almost always ferment the sugar-alcohol, mannitol, and this is helpful in distinguishing them from other staphylococci and micrococci.

Staphylococcus aureus has an extraordinary array of unpleasant attributes. All strains cause blood plasma to clot (coagulate) even in the presence of anticoagulants. This is achieved by enzymes called *coagulases*. A test for coagulase production is the most important single procedure in identifying bacterial colonies suspected of being *S. aureus*. Production of coagulase correlates better with virulence than any other single factor when considering the staphylococci. However, at present there is little evidence that coagulase by itself is responsible for virulence.

COAGULASE

Staphylococcus aureus also produces one or more toxins which hemolyse red blood cells (Fig. 23-3). One such hemolysin, the alpha toxin, acts *in vitro* on the red blood cells of nonhuman species and correlates well with virulence. However, antibody to alpha toxin does not give protection from infection. Another hemolysin, the delta hemolysin, is toxic to human white blood cells, and therefore is called a *leukocidin*. In laboratory animals injection of bacteria-free filtrates from broth cultures of *S. aureus* into the skin may produce an area of skin death, or with injection intravenously the animal dies. There is evidence that these two effects sometimes play a role in human disease also.

Table 19-3
Page 365

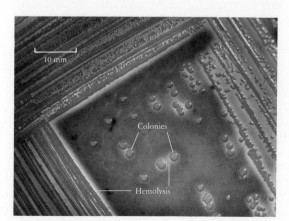

FIGURE 23-3
Staphylococcus aureus colonies growing on sheep blood agar medium. Note the zone of clearing around the colonies as a result of destruction of red blood cells in the medium. (Courtesy of F. Schoenknecht)

As mentioned in Chapter 19, other extracellular products may play a role in staphylococcal infections. These include hyaluronidase, staphylokinase (an enzyme that converts an inactive clot-dissolving substance in blood to an active form), proteinases, lipases, and penicillinase.

A large percentage of strains of *S. aureus* produce a lysozyme-like substance which destroys some organisms among the normal flora. Some strains also produce another bactericidal substance which acts on other gram-positive organisms. These substances may assist some staphylococci in colonizing areas of the host which have a large existing microbial population.

Impetigo

Impetigo, another bacterial skin disease, is not uncommon among schoolchildren among whom it can spread in epidemic fashion. Unlike boils the infection is very superficial, involving patches of epidermis just beneath the dead scaly layer. The growth of the bacteria typically results in a thin-walled blister which breaks and is replaced by crust and areas of weeping of plasma. Usually there is no accompanying fever or pain, although lymph nodes near the involved area often enlarge, indicating that bacterial products often enter the lymphatic and blood vascular systems.

Most commonly, *S. aureus* in large numbers is isolated in pure culture from the blister fluid of persons with impetigo. In such cases the staphylococcus has normally been of bacteriophage type 71, rather than a variety of the more common phage types. This indicates that organisms which cause impetigo may have some peculiar properties which allow them to infect the intact superficial skin layers. So far scientists have not been able to define these traits, perhaps because expression of them occurs only in the human "medium" and not in laboratory media. However, it is known that the staphylococcal strains causing impetigo produce larger amounts of hyaluronidase than others, and also a diffusible bactericidal substance which acts against other gram-positive organisms, including the impetigo-causing streptococci discussed later. These factors may aid colonization and spread within the skin although they are not solely responsible for virulence.

Streptococcus
Page 256

In many other persons with impetigo, the blister fluid contains the bacterium *Streptococcus pyogenes* instead. This organism (Table 23-3) can be recovered

TABLE 23-3
Characteristics of *Streptococcus pyogenes*

Identification	Extracellular Products	Pathogenic Potential
Gram-positive cocci in chains or pairs. Colonies beta-hemolytic, small. Catalase negative. Cell wall contains group A polysaccharide. Strains distinguished by cell wall proteins (M and T antigens).	Hemolysins (streptolysins) O and S, streptokinase, DNAase, hyaluronidase, and others.	Causes impetigo, pharyngitis, wound infections, puerperal fever. Late complications of infection include glomerulonephritis and rheumatic fever.

in pure culture on laboratory media or, later in the disease, mixed with a *S. aureus* strain which can be of various bacteriophage types. Although material from streptococcal infections usually shows the organisms as small gram-positive cocci occurring in pairs, in culture the organisms produce the chains of cocci typical of other streptococci. The organism is more fastidious than staphylococci in its nutritional requirements, and even on blood agar the colonies are smaller than those of staphylococci after 24 hours incubation. The vast majority of strains hemolyse red blood cells of various species and thus there is an area of clearing of the medium surrounding colonies growing on blood agar. As mentioned previously this effect on the blood agar is called "beta hemolysis" and also occurs with several other species of bacteria including other streptococci and most *S. aureus*. Despite the fact that several other species produce it, beta hemolysis is, however, very useful in spotting possible colonies of *S. pyogenes* in cultures of material from people or their environment. One need select only those colonies which are beta hemolytic for further identification procedures. The often used term, "beta hemolytic streptococcus," includes many different kinds of streptococci, some requiring anaerobic growth conditions, some with little or no virulence, as well as the highly pathogenic *S. pyogenes*. The latter organism is distinguished from the others by using antiserum to identify its specific cell wall antigen, called the group A polysaccharide.

Like *Staphylococcus aureus, Streptococcus pyogenes* produces a wide variety of extracellular products. For example, two hemolysins may participate in producing the changes in blood mentioned above. The first of these is streptolysin S, an oxygen stable, nonantigenic substance. The second hemolysin is streptolysin O which becomes inactive on exposure to air and is highly antigenic. Thus the beta hemolysis seen on aerobically incubated culture plates is caused mostly by streptolysin S, and optimum hemolysis occurs under atmospheres of reduced oxygen concentration because of the added effect of streptolysin O. In infections with *S. pyogenes*, streptolysin O is released by the organisms and is absorbed by the body. Since it is antigenic, antistreptolysin O antibodies often appear in the serum during the course of an infection and can be detected and measured in the laboratory. An increase in the amount of antistreptolysin O in a blood specimen taken one to two weeks after the infection starts, compared with the amount of antistreptolysin O at the beginning, is good evidence that the infection was caused by *S. pyogenes*.

A number of extracellular products may contribute to virulence. These include enzymes which break down protein, nucleic acid, and the hyaluronic acid of host tissues. As with staphylococci it is doubtful whether any one of these factors plays an essential role in pathogenicity. However, the role of surface antigens of *S. pyogenes* is quite well defined, at least in experimental infections in laboratory animals. The most important substance is a heat and acid resistant protein, called *M-protein,* which interferes with phagocytosis of the organism by white blood cells of the host, thus preventing its destruction and permitting it to continue to multiply. M-protein is thus very important in virulence, and antibody to it plays a big role in immunity by opsonizing the organism. However, since

Streptococcal hemolysis
Figure 5-8; Page 116, Page 257

STREPTOLYSIN O

M-PROTEIN

there are many different strains of *S. pyogenes* and a number of different M-proteins, immunity to one strain does not necessarily provide immunity to another. In some strains, hyaluronic acid (the same substance normally present in body tissue) is produced by the streptococcus in the form of a capsule. This substance also interferes with phagocytosis and results in increased virulence.

Although all strains of *S. pyogenes* possess the same species specific carbohydrate antigen (the Lancefield group A polysaccharide) enmeshed in the mucocomplex of their cell walls, they have in addition a variety of other cell wall antigens. The M-proteins can be identified by precipitin reactions of streptococcal extracts with specific antibodies produced in experimental animals. Unfortunately only about half of the *S. pyogenes* isolates can usually be typed by identifying their M-protein; the remainder fail to react with any of the available antisera. The problem of typing strains which do not yield to M-protein typing is partly solved by identification of other cell wall proteins, the T-proteins. In contrast with M-proteins the T-antigens have nothing to do with virulence, but using specific antisera to identify T-antigens by agglutination and M-proteins by precipitation makes it possible to identify most strains of group A beta hemolytic streptococci.

TYPING *S. PYOGENES*

For many years it has been noticed that persons with infections caused by *S. pyogenes* may suddenly develop fever, high blood pressure, and blood and protein in their urine during convalescence. A few may become so sick that death follows. This complication of streptococcal infection is caused by inflammation of small tufts of tiny blood vessels (glomeruli) within the kidney (nephros), and is called acute *glomerulonephritis*. This complication occurs only after infection by certain types of *S. pyogenes* and not others. Another late complication, rheumatic fever, occurs much more commonly as a result of throat rather than skin infections, whereas in recent years the reverse has generally been true of glomerulonephritis. In both of these complications it is unusual to find *S. pyogenes* or other bacteria in the diseased tissues, and both conditions probably have an immunological basis.

GLOMERULONEPHRITIS

DISEASES OF THE SKIN CAUSED BY VIRUSES

Smallpox (Variola)

In January 1962 a man from Pakistan arrived in London by airplane and traveled to Wales. He had been vaccinated the previous month but developed no skin reaction, indicating the possibility that the vaccine was no good or improperly administered. About the time he arrived in Wales, he became ill and was admitted to the local hospital. His symptoms were only fever and rash, and no definite diagnosis was made. A pregnant housewife living a short distance from the hospital became ill about three weeks later and suffered a miscarriage at home. A neighbor helped her with the delivery, but the sick woman failed to improve and required hospitalization because of progressing illness. She died shortly after

SMALLPOX EPIDEMIC

entering the hospital. An autopsy revealed no definite cause for her illness. However, within two weeks of her death the diagnosis became clear when the friend who had helped with the home delivery of her dead child developed the typical pus-filled blisters of smallpox and died shortly thereafter. In fact all three of these people had smallpox, but in the first two it was in a form which was not easily recognizable as such.

Subsequently the dead mother's sister, brother, husband, son, and physician-consultant also developed smallpox, and the mother of one of them also developed smallpox. Two of these people died.

By now the diagnosis was well known and public health authorities began a systematic campaign of vaccinating people in the area. However, secondary cases began developing among the people who were exposed to the friend who had helped during the miscarriage and who subsequently died of smallpox. Seven cases of smallpox developed among these contacts; her husband, father, mother, brother, and three sisters. Two cases also developed among the patients of a doctor who had visited the friend before the diagnosis was known. All had been vaccinated, but too late to prevent the illness. A number of other cases developed before the epidemic was contained.

In summary, this modern epidemic, which apparently started by introduction of the smallpox agent by a patient who arrived by airplane with an atypical illness, involved at least 50 cases and caused 18 deaths. Four hospitals were involved. Very few of the victims had been vaccinated since childhood, and some had never been vaccinated. In several instances there was definite evidence of transmission of the virus by healthy persons or inanimate objects. A similar epidemic in Yugloslavia in the spring of 1972 resulted in 173 cases and claimed 34 lives.

PATHOGENESIS OF
VARIOLA

Smallpox virus enters the body of the victim through the respiratory tract, carried in with air contaminated by a patient with smallpox. The virus is taken up by cells in the respiratory tract, multiplies, and is released into the lymphatic vessels. Further multiplication probably occurs in cells of the reticuloendothelial system. Large amounts of virus later enter the bloodstream. From there it is disseminated throughout the body, but localizes in cells of body tissues as well as in the living cells of the epidermis, where it produces the skin changes resembling those seen after successful vaccination. Death from smallpox presumably results from the damage produced throughout the body organs. There are actually two variants of the virus: one produces mortality rates of 20 to 50 percent (variola major) and the other produces death in about 1 percent of the victims (variola minor or alastrim).

LABORATORY
DIAGNOSIS

The rash of a typical case of smallpox begins as small red spots, which quickly develop into small, firm pimples, and finally the pimples become small blisters as the infected cells break down. Initially the blisters contain clear fluid, but later they become filled with pus. In the earliest stages before pus forms, scrapings of the involved skin can be smeared onto glass microscope slides, stained, and inspected with the light microscope. The virus particles are often visible in large numbers as small bodies near the limits of resolution of the microscope. Skin cells, if present, may show poorly defined cytoplasmic inclusion bodies. Electron microscopic examination can also be made of scrapings, or even the pus and crusts of later stages. This approach has been used in the rapid diagnosis of skin diseases which might possibly be smallpox, although it can only identify or rule out viruses of the pox group and not distinguish among them. Smallpox virus antigen can also be detected in material from the skin rash by preparing smears on microscope slides and staining with a fluorescent antibody prepared by immunizing a laboratory animal with a related virus. This method is fast and accurate, but failures can occur, as happened in the United States in 1965 when a false diagnosis led to vaccinating numerous contacts to prevent a feared smallpox epidemic from a foreign visitor who later proved to have only chicken pox. Antigen can also be identified serologically by precipitin and complement fixation tests.

Serological tests
Pages 378, 380

Smallpox epidemics have been a scourge of mankind for many centuries, and nowhere have they been more severe than on the North American continent following the first arrivals of Europeans. An epidemic swept the Massachusetts coast during 1617–1619, killing about nine-tenths of the native American inhabitants. A visitor to the friendly Indian chief Massassoit in 1621 found many skeletons lying on the ground because not enough people remained alive to bury the many dead.

There was enormous fear of the persistence of this invisible killer, so that when burials were performed, they were made in "pox acres" remote from usual habitations. Houses in which smallpox victims were cared for were burned to the ground when the person died or recovered. That such fear was to a certain extent justified is dramatized by some recent studies showing that crusts from the skin of smallpox victims, stored in envelopes at room temperature, showed more than 50,000 viable virus particles per crust after three years. After ten years more than 1000 per crust could still be demonstrated!

Early attempts at immunologic control made use of the interesting principle that introduction of an infectious agent into the host via an *abnormal route* may often result in a milder form of the disease. Thus ancient Chinese and Indian people rubbed material from smallpox cases into their nostrils or skin (variolation) in an effort to contract mild disease. If successful, these people would recover from the resulting illness with a lasting immunity. Unfortunately the disease produced was sometimes severe, or even worse started an epidemic among the members of their communities, as was shown as recently as 1970 in Africa where variolation was still occasionally employed.

VARIOLATION

Jenner's observation that the pox virus of cows could infect humans, producing a mild illness but substantial immunity to smallpox, was, of course, a landmark in control of infectious diseases. This "experiment of nature" demonstrated clearly that an organism of little or no virulence could confer immunity against a related organism of dangerous potential. Artificial inoculation of people with cowpox became a widespread practice which was highly effective in controlling the spread of serious smallpox.

CROSS-IMMUNITY

Present-day vaccination against smallpox, however, utilizes vaccinia—a different virus—the origin of which is buried in the obscurity of time. Strangely enough present-day laboratory studies show the vaccinia virus to be more closely related to the smallpox viruses than to cowpox.

The immunity resulting from vaccination is substantial, although it wanes with time and often allows mild and atypical smallpox illness which is, nevertheless, transmissible. Smallpox epidemics in Western countries, such as the one described at the beginning of this section, have shown that 70 to 95 percent of the cases occur in people who have never been vaccinated. Those who contract smallpox even many years after vaccination usually have a milder illness. For example, in a recent epidemic the mortality was 53 percent in those never vaccinated, but only 15 percent in those last vaccinated 15 years previously. Finally, vaccinated persons who get smallpox transmit the disease to others only one-third as readily as those who have never been vaccinated.

Although recent studies confirm that there is no natural host other than man for smallpox viruses, eliminating smallpox from the world by vaccination has proved a difficult task. During the first half of the twentieth century smallpox was eliminated from Europe and North America as an endemic disease. Since then intensive efforts by the World Health Organization have led to the administration of hundreds of millions of doses of vaccine, predominantly in Asia, Africa, and South America. Despite this almost 118,000 cases were reported in 1967. Intensified programs undertaken at that time have resulted in dramatic declines in the incidence of smallpox, and sustained international cooperation appears to be finally winning the battle against this disease.

Paradoxically while smallpox vaccination is being pushed extensively in areas of endemic disease, routine vaccination of all children has been discontinued in the United States. Routine smallpox vaccination is a relatively safe procedure, the few fatalities most commonly resulting when a person with eczema, (the weepy allergic skin disease), is exposed to someone who has recently been inoculated with smallpox vaccine. Only about seven people die each year from vaccination among the 14 million people in the United States who receive their first vaccination. However this would mean at least 210 deaths from vaccination over the next 30 years. As of 1972 the United States had not had a smallpox epidemic in 23 years, and the chances of imported smallpox will decrease further with the decreasing incidence of the disease world-wide. Even so, we could still probably have two smallpox importations every three years without exceeding a total of 210 deaths, because modern methods can quickly limit spread of any epidemics which might start. There is even a medication, a thiosemicarbazone derivative, which can prevent smallpox if given soon after exposure.

At present, memories of epidemics of the recent past are still too vivid for anyone seriously to suggest abandoning smallpox vaccination altogether. It is interesting, however, to consider at what point in the battles against infectious diseases the human species can safely afford to neglect exposing its immune mechanisms to the antigens of pathogenic microbes.

A few years ago in a large suburb of Seattle, Washington, a 29-year-old housewife complained to her physician of a painful rash which had appeared on one side of her chest. She had had chicken pox years ago while in grammar school, and it had cleared up in the usual fashion. The doctor examined the rash on her chest and noticed that it consisted of small red bumps, blisters, and scabs, and that it seemed to follow the branches of one of the nerves of skin sensation. He informed the housewife that she had shingles (herpes zoster), and that it represented a reactivation of the chicken pox infection she had had many years ago. This seemed somewhat unlikely to the housewife, but her skepticism lessened appreciably when the first of her four children developed chicken pox two weeks later, followed by the other three in succession, and then their friends, until half of the younger children in the immediate community had become involved.

PATHOGENESIS OF
VARICELLA

The chicken pox virus enters the body of a person with air contaminated by a patient with chicken pox. The virus is thought to multiply in cells lining the respiratory tract and from there infects cells of the body tissues. From the latter it is released into the bloodstream and is carried to the cells of the epidermis. The infected skin cells initially enlarge, then break down leaving a fluid-filled blister beneath the skin. This later may become filled with pus, rupture, and form a crust or scab before finally healing. In short the development of varicella and smallpox are quite similar, although the former is considerably less severe and deaths are unusual in normal people. The virus is, however, a threat to babies of mothers who contract the disease near the time of birth, and to persons whose immune mechanisms are impaired. In such instances the virus causes extensive damage to internal organs which often results in death. In older children and adults the virus sometimes causes extensive damage to the lung and in past years as many as one out of five such cases have died.

So far we do not fully understand the development of herpes zoster. However, it is quite clear that in some persons who develop chicken pox, the virus is not entirely eliminated from the body at the time of recovery. This surviving virus is thought to enter the cells of some nerves of sensation. These nerves, responsible for transmission of sensation from the skin, are very complex and interesting structures. The nuclei of the nerve cells themselves are located in small bodies called *ganglia* lying near the spinal column. From the nucleus the cell stretches out an extremely long process all the way to the skin, as well as a shorter process to the spinal cord. A number of these cell processes are enclosed together in a sheath of cells, and each nerve fiber within the sheath is also enclosed by a layer of specialized cells. Lymphatic channels are also present within these nerve bundles, as well as connections with the space containing the cerebrospinal fluid.

It is not known yet where in this system of ganglia and nerve bundles the chicken pox virus stays during the many years the person is without evidence of infection. It is also not known in what form the virus exists, whether incomplete (noninfectious) or mature virus. However, when herpes zoster begins, the first change is an inflammatory process in the ganglion, which is then followed by the appearance of infectious virus in the skin supplied by the corresponding nerve cells, the cerebrospinal fluid in some cases, and possibly the blood. High levels of antibody appear within a few days, but despite this, skin blisters containing infectious virus are present for a much longer time than in patients with chicken pox, where antibody is slower in appearing.

It seems reasonable to postulate that the hidden virus resides in the nerve cell and in producing shingles it begins multiplying in the nucleus of the cell. Infectious virus is then probably carried along the processes of the nerve cell in channels inaccessible to the action of neutralizing antibody, and is thereby carried to the preferred host cells, the cells of the skin.

The mechanism of persistence of the virus may represent an important adaptive solution to the problem of survival in small isolated populations. It has been noticed, for example, that when a highly infectious virus like measles is introduced into such a population, it spreads quickly and infects most of the susceptible individuals, who then become immune or succumb to the infection. The measles virus then disappears from the community. By contrast, when chicken pox is introduced, it persists indefinitely as recurring epidemics whenever sufficient susceptible children have been born. The sources of these recurring epidemics could quite possibly be cases of shingles.

The chicken pox virus is a member of the herpes group, which includes the virus of herpes simplex (cold sores). Like other herpes viruses, the varicella-zoster virus has an icosahedral shape with a DNA core and a lipid-containing envelope. If material is scraped from the base of the blisters, smeared on microscope slides, and stained, an important finding can often be made with a light microscope. Although the individual virus particles themselves cannot be clearly identified, they cause the host to produce large, multinucleated cells, and finding them is very useful in identifying infections with herpes viruses and in ruling out variola. Also sometimes visible under the light microscope are infected host cells with inclusion bodies within the nucleus, the site of viral assembly. Specific antiserum can also be used to identify viral antigen in scrapings, and this is most quickly and conveniently used as a fluorescent antibody.

Herpes viruses
Table 17-1; Page 335, Page 676

Since chicken pox and shingles have only rarely been fatal for normal children, and most people have been infected by the virus by the time they reach adulthood, there has been little effort at developing control measures. Indeed since chicken pox is so much more severe in adults, it may be wise to promote childhood infection. Attempts at protecting very susceptible people (such as patients with leukemia) have been made by injecting them with gamma globulin pooled from normal people. These attempts have not usually been successful, probably because of the low antibody levels. On the other hand, use of globulin obtained from patients having recovered from shingles was highly effective in preventing chicken pox in normal children. Administration of chemicals which interfere with DNA synthesis has also been of benefit when given early in the illness to some patients.

BACTERIAL
SUPERINFECTION IN
RUBEOLA

RUBELLA, A CAUSE OF
BIRTH DEFECTS

Measles (rubeola) develops in a similar way as smallpox or chicken pox, but the major effects are on the respiratory tract as well as the skin. Serious consequences of measles are quite common and result from secondary infection by opportunistic bacteria such as *S. pyogenes, Haemophilus influenzae,* and *Streptococcus pneumoniae.* Effective control of measles is now accomplished by using a living vaccine derived from measles virus considerably lessened in virulence (attenuated) by prolonged growth in the laboratory. Measles can apparently be followed years later by a very rare disease marked by progressive degeneration of the brain lasting months or years. Measles virus antigen can be demonstrated in the brain of such patients, and high levels of measles antibody are present in their blood. It is as yet uncertain whether the viral antigen represents measles virus or a closely related unknown form.

    German measles (rubella) also develops in a similar way, but the effects on both skin and respiratory tract are mild, and it is less contagious. For this reason many people escape childhood without developing the infection and are therefore susceptible as adults. Epidemics may therefore occur among college students and military recruits. Unfortunately in infected pregnant women during the stage when the virus is circulating, the fetus contracts the infection too. In fact if the mother contracts rubella during the first eight weeks of pregnancy, at least 90 percent of the fetuses will become infected. During this stage of development the tissues of the fetus are easily damaged by the virus, which arrests tissue cell multiplication and produces chromosome abnormalities. Miscarriages often result, but more commonly the child lives and is born with deafness, defective heart, poor vision, or mental retardation. These defects, however, do not always follow infection of the fetus. The risk of serious defects among live-borne children has been 30 to 50 percent when rubella was contracted during the first

TABLE 23-4
Viral Diseases Involving the Skin

	Variola (Smallpox)	Varicella (Chicken pox)	Rubeola (Measles)	Rubella (German Measles)
Causative agent	Variola virus, a pox virus.	Varicella-zoster virus, a herpes virus.	Rubeola virus, a paramyxovirus.	Rubella virus, (uncertain classification).
Importance	Very contagious; variola major has a high mortality.	After recovery from chicken pox, the infection may become latent: recurs as herpes zoster.	Damages the respiratory mucosa; thereby predisposes to serious bacterial infections of respiratory system.	Infections that develop during pregnancy commonly cause fetal damage; virus excretion continues after birth for extended periods of time.
Control	Vaccination with vaccinia virus; medications given shortly after exposure can prevent the disease.	No vaccines available; passive immunization for highly susceptible persons such as leukemics.	Vaccine: a live, attenuated rubeola virus.	Vaccine: a live, attenuated rubella virus.

month of pregnancy, with the risk declining with infections occurring later in pregnancy. Some of these children are unable to develop neutralizing antibody to rubella virus and continue to excrete the virus for many months after birth (a possible example of immune tolerance). The virus can also persist for years in tissues inaccessible to antibody, such as the lens of the eye.

To reduce the possibility of infections in pregnant women, attempts at control of rubella are now underway with the widespread use of a live vaccine for immunizing children. This vaccine was derived from German measles virus by many subcultures in tissue cell cultures. It produces milder disease than the "wild-type" rubella virus. It should not be administered to women who might be pregnant since it is remotely possible that the vaccine virus might also damage the fetus.

The principal features of these viral skin diseases are given in Table 23-4.

FUNGUS DISEASES OF THE SKIN

Earlier in this section we mentioned the possible role of one of the members of the normal flora, the yeast *Pityrosporum,* in causing a mild skin disease, tinea versicolor. Other fungi are responsible for more clear-cut infections of the skin, although even in these cases there are strong host factors which play a role. The yeast *Candida albicans* is commonly found among the normal flora, yet in some people it invades the nails, skin, and subcutaneous tissues. In many people with candida skin infections, no precise cause for the invasion can be determined. Similarly, a variety of mold-type fungi may invade the skin, hair, and nails, producing conditions with such colorful names as athlete's foot, jock itch, and ringworm, but only in a minority of all persons colonized by the organisms. These skin-invading molds belong to the genera *Trichophyton, Microsporum,* and *Epidermophyton,* and are collectively called *dermatophytes.* They are peculiar in that they are unable to grow in living tissues, but instead multiply in structures containing keratin. Most of their unpleasant effects result from secondary bacterial invasion or allergic reactions to the presence of the fungi, but thickening of nails and loss of hair give rise to a few complaints, too. Some of these agents have major reservoirs in soil or on the skins of animals, while some are only known to attack man.

Table 14-3
Page 291

SUMMARY

The normal skin is scaley and uneven at the microscopic level. Oily secretions and sweat are produced by glandular elements and support the growth of a rich aerobic and anaerobic normal flora. The most important microbial groups are diphtheroids (including *Propionibacterium acnes*), micrococci and *Staphylococcus epidermidis,* and yeasts of the genus *Pityrosporum.*

Bacterial pathogens most frequently invade skin damaged by trauma or other factors. *Staphylococcus aureus* is one of the most important, producing furuncles, carbuncles, and impetigo. It characteristically produces tissue destruction and abscesses if it escapes localization in the skin and gains access to other parts of the body. *Streptococcus pyogenes* (group A beta hemolytic streptococci) also causes skin infections such as impetigo. Infections by certain strains of this bacterium sometimes result in glomerulonephritis, an inflammatory condition of the kidneys, developing during healing of the skin disease.

A number of viral diseases involve the skin, although the mode of entry to the body is usually respiratory or gastrointestinal. Smallpox (variola) has until recent years been the most dreaded of these diseases because of its severity and extremely infectious nature. Vaccination with the related virus, vaccinia, has now almost conquered this disease. Chicken pox (varicella) is caused by the varicella-zoster virus, a member of the herpes virus group, to which the herpes simplex (cold sore) virus also belongs. These viruses can exist in tissues in an inapparent form, only to produce active disease years after the initial infection. Measles (rubeola) characteristically produces its most severe damage to the respiratory system, resulting in superinfection by bacteria. An unusual degenerative brain disease is now thought to be a late complication of measles virus infection or a close relative. German measles (rubella) is a mild disease, chiefly of importance because of its ability to infect the fetus and produce birth defects.

A number of fungi infect the skin. *Candida albicans* is one of the most invasive. Others of the genera *Trichophyton, Microsporum,* and *Epidermophyton* can attack only keratinized structures such as the outer layers of epidermis, hair, and nails. Inflammatory reactions may result from allergy to fungal products or superinfection by bacteria.

QUESTIONS

1. What are the principal attributes of skin that make it resistant to infection?
2. Discuss the pathogensis of a carbuncle and the responsible organism.
3. Discuss the pros and cons of routinely immunizing all United States children against smallpox.
4. What is the relationship between varicella and herpes zoster?
5. Compare and contrast the complications of measles (rubeola) and German measles (rubella).

FURTHER READING

MAIBACH, H. I. and G. HILDICK-SMITH (eds.), *Skin Bacteria and Their Role in Infection.* New York: McGraw-Hill, Inc., 1965.

MARPLES, M. J., *The Ecology of the Human Skin.* Springfield, Ill.: Charles C Thomas, 1965.

———, "Life on the Human Skin," *Scientific American,* 220 (January 1969), 108–115.

ROGERS, F. B., " 'Pox Acres' on Old Cape Cod," *New England Journal of Medicine,* 278 (1968), 21–23.

ROTHMAN, S. and A. L. LORINCZ, "Defense Mechanisms of the Skin," *Annual Review of Medicine,* 14 (1963), 215–242.

Chapter 24

Infections of the Upper Respiratory System

Infectious diseases of the upper respiratory system are by far the commonest afflictions of mankind, and far outweigh any other in cumulated misery and loss of productivity. The following sections illustrate features of microbial interactions with the upper respiratory tract and its appendages.

ANATOMICAL AND PHYSIOLOGICAL CONSIDERATIONS

A person normally breathes about 16 times a minute, 500 cc of air with each breath, or over 11,500 liters of air per day. This enormous volume of air with its accompanying microbes flows into the respiratory system (Fig. 24-1) as a result of the vacuum produced in the lung by chest and abdominal muscles. The air enters at the nostrils (a common site of colonization by *Staphylococcus aureus*), flows into the nasal cavity, and curves downward through the throat. It enters the lower respiratory tract below the epiglottis, a muscular fold of tissue which closes off the lower tract during swallowing (and site of a serious infection of children by *Haemophilus influenzae*). The nasal cavity is a chamber above the roof of the mouth, incompletely divided into right and left halves by a vertical wall extending almost back to the throat. Spongy masses of tissue bulge into each half of the chamber from the outside walls. These tissues are similar to the erectile tissues of the genitalia in that they expand and contract with alterations in blood flow controlled by nervous reflexes. With extreme enlargement they contribute to "nasal congestion" or obstruction of the airways. The stimuli for these changes are not all well defined, but probably include temperature, humidity, and emotional factors, as well as the irritating effect of infections.

443

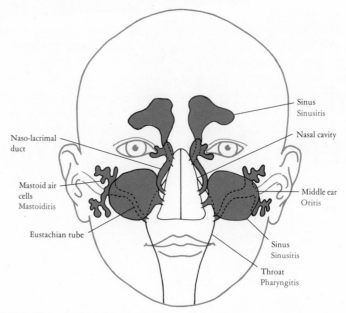

FIGURE 24-1 Upper respiratory system.

Openings of the tear ducts, sinuses, and eustachian tubes from the ears enter the chamber on each side. Thus the conjunctiva of the eye, middle ear, and sinuses of the skull (including the mastoids) are connected to the nasal chamber. Collections of lymphoid tissue, the tonsils and adenoids, are located where the mouth and nasal chamber join the throat. Such tissue is important in the production of immunity to infectious agents, but paradoxically it can be the site of certain infections, and with enlargement can contribute to infections of the ears by interfering with normal drainage through the eustachian tubes. The lining cells of the nasal chamber and its appendages (including the middle ear) have tiny hairlike projections (cilia) along their free border. The cilia beat synchronously and continually propel a film of mucus (secreted by other cells) to the throat and nose.

MUCUS AND CILIARY ACTION DEFEND AGAINST INFECTION

The rich and varied normal microbial flora of the nasal passage and throat often includes potential pathogens such as *Streptococcus pneumoniae* (a gram-positive diplococcus) and *Haemophilus influenzae* (a small gram-negative rod). The ears and sinuses normally are kept free of microorganisms by the ciliary action of the lining cells which continuously move mucus outward through small passages into the nasal cavity. The pH of the nasal mucus is usually about 6.5; breathing cool air has been found to shift the pH toward alkaline, whereas hot air results in more acid mucus. The slight acidity favors the growth of some respiratory pathogens.

The primary functions of the upper respiratory tract are to regulate the temperature and humidity of inspired air and to remove or destroy undesirable microbes and other foreign material. Nervous reflex mechanisms immediately increase the blood flow when cold air arrives, or decrease it with warm air. Transfer of heat between the blood and the air usually adjusts the temperature to within two or three degrees of body temperature by the time the air reaches the lung. The air is also saturated with water vapor, the nasal passage having the ability to give up more than a quart of water per day to inspired air. Expired air is cooled on passing out, giving back much of the water vapor to the mucous membrane. Trapping of inspired particles is also very efficient. Turbulence produced by nasal hairs causes larger particles to impinge on the mucous film and become trapped. Even with small particles, 1 to 5 μm in size, only 50 percent escape trapping in the upper airways. It seems likely that infections of the upper respiratory tract interfere with these primary functions and thereby promote lower respiratory tract infection.

NORMAL FLORA OF THE UPPER RESPIRATORY SYSTEM

Even though the eye is exposed to a multitude of microbes, about half of all normal healthy people yield negative cultures of the conjunctiva. Presumably this sparcity of microbes is the result of frequent mechanical washing with lysozyme-rich tears and the blinking reflex of the eyelid. Organisms impinging on the moist membrane covering the eye are almost immediately swept into the naso-pharynx. When organisms are recovered in cultures, they usually are scant and consist of species found among the flora of the nasal chamber or skin (Table 24-1). Micrococci, staphylococci, diphtheroids, nonpathogenic *Neisseria* species, and *Haemophilus* species are the groups most commonly encountered.

Neisseria
Page 256;
Haemophilus
Pages 114, 253

The secretions of the nasal entrance usually contain large numbers of diphtheroids, micrococci and staphylococci, and smaller numbers of environmental microbes such as *Bacillus* species. About a third of normal people carry *S. aureus.*

TABLE 24-1
Normal Flora of the Upper Respiratory System

Name	Appearance	Comments
Micrococci and staphylococci	Gram-positive cocci	May include the pathogen, *S. aureus*
Corynebacterium	Pleomorphic gram-positive rods, non-motile, non-spore-forming	Nonpathogens collectively referred to as "diphtheroids;" the pathogen is *C. diphtheriae*
Neisseria	Gram-negative diplococci	May include the pathogen, *N. meningitidis*
Haemophilus	Small pleomorphic gram-negative rods	May include the pathogen, *H. influenzae*
Bacteroides	Small pleomorphic gram-negative rods	Strict anaerobes
Streptococcus	Gram-positive cocci in chains	Alpha (especially viridans streptococci), beta, and gamma types; may include the pathogen, *S. pneumoniae*

The nasal passages have a relatively sparse aerobic flora, the species being those found in the nasal entrance or in the nasopharynx. The nasopharynx contains large numbers of microorganisms, mostly alpha streptococci of the viridans group, gamma streptococci, *Neisseria* species, and diphtheroids. Anaerobic gram-negative bacteria including the genus *Bacteroides* are also present in high concentrations. In addition, pathogens such as *S. pneumoniae, H. influenzae,* and *N. meningitidis* are commonly found, especially during the cooler seasons of the year. Carriage of *Streptococcus pyogenes, Mycoplasma pneumoniae, Corynebacterium diphtheriae, Bordetella pertussis,* and other respiratory pathogens occurs less commonly.

THE COMMON COLD
(ACUTE AFEBRILE INFECTIOUS CORYZA)

The all too familiar symptoms of this disease are the result of upper respiratory tract infections by any of about 80 known viruses and a few bacteria. Myxoviruses (influenza A, B, and C; parainfluenza and respiratory syncytial), adenoviruses (eight types), and enteroviruses (Coxsackie A and B, four types; seven ECHO viruses) have all been recovered from persons with colds. However, the vast majority of cases are now known to be due to rhinoviruses, small acid-sensitive RNA viruses of the picorna group. These viruses were initially very difficult to cultivate in the laboratory since they failed to infect laboratory animals or tissue cell cultures. Workers in England then discovered that if cell cultures of monkey or human origin were incubated at 33°C instead of body temperature, and at a slightly acid pH instead of the alkaline pH of body tissues, in many cases positive cultures were obtained. It is noteworthy that the lower temperature and pH are conditions which normally exist in the upper respiratory tract.

Human beings are probably the only significant reservoir of the rhinoviruses which produce colds. The viruses are very likely contracted by inhalation, and close contact with an infected person appears to be necessary to transmit the agents. After lodging on the respiratory lining cells, infection of a few of the cells is established, virus is replicated intracellularly, is discharged, and infects adjacent cells. The irritation stimulates nervous reflexes which cause an increase in nasal secretions, sneezing, and swelling of the mucosa and erectile tissue so that the airway is partially or completely obstructed. Later, an inflammatory reaction occurs, with dilation of blood vessels, oozing of plasma, and congregation of polymorphonuclear white blood cells. The secretions may then contain pus and blood. The infection may be limited, probably by interferon release, but it may extend into the ears, sinuses, or lower respiratory tract before it is halted. Recovery is associated with the appearance of specific antibody which generally gives immunity to the infecting agent for two or more years. Locally secreted IgA antibody is thought to be responsible for the immunity since reinfection can occur even though plasma antibodies are present. Persons with colds have virus in their nasal secretions from one or two days before the symptoms begin to eight

COLDS HAVE MANY
CAUSES, PRINCIPALLY
RHINOVIRUSES

Rhinovirus
Page 679

PATHOGENESIS

Interferon
Page 337

or more days after. Some persons become infected and have infectious virus in their nasal secretions without developing a cold. Such people may be more likely than ill persons to transmit the virus since no one would suspect they had the infection.

There is as yet little evidence that the scientific proof of the 1960s has convinced the average person that colds are caused by viruses. The conviction that colds are caused by cold is deeply ingrained, and scientific studies have been repeatedly undertaken to define the relationship between exposure to cold and the common cold. Studies as long ago as the 1920s showed that the incidence of colds increases dramatically and simultaneously in all parts of a region with the onset of cold weather. On the other hand, another important study, reported in 1933, showed that colds disappeared from the Arctic island of Spitzbergen during the long winter when no ships came. Colds reappeared in Spitzbergen when ships arrived in the late spring. This indicated that new sources of infectious virus were necessary to produce colds, regardless of the temperature.

ROLE OF WEATHER UNCLEAR

Very careful recent studies have shown that the incidence of colds was the same when a rhinovirus was administered to nonimmune volunteers exposed to chilling as in those without chilling. Thus there is little evidence that exposure to cold increases susceptibility to the virus. Finally, it has been shown that in semitropical areas a sharp increase in colds occurs with the onset of the rainy season, even though the mean temperature stays about the same. The influence of season on the incidence of colds and other respiratory disease has been clearly demonstrated, but the mechanism for this phenomenon has not. It seems likely, however, that the physiological state of the nasal chamber (including temperature, pH, air velocity, and flow of mucus) is influenced by meteorological conditions, and might be important in aiding growth and dissemination of virus, in increasing the severity of symptoms produced by infection, or in decreasing the number of viral particles necessary to produce illness. Moreover, people tend to congregate inside when the weather is bad, perhaps increasing opportunities for the viruses to spread.

Specific control measures are not available for colds. The large number of different causative agents make the development and use of vaccines impractical. Rhinoviruses, like other viruses, are unresponsive to antibiotics and other medications that control bacterial infections. Of course, avoiding close contact with persons with colds helps limit their spread.

THROAT INFECTIONS

Adenovirus Infections

Children and young adults may develop an illness resembling a cold, but with high fever, very sore throat and severe cough, and swelling of the lymph nodes of the neck. A whitish-gray exudate may spread over the tonsils and throat. Some persons develop pneumonia as evidenced by X-ray films, or infection of the eye (conjunctivitis). Recovery occurs spontaneously over one to three weeks.

Adenoviruses
Table 17-1; Page 335, Page 677

The cause of this illness in most cases is an adenovirus, a moderate size (75 nm diameter) icosahedral virus with a DNA core. More than 30 strains of adenoviruses infect man, and all share a common antigen that can be demonstrated by complement fixation tests. The individual strains are differentiated by hemagglutination-inhibition or neutralization tests with specific antisera. The viruses can be cultured from respiratory secretions and feces by using animal cell cultures.

Little is known about the pathogenesis of this disease. The source of the agent is always another human being; related viruses infect animals, but not man. Once inside the cells of the host, the virus grows in the nucleus, producing inclusion bodies. In severe infections extensive cell destruction and inflammation occur. Different types of adenoviruses vary in their predilection for different tissues. For example, the acute respiratory illness described above is likely to be caused by adenoviruses 4, 7, or 21, whereas type 8 is likely to cause extensive eye infection with few other symptoms. Adenoviruses 1 and 2 produce a mild throat infection in young children and then become latent in the lymphoid tissues where they remain for years.

TISSUE SPECIFICITY

Adenoviral throat infections may resemble infectious mononucleosis and "strep throat." Various other agents (including the bacteria *S. pneumoniae* and *Haemophilus species*) may also cause conjunctivitis and infection of the nose and throat.

Killed vaccines have been used successfully to prevent acute respiratory disease by adenoviruses 3, 4, and 7. These vaccines have been used mainly for groups of military recruits, because these kinds of adenovirus infections are common with them, but are not frequent enough in the general population to warrant routine use of vaccines. Antibiotic treatment is of no value for adenovirus infections and may do harm by suppressing some of the normal bacterial flora and allowing growth of resistant opportunists.

"Strep Throat"

The most important bacterial infection of the throat is due to *Streptococcus pyogenes*, the group A beta hemolytic streptococcus described previously as being a cause of impetigo. Streptococcal infections of the throat often resemble adenoviral infections very closely, but in general cause greater enlargement and tenderness of the lymph nodes in the neck. Conjunctivitis and pneumonia are not usually present. Some strains produce a toxin (*erythrogenic toxin*), which is absorbed and carried by the bloodstream to the skin, resulting in a red rash. When this happens, the disease is called scarlet fever. Production of erythrogenic toxin is under the genetic control of temperate bacteriophage carried by the bacterium, an example of phage conversion. Although abscess formation and other local complications may prolong the illness, most patients with streptococcal sore throat recover spontaneously after about a week. In fact many infected persons have only the symptoms of a mild cold. Streptococcal throat infections are spread both by the respiratory route and by contaminated food. The incidence of *S. pyogenes* as a cause of sore throat varies greatly with age, time of year, and geo-

Impetigo
Page 432

Lysogenic conversion
Page 322

graphical location. Among students with sore throats at a large West Coast university, *S. pyogenes* was isolated in less than 5 percent; however, among some groups of military recruits, the incidence has been 25 percent.

Throat infections, like skin infections, may lead to glomerulonephritis as a result of antibody reacting with streptococcal products. In addition, persons with untreated throat infections with *S. pyogenes* carry a risk of about 0.5 percent of developing acute *rheumatic fever*. This complication usually occurs about three weeks after the onset of the sore throat, and is characterized by inflammatory changes in the joints, heart, skin, and other tissues. Heart failure and death may ensue, but usually the process subsides with the help of medications. Unfortunately, the heart valves may be left permanently damaged, and if the person develops subsequent infections with *S. pyogenes,* rheumatic fever promptly recurs and produces more damage to the heart. In contrast to acute glomerulonephritis which is caused by only a few types of *S. pyogenes,* any M-protein type may result in rheumatic fever. Rheumatic fever is far more common following throat infections than infections of the skin or other body sites. In many cases, at the time of the development of rheumatic fever, *S. pyogenes* can no longer be cultured from the throat. Cultures of the blood, heart, and joint tissue are also usually negative. The mechanism by which *S. pyogenes* produces acute rheumatic fever is still a mystery, but most likely a reaction between antibody and some streptococcal product is responsible for the tissue damage. Alternatively, antibody rising in response to some streptococcal antigen may cross-react with the person's tissues. Indeed some streptococci have been shown to have a cytoplasmic membrane antigen in common with human heart muscle, but the significance of this observation is still in doubt. The incidence of rheumatic fever in the United States has been dropping steadily for many years, perhaps due in part to the widespread practice of giving penicillin treatment for sore throats. Only about 3000 cases of rheumatic fever are reported in the United States each year.

Adequate ventilation and avoidance of crowding helps to control the spread of streptococcal infections. Persons suspected of having streptococcal sore throats have their throats cultured, and if the diagnosis is confirmed, they are treated with penicillin. The organisms are eliminated by treatment in about 90 percent of the cases and the risk of rheumatic fever becomes infinitesimal. The remaining 10 percent may become permanent carriers, but in them, avirulent mutants lacking M-protein are gradually selected. The reasons for this are unknown, but presumably involve acquired immunity to the M-protein. Persons with rheumatic fever are usually advised to take penicillin for at least five years, and sometimes for life, to prevent reinfection and the high risk of recurrent heart disease.

Diphtheria

Diphtheria usually begins as a mild sore throat and slight fever, with a disproportionately great amount of fatigue and malaise. Swelling of the neck is often dramatic. A whitish-gray membrane forms on the tonsils, throat, or in the nasal cavity. Heart and kidney failure and paralysis may follow these symptoms.

The cause of diphtheria is *C. diphtheriae,* a nonmotile gram-positive rod of

Glomerulonephritis
Page 434
M-protein
Page 433

RHEUMATIC FEVER

Corynebacterium
Page 258

variable shape. The organisms grow aerobically on most enriched laboratory media, but a coagulated serum medium (Loeffler's), and an agar medium containing potassium tellurite and enriched with blood or serum are usually employed. If *C. diphtheriae* bacilli grown on Loeffler's medium are stained with methylene blue, they show a characteristic appearance, with irregularities of staining and metachromatic granules. The tellurite medium prevents the growth of most members of the normal flora, yet allows *C. diphtheriae* to grow, giving characteristic colonies in one or two days. Biochemical tests, to demonstrate the ability to ferment glucose, maltose, and starch, and to reduce nitrate or break down urea, are used to identify the species. Most strains of *C. diphtheriae* produce a very powerful exotoxin which can be identified using specific antiserum. Infection with a temperate bacteriophage is required for toxin production, another example of phage conversion.

Exotoxin
Page 364

The source for *C. diphtheriae* is another human being, either a carrier or a case of diphtheria. The organisms are inhaled and establish infection in the upper respiratory system. They have very little invasive ability, but the powerful toxin which they elaborate is absorbed by the bloodstream. The gray-white membrane which forms is made up of host mucous membrane and inflammatory cells which have been killed by the growing organisms. This membrane may come loose and obstruct the airways so that sometimes the patient may smother to death. Absorption of the toxin by body cells results in cessation of cellular protein synthesis, the effect being at the level of transfer of activated amino acids from tRNA to the growing peptide chain. Studies using radioactive tracers have shown that in the presence of toxin, a fragment of NAD is bound irreversibly to one of the enzymes required to carry out the amino acid transfers in protein synthesis. The enzyme is then unable to carry out its vital function in protein synthesis, and injury or death of host cells results. Toxoid, prepared by formalin treatment of the toxin, is used to immunize against diphtheria.

**PATHOGENESIS OF
DIPHTHERIA**

Antibodies arising in response to toxoid administration specifically neutralize toxin. Since diphtheria results from toxin absorption rather than microbial invasion, control of diphtheria can be effectively accomplished by immunization with toxoid. The well-known childhood "shots" (DPT) consists of diphtheria and tetanus toxoids, and pertussis vaccine, all three generally given together. Unfortunately, immunizations are often neglected, and especially among socioeconomically disadvantaged groups serious epidemics periodically occur. For example, 66 cases of diphtheria with three deaths occurred in San Antonio during the first eight months of 1970. Seventy-five percent of the cases occurred in persons who had had no previous immunization.

PREVENTION

EARACHE (OTITIS MEDIA) AND SINUS INFECTIONS (SINUSITIS)

Infections of the middle ear (the space just behind the eardrum) result when infectious agents from the nasal passages and throat spread upward through the eustachian tube. This may occur simply by cell-to-cell spread of the infection, undoubtedly assisted at times by changes in pressure forcing infected secretions upward in the tube. Most people have experienced the sensation of pressure when driving down a steep hill or descending in an aircraft. Because of the infection, damage to the ciliated cells, and resulting inflammation, there is build up of pressure from fluid and pus collecting behind the eardrum. The throbbing ache is produced by pressure on the nerves supplying the middle ear. Normal drainage through the eustachian tube is impaired by poor ciliary action of the mucous membrane and inflammation which tends to decrease effective diameter of the tube.

As one would expect, the organisms causing otitis media are those which infect the upper respiratory system. The majority of bacterial middle ear infections are due to *S. pneumoniae,* and a large additional share are due to *H. influenzae. Streptococcus pyogenes* may cause severe otitis media, but is less common than the other two species. Cultures for bacteria are negative in about half the cases and such infections are presumed to be caused by respiratory viruses. Also, some are due to species of *Mycoplasma,* bacteria which lack a cell wall and require special media for culture.

The bone of the skull behind the ear is honeycombed with small air cells called mastoid cells, and these connect with the middle ear and can be infected through the connections with the nasal chamber. The respiratory viruses associated with the common cold frequently involve the sinuses, and bacterial infections with *S. pneumoniae* and *H. influenzae* are not uncommon. Bacterial infections of the ear may result in perforation of the eardrum, and rarely, those of the ear and mastoids may extend into the bone to involve nerves passing through the skull or produce infection of the brain and its covering membranes (meningitis). Occasionally, bacterial infections of the sinuses spread in a similar fashion.

Scientific studies of the relationship between viral and bacterial upper respiratory diseases are as yet incomplete. However, upper respiratory viral infections probably often result in superinfections with *S. pneumoniae, H. influenzae, S. pyogenes, N. meningitidis,* and other bacterial respiratory pathogens. Indeed, as mentioned in Chapter 23, one of the greatest hazards of measles (rubeola) is bacterial superinfection of the respiratory system, since the measles virus infects the mucous membranes just as it does the skin. Damage to the membrane caused by the virus is conducive to bacterial invasion.

BACTERIAL
SUPERINFECTIONS

SUMMARY

The nasal corner of the eyes, middle ears, mastoids, and sinuses all connect with the nasal passages and throat to make up the upper respiratory system. Inhaled organisms establish infections in the lining cells of the nasal airways, and those infections may spread along common membranes to other parts of the upper respiratory system and to the lung. Rhinoviruses, adenoviruses, and enteroviruses as well as the bacterial agents *Streptococcus pneumoniae, Haemophilus influenzae,* and *Streptococcus pyogenes* are frequent causes of infections in this area. The principal complications are structural damage to the eardrums or drainage channels, or extension to the bone of the skull and the nervous system. There appears to be an increased hazard of bacterial infections after viral infections of the upper respiratory system.

QUESTIONS

1. What are the primary functions of the nasal passages and how might infection interfere with them?
2. What portions of the upper respiratory system are normally free of microorganisms?
3. The common cold probably causes a greater loss of human productivity than almost any other infectious disease. Why has so little progress been made in controlling colds?
4. What is rheumatic fever and how does it relate to infections by *Streptococcus pyogenes* (Group A beta hemolytic streptococci)?
5. Why is diphtheria such a dangerous disease when *Corynebacterium diphtheriae* rarely invades beyond the respiratory mucosa?

FURTHER READING

ANDREWES, C. H., "The Viruses of the Common Cold," *Scientific American,* 203 (December 1960), 88.

FREIMER, E. H. and M. McCARTY, "Rheumatic Fever," *Scientific American,* (December 1965).

WOOD, B. W., *From Miasmas to Molecules.* New York: Columbia University Press, 1961.

Lower Respiratory
Tract Infections

The lower respiratory system (Fig. 25-1) consists of the windpipe (trachea) and its various branching divisions and subdivisions (bronchi and bronchioles), ending in the tiny, thin-walled air sacs (alveoli) that make up the lungs. The lungs are surrounded by two membranes, one adheres to the lung and the other to the wall of the chest and diaphragm. These membranes (pleura) normally slide to and fro against each other as the lung expands and contracts. There is thus a potential space between the two membranes where products of infection can accumulate and thereby compress the lungs.

In contrast to portions of the upper respiratory system, microorganisms are normally absent in the lower tract. As with the upper tract much of the lower respiratory system is lined with ciliated cells and a film of mucus. This film is constantly swept upward from the bronchioles and bronchi toward the throat at the rate of about an inch per minute under normal conditions, and its function is trapping and removing microorganisms and other foreign material. Extraneous factors such as tobacco smoke may decrease ciliary action, and prolonged exposure to irritants leads to loss of cilia and flattening of the lining cells. The cough reflex, which also aids in expelling foreign materials, is activated by irritants and excessive secretions, and can be depressed by external factors, such as alcohol and narcotics. Finally, phagocytic macrophages are numerous in the lung tissue and readily move into the alveoli and airways in response to foreign substances, such as microbes. These protective mechanisms are very efficient, especially against bacterial pathogens, and it is unusual to see bacterial lung infections, except in those who have impaired mechanisms.

DEFENSE MECHANISMS

453

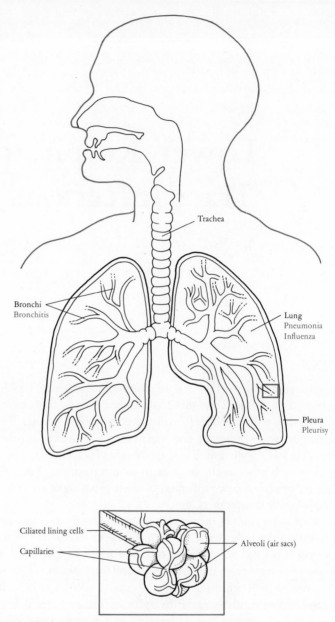

Trachea

Bronchi
Bronchitis

Lung
Pneumonia
Influenza

Pleura
Pleurisy

Ciliated lining cells

Capillaries

Alveoli (air sacs)

FIGURE 25-1   The lower respiratory system.

PNEUMONIA

Pneumonia results from inflammation of the lung and is usually manifested by fever, cough, chest pain, and production of sputum. Pneumonia due to microbial infection may result when inhaled pathogenic organisms escape the trapping mechanisms of the upper and lower respiratory system and multiply in the airways, alveolar spaces, or lung tissue itself. Perhaps more commonly, such microbes first establish themselves in the upper respiratory system and then a large inoculum is accidentally carried into the lung with a ball of mucus during a deep breath or when the cough reflex is depressed. Thus upper respiratory infection is often a prime factor in the development of lower tract infection. Serious pneumonias in babies and adults have also resulted when organisms of little virulence were inhaled into the lung in high concentrations from air conditioning or inhalation equipment containing contaminated water.

Pneumococcal Pneumonia

One of the most common bacterial causes of lung infection is the gram-positive coccus, *Streptococcus pneumoniae* (also known as *Diplococcus pneumoniae,* or the pneumococcus, Table 25-1). As the generic name *Diplococcus* implied, pneumococci tend to occur in pairs, although chains are often seen, especially in fluid cultures. Most strains of pneumococci are not completely spherical, but the adjacent surfaces of the two organisms are somewhat flattened, and the opposite surface is somewhat pointed (Fig. 25-2). As mentioned in earlier chapters the virulence of these organisms is due to the presence of a large polysaccharide capsule which interferes with their ingestion by phagocytic host cells. In Chapter 9 we mentioned that production of capsules is under genetic control, and a large number of different antigenic types of capsule are known.

Transformation
Pages 194–197

The organisms are readily recovered in cultures of sputum and blood from patients with pneumonia. Growth is enhanced by the presence of blood or serum and glucose in the medium and an atmosphere of 5 to 10 percent carbon dioxide. The colonies produce hemolysis and a green discoloration on blood agar medium, and may closely resemble organisms of the viridans group of streptococci, universally present among the upper respiratory tract normal flora. A feature that distinguishes pneumococci from the viridans streptococci, however, is the presence in pneumococci of an autolytic enzyme which characteristically destroys the organisms in the center of colonies. The older colony thus has an umbilicated appearance (Fig. 25-3). Lysis of pneumococci also occurs in the presence of bile,

CHARACTERISTICS
OF PNEUMOCOCCI

TABLE 25-1
Characteristics of *Streptococcus* (*Diplococcus*) *pneumoniae*

Identification	Normal Habitat	Pathogenic Potential
Gram positive cocci in pairs. Colonies umbilicated, show alpha hemolysis. Bile soluble.	A common inhabitant of the upper respiratory tract of human beings.	Encapsulated strains are opportunistic pathogens. Cause eye, sinus and ear infections, and pneumonia. Infections sometimes spread to heart or brain. Immunity results from antibody to the capsule, but there are a large number of capsular types.

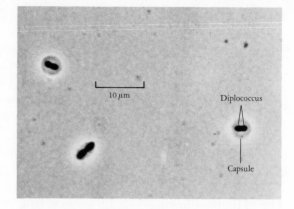

FIGURE 25-2
Streptococcus pneumoniae. Note that the organisms are in pairs, accounting for the generic name *Diplococcus* formerly given to this species. The large capsule surrounding the organism is made more distinct because of specific antiserum added to the surrounding fluid. (Courtesy of J. T. Staley and J. P. Dalmasso)

and bile solubility is a key test in identifying these organisms and differentiation from viridans streptococci.

Pneumococci normally inhabit only human beings, and encapsulated or nonencapsulated variants may commonly be present among the upper respiratory flora in normal people. When the encapsulated (virulent) pneumococci enter the alveoli of a host lacking antibody, they may multiply rapidly and elicit an inflammatory response. Serum and phagocytic cells pour into the air sacs of the lung, causing difficulty in breathing and production of sputum. This increase in fluid is what produces the abnormal shadows on X-ray films of the chest in patients with pneumonia. The inflammation may involve nerve endings, giving pain, and when this pain arises from the pleura, the sick person is said to have *pleurisy*. Pneumococci commonly enter the bloodstream from the inflamed lung and occasionally produce *septicemia* (infection of the blood stream) or *endocarditis* (infection of the heart lining), or establish infections, such as meningitis, elsewhere

PATHOGENESIS OF
PNEUMOCOCCAL
INFECTIONS

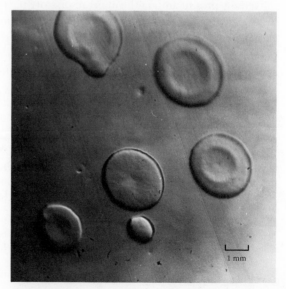

FIGURE 25-3
Streptococcus pneumoniae, umbilicated colonies. Note varying stages of collapse of the central portion of the colonies. (Courtesy of J. P. Dalmasso)

in the body. If infected persons fail to develop such complications, then sufficient specific antibody usually arises within a week or two to allow trapping and destruction of the organisms by lung phagocytes, and complete recovery usually results. Most pneumococcal strains have little ability to destroy lung tissue.

Treatment of most pneumococcal infections is very effective if given soon enough. Pneumococci are almost uniformly susceptible to penicillin, erythromycin, and several other commonly used antibiotics, but strains resistant to tetracycline are now occasionally encountered.

As with several other common bacterial infections there has been renewed interest in developing preventive measures especially for high risk patients, such as those suffering from chronic lung diseases or alcoholism. Vaccines can be prepared to stimulate anticapsular antibodies, giving immunity to pneumococcal disease. To be effective the vaccines must include pneumococci of each of the capsular types prevalent in the community. An estimated 20,000 deaths from pneumococci could be prevented by immunization each year in the United States alone, and innumerable other deaths in countries with less well-developed health facilities.

Klebsiella Pneumonia

Among the less common but more serious causes of pneumonia are enterobacteria of the genus *Klebsiella*. The pathogenesis (sequence of events leading to disease) is thought to be similar to that with pneumococci, except that permanent damage to the lung usually results and complications are more frequent. Most cases occur in persons who drink heavily, for alcohol is known to interfere with lung defense mechanisms at several levels (ciliary action, cough reflex, and phagocytosis). *Klebsiella* are differentiated from other enterobacteria by the presence of a large capsule, lack of motility, and their inability to decarboxylate the amino acid ornithine. A half dozen species have been defined largely on the basis of differences in biochemical capability. There are numerous capsular types, but neither species designation nor capsular type correlates well with virulence for humans, perhaps because the state of host defenses is the controlling factor in pathogenesis.

The organisms persist in the environment abundantly relative to other enterobacteria, but are susceptible to strong disinfectants. They are susceptible *in vitro* to some antibiotics, although the response of infections to therapy is often slow. *Klebsiella* is also prone to develop resistance to some antimicrobial medications very quickly because of high mutation frequencies to antibiotic resistance. They also commonly contain R-factors, responsible for drug resistance. In some hospitals they are now considered the principal source of R-factors transferrable to other pathogens. Virulence is due partly to the antiphagocytic property of the capsules, and endotoxin may well play a role in the tissue destruction characteristic of this disease. Laboratory diagnosis is complicated by the fact that *Klebsiella* may be present in the oropharynx of normal people, and thus contaminate sputum specimens.

Carboxyl group
Table 2-2, Page 24

R-factors
Page 201

Chapter Twenty-Five
Lower Respiratory
Tract Infections

Mycoplasma Pneumonia

Mycoplasma
Page 262

Pneumonia due to *Mycoplasma pneumoniae,* frequently referred to as *primary atypical pneumonia,* may resemble pneumococcal pneumonia very closely. However, patients are usually not so acutely ill and serious complications are not so common. *Mycoplasma pneumoniae* lacks a cell wall and is thus easily deformed allowing it to pass through filters that retain other bacteria. Since it stains poorly and has a plastic shape it is difficult to detect microscopically. *Mycoplasma pneumoniae* is extremely fastidious in its growth requirements, and a medium supplemented with fresh yeast extract and horse serum is usually employed for cultures of this species. Growth is slow, with colonies often requiring two to four weeks to appear. Differentiation from other species of *Mycoplasma* among the normal upper respiratory flora is based on the ability of *M. pneumoniae* to produce acid from glucose and hemolyze red blood cells. Specific antibody inhibits growth of these bacteria *in vitro,* and thus precise identification can be made by a serological method.

The mechanism by which these organisms produce pneumonia is unknown. One interesting aspect is the appearance in the bloodstream of antibodies called "cold agglutinins" which cause the clumping of red blood cells at low temperatures. The clumps disaggregate when the suspension is warmed. The reason for the appearance of these antibodies is unclear; they, however, provide a useful diagnostic test, since approximately 80 percent of hospitalized persons with mycoplasma pneumonia have them. A more specific diagnostic aid is a complement-fixation test using mycoplasma antigen.

Action of penicillin
Page 83

Not much is known about the survival of these organisms outside the host. *Mycoplasma pneumoniae* is presumed to spread from person to person by droplet infections. The organisms are not susceptible to penicillin, because they lack a cell wall, but are susceptible to several other antibiotics, such as erythromycin and tetracyclines. Many viruses are capable of producing pneumonia resembling mycoplasma pneumonia. Antibiotic therapy is of no value in viral pneumonia.

INFLUENZA

Myxoviruses
Page 680; Table 17-1,
Page 335

Respiratory infection due to myxoviruses of the influenza group is called *influenza,* and differs from bacterial pneumonia in its tendency to spread explosively and to produce more fatigue, aches, and pains. There is also a tendency to involve the bronchi and bronchioles and their supporting tissue to a greater extent than the alveoli. Even though the vast majority of people recover completely from influenza without any treatment, the number of deaths is high because so many people become infected. For example, in the Asian influenza epidemic of 1957–1958, about 26,000 deaths resulted in the United States among more than 50 million people infected. Mortality tends to be highest among the aged, the pregnant, and those with heart diseases. Death may be due to influenza itself, although more commonly it is due to secondary invasion of the damaged lung by pathogenic bacteria. *Staphylococcus aureus, Haemophilus influenzae, Streptococcus pyogenes,* and *Streptococcus pneumoniae* are the chief offenders.

The influenza viruses can be isolated in a variety of tissue cell cultures in which they may produce cytopathic effects or hemadsorption (the adherence of red blood cells to tissue cells infected by the virus). The viruses hemagglutinate red blood cells from several animal species. Antisera specific for the viral nucleoprotein can be used to classify influenza viruses into three major groups, A, B, and C; several subgroups and types are identifiable using antibodies specific for hemagglutinins (envelope antigens). For example, influenza A viruses have a common nucleoprotein antigen which differs from that of groups B and C; A-2 (Asian influenza) virus is distinguished from other group A viruses by its specific hemagglutinating antigen.

The influenza virus enters the respiratory system by inhalation of air contaminated by virus from another infected person. Penetration of the mucous blanket is presumably aided by a viral enzyme, neuraminidase, attached to the envelope. Infection is established in cells lining the air passages and large amounts of virus are released to infect other cells. Death of the ciliated cells, inflammation, and leakage of plasma are prominent features.

More than 30 pandemics (epidemics of worldwide scope) of influenza have swept the world in the last 450 years, one of the most devastating occurring in 1918–1919. Even today the inability of public health measures to restrain this disease has been startlingly evident. Figure 25-4 shows the effect of the Hong Kong strain of mutant A-2 influenza virus on death rates in the United States. Even though the "Hong Kong" virus was known to be threatening six months before the epidemic reached the United States, there were an estimated 19,700 total deaths related to the 1968–1969 epidemic. The rapid spread of the Hong Kong virus was facilitated by the fact that it was a mutant with a modified envelope protein against which antibodies resulting from the earlier Asian epidemic were relatively ineffective. In fact frequent mutation is an important characteristic of influenza viruses, and minor variations are detected almost yearly. Mutants with major antigen differences from previous strains are likely to produce pandemics if they also infect easily and produce high titers of virus in the respiratory secretions of the host.

The reason for the apparent genetic instability of influenza viruses is not yet completely understood, but scientific studies suggest that the viral RNA consists of several distinct molecules. Experimentally, when infections are established in a cell by two strains simultaneously, virus particles are produced with some characteristics of each of the infecting strains, presumably because of inclusion of nucleic acid molecules from each strain within a single capsid. This finding suggests the possibility that natural strains with new characteristics might arise by this mechanism as well as by mutation. The source for new genetic information might be influenza viruses which normally infect animals such as horses and pigs, but this is purely conjectural. With the possible exception of the virus causing the 1918–1919 pandemic, even the most virulent influenza viruses so far known have generally produced only mild illness in the majority of healthy young peo-

Hemagglutination- inhibition
Page 385

PANDEMIC INFLUENZA

GENETIC INSTABILITY

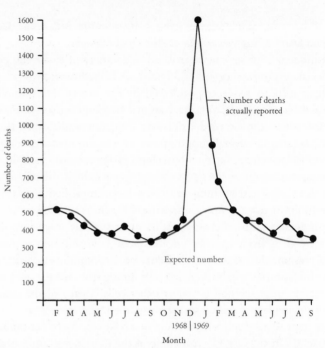

FIGURE 25-4   Increased number of deaths from respiratory disease during an influenza epidemic. (From Morbidity and Mortality Weekly Report Annual Supplement Summary 1968, v. 17, no. 53, December 1969)

ple. Nevertheless increasing population density and speed of travel make the prospect of a future, more lethal variant a real concern.

Control of influenza presently depends on immunization with a vaccine prepared from a threatening epidemic strain or a close relative. Greater than 60 percent protection is provided against the homologous strain, substantially less with antigenically related variants, and none at all against other strains. A chemotherapeutic agent, amantadine hydrochloride, is effective in preventing influenza from type A viruses, and acts by preventing viral penetration into host cells.

WHOOPING COUGH (PERTUSSIS)

Whooping cough is primarily a childhood disease and is manifested by nasal congestion and a mild cough at onset. After a week or two, these symptoms are replaced by spasms of violent coughing followed by a loud gasping noise as the patient draws a breath. Vomiting and convulsions frequently occur.

Bordetella
Pages 251, 364

The causative agent of whooping cough is *Bordetella pertussis,* a small, strictly aerobic gram-negative rod bacterium. Early in the illness it is present in enormous numbers in the respiratory secretions. The organism is highly fastidious, and can be isolated only on special media such as one containing potato extract, glycerol, and 20 percent blood. The organism is encapsulated and produces a protein toxin. Specific identification of the organism is made using antiserum; indeed fluorescent antibody can reliably identify the organisms in smears of naso-pharyngeal secretions.

Infection is confined entirely to the surface of the tracheobronchial system in which the organisms grow in dense masses. The mucus becomes thick and tenacious and ciliary action is impeded. These factors result in the characteristically violent but relatively ineffective coughing. Some of the organisms disintegrate, releasing their toxic substance, causing death of some of the lining cells and stimulating a rise in the numbers of lymphocytes in the bloodstream of the host. About 10 percent of infants with pertussis die of the disease.

The organisms do not tolerate drying or sunlight and die quickly outside the host. They also are quite susceptible to several antibiotics and to disinfectants. Spread is entirely by inhalation of droplets from the respiratory system of an infected person. Control is achieved by intensive vaccination of infants with a killed vaccine. Effectiveness of the vaccine is about 70 percent and its widespread use early in life is probably responsible for the drop from 265,269 cases in the United States in 1934 to 4810 cases in 1968. Pertussis continues to be a major problem in poverty areas of the Southeast and the big city slums because of the low percentage of infants who receive vaccine in these areas.

TUBERCULOSIS

Tuberculosis is characteristically a chronic infection of the lungs manifested by fever, weight loss, cough, and sputum production, lasting months or years. In some patients the infection spreads to bone, meninges, kidneys, or other parts of the body. In rare instances tuberculosis begins as an infection of the skin or gastrointestinal tract.

Early in the twentieth century the death rate from tuberculosis in the United States was about 200 per 100,000 population, and almost everyone in larger urban areas had contracted the infection by adulthood. Although the initial infection was usually arrested by body defenses and went unnoticed, reactivation during later life was very common; an estimated one out of ten infected persons eventually died of the disease. Today new cases of tuberculosis occur at the rate of about 20 per 100,000 population per year, with most permanently arrested by modern treatment. Schoolchildren have a very low infection rate and fewer than 5 percent of young adults have been infected. The reasons for the dramatic decline are incompletely understood, since it began before the availability of specific therapy and modern public health measures. Nevertheless tuberculosis is far from conquered, especially among the economically disadvantaged. Overcrowding and malnutrition are important in the spread of tuberculosis, and worldwide it is still the leading cause of death.

Causative Organism

Tuberculosis is caused by *Mycobacterium tuberculosis* (Table 25-2), the "tubercle bacillus," a slender rod-shaped bacterium closely related to a large group of

Mycobacterium
Page 259

TABLE 25-2
Characteristics of *Mycobacterium tuberculosis*

Identification	Source	Pathogenic Potential	Control
Acid fast rods. Aerobic Require special growth media. Slow growing. Forms cords.	Human beings with tuberculosis	Highly infectious. Initial infection often asymptomatic, is confined to small area by host defenses and becomes inactive. Infection commonly reactivates later in life.	Relatively resistant to disinfectants. Cases detected by tuberculin tests or chest x-rays and treated with anti-microbial drugs. Prevention by BCG, a live vaccine.

Acid-fast staining
Page 72

branching bacteria. Mycobacteria contain a waxy material in their cell walls and show the acid-fast staining property. Unfortunately the simple acid-fast staining procedure does not give a definite identification of *M. tuberculosis* because other pathogenic and nonpathogenic mycobacteria are common. Some of these other acid-fast bacteria are present in the secretions around the urinary opening, in wax of the ear, and in grain products and other foodstuffs. They are frequently present in tap water and air, and thus can contaminate stains and other reagents used in microbiological procedures. Moreover acid-fast bacteria are difficult to remove from glassware, and washing and autoclaving often do not change their shape or acid-fast staining property.

SLOW GROWTH OF
M. TUBERCULOSIS

Growth of *M. tuberculosis* fails on most laboratory media probably because of traces of toxic fatty acids. Specialized media, such as one consisting of coagulated egg yolk, with potato starch, glycerol, and asparagin, have been used successfully for many years. Growth occurs only under aerobic conditions and is enhanced by 5 to 10 percent carbon dioxide, but even under ideal conditions small inocula require at least two weeks to appear as colonies, and six weeks are not infrequently required. The generation time for *M. tuberculosis* is 12 hours or more, in contrast to 20 minutes for many eubacteria. Because of such slow growth, special techniques must be used to cultivate them from materials containing other faster-growing bacteria. The latter must be killed or their colonies will cover the medium and prevent the appearance of the slow-growing tubercle bacillus.

RESISTANCE TO
DISINFECTANTS

Mycobacteria show about the same susceptibility to heat as most other vegetative bacteria and small inocula are readily killed by pasteurization. On the other hand, most mycobacteria are relatively resistant to drying, ultraviolet light, disinfectants, and many of the medications useful in treating other bacterial diseases. This provides a way of recovering mycobacteria from mixtures of faster-growing microbes, thus allowing laboratory identification of the cause of tuberculosis and tests for susceptibility to medications. Strong alkali, for example, or even the antibiotics, penicillin and polymyxin, can be used to rid a material of other bacteria while leaving viable mycobacteria that can then be cultivated.

Differentiation of *M. tuberculosis* from other acid-fast bacteria is carried out by determination of growth rate at different temperatures, colonial appearance, and biochemical tests. The peculiar arrangement of the bacteria into thick ropes in some fluid media is highly characteristic of tubercle bacilli. Tests of pathogenicity for different species of laboratory animals are also sometimes needed. A number of species of mycobacteria other than *M. tuberculosis* can cause infections in human beings. Except for *M. bovis* they are, however, generally less likely to be pathogenic than *M. tuberculosis*.

Development of Tuberculosis

The organisms are inhaled in air which has been contaminated by a person with tuberculosis. The microbes lodge in the alveoli of the lung and produce an inflammatory response. The organisms are then ingested by macrophages, where they survive destruction and may multiply, being carried to lymph nodes in the region and often to other parts of the body. As mentioned in Chapter 20, after one or two weeks a delayed-type hypersensitivity to the organisms develops. A brisk reaction then occurs at the sites where the bacilli have lodged. Macrophages collect around the bacteria and some coalesce to form large multinucleated cells (Langhans giant cells). Lymphocytes collect surrounding these cells. This kind of inflammatory focus is called a *granuloma,* and is the characteristic response of the body to microbes and other foreign substances that resist digestion and removal. The granulomas of tuberculosis are called tubercles. The mycobacteria cease to multiply in the tubercles, but a few may remain viable for many years. In large collections of tubercles the blood supply may be so poor that the tissue cells die and autolyse. If this process involves a bronchus, the dead material may discharge into the airways, causing a cavity and spreading the organisms to other parts of the lung. Coughing and spitting transmits the organisms to other people as well. A persistent lung cavity may remain with prolonged shedding of bacteria into the bronchus. Involvement of blood and lymphatic vessels or the covering membranes of the brain may lead to dissemination into the vascular system, or to the cerebrospinal fluid.

Tuberculin hypersensitivity
Page 398

Hypersensitivity to the bacilli is detected by injecting tuberculin, the supernatant from a culture of *M. tuberculosis,* or a purified fraction of the supernatant (purified protein derivative, PPD) into the skin. In persons who are sensitive, redness and swelling develop at the injection site, reaching a peak intensity in 48 to 72 hours. This test is called a tuberculin test. A strongly positive reaction is thought to indicate the presence of living bacilli somewhere in the body. This is associated with the possibility that impaired general health or certain treatments that suppress the immune response may result in renewed multiplication, enlargement of the tubercles, and actively progressing disease. This may happen even years after the initial infection.

In the United States today control of tuberculosis is carried out by identifying infected persons by chest X-rays and skin tests, and by giving antimicrobial medications to those who have active disease, since they are now the only important sources of infection. Milk contaminated with *M. bovis* (species similar to *M. tuberculosis* but prevalent in cattle) was formerly a major cause of human tubercu-

CONTROL MEASURES

losis, principally involving the lymph nodes of the neck and the gastrointestinal tract. Milk, however, is no longer a significant source because of the widespread use of pasteurization and surveillance of dairy cows.

Drugs useful in the treatment of infections due to *M. tuberculosis* and *M. bovis* are isonicotinic acid hydrazide (INH), para-amino salicylic acid (PAS), the aminoglycoside antibiotics (such as streptomycin), and rifampin. Other species of mycobacteria are often highly resistant to these and other medications. Even with *M. tuberculosis* and *M. bovis,* resistant mutants occur with high frequency. Since mutants simultaneously resistant to more than one drug have a much lower fre-

quency, two or more drugs are usually given together in treating tuberculosis. Because of the very slow generation times of most pathogenic mycobacteria and their resistance to destruction by body defenses, treatment must generally be continued for one or more years to obtain permanent arrest of the disease. Even so, nonmultiplying bacilli enclosed within old tubercles may not be killed by treatment.

Vaccination against tuberculosis has been widely used in Scandinavia and elsewhere and is of proven value. The vaccinating agent, a living attenuated

mycobacterium known as Bacillus Calmette-Guerin (BCG), is probably derived from a human or bovine mycobacterium. Repeated subculture in the laboratory over many years resulted in selection of a mutant which has little virulence for humans and causes the development of some immunity. The protective effect is of value to nurses and tuberculosis sanitorium personnel, has been about 80 percent effective (incidence of tuberculosis being one-fifth that occurring in unvaccinated persons), and lasts for several years. Also persons receiving the vaccine develop delayed hypersensitivity to tuberculin. In some populations the tuberculin reaction has been slight or negative a year after vaccination, while in others the positive reaction has persisted. Public health authorities have discouraged its routine use in the United States because most people who now develop active tuberculosis do so as a result of activation of an infection acquired years before. Many physicians prefer that nurses and other high risk personnel be checked for tuberculosis by periodic tuberculin tests and treated if a positive result occurs. Vaccination is of no value and occasionally may be harmful to persons already infected by *M. tuberculosis,* since a severe reaction may occur at the vaccination site with possible activation of latent tuberculosis. Scars at the site of injection, production of a positive skin test, and (rarely) disease in a peculiarly susceptible person are the other undesirable results of routine BCG vaccination.

FUNGOUS INFECTIONS OF THE LUNG

Coccidioidomycosis

Valley fever and desert rheumatism are the other names for this disease occurring commonly in certain hot, dry, dusty areas of the Americas. Fever, cough, chest pain, and loss of appetite and weight are common features. About one out of ten suffering the illness develop hypersensitivity manifested by a rash on their shins or other parts of the body and pain in their joints. The vast majority of persons recover spontaneously within a month and, in contrast to tuberculosis, have little risk of later reactivation. In a few the disease closely resembles tuberculosis. Death of lung tissue may lead to the formation of cavities and, more rarely, the infection spreads throughout the body so that the skin, mucous membranes, brain, and internal organs become involved. About half of those with disseminated infection die unless proper treatment is given (see the following).

The causative agent, *Coccidioides immitis,* is a dimorphic fungus living in the soil. It grows only in the Western Hemisphere, in parts of Argentina, Paraguay, Venezuela, Mexico, and in California, Arizona, New Mexico, Texas, and Utah. Infectious spores have unknowingly been transported to other areas (such as southeastern United States), but there is no evidence that the fungus can establish itself in other climates. In the endemic areas infections occur only during the dusty seasons and disappear during rains. Persons have become infected while traveling through contaminated areas.

The tissue phase of the organism is the form present in infected tissues and can be identified by microscopic examination of sputum or pus. It is a thick-walled sphere ranging in size usually from 20 to 80 μm in diameter; in contrast to most other fungi growing in tissue, it never buds. The larger spheres contain several hundred small cells called "endospores," although they have little resemblance to the endospores of bacteria (Fig. 25-5a). The mold form of the organism grows readily on most laboratory media at room temperatures and is the form which grows in soil. On Sabouraud's medium (a specialized medium for growing fungi), growth usually occurs in three to five days. Such cultures are extremely infectious since most of the hyphae develop numerous very light barrel-shaped arthrospores (Fig. 25-5b). These separate easily from the hyphae and become airborne. Since other molds may resemble *C. immitis,* cultures may need to be converted to the spherule form to complete the identification. Suspensions of the arthrospores are prepared in saline and injected into laboratory mice or into a special liquid medium incubated at 40°C. Spherules usually develop within a week and can be identified microscopically.

Coccidioides
Table 14-3, Page 291

Dimorphic fungi
Page 283

MOLD FORM
EXTREMELY INFECTIOUS

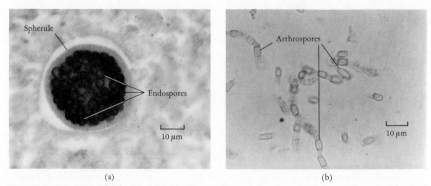

(a)   (b)

FIGURE 25-5 *Coccidioides immitis.* (a) Tissue phase. Spherule containing endospores (stained preparation). (b) Mold phase. The barrel-shaped arthrospores are characteristic of this species.

Coccidioidomycosis is initiated by inhaling arthrospores from fungi growing in the soil. Infectious arthrospores do not develop in man or animals so that transmission from animal to animal does not occur. After lodging in the lung, the spores develop into spherules which mature and discharge their endospores, each of which can develop into another spherule. The pathogenesis is similar to tuberculosis. The immune response can be measured by skin testing with coccidioidin, a culture supernatant analogous to tuberculin. Most persons show a positive coccidioidin skin test within three weeks after inhaling the fungal spores and retain the capacity to give a positive skin test for life if they remain in the area where the fungus occurs. Those who move away for several years and those with disseminated disease often lose skin reactivity to coccidioidin. Within the first month of illness, most persons with coccidioidomycosis will develop measurable precipitating antibodies, but these generally drop to low titers by the third month regardless of the progress of the disease. Complement-fixing antibodies arise later than precipitating antibodies and often remain detectable for years, although the titer falls when recovery occurs. Titers of complement-fixing antibodies continue to rise in disseminated infections, and this laboratory finding means that treatment must be given to save the person's life.

Complement-fixation
Page 380

The only effective control of the fungus is attained by measures which decrease dust. Intensive efforts at producing vaccines have so far not been successful. Serious infections can often be arrested by intravenous administration of the antibiotic, amphotericin B. Treatment must be continued for months, but reactivation of the disease can occur months or years after therapy is discontinued.

Figure 32-8
Page 563

Histoplasmosis

This disease is very similar in many respects to coccidioidomycosis, except for its spotty occurrence in tropical and temperate zones around the world. The causative organism is the dimorphic fungus, *Histoplasma capsulatum,* a fungus preferring soils contaminated by bat or bird droppings. Caves and chicken coops are

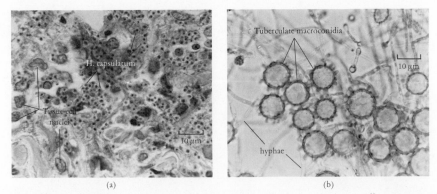

FIGURE 25-6 *Histoplasma capsulatum.* (a) Tissue phase. Numerous tiny yeast cells are present within the cytoplasm of macrophages. (b) Mold phase. The tuberculate macroconidia appear in cultures grown at 25°C.

notorious sources of the infection. In tissue phase it is an oval yeast 2 to 5 μm in size, often growing within macrophage cells of the host (Fig. 25-6a). The mycelial phase shows chlamydospores characterized by numerous projecting knobs (Fig. 25-6b). Delayed hypersensitivity is detected by skin testing with histoplasmin, an antigen analogous to coccidioidin and tuberculin. Precipitating and complement-fixing antibodies develop as with coccidioidomycosis. Because cross reactions often occur with antigens of *Coccidioides* and another pathogenic fungus, *Blastomyces dermatitidis,* skin tests and antibody determinations are helpful in establishing the cause of the infection only if antigens from all three fungi are tested at the same time. The antigen corresponding to the infecting fungus usually shows the larger skin test and antibody response. Probably half of the reported cases of histoplasmosis have occurred in the United States, primarily the Mississippi River drainage and the South Atlantic states. Skin tests reveal that millions of people living in those areas have been infected. As with *Coccidioides* only a tiny fraction of those infected develop serious illness. The antibiotic amphotericin B is helpful in arresting serious infections.

OTHER LUNG DISEASES CAUSED BY MICROBES

Numerous additional bacteria, viruses, and fungi produce infections of the lung. Other agents are unable to establish infection, but repeated exposure to them from inhaled dust may nevertheless result in serious damage from the inflammatory reaction they elicit. For example, repeated inhalation of hay dust containing certain thermophilic branching bacteria may produce an allergic reaction to the bacterial antigens. It is characterized by severe inflammation and granuloma formation in the airways and alveoli. Finally, spores of saprophytic fungi may be responsible for some cases of asthma. In this case there is little damage to lung tissue, but immediate hypersensitivity to the foreign antigen causes spasm of the airways and difficulty in breathing.

Thermoactinomyces
Page 262

SUMMARY

Microorganisms enter the lung with inspired air, or in infected material from the upper respiratory tract. They may be removed by defense mechanisms, or produce a variety of diseases depending on the nature of the infecting organism. *Streptococcus (Diplococcus) pneumoniae, Klebsiella* species, and other eubacteria can produce lung infections, as can viruses such as those causing influenza. Tuberculosis results from infection with *Mycobacterium tuberculosis,* and related organisms may sometimes cause a similar disease. Chronic illnesses resembling tuberculosis may also result from infection by certain fungi such as *Coccidioides immitis* and *Histoplasma capsulatum.* Severe lung disease can result from inhaling saprophytic bacteria because of hypersensitivity of lung tissue to their antigens. Vaccines can help in preventing some lower respiratory infections, including pneumococcal pneumonia, influenza, whooping cough, and tuberculosis.

QUESTIONS

1. What are the mechanisms by which the lower respiratory system is protected from infection?
2. On what basis does specific immunity to pneumococcal disease depend? Does immunity to one strain of *Streptococcus pneumoniae* give immunity to all strains of this species? Explain.
3. Give reasons why the total number of deaths from an influenza epidemic is high despite the fact that influenza itself is usually a mild disease.
4. If some means were devised completely to prevent transmission of *Mycobacterium tuberculosis,* how long would it take tuberculosis to disappear?
5. Which would you expect to be a more sensitive test for tuberculosis, a chest X-ray or a tuberculin test for delayed hypersensitivity? Why?

FURTHER READING

COMROE, J. H., "The Lung," *Scientific American,* 214 (February 1966), 56–58.

DUBOS, RENE and JEAN, *The White Plague.* Boston: Little, Brown and Co., 1952.

LONG, E. R., "The Germ of Tuberculosis," *Scientific American,* 192 (1955), 102.

ROUECHE, BERTON, "The Liberace Room," in *Annals of Epidemiology.* Boston: Little, Brown and Co., 1967.

Chapter 26

Alimentary Tract Infections and Food Poisonings

The alimentary tract, like the skin, is one of the body's boundaries with the environment and is continually exposed to microbes of the endogenous flora as well as extraneous organisms. It is, so to speak, our "inside outside," and it is one of the major routes of access for invading germs.

ANATOMIC AND PHYSIOLOGIC CONSIDERATIONS

The alimentary tract includes the mouth, esophagus, stomach, small and large intestines. The system also includes some very important appendages: the salivary glands, the liver and gall bladder, and the pancreas. These are attached to the alimentary tract by tubes through which pass the fluids produced by the appendages. Figure 26-1 illustrates the relationships of the various parts of the alimentary system.

The Mouth

The teeth are made up largely of calcium phosphate with some protein matrix. The outer portion, or enamel, is especially dense, and yet certain bacteria are able to penetrate and produce cavities and tooth destruction. Both microscopic and large crevices in the surfaces of the teeth tend to collect food particles that are sites of microbial colonization. Saliva is secreted into the mouth from the various salivary glands at a rate of about 1500 ml per day. It serves to keep the mouth clean and lubricated, helps maintain a neutral pH, and since it is saturated with

PROTECTIVE ROLES OF SALIVA

469

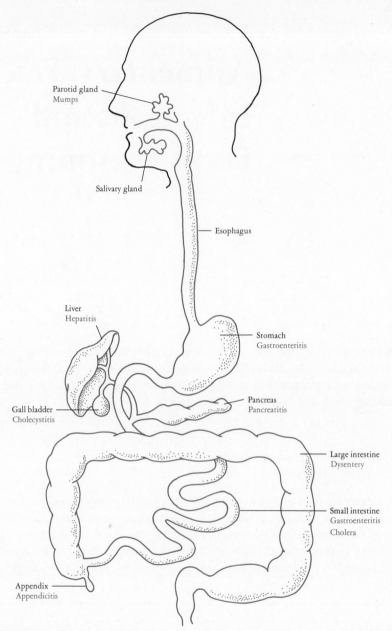

FIGURE 26-1   The alimentary system.

calcium, tends to prevent the calcium phosphate of the teeth from dissolving. It has readily demonstrable inhibitory and killing powers against various groups of microorganisms, both normal microbial flora and pathogens, but the source and significance of these effects are still an unsettled matter.

The Stomach

The esophagus connects the mouth with the stomach, a distensible sac with a muscular wall and lining cells, some of which produce hydrochloric acid and pepsinogen, a precursor of the protein splitting enzyme, pepsin. Nerves and hormones cause the stomach to react in response to changes in other parts of the body, such as the brain or small intestine, as well as to local factors, such as irritation arising from infections. In response to food the stomach begins a mixing action, bringing acid and enzyme into contact with the ingested material. Many types of microbes are destroyed by this action.

The Small Intestine

A valve at the end of the stomach controls the passage of the well-mixed gastric contents into the intestine. For example, the rate of passage slows down if there is considerable fat in the food or if it is markedly hypertonic or hypotonic. The small intestine in an adult is about 8 or 9 feet long and has an enormous absorptive and secretory area, about 30 square feet. This large area results from the presence of many small fingerlike projections (villi), 20 to 40 per square millimeter, each 0.5 to 1 mm long. Microbial diseases may markedly interfere with functions such as absorption and secretion by effects on the cells lining the intestine and on the transit of intestinal contents. Small patches of lymphoid tissue spot the walls of the small intestine and may be selectively invaded by certain pathogens.

The liver has many functions, one of which is to remove breakdown products of hemoglobin from the bloodstream. These yellowish-green materials are then excreted with the bile, the fluid produced by the liver. Interference with this process by infection or other factors may produce jaundice (a yellow color of the skin and eyes). The gall bladder is simply a storage area for the bile, where it is concentrated and its acidity neutralized. Infection of the gall bladder, *cholecystitis,* is often a life-threatening disease, but in persons who are carriers pathogenic organisms can sometimes grow in the gall bladder for years without causing symptoms. Bile helps emulsify fats in the intestine and has antibacterial action against some microbes. The pancreas also has multiple functions, including the production of several digestive enzymes. About 500 ml bile per day pour into the upper portion of the small intestine from the liver and gall bladder, and another 2000 ml of digestive juices are added by the pancreas. These fluids as well as digestive juices of the small intestine itself are alkaline and neutralize the stomach acid. They also enzymatically break down foodstuffs into absorbable amino acids, fats, and simple sugars. These smaller molecules are selectively taken up by the lining cells of the small intestine.

ANTIMICROBIAL ROLE OF STOMACH

SITES OF VULNERABILITY TO MICROBIAL PATHOGENS

APPENDAGES OF THE SMALL INTESTINE

Because of the great absorptive ability of the small intestine, only about 300 to 500 ml of fluid normally reach the large intestine per day. The large intestine also can absorb some water, electrolytes, vitamins and amino acids. After this further absorption, the normal semisolid feces remain.

NORMAL FLORA OF THE ALIMENTARY TRACT

Because of the abundance of food and moisture, the total numbers of microbes in some areas of the alimentary tract are considerably greater than on the skin (Table 26-1). The variety of different species and unnamed organisms is very high and there may be frequent variations in relative numbers. These changes depend on factors such as diet, antibiotic treatment, and mechanical alterations, including dental work which eliminates or creates an ecological niche for certain groups.

MOUTH FLORA AND DENTAL CARIES

Among the many varieties of bacteria, yeast, and protozoa normally present in the mouth, streptococci such as *Streptococcus mitis,* peptostreptococci, and diphtheroids usually predominate. Also commonly present in large numbers are small gram-negative cocci of the genus *Veillonella,* strict anaerobes that can metabolize

Lactic acid bacteria
Pages 257, 258

the lactic acid produced by streptococci and thereby help reduce the acidity. Aerobic gram-negative diplococci of the genus *Neisseria* are present in smaller numbers, as are various anaerobic gram-negative rods such as *Bacteroides,* spirochetes, and actinomycetes. The near-neutral conditions of the mouth maintained by saliva and alkali-producing bacteria discourage the growth of lactobacilli, except in areas where other organisms reduce the pH.

Tenacious collections of bacteria that form in crevices of the teeth and even on smooth surfaces (dental plaques) are responsible for tooth destruction. The density of microbes in plaques may be very high, exceeding 10^{11} bacteria per

TABLE 26-1
Predominant Microbial Flora of the Alimentary Tract

Mouth	Stomach	Small Intestine	Large Intestine
Streptococci, diphtheroids, species of *Neisseria, Veillonella,* and *Bacteroides.*	Scant or absent.	Enterobacteria, lactobacilli, streptococci and yeasts. Counts low; range from about 1000 (upper) to 100,000 (lower intestine).	Anaerobes of genera *Bacteroides* and *Lactobacillus* predominate. Enterobacteria also numerous. Count high; about 10^{11} per gram of feces.

gram. Although many kinds of bacteria and yeast have been identified in plaques, the destructive action is mainly by lactic acid bacteria, particularly certain streptococci. Scientists have placed tiny electrodes into plaques and studied what happens to the pH when sugar and other carbohydrates enter the mouth. In certain instances the response to carbohydrate is dramatic and the pH drops from normal (7.0) to about 5.0 within minutes. This means a hundredfold increase in acidity, to a level where calcium phosphate begins to dissolve. The duration of the acid state is directly dependent on the duration of exposure to sugars and their concentration. Thus people who eat frequent snacks are more prone to dental caries than those who consume large meals less often. After the food leaves the mouth, the pH slowly rises to neutral. The effect is often delayed by the ability of some bacteria to store a portion of the foodstuff as carbohydrate, which is then metabolized to acid.

Studies with animals raised in a germ-free state have shown that with a diet suitably rich in sugar, the introduction of either streptococci or lactobacilli can cause dental caries. The production of caries is determined not only by diet but also by properties of the strain of microbe and the genetic background of the animal. However, no caries develop in the teeth unless the organisms are introduced, regardless of dietary or other factors. Indeed even animals given antibiotics to render them free of bacterial strains able to produce caries will remain free of caries until they have contact with animals carrying the caries-producing organisms.

Strains of streptococci able to produce dental caries have been found to produce sticky gluelike substances (dextrans) by polymerizing sucrose. This enables the bacteria to adhere to tooth surfaces despite the cleansing action of saliva and the musculature of the mouth. Populations whose diet contains little sucrose are virtually free of caries. For example, a recent survey of over 500 schoolchildren in villages of northern Thailand failed to turn up a single case of dental caries even though their rice diet is high in carbohydrate. When sweets are introduced into the diets of such populations, caries quickly become common. Although a low-sugar diet helps to prevent caries, unfortunately it does not ensure against other dental disease. Formation of bacterial deposits near the junction of the teeth and gums may lead to softening of the tissues and loss of teeth (periodontal disease). Although bacteria contribute, this process occurs by a different mechanism from caries and is not yet well understood.

MICROBES PRESENT IN THE GASTROINTESTINAL TRACT

The rich microbial flora of the mouth is absent in the fasting stomach. Both hydrochloric acid and other antimicrobial secretions play a role in microbial killing, depending on the species of microbe. The small intestine is also relatively free of microbes. Studies of healthy people usually show no more than 1000 bacteria per milliliter in the upper small intestine and a 100,000 or less per milliliter

PLAQUE BACTERIA PRODUCE ACID FROM SUGAR, CAUSING CARIES

Dextrans
Page 45;
Figure 2-22, Page 46

PERIODONTAL DISEASE

in the lower small intestine. The predominant organisms have usually been bacteria able to grow in air, including facultatively anaerobic gram-negative rods and various streptococci. Lactobacilli are fairly commonly found in small numbers, as are small numbers of yeast such as *Candida albicans*. The main reason for such low concentrations of microorganisms is thought to be the rapid passage of the various digestive juices through the small intestine.

Bacteroides
Page 254

In contrast to the scanty numbers of organisms in the stomach and small intestine, the colon contains very high counts, usually in the neighborhood of 100,000,000,000 per gram of feces. Anaerobic organisms of the genera *Lactobacillus* and *Bacteroides* generally exceed the others by about a hundredfold. Many of these organisms require special methods for cultivation in the laboratory, and some have not been successfully grown in pure culture. Many fail to correspond to known species and have simply been assigned numbers. Facultatively anaerobic gram-negative rods, particularly *Escherichia coli,* and members of the genera *Enterobacter, Klebsiella,* and *Proteus,* predominate among fecal microbes able to grow aerobically. In contrast to the anaerobes they are easy to cultivate and some have been studied in exhaustive detail.

Enterobacteria
Page 252

BIOCHEMICAL ACTIVITIES OF INTESTINAL MICROBES

Because of the many different kinds of bacteria living together in the alimentary tract, some of which depend on each other for growth, it is very difficult to predict which metabolic effects observed in the test tube are important to the animal host. It is, however, clear that some intestinal bacteria can use up vitamin C (ascorbic acid) and several B vitamins (choline, folic acid, cyanocobalamine, and thiamine). Thus under certain conditions, intestinal bacteria can contribute to vitamin deficiency. They also produce ammonia, organic acids from carbohydrates, and degrade bile and digestive enzymes.

BLIND LOOP SYNDROME

Some of these metabolic effects undoubtedly can play a role in human disease. One example is an unusual condition called the "blind loop syndrome" in which, due to disease or surgery, a pocket of the small intestine becomes relatively isolated from the rest. In the pocket, flow of intestinal juices is lessened and large populations of the normal flora build up. Affected individuals may become severely anemic and lose weight because the large numbers of bacteria use up certain vitamins and degrade bile. This condition may be alleviated by correcting the structural abnormality or by giving antibiotics to reduce the numbers of organisms in the loop.

Although some intestinal microbes may perform potentially harmful metabolic actions, other common intestinal bacteria synthesize an excess of useful vitamins. These include niacin, thiamine, riboflavin, pyridoxine, cyanocobalamine (vitamin B12), folic acid, pantothenic acid, biotin, and vitamin K. These syntheses are of enormous importance for the nutrition of some animals. For man they are of doubtful importance except when there is an inadequate diet. With poor nutrition oral antibiotics sometimes bring about vitamin deficiency presumably by causing a decrease in vitamin production by colon microbes.

SOME VIRAL DISEASES OF THE ALIMENTARY SYSTEM

Mumps

Mumps is an illness characterized by fever and painful swelling beneath one or both ears. This condition is due to an infection of the parotid salivary glands which are located between the angle of the jaw and the ear. The agent responsible is a para-myxovirus known as mumps virus. The agent is transmitted by saliva, perhaps sprayed into the air while the infected person is talking. The virus is known to occur in the saliva of infected persons as much as a week before they become sick. There is also recent evidence that a "carrier" state may occur, in which seemingly healthy persons shed virus in their saliva for weeks or months. In a few infected persons no illness occurs, and in others the major effects appear in the central nervous system, the testes, ovaries, pancreas, or thyroid gland with or without involvement of the parotid glands.

Paramyxovirus
Table 17-1, Page 335;
Page 680

Since other infectious agents may sometimes involve the parotid glands, not every patient with fever and painful swellings can be assumed to have mumps. Identifying mumps virus infection is most commonly done by testing the person's blood serum for development of complement-fixing antibodies against known mumps virus antigen prepared in the laboratory. In addition the virus can be isolated by injecting saliva or urine from infected persons into chicken embryos. Mumps viruses are related antigenically to some viruses causing respiratory diseases in children and to Newcastle disease virus (which infects chickens). A vaccine can give partial protection against mumps, but no useful antimicrobial therapy is available for prevention or treatment.

Complement fixation
Page 380

Hepatitis, an Infection of the Liver

In most young people hepatitis is the result of a viral infection of the liver. Either of two distinct viruses are involved, infectious hepatitis (IH) virus and serum hepatitis (SH) virus, both extremely difficult to study because of their small size and our inability to propagate them outside the human host (Table 26-2). These viruses infect the cells of the liver and interfere with their normal function, resulting in jaundice.

Infectious hepatitis spreads in epidemic fashion and has an incubation period averaging about 25 days, but with a wide range. Serum hepatitis generally has a longer incubation period, but its range of 40 to 150 days overlaps that of infectious hepatitis. Infection with the first virus does not produce immunity to the second, and vice versa. Both viruses are relatively resistant to heat and to

RESISTANCE TO HEAT
AND DISINFECTANTS

TABLE 26-2
Hepatitis Viruses

Characteristics	Pathogenic Potential	Epidemiology
Very small viruses of uncertain classification. Not cultivable *in vitro*. Resist heat and disinfectants.	Preferentially damage the liver and often produce jaundice.	Source: human cases and carriers. Incubation period: average 25 days for IH, much longer for SH. Transmission: fecal oral route for IH, blood or blood products for SH- as little as 0.0001 ml. of blood may be infective.

chemicals. Neither boiling nor the usual disinfectants give reliable killing of the viruses—only gas sterilization or autoclaving is effective.

Gamma globulin
Pages 371, 422

Pooled human gamma globulin protects against IH virus, but generally not against SH. The SH virus is usually transmitted by human blood or blood products. Persons receiving blood transfusion are vulnerable (about 3 percent get hepatitis), and also persons who wash hospital laboratory glassware, and drug addicts who share injection syringes. Formerly medical or dental instruments transmitted the infection to many because of a mistaken reliance on ineffective chemical "sterilization." At least 28,000 cases were caused during World War II when American troops were given yellow fever vaccine to which human serum had been added as a stabilizer. Some blood plasma samples were shown to contain 10,000 infective units per milliliter!

SPREAD OF SERUM HEPATITIS

SPREAD OF INFECTIOUS HEPATITIS

The IH virus is excreted in the feces of infected persons and is usually spread to others by the fecal-oral route. This often arises with inadequate hand washing by food handlers, but shellfish can concentrate the virus from fecally polluted waters and when eaten raw cause hepatitis. Transmission by blood products can also occur, but is much less important.

The presence of so many people in the general population with SH virus has been hard to explain since its principal mode of transmission is by blood products. However, a series of interesting developments began with a 1966 report of a previously unidentified antigen in blood from a native Australian (and therefore called "Australia antigen"). This antigen was subsequently found in high incidence in certain groups of people, especially those with hepatitis. In 1967 studies of hepatitis in an institution for the mentally retarded indicated strongly that SH and IH were distinct viruses but that both could be transmitted either by injection or by mouth.

In 1968 another report revealed that blood from those having had multiple transfusions (such as patients with hemophilia) often had antibodies which reacted with something in the blood of patients with SH. By 1969 it was shown that this substance was Australia antigen. It was separated, examined by electron microscopy, and shown to include a small particle about 20 nm in diameter. Presumably these particles are the long-sought SH virus. After exposure of a normal person to blood from someone with SH infection, Australia antigen appears in the blood of the recipient as early as 27 days later. Liver injury usually develops one to seven weeks after the first appearance of the antigen. In a few instances this antigen has been shown to circulate for more than eight years, although it generally disappears soon after recovery from hepatitis. There is now a good deal of conjecture over what harmful effects the long-standing circulation of Australia antigen might have on the infected person, his offspring, and his associates. Sensitive tests for detecting Australia antigen have been developed to identify carriers of SH virus so their blood will not be used inadvertently for transfusions. The natural modes of spread are not yet defined, but it is likely that using things such as towels or toothbrushes in common could transfer the virus from one person to another under some circumstances. There are also reports of its detection in human excretions, indicating the possibility of transmission by this mode.

SOME BACTERIAL INFECTIONS OF THE GASTROINTESTINAL TRACT

Diarrhea and Vomiting (Gastroenteritis)

Diarrhea results from an outpouring of fluid through the gastrointestinal tract. When caused by an infectious agent, it is often associated with fever. Vomiting and cramps often accompany the excessive irritation. When pus and blood are found in the diarrheal fluid, the condition is often called *dysentery.*

Gram-negative rods of the genus *Salmonella* (Table 26-3) are a prominent cause of gastroenteritis, infecting over a million people in the United States every year. The organisms generally enter the gastrointestinal tract with food, although contaminated water, fingers, and other objects may be sources. The food products most commonly contaminated are eggs or egg products, or poultry. Since 1963 epidemics have also resulted from contaminated brewers' yeast, protein supplements, dry milk, and even a medication used to help diagnose intestinal disease! Why such a large variety of substances may be contaminated with *Salmonella* is not hard to explain. They infect a wide variety of animals—from cows and chickens to pet turtles—and it is primarily the species that originate from animals that produce diarrhea in people.

Enterobacteria
Page 252 and Figure 12-15,
Page 253

The mechanisms by which *Salmonella* produce intestinal infection are not known. They do, like all gram-negative rods, possess endotoxin, but this alone cannot explain their effect. Some *Salmonella* are able to penetrate the cells lining the intestine, and the resulting damage is undoubtedly an important factor in their pathogenicity. In patients with diarrhea caused by *Salmonella,* the causative organisms are usually present in the feces in large numbers. Since many species of *Salmonella* are relatively resistant to toxic chemicals, selective media containing such substances can be used in culturing to suppress the normal flora and allow the *Salmonella* to grow. In contrast to the majority of strains of *Escherichia, Enterobacter,* and *Klebsiella,* only very rare strains of *Salmonella* ferment lactose. *Salmonella* are identified by a series of biochemical tests and by reactions with specific antisera against somatic and flagellar antigens. One of the anomalies of history is that a different species name has been assigned for each set of antigens, and be-

Selective media
Page 116

cause of this, over 700 species of *Salmonella* have been recognized! Present feeling is that only three are justified: *S. typhi, S. enteritidis* and *S. cholerae-suis*.

Control measures for these pathogenic bacteria have concentrated on establishing reporting systems for epidemics, tracing of sources by careful identification of individual strains, and routine sampling of animal products for contamination. Unfortunately the incidence of reported cases has tended to rise, and the organisms have also shown increasing resistance to antimicrobial medicines. This is partly due to selection of resistant strains by the widespread use of such medicines in animal feeds in an attempt to promote more rapid growth.

Dysentery

Another frequent source of intestinal infection is the genus *Shigella* (Table 26-3). One of the most severe and extensive shigella epidemics of modern times began in Guatemala early in 1969 with antibiotic-resistant *Shigella dysenteriae* striking almost simultaneously at widely separated villages. Characteristically the infected persons had fever, diarrhea, and vomiting, with pus and blood appearing in the feces 12 to 72 hours after onset. The condition subsided spontaneously in five to seven days in most instances. The causative organism was not immediately determined since the epidemic strain was unusually fastidious in its growth requirements and most cultural attempts failed to yield *Shigella*. State public health laboratories in the United States began to isolate the agent from travelers returning during the same year, but spread of the agent from them to those living in the United States occurred only rarely.

PATHOGENESIS OF
DYSENTERY

Like *Salmonella* these organisms enter the mouth on materials contaminated directly or indirectly by feces, but the source is essentially always another human being since these organisms usually infect only man and other primates. The attack on the gut would appear to differ from *Salmonella* in that the organisms invade and multiply in the lining cells. Death of the cells results in intense inflammation and formation of small abscesses. Severe dysentery is thus the common result of *Shigella* infections. Children infected with some *Shigella* strains often have headache, a stiff neck, and convulsions. Adults commonly have painful joints for weeks or months after recovery. Since these bacteria are not commonly found in the general blood circulation during dysentery, many feel some

Protein toxins
Page 364

TABLE 26-3
The Genera *Salmonella* and *Shigella*

	Characteristics	Source	Pathogenic Potential
Salmonella	Lactose non-fermenting enterobacteria, usually motile.	Cold and warm blooded animals, including man.	Gastroenteritis, sometimes invasion of blood stream; *S typhi* causes typhoid.
Shigella	Lactose non-fermenting enterobacteria, non-motile.	Man.	Dysentery, Gastroenteritis.

bacterial toxin may be absorbed to explain the extra-intestinal symptoms. However only in strains of *Shigella dysenteriae* has presence of an exotoxin been established.

Most strains of *Shigella* grow readily on many laboratory media and can be isolated by culturing the feces of infected persons. They are, however, generally less resistant to inhibitory agents used in selective media for *Salmonella* species and many strains will not grow on these media. They also enter the feces with particles of dead intestinal lining cells, pus, and blood, and are thus not uniformly present in the feces. The sample selected for culture may not even contain any *Shigella.* Thus, as in the Central American epidemic, multiple cultures on special media and antibody tests on the serum of patients may be necessary to demonstrate the infection.

Shigellae, like salmonellae, are lactose nonfermenting enterobacteria. They are, however, nonmotile (they lack flagella) and are relatively restricted in their biochemical capabilities in comparison with most others of this large family. Separation of the various species within the genus *Shigella* is done using antisera against the somatic (cell wall) antigens.

Control of the spread of shigellae is almost entirely accomplished by sanitary measures and surveillance of food handlers and water supplies. It is important to note in this regard that the Guatemalan epidemic mentioned above followed a severe drought and subsequent flood during which sanitation suffered.

Cholera

Cholera, an ancient disease, is a unique form of severe epidemic diarrhea which had been limited largely to India and Southeast Asia in recent times. However, during 1961 to 1972, extensions into the Middle East, USSR, and Africa have occurred (Fig. 26-2). This disease is characterized by profuse outpouring of fluid from the intestine, often amounting to 15 percent of body weight within a few hours. So much water and electrolytes leave the body that the blood becomes thickened and reduced in volume so that there is insufficient blood flow to keep vital organs such as the kidney working properly. Many people die unless the lost fluid can be replaced promptly.

An early physician observer described the cholera patient as follows: "The face was sunken as if wasted by lingering consumption, perfectly angular, and rendered peculiarly ghastly by the complete removal of all the soft solids, in their places supplied by dark lead-colored lines; the hands and feet were bluish-white, wrinkled as when long macerated in cold water; the eyes had fallen to the bottom of their orbes, and evinced a glaring vitality, but without mobility; and the surface of the body was cold." Microscopic inspection of the feces typically reveals the causative organism, *Vibrio cholerae,* which are slightly curved gram-negative rods with the darting motility characteristic of bacteria with polar flagella. Since the normal flora are commonly present in large numbers, an enrichment procedure is often helpful in culturing the pathogen from feces. The medium used for this is a highly alkaline broth in which the vibrios grow well but other organisms

SHIGELLAE AND SALMONELLAE ARE LACTOSE NON-FERMENTERS

V. cholerae
Figure 12-11; Page 248
Enrichment
Page 115

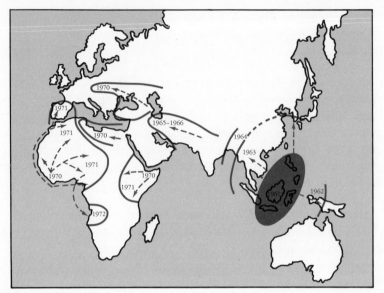

FIGURE 26-2 The spread of cholera 1961–1972. (From Morbidity and Mortality Weekly Report, 20 May 1972, p. 171.)

are inhibited. The actively motile *Vibrio cholerae* congregate at the surface of the fluid where the oxygen concentration is greater and subcultures are made from the surface to solid media for isolation. Confirmation and identification of individual strains are made by specific antisera and susceptibility to certain bacteriophages.

PATHOGENESIS
OF CHOLERA

Much research has been done to determine how *V. cholerae* produces cholera. Because there is no visible damage to the lining cells of the small intestine, it has long been assumed that the bacterium produced a toxin which interfered with the normal transfers of fluid by the intestinal cells. Indeed such a toxin has now been isolated and characterized. The toxin acts by causing excessive secretion of water and electrolyte by the intestinal cells without greatly altering their absorptive quality. The exact mechanism is not yet understood although the toxin appears to cause increased levels of cyclic AMP in the intestinal cells. Evaluation of a toxoid for immunizing people is now underway, but toxoids prepared from toxins of other gram-negative pathogens have not generally been successful immunizing agents. Some strains of *E. coli* have recently been found to produce a mild choleralike illness by a toxin which acts in a similar manner.

Cyclic AMP
Page 219

Although lower animals associated with man may harbor *V. cholerae,* man himself is thought to be the only important reservoir. Between epidemics the organisms persist in the intestines of human carriers, who have been shown to excrete the organisms for as long as six years. Control of cholera is aided by administering a vaccine of killed *V. cholerae* which stimulates bactericidal and antitoxic antibodies in the recipients, but protection lasts only a few months. Well-nourished individuals living under sanitary conditions appear to be quite resistant to cholera. Indeed, over the decade ending in 1970, only six cholera cases and no deaths were reported among the millions of American travelers in areas of the world where cholera existed. The requirement of cholera vaccination for travelers entering the United States has recently been dropped.

Typhoid Fever

People sick with typhoid typically develop a fever, which goes up over a three-day period, followed by severe headache and abdominal pain. With some, rupture of the intestine and shock due to loss of blood occurs. If treatment is not given, about one out of five persons die. The causative organism, *Salmonella typhi,* infects only man. It usually enters the gastrointestinal tract with contaminated food or water as do other salmonellae, and like them readily penetrates the intestinal lining. Phagocytic defense cells appear in the gut wall to ingest the invaders but the typhoid bacilli multiply inside the cells and are carried to other parts of the body by the bloodstream. *Salmonella typhi* also localizes in the collections of lymphoid cells of the gut wall, and destruction of this tissue can lead to intestinal rupture and hemorrhage.

The organisms can be recovered from the blood or bone marrow of an infected person, and later, by using selective and enrichment media, from the feces. A rise in antibody titer over a one- to two-week period against the organisms can often be demonstrated in the blood of the infected person and may be helpful in identifying the cause of his illness.

Cultural methods
Page 116

This strictly human disease is maintained in nature by carriers, people who appear perfectly well but who may excrete as many as 10,000,000,000 typhoid bacilli per gram of their feces. Since far fewer organisms are required to infect under natural conditions, it is easy to see how dangerous carriers can be. One of the most notorious carriers was Typhoid Mary, a young Irish cook living in New York State in the early 1900s. She is known to have been responsible for at least 53 cases of typhoid fever transmitted over a 15-year period. At that time, about 350,000 cases of typhoid occurred in the United States each year. Now with improved sanitation and public health surveillance measures the reported incidence of typhoid fever in the United States is about 300 cases per year.

A vaccine against *S. typhi* and related organisms is widely used by the military and in countries of high prevalence and is effective in reducing the incidence and severity of the disease; however, it does not offer complete protection. Searching out and treating carriers is another control measure. Carriers can be detected by isolating the organism from stool cultures and by serological tests for antibodies against the Vi (capsular) antigen of *S. typhi.*

Page 422

PROTOZOAN INFECTIONS OF THE GASTROINTESTINAL TRACT

Several species of protozoa produce intestinal infections and may be important in causing human disease. All the major groups (Chapter 15) are represented, as,

Pages 301–306

Figure 15-6
Page 300

for example, *Giardia lamblia* (flagellate), *Balantidium coli* (ciliate), *Isospora belli* (sporozoon), and *Entamoeba histolytica* (ameba).

Perhaps the most important of the protozoa producing intestinal infections of people is *E. histolytica*. This ameba, one of the causes of dysentery, is found world-wide, being especially prevalent in areas of poor sanitation. Encysted organisms usually enter the alimentary tract with contaminated food or water and, once in the intestine, liberate trophozoites. Upon reaching the upper portion of the large bowel, the trophozoites begin feeding on mucus and cells lining the intestine. Several digestive enzymes produced by the amebae aid them in penetrating through the lining cells into the intestinal wall. In fact they may burrow into blood vessels and be carried to the liver or other body organs. Continued multiplication and tissue destruction at intestinal and extra-intestinal sites result in the formation of abscesses. The irritant effect of amebae on the lining cells results in increased movements of the bowel and production of fluid. Thus diarrhea, often tinged with blood, develops. Direct examination of diarrheal fluid on a warm microscope slide may reveal the motile trophozoites, which generally range from about 20 to 40 μm in diameter. Many persons develop a chronic infection which may be asymptomatic and therefore unnoticed. Trophozoites are usually not seen in examination of the feces of such patients, but cysts are commonly present. On passage through the large bowel the cysts typically develop four nuclei, and in this form are infectious for the next host. Cysts can be identified by microscopic examination of feces, and methods are available for concentrating them from other fecal material to aid their detection.

Entamoeba histolytica can be grown anaerobically in pure cultures or in aerobic fluid cultures containing bacteria. They grow better when certain bacteria are added to the medium, and a mixture of bacteria gives better growth than a single bacterial species. In studies with some germ-free animals, *E. histolytica* has caused little or no injury to the intestinal lining unless bacteria were also introduced. Indeed in the treatment of persons with amebic dysentery, antibiotics active against bacteria are very effective even though they have no direct activity against the amebae. This indicates that a synergistic action of amebae and bacteria is required to produce the intestinal disease.

FOOD POISONINGS

Illness resulting from microbial growth in food results chiefly from two different mechanisms: (1) the microorganisms may infect the person who ingests the food or (2) products of microbial growth in the food may be poisonous. Gastroenteritis, dysentery, and typhoid fever due to bacteria of the genera *Salmonella* and *Shigella* may arise by the first mechanism, and a wide variety of common microbial species (including members of the bacterial genera *Clostridium, Bacillus, Streptococcus,* and *Staphylococcus*) may be responsible for illnesses by the second mechanism. Although all such illnesses arising from microbial growth in food are sometimes considered "food poisoning," the following discussion deals with illnesses caused by microbial products rather than from infection.

Staphylococcal Food Poisoning

One of the most common forms of food poisoning is due to *Staphylococcus aureus,* the same species of bacterium responsible for boils and other infections. The illness ensues a few hours after ingesting contaminated food, is manifested by nausea, vomiting, and diarrhea, and is followed by recovery within another few hours. The demoralizing but rarely fatal illness is the result of an exotoxin produced when certain strains of *S. aureus* grow in a suitable foodstuff, usually one high in carbohydrates. This exotoxin is called an enterotoxin ("entero-" referring to intestine, not to be confused with endotoxin of the cell walls of gram-negative rods) and may be produced quickly at room temperature, often within only 1 or 2 hours. Moreover the toxin is relatively heat stable, so that subsequent cooking may serve to kill the staphylococci, but toxic activity may remain. Several antigenically distinct varieties of enterotoxins are known to be produced by the various food-poisoning strains of *S. aureus,* and they can be identified in extracts of contaminated foods by using specific antisera. *Clostridium perfringens,* which produces enterotoxin upon sporulation, is another common cause of food poisoning. The pathogenesis of poisoning by species of *Bacillus* and *Streptococcus* is unclear.

S. aureus
Pages 430–432, 516

ENTEROTOXIN IS HEAT STABLE

Poisonings Due to Fungi

In former times, ergot poisoning ("St. Anthony's Fire") followed the ingestion of grain contaminated with certain fungi. Afflicted persons had agonizingly painful convulsions, and some developed gangrene of their hands and feet.

Proper standards of agricultural practice and public health surveillance of food grains now make it unlikely that ergot-producing fungi will contaminate food and give ergot toxicity. However, another kind of fungal toxin is of great current interest. This fungal poison, called *aflatoxin,* is produced by common molds of the genus *Aspergillus.* Aflatoxin can cause acute poisonings in many species of animal, affecting chiefly the young. A few milligrams, for example, may cause severe liver damage and death of a dog within 72 hours. Much interest in aflatoxin arose with the observation that tiny traces of the poison in the foodstuffs of certain animals caused tumors. In fact, for trout, aflatoxin is one of the most potent inducers of tumors known. Aflatoxins have been demonstrated in many human foods, and high levels of a metabolite may appear in cow's milk. Aflatoxins produced by various strains of molds are not always the same and a family of related compounds is now known. Their possible role in human disease is still under study.

Further Reading
Page 294

AFLATOXINS

Botulism

References were made in Chapter 13 to paralysis of human beings and other animals who eat shellfish containing red-tide algae of the genus *Gonyaulax.* Botulism, another microbial poisoning that is characterized by paralysis, is one of the most feared diseases of mankind. The cause is the gram-positive, rod-shaped,

Gonyaulax
Page 275
Clostridium
Page 257

strictly anaerobic eubacterium, *Clostridium botulinum,* widely distributed in soils around the world. The name *botulinum* comes from the Latin word for sausage, and was chosen because some of the earliest recognized cases of botulism occurred in people who had eaten contaminated sausage. However, many other foods have been sources of the poisoning, including vegetables, fruits, meats, and cheese. Like other clostridia, *Cl. botulinum* produces endospores which may be greatly resistant to heat, and thus persist in foodstuffs despite cooking and canning processes. These spores can later germinate and growth of the bacterium can be associated with release of a powerful exotoxin. When someone eats the contaminated food, the toxin is absorbed into the bloodstream and may circulate for as long as three weeks. This circulating toxin is then carried to the various nerves of the body, where it acts by blocking transmission of nerve signals to muscle, thus producing paralysis. Twelve to 36 hours after eating toxin-containing food, blurred or double vision gives the first indication of paralysis. All muscles may be affected, but respiratory paralysis is the most common cause of death, which ensues in more than half the victims of botulism. There is no effect on sensation or most other functions of the brain. The toxin is one of the most powerful poisons known, a few milligrams being sufficient to kill the entire population of New York City. Botulism has resulted from eating a single contaminated string bean, or even licking a finger contaminated by toxin. Fortunately the toxin is highly heat labile, and even high concentrations are completely inactivated by boiling the food for 15 minutes.

Clostridium botulinum grows readily under anaerobic conditions on most nonselective bacteriologic media. The optimum temperature for growth is between 30 to 35°C. Final identification depends on demonstration of botulinum toxin. There are six types recognized within the species—A, B, C, D, E, and F—of which A, B, and E account for essentially all of human botulism. The division into types is based on differences in toxins as shown by using specific antisera prepared in laboratory animals. Antibody produced against one toxin will neutralize only that toxin and not the others.

Toxin neutralization
Page 386

In the five years preceding 1922 there were 83 outbreaks of botulism in the United States, many tracable to commercially canned foods. Because of the work of the distinguished microbiologist, Dr. Karl F. Meyer and his colleagues, it was shown that canning methods then in use were inadequate to kill the heat-resistant spores of *Cl. botulinum*. Strict controls were placed on commercial canners to ensure adequate sterilizing methods. Following this, outbreaks caused by commercially canned foods almost disappeared. However, lapses in proper canning techniques led to death from botulism of a man in Westchester County, New York, in the summer of 1971, the first case there since 1934. The lethal poison was traced to vichyssoise soup canned by a large and well-known firm. Prompt action by physicians and public health officials led to tracking down most of the over 6,000 cans from the same lot of soup that had been distributed to more than a dozen companies. As a result only one additional case of botulism developed. The episode, however, points up the potential for disaster when defective processing occurs in large, food-producing and distributing firms.

Most cases of botulism from inadequately processed canned foods have been caused by type A or B. Awareness of major outbreaks of type E botulism occurred in the early 1960s, primarily in Japan and the United States. Like other types of *Cl. botulinum,* type E strains are soil organisms, but they are more readily taken up by fish and sea mammals. Type E strains also differ in that toxin can be produced at lower temperatures and at greater exposure to aerobic conditions. For these reasons type E botulism is generally associated with ingestion of contaminated seafoods.

Botulism is treated by administering the appropriate antitoxin. This neutralizes toxin circulating in the bloodstream, but recovery of the nerves is slow, requiring weeks or months. Control of the disease depends on proper sterilization and sealing of food at the time of canning and heating it adequately preparatory to serving. One cannot rely on a "spoiled" taste or appearance to detect contamination, because such changes may not always be present. Modern research on botulism has focused on ways to achieve more rapid identification of the type of toxin and on finding medications to neutralize its toxic effects.

PREVENTION AND
TREATMENT

SUMMARY

One of the major routes of microbial invasion is the alimentary tract, and the major manifestations of resulting diseases may involve any part of the body. This chapter deals with microbial diseases in which the alimentary system itself is heavily involved by the disease process. Members of the normal flora with little invasive ability can produce dental caries or the blind loop syndrome. Certain viral infections have specific tissue affinities, such as the parotid glands for mumps and the liver for hepatitis viruses. The genus *Salmonella* is responsible for a wide spectrum of human diseases, ranging from gastroenteritis (generally from species colonizing other animals) to the more serious typhoid fever from *Salmonella typhi,* a strain infecting only human beings. *Shigella,* another facultatively anaerobic gram-negative rod, is a frequent cause of dysentery as is the ameba, *Entamoeba histolytica.* In cholera there is no apparent structural damage to the intestine, yet severe alteration of intestinal function results. The toxin of *Vibrio cholerae* greatly increases intestinal fluid excretion, producing watery diarrhea and dehydration of the body.

Microbial food poisonings occur as the result of eating foods contaminated by certain toxic microbes or the by-products of their growth in the food. Bacteria of the genera *Bacillus, Clostridium, Streptococcus,* and *Staphylococcus* are commonly responsible for poisoning manifested by nausea, vomiting, and diarrhea. In the case of *Staphylococcus aureus,* a heat-stable enterotoxin (a type of exotoxin) is responsible. Fungal toxins include aflatoxins of *Aspergillus* species which can produce liver damage and tumors in some animals. One of the most feared types of poisonings is botulism, a severe, often fatal paralysis caused by the powerful exotoxins of *Clostridium botulinum.*

1. Discuss the various mechanisms by which microorganisms can interfere with normal function of the gastrointestinal tract. Why is cholera such a devastating disease when *Vibrio cholerae* neither invades the body nor destroys the intestinal mucosa?

2. Discuss the relative importance of microbial and dietary factors in dental caries.

3. Should one consider blood agar prepared from human blood to be sterile? Explain. What danger is there in washing a test tube containing human serum?

4. Explain how a foodstuff responsible for staphylococcal food poisoning may fail to yield growth of *Staphylococcus aureus* when inoculated onto appropriate culture media.

5. What simple procedure that can be carried out in every household kitchen will assure that no one gets botulism from canned food?

FURTHER READING

BURNETT, G. W. and H. W. SCHERP, *Oral Microbiology and Infectious Disease.* Baltimore: Williams and Wilkins, 1968.

HIRSCHHORN, N. and W. B. GREENOUGH, "Cholera," *Scientific American,* 225 (August 1971), 15.

NOLTE, W. A. (ed.), *Oral Microbiology.* St. Louis: C. V. Mosby and Co., 1968.

RENSBERGER, B., "Grim Detective Case: Search for Vichyssoise," *The New York Times,* (July 18, 1971), 1.

ROUECHE, BERTON, "Family Reunion," in *Annals of Epidemiology.* Boston: Little, Brown and Co., 1967.

———, "A Game of Wild Indians." *ibid.*

———, "S. Miami." *ibid.*

———, "The Simpsons and the Hepatitides," *The New Yorker Magazine,* (August 21, 1971), 72–86.

SULLIVAN, W., "Hepatitis Traced to an Oysterman," *The New York Times,* (November 19, 1961).

Chapter 27

Infections of the Genitourinary Tract

The human body is efficiently constructed to prevent microbial invasion, a fact which is particularly evident in studying the anatomy of the urinary tract. However, under certain circumstances the genitourinary tract can be invaded by a variety of organisms from the normal flora, which act as opportunists, and by pathogenic species as well.

ANATOMICAL AND PHYSIOLOGICAL CONSIDERATIONS

The urinary tract consists of the kidneys (in the upper urinary tract), the bladder (lower urinary tract), and accessory structures. The kidneys act as a specialized filtering system to cleanse the blood of many waste materials, selectively reabsorbing other substances that can be reutilized. Figure 27-1 shows that each kidney is drained by a tube, the ureter, which connects it with the urinary bladder. The bladder empties through the urethra. Special groups of muscles near the urethra act as sphincters to keep the system closed most of the time and help prevent infection. The downward flow of urine also helps by flushing out microorganisms that may have invaded, before they have a chance to multiply and infect.

Even when infection occurs in the lower urinary tract, the kidneys are still protected to some extent by the action of other sphincter muscles at the junctions of the bladder and the ureters. It is fortunate that these mechanisms usually protect against kidney infection, because such infections can be very difficult to eradicate, are often chronic, and can lead to marked destruction of the kidneys. Death promptly follows kidney failure unless the patient is lucky enough to be able to

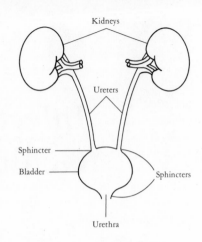

FIGURE 27-1
The anatomy of the urinary tract. Urine flows from the kidneys down the ureters into the bladder, which empties through the urethra. Sphincter muscles help to prevent any contamination from ascending.

use one of the all too scarce artificial kidneys, or perhaps to receive a kidney transplant.

The urinary tract is protected from infection by a number of mechanisms besides its anatomy. The urine normally contains certain organic acids and other antimicrobial substances, and small quantities of antibodies. During urinary infections larger quantities of specific antibodies can be found in the urine. There is evidence that antibody-forming lymphoid cells infiltrate the infected kidneys or bladder and form protective antibodies locally at the site where they are needed. Also, during infection an inflammatory response ensues in which phagocytes are of utmost importance in engulfing and destroying invading microorganisms.

The anatomy of the male and female genital tracts is shown in Figure 27-2. When an ovum is extruded from an ovary, it is swept into the adjacent fallopian tube, and then into the uterus, by the action of ciliated cells (Fig. 27-2b). Normally the uterus is protected from infection by sphincter muscles at the cervical canal, which connects the cervix (lower third of the uterus) with the vagina. Because the vagina is the portion of the female genital tract connecting to the exterior, it is the most frequent site of infections of this system. Lining cells of the vaginal mucosa are influenced by the hormone estrogen. When estrogen is present, glycogen is deposited in the lining cells. The carbohydrate is converted to lactic acid by lactobacilli, resulting in an acid pH. Thus the normal flora and resistance to infections of the female genital tract vary considerably with the hormonal status of the female.

NORMAL FLORA OF THE GENITOURINARY TRACT

Normally the urine and the urinary tract above the entrance to the bladder are completely free of microorganisms; however, the urethra has a normal flora. Species of *Streptococcus* (alpha-hemolytic or enterococci), *Neisseria, Mycobacterium,*

Bacteroides, and some of the enterobacteria are commonly isolated from the normal urethra.

As mentioned the normal flora of the female genital tract is influenced by the action of estrogen hormone. For several weeks after birth the vagina of newborn females remains under the influence of maternal hormones and has an acid pH. During this time lactobacilli are predominant in the normal flora. As the influence of maternal hormones wanes, the pH of the vagina increases to neutrality and remains neutral until puberty. During this childhood period the normal flora consists of a variety of cocci and rod-shaped organisms, including species of *Streptococcus* and many other genera. At puberty lactobacilli again become predominant, although smaller numbers of yeasts and other bacterial species are also present. After the menopause, there is a return to a neutral pH and the mixed flora typical of childhood.

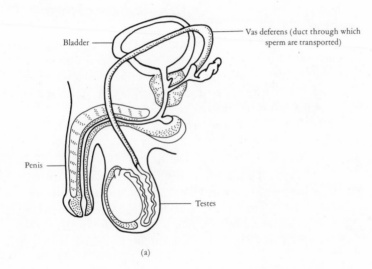

(a)

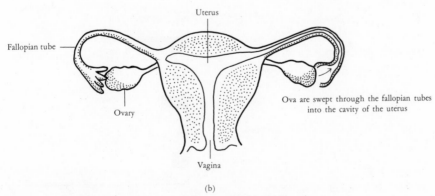

(b)

FIGURE 27-2 The anatomy of the genital tract. (a) Male (b) Female

A number of factors can predispose to urinary tract infections. Any situation that leads to stasis of the urine increases the chances of infection. For example, after anesthesia and major surgery, the reflex ability to void urine may be inhibited for a time. In this circumstance urine accumulates and distends the bladder, which is a very elastic organ. Even a few bacteria which have managed to evade defenses and enter the bladder can multiply during stasis, causing infection of the bladder. A similar problem commonly occurs in paraplegics or persons who have paralysis of the lower part of the body. Because they lack nervous control, paraplegics are unable to void normally and almost invariably develop urinary tract infections.

In the female the vagina is closely adjacent to the urethra (Fig. 27-2b) and the pressure associated with coitus may permit the introduction of organisms into the urinary tract. The bladder infections which result are often referred to as "honeymoon cystitis," a not infrequent disease accompanying initial sexual experience.

Unless infection of the lower urinary tract is overcome, either by successful treatment or by immune mechanisms, it can ascend through the ureters to involve the kidneys. Thus most kidney infections develop subsequent to bladder infection. Bacteria carried by the bloodstream can infect the kidneys, but this occurs much less often than ascending infections.

Infections of the urinary tract occur far more frequently in females than in males because the female urethra is short (about 1.5 inches, compared with 8 inches in the male), and closely adjacent to the genital and intestinal tracts. About 90 percent of urinary tract infections are caused by bacterial species which are part of the normal intestinal flora and which consequently are readily available to invade when defenses are interrupted (Fig. 27-3). Serological studies of infecting organisms have shown that they usually are derived from the patient's intestinal bacteria. For example, in these studies *Escherichia coli* isolated from urine

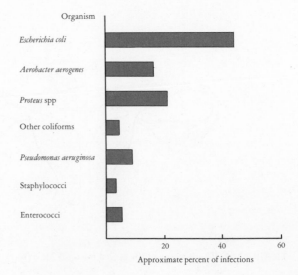

FIGURE 27-3
Common causes of urinary tract infections. The percentages vary considerably from one study to another. Often infections are caused by more than one kind of organism.

during urinary tract infections was almost always the same serotype as *E. coli* from the intestine of the same patient.

Figure 27-3 shows that *E. coli* accounts for about 50 percent of urinary tract infections, but other gram-negative rods of different genera, such as *Proteus* and *Enterobacter,* are also often found. *Pseudomonas aeruginosa,* a strictly aerobic gram-negative rod, is a particularly troublesome urinary pathogen because it is difficult to treat successfully. *Streptococcus faecalis* (enterococcus), found in normal bowel flora, is the most frequently isolated gram-positive organism responsible for urinary infections.

Pseudomonas aeruginosa in wound infections
Pages 517, 518

Causative organisms can usually be recovered easily and in large numbers from the urine of infected patients. It is necessary to collect voided urine specimens carefully to minimize external contamination from the genitalia or intestinal tract. This is more difficult in the female than the male and is best accomplished by collecting a midvoiding sample after carefully cleansing the external genitalia. Normal urine collected in this way usually contains less than 10,000 bacteria per milliliter. Concentrations of bacteria greater than 100,000 indicate infection. Colony counts are done to determine the number of bacteria per milliliter of urine. It is important to remember that bacteria multiply exponentially; therefore cultures for colony counts must be prepared promptly after the urine is collected to obtain accurate counts of bacteria.

Urine specimens are cultured on media especially suitable for the growth of enterobacteria and on blood agar which will also support the growth of enterococci and other gram-positive pathogens. *Escherichia coli,* the organism most often responsible for urinary tract infections, is the prototype of the enterobacteria.

Enterobacteria
Pages 252, 253

During some diseases in which bacteria primarily attack areas of the body other than the urinary tract, the causative organisms may be found in the urine. Thus in typhoid fever, *Salmonella typhi* organisms become disseminated throughout the body and are excreted in the urine beginning about one to two weeks after infection. In like manner slender spirochetes of the genus *Leptospira* (the cause of leptospirosis) are commonly found in the urine. Contamination of water with urine from infected animals or human beings is the major means of transmission of these bacteria. The organisms may continue to be excreted in the urine after recovery from these illnesses.

Occasionally another type of kidney disease results as a sequel to infection elsewhere in the body, even though the kidneys are not actually infected. An example is the disease called post-streptococcal glomerulonephritis which occurs rarely and which follows streptococcal infection in the throat or elsewhere. This disease represents a hypersensitivity reaction to soluble antigens of only a few M-types of *Streptococcus pyogenes.* The antigens combine with specific antibodies to form antigen-antibody complexes which interact with complement. These immune complexes are deposited in the kidney, where they cause tissue damage. Although this kidney disease is associated with streptococcal infection, the bacteria are not found in the urinary tract and in fact the infection usually has cleared before hypersensitivity reactions become apparent.

Streptococcus pyogenes
Pages 433, 434

BACTERIAL INFECTIONS OF THE GENITAL TRACT

Puerperal Fever

Group A beta hemolytic streptococci (*S. pyogenes*) are a major cause of puerperal fever (childbirth fever). They are transmitted by direct contact at childbirth with contaminated instruments or hands. In addition to *S. pyogenes,* the causative bacteria may be some of the anaerobic species (clostridia, bacterioides, and anaerobic streptococci) that infect wounds and find a fertile area for growth in tissues injured by trauma.

The history of puerperal fever illustrates the general tendency of man to resist change. For centuries it had been an accepted fact that many mothers developed fever and died following childbirth. About the middle of the nineteenth century, in the hospitals of Vienna (the major medical center of the world at that time), about one out of every eight women died of puerperal fever following childbirth. A Hungarian physician named Semmelweiss, who had come to Vienna to practice, studied this abhorrent situation and concluded that the disease was transmitted by the attending physicians or their instruments. He instituted a program of simple disinfection of all instruments used during delivery and was able to cut the maternal death rate at childbirth from about 12 percent to less than 1 percent. Instead of accepting these findings and the new techniques of disinfection Semmelweiss used, his colleagues at Vienna refused to face the fact that they had been responsible, even unknowingly, for the death of so many patients. The work of Semmelweiss was attacked, and he was forced to leave Vienna and return to his native Hungary. There he was again able to use disinfection techniques and to achieve a remarkable reduction in the number of deaths from puerperal fever. The famous American physician, Dr. Oliver Wendell Holmes, independently made the same findings at about the same time.

Over the years, with increasing improvement (and acceptance) of disinfection and sterilization techniques, puerperal fever has become rare in countries with adequate medical facilities. However, even now, outbreaks occur occasionally and it may be very difficult to identify the source. Recently in the United States a number of cases occurred despite careful precautions and use of proper techniques. In some instances it was found that a medical attendant was an anal carrier of *S. pyogenes,* which was disseminated by movement of his clothing.

Venereal Diseases

GONORRHEA

Although several venereal diseases are known, by far the most prevalent is gonorrhea. At present in the United States gonorrhea has the highest incidence of any bacterial disease, and it is increasing at an alarming rate (Fig. 27-4). The causative organism, *Neisseria gonorrhoeae* (gonococcus), is a gram-negative diplococcus, identical in appearance to the closely related *N. meningitidis* (the cause of epidemic meningitis). The cocci are flattened on adjacent sides, and are frequently found within certain white blood cells (PMNs) in pus, as indicated in Figure

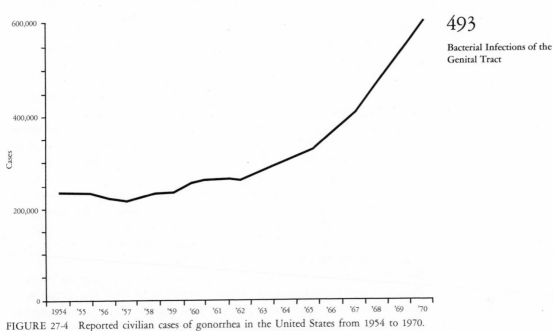

FIGURE 27-4   Reported civilian cases of gonorrhea in the United States from 1954 to 1970. (Morbidity and Mortality Report, United States Public Health Service **19**:59, 1971).

27-5. Gonococci are parasites of man only, preferring to live on mucous membranes of the host. They are susceptible to cold and drying, and hence do not survive well outside their host. For this reason gonorrhea is transmitted primarily by direct contact, and as the bacteria live in the genital tract, the contact is almost always venereal. There are certain exceptions to this general rule. For instance, one form of gonococcal disease is an eye infection (opthalmia neonatorum) that can be transmitted from mother to infant at the time of birth by passage through an infected birth canal. This form of gonococcal infection has been controlled in the United States by laws requiring the use of silver nitrate or an antibiotic instilled directly into the eyes of all newborn infants.

Gonococci grow in the genitourinary tract only in areas where a certain kind of epithelial (lining) cells are found (the columnar epithelial cells). Con-

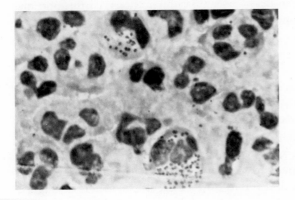

FIGURE 27-5
Neisseria gonorrhoeae in gram-stained material from a patient with gonorrhea. The diplococci are abundant within polymorphonuclear leukocytes in the pus.

sequently gonorrhea differs in males and females because of the differences in location of these particular cells. In males the disease is characterized by inflammation of the urethra, with pain on urination and a thick, pus-containing discharge which becomes apparent a few days after infection. Usually the disease is self-limiting and clears of its own accord, but the infection may extend in the genital tract, resulting in complications. For example, extensive inflammatory reaction to the infection can lead to the formation of fibrous tissue which obstructs the urethra, predisposing to future urinary tract infection. If fibrous tissue occludes the tubes that carry the sperm (vas deferens) or if tissue in the testes is destroyed by inflammation, sterility may result.

Gonorrhea follows a different course in women. Gonococci thrive in the cervix and fallopian tubes (Fig. 27-2b) as well as other areas of the female genital tract. The early stages of infection, involving the cervix, are commonly mild and the woman may be unaware of the disease. The infection, however, frequently progresses upward into the uterus and the fallopian tubes. Scar and fibrous tissue formed in response to the infection may block normal passage of the ovum through the tube, and is thus a cause of sterility in females. This situation can also lead to abnormal pregnancy during which the ovum is fertilized and develops in the fallopian tube or in the abdominal cavity outside of the uterus (ectopic pregnancy). Another common complication of gonorrhea, in both females and males, is gonococcal arthritis caused by growth of the cocci within joint spaces. Any joint may be affected, especially the larger ones.

About 90 percent of females with gonorrhea have no symptoms and they often unwittingly harbor and transmit gonococci for months or even years. Late in the disease they may develop some of the complications mentioned previously. These female carriers are the source of infection responsible for a large proportion of the current epidemic of gonorrhea.

New problems have arisen as a result of the widespread use of birth control pills, often referred to as the "Pill." It has been reported that use of the Pill promotes several kinds of venereal diseases by increasing both the pH and the moisture content of the vagina. The alkaline environment produced is particularly favorable to the growth of gonococci and a number of other pathogens. It has been observed that women on the Pill are especially vulnerable to gonorrheal infection and more likely to develop serious complications. Furthermore the complications usually become apparent early during the disease in these women, often within a few days after infection.

Gonococci can be recovered from purulent discharge or from secretions collected from the infected areas. Gram-stained smears usually reveal the characteristic gram-negative diplococci within PMN cells in pus, but in order to identify the organisms as *N. gonorrhoeae* it is necessary to culture them. They grow on rich medium, such as chocolate agar, at 35 to 37°C, in 5 to 10 percent carbon dioxide, forming small colonies. All members of the genus *Neisseria* synthesize oxidase that gives a visible reaction with a particular dye; hence they are called oxidase-positive. This reaction can be observed readily by touching a portion of the colony to filter paper soaked with the dye. It can also be used to distinguish colonies of *Neisseria* by flooding a petri dish culture with the dye; oxidase-positive colonies will turn pink and then black, and while at the pink stage they can be subcultured to obtain a pure culture. Biochemical tests (sugar fermentations) must be done in order to differentiate *N. gonorrhoeae* from other neisseriae. Fluorescent antibody techniques are also helpful for identifying gonococci.

Although gonococci are usually susceptible to penicillin, quite resistant strains are being selected. Thus, 15 years ago 300,000 units of penicillin were usually sufficient to cure a case of gonorrhea, but now a dose of 4.8 million units is often necessary, and even this dose is not uniformly successful.

Several factors are involved in the increasing incidence of gonorrhea. One is the change in sexual mores, with the new sexual freedom leading to a greater chance of infection, coupled with the fact that the female carrier of gonococci can unknowingly transmit the bacteria over months or even years. There is no lasting immunity after infection, and the individual who has recovered from gonorrhea soon becomes susceptible to reinfection. A great effort, so far unsuccessful, is being made to develop a vaccine that will provide effective protection against gonococci. Attempts to identify female carriers so that they can be treated appear to offer more promise.

Syphilis, another venereal disease of major importance, is caused by a spirochete, *Treponema pallidum.* The organisms (treponemes) of syphilis are motile, extremely thin, tightly coiled spirals. They cannot be seen on gram-stained smears because they are so thin; thus it is necessary to use special methods such as dark-field examination or impregnation with silver stains to visualize them. These treponemes cannot be grown in culture, and under natural conditions are strictly parasites of man. They can be grown in limited numbers in the testes of rabbits. Like the gonococcus, *T. pallidum* is exquisitely sensitive to destruction by drying and to temperature changes, and for this reason it is almost exclusively transmitted by sexual contact.

The disease, syphilis, occurs in three stages. As a rule, in *primary syphilis,* treponemes transmitted by sexual contact grow and multiply in local areas of the genitalia, spreading from there to the lymph nodes and bloodstream. About three weeks after infection an ulcer, called a hard chancre, appears at the site of infection. This represents a cellular response to the bacterial invasion, and examination of a drop of fluid expressed from the chancre reveals that it is teeming with infectious *T. pallidum.* Whether treated or not, the chancre disappears within a period of four to six weeks and the patient may mistakenly believe that the disease is cured. However, about two to ten weeks later a rash and other manifestations of *secondary syphilis* may appear. By this time the spirochetes have spread throughout the body and infectious lesions occur on the skin and mucous membranes in various locations, especially in the mouth; therefore syphilis can be transmitted by kissing during this stage. The secondary stage lasts for weeks to months, sometimes as long as a year, and gradually subsides. About 50 percent of untreated cases never progress past the secondary stage; however, some cases continue to the tertiary stage, which occurs after a latent period of 5 to 20 years, or even longer.

Tertiary syphilis represents a hypersensitivity reaction to small numbers of

T. pallidum that grow and persist in the tissues. The treponemes may be found in almost any part of the body, and the symptoms of tertiary syphilis depend on where the hypersensitivity reactions occur. If they are in the skin, bones, or other areas not vital to existence, the disease is not life-threatening. However, if they occur within the walls of a major blood vessel, they may lead to rupture of the vessel and instant death; in the eyes they cause blindness; in the central nervous system they can cause insanity, characteristic paralysis, and other manifestations.

Primary or secondary syphilis can be diagnosed by finding the motile spirochetes in fresh material from lesions on the skin or mucous membranes of the genitalia or the mouth. Darkfield illumination permits the thin spirals to be seen. Dried specimens can be examined by using fluorescent antibody techniques. In tertiary syphilis the patient is not infectious and the internal lesions contain so few organisms that they cannot be detected by these means.

During all the stages of syphilis, serological tests are helpful in diagnosis. Although *T. pallidum* cannot be grown in culture and thus is not readily available as a source of antigen, it was found quite by accident that antibodies formed during syphilis will react with certain substances extracted from beef heart. This fortuitous discovery allowed the development of tests for serum antibodies against *T. pallidum,* using the beef-heart substance as "antigen." At first, complement-fixation reactions, such as the Wassermann test, were widely used, but now most laboratories use a form of precipitation reaction called the flocculation test; examples are the Kahn test and the VDRL test (Venereal Disease Research Laboratory). In these tests a fluffy "antigen"-antibody precipitate, or flocculation, is formed. False positive reactions occur fairly frequently, so that flocculation tests are useful for preliminary testing of large numbers of people, but positive tests do not always indicate that the subject has syphilis. Other methods must be employed to confirm the suspected diagnosis, using specific antigens of *T. pallidum.* The fluorescent treponemal antibody test (FTA-ABS) is especially valuable (Fig. 27-6). In this indirect immunofluorescence test, the patient's serum is allowed to interact under carefully controlled conditions with killed *T. pallidum* organisms that have been grown in the testes of rabbits. If specific antitreponemal antibodies are present in the serum, they combine with the treponemes and excess serum proteins are washed from the slide. In order to detect the reaction fluorescent dye-tagged antibodies against human gamma globulin are added. When specific antitreponemal antibodies have combined, the treponemes will fluoresce when examined under ultraviolet light.

Sometimes it is difficult to establish a diagnosis of syphilis. A chancre may fail to develop, so that the primary stage goes unnoticed, or the symptoms that are observed may be confused with symptoms of another disease. *Syphilis has been called "the great imitator" because it resembles or imitates many other diseases as it progresses.* Also, congenital syphilis may go unrecognized if the disease is not apparent in the mother. It is possible for mothers who have passed the infectious stages of secondary syphilis and have no symptoms of the disease to transmit the treponemes across the placenta to the fetus.

Treponema pallidum is always susceptible to low doses of penicillin, and thus

a. Killed *Treponema pallidum* spirochetes, fixed to the slide, are incubated with the subject's serum. If specific antibodies are present they combine with the spirochetes.

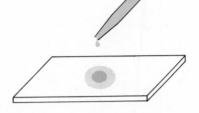

b. The slide is washed to remove excess serum, and fluorescent-labeled anti-human-gamma-globulin (anti-HGG) antiserum is added. The anti-HGG antibodies combine with any human antibodies already on the spirochetes.

c. Following incubation, the slide is washed and examined microscopically using ultraviolet light. The *T. pallidum* organisms coated with antibodies fluoresce and can be visualized readily.

T. pallidum +
specific antibodies

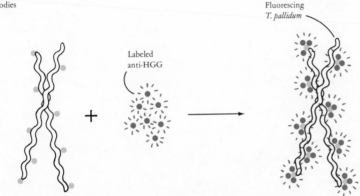

FIGURE 27-6   The fluorescent treponemal antibody (FTA-ABS) test. This test employs indirect immunofluorescence to demonstrate humoral antibodies against *Treponema pallidum*.

penicillin is usually the treatment of choice. Of course, other antibiotics must be used to treat those persons who are allergic to penicillin. Treatment must be continued for a longer period during tertiary than during primary or secondary syphilis. There is no lasting immunity to syphilis either after spontaneous recovery or treatment early in the disease, nor have effective methods of immunization been developed.

In general syphilis is decreasing in incidence because of the use of effective antibiotics and an intensive program for tracing and treating contacts of infected individuals. However, occasional increases occur whenever precautions are relaxed. For example, the number of cases of syphilis in the United States each year dropped precipitously following the introduction of penicillin, as shown in Figure 27-7; there was, however, an increase in 1962. When more stringent precautionary measures were taken, the rate decreased again.

Several factors explain the frequent decreases in incidence of syphilis as contrasted with the tremendous increase in the number of cases of gonorrhea. For

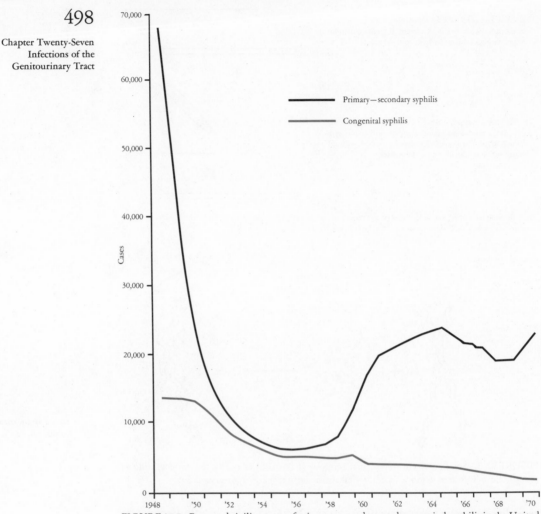

FIGURE 27-7 Reported civilian cases of primary, secondary, and congenital syphilis in the United States from 1948 to 1970. (Morbidity and Mortality Report, United States Public Health Service **19**:58, 1971).

one thing case workers carefully trace all reported cases of both syphilis and gonorrhea, attempting to reach and treat all sexual contacts of these patients; this approach usually allows contacts to be treated prophylactically during the long (one to three weeks) incubation period before syphilis develops. However, the incubation period of gonorrhea is so brief (one to three days) that the disease has usually developed in sexual contacts before they can be traced and treated. Frequently, a number of other contacts will also have been infected. A second important factor is that serological tests will identify subjects with unsuspected syphilis, but no serological tests for gonorrhea are yet available for screening large numbers of people. Another factor of major importance is the presence of large numbers of female carriers of gonococci—often women who have never had symptoms of gonorrhea. These carriers are the major reservoir of gonococci, and to date there is no easy way to identify them. It is necessary to culture secretions from the genital tract and to identify *N. gonorrhoeae* from the cultures—a laborious and expensive procedure. Thus it appears that the incidence of gonorrhea will continue to increase even as syphilis declines.

Although gonorrhea and syphilis account for a majority of cases, other venereal diseases caused by bacteria occur in the United States. One which is not uncommon in areas of the South is lymphogranuloma venereum, caused by obligate intracellular organisms of the genus *Chlamydia* (*Bedsonia*). Another is chancroid, which results from infection with small gram-negative rods, often seen in chains, called *Haemophilus ducreyi*.

NONBACTERIAL INFECTIONS OF THE GENITOURINARY TRACT

One of the more common fungal causes of genitourinary disease is Candida albicans (Fig. 27-8). This budding yeast is part of the normal flora of the mucous membranes of about 35 to 40 percent of people, and usually it is nonpathogenic. The interaction between large numbers of bacteria and small numbers of fungi results in an ecological balance between the groups. When this balance is upset, disease may occur. This happens most often after intensive antibacterial treatment, particularly with some of the more effective broad-spectrum antibiotics which suppress much of the normal bacterial flora and allow fungi to grow. The large numbers of *C. albicans* which are found in such a situation can cause extremely persistent and severe infections of mucous membranes in the genitourinary tract and also in the gastrointestinal tract.

Candida albicans is an oval, budding yeast, about 6 μm in length (Fig. 27-8a). When grown at 37°C on rich medium, it forms smooth, creamy colonies of budding yeasts. However, under less favorable conditions such as a less nutritious medium at room temperature pseudohyphae are formed along with the budding cells. Pseudohyphae consist of very elongated budding cells in chains,

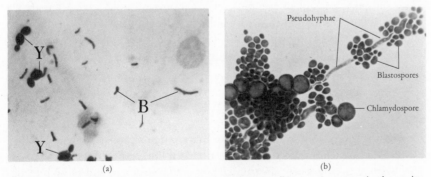

FIGURE 27-8 *Candida albicans* (a) The morphology of *C. albicans* as it appears in the genitourinary tract. The budding years (Y) are large (6 μm) and readily distinguished from bacteria (B). An epithelial cell from the genitourinary tract is seen in the upper right-hand corner (b) *C. albicans* grown in culture. (Courtesy of S. Eng)

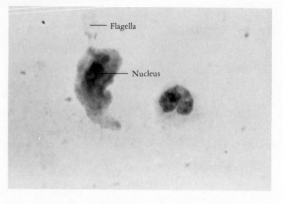

FIGURE 27-9
Flagellate of the genus *Trichomonas* in gram-stained material from the human genitourinary tract. The protozoan is on the left and a leucocyte on the right. Identification of *Trichomonas* sp. is easier in a fresh specimen than in a stained smear because the living organisms have a characteristic motility. (Courtesy of S. Eng)

resembling hyphae in a mycelium. Frequently round, thickened spores (chlamydospores) form at the end of pseudohyphae of *C. albicans* (Fig. 27-8b). These morphological characteristics help to distinguish *Candida* from other yeasts. The carbohydrate fermentation pattern may also be used for identification of *C. albicans*.

Certain protozoa of the genus Trichomonas, often members of the normal flora of the genitalia, may invade the genitourinary tract to cause disease. They frequently cause a mild infection of the vagina in females, and less often an infection of the prostate in males, or in the bladder in either sex. These flagellate, unicellular protozoa are easily identified by microscopic examination of the urine or infected exudate (Fig. 27-9).

A viral infection, genital herpes, is of particular interest because of its possible, or suspected, relation to cancer of the genital tract. It is caused by herpes virus, type 2, which is transmitted by venereal contact. (The virus closely resembles herpes virus, type 1, the cause of herpes simplex, commonly known as fever blisters or cold sores.) Serological studies have shown that women with carcinoma of the cervix are much more likely than normal women to have serum antibodies against herpes virus, type 2. This finding suggests that the virus could be involved as a predisposing factor or as a cause of carcinoma of the cervix. Alternatively, it is possible that some other factor predisposes to both carcinoma of the cervix and genital herpes. At present extensive investigations are under way to attempt to clarify the relationship of herpes virus, type 2, to human cancer.

SUMMARY

Urinary tract infections, much more common in females than in males, are usually caused by members of the normal intestinal flora, especially *Escherichia coli*. The bacteria invade when normal defenses are interrupted, and cause infection of the bladder or kidney or both. As a rule infections occur first in the bladder and reach the kidneys by ascending through the ureters.

Historically important, but uncommon in the United States at present, is puerperal fever (childbirth fever), often caused by *Streptococcus pyogenes*. Among bacterial infections of the genital tract that are extremely common at present are some of the venereal diseases. Gonorrhea results from infection with *Neisseria gonorrhoeae*, and syphilis is caused by the spirochete, *Treponema pallidum*.

Nonbacterial infections of the genitourinary tract occur frequently. Fungi such as *Candida albicans*, protozoa such as *Trichomonas vaginalis*, and viruses such as herpes virus, type 2, are common causes of disease in the lower genital tract.

QUESTIONS

1. Describe the normal flora of the genitourinary tract.
2. Why are gram-negative rods and enterococci the most common causes of urinary tract infection?
3. Gonorrhea can be treated effectively with antibiotics and yet this disease is epidemic and increasing in incidence. Why?
4. Check the appropriate blanks below:
   Syphilis may be transmitted by the following:

	Primary syphilis	Secondary syphilis	Tertiary syphilis
Sexual intercourse	_____	_____	_____
Kissing	_____	_____	_____
Blood transfusions	_____	_____	_____
Mother to fetus across the placenta	_____	_____	_____

5. What are some possible reasons that there is no effective vaccine against syphilis?

FURTHER READING

MUDD, S. (ed.), *Infectious Agents and Host Reactions.* Philadelphia: W. B. Saunders Co., 1970.

ROSEBURY, T., *Microbes and Morals.* New York: Viking Press, 1971.

WILSON, J. W. and O. A. PLUNKETT, *The Fungous Diseases of Man.* Berkeley: University of California Press, 1967.

Chapter 28

Infections of the Nervous System

Infections of the nervous system are relatively uncommon, but are apt to be devastating when they occur. This is because they strike at our ability to move, to feel, or to think. Nervous system infections may be caused by bacteria, viruses, fungi, and even some protozoa.

ANATOMIC AND PHYSIOLOGIC CONSIDERATIONS

The brain and spinal cord make up the central nervous system and are enclosed in bony cavities, the brain in the skull, and the spinal cord inside the spinal canal (Fig. 28-1). The network of nerves throughout the body is connected with the central nervous system by larger nerve trunks which penetrate the cavities. These larger nerves can therefore be damaged by infections of certain bones. Two kinds of nerves are involved in the network: motor nerves which activate different parts of the body and sensory nerves which transmit feelings, such as heat, pain, light, and sound. All the nerves are made up of special cells that often have very long extensions that transmit electrical impulses. These long extensions can sometimes regenerate if damaged, but there can be no repair if the nerve cell itself is killed, as may occur, for example, in poliomyelitis.

CEREBROSPINAL FLUID

Deep inside the brain are several cavities filled with a clear fluid called *cerebrospinal fluid*. It is continually produced in the cavities and then flows out through small openings to spread over the surface of the brain and spinal cord and out along the nerve trunks which penetrate the bony cavities. This fluid flows into lymph channels and eventually enters the bloodstream. In persons with sus-

502

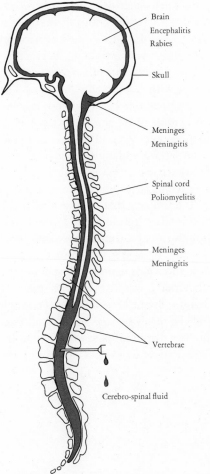

Brain
Encephalitis
Rabies

Skull

Meninges
Meningitis

Spinal cord
Poliomyelitis

Meninges
Meningitis

Vertebrae

Cerebro-spinal fluid

FIGURE 28-1
The central nervous system.

pected infections of the central nervous system, a needle can be safely inserted between the spinal vertebrae in the small of the back, and a sample of cerebrospinal fluid can be removed from the spinal canal for examination (Fig. 28-1). In this manner the causative agent can often be found and identified and appropriate treatment administered.

Covering the surface of the brain and spinal cord are two membranes called *meninges,* and it is between these that the spinal fluid flows. Inflammation of these membranes is called *meningitis.* Obviously both blood vessels and nerves must pass through both these membranes and the film of cerebrospinal fluid. Meningitis may therefore involve these vessels and nerves and cause their malfunction.

The central nervous system is in a well-protected environment, and infectious agents cannot readily get to it. When they do, it seems that it must have happened almost by accident, since almost all microbes that infect the nervous system infect other parts of the body vastly more frequently. The routes by which microbes reach the brain and spinal cord are as follows:

MENINGES

1. *The bloodstream.* This is probably the chief source of central nervous system infections, but indeed it is very difficult for infectious agents to cross from the bloodstream to the brain. This block to passage of harmful agents is often called the "blood-brain barrier." Although its exact nature is unknown, it is effective in preventing microbes from entering nervous tissue in all but a few of those with bloodstream infections. The reasons for these few failures of the blood-brain barrier are not known, but probably relate to the concentration of organisms in the bloodstream and the length of time they circulate. Veins in the face may connect directly with those on the brain surface, and upper facial infections may therefore sometimes spread to the brain.

2. *Entrance by means of the nerves.* Some agents penetrate the central nervous system by traveling up the nerves. It is unlikely that they pass inside the processes of the nerve cells themselves, but instead move through passages between the cells which surround and protect the nerve cell processes. In some instances the agent grows in these surrounding cells.

3. *Direct extension of infections elsewhere.* Infections in bone surrounding the central nervous system may rarely erode inward to reach the brain or spinal cord. They must penetrate not only the bone but also the tough outer membranes which surround the meninges and bone. Skull fractures may produce nonhealing injuries that predispose to infection. Infections may also extend from the respiratory sinuses or, more commonly, from the middle ear.

SOME BACTERIAL INFECTIONS OF THE NERVOUS SYSTEM

Meningococcal Meningitis

Meningococcal meningitis is often called epidemic meningitis although it typically appears at widely separated locations throughout the year. With the usual case there are symptoms of a mild cold, followed by sudden onset of severe throbbing headache and fever, and marked pain and stiffness of the neck and back. Purplish spots may appear on the skin. The sick person may develop shock and die within 24 hours of feeling and appearing completely well.

Neisseria
Page 256

The causative organism, *Neisseria meningitidis* (meningococcus), can usually be found on smears made from cerebrospinal fluid or scrapings of the purple skin spots. They are gram-negative cocci occurring in pairs (diplococci), each member of which has a flattened side where it opposes its neighbor. Most of the organisms seen in cerebrospinal fluid are within white blood cells which enter the fluid in response to the infection.

The organisms readily grow on medium enriched with blood, especially if incubated in an atmosphere with increased carbon dioxide concentration. The meningococcus, like other *Neisseria* species, gives a positive oxidase test, a portion of a colony immediately producing a color reaction when mixed with a reagent to detect cytochrome C oxidase activity. The fermentation pattern resulting when the organism grows in different sugars helps separate it from *N. gonorrhoeae* and other *Neisseria* species which may rarely cause meningitis. Some meningococcal strains are quickly killed by moderate chilling, and this may account for some failures to recover the organism in cultures of material from patients.

For epidemiological purposes meningococci can be further characterized by identifying their surface antigens using specific antisera produced in animals. On this basis *N. meningitidis* is divided into several types designated by capital letters. Most infections are caused by types A, B, or C. Even strains within each type are heterogeneous and bacteriocin typing may be useful in studying the ecological relationships of a single strain. Bacteria actually have a remarkable degree of individuality when carefully studied.

Under natural conditions *N. meningitidis* is strictly a human parasite, and infection is virtually always transmitted by exposure to a person carrying this species. The vast majority of infections pass unnoticed and are only detected by the chance recovery on culture of a heavy growth of the organisms from the throat of an infected person. If a selective medium is used to suppress the growth of other throat flora, meningococci can often be recovered in about 15 percent of normal persons, and in closely crowded conditions, such as military camps, the organisms probably spread to almost all persons present. Even so the rates of cerebrospinal meningitis remain low. Figure 28-2 shows how susceptibility decreases dramatically with age. The rise in incidence in the age range of 15 to 25 years is due to cases in the armed forces. Factors such as fatigue and crowding that are present during military training apparently overcome the natural resistances developing with age.

A most fascinating shift in the interrelationships between man and microbes is now being documented with the meningococcus. Until recent years *N. meningitidis* type A was responsible for most outbreaks of cerebrospinal meningitis, while types B and C generally caused the isolated cases appearing from time to time. After a peak number of cases (5077) reported in 1953, there was a fall in incidence until about 1962 when the rates began to climb again, and with the rise, type B strains began to appear in increasing numbers. By the early part of 1965 almost all the reported cases were due to type B and a few type C organisms. But a further shift then began to appear, so that by 1970 more than two-thirds of all the reported cases were due to type C organisms! Moreover, before 1953 virtually all meningococci were exquisitely susceptible to sulfa drugs. Coincidentally with the rise in types B and C, there appeared a dramatic rise in resistance to these drugs. By the first half of 1970 more than two-thirds of the strains were resistant to sulfa drugs. Factors which may have played a role in these shifts include relative immunity of the population to type A strains developed during previous epidemics, widespread use of sulfa drugs which resulted in the selection of rare sulfa-resistant strains known to be present even before the introduction of sulfa drugs, crowding of susceptible populations, and genetic interchange among different strains. With regard to the latter, transforming DNA has been

N. MENINGITIDIS
TYPES

Bacteriocin typing
Figure 22-2, Page 416

Transformation
Pages 194–197

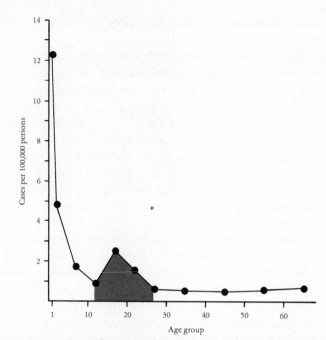

FIGURE 28-2   Meningococcal disease rates by age group in the United States in 1969. The rise in incidence indicated by the colored area is due to cases in the armed forces. (Morbidity and Mortality Weekly Report, 24 October 1970, p. 415)

demonstrated in cultures of *N. meningitidis* and could possibly play a role in nature by transferring genetic information among different strains of meningococci.

The mechanisms by which meningococci produce cerebrospinal meningitis have been the subject of considerable study. The organisms probably infect when the victim inhales droplets from another person, almost always a carrier, or one with an unnoticed infection. They establish an infection in the upper respiratory tract and from this site are seeded into the bloodstream. The blood carries the organisms to the meninges and spinal fluid where they multiply faster than they can be engulfed and destroyed by the white blood cells. The inflammatory response, with the formation of pus and clots, may obstruct the normal outflow of cerebrospinal fluid, so that the brain is squeezed flat against the skull from internal pressure. Infection of brain tissue, and later scar formation, may damage motor nerves and produce paralysis. Moreover *N. meningitidis* circulating in the bloodstream injures the blood vessels so that they cannot maintain normal blood pressure, and shock results. The smaller blood vessels of the skin are affected (small hemorrhages producing the purple spots). Vital organs are also often damaged.

Blood vessel damage very likely results partly from endotoxin. Not only do endotoxins cause shock but they also produce fever and heightened sensitivity to further exposure to endotoxin. This increased sensitivity, called the *Shwartzman phenomenon,* can be demonstrated in animals by using meningococci. It may

Endotoxin
Page 365

well be responsible for the damage to the small blood vessels seen in people. The tendency of meningococci to autolyse (spontaneously rupture) may explain the release of endotoxin.

Available evidence indicates that type-specific opsonins can protect against meningococcal disease. In fact vaccines have been prepared against types A and C which protect animals from experimental infections by *N. meningitidis*. These are currently being used to immunize certain high risk human populations in an effort to control meningitis.

Other Bacterial Causes of Meningitis

Many other bacteria may cause infection of the meninges, and in most cases, just as with *N. meningitidis,* the organisms are relatively commonly carried by healthy people, are transmitted by inhalation, and only rarely produce meningitis. Organisms in this category include the respiratory pathogens *Streptococcus pneumoniae* and *Haemophilus influenzae.* On the other hand, these common organisms of the respiratory tract seldom cause meningitis in newborn babies, presumably because antibody against them is transferred across the placenta from the mother. Instead gram-negative rods such as *Escherichia coli* from the mother's intestinal tract, or even *Flavobacterium meningosepticum* from the nursery water faucet, are more likely to cause meningitis in newborn babies. The current view is that the newborn lacks protection against gram-negative rods such as these because the protective antibodies are bactericidal IgM antibodies that are unable to pass across the placenta. Other causes of meningitis of newborn babies are beta hemolytic streptococci of serological group B, and *Listeria monocytogenes* (a small gram-positive rod that resembles diphtheroids but unlike them, is motile). These two species of bacteria are carried among the vaginal flora of some women, but why they occasionally cause infections in babies is unexplained.

S. pneumoniae
Page 455
H. influenzae
Pages 114, 253, 443

IgM
Table 20-3, Page 372

LEPROSY, A BACTERIAL DISEASE OF THE PERIPHERAL NERVES

A few years ago a 40-year-old Costa Rican domestic worker living in New Jersey consulted a doctor because of some thickened plaquelike areas on her face, earlobes, and extremities. She also complained of loss of sensation over areas of her arm. She had seen several doctors concerning this problem over the previous seven years, but no definite diagnosis or treatment had been made. This time, however, the doctor cut a small piece (biopsy) of skin from one of the plaques and stained a very thin slice of it for acid-fast bacteria. On inspection under the microscope, acid-fast rodlike bacteria were seen, and the diagnosis of leprosy was established. Her story is similar to cases encountered every year in the United States (Fig. 28-3).

Leprosy was formerly common in Europe and America, and in 1872 Hansen of Norway demonstrated the causative bacteria in the tissues of leprosy patients.

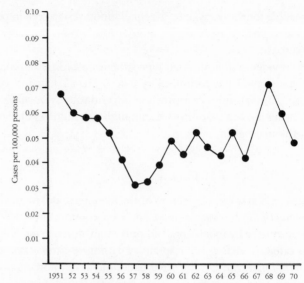

FIGURE 28-3 Incidence of leprosy in the United States from 1951 to 1970. One hundred twenty-nine new cases were reported in 1970. (Morbidity and Mortality Weekly Report Annual Supplement-Summary, 1970, United States Public Health Service)

This is said to be the first time a bacterium was causally linked to a human disease. Like tuberculosis, leprosy began to decrease in Europe and America for unknown reasons, and today it is chiefly a disease of warmer and economically under-developed countries, with an estimated (1967) 7 million victims in tropical Africa and the Indian subcontinent alone. It has in many respects been a baffling disease, and its causative organism has yet to be convincingly grown on artificial medium.

Because of the frustrations of trying to grow the causative organism *in vitro,* there have been many attempts to infect laboratory animals. A tremendous ad-vance was made in 1960 when it was demonstrated that the organisms would multiply in the footpads of mice. It has been found more recently that the disease in mice may slowly progress, so that after two years, it closely resembles human leprosy.

Mycobacterium
Page 259

From studies of infected mice it was possible to find some interesting infor-mation about the causative organism of leprosy, *Mycobacterium leprae.* One of the more striking findings was that these bacteria have a very long generation time estimated to be about twelve days! This finding helps account for the very long incubation period of human disease, probably a minimum of two years and often ten years or longer. In addition, growth in the mice permitted studies aimed at finding new medicines for the treatment of leprosy in human beings.

The earliest detectable finding in infection with *M. leprae* is invasion of the small nerves of the skin. Indeed this is the only known human pathogen which preferentially attacks the peripheral nerves. Thereafter the course of the infection is thought to depend on the immune response of the host. In most cases cell mediated immunity and delayed hypersensitivity to the invading bacteria develop. The bacteria remain very scanty, but nerve damage may progress and lead to disabling deformity. In the majority of instances, however, the disease arrests itself after a few years and the nerve damage, although permanent, does not progress further. This kind of leprosy is called the *tuberculoid* type. Such persons rarely, if ever, transmit the disease to others.

In other infected persons cellular immunity and delayed hypersensitivity either fail to develop, or are lost, and unrestricted growth (rather than destruction) of *M. leprae* occurs in macrophages. This relatively uncommon form of leprosy is referred to as the *lepromatous* type. In addition to the nerves, most tissues of the body become involved. The tissues and mucous membranes swarm with the leprosy bacteria which can be transmitted to others by intimate direct contact. A similar disease can be produced in mice by removing their thymus and giving them X-ray treatment or antibodies against their lymphocytes to destroy their capability for cellular immunity. With lepromatous leprosy, there may be a general impairment of delayed hypersensitivity responses, not only to leprosy bacilli, but also to many other antigens. Even administration of normal lymphocytes often fails to restore delayed hypersensitivity and cellular immunity in these patients. On the other hand, a return toward normal immune function occurs when the disease is controlled by medications.

It is now widely known that leprosy is one of the least contagious of all infectious diseases, and the morbid days when lepers were forced to carry a bell or horn to warn others of their presence are fortunately past. Today persons with leprosy can expect a normal lifespan provided ill-grounded fears and social pressures do not cause them to avoid early medical consultation. In 1968 there were 78 new cases of lepromatous leprosy diagnosed in the United States and Puerto Rico, most contracted during residence in other parts of the world. The risk to persons in the armed services abroad appears to be small, perhaps 30 cases having been detected between 1940 and 1970. However, one of these, living in a northern state, is known to have transmitted the disease to his wife and two of his children. In Vietnam and neighboring countries the incidence of leprosy among native people has been estimated at about 5 per 1000. It will take a number of years before the impact of the Southeast Asian war on leprosy incidence in the United States and other countries is certain. In the past the average time between exposure and discovery of the disease in armed forces veterans has been about ten years. The disease is detected by microscopic inspection of scrapings or biopsies of infected tissues. Arrest of the disease can usually be achieved by long-term treatment with sulfones.

SOME VIRAL DISEASES OF THE NERVOUS SYSTEM

Poliomyelitis (Infantile Paralysis)

A person developing poliomyelitis usually first suffers the symptoms of meningitis: headache, fever, stiff neck, and nausea. In addition, pain and spasm of some

Cellular immunity
Page 386
Delayed hypersensitivity
Page 398

Sulfones
Page 566

muscles generally occur, later followed by paralysis, and finally a relative shrinking of muscle and failure of normal bone development in the affected area. In the more severe cases the muscles controlling respiration are involved and the person requires a machine to pump air in and out of his lungs. Some recovery of function is the rule if the life of the person can be maintained. Sensation is not affected.

The infectious agent, poliovirus, can be recovered from the throat and feces early in the development of the illness, later in the blood, and finally, in some patients, in the cerebrospinal fluid. Excretion in the feces continues for weeks or months, but is transitory in the other sites. Introduced into tissue cell cultures (usually consisting of monkey kidney cells) the virus produces a cytopathic effect which can readily be seen under the low power of the microscope. Poliomyelitis is generally caused by one of three types of poliovirus, but occasionally may be caused by other enteroviruses. The three polioviruses are designated by Roman numerals I, II, and III, and have only slight antigenic relationship to one another. The antiserum which prevents cytopathic effect identifies the virus.

Virus neutralization
Page 384

Picornaviruses
Page 678

Polioviruses, like other enteroviruses, are small, RNA-containing viruses which are quite stable under natural conditions, but are inactivated by pasteurization and proper chlorination of drinking water. Polioviruses can infect a cell under natural conditions only if that cell has certain receptors on its surface to which the virus can attach. This helps explain the remarkable selectivity of these viruses for motor nerve cells of the brain and spinal cord, whereas most other kinds of nerve cells are spared. Destruction of the infected cell occurs upon release of the mature virus.

Up until the late 1950s poliomyelitis was a terrifying threat with a greater impact on economically advanced countries. Fear of the disease was widespread, and hundreds of thousands of people in the United States contributed dimes to the National Foundation for Infantile Paralysis to help victims and to support research to solve this problem. President Roosevelt, himself a polio victim, helped dramatize the need for this support.

Efforts in studying the disease were thwarted because of its very limited host range (only humans and a few other primates). Therefore one of the most important developments in modern medical microbiology was the demonstration (1949) by Dr. John Enders and his colleagues at Harvard University that polioviruses grow readily in tissue cultures, producing clear-cut cytopathic effects. Details of pathogenesis and ecological relationships of the viruses were then worked out. By 1954 Dr. Jonas Salk had perfected a formalin-inactivated virus vaccine which was widely employed and soon dramatically lowered the incidence of paralytic poliomyelitis. Meanwhile tissue culture techniques allowed the selection of mutant poliovirus strains which lacked ability to attack nervous tissue, but could nevertheless stimulate antibody formation against wild-type virulent virus. Living vaccines consisting of such attenuated viruses were thus quickly developed, and those prepared by Dr. Albert Sabin were chosen for use. As a result of these historic research developments, we now generally have less than 50 cases a year of paralytic poliomyelitis in the United States as opposed to the many thousands in earlier years.

Reference was made to the seemingly greater impact of poliomyelitis in economically advanced countries. Poliomyelitis occurs in poorer crowded countries, but epidemics of paralytic disease generally appear to affect small groups of geographically or culturally isolated people in such countries. Among the remaining population the virus is widespread, and very few escape childhood without infection. Therefore mothers have antibody in their bloodstream and some of this antibody crosses the placenta and enters the baby's circulation. Babies are thus partially protected against bloodstream infection and nervous invasion for as long as the mother's antibody lasts—usually about two or three months. During this time, because of widespread exposure to polioviruses through crowding and unsanitary conditions, the infant is likely to develop a mild infection of his throat and intestine, and thereby achieve life-long immunity. Even if exposure is delayed beyond two or three months, it is likely to occur in the earlier years when the disease is less likely to be severe.

By contrast in areas such as a suburban American community the virus sometimes cannot spread fast enough to sustain itself, and may die out. If reintroduction from another community is delayed sufficiently, then some people of all ages may lack antibody and be susceptible, including older children and adults. When the virus is finally reintroduced, a high incidence of paralysis results.

We have only limited information so far on the duration of immunity following vaccination. If it is only a few years, there is a very real danger of a large buildup of an adult population with susceptibility to polioviruses. The Sabin vaccine virus is excreted and hence transmissible to others. However, there is disagreement as to whether ongoing programs of vaccinating children with Sabin vaccine in areas with good sanitation will allow adequate spread of the attenuated virus to adults to maintain their immunity. So far the intensive use of living vaccine given by mouth has produced dramatic reductions in the incidence of paralytic poliomyelitis in several countries. It will be interesting to see whether the wild-type (virulent) virus can gradually be displaced from human beings and whether the man-chosen poliovirus (vaccine) mutants can maintain themselves in nature.

The production and delivery of adequate vaccine require the most highly sophisticated technology. A susceptible adult population could become highly dependent on this technology for protection from poliomyelitis and other diseases. On the other hand, two incidents in the development of inactivated poliomyelitis vaccine point out dangers which can happen despite the extreme caution with which mass administration of biological agents is treated. In the first episode, formalin inactivation was inadequate and vaccine recipients received some living wild virulent virus which caused cases of paralysis. The greatest care was used in preparation of the vaccine, but a change in the nature of the viral suspension produced an unexpected delay in inactivation by formalin. The second finding was that the poliovirus used for the vaccine was unknowingly contaminated by another virus, called simian virus 40 (SV_{40}), growing in the monkey tissue cultures. As it happened this virus withstood formalin exposures which killed the polioviruses, so that many people received the living monkey virus.

SV_{40}
Page 352

Although SV$_{40}$ has been shown to induce tumors in laboratory animals (Chap. 18), many years have now gone by with no ill effects appearing in the vaccine recipients. This episode, however, demonstrated that the risks of large-scale tamperings with our species cannot always be adequately foreseen.

Rabies

Rabies, or hydrophobia, is one of the most feared of all human diseases because it is agonizing in its manifestations and almost always fatal. It is an acute infectious disease involving the central nervous system and is usually transmitted to warm-blooded animals by the saliva of biting animals infected with rabies virus. In human beings the disease usually develops after an incubation period of 30 to 60 days, although the range is from ten days to more than a year. A prominent feature of the illness is spasms of the muscles of the mouth and throat at the sight (or even the thought) of water, and thus the popular name "hydrophobia."

Identification of the causative agent is usually made most quickly by preparing smears of the brain of the dead person or animal. These smears are stained with a combination of dyes called Seller's stain, or with fluorescent antirabies antibody and inspected for nerve cells containing viral inclusions (*Negri bodies*). Rabies virus can be cultivated in animals, such as laboratory mice, or in tissue culture. As mentioned elsewhere the virus is large and of an unusual "bullet" shape. Its relationship with other viruses is not yet settled, although it is known to be lipid-containing with single-stranded RNA and it buds from the surface of infected cells. These features indicate that it is an enveloped virus when naturally transmitted. At present only one antigenic type is known, but differences in virulence and ability to produce Negri bodies exist from isolate to isolate.

Rhabdoviruses
Page 681

Although infection has been demonstrated by both the respiratory and oral routes under research conditions, the principal mode of transmission to man is passage of infected saliva through the skin barrier by way of a bite or other break in the skin. Once through the skin, virus passes slowly by an unknown mechanism along the course of nerves to the central nervous system where it multiplies in some of the motor nerve cells without causing cell destruction. The slow travel via the nerve is thought to account for the very long incubation periods often observed with this disease. As a matter of fact, with laboratory animals, rabies deaths were prevented by cutting the nerve coming from the site of injection of virus. The mode of spread of virus through the nerves and central nervous system is still poorly understood, and the explanation of how virus gets to saliva-producing glands and other tissues (such as fat) is obscure.

Rabies infections are widespread in wild animals and they serve as an enormous reservoir from which human infection can be transmitted (Fig. 28-4). In South America the vampire bat is often infected, and 700 deaths of cattle from this source were reported in 1969 from Brazil alone. Foxes and squirrels are slowly spreading the virus across Europe, and in the United States raccoons, skunks, foxes, and bats constitute the chief reservoir hosts. The vast majority of human cases in the United States have been caused by dog bites, and this is true in most

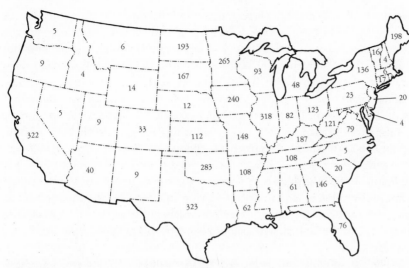

FIGURE 28-4   Number of cases of animal rabies reported by state in 1971. (Morbidity and Mortality Weekly Report, 6 May 1972, p. 155)

areas of the world. The dog population in the United States is about 25 million, and an estimated 1 million people get bitten each year. Fortunately, since World War II the incidence of dog rabies (and human rabies) has dropped dramatically and now almost three-quarters of reported rabies infections are in wild animals. Unfortunately there are incomplete data on the natural history of rabies in wild animals. In dogs anywhere from 10 to 100 percent will die of rabies following injection with wild rabies viruses from different sources. About three-quarters of those that develop rabies excrete virus in their saliva and about one-third of these begin excreting it one to three days before they get sick. Some dogs become irritable and hyperactive with the onset of rabies, salivate copiously, and attack people, animals, and inanimate objects. Perhaps more common is the "dumb" form of rabies, where the dog simply stops eating, becomes inactive, and suffers paralysis of throat and leg muscles. Excretion of virus in the saliva tends to be less frequent in the latter group.

RABIES IN DOGS

    A most dramatic decline in rabies incidence in people has resulted from immunizing dogs and other pets against rabies infection. This in effect places a partial barrier to the spread of virus from wild animal reservoirs to human beings. Since pets vary in their tolerance of rabies vaccines depending on age and species, several kinds of vaccines are available. In recent years these have consisted of both attenuated and inactivated rabies viruses.

    The risk of rabies developing in people bitten by dogs having rabies virus in their saliva may be as high as 30 percent, and with bites by other wild animals, the risk may be higher still. Louis Pasteur discovered that this risk can be lowered considerably by administering rabies vaccine as soon as possible after exposure to the virus. Presumably the effectiveness of this measure is based on accessibility of the rabies virus and susceptibility to inactivation by immune mechanisms at some time during its long sojourn to the motor nerve cells of the brain. By giving

RABIES VACCINE

large doses of vaccine immediately after the bite, sufficient immunity can be developed in most cases in time to prevent infectious virus from reaching the central nervous system. In fact the risk of rabies can thereby be lowered by about 85 to 90 percent. Unfortunately, however, there is a risk of central nervous system disease from the vaccine itself. This arises because the vaccine virus is cultivated in animal brain tissue containing an antigen held in common by both the animal and the human brain. The person receiving vaccine thus may develop an immune response to antigens of his own brain, producing "allergic encephalitis" in about one out of 8000 instances. This risk has been lowered to about one in 25,000 since 1957, with the availability of vaccine cultivated in duck embryos instead of animal brains. In fact the duck embryo vaccine is safe enough to be used routinely to immunize veterinarians and people living in areas of high rabies incidence, before they are bitten by a rabid animal, or by laboratory workers who work with rabies virus. In the United States about 30,000 people get post-exposure vaccine each year.

The risk of rabies can be lowered further still by injecting rabies antibody into the exposed person. This antibody is produced by injecting vaccine into an animal such as a horse. The horse antiserum must be given to the exposed person in large doses, and because it is a foreign protein, he soon develops antibodies against the antiserum. This may result in serum sickness, or anaphylaxis. If antibodies obtained from human volunteers who have been immunized against rabies are available, the dangers of using horse serum are avoided.

Serum sickness
Page 394
Anaphylaxis
Page 395

A synthetic double-stranded RNA (polyriboinosinic-polyribocytidylic acid, or "poly I:C") has been shown effective in preventing rabies in experimentally infected rabbits. This material induces interferon production of host cells without interfering with antibody production. It is possible that a substance such as this will be found that is more effective than rabies antiserum in protecting humans against rabies until sufficient host antibody is developed from administered vaccine.

SUMMARY

In comparison with other tissues the brain and spinal cord are relatively protected from infections, and central nervous system involvement generally occurs in only a fraction of the total infections produced by a microbial agent. In many cases examination of cerebrospinal fluid is helpful in recovery and identification of the infecting microbe. The respiratory pathogens, *Neisseria meningitidis, Haemophilus influenzae,* and *Streptococcus pneumoniae* are the most frequent causes of bacterial meningitis, although in newborn babies the cause is more likely to be *Escherichia coli* or even a gram-negative rod from the environment. *Mycobacterium leprae,* the cause of leprosy, is the only pathogen which preferentially attacks peripheral nerves. The disease exists in two main forms, tuberculoid and lepromatous, the latter associated with defective delayed hypersensitivity. Poliomyelitis and rabies are viral diseases of the nervous system. The former is generally caused by three enteroviruses, designated poliovirus I, II, and III. The viruses preferentially attack the cells of motor nerves. Control of the disease has been dramatically achieved by development of vaccines. A huge reservoir of rabies virus exists in wild animals, but the main source of human infections is the domestic dog. Human rabies control rests largely on vaccination of dogs against rabies virus. Routine vaccination of human beings is avoided because of a small danger of developing allergic encephalitis. Human vaccination is justified for persons with possible exposure to the virus.

QUESTIONS

1. The continuing interaction between *Neisseria meningitidis* and man has been accompanied by shifting patterns in serological types responsible for meningococcal disease. Discuss possible explanations for these changes.
2. Explain why patients sometimes die from meningococcal infections despite intensive treatment with penicillin, a medication which is highly effective in killing the organism.
3. What defect in host defenses is associated with the development of lepromatous (as opposed to tuberculoid) leprosy?
4. Explain the apparent paradox of a higher incidence of paralytic poliomyelitis in countries with good standards of sanitation.
5. Give reasons why only a few cases of rabies occur in human beings in the United States despite the widespread distribution of rabies in animals.

FURTHER READING

BRUBAKER, M. L., C. H. BINFORD and J. R. TRAUTMAN, "Occurrence of Leprosy in U.S. Veterans After Service in Endemic Areas Abroad," *Public Health Reports,* 84 (December 1969), 1051–1058.

MARTIN, BETTY, *Miracle at Carville.* New York: Doubleday and Co., 1950.

MELNICK, J., "Enteroviruses," *Scientific American,* 200 (February 1959), 89.

ROUECHE, BERTON, "A Lonely Road," in *Eleven Blue Men and Other Narratives of Medical Detection.* New York: Berkeley Medallion, 1965.

————, "The Incurable Wound," in *Annals of Epidemiology.* Boston: Little, Brown and Co., 1967.

Chapter 29

Wound Infections

Most infections can be compared to the invasion of Troy by the Greeks with their wooden horse, in that microorganisms use devious methods to overcome host defenses and invade. Infections of wounds, however, are perhaps better compared to the capture of a city where the fortifications have been broken down, giving invaders relatively free access.

Whether a wound is caused accidentally by trauma or intentionally by surgery, the open area is extremely susceptible to microbial invasion. Once a wound has become infected, pus often forms in the injured area resulting in a wound abscess. An abscess is a collection of pus (including white blood cells and any infecting organisms that may be present) which often becomes surrounded by clotted blood, separating the infected area from normal tissue. Consequently abscess formation tends to localize an infection and help prevent its spread.

ORGANISMS OFTEN RESPONSIBLE FOR WOUND INFECTIONS

Virtually any kind of pathogenic or opportunistic organism can cause wound infections under appropriate conditions, but certain types of wounds are subject to infection by particular organisms. Staphylococci are by far the most frequent cause of wound infections. Aerobic or facultative bacteria also common in wounds include streptococci, enterobacteria, and pseudomonads. Other microorganisms that less frequently infect wounds include a variety of fungi that invade wounds where conditions favor their growth and some animal pathogens that are transmitted by animal bites or scratches.

Anaerobic microorganisms are responsible for about 10 percent of wound infections. Therefore, it is necessary to use special culture media and methods for obtaining anaerobic conditions, along with the usual aerobic techniques, when examining material from wounds. One reason for the relatively frequent occurrence of anaerobe-induced wound infections is that anaerobic conditions often develop in tissues following wound injury; thus, anaerobic microorganisms may find an excellent environment for growth there. Puncture wounds caused by contaminated nails, thorns, splinters, and other sharp objects introduce foreign material and microorganisms deep into the body. Bullets and other projectiles that enter at high velocity can carry into the tissues with them contaminated fragments of skin or cloth. Because of the force with which they enter, they cause relatively small breaks in the skin which may close quickly, masking areas of extensive tissue damage.

Another major reason why anaerobes often infect wounds is that they are present in tremendous numbers among the normal flora and therefore are readily available to invade wounded areas. In fact, all of the anaerobic bacteria considered in this chapter as causes of wound infections are included among the normal flora of man, and may be the dominant organisms numerically.

AEROBIC INFECTIONS
OF SURGICAL WOUNDS

Even when the most careful precautions are taken during surgery, the surgical wound may become infected. *Staphylococcus aureus* is the most common cause of infections of surgical wounds as well as of those resulting from accidental trauma. The presence of sutures in surgical wounds is an important predisposing factor. Experiments showed that hundreds of thousands of *S. aureus* could be injected under the skin without causing serious damage in a normal individual; however, less than a hundred of the same organisms caused abscess formation when introduced on a suture. Tissue reactions to sutures and to other foreign bodies strongly favor the establishment of staphylococcal infection.

Staphylococcus aureus
Pages 430–432

Many strains of *S. aureus* produce penicillinase; therefore penicillinase-resistant penicillins (such as methicillin) are generally used to treat staphylococcal infections. Some strains with resistance to methicillin have emerged, but infections with these strains can be successfully treated with other antibiotics.

Streptococcus pyogenes and enterobacteria are also common causes of surgical wound infections (Table 29-1).

Streptococcus pyogenes
Pages 433–434
Enterobacteria
Pages 252–253

INFECTIONS OF BURNS

Burned areas, with their damaged skin, offer ideal sites for infections by bacteria of the environment or of the normal flora. Almost any potential pathogen can infect burns, but one of the most prevalent and hardest to treat is *Pseudomonas*

TABLE 29-1
Important Causes of Wound Infections

Type of Wound	Causative Agent
Wound resulting from accidental trauma	*Staphylococcus aureus*
	Streptococcus pyogenes
	Clostridium sp.
	Sporotrichum schenkii
Surgical	*S. aureus*
	S. pyogenes
	Pseudomonas aeruginosa
	Enterobacteria
Burn	*P. aeruginosa*
	S. aureus
	S. pyogenes
	Enterobacteria
Animal-inflicted	Rabies virus
	Pasteurella multocida

aeruginosa. This gram-negative rod forms a bluegreen pigment that diffuses into the medium, and that may give a bluegreen color to pus. The colonies are oxidase-positive. *Pseudomonas aeruginosa* is not fastidious, and grows in many unlikly media, including some disinfectant solutions. Pseudomonas is frequently found among the normal intestinal flora, and as an opportunist in urinary tract infections. It is especially dreaded because of its resistance to antibiotics. In fact, infections with pseudomonas are a frequent cause of death in burn patients; however, a new and promising method has recently been introduced for combatting these infections. Although more evidence is needed as to its efficiency, it offers hope that the death rate from infected burns may soon be lowered. Briefly, the method involves immunization of badly burned patients with killed pseudomonas vaccines within the first few days after injury and at intervals during the first two weeks thereafter. Antibodies against the bacteria are produced, usually before the wounds have time to become infected, and the antibodies protect against pseudomonas invasion.

Pseudomonas aeruginosa is also of great importance as a cause of other kinds of infections, including those of surgical wounds. Treatment with antibiotics to control other invading bacteria predisposes to infection with antibiotic-resistant pseudomonads.

ANAEROBIC WOUND INFECTIONS CAUSED BY CLOSTRIDIA

Tetanus (Lockjaw)

One of the most dangerous diseases resulting from anaerobic infection of wounds is tetanus, caused solely by the action of a powerful exotoxin produced by *Clostridium tetani.* Even though this disease is easily prevented by immunization with tetanus toxoid, hundreds of people in the United States contract the disease and many die from it each year. For example, during 1965 and 1966, 507 cases of tetanus were reported to the Public Health Service, of which 345 were fatal. It was found that 97 percent of these people who developed tetanus had never been immunized with toxoid. Adequate immunization, a very simple procedure, would have prevented the disease.

On a global basis, tetanus occurs much more frequently than in the United States. In many parts of the world, it is a common cause of death in newborn babies, as a result of cutting the umbilical cord with instruments contaminated with *Cl. tetani*.

Although tetanus was accurately described by Hippocrates in the fourth century B.C., it was not until the late 1800's that the causative organism was identified and its toxin discovered. It was recognized that tetanus bacteria do not invade past the wound area, and that the wide-spread symptoms of tetanus result from the action of an exotoxin which is absorbed from the infected wound. At that time, the disease was studied experimentally in animals which were wounded and infected with *Cl. tetani* by rubbing soil into the wounds. This was effective in producing the infection because the tetanus organisms are widely distributed in soil. These bacteria are also commonly found in the gastrointestinal tracts of man and other animals; hence, manured soil is apt to be especially rich in *Cl. tetani*.

Clostridium tetani is a strictly anaerobic, motile, gram-positive, spore-forming rod. The organisms may rapidly become gram-negative, especially if they are exposed to air or placed in other unfavorable conditions. Nevertheless, the presence of characteristic terminal round spores (Fig. 29-1) is presumptive evidence that an anaerobic organism is the tetanus bacillus. The spores are extremely resistant to unfavorable conditions. They survive heating at 100°C for 30 minutes, and it is necessary to autoclave material at 121°C for 15 or 20 minutes in order to assure that all tetanus spores have been killed. This property permits the isolation of *Cl. tetani* from a mixture of bacteria by heating to 80°C for 10 minutes to kill vegetative bacteria. Tetanus spores will survive this heat, and can subsequently germinate and grow when placed in favorable conditions.

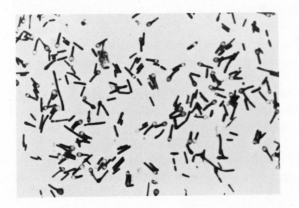

FIGURE 29-1
The terminal endospores of *Clostridium tetani*. (Courtesy of CCM: General Biological, Inc.)

Cultures of *Cl. tetani,* incubated under strictly anaerobic conditions, will show visible growth in 4 days to one week. The motile bacteria tend to spread and grow on solid media as a thin, translucent surface film which is easily overlooked by an inexperienced microbiologist. Often, it is necessary to inject cultures into mice to determine whether the lethal tetanus toxin is being produced. Mice protected with specific antitoxin will survive, while unprotected mice are killed by the toxin.

The isolation of *Cl. tetani* from wounds is not definite proof that a person has tetanus; conversely, failure to find the organisms does not eliminate the possibility of tetanus. The reasons for these observations are that tetanus spores may be isolated from wounds in which the organisms are not producing toxin; also, spores may remain dormant in the tissues for long periods of time, germinating and producing toxin after the wound has healed. In a recent study in the United States, it was found that the organisms could not be isolated from 7 percent of the cases studied, even though the symptoms of tetanus were undeniable.

The wounds that lead to tetanus may be very small and seemingly insignificant because the toxin is effective in extremely small amounts. The important thing is that the wound must provide the anaerobic conditions necessary for toxin production.

In contrast to the spores, the toxin of *Cl. tetani* is heat-labile, and readily destroyed by heating at 65°C for 5 minutes. It is also destroyed by enzymes of the gastrointestinal tract. Tetanus toxin is one of the most lethal microbial products known. It has been estimated that one milligram of the purified exotoxin, properly administered, could kill over 30 million mice.

Tetanus toxin acts on nerve cells; however, the mechanism of its action is not fully understood. It is a protein (M.W. 67,000), which diffuses away from the infected wound and may enter the circulation; however, it can also travel along the nerves to the spinal cord. In the central nervous system it interferes with control of certain reflex activities, resulting in violent spasmodic contractions of muscles that counteract each other. These contractions can be triggered by any minor stimulus. The muscles of the jaw are particularly susceptible, hence the common name of lockjaw has been given to tetanus.

The incubation time of the disease may be as short as four days after infection, or as long as many weeks if dormant spores germinate in the tissues. Tetanus begins with restlessness, irritability, stiffness of the neck, a tight jaw, and sometimes convulsions, particularly in children. As more muscles tense, the pain grows more severe, breathing is labored, and after a period of almost unbearable pain, the patient often dies of respiratory failure.

Tetanus is treated by administering antitoxin. Since there is only one antigenic type of toxin, only one type of antitoxin is necessary. Antitoxin prepared by immunizing horses used to be given, but it frequently gave rise to the complication of serum sickness. Recently, gamma globulin from human beings immunized with tetanus toxoid has become available, and this is now the treatment of choice. This antitoxin does not cause hypersensitivity reactions and is much more effective than horse antitoxin in man. In addition to antitoxin treatment the wound must be cleansed and all dead tissue or other material that would support the growth of tetanus bacteria removed. Penicillin can prevent the growth of any *Cl. tetani* it reaches, thus preventing the formation of more exotoxin.

Prophylactic immunization with tetanus toxoid is by far the best weapon against tetanus. Three injections of the toxoid are given, usually beginning during the first year of life. A "booster" dose is given after a year, and another when the child enters school. Once immunity has been established by this regime, "booster" doses at about 10-year intervals will maintain an adequate level of protection. More frequent doses of toxoid are not recommended because of the danger of an allergic reaction following intensive immunization. This sort of hypersensitivity reaction can result when antigen (the toxoid) is introduced into an individual who already has large quantities of humoral antibodies against the antigen.

Gas Gangrene

In addition to tetanus, bacteria of the gas-gangrene-producing group of clostridia can cause serious wound infections. This group of gas-forming organisms includes *Cl. perfringens* (also known as *Cl. welchii*) and several other species. At present, gas gangrene is rare except in battlefield situations where wounds cannot be promptly treated; however, it occurs occasionally as a result of illegal abortions and accidents.

NONCLOSTRIDIAL ANAEROBIC INFECTIONS OF WOUNDS

With increasing progress in techniques for achieving anaerobic conditions, it became apparent that many wound infections could be caused by anaerobes other than the clostridia. Although much remains to be learned about such infections, they appear to be caused by certain organisms of the normal flora: namely, the gram-negative non-spore-forming rods of the genera *Bacteroides* and *Fusobacterium*; and gram-positive cocci, principally streptococci.

In addition, certain funguslike bacteria can cause wound infections. Among these is *Actinomyces israelii,* gram-positive, anaerobic, branching bacteria which are occasional members of the normal flora of the upper alimentary and respiratory tracts. Within pus from infected areas these actinomycetes form microcolonies known as "sulfur granules" because of their yellow color. When a "sulfur granule" is examined microscopically, it is seen to consist of filaments of the bacteria arranged in radial fashion around the periphery of each granule (Fig. 29-2). As might be expected, wounds caused by dental procedures sometimes become infected with actinomyces that are part of the normal flora of the mouth.

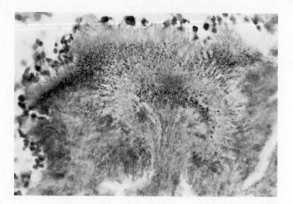

FIGURE 29-2

Actinomyces israelii. These fungus-like bacteria are part of the normal flora of the oral cavity of man, but under unusual circumstances they can invade and cause an infection. This specimen is part of a colony of *A. israelii* in pus from an infected patient. The colony appears as a yellow mass; therefore it is called a "sulfur granule." In this micrograph filaments of bacteria can be seen radiating from the edge of the colony. (Courtesy of CCM: General Biological, Inc.)

ANIMAL BITES

Normal flora and pathogens of animals can be transmitted to man via animal bites. The virus disease, rabies, is one of the most feared infections transmitted in this manner, but less serious bacterial infections also occur. For example, the bites of cats, dogs, and some other mammals are sometimes contaminated with *Pasteurella multocida*. This small gram-negative rod is similar to the causative agent of the plague, *Yersinia pestis,* and also resembles the bacteria that cause tularemia, *Francisella tularensis. Pasteurella multocida* organisms are fastidious, aerobic, and non-spore forming. Inhabitants of the normal flora of mouth and the upper respiratory tract of the natural animal hosts, they can cause severe infection of the natural host when defenses are suppressed or stressed (as by a severe viral infection). In man, infections with *P. multocida* occur following animal bites, and the infections may be severe.

Pasteurella
Page 253

Rat bites can also lead to infections. A fever-producing disease (rat-bite fever), caused by *Spirillum minus,* is transmitted by the bites of contaminated rats. This disease is rare in the United States. More common here (but still not frequent) is the rat-bite fever caused by *Streptobacillus moniliformis*. Whereas *Spirillum minus* has not yet been cultivated in the laboratory on artificial media, *Streptobacillus moniliformis* can be grown on media containing blood. This organism shows unusual variation in morphology when examined in the gram stain. Long filamentous forms with bulbous ends and distortions are common, and many of these unusual shapes and forms resemble the L-forms of other bacteria. Nonetheless, infections with *S. moniliformis* are effectively treated with penicillin, indicating that the infecting form has a cell wall.

Wounds from cat scratches and bites occasionally lead to a disease known commonly as cat-scratch fever. The disease affects lymph nodes draining the area of injury. Typically, it is a mild illness lasting a few weeks to a few months. The causative agent has not been definitely identified, but appears to be a chlamydia. The disease affects man only, and cats which transmit it give no evidence of illness.

FUNGAL INFECTIONS OF WOUNDS

Over the twenty-year period from about 1925 to 1945, over 3,000 cases of an unusual type of wound infection were reported in South African mine workers. Typically, wounds caused by splinters from mine timbers became ulcerated, often without causing pain. After about a week, nodules developed in a chain along the lymphatic vessel draining the wound and these nodules also developed into ulcers. The lymph node draining the area usually became enlarged, and the lesions persisted for long periods of time even though the patients were usually not very ill. These miners were infected with a common fungus, *Sporotrichum schenckii,* which grows especially well on timbers in mines where both the temperature and the humidity are consistently high. It is also common on vegetation, and frequently infects florists, gardeners, and other workers who suffer wounds from contaminated thorns or splinters. Rarely, the disease becomes systemic, but usually it follows the chronic, persistent course described above. The fungus may also spread in the skin, causing scaly, flat plaques or wart-like lesions, but inoculation by a puncture wound more often produces lymphatic involvement.

The causative fungus, *S. schenckii* is dimorphic. In nature, or in cultures incubated at room temperature, it forms a fluffy mold mycelium, tan to brown in color. *In vivo* and *in vitro* at 37°C, it grows as a budding yeast. In the mold form, hyphae are much thinner than those of most fungi and spores are formed along the hair-like hyphae. Hence the name sporotrichum (-trichum = hair-like). The spores (conidia) are oval or rounded, about 2 to 5 μm in length, and occur in clusters on branches (conidiophores) at right angles to the hyphae, producing a flower-like appearance (Fig. 29-3a). The yeast phase at 37°C and in tissues is characterized by elongated cigar-shaped cells (up to 5 μm long) with from one to three buds at both ends of the cell (Fig. 29-3b).

Other fungi can invade and infect wounds, but none produces such a characteristic chronic disease as that resulting from *S. schenckii*.

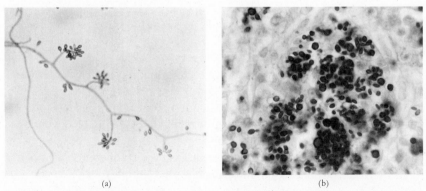

(a)   (b)

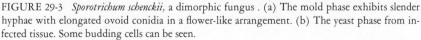

FIGURE 29-3 *Sporotrichum schenckii,* a dimorphic fungus . (a) The mold phase exhibits slender hyphae with elongated ovoid conidia in a flower-like arrangement. (b) The yeast phase from infected tissue. Some budding cells can be seen.

Wounds interfere with normal defenses and therefore predispose to infection. *Staphylococcus aureus* is the most frequent cause of wound infections. Other aerobic microorganisms often responsible include streptococci, enterobacteria, and pseudomonas. Wounding often leads to the development of anaerobic conditions in the tissue and subsequent invasion by anaerobic bacteria such as clostridia, streptococci, bacteroides, fusobacteria, and actinomycetes.

Surgical wounds often become infected with *S. aureus* that are resistant to penicillin. The presence of sutures in the wound predisposes to staphylococcal infection. *Streptococcus pyogenes* and enteric bacteria are also frequent causes of surgical wound infections.

Burns are particularly susceptible to infections. *Pseudomonas aeruginosa* is one of the most prevalent agents in burn infections and a frequent cause of death in burn patients, because of its resistance to antibiotics.

Tetanus is an important disease resulting from anaerobic infections of wounds with *Clostridium tetani*. The disease is caused by the potent tetanus nerve toxin; therefore it is treated with tetanus antitoxin and prevented by immunization with tetanus toxoid. Gas gangrene is also produced by members of the genus *Clostridium*.

In addition, anaerobic infections of wounds are caused by *Actinomyces israelii*, an anaerobic actinomycete that is part of the normal flora. It forms small yellow colonies in pus, called "sulfur granules."

Animal bites and scratches may transmit animal pathogens or normal flora to the unnatural host, man, resulting in serious disease. Among the infectious agents transmitted in this manner are rabies virus, *Pasteurella multocida*, *Spirillum minus*, *Streptobacillus moniliformis*, some of the chlamydiae, and others.

Various fungi can infect wounds. One of the most readily recognized fungus infections is caused by *Sporotrichum schenckii*, which invades small wounds and causes a characteristic lymph node involvement.

QUESTIONS

1. List some of the aerobic or facultative bacteria that most often cause wound infections.
2. Why is *Pseudomonas aeruginosa* likely to infect severe burn wounds? Since this organism is often considered "nonpathogenic," is a pseudomonas wound infection cause for concern? Why?
3. In theory, the disease tetanus can be eradicated, yet it still occurs. Why is this and what can be done to eliminate tetanus?
4. Can tetanus be treated effectively? How?
5. List a few kinds of infection that can be transmitted by animal bites or scratches.

Infections Involving
the Blood Vascular
and Lymphatic Systems

ANATOMICAL AND PHYSIOLOGICAL CONSIDERATIONS

Almost everyone today is familiar with the general structure and function of the blood vessel system of the body, although few people had ever considered the possibility of blood circulation before William Harvey's convincing argument in 1628. Force for the movement of blood is supplied by the heart, a muscular pump enclosed in a sac called the pericardium. As shown schematically in Figure 30-1, the heart is divided into a right and left side, separated by a wall of tissue through which blood normally does not pass in adult life. The right and left parts of the heart are each divided into two chambers, one which receives blood (atrium) and the other which discharges it (ventricle). A valve is located at the entrance and exit of each ventricle. Although not common, infections of the heart valves, muscle, and pericardium may obviously be disastrous because of their effect on the vital functions of the circulatory system. Blood from the right ventricle flows through the lungs and into the atrium of the left part of the heart. Blood then passes into the left ventricle and is pumped through the aorta to arteries and capillaries supplying almost all tissues of the body.

A system of veins collects the blood from the tissue capillaries and carries it to the atrium of the right heart. Since pressure in the veins is low, one-way valves help keep the blood flowing in the right direction. Thus as the blood flows around the circuit, it alternately passes through the lung and the tissue capillaries. Also, during each circuit a portion of the blood passes through other organs such as the spleen, liver, and lymph nodes, all of which, like the lung, contain fixed phagocytic cells of the reticulo-endothelial system. Foreign material, such as microbes, may be removed by the phagocytes on passage through such tissues.

Reticuloendothelial system
Page 362

525

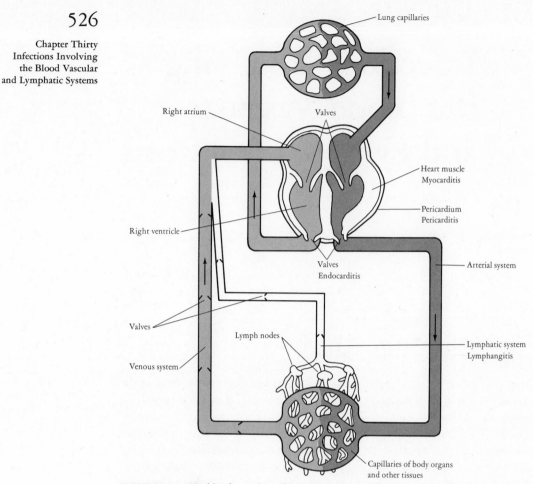

FIGURE 30-1   The blood vascular and lymphatic systems. The lymphatic drainage from the lung is omitted for clarity.

LYMPHATIC SYSTEM

The system of lymphatic vessels is less well understood than that of the blood vessels. This system begins in tissues with tiny tubes, which differ from blood capillaries in having closed distal ends and being somewhat larger. Inside these lymphatic vessels is lymph, an almost colorless fluid. This fluid comes from plasma which has oozed through the blood capillaries to bathe the tissue cells and which is then taken up by the lymphatics. Unlike the blood capillaries the lymphatics readily take up foreign material such as invading microbes or their products. The tiny lymphatic capillaries join progressively larger vessels in a manner similar to the roots of a plant. Many one-way valves in the lymphatic vessels keep the flow moving in the general direction away from the capillaries. Both contraction of the vessel walls and compression by movements of muscles force the fluid along.

At many points in the system, the lymphatics drain into small, roughly bean-shaped bodies called lymph nodes. These nodes are so constructed that for-

Lymphoid tissues
Page 375

eign material such as bacteria is trapped; they also contain cell types that are involved in phagocytosis and antibody production. Lymph flows out of the nodes through other vessels that eventually unite into one or more large tubes that discharge lymph into the blood system through connections with large veins in the chest. An infection of a hand or a foot is sometimes evidenced by the spread of a red streak up the limb from the infection site. This streak represents the course of the lymphatic vessels which have become inflamed due to the infectious agent, and the condition is called lymphangitis. It may stop abruptly at a swollen and tender lymph node, and later continue to yet another lymph node. This demonstrates the ability (even though sometimes temporary) of the lymph nodes to clear the lymph of an inflammatory agent. As indicated earlier the blood and lymph carry such antimicrobial agents as white blood cells, antibodies, complement, lysozyme, beta lysin, and interferon. Lymph and blood may clot in regional vessels as a result of inflammation arising from infection or antibody-antigen reactions.

Inflammation
Page 361

Lysozyme and beta lysin
Page 362

BACTERIAL DISEASES OF THE BLOOD VASCULAR SYSTEM

Subacute Bacterial Endocarditis

Table 30-1 outlines the important features of subacute bacterial endocarditis. This is an infection of the inner lining of the heart and is usually localized on one of the valves. It usually occurs in abnormal hearts, perhaps as a result of rheumatic fever or some other disease, or a birth defect. Afflicted persons typically become ill very gradually, developing fever and slowly losing pep and vigor over a period of weeks or months. The causative organisms of endocarditis are usually shed from the infected valve into the circulation and can often be identified by culturing samples of blood drawn from an arm vein. They are almost always organisms frequently present among the normal flora of the infected person, most commonly streptococci of the viridans group or other streptococci of low virulence, or *Staphylococcus epidermidis*.

Gram-positive cocci
Page 256

TABLE 30-1
Features of Subacute Bacterial Endocarditis

Host Factors	Microbial Cause	Pathogenesis

Structural abnormality of heart:
—birth defect
—rheumatic fever
—other diseases
	Normal flora:
	—*Staphylococcus epidermidis*
	—viridans streptococci
	—other
		Turbulent blood flow
		Clot formation
		Colonization of clot
		Enlargement of infected clot
		Release of fragments:
		  —plugging of vessels
		  —damage due to immune complexes

PATHOGENESIS OF
BACTERIAL
ENDOCARDITIS

Immune complexes
Pages 394–395

S. aureus
Pages 430–432
S. pneumoniae
Pages 455–456

A very few organisms of the normal flora frequently gain entrance to the bloodstream during dental procedures, brushing of teeth, or other trauma, but are normally quickly eliminated by body defense mechanisms. On the other hand, in an abnormal heart, turbulent blood flow fosters the formation of a film of blood clot and traps such organisms in areas where phagocytes have difficulty acting. Thus the high levels of antibody often present in bacterial endocarditis are of little value and may even foster the progress of the infection by clumping the bacteria and depositing them on the heart. The organisms multiply extensively and more clot is progressively deposited around them, gradually building up a fragile mass. Bacteria and pieces of infected clot continually break off and, if the latter are large enough, may block important blood vessels and lead to tissue death, or weakening and ballooning out of larger vessels. Circulating immune complexes may lodge in the kidney, producing a form of glomerulonephritis. Even though the organisms normally have little invasive ability, great masses of them growing in the heart are sometimes able to burrow into the heart or valve tissue, even resulting in valve rupture.

In some instances of infective endocarditis culturing the blood from an arm vein fails to yield the causative agent. This is especially likely when the infection is on the right side of the heart and blood from the infected site must pass through both the lung and the tissue capillaries (with their phagocytic cells) before reaching the arm vein. In other instances the organism may be too fastidious to grow in usual bacteriologic media, and may even be a rickettsia or virus.

Infections of the heart can also occur with more virulent microbes such as *Staphylococcus aureus* or *Streptococcus pneumoniae*, usually causing a rapidly progressing illness called acute bacterial endocarditis. Such organisms are much more likely than those causing subacute bacterial endocarditis to invade the heart and its valves and produce permanent tissue destruction, since they have means of overcoming tissue defenses.

Septicemia

"Blood poisoning" (septicemia) is illness associated with microbes or their products circulating in the bloodstream (bacteria sometimes enter the bloodstream without causing illness, and this is referred to as *bacteremia*). The main findings in septicemia are fever and a drop in blood pressure, probably due at first to impaired strength of the heart and walls of blood vessels. In severe septicemia the blood vessel damage may be irreversible. If the drop in pressure is marked and prolonged, there is insufficient flow of blood to supply adequate oxygen to the organs and other tissues, and shock results. Septicemia is almost always the result of an infection somewhere else in the body, and thus is commonly caused by one of the usual bacterial pathogens. In addition, in recent years alterations in the normal body defenses as the result of medical treatments have resulted in many septicemias from microbes that normally have little invasive ability. Indeed they typically originate from among the normal flora or the environment.

Although gram-positive bacteria and fungi can cause septicemia, gram-negative rod shaped bacteria are most often responsible. Septicemia due to gram-negative rods can be devastating. Shock is common and only about half of people with this kind of infection survive. Cultures of the blood in these cases most commonly reveal facultatively anaerobic gram-negative rods such as *Escherichia coli, Enterobacter aerogenes, Serratia marcescens,* or *Proteus mirabilis.* Among the aerobic gram-negative rod bacteria encountered are organisms commonly found in nature, such as *Pseudomonas aeruginosa. Pseudomonas aeruginosa* has extraordinary biochemical capabilities, is very resistant to antibiotics, and is able to grow under a variety of conditions unfavorable to the usual bacterial pathogens.

Not all the gram-negative organisms causing septicemia will grow in air. As mentioned in Chapter 26 anaerobic gram-negative rods of the genus *Bacteroides* make up a major percentage of the normal flora of the large intestine, and these not infrequently cause septicemia. This group of organisms has, until recently, been very poorly classified, and many strains have been isolated that fail to fit into existing species. Recently gas chromatographic examination has shown promise as a tool for identifying relationships among this group of bacteria. In this procedure the bacterial cells are degraded to simpler components, and the relative amounts of each are measured.

How gram-negative bacteria produce septicemia and shock is still not completely understood, but the probable sequence of events is shown in Table 30-2. Gram-negative bacteria contain endotoxin, the poisonous material making up part of their cell walls. This type of toxic substance has been discussed in Chapter 19. It is a high molecular weight lipopolysaccharide and relatively heat stable. When injected into the bloodstream of laboratory animals in milligram doses, it produces fever and shock, often leading to death.

Despite these effects of endotoxins defined under laboratory conditions, there is as yet no agreement about the extent to which endotoxins are released in infections and the exact role they play in specific diseases caused by gram-negative bacteria. Nevertheless it is very likely that endotoxin causes the principal damage to the host in many septicemias, especially those resulting from relatively avirulent bacteria of the normal flora and the environment. Other bacterial products, including protein toxins, probably play a role in some cases of septicemia.

Enterobacteria
Page 252

Pseudomonas
Page 250

Page 365

TABLE 30-2
Probable Sequence of Events in Septicemia Caused by Gram-Negative Rod Bacteria

1. Bacteria enter the bloodstream from a focus of infection.
2. Endotoxin is released.
3. Endotoxin acts on the heart and blood vessels causing impaired contractile strength which becomes irreversible if prolonged.
4. Decreased force of heart contractions and dilatation of blood vessels result in a drop in blood pressure.
5. Insufficient pressure and maldistribution of blood result in an inadequate supply of oxygen to vital organs.
6. Death results in many cases even though the infection is controlled because of damage to the circulatory system from endotoxin.

A VIRAL INFECTION OF THE HEART MUSCLE

Myocarditis, or inflammatory disease of the heart muscle, may be caused by bacteria, fungi, or protozoa, but more commonly is due to a viral infection. Sometimes the inflammation involves primarily the sac around the heart (pericardium) and is called *pericarditis*. Involvement of the heart muscle may produce only pain in the chest and abnormalities of the electrical impulses of the heart, but if the infection is extensive, the heart is unable to contract effectively, and heart failure may result. This problem is most likely to occur in infants, but can arise at any age.

Identification of the causative microbe can sometimes be made from cultures of fluid from beneath the pericardial sac, or from cultures of feces or throat secretions plus demonstration of a rise in neutralizing antibody during the course of the illness. In most of the instances where an agent has been recovered, a Coxsackie virus has been found.

Picornaviruses
Page 678

The virus presumably reaches the heart by the bloodstream or lymphatic system after infecting the respiratory or gastrointestinal tract. However, the reason why only a few of the infected persons during any given epidemic manifest myocarditis is unknown. Once in the heart the virus extensively damages the muscle cells in which it replicates, and some of these cells die and are replaced by scar tissue. In some terminal cases of myocarditis after an illness of many months, viral antigen can still be demonstrated in the heart muscle by using fluorescent antibody against Coxsackie viruses. Mysteriously enough the virus cannot usually be recovered in cultures even though large amounts of viral antigen are demonstrable.

Coxsackie viruses are able to grow and produce disease in mice, but susceptibility of the mice is markedly age-dependent. In fact only newborn mice are consistently affected. Moreover Coxsackie viruses differ from each other in the types of mouse tissue cells they will attack. Group A viruses characteristically damage the muscles of the mouse, and group B attack the central nervous system, liver, pancreas, and fatty tissues. Different Coxsackie viruses show differing tissue affinities in man also. Most persons with viral myocarditis have group B Coxsackie infections.

Because only a small fraction of persons infected with Coxsackie viruses develop serious illness, and very few of these die, there has not been much effort to establish control measures. No vaccines are available. As with other enteroviruses, sanitary practices are unlikely to limit spread of the agent, although they should help decrease the infecting dose of virus received by the victims. Excretion of the virus in the feces of an infected person may continue for weeks, even after mild, unnoticed infections. The presence of healthy disseminators in a community undoubtedly aids in the spread of Coxsackie viruses.

A VIRAL DISEASE OF LYMPH NODES: INFECTIOUS MONONUCLEOSIS

Infectious mononucleosis ("mono") is a disease familiar to many students because of its highest incidence among economically privileged persons between the ages of 15 to 24 years. The term "mononucleosis" refers to the fact that infected persons have an increased number of mononuclear white blood cells (white blood cells with unsegmented nuclei) in their bloodstreams. The most dramatic symptoms of the disease are fever, sore throat, and enlargement of the lymph nodes and spleen. In some instances complications such as myocarditis, meningitis, hepatitis, or paralysis develop. One very interesting aspect of the illness is the development of an antibody in the bloodstream of the sick person which will react with an antigen on the surface of red blood cells from sheep. In fact, until recently this was the only practical way to separate patients with infectious mononucleosis from those with other infectious conditions closely resembling it. This was because infectious mononucleosis, although long presumed to be a viral disease, did not have an identifiable causative agent.

Lymphocytes
Figure 20–6, Page 377

In 1968 the first reports appeared with some very interesting developments relating to infectious mononucleosis. These derived from work extending back to 1964 on a peculiar virus seen in cell cultures prepared from a malignant tumor of lymphatic tissues. The virus was shown to be unrelated to any common known virus although its appearance was similar to herpes viruses. The virus is now popularly called Epstein-Barr virus (or EB virus) after its discoverers. Using EB virus grown in tissue cell cultures, it was possible to detect and quantify anti-EB virus antibody in the bloodstreams of various people. The highest titers of antibody were found in persons with the type of malignant lymphatic tumor which originally yielded the virus, but many others were also found to have antibody. Of special interest, however, was the finding that those who contracted infectious mononucleosis consistently lacked antibody to EB virus before the disease started, and then developed high titers of antibody during the illness. This antibody persists for years afterward. Moreover, EB virus (or possibly a close relative) could be cultured from their mononuclear blood cells in a manner similar to the malignant tumor.

EB virus
Page 355

Present evidence indicates that EB virus is widespread and infects most economically disadvantaged groups at an early age without producing significant illness. In such populations infectious mononucleosis is quite rare. On the other hand, more than 50 percent of students entering some United States colleges have shown no antibody to EB virus and presumably have escaped past infection with the agent. Infectious mononucleosis occurs exclusively in antibody-negative subjects. This information is consistent with the hypothesis that EB virus causes infectious mononucleosis. Epidemiological evidence indicates that kissing may be one mode of transmission of infectious mononucleosis.

A BACTERIAL DISEASE OF LYMPH NODES: TULAREMIA

Not long ago an outbreak of disease characterized by fever and enlargement of lymph nodes appeared in Vermont. Altogether 72 cases were identified, and all occurred among persons who had been in contact with fresh muskrat pelts or skinned muskrats. About the same period, almost 3000 miles away, a similar disease developed in a young man in the state of Washington who had recently skinned a bobcat. All had tularemia, caused by an agent widespread among wild animals and arthropods—an agent which was often fatal for man in the days before antibiotics were available for treatment.

Francisella
Page 251

The causative microbe, *Francisella tularensis* (formerly called *Pasteurella tularensis*), is a pleomorphic, nonmotile, aerobic gram-negative rod that derives its name from Tulare County, California, where it was first studied. The organism can be cultured from blood and other materials from infected persons and animals. This pathogen does not grow well unless 0.1 percent cystine and 1 percent glucose are added to rich medium. Good growth is apparent within 48 hours on such media. Unfortunately contaminating microbes from the environment or normal flora, if present in the inoculum, may obscure the growth of *F. tularensis*.

ASCOLI TEST

If this is likely, the material is injected into a laboratory animal such as a white mouse. The defense mechanisms of the mouse quickly destroy the contaminating flora, while the pathogen invades its tissues. There *F. tularensis* produces a soluble antigen which can be extracted and identified with specific antiserum (*Ascoli test*).

Tularemia occurs in many areas of the Northern Hemisphere, including all the states of the United States except Hawaii. In the eastern United States, human infections usually occur in the winter months, as a result of skinning rabbits (thus the common name, "rabbit fever"). In the West infections result primarily from bites of ticks and flies, and thus occur during the summer. *Francisella tularensis* probably cannot penetrate the unbroken skin, but enters small cuts or scratches which may not be visible, or through mucous membranes of the eye or mouth. It typically causes an area of infection, an ulcer formation, at the site of entry,

Ticks
Page 544

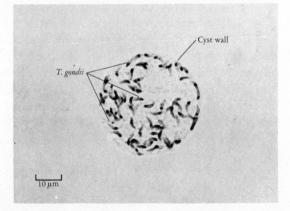

FIGURE 30-2
Toxoplasma gondii cyst. (Courtesy of J. Remington)

and in the lymph nodes of that area of the body, which may become very large and filled with pus. The organisms frequently spread to other parts of the body. These bacteria are of interest because (like *M. tuberculosis*) they are ingested but not killed by phagocytic cells, and readily grow within them. This may explain why the infection persists in some persons even though they have high titers of antibody in their blood. Before the days of effective treatment, about one out of ten persons with the disease died. The mechanisms of cellular immunity which arise in the host to get rid of this and other organisms which persist intracellularly are still not fully understood. Both delayed-type hypersensitivity (demonstrated by injecting antigen from the bacterium into the skin of the infected person) and humoral antibody quickly arise during infection, and their demonstration can be used to help identify the disease when positive cultures cannot be obtained.

Delayed hypersensitivity
Page 398

TOXOPLASMOSIS: A DISEASE CAUSED BY PROTOZOA

Toxoplasmosis is a disease caused by the small crescent-shaped protozoon, *Toxoplasma gondii*. This microbe (Fig. 30-2) can infect a broad range of hosts, including both birds and mammals, and disastrous epidemics among commercially valuable animals have occasionally occurred. Overt disease in human beings as a result of this organism is unusual, although infection is common. Three main illness patterns are seen: (1) infections of the fetus resulting from asymptomatic infections of expectant mothers—in this instance the baby may be born mentally defective or merely have slight damage to the retina of the eye; (2) infections of adults, especially those with serious underlying diseases, may be manifest as myocarditis or generalized systemic involvement involving most body tissues; and (3) a mild illness resembling infectious mononucleosis usually seen in young adults. Studies of college students at the University of California and at Harvard College have shown about one out of five have been infected with *Toxoplasma*. Infection is thought to occur mainly from eating improperly cooked meat, and an outbreak of five cases recently occurred among a group of New York medical students who ate inadequately cooked hamburger. Their illness resembled infectious mononucleosis, and they had lymph node enlargement lasting about three months.

The life cycle of *T. gondii* has long been a mystery, but it is now known that the organism is a coccidium of cats, infecting the cells lining the cat intestine (coccidia are sporozoa—distant relatives of malaria parasites). In one American city about half the cats tested had evidence of infection. Other animals become infected when they ingest food or water contaminated by feces from infected cats. The organisms enter the tissues of the new host—perhaps a cow—where they develop into a cystic form containing many infectious progeny. These persist alive for an indefinite period and can transmit the infection when another animal such as man ingests the inadequately cooked flesh of the cow. However, the complete cycle of growth can only develop with infection of the intestinal-lining cells

Sporozoa
Page 302

of a cat or closely related species. It is not yet known to what extent cats are directly responsible for human infections, but infected meat appears to be a more important source.

Toxoplasma gondii is of general interest because of its wide host range. The organisms are infectious for almost all cells of warm-blooded animals except nonnucleated red blood cells. It was also one of the first infectious agents shown to produce general enhancement of resistance to intracellular infections by other species of microorganisms. The practical significance of this observation is as yet undefined, but in view of the widespread and benign nature of most human infections, it is interesting to speculate on the possible role of this coccidium of cats in ameliorating human disease. Nevertheless expectant mothers would do well to avoid cats and rare meat.

SUMMARY

There are two networks of vessels in the body, the blood vascular system and the lymphatic system. As the heart pumps blood around the blood vascular circuit, any microbes present are likely to be removed by the phagocytic cells of the spleen, lung, and other organs. Unlike the blood vessels the lymphatics begin as blind-ended capillaries which are readily permeable to microbes. Lymph is subject to the decontaminating action of phagocytes in lymph nodes. Failure of body defense mechanisms may result in infection of the heart valves and lining (endocarditis), the muscle and outside covering of the heart (myocarditis and pericarditis), and the lymphatic vessels (lymphangitis). Microorganisms circulating in the bloodstream may release toxins and produce septicemia. Tularemia and infectious mononucleosis are diseases in which the lymph nodes are often prominently involved. Toxoplasmosis is an unusual manifestation of widespread infections by *Toxoplasma gondii,* a protozoan parasite of cats.

QUESTIONS

1. How is it that avirulent bacteria of the normal flora can cause the serious disease, bacterial endocarditis?
2. Why is it that infections by Coxsackie viruses are manifest in different ways, sometimes by heart disease and sometimes by disease of other parts of the body?
3. Discuss the relationship between EB virus and infectious mononucleosis.
4. Tularemia can be identified as the cause of death of an animal in less time than it takes to recover and identify *Francisella tularensis* in culture. Explain how.
5. How would the elimination of cats from the world jeopardize the existence of *Toxoplasma gondii* (if at all)?

BRAUDE, A. I., "Bacterial Endotoxins," *Scientific American,* 210 (March 1964), 36.

DAVIS, B. D., R. DULBECCO, H. S. GINSBERG, and W. B. WOOD, *Microbiology.* New York: Hoeber Medical Division, Harper and Row, 1969.

FRENKEL, J. K., "Pursuing Toxoplasma," *Journal of Infectious Diseases,* 122 (December 1970), 553–559.

HALL, J. G., "The Flow of Lymph," *New England Journal of Medicine,* 201 (September 1969), 720–722.

MAYERSON, H. S., "The Lymphatic System," *Scientific American,* 208 (June 1963), 80.

NIEDERMAN, J. C., A. S. EVANS, L. SUBRAHMANYAN, and R. W. McCOLLUM, "Prevalence Incidence and Persistence of EB virus Antibody in Young Adults," *New England Journal of Medicine,* 282 (February 1970), 361–365.

WIGGER, C. J., "The Heart," *Scientific American,* 196 (May 1957) 74.

WOOD, J. E., "The Venous System," *Scientific American,* 218 (January 1968), 86–96.

ZWEIFACH, B. W., "The Microcirculation," *Scientific American,* 200 (January 1959), 54.

Chapter 31

Microbes, Men and Multicellular Parasites

For the most part the focus of this book is on unicellular microbes (bacteria, protozoa, algae, fungi) or subcellular agents (viruses). However, parasitism of man and other animals involves multicellular forms as well, and the principles of host response and methods of microscopic identification of these organisms are similar to those of the smaller forms. Of particular importance here is the fact that some of these multicellular parasites are themselves hosts for sub-, uni- and multicellular agents, some of which they can transmit to man with devastating results. We shall explore some interesting examples of multicellular parasitism and its consequences. Comprehensive treatment of this subject is given in textbooks on parasitology and entomology.

CLASSIFICATION

The animal subkingdom Metazoa includes all multicellular forms except sponges and possibly some primitive wormlike creatures of the sea. The majority of forms parasitic for man are represented in three phyla: Platyhelminthes, Nematoda, and Arthropoda. Organisms of the first two phyla are primitive wormlike creatures, but those of the third phylum, Arthropoda, including insects, arachnids, and crustaceans, are highly advanced on the evolutionary scale. Arthropods almost certainly arose from segmented wormlike animals with legs on each segment. In the course of evolutionary development the segments have tended to fuse, and the appendages have become mouth parts, specialized organs of locomotion, respiratory or mating apparatus, and so on. Thus unlike members of the more primitive phyla, the arthropod parasite often moves easily from host to host, and with the piercing apparatus of his mouth feeding directly from the bloodstream of a vertebrate host. Members of the phyla Platyhelminthes and Nematoda are known to carry infectious microbes in a few instances involving other animals, but not man. By contrast Arthropoda includes some extremely important vectors of organisms pathogenic for man.

536

TABLE 31-1
Examples of Arthropod-Borne Diseases

Disease	Causative Agent	Vector
Malaria	*Plasmodium* sp.	Mosquitoes (*Anopheles* species)
Equine encephalitis	Arboviruses	Mosquitoes (*Culex* species)
Plague	*Yersinia pestis*	Flea (*Xenopsylla cheopis*)
Typhus	*Rickettsia prowazekii*	Louse (*Pediculus humanus*)
Rocky Mountain spotted fever	*Rickettsia rickettsii*	Ticks (*Dermacenter* species)

ARTHROPODS AND ARTHROPOD-BORNE INFECTIONS

Examples of some arthropod-borne infections are shown in Table 31-1 with causative agents and principal vectors.

The Mosquito

One of the most medically important arthropods is an insect, the mosquito. The mosquito mouth parts consist of sharp stylets which are forced through the skin to the subcutaneous capillaries. One of these needlelike stylets is hollow, and salivary secretions of the mosquito are pumped through it. These secretions help keep the victim's blood from clotting as it is drawn into a tube formed by the other mouth parts, and they may also act on the blood vessels to increase blood flow. The injected mosquito saliva contains antigens to which some individuals are hypersensitive, and it may also contain agents causing serious infectious diseases. Only in the female are the mouth parts adapted to piercing animal skin; the male mosquito feeds on plant juices instead. Ovarian development and egg production of mosquitoes are markedly enhanced by a blood meal, indicating that blood is the source of important hormones or nutrients which contribute to reproductive functions.

Some important features of mosquito anatomy are shown in Figure 31-1. On either side of the mouth parts, note two types of sense organs, the palps and the antennae. The former probe the victim's skin and are thought to help the mosquito find a suitable place to make her "bite." The antennae pick up more subtle sensations. For example, in the male they respond to the buzzing sound characteristic of females of the same species, and thus aid him in finding a mate.

The muscles of the throat suck blood into the digestive system and into the large storage areas. The latter allow the mosquito to take in as much as twice its body weight in blood, giving a relatively good chance of picking up microbes circulating in the host's capillaries. Other parts of the digestive tract are the stomach, in the wall of which the malaria parasite develops, and the midgut through which the parasite penetrates into the mosquito's body cavity to infect the salivary glands.

The more important groups of mosquitoes include *Aedes, Culex,* and *Anopheles,* distinguished largely by microscopic examination of their palps, antennae, wings, claws, mating apparatus, and other features.

IDENTIFICATION

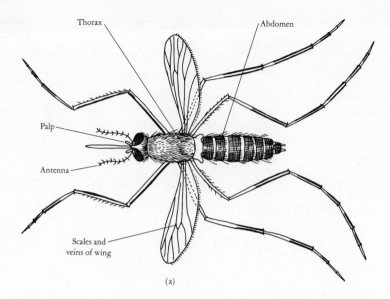

(a)

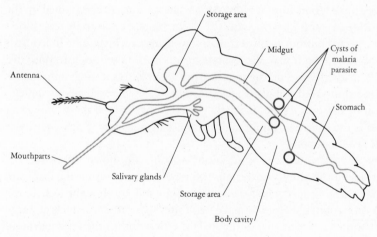

(b)

FIGURE 31-1 Mosquito anatomy. (a) Features important in identification. (b) Internal anatomy.

Although mosquitoes may act as vectors for metazoan agents such as certain worms, and for bacterial agents such as *Francisella tularensis,* the protozoan and viral agents transmitted by mosquitoes are numerically far more important. A few of the infections of mosquitoes that are important to man are considered in the following sections.

MALARIA

Malaria, an ancient scourge of mankind, was suspected of being mosquito-borne at least as early as the second decade of the eighteenth century. This suspicion was dramatically confirmed in the late 1800s when mosquito control meas-

ures in Havana led to a drop in the incidence of malaria from about nine out of
ten persons to one out of fifty. In the early 1900s, effective mosquito control
reduced the incidence of malaria and yellow fever and allowed completion of the
Panama Canal.

The most characteristic feature of malaria in man is paroxysms of fever
occurring at one-to-three-day intervals, with relative well-being in between. The
protozoan parasite, a species of *Plasmodium,* grows and divides in the red blood
cells of the host (Fig. 31-2). After a time the offspring of this division rupture
the red blood cells and are released into the plasma. The patients' fevers result
from this periodic release of the plasmodia. Most of the young parasites then

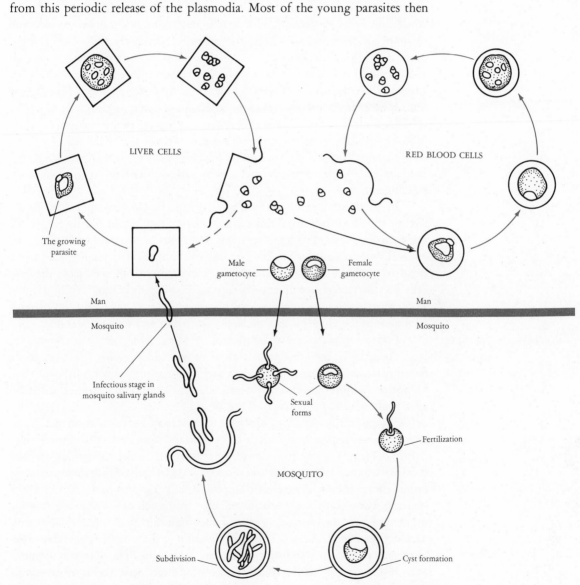

FIGURE 31-2 Life cycle of *Plasmodium* species.

enter new red blood cells and again multiply, and the cycle is repeated. Some of them may develop into specialized sexual forms (*gametocytes*), different in appearance and in susceptibility to antimalarial medicines. These forms cannot develop further in the human host and are not important in causing his symptoms. However, they are infectious for certain species of mosquitoes (especially certain species of *Anopheles*) and thus ultimately for transmission of malaria from one human being to another.

DEVELOPMENT IN
THE MOSQUITO

When a patient with malaria is bitten by a suitable species of mosquito, it may ingest the sexual forms of the malarial parasite. Shortly after entering the intestine of the mosquito, the male gametocyte produces tiny whiplike bodies which unite with the female gametocyte in much the same way as the sperm and ovum of higher animals. The resulting organism enters the wall of the midgut of the mosquito and forms a cyst, which enlarges as the zygote undergoes meiosis and division into numerous asexual forms. These rupture into the body cavity and find their way to the salivary glands where they may be injected into a new human host. It is of interest that the infection of the mosquito fails to impair its normal function or longevity. This may imply long evolutionary association of parasite and the mosquito host, with selection of those variants most favorable to coexistence with the host mosquito. Unhappily the relationship between plasmodium and man is not so advanced, and malaria is often a severe and sometimes fatal illness.

Injected into a new host by the mosquito, the parasites are carried to the liver by the bloodstream and leave the circulation to infect liver cells. Each parasite enlarges and subdivides, producing thousands of daughter cells. These are released into the bloodstream and establish the cycle of infection described above involving red blood cells. With some species organisms released from the liver cells may also infect new liver cells, thus establishing a cycle of infection in the liver. A sustained liver infection may result, which may persist despite effective treatment of the bloodstream infection and malaria may reoccur later on.

CONTROL OF MALARIA

Control of malaria is accomplished by using insecticides and eliminating breeding areas for the principal mosquito vectors and in giving treatment to infected patients. The patients are identified by preparing stained smears of their blood and examining them microscopically for the malarial parasites. Intensive efforts, often with assistance from the World Health Organization, have markedly reduced or eliminated mosquito-borne malaria from many countries. Complete eradication of human malaria is theoretically possible, but has proved difficult to achieve. Ceylon, for example, was formerly subject to serious malarial epidemics, one in 1935 causing an estimated 80,000 deaths. With intensive eradication efforts the incidence had dropped to only 17 cases in 1963, and DDT spraying was discontinued and surveillance efforts relaxed. Following this the incidence promptly rose again, so that in the single month of February 1968, over 42,000 cases were diagnosed. Formerly common in the United States malaria became very rare with establishment of control measures. The dramatic reappearance of malaria associated with the return of military personnel from Southeast Asia (each of the 50 states reported cases in 1970, with almost 4000 cases total) was mostly due to infections acquired outside the country. In some instances transmission to civilians has resulted from blood transfusions. Transmission can also occur when heroin addicts share syringes; one such episode in 1971 resulted in 48 infections.

Mosquitoes may also bear viruses which cause *encephalitis*, an inflammatory process of the brain. Human and domestic animal encephalitis caused by arboviruses occurs in endemic or epidemic fashion every year in the United States. During 1969, for example, cases were identified in California, Florida, Georgia, Idaho, Minnesota, New Jersey, New Mexico, North Carolina, Ohio, Oregon, South Carolina, Texas, Washington, and Wisconsin. Among these are the equine encephalitis viruses, which characteristically infect horses, but which can cause death, mental disorders, or motor nerve impairment in people.

Arboviruses
Page 679

Normally the viruses pass from one mosquito to another by common exposure to another host, typically one of the birds living in a swamp. The viruses live in the infected mosquito for the life of the mosquito, but cannot spread directly from one mosquito to another nor from the mosquito to its egg (transovarially). The fact that under proper circumstances a mosquito may be infected with the virus for its whole lifespan indicates how well adapted the parasitic agent is to its host. Every now and then, when the mosquitoes are unusually plentiful and the number of infected birds is high, the virus may be spread to other areas and to less adapted hosts. Mosquitoes may then transmit the virus to domestic fowl, to horses, or to man. The transfer of virus from mosquito to swamp bird to mosquito to bird, and so on, is often referred to as a "cycle," although it is more like a never-ending chain with repeating links. The spread to other hosts is usually a dead end because of a high mortality of the new host and because the species of mosquitoes that usually bite these hosts are less able to sustain infection and transmit the virus. It is likely, however, that on occasion mutants among the infecting viruses may multiply and persist in the new vector and cause less serious disease in the new host, so that a new cycle develops.

EQUINE ENCEPHALITIS VIRUSES

Control of equine encephalitis is achieved chiefly through mosquito control, although vaccines given to animals may be helpful in some instances. Control of a major epidemic of Venezuelan equine encephalitis in the United States in the summer of 1971 involved malathion spraying of 8 million acres and vaccination of approximately 1.3 million horses. About 1500 horses died and at least 84 human cases developed before the epidemic came under control.

CONTROL OF EQUINE ENCEPHALITIS

The Flea

In contrast to mosquitoes, fleas, also classed as insects, have no wings, and depend on their powerful hind legs to jump quickly from place to place.

The anatomy of a flea is diagrammed in Figure 31-3. Points of special note are mouth parts, spines or "combs" about the head and first chest segment, and the digestive tract. Note especially the muscular pharynx (for sucking blood from the host) and the long esophagus followed by the spiny valve composed of rows of teethlike cells.

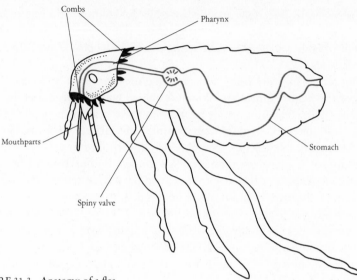

FIGURE 31-3   Anatomy of a flea.

Enterobacteria
Page 252

PLAGUE

Plague is primarily a disease of rats and other rodents, and their fleas. The flea contracts the plague bacillus *Yersinia pestis* (formerly called *Pasteurella pestis*) by feeding on an infected rodent. So many of the bacteria may be present in the blood of the rat that the flea may ingest 25,000 bacteria in the course of a single feeding. The organisms pass into the stomach of the flea and multiply there, producing sticky masses of bacteria which can cause obstruction by adhering to the spiny valve. The blood sucked into the esophagus then cannot pass into the stomach, but regurgitates into the puncture wound in the host's skin carrying the plague bacilli with it. Moreover, the hungry flea tries again and again to feed and may infect several other hosts with *Y. pestis*. Infected fleas may live for long periods, since the obstruction often resolves spontaneously, but the infection persists and *Y. pestis* is excreted with the feces. This material may also infect man or lower animal hosts when rubbed or scratched into flea bites.

Even when many susceptible rodents are present, the mortality rate due to *Y. pestis* infection may be so high that the fleas find a rapidly diminishing supply of natural hosts. In this case they will turn to human beings, who thereby become infected with *Y. pestis* and develop *bubonic plague*. Carried to the nearby lymph nodes of men by tissue fluids, the bacteria cause enlargement of the nodes ("bubo"—thus "bubonic" plague). Then large numbers of the bacilli are released into the bloodstream and distributed to all areas of the body. In some cases bleeding beneath the skin is a prominent feature; the darkened blood may have suggested the name "Black Death" for the European epidemic of the fourteenth century.

Yersinia pestis bacilli in the blood are carried to the lungs, where they may produce a form of the disease which spreads easily from person to person without the mediation of fleas. The bacilli enter the bronchial secretions and with coughing are sprayed into the air, and can thus be inhaled by others. The disease resulting from inhalation rather than flea bite is called "pneumonic plague" and accounts for most epidemic spread among human beings. In both bubonic and pneumonic plague the bacteria can be seen in infected material and readily grown on laboratory media. They are gram-negative enterobacteria which stain more intensely at each end, looking somewhat like a safety pin.

In early 1968, 90 cases of plague with 36 deaths occurred in villages of Central Java. Indonesian and American health officers imposed strict quarantines to prevent person-to-person spread of the disease, and gave plague vaccine to those living in areas adjacent to the epidemic. Rats were trapped, and the most frequent parasite found on them was the rat flea, *Xenopsylla cheopis*. Villages were heavily sprayed with DDT to reduce the numbers of fleas. The epidemic was quickly controlled, and cases ceased to appear within about two months. Plague in human beings is an unusual disease in the United States today, although it has been identified in the wild rodent population of at least 15 states and thus still remains a threat. A case in the summer of 1971 in a Navajo woman from New Mexico was traced to plague-infected prairie dogs.

CONTROL OF PLAGUE

Lice

Like fleas lice are small wingless insects which prey on warm-blooded animals by piercing their skin and sucking blood. The legs and claws of lice are adapted for holding on to body hair and clothing, rather than for jumping from place to place. Like fleas and mosquitoes, lice are susceptible to microbial infections by agents which can be transmitted to man.

The most notorious of the lice, *Pediculus humanus,* or "cootie," is 2 to 3 mm long, with a characteristically small head and thorax and large abdomen (Fig. 31-4). The louse has a membranelike lip with tiny teeth to anchor himself firmly to the skin of the host. Within the floor of the mouth is a piercing apparatus somewhat similar to that of fleas and mosquitoes. *Pediculus humanus* has only one host, man, but easily spreads by direct contact or by contact with personal items, especially in situations of crowding and poor sanitation. A similar organism, *Phthirus pubis* ("crab louse"), is not uncommon among young adults, most frequently transmitted by sexual intercourse. It is not known to be a vector of infectious diseases.

Epidemic typhus (not to be confused with typhoid, caused by *Salmonella typhi*) is a louse-borne disease that tends to occur under crowded and unsanitary

TYPHUS

FIGURE 31-4 Lice. (a) *Pediculus humanus.* (b) *Phthirus pubis.* (Courtesy of F. Schoenknecht)

living conditions, especially during the social disruption of wartime. In fact, typhus has been responsible for many more deaths than have the armaments of war. Patients with this disease develop fever abruptly, followed in a few days by rash and confusion (typhus means "hazy"). Damage to small blood vessels then becomes prominent, resulting in gangrene and hemorrhages beneath the skin. Involvement of heart, brain, and kidneys may contribute to the illness, and mortality ranges from about 10 to 40 percent in different epidemics. This disease is caused by *Rickettsia prowazekii,* a tiny obligate intracellular bacterium transmitted by *P. humanus.*

Rickettsia
Page 263

Lice feeding on typhus-infected persons ingest the rickettsiae. The microorganisms are taken up by the cells which line the gut of the louse and multiply within them. The infected cells die and liberate rickettsiae, which then pass through the louse intestine and are excreted in the feces. Scratching the itchy lice bites causes the new victim to rub some of the excreta into the wound, thereby inoculating himself with *R. prowazeckii.* Infection may at times also occur by inhalation of the dried louse feces or by rubbing it accidentally into the eyes.

Like most rickettsiae, *R. prowazekii* does not grow on cell-free media, but must be grown in susceptible living cells. The organisms can be recovered from human patients or lice by inoculating blood or other material into laboratory animals, embryonated eggs, or tissue cell cultures. However, it is easier and safer for laboratory workers to look for a rise in titer of antibody to the rickettsia in the patient's blood. This is done by making use of a dramatic example of antigenic sharing by members of two greatly different genera of bacteria, *Rickettsia* and *Proteus.* Certain strains of *Proteus vulgaris* and *mirabilis* have cell wall antigens that react with antibodies formed against *R. prowazekii* and some other rickettsiae. The resulting antibody-antigen reaction produces agglutination of the *Proteus.* This procedure employing strains of *Proteus* to diagnose rickettsial disease is known as the *Weil-Felix test.*

WEIL-FELIX TEST

Control of epidemic typhus can readily be achieved by using insecticides to kill lice. For all practical purposes transmission of the disease from person to person can result only from *P. humanus* infestation, and man is the only reservoir for *R. prowazekii.* Unfortunately, complete eradication of typhus is complicated by the fact that persons who recover from the disease may continue to have living rickettsiae in their body tissues for many years without apparent ill effects. Under conditions of stress, fading immunity, and other unknown factors, these organisms may multiply and again cause typhus. If *P. humanus* is present, the recrudescent disease may be transmitted to others, producing a new wave of epidemic typhus. *Rickettsia prowazekii* is thought to persist in the tissues through a delicate balance between defensive factors of the host and attributes of the microbe which protect against them, but the exact mechanisms involved are unknown.

Latent infection
Page 366

Ticks

Ticks are not insects but arachnids. Arachnids differ from insects by their lack of wings and antennae, and the fact that their thorax and abdomen are fused. Adults have four pairs of legs as compared to the three pairs insects have. Most ticks that attack humans are intermittent parasites, attaching from time to time to feed on blood, then dropping off to rest, grow, or deposit eggs. Their bite is usually completely undetected and one discovers the tick with surprise, firmly attached and getting swollen with ingested blood.

Dermacentor andersoni, a common American tick, is a saclike creature about 6 mm long with eight legs attached toward the front (Fig. 31-5). The anterior end has conspicuous mouth parts, but no distinct head.

Like many other ticks *D. andersoni* has a broad host range—more than 25 animals. The attachment of ticks to the skin is very secure. In fact, if one tries to pull off a tick, the mouth parts may separate from the tick rather than tearing from the skin of the host. How the tick attaches itself without causing pain is unclear, but it is known that some produce a toxin powerful enough to cause paralysis. Paralyzed men and animals usually recover rapidly following removal of the tick.

Ticks are susceptible to microbial infections, including protozoan, viral, and bacterial agents, some of which are infectious for man and domestic animals. For example, *D. andersoni* may be infected with *R. rickettsii* which may be passed between ticks and wild rodents without producing serious disease in either vector or reservoir host. However, in man, the rickettsia produces *Rocky Mountain spotted fever,* a serious illness somewhat resembling typhus. The observation that the rickettsiae can pass from generation to generation of tick through the eggs (transovarial passage) is of great interest, demonstrating the ability of another microbial species to coexist with its arthropod host for long periods without harming it significantly.

ROCKY MOUNTAIN SPOTTED FEVER

Dermacentor andersoni may also harbor the pathogenic rickettsia, *Coxsiella burnetii,* the cause of *Q fever,* usually contracted by man through inhalation of infected tick feces or by drinking milk from an infected cow. Another agent of importance is the arborvirus of *Colorado tick fever.* This virus is endemic in wild rodents of the Rocky Mountain region of the United States. In man it typically gives illness with fever, but no rash. Finally, these ticks may be infected with the bacterium *Francisella tularensis,* the gram-negative rod which causes tularemia. Like *R. rickettsii, F. tularensis* can pass transovarially.

Q FEVER

Tularemia
Page 532

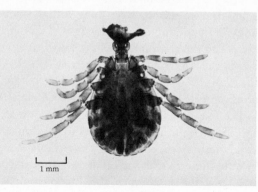

1 mm

FIGURE 31-5
The tick, *Dermacenter andersoni.* The mouth parts are imbedded in a piece of the host's skin torn away when the tick was removed. Note that four pairs of legs are present as opposed to the three pairs of insects. (Courtesy of C. Mecklenberg)

These are just a few of the multicellular parasites attacking the skin of man. There are many more, such as the buffalo gnat (a vector of a parasitic worm), tsetse flies (vectors of the protozoan agent of sleeping sickness, *Trypanosoma*), and sandflies (vectors of the bacterium, *Bartonella,* the cause of an infection of the red blood cells). Other arthropods are relatively permanent residents of the skin, such as *Sarcoptes scabiei,* the cause of an itchy rash, and *Demodex folliculorum,* which lives in the pilosebaceous glands, usually without producing symptoms.

NEMATODES (ROUNDWORMS)

PINWORMS

The roundworms are a very large group of parasites, some of which like *Enterobius vermicularis,* the pinworm of children, are widely known. *Enterobius* worms live in the large intestine, migrating to the anus where they discharge their eggs. The inflammation and itching which result often produce sleeplessness and behavior disorders. Another intestinal roundworm, of greater health importance, is the hookworm.

Hookworms

Hookworms are about 10 mm long and live in the small intestine of man, attaching by small hooks or plates located about their mouth. They feed by sucking the blood of the host. One authority, in 1962, estimated that 630 million people were infested world-wide, and the daily loss of blood was greater than all the blood transfusions given in the United States in one year. More than 1000 worms may be found in a single person, and the constant loss of blood frequently produces anemia. The lifespan of the worm is about six years, but medicine can reduce or eliminate the infestation.

LIFE CYCLE OF
HOOKWORMS

The female worm releases her eggs—perhaps 25 million in the course of her lifetime—and they are discharged with the feces. Thus hookworm eggs may contaminate the soil in those warm climates of the world where people defecate on the ground. The eggs hatch, releasing a tiny larva that develops in the soil and becomes infectious. These larva infect by penetration through the skin, generally of the feet. They then enter the bloodstream and are swept along with the blood to the heart and lungs. In the lungs they push through the blood vessel walls into the alveoli, and thence gain entrance to the gastrointestinal system when lower respiratory tract secretions are swallowed. Once in the small intestine they attach and mature to complete their life cycle.

Infestation with hookworms can be detected by examining fecal material for the eggs (ova) (Fig. 31-6). The four-celled ovum identifies the worm as a hookworm, and the species may be determined by allowing it to develop and hatch in a petri dish. The species *Necator americanus* and *Ancylostoma duodenale* parasitize humans. Other hookworms infest animals such as the dog and cat, and although the latter are able to penetrate the skin of man, they cannot complete their life cycle in man. They wander around through the deeper layers of the skin

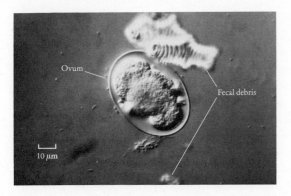

FIGURE 31-6
Hookworm ovum. The egg is in the four celled stage characteristically found in the feces. (Courtesy of Dr. F. Schoenknecht)

for a while and then die, often producing dramatic trails of inflammation and itching. Children may contract infestation of this kind by playing in sandboxes where dog or cat feces have been deposited.

Ascaris

Ascaris lumbricoides is a human parasite which, like the hookworm, has relatives that infest other animals, including the dog and the pig. Ascaris infestation is common in areas of poor sanitation or where human feces are used as fertilizer. They live free in the human gut, sucking in intestinal juices for nourishment, and often causing little trouble. Individuals with massive infestations may, however, suffer severe consequences, as, for example, when the worms get tangled into a ball which completely obstructs the intestine. At other times the worms wander into parts of the body they do not usually infest, like the liver. Fortunately, they only live about one year.

The female ascaris can produce about 200,000 eggs a day. These pass through the intestinal tract and are deposited on the soil with feces. Since some people have several hundred worms, the eggs can be spread in enormous numbers, and easily contaminate fingers, foodstuffs, or drinking water which have come into contact with the soil. After a period of development, the eggs are infectious by ingestion. When the eggs enter the gastrointestinal tract of a human being, they hatch and the tiny larva burrows through the intestinal wall and enters intestinal veins. Then, just as with hookworm larva, they are carried to the lung, enter the air passages, and return to the intestine upon being swallowed. Why they take the long journey through the blood vessels is obscure. Eggs of ascarid worms from dogs and cats may infect children who ingest contaminated food or other materials, but just as with hookworms the larvae are unable to complete their life cycle. Their wanderings through tissues such as the eye and liver may produce serious illness.

ASCARID WORMS OF DOGS AND CATS

Trichinella and Trichinosis

In the spring of 1968 an Ohio family of seven ate meals containing pork. Between two and five weeks later, all but one of them became sick with fever, muscle pain, and swelling around the eyes. Some were also nauseated and had a rash.

Trichinosis was suspected. Epidemiologists investigated and the diagnosis was confirmed when it was found that some of the pork had been eaten raw as sausage, and tests on some of the remaining sausage showed three larva of trichina worms in 50 grams (less than 2 ounces) of meat.

This episode is not atypical of many that occur all over the world every year. The cause of the disease trichinosis is *Trichinella spiralis,* a tiny worm 1 to 4 mm long which lives in the small intestine of meat-eating animals, especially rats, pigs, bears, dogs, and human beings. The female worm discharges living young into the lymph and blood vessels of the intestine without an intervening egg stage, and these larvae are carried to all parts of the body. They cause an inflammatory reaction and many of them die and are dissolved by body defense mechanisms. Those which arrive alive in the muscles often lodge there. The muscle tissues react to form a little scar tissue and inflammatory cells around them, and they stay alive for months or years. Then if the flesh of the host is eaten by man or another animal, the digestive juices of the new host release the worm from its case, it burrows into the intestinal lining, matures, and the female worms begin producing larvae in the new host to complete their life cycle. Each female may live four months or more and produce 1500 young.

The Ohio family mentioned was lucky in that the number of larvae in the pork was small. In some instances the same quantity of meat has contained 1 million encysted worms! Since about half would be female, and each one produces more than 1000 young, it is easy to see why trichinosis is sometimes a serious, even fatal disease.

Laboratory diagnosis of trichinosis in humans is carried out in two ways. The simplest method is to draw some blood from the patient and test for complement-fixing or flocculating antibodies to worm antigen. The other way is to remove a small piece of muscle (biopsy), cut it into ultra-thin slices, and inspect under the low-power lens of the microscope to find the worm larvae. As with infections caused by many other species of worm, there is a high percentage of a particular kind of white blood cell (eosinophile).

Complement-fixation
Page 380

Prevention of trichinosis in people depends on cooking meat adequately throughout, that is, at least 140°F for 30 minutes per pound. Pork has been the chief offender, presumably because of the practice of feeding uncooked garbage containing meat scraps to pigs. However, beef ground in a machine previously used for pork has also resulted in cases of trichinosis, and bear meat is a notorious cause. Government inspection of meats does not detect *Trichinella* infested meats, and adequate destruction of worm larvae by freezing is often not practical because of the low temperatures and long time required.

PLATYHELMINTHES (FLATWORMS)

A "third phylum" containing multicellular parasites of human beings is Platyhelminthes, the "flatworms." Relative to arthropods and nematodes, flatworms are very primitive forms of life. Their digestive tract is usually incomplete or absent, and there is no blood circulatory system. Most are hermaphroditic, but some forms such as human blood flukes have separate male and female worms. Sexual reproduction is usually followed at some point in the life cycle by a mode of reproduction that resembles budding. Flatworms parasitic for man are divided into two major groups: flukes, which are relatively short, flat, bilaterally symmetrical worms that generally attach by one or more sucking discs, and tapeworms, usually longer and ribbonlike in appearance.

Flukes

The adult flukes may inhabit blood vessels, organs such as lung or liver, or the intestinal tract. Their life cycle is often very complicated, involving several hosts.

Schistosomiasis is a disease caused by flukes of the genus *Schistosoma*. It is often a chronic, slowly progressive illness resulting in damage and loss of function of the liver, with resulting malnutrition, weakness, and accumulation of excess fluid in the abdominal cavity. World-wide incidence of schistosomiasis has been estimated at 100 million cases. In the Western Hemisphere, the disease occurs chiefly in the Caribbean area and South America, and many thousands of cases have been seen in cities of the United States to which Latin Americans have emigrated.

Schistosoma mansoni, a common cause of schistosomiasis, is about 10 mm long and lives in the small veins of the intestinal wall. Its life cycle is depicted in Figure 31-7. The female worm discharges eggs, some of which are swept into the liver by the flow of blood, and others rupture through the blood vessels and adjacent intestinal lining to enter the feces. The major symptoms of the disease are caused by the eggs, many of which lodge in the liver, producing inflammation and scarring and eventual loss of liver function.

The cercarial form infects human beings by burrowing through the skin of persons wading in infested waters. They penetrate to the blood vessels beneath the skin, leaving their tails at the body surface, and are carried by the blood flow to the intestinal veins where they become mature and complete their life cycle. The worms can live for more than 25 years, and continue to produce eggs and consequent liver damage.

Measures used in the control of schistosomiasis include treatment of infected cases, sanitary disposal of feces, chemical and biological control of snail hosts, and anticercarial measures. The expanding use of irrigation rice farming, as well as the existence of animal hosts for some species of schistosomes, has made control difficult.

The disease "swimmers itch," common in parts of North America and other areas of the world, is caused by the cercariae of schistosomes of birds or other animals. These cercariae penetrate the skin, but just as with animal hookworms and ascarids, they are unable to complete their life cycle in human beings.

Tapeworms

Tapeworms, like other flatworms, are primitive creatures somewhat more advanced than jellyfish. However, because of their parasitic adaptation, tapeworms have lost an important evolutionary feature, their intestinal tract. In gross ap-

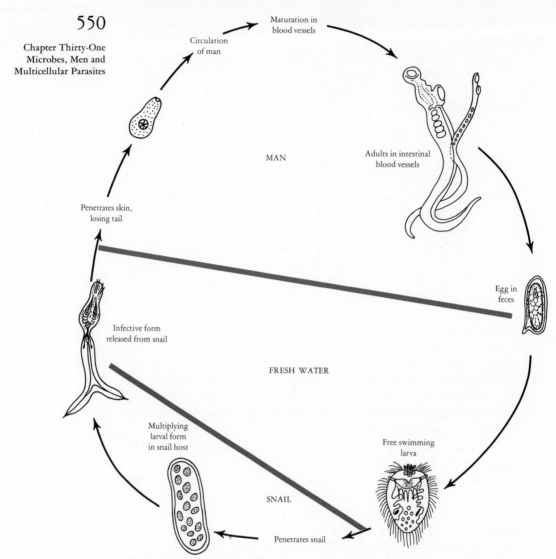

Maturation in
blood vessels

Circulation
of man

MAN

Adults in intestinal
blood vessels

Penetrates skin,
losing tail

Egg in
feces

Infective form
released from snail

FRESH WATER

Multiplying
larval form
in snail host

Free swimming
larva

SNAIL

Penetrates snail

FIGURE 31-7   Life cycle of *Schistosoma mansoni* (developmental forms not to scale).

pearance they resemble a white ribbon made up of more or less well-defined segments. Their flatness ensures that all the cells constituting the organism are close
to the outside environment from which nutrients must diffuse.

FISH TAPEWORM   *Diphyllobothrium latum,* a parasite found in human beings, has the general
structure of most tapeworms. It has a head approximately 1×2 mm in size
designed for attaching the worm to the intestinal lining. At the end of the slender
neck is a small flattened segment, which is followed by progressively wider and
better defined segments as one moves away from the head. New segments are
continually formed at the neck and mature as they are pushed away from the head
by succeeding segments. These worms may live for more than ten years, and thus

attain enormous lengths of as much as 40 feet or more folded back and forth in the host intestine.

Microscopic inspection of a mature segment reveals that it contains little more than an egg-producing apparatus and (in the same segment) an organ which produces sperm. The worm might then be viewed as a string of hermaph-

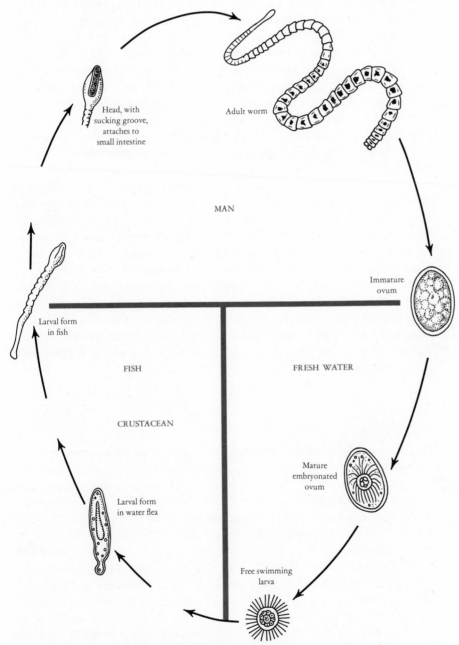

FIGURE 31-8 Life cycle of *Diphyllobothrium latum* (developmental forms not to scale).

roditic offspring arising from a neck and a head which serves mainly for attachment of the string to the wall of the host's intestine.

The number of eggs produced by a large worm may be great, perhaps a million a day. These eggs pass out of the segments through an opening near the center of one flat side, and are discharged to the outside world to begin a life cycle even more complex than that of *Schistosoma* (Fig. 31-8). If carried into freshwater, a round ciliated embryo hatches out and swims freely until eaten by a crustacean such as a water flea, the first intermediate host. It burrows into the body cavity of the new host, where it develops into a wormlike creature. The crustacean may then be eaten by a fish (second intermediate host), which in turn becomes infested by the growing worm. The parasite burrows into the flesh of the fish where it may grow as large as 15 mm, although in achieving this size its host may be eaten by a larger fish and pass along the growing worm. The worm can only reach adulthood if the raw or incompletely cooked fish is eaten by mammals such as man, dog, cat, or bear. Man, however, seems to be the preferred host, and develops the finest and most fertile specimens of *Diphyllobothrium*.

Note that, as with many other multicellular parasites, the earlier stages of development of tapeworms occur in intermediate hosts (for example, the snail and fish with *D. latum*), and adulthood with egg production in the definitive host (for example, man). The developing stages are invasive, and grow within the tissues of the intermediate host, while the adult normally simply attaches to the intestinal lining. Unfortunately, man can sometimes become infected by intermediate developmental forms of tapeworms, producing serious, even fatal, disease. This usually occurs when man ingests eggs of certain tapeworm species rather than the larvae, but may also come about with the practice in some locales of applying raw meat to open wounds in the mistaken belief the meat will aid recovery. If tapeworm larvae are present in the meat, they may penetrate the tissues via the wound.

SUMMARY

In addition to the unicellular and subcellular agents discussed earlier in this book, multicellular organisms may also parasitize man. Three phyla are involved: Arthropoda, Nematoda, and Platyhelminthes. The importance of arthropod parasites lies mainly in their susceptibility to microbes which are pathogenic for man. Mosquitoes may be infected with such agents as *Plasmodium* species (causes of malaria) and the viruses responsible for equine encephalitis. Fleas may carry the enterobacterium, *Yersinia pestis,* cause of plague; and lice, *Rickettsia prowazekii,* cause of typhus. Arachnids such as ticks may also be hosts for microbial pathogens, including *Rickettsia rickettsii* (cause of Rocky Mountain spotted fever), *Coxsiella burnetii* (cause of Q fever), and *Francisella tularensis* (cause of tularemia). Insects and arachnids infect man with such pathogenic microbes when they puncture the skin for a blood meal. By contrast, parasitic forms of the phyla Nematoda and Platyhelminthes are not important carriers of microbes pathogenic for man, but many themselves invade the human body. In some cases the adult worms live in man, and in others disease is caused by larval forms whose adult stage may be in other animals. Some examples are hookworms, ascarids, and *Trichinella* (phylum Nematoda); and the fluke *Schistosoma* and the tapeworm *Diphyllobothrium* (phylum Platyhelminthes).

QUESTIONS

1. Since mosquitoes feed on both human beings and plants, what do you think about the possibility of mosquitoes transmitting agents of human diseases to plants, and vice versa?
2. Since parasitic arthropods may feed on more than one mammalian host, it seems likely that simultaneous infection with agents human and animal diseases sometimes occurs. From knowledge of earlier chapters, would genetic interchange be possible?
3. Give the principal features of diseases acquired by man from other mammals.
4. In what ways might precise laboratory identification of the arthropod responsible for transmitting an epidemic disease help to influence control strategy?
5. Because of the host-range specificity of many microbes, it would seem possible to find microbes which would infect parasitic worms but not their hosts. What difficulties might you anticipate in applying such a biological control measure to the parasitic worms discussed in this chapter?

FURTHER READING

BELDING, D. L., *Textbook of Parasitology.* New York: Appleton-Century-Crofts, 1965.

HAWKING, FRANK, "The Clock of the Malaria Parasite," *Scientific American,* 222 (June 1970), 123.

HERMS, W. B. and M. T. JAMES, *Medical Entomology.* New York: Macmillan Co., 1961.

JOHNSON, W. H., L. E. DeLANNEY, T. A. COLE, and A. E. BROOKS, *Biology,* 4th ed. New York: Holt, Rinehart and Winston, Inc., 1972.

JONES, J. C., "The Sexual Life of a Mosquito," *Scientific American,* 218 (April 1968), 108.

KADIS, S., T. C. MONTIE, and S. J. AJL, "Plague Toxin," *Scientific American,* 220 (March 1969), 92–100.

LANGER, W. L., "The Black Death," *Scientific American,* 210 (February 1964), 114.

ROTHSCHILD, MIRIAM, "Fleas," *Scientific American,* 213 (December 1965), 44.

ROUECHE, BERTON, "A Swim in the Nile," in *A Man Named Hoffman.* Boston: Little, Brown and Co., 1965.

———, "Alerting Mr. Pomerantz," in *Eleven Blue Men.* New York: Berkeley Medallion, 1965.

YOELI, M., "Animal Infections and Human Disease," *Scientific American,* 202 (May 1960), 161.

ZINSSER, HANS, *Rats, Lice and History.* Boston: Little, Brown and Co., 1935.

Chapter 32

Antibiotics and Chemotherapeutic Agents

Similarities in the biosynthetic pathways of all living cells have been discussed previously. Fortunately differences do exist, and substances can be found which interfere with vital biochemical pathways of microbial cells in concentrations which do little or no harm to mammalian cells. Such substances are said to have *selective toxicity;* they poison one type of cell, but spare others. The search for such substances began in the early part of the twentieth century with the observation that some dyes had great ability to stain one kind of bacterial cell but little affinity for another. It was reasoned that dyes having selective affinity for a group of cells would selectively kill them as well. Although this idea was only partly true, it led to the discovery of sulfa drugs, the first of a long series of medicines demonstrating substantial activity against pathogenic microbes when given to people in doses producing little or no toxic effect. The ratio of maximum tolerated dose

to the minimum dose required to cure infections is called the *therapeutic index.* The quest for medicines of high therapeutic index represents one of the greatest success stories of modern science.

It has been customary to divide the drugs which have emerged into two large classes: *chemotherapeutic agents,* synthesized from known chemicals *in vitro* without the aid of living cells or extracted from plants; and *antibiotics,* produced by microbial cells. As shown in Table 32-1, these antimicrobial medicines can be grouped according to their effectiveness against different kinds of microbes. For example, the *macrolides* are chiefly active against gram-positive bacteria, whereas the *polymyxins* are useful against gram-negative bacteria. Antibiotics and chemotherapeutic agents may also be classified according to whether their principal action results in killing microbes or merely inhibiting them. Thus penicillin is classed as *bactericidal,* and tetracycline is *bacteriostatic.*

554

TABLE 32-1
Spectra of Activity of Antimicrobial Medicines

Primarily active against gram-positive bacteria

Penicillins	Vancomycin
Bacitracin	Macrolides

Primarily active against gram-negative bacteria

Polymyxins

Activity against many gram-positive and gram-negative bacteria

Tetracyclines	Cephalosporins
Chloramphenicol	Sulfa drugs
Aminoglycosides	

Active against mycobacteria

Aminoglycosides	Isonicotinic acid hydrazide (INH)
Cycloserine	Para-aminosalicylic acid (PASA)
Rifampin	Pyrazinamide
	Ethionamide

Active against protozoa

Emetine	Sulfa drugs
Quinine	Primaquine
Chloroquine	Chloroguanide

Active against fungi

Amphotericin	Nystatin
Griseofulvin	

ANTIBIOTICS

As shown in Table 32-2, only three groups of microbes have yielded useful antibiotics: actinomycetes (filamentous branching bacteria), eubacteria of the genus *Bacillus,* and saprophytic molds. Almost all useful antibiotics have come from four genera: *Streptomyces, Bacillus, Penicillium,* and *Cephalosporium.* To obtain a desired antibiotic a carefully selected strain of the appropriate producing species is inoculated into medium and incubated. As soon as the maximum antibiotic concentration is reached, the drug is extracted from the medium and extensively purified. Its activity is then standardized against a reference sample of pure anti-

Actinomyces
Page 261, 262
Bacillus
Page 257
Molds
Pages 282, 283

TABLE 32-2
Sources of Antibiotics used for Systemic[a] Administration

Actinomycetes	Bacillus Species	Molds
Tetracyclines	Polymyxins	Penicillins
Chloramphenicol	Bacitracin	Cephalosporins
Macrolides		Griseofulvin
Aminoglycosides		
Polyenes		
Vancomycin		
Cycloserine		

[a]Systemic means given by injection or by mouth for absorption into the circulatory systems, as opposed to use on body surfaces.

biotic, and it is put into a form suitable for administration (such as flavored, buffered, coated to protect against stomach acid, put into capsules, pressed into tablets, or mixed with substances to prolong absorption). New antibiotics must receive extensive evaluation of their properties: spectrum of activity against pathogens; toxicity (to define permissible doses); identification of pathways for their metabolism and excretion by the host; measurement of concentrations obtained in tissues, blood, and other host body fluids; effect of protein-binding and pH on their activity against microbes; and stability under various conditions. In many instances the microorganism which produces the antibiotic yields several closely related substances as well, and mutants of the strain are selected which give the best production of the most desirable analogs. For example, a mutant of superior ability for penicillin production was obtained from mold cultures exposed to X-rays to induce mutations.

MAJOR GROUPS OF ANTIBIOTICS

Penicillins

Penicillin was discovered accidentally in 1928 when Sir Alexander Fleming, working at St. Mary's Hospital in London, noticed that a mold contaminant on a culture plate containing staphylococci resulted in a zone of killing of the bacteria. The mold was a species of *Penicillium*—thus the name of the antibacterial substance, *penicillin*. Filtrates of broth cultures of the mold proved to have insufficient activity for treating infected wounds and it was not until purification attempts at Oxford University a decade later that sufficiently potent material became available for treating infections. The intervention of World War II spurred British and American workers to develop means for large-scale production and to determine the chemical structure. Several different analogs of penicillin were found to be present in the mold cultures, and these were designated F, G, X, K, O, and so on. Of these, penicillin G (or benzyl penicillin), (Fig. 32-1) was found to be most suitable, being relatively stable and possessing high antibacterial activity.

Penicillin was the first safe and effective antibiotic for systemic therapy, and it remains today the drug with the greatest therapeutic index. However, penicillin G has the drawback that it is unstable in acid solutions, and therefore cannot survive stomach acid, and be absorbed consistently after oral administration. Moreover its spectrum of activity is limited to gram-positive bacteria (such as *Streptococcus pyogenes*) and a few gram-negatives (such as *Neisseria meningitidis*). It

Penicillinase
Page 200

is completely ineffective against bacteria which produce penicillinase. The development of penicillin V, a relatively acid stable compound, only partially answered these objections.

In 1959 English scientists discovered that the main portion of the penicillin molecule—6-amino penicillanic acid—could be produced by appropriate mold cultures and then modified chemically by addition of side chains. From the many

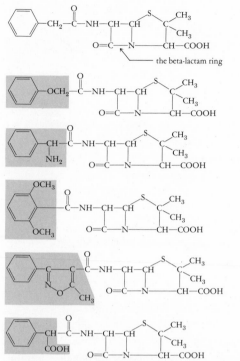

Penicillin G

Penicillin V
(acid resistant)

Ampicillin
(broadened spectrum)
acid resistant

Methicillin
(penicillinase resistant)

Oxacillin
(acid and penicillinase resistant)

Carbenicillin
(broadened activity against gram negative rods)

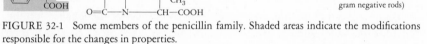

FIGURE 32-1 Some members of the penicillin family. Shaded areas indicate the modifications responsible for the changes in properties.

resulting compounds, several were selected for use in treating infections (Fig. 32-1). Further alterations have yielded a large family of penicillins with differing properties for use in various kinds of infections. For example, carbenicillin is effective against many strains of *Pseudomonas aeruginosa,* a species generally resistant to other penicillins.

Although nontoxic, penicillins occasionally cause death when administered to people who are allergic to them (an estimated 300 to 500 deaths per year in the United States); as mentioned in Chapter 21 the penicillins themselves are nonantigenic, but can act as haptens when they or their breakdown products attach to body proteins. This is why doctors and nurses always ask about penicillin allergy before giving the drugs. In the past some people probably became hypersensitive to penicillin because of unknowing exposure to it in vaccines, or milk from cows under treatment with the drug, but now strict controls make this possibility less likely.

Another drawback of the penicillins is their instability in aqueous solutions, so that they must be used promptly after preparation to avoid loss of antibacterial activity. Even water condensing from the air may result in decreased activity of penicillins which have been refrigerated. Also widespread and often inappropriate use of penicillins has resulted in a selective increase in resistant strains as the sensitive ones were killed. Some resistant bacteria even grow in penicillin

Haptens
Page 375

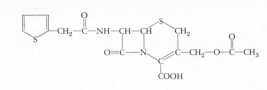

FIGURE 32-2   Cephalosporin derivatives.

solutions and can cause infections when such contaminated medicine is administered for treatment. Finally, as with most medicines taken orally, food may interfere greatly with absorption of penicillins.

Cephalosporins

A strain of mold of the genus *Cephalosporium,* obtained from the sea near a sewage
outlet, was found to produce antibiotic substances somewhat similar in structure
to penicillin (Fig. 32-2). One of these, cephalosporin C, could be modified
chemically to produce four useful antibiotics. Two of these are susceptible to
stomach acid and must be given by injection; the other two can be given orally.
Like penicillins they are bactericidal and have a high therapeutic index. They are
active against many gram-positive and gram-negative pathogens. Their difference
in chemical structure is sufficient to make them relatively resistant to penicillinase
and to permit administration to some people allergic to penicillins. They are,
however, susceptible to cephalosporinases produced by a variety of bacteria.

Tetracyclines

Unlike penicillins and cephalosporins, that are of fungal origin, tetracyclines
(Fig. 32-3) are produced by certain strains of *Streptomyces.* A family of useful tetracyclines has been produced by selection of appropriate producing strains and
by chemical alteration. This family includes oxytetracycline, chlortetracycline,
demethylchlortetracycline, minocycline, and tetracycline. All have a very similar

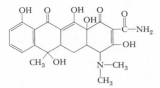

FIGURE 32-3

Tetracycline. Modifications of this structure yield oxytetracycline, chlortetracycline, demethylchlortetracycline, minocycline, and others.

spectrum of activity, but vary in their stability, toxicity, and affinity for blood proteins. They are often referred to as "broad spectrum" antibiotics because they show activity against strains from many different genera, both gram-positive and gram-negative. This term, however, may be misleading because of the high incidence of tetracycline resistance among various genera of bacterial pathogens. The tetracyclines are usually given orally and have a low therapeutic index, producing relatively low levels of antimicrobial activity in body fluids. They are bacteriostatic, and are easily bound and inactivated by foods and metallic ions.

Tetracyclines, like many other drugs, bind reversibly to serum proteins. The binding is an equilibrium reaction, so that a fraction of the antibiotic is free to react with microbes and to diffuse into tissue fluids. When the fraction of free drug is small, as with tetracyclines, concentrations entering body fluids are relatively low. Excretion is delayed, since protein-bound drug does not normally enter the urine.

In the past, tetracyclines have been widely and indiscriminately used: for example in animal feeds; in infectious illnesses of man when the etiologic agent could not be readily determined; and in efforts to prevent secondary bacterial infections in persons suffering from colds, influenza, measles, heart failure, or surgical wounds. It is now known that this latter practice may actually increase the rate of infection, probably because the tetracyclines interfere with the protective effect of the normal bacterial flora and allow colonization or overgrowth by resistant microbes, including *Staphylococcus aureus, Candida albicans, Pseudomonas aeruginosa,* and species of *Proteus* and *Klebsiella.* Extensive use of tetracyclines in former times has resulted in an increasing percentage of resistant strains of *Streptococcus pyogenes, Neisseria gonorrhoeae,* and *Bacteroides* species, all of which were previously susceptible. Toxic effects of tetracyclines range from diarrhea and discoloration of teeth to growth retardation of infants. Severe malfunction of the liver can result from large doses, and tetracyclines administered after prolonged storage (such as in the family medicine chest) can cause serious derangement of kidney function due to a deterioration product of the drug.

UNTOWARD EFFECTS OF TETRACYCLINES

Chloramphenicol

Originally isolated from *Streptomyces venezuelae,* chloramphenicol (Fig. 32-4) is now synthesized more cheaply by chemical methods without the need for biosynthetic steps. Like tetracyclines it has a broad spectrum of antimicrobial activity and is generally bacteriostatic in concentrations reached in body fluids. It has a higher therapeutic index than the tetracyclines, is less bound to serum protein, and readily diffuses into cells and spinal fluid to act on bacterial pathogens out

FIGURE 32-4

Chloramphenicol. The structure of the side chain (shaded portion of the molecule) is of critical importance to its antimicrobial activity, but the part of the molecule responsible for toxic effects is not known.

$$NO_2 - \text{(benzene ring)} - \overset{\overset{OH}{|}}{CH} - \overset{\overset{CH_2OH}{|}}{CH} - NH - \overset{\overset{O}{||}}{C} - CHCl_2$$

UNTOWARD EFFECTS OF CHLORAMPHENICOL

of the reach of other antibiotics. Toxic effects result with high doses—mainly interference with the normal development of red blood cells. Much more serious, however, is a reaction occurring in about 1 out of every 40,000 patients who receive the drug. This complication (called aplastic anemia) is characterized by inability to form white and red blood cells, and often ends in death from infection or the development of leukemia. For this reason use of chloramphenicol is generally restricted to life-threatening infections where equally effective alternate treatment is not available. Nevertheless chloramphenicol is extensively used in economically poor countries because of its low price. Although it has not yet been proven, persons who develop aplastic anemia after receiving chloramphenicol are thought to have a genetically determined defect in metabolizing the drug. During the years of peak use of chloramphenicol in the United States, fewer than half the persons developing aplastic anemia had received the antibiotic, which shows that there are other causes of aplastic anemia besides chloramphenicol.

Aminoglycosides

Glycosidic bond
Page 44

The aminoglycosides are classified together because all have two or three chemical components linked glycosidically (Fig. 32-5). They share many properties, but

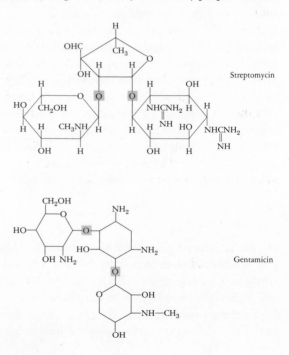

Streptomycin

Gentamicin

FIGURE 32-5
Aminoglycoside antibiotics. The shaded areas indicate the glycosidic linkages.

do not necessarily cover the same spectrum of antimicrobial activity. Strepto-mycin, the first of the aminoglycosides, was found after screening 10,000 cultures of soil bacteria; the producing strain, *Streptomyces griseus,* was originally dis-covered in the throat of a chicken! It is a bactericidal antibiotic, active against many gram-positive and gram-negative bacteria, and one of the few antibiotics that are active against *Mycobacterium tuberculosis.* It is poorly absorbed from the gastrointestinal tract and must be given by injection. Unfortunately, usefulness of this promising drug was soon impaired because of the high frequency of very resistant mutants among many bacterial pathogens. Toxicity involves damage to the kidney and the nervous apparatus for equilibrium, but the most feared effect is irreversible deafness. The toxic effects result when prolonged high concen-trations of the drug occur in the bloodstream, a situation that is especially likely to occur in persons with kidney diseases that interfere with normal excretion in the urine. Streptomycin has little activity against anaerobic bacteria or under acid conditions.

Neomycin and kanamycin have been widely used orally to reduce intestinal enterobacteria prior to surgery, but there is evidence that this practice may lead to superinfection with resistant organisms, including anaerobic bacteria such as *Bacteroides* species. Gentamicin, a newer aminoglycoside, has a special role in treating infections with *Pseudomonas.*

Miscellaneous Chemical Types

The macrolide antibiotics (Fig. 32-6), erythromycin and oleandomycin, share the properties of being active primarily against gram-positive bacteria (a spectrum similar to penicillin G), usually bacteriostatic, and readily absorbed after oral ad-ministration. They have a high therapeutic index. Their main use is in treatment of infections in persons allergic to penicillins. Unfortunately resistant variants sometimes arise during treatment of staphylococcal infections.

Lincomycin is chemically distinct from other antibiotics, but has an anti-microbial spectrum similar to the macrolides. The therapeutic index is high, and its action is generally bacteriostatic in the concentrations reached in body fluids. A number of chemical modifications of lincomycin has now been developed, some of which are better absorbed and show greater *in vitro* activity than the

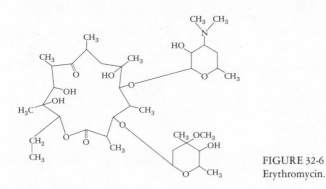

FIGURE 32-6
Erythromycin.

parent compound. Although staphylococci resistant to the macrolides may be sensitive to lincomycins, resistant forms quickly arise on exposure to the latter drugs.

Polypeptide antibiotics (Fig. 32-7) (such as bacitracin and polymyxin) generally are toxic and poorly absorbed after oral administration. Bacitracin is a bactericidal antibiotic produced by a strain of *Bacillus subtilis* originally obtained from dirt in a wound from a girl named Tracy; hence the name bacitracin. It is effective in treating infections by gram-positive bacteria such as *Staphylococcus*

Polypeptides
Page 30

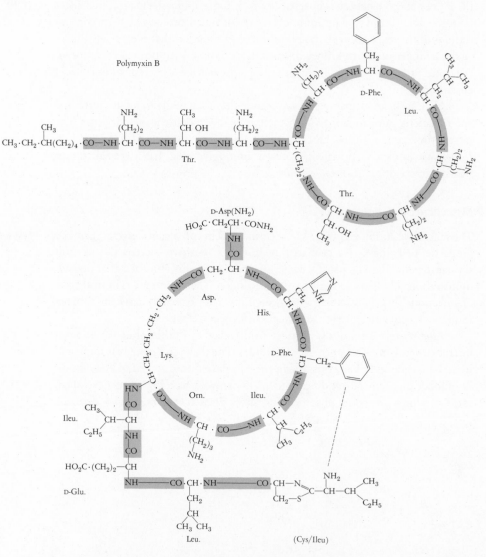

FIGURE 32-7 Polypeptide antibiotics. Shaded areas indicate peptide bonds.

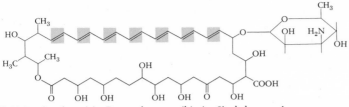

FIGURE 32-8   Amphotericin B, a polyene antibiotic. Shaded areas show unsaturated (ene) groups.

aureus. However, it is relatively toxic when given by injection and is therefore used mainly on body surfaces. Polymyxins are produced from another species of *Bacillus* and, in contrast to bacitracin, are active against gram-negative organisms. They are useful primarily in treating serious infections due to *Pseudomonas aeruginosa,* but share with bacitracin the property of poor diffusion into infected tissue. In addition, polymyxins are readily inactivated by pus and dead tissue probably because they react with free nucleic acids.

Polyene antibiotics (Fig. 32-8) (such as nystatin and amphotericin) have activity against some fungi and protozoa and, although toxic, are among the few available agents for treatment of infections by eucaryotic forms.

Uses of amphotericin
Pages 466, 467

The *rifamycins,* discovered in 1957 in cultures of *Streptomyces mediterranei,* were found unsuitable for therapeutic use, but a derivative, rifampin (Fig. 32-9) has shown great activity against many gram-positive and gram-negative organisms as well as mycobacteria. Even more exciting is its apparent action against certain viruses. Rifampin can be administered orally and little toxicity has been noted so far. Its main defect has been that highly resistant mutants are readily selected from several bacterial genera. The rifamycins are chemically unrelated to other antibiotic families.

Finally, a few antibiotics have selective activity against the cells of malignant tumors. Unfortunately the therapeutic index is very low, and some toxic effects on the host are the rule. Actinomycin D (Fig. 32-10) and mitomycin C are examples of such agents. They have given temporary arrest of some tumors lasting several years.

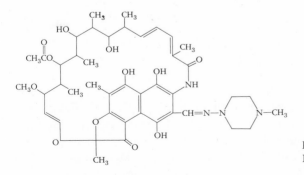

FIGURE 32-9
Rifampin, which shows both antibacterial and antiviral activity.

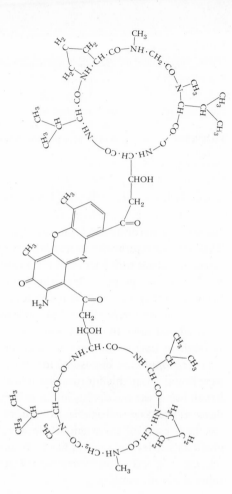

FIGURE 32-10
Actinomycin D, an anti-tumor antibiotic.

MODES OF ACTION OF ANTIBIOTICS

Some of the modes of action of antibiotics, covered in Chapters 4 and 8, are summarized in Table 32-3. Several different biochemical steps in the synthesis of bacterial cell wall can be blocked as can various portions of the protein and nucleic acid biosynthetic pathways. Finally, some antibiotics cause direct damage to cytoplasmic membranes.

Modes of action
Pages 83, 177, 178

CHEMOTHERAPEUTIC AGENTS

Sulfonamides

The discovery of sulfa drugs was the result of the systematic and tedious examination of hundreds of dyes carried out at the I.G. Farben Company in Germany during the early 1930s. These dyes were tested for activity against streptococcal infections, and one, prontosil, was found to be effective. Mysteriously the drug was effective *in vivo* but not *in vitro*. British and French scientists shortly solved

TABLE 32-3

565

Mechanisms of Action of Some Antibacterial Agents

Type of Drug	Known Mechanism of Action
Penicillin	Blocks normal cell wall production
Chloramphenicol	Blocks protein synthesis by interfering with incorporation of amino acids
Tetracycline	Blocks protein synthesis by interfering with tRNA-amino acid on the ribosome
Aminoglycoside	Affects the ribosome, giving abnormalities of protein synthesis
Polymyxin	Damages the cytoplasmic membrane
Rifamycin	Prevents microbial RNA synthesis by reacting with RNA polymerase

this puzzle by showing that the activity of the dye actually resulted from sulfanilamide formed in the animal body by splitting off a chemical side chain of the dye. Sulfanilamide was found to be readily absorbed and to have a high therapeutic index, although insoluble at acid pH and therefore prone to precipitate in the kidney and produce damage. Since it was a small molecule not greatly bound by serum protein, it diffused readily into tissues and spinal fluid. It was readily modified chemically so that a whole family of sulfa drugs with different properties is now available. Unfortunately widespread and indiscriminate use of sulfa drugs has led to selection of resistant mutants in many genera of pathogenic organisms so that the drugs are limited in their usefulness today. In addition, certain forms of sulfa drug result in severe allergic reactions sufficiently often to make other antimicrobial medications preferable.

Resistance of *N. meningitidis*
Page 505

Sulfonamides are one of the few classes of chemotherapeutic drugs for which the mode of action against microbes is understood (Chapter 7). In contrast to mammalian cells, some microbes (including some protozoa, fungi, and many bacteria) are incapable of using various forms of the vitamin folic acid found in plant or animal foods. Such microbes synthesize their folic acid from para-aminobenzoic acid (PABA). Sulfa drugs by competitive inhibition cause blockade of folic acid synthesis. Folic acid is essential in the metabolism of cells because it is a required component of an important enzyme which has essential roles in the production of purine and pyrimidine bases and some amino acids. Such substances may be considered the *end products* of the folic acid containing enzyme. The effect of sulfonamides on susceptible microbes is not seen until several cell divisions have occurred, since folic acid formed prior to the addition of sulfonamide must first be depleted. The action of sulfa drugs can readily be reversed by adding PABA or the end products. Sulfa drugs may act on some microbes in ways different from that described above. In fact pathogenic effects of some rickettsiae may actually be enhanced by sulfonamides.

Pages 145–146

Purines and pyrimidines
Page 40

Other Chemotherapeutic Agents

Several other small molecular weight chemicals (Fig. 32-11) selectively inhibit metabolic processes of certain microbes. *Trimethoprim* interferes with a later stage in folic acid metabolism, and markedly potentiates the chemotherapeutic

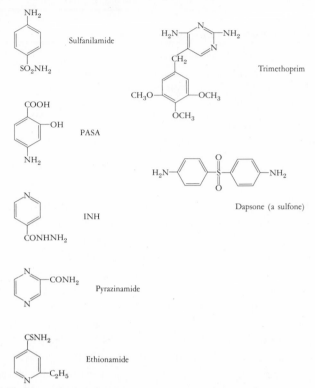

FIGURE 32-11 Chemotherapeutic agents.

Leprosy
Page 507

action of sulfonamides. *Sulfones,* useful in treating leprosy, probably have a similar mode of action to the sulfonamides, since they are antagonized by PABA. *Para-amino-salicylic-acid* (PASA), a drug useful only in treating tuberculosis, is also antagonized by PABA. Other simple cyclic compounds including *isonicotinic acid hydrazide* (INH) (Fig. 32-11 and Table 32-1) are also useful in treating tuberculosis. Their mechanisms of action are not yet well defined, but they differ from sulfonamides since PABA is not antagonistic.

QUININE

Ancient remedies sometimes remain useful. For example, over 300 years ago Spanish monks in Peru discovered that the chinchona bark used medically by the native people was effective against malaria. It is now known that *quinine* (Fig. 32-12) is the chief antimalarial ingredient in chinchona bark, and it is still useful in treating persons infected with *Plasmodium falciparum* strains resistant to more modern medications. Little is known of the mode of action of quinine except that it has diverse effects on specific microbial and mammalian cells. Only certain developmental forms of the malarial parasite are susceptible. Other agents useful

FIGURE 32-12
Quinine, an ancient antimalarial medication still lifesaving in some cases.

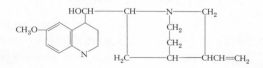

in treating diseases caused by protozoa are given in Figure 32-11 and Table 32-1. Although some, like emetine, are ancient remedies, their modes of action generally are still poorly defined.

IN VITRO DETERMINATION OF SUSCEPTIBILITY OF MICROORGANISMS TO ANTIMICROBIAL AGENTS

For some pathogenic species, susceptibility to therapeutic agents is predictable (for example, penicillin G always kills pneumococci, and chloroquine is generally active against *P. vivax* causing malaria). With others (such as *S. aureus*) there is no way of knowing reliably which drug is likely to be effective in a given case. In treating infections it has often been the practice to try one antimicrobial after another on the patient until a favorable response is observed, or if the infection is very serious, to give several antimicrobial agents together. Both approaches are undesirable, because with each unnecessary drug goes unnecessary risk of toxic or allergic effects, and undesirable alterations of the normal flora. *The cornerstone of rational treatment of infectious diseases thus rests with choosing the antimicrobial agent most likely to act against the offending pathogen and as few other cells as possible.* Practical methods exist for determining the susceptibility of the more rapidly growing bacterial pathogens. Some of the basic concepts involved in *in vitro* determinations of bacterial susceptibility are given below.

Quantification of Susceptibility

In principle, a series of decreasing concentrations of the antimicrobial agent is prepared in a suitable growth medium, and a suspension of the infecting organism is added to each concentration. After a period of incubation, the concentrations which have inhibited visible growth are determined (Fig. 32-13). The lowest concentration capable of preventing growth is called the minimal inhibitory concentration (MIC). Cultures containing concentrations of antimicrobial agent which show no growth of the microbe can be subcultured to medium lacking antibiotics to see whether viable organisms remain. In this way the minimal bactericidal concentration (MBC) can also be determined. The organism is said to be "sensitive" to the least concentration that inhibits growth, (1 microgram per ml.) (Fig. 32-13). If the concentration required to *kill* the organisms is only two to four times as much as the inhibitory concentration, the antimicrobial agent is said to be bactericidal; if higher concentrations are required, it is bacteriostatic.

MINIMAL INHIBITORY
AND MINIMAL
BACTERICIDAL
CONCENTRATIONS

Assays of Antimicrobial Agents

Merely knowing the MIC and MBC is not enough to tell whether a drug will be effective in treating an infection. It is also necessary to know whether these

(a)   A series of tubes containing decreasing concentrations of antimicrobial agent.

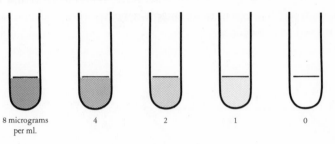

8 micrograms   4   2   1   0
per ml.

(b)   Addition of an invisible inoculum doubles the volume.

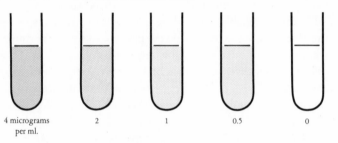

4 micrograms   2   1   0.5   0
per ml.

(c)   Appearance of growth in the more dilute solutions following incubation.

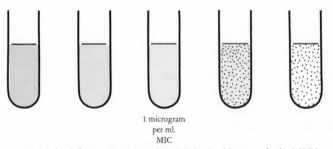

1 microgram
per ml.
MIC

FIGURE 32-13   Minimal Inhibitory Concentration (MIC). In this example the MIC is 1.0 microgram per ml. The tube containing medium without antibiotic is a growth control. In actual practice, an antibiotic control is also included (organism of known MIC).

concentrations are likely to occur in infected tissues. To do this each new drug is given to human subjects in doses known to be safe and nontoxic. Samples of their blood, urine, and other body fluids are then collected at different time intervals following administration of the drug. A very sensitive organism can then be used to measure the amount of antimicrobial drug present in the fluids. To do this a culture of the organism is spread over the surface of agar medium. Holes are then punched out of the agar, some of which are filled with fluid containing known concentrations of the drug to be assayed, while others are filled with the body fluid being tested. Following incubation, zones of inhibition are formed around the agar wells, their diameter depending on the concentration of antibiotic present. By measuring the zone sizes and plotting them against the corresponding concentrations, a curve relating zone size and concentration is obtained, from which the concentration of antimicrobial agent in the body fluid can be read (Fig. 32-14). Thus concentrations present at different times following administration of a drug can be determined, as in Figure 32-15 for serum levels of a penicillin.

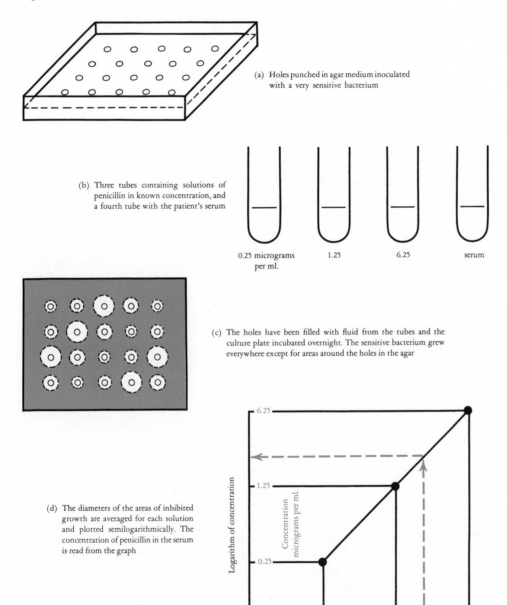

(a) Holes punched in agar medium inoculated with a very sensitive bacterium

(b) Three tubes containing solutions of penicillin in known concentration, and a fourth tube with the patient's serum

0.25 micrograms per ml.   1.25   6.25   serum

(c) The holes have been filled with fluid from the tubes and the culture plate incubated overnight. The sensitive bacterium grew everywhere except for areas around the holes in the agar

(d) The diameters of the areas of inhibited growth are averaged for each solution and plotted semilogarithmically. The concentration of penicillin in the serum is read from the graph

Logarithm of concentration

Concentration micrograms per ml.

6.25

1.25

0.25

10   20   30

Zone diameter, mm.

FIGURE 32-14 Assay of penicillin in a patient's serum.

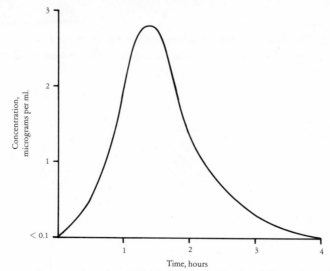

FIGURE 32-15   Concentration of penicillin in a patient's blood at different times after an oral dose.

The Meaning of "Sensitive"

Although often used in the quantitative sense, the word "sensitive" is also commonly used in a qualitative sense to describe organisms susceptible to concentrations of antimicrobial drug present in the blood of patients under treatment. The word "resistant" is used for microorganisms requiring substantially higher concentrations. Thus an organism with an MIC of only 20 micrograms per ml. of polymyxin would nevertheless be called "resistant" since blood levels of this drug are usually lower than 5 micrograms per ml. Microorganisms requiring inhibitory concentrations on the borderline between sensitive and resistant are often called "intermediate."

Distinguishing Sensitive and Resistant Organisms

ANTIBIOTIC DISCS FOR
SENSITIVITY TESTING

Since, in practice, it is technically difficult, expensive, and time consuming to determine MICs, qualitative susceptibilities of bacterial pathogens are determined using filter paper discs impregnated with antimicrobial agents. Thus to determine whether the *S. aureus* strain isolated from a boil is sensitive to penicillin or not, a culture of the organism is spread over the surface of agar medium and an antibiotic disc containing a precise amount of penicillin is placed on top of it. Following incubation, the zone of inhibition of growth is measured and, if large enough, shows that the organism may be called *sensitive*. To determine whether the inhibition zone is large enough, it is necessary to refer to tables of values for each drug. These tables were prepared by determining MICs and inhibition zone sizes simultaneously on bacteria of a wide range of susceptibility. In this way the inhibition zone diameters were correlated with the MICs, and knowing the expected blood levels of each drug, correlations between zone diameters and sensitivity could also be determined. Since antimicrobial agents diffuse through agar at different rates, there are different standards for each drug.

Standardization of Sensitivity Determinations

The medium used for determining sensitivity should be roughly comparable to the tissue fluids of the body in certain constituents, or erroneous information may result. For example, if the growth medium contains significant amounts of PABA (as many do), organisms will appear to be resistant to sulfa drugs even though quite susceptible to sulfonamides in the blood of a patient where there is little PABA. A hypothetical organism which is able to utilize preformed folic acid and synthesize it would falsely appear sensitive on a medium lacking folic acid. Other drugs may also show altered activity in certain media; for example, tetracyclines are inactivated by di- and trivalent metallic ions, and show enhanced activity at low pH, and aminoglycosides have decreased activity at low pH and increased activity at high pH. Because of such variations, a committee of the World Health Organization has developed detailed recommendations for standardization of methods used in determining bacterial susceptibility. Little work has been done yet in standardizing procedures for determining fungal, protozoan, viral, or metazoan susceptibilities.

LIMITATIONS OF THE VALUE OF ANTIMICROBIAL AGENTS

Selection of Resistant Variants

During the first few years after the introduction of sulfonamides and penicillin, there was great hope that such agents would soon do away with most infectious diseases. Yet today resistance limits the usefulness of all known antimicrobial drugs. For some organisms (such as *S. aureus*) penicillin-resistant strains were well established in nature before the introduction of penicillin, although at a low frequency. Heavy use of penicillin (now measured in many tons per year) soon fostered their selection, so that by 1950 more than 50 percent of strains of staphylococci causing infections were resistant. Such strains do not readily disappear when usage of an antibiotic is discontinued since they are well adapted for survival even without the competitive advantage of widespread drug usage.

Clones of sensitive bacteria contain mutants resistant to most antimicrobial agents ranging in frequency from about 10^{-3} to 10^{-9}. Under certain conditions of sustained heavy usage of antimicrobials, selection of these mutants may also occur, giving rise to resistant strains. Generally these mutants can maintain their dominance only with the help of heavy antibiotic usage. As soon as the drug is withdrawn from use, they lose their competitive advantage and again become a minor part of the total population of that species. This is true of chloramphenicol-resistant *S. aureus*, for example.

Transfer of genetic information from resistant to sensitive strains of bacteria by transduction and conjugation can also markedly increase the incidence of resistance to antimicrobial medications. Transfer of the genetic information necessary for resistance occurs very infrequently, but the presence of the appropriate antimicrobial drug gives a marked selective advantage to the recipient bacterium. Moreover, resistance factors (R factors), transferred by conjugation, may contain genes for resistance to several medications. Thus a person receiving a single antimicrobial agent runs the risk of having a sensitive pathogen acquire resistance to several drugs.

Finally, the establishment of resistant strains in a community is markedly influenced by conditions which allow their spread from person to person and their ability to colonize. Thus overcrowding, mobility, ventilation, sanitation, and antibiotic usage all may play major roles. The administration of an antibiotic to a person markedly increases the probability that he will be colonized by extraneous antibiotic-resistant strains, presumably because of suppression of sensitive strains in the normal flora which would otherwise compete with the resistant forms.

Some of the mechanisms by which microbes resist the action of antibiotics and chemotherapeutic agents are now known, and many studies of these mechanisms are being carried out in the hope of devising means to circumvent them.

MECHANISMS OF
RESISTANCE TO
ANTIMICROBIAL DRUGS

Some microbes resist the action of antimicrobial drugs by degrading them. For example, references have been made earlier to penicillinases, microbial enzymes which can attack the penicillin molecule and destroy its activity. A number of penicillinases are known, differing in inducibility, kinetics, and site of action against penicillins. Most act by opening the beta lactam ring (Fig. 32-1), but other penicillinases attack other sites on the penicillin molecule. Resistance to other antibiotics occurs through microbial enzymes which add a chemical group (such as phosphate) and thus cover up sites cn the molecule which normally would react with vital structures in the microbes. Resistance to other drugs is achieved by changes in permeability of the cell wall or cell membrane to prevent entry of the antimicrobial agent. Finally, with sulfa drugs some resistant bacteria have been shown to have increased production of the metabolic analog (PABA) and thus prevent the competitive biosynthetic blockade produced in sensitive cells.

Other Limitations

Drugs which act against microbes only prevent their growth or increase the rate of their killing. They have no effect on microbial toxins that have already been released and, in some instances, by killing microbes, they may enhance release of toxins with life-threatening consequences. Moreover the effectiveness of antibiotics is influenced by their ability to diffuse to the site of the infection. For example, aminoglycosides (such as streptomycin) may fail against meningitis because they cross the blood-brain barrier poorly. Conditions at the site of infection may not be favorable for antibiotic action. Penicillin kills streptococci in a few hours in broth medium, but many days are required for the same concentration to kill them if the cocci are sequestered in a blood clot. The acid of urine interferes with the activity of streptomycin, and polymyxins are neutralized by nucleic acids in pus. For the most part antimicrobial agents are of little value in eliminating infectious agents in abscesses or other sites where the organisms are not actively dividing. Antimicrobial medicines often fail to eliminate infections with sensitive organisms in persons with impaired host defenses, particularly those with impaired phagocytic function. Finally, antibiotics may prevent development of immunity to an infecting agent, presumably by allowing elimination of the organism before an adequate primary immunological response can occur.

SUMMARY

Antibiotics and chemotherapeutic agents are of value in treating infections because they have the property of *selective toxicity*. The ratio of antimicrobial activity to unfavorable effects on the host is referred to as the *therapeutic index*. Antibiotics are derived from only a few genera of microbes, and exhaustive searches of many strains are required to find ones suitable for use. The major groups of antibiotics consist of families of chemically related forms with varying properties, some resulting from manipulations of producing microbes and some from chemical alterations of products of biosynthesis. The sites of action include either the cell wall, protein synthetic mechanisms, the cytoplasmic membrane, or nucleic acid synthesis. Sulfonamides and some other chemotherapeutic agents act by interfering with synthesis of folic acids, necessary for coenzymes that mediate transfers of one carbon fragments.

The value of antibiotics and chemotherapeutic agents has been markedly impaired by the emergence and dissemination of resistant microbes. Studies of the mechanisms of action and the mechanisms of resistance have aided in finding new antimicrobial medicines. Antibiotics and chemotherapeutic agents have been of enormous value in controlling many serious infections, but their continued value depends largely on judicious use to minimize the incidence of resistant forms.

QUESTIONS

1. How do you interpret the fact that so few strains of microbes produce antibiotics?
2. Why is it much harder to find an antibiotic useful against viral diseases as opposed to diseases caused by bacteria?
3. How is it that bacteriostatic antimicrobial agents can cure infectious diseases?
4. Why might it be inadvisable to use antimicrobial agents to treat minor infections which would be overcome by body defenses anyway?
5. What is meant by sensitivity of a microbe to an antimicrobial agent?

BALDRY, P. E., *The Battle Against Bacteria.* New York: Cambridge University Press, 1965.

GARROD, L. P. and F. O'GRADY, *Antibiotics and Chemotherapy,* 3d ed. Baltimore: Williams and Wilkins Co., 1971.

GAUSE, G. F., *The Search for New Antibiotics—Problems and Perspectives.* New Haven: Yale University Press, 1960.

GOODMAN, L. S. and A. GILMAN (eds.), *The Pharmacological Basis of Medical Therapeutics,* 4th ed. New York: Macmillan Co., 1970.

HEDGECOCK, L. W., *Antimicrobial Agents.* Philadelphia: Lea and Febiger, 1967.

ROUECHE, BERTON, "Something Extraordinary," in *Eleven Blue Men and Other Narratives of Medical Detection.* New York: Berkeley Medallion, 1970.

WATANABE, I., "Infections and Drug Resistance," *Scientific American,* 217 (December 1967), 19.

IV

Microbes and the Environment

Nodules formed on root system of soybean plant (courtesy of Dr. F. J. Bergersen; Bergersen, F. J., 9th Internat. Congress of Soil Sci. Trans., Vol. II, Paper 6, p. 49, 1968)

Sterilization, Disinfection, Decontamination

To a large extent routine control of undesirable microbes is achieved by mechanical removal (washing and scrubbing), by drying, and by exposure to the competitive action of other microbes. However, in addition, it is frequently important to use special methods for decontaminating materials and keeping them free of unwanted microbes. Some of the chief approaches to accomplishing these goals are discussed.

DEFINITIONS

Sterilization

Removing or killing *all* microbes on an object or in any material is called *sterilization*. The elusive nature of this definition is exemplified by an incident that occurred during the controversy over spontaneous generation. In those days urine was commonly used as a growth medium, because it is nutritious for microbes, easily obtained, and almost free of microorganisms. Pasteur first showed that if urine were heated to a high temperature for a period and then kept from exposure to airborne contaminating microbes, it would remain free of microbial growth. Others, in repeating these experiments, found that such heated and presumably sterile urine would sometimes show luxuriant growth of bacteria upon the addition of sterile alkali. They interpreted this as "spontaneous generation"; in fact however, airborne spores which had contaminated the urine before heating were not killed by the heat and subsequently grew. The spores were only injured and rendered incapable of growth by the normal acidity of urine. When

Spontaneous generation
Page 6

575

the acidity was neutralized by sterile alkali, the spores were able to germinate. Therefore *"killing" means making microbes unable to reproduce even under the most favorable growth conditions.*

Disinfection

The reduction or elimination of pathogenic microbes in or on materials so that they are no longer a hazard is the process of *disinfection.* Unlike sterilization disinfection implies that some living microbes may persist. The term *disinfectant* is generally used for chemical agents employed to disinfect inanimate objects. In the United States the term *antiseptic* is generally used to indicate a suitably nontoxic disinfectant for use on animal tissues. *Decontamination* is often used interchangeably with disinfection, but implies a broader role including inactivation or removal of microbial toxins as well as microbial pathogens.

Germicide

An agent capable of killing most microbes (germs) rapidly is a *germicide.* Special terms (such as bactericide, fungicide, and viricide) are used to indicate killing action against specific microbial groups (such as bacteria, fungi, and viruses). Related terms (such as bacteriostatic and fungistatic) indicate that an antimicrobial agent is primarily inhibitory in its action, preventing growth without substantial killing. *These terms refer to generalities, since an agent may be -cidal against one sepcies of microbe and -static against another.* Also the action may change depending on concentration, pH, temperature, and other factors.

HEAT AS AN ANTIMICROBIAL AGENT

The value of cooking to prevent disease was recognized centuries before the discovery of microbes. Cooking times and temperatures designed to satisfy taste often are adequate to kill microbes. Unfortunately, this technique sometimes fails because of a discrepancy between observable changes on the surface and those inside the food. Temperatures adequate to kill the microbes in question must be maintained for an adequate time throughout the material being heated, and not just on the surface.

Dry Heat

Heating of materials other than food is frequently used to kill microorganisms. Sterilization of laboratory equipment and glassware is a familiar application, but similar techniques are used on landing modules in the space program. Dry heat is generally less effective than wet heat in killing microbes. Preparation of sterile laboratory glassware usually requires temperatures of 160 to 180°C for 1½ to 2 hours with static air. Circulating hot air sterilizes in half the time because of more efficient transfer of heat to the glassware.

Boiling

The use of boiling to prevent disease extends back at least to the time of Aristotle, who advised Alexander the Great to have his armies boil their drinking water. Under ordinary circumstances, with concentrations of microorganisms of less than 1 million per milliliter, suspensions of vegetative cells and eucaryotic spores can be sterilized in about 10 minutes. Most viruses also succumb, although large concentrations of hepatitis virus and scrapie agent may partly survive boiling for longer times. Some bacterial endospores including certain strains of *Clostridium perfringens* and *Clostridium botulinum* survive boiling for hours.

In boiling, heat is added to the fluid in a vessel with a progressive rise in temperature to a certain point. With water at sea level the temperature rises until it reaches 100°C (212°F). An additional amount of heat energy (heat of vaporization) is then taken up by the water without a change in temperature, whereupon boiling begins to occur. Practically speaking, the temperature does not go above 100°C, and at high altitudes water even boils several degrees lower than this because of the reduced atmospheric pressure. Alkali is sometimes added to boiling water to increase the killing power against spores. However, to kill the more resistant strains of spores, it is necessary to raise the temperature of boiling by increasing the pressure.

Slow viruses
Page 681
Clostridial diseases
Pages 482, 483, 518, 521

Pressure Cooking and Autoclaves

The first pressure cookers were probably used in the early 1800s to produce canned food for Napoleon's armies. These devices simply enclosed the vessel containing heated water so that pressures above those of the atmosphere might thus be attained. A suitable safety valve was included to prevent an explosion. Water at temperatures above 100°C could thus be employed, and even the most heat-resistant living microbes could be killed in a few minutes (penetration of the heat and steam may take longer if large objects are being sterilized, and this penetration time must be added to the killing time).

The modern autoclave is, in principle, a sophisticated pressure cooker with mechanisms for regulating the steam pressure and ensuring complete evacuation of air from the chamber (Fig. 33-1). The presence of air would allow an increase in pressure within the chamber without a corresponding increase in temperature. Thus it is important always to check the temperature as well as the pressure, since the latter is of little or no significance in killing microbes. Superheated steam (dry steam) is likewise unsatisfactory and kills at slow rates, similar to hot air. Pressures and temperatures inside an autoclave should correspond, as illustrated in Figure 33-2. The conditions usually employed are 15 pounds pressure above atmospheric and 121°C (250°F), although lower temperatures are used for some heat-labile bacteriologic media.

The main reason for the great effectiveness of the autoclave as compared to a hot air oven relates to the ready heat coagulation of essential enzymes which occurs under moist conditions. Release of the latent heat of vaporization by

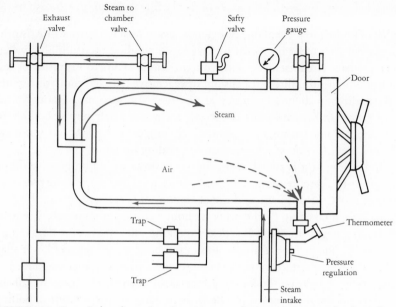

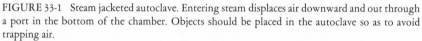

FIGURE 33-1 Steam jacketed autoclave. Entering steam displaces air downward and out through a port in the bottom of the chamber. Objects should be placed in the autoclave so as to avoid trapping air.

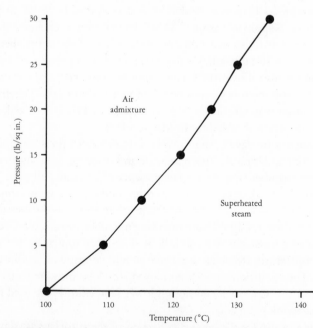

FIGURE 33-2 Relationship between temperature and pressure of steam. Operating conditions falling to the left or right of the curve are unsatisfactory for autoclaving.

steam condensing on organisms may also play a role in the rapid killing. On the other hand, the killing of organisms with dry heat occurs by unknown mechanisms. It appears to take place almost equally well whether or not oxygen is present, indicating that oxidation by atmospheric oxygen is not the primary cause of death.

The autoclave is the simplest and the most consistently effective means of sterilizing objects. The following are some practical aspects of its use.

1. *Air trapping.* The importance of removing air from the chamber has already been mentioned. Since steam displaces air downward in the chamber, long thin containers should not be inserted into the autoclave in an upright position.

2. *Heat penetration.* A cold object placed in an autoclave takes time to warm up, over and above the time at which the chamber temperature reaches 121°C. For example, a liter container of fluid must be autoclaved for 25 minutes after the chamber reaches 121°C, but 4 liters in a container requires an hour if all microorganisms are to be killed.

3. *Contact with steam.* Organisms protected from contact with steam (for example, within oil, occlusive wrappings, or containers of talcum powder) cannot reliably be killed by autoclaving.

4. *Heat indicators.* Test tubes and tapes containing a heat-sensitive chemical indicator are often used with objects when they are autoclaved. The indicator changes color during the autoclaving and thus gives a visual means of checking whether the objects have been subjected to adequate heat. A changed indicator does not, of course, always indicate that the object is free of living microbes, because even wrapped objects may have organisms reintroduced by insects, water, air, and so on. Also, some indicators will turn color under circumstances which do not guarantee complete microbial killing, such as standing overnight in an open autoclave against a hot chamber wall. Tubes or envelopes containing large numbers of heat-resistant spores of the nonpathogenic bacterium *Bacillus stearothermophilus* are also frequently used in packs of materials being autoclaved. Death of these spores indicates adequate killing at the point where the tube or envelope was placed, preferably near the center of the pack.

5. *Elevated boiling points under pressure.* Fluids are prevented from boiling in the autoclave by the elevated pressure even though well above their normal boiling point. If at the end of the period of autoclaving, valves are opened to release the pressure, these fluids will immediately boil —and may even explode their containers. Pressure must be maintained to prevent them from boiling until temperatures have dropped below the boiling point at the ambient pressure.

6. *Some materials cannot withstand 121°C without deleterious effects.* Most autoclaves have a mechanism allowing operation at lower temperatures.

USE OF THE AUTOCLAVE

Chapter Thirty-Three
Sterilization,
Disinfection,
Decontamination

Pasteurization of food
Page 652

Page 8

In the 1800s the French wine industry was simultaneously threatened by foreign competition and a microbial invasion from within. Good wine is the result of properly controlled conversion of fruit sugars to ethyl alcohol by yeast. Bacteria and molds from the environment may become established in the wine, performing unfortunate degradations of the alcohol and producing various ill-flavored metabolites. Pasteur found that with moderate heating of wine at just the right conditions of time and temperature, the spoilage microbes could be killed without significantly changing the taste of the wine. He verified his findings by supplying heated and unheated wines to ships of the French Navy, and the heated wines were consistently found to be superior. This process of controlled heating at temperatures below boiling (now called *pasteurization*) helped save the French wine industry and is widely used today for ridding milk and other foods of disease-causing bacteria. Certain items of hospital equipment, such as oxygen masks, which may not withstand the higher temperatures of the autoclave can also be pasteurized. The temperatures and times used vary according to the organisms present and the heat stability of the material. However, practically speaking, pasteurization always causes a reduction in the numbers of microbes present and thereby retards spoilage in foodstuffs.

As indicated in Chapter 1 repeated controlled heating at relatively low temperatures (tyndallization) can sterilize, since it will even kill spore-forming bacteria, provided the suspending medium enables their spores to germinate.

FILTRATION

EARTHEN AND ASBESTOS
FILTERS

Filters capable of removing bacteria from fluids were devised during the last decade of the nineteenth century. They were made from metallic compounds of silica, including porcelain, diatomaceous earth, or asbestos, partially fused by intense heat. Filters of this type are still used today. Their action is often thought of as a microscopic sieve, holding back microbes but letting the suspending fluid pour through the small holes. In fact the passages through such filters are very tortuous, with some large open areas and many blind passages. The size of the passageways is often considerably larger than the microbes it retains, and trapping of microbes by electrical forces may be involved since there are charges on the walls of the filter passages.

From these considerations it is easy to see that the effectiveness of these filters varies not only as the pore size but also with the chemical nature of the fluid and the amount of pressure used to transfer it across the filter. These filters are often used satisfactorily to remove microbes larger than viruses from aqueous fluids. They are relatively inexpensive and unlikely to clog. However they are not suitable for some other purposes because of the relatively large amount of filtrate absorbed by the filter, the possible introduction of metallic ions into the filtrate, or the adsorption of biological substances such as enzymes from the filtrate.

MEMBRANE FILTERS

In recent years membrane filters composed of compounds such as cellulose acetate have been widely used in the laboratory and in industry. Many beers now are filtered by this means rather than pasteurized. The filters are paper thin and are produced with graded pore sizes extending below those of the smallest known viruses. They are relatively inert chemically and adsorb very little of the fluid or its biologically important constituents. They become semitransparent when immersion oil is applied, thus allowing microscopic inspection for any microbes that adhere to their surface. Their chief disadvantages are that they are expensive and have a tendency to clog.

Filters also figure in plans for space exploration, since scientists want to avoid introducing terrestrial microbes to other planets. In fact the International Committee on Space Research has specified that when a vehicle is landed on another planet, the probability of it carrying a single microbe must be less than 1 in 10,000. As indicated previously heat is a highly reliable method of ridding the metallic surfaces of space vehicles of living microbes. But what about the reliability of filters needed to free heat-labile fluids of microbes? Recent studies comparing different kinds of filters have indicated that the older types—diatomaceous earth, porcelain, and asbestos—were not sufficiently reliable. Only membrane filters gave satisfactory results for use in sterilizing heat-susceptible fluids in space vehicles. Even so it is very difficult to guarantee the complete absence of every microorganism, although one can usually state with some precision what the probability is of a microorganism escaping the procedure designed to kill or remove it.

RELIABILITY OF FILTERS

Moreover aside from the smallest pore size membranes, filters will not sterilize fluids containing viruses. Bacteriophages and other viruses readily pass through filters and, if present in the original suspension, will be present in the filtrate.

CHEMICAL AGENTS

Most chemical methods are unreliable for sterilization, and experience has repeatedly shown that it is dangerous to assume that they are effective for this purpose. Chemical agents do, however, have value in disinfection, that is, reducing the numbers of and eliminating certain dangerous microbes. Satisfactory results are more likely to be obtained if microbial counts are low and if the surfaces to be disinfected are clean and free of interfering substances, such as protein. Heat generally enhances the action of disinfectants. Chemical disinfectants work very slowly against bacterial endospores, the bacteria causing tuberculosis (genus *Mycobacterium*) and some viruses. A very large number of disinfecting agents are in use. The major groups of chemical agents and some of their characteristics are shown in Table 33-1.

LIMITATIONS OF
DISINFECTANTS

Alcohols

Ethyl and isopropyl alcohol are rapidly effective in killing vegetative bacteria and fungi, but are of little value against spores and some medically important viruses.

TABLE 33-1
The Major Groups of Chemical Agents Useful in Disinfection

Class of Disinfectant	Most Vegetative Bacteria	Mycobacterium Tuberculosis	Bacterial Endospores	Fungi	Viruses	Some Trade Names
Alcohols						
Ethyl and isopropyl	+	+	−	+	+ or −	
Chlorine						
Sodium hypochlorite and						Clorox,
chloramines	+	+	+	+	+	Dichloramine T, Halazone
Formaldehyde						
Formalin and						
glutaraldehyde	+	+	+	+	+	Cidex
Iodine						
Tincture of iodine and						Isodyne,
iodophores	+	+	+	+	+	Wescodyne, Betadine
Phenolics	+	+	−	+	+ or −	Lysol, Dettol, Phisohex
Quaternary ammonium compounds	+	−	−	+ or −	+ or −	Zephiran, Cetrimide, Diaparene

+ active
− little or no activity
+ or − active against some, but not others

Hepatitis viruses
Page 475
Picornaviruses
Page 678

In former days some doctors, dentists, and other medical personnel prepared their instruments by soaking them in alcohol. This sometimes resulted in serious illness and deaths because of the failure of alcohol to eliminate viable hepatitis viruses from the instruments. Certain picornaviruses are also relatively resistant, especially to isopropyl alcohol. Alcohols act by coagulation of essential proteins, and thus are only effective when water is present. The final concentration of alcohol in the material that is being disinfected should be about 75 percent by volume for optimal results. Because of the difficulty in hydrating spores, they are frequently quite resistant to alcohols. Valuable as disinfecting agents by themselves, alcohols can also enhance the activity of other chemical agents, such as iodine, chlorhexidine, and quaternary ammonium compounds. Use of alcohols is, of course, limited to materials resistant to their solvent action.

Chlorine

Chlorine is widely used in disinfection in the form of free gas, in chlorine-releasing organic compounds such as chloramines, and as hypochlorite ion. Chlorine gas reacts with water to produce hypochlorite, which is thought to act on microbes by oxidation of essential proteins. Household bleaches (such as Clorox) consist of about 5.25 percent sodium hypochlorite solution, and a tablespoon of such preparations in a gallon of water results in a solution of about 200 parts per million (ppm) of available chlorine. This is about 100 times the amount lethal to most pathogenic microbes, but such high concentrations are usually necessary to speed action and also because organic material that is often present greatly interferes with the disinfectant activity. Use of chlorine disinfectants is limited to materials resistant to the corrosive action of these compounds. For example, certain kinds of rubber and metals are broken down by such disinfectants.

Iodine

Iodine, like chlorine, is very active against most microbial species. Its mode of action, however, is thought to relate to iodination of the amino acid tyrosine present in the enzymes and structural proteins of microorganisms. Iodine, like several other disinfectants, has enhanced activity when dissolved in alcohol to produce a tincture. However, simply swabbing the skin with tincture of iodine may not kill all the spores of *Clostridium perfringens* when they have been experimentally applied. Compounds (called iodophores) of iodine with surface-active agents have been very popular because they do not sting as much when applied to wounds and are not as likely to stain. Careful evaluation shows that these benefits are accompanied by a substantial reduction of antimicrobial activity.

Formaldehyde

Eight percent formaldehyde is an extremely active disinfectant, killing all forms of microbial life in minutes. Lower concentrations, however, are only slowly active against some bacterial endospores and some viruses. Failure to recognize this fact has caused several near disasters because of the use of formaldehyde to kill living agents in vaccines. Solutions of formaldehyde have limited use because of their irritating vapors, and in recent years other aldehydes (such as glutaraldehyde) have been used instead. Like formaldehyde, they appear to act by combining with amino groups in microbial proteins.

Poliomyelitis vaccine
Pages 511, 668

Phenolics

Phenolics are a very large group of compounds more or less chemically related to phenol (carbolic acid). The group includes cresols and xylenols (Fig. 33-3).

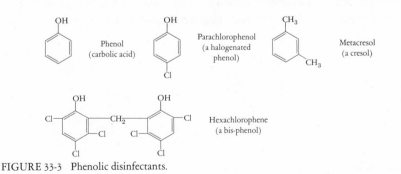

FIGURE 33-3 Phenolic disinfectants.

In high concentration (5 to 10 percent) they kill most bacteria including the tubercle bacilli (*Mycobacterium tuberculosis*). Five percent phenol can also be used against all the major groups of viruses including picorna viruses, but some of the more commonly used derivatives of phenol lack sufficient activity. Lower concentrations of phenol (or its derivatives) may be ineffective. In fact some gram-positive cocci and gram-negative rod bacteria not only grow in the presence of 0.1 percent phenol but are actually able to utilize it. The major advantage of phenolics is their ability to remain active when mixed with soaps and detergents. An exception to this rule occurs with some of the halogenated xylenols and cresols, which are inactivated by organic materials including the oily material in some lotions. Phenolics are thought mainly to act against the cytoplasmic membrane of cellular microbes. Their mode of action against viruses is not yet clear.

One phenol derivative, hexachlorophene, deserves special mention because of its extremely widespread use in household hand soaps, deodorants, and the antiseptic scrubs of surgeons. It has substantial bacteriostatic and bactericidal activity against *Staphylococcus aureus* and tends to be retained by the skin, so that antimicrobial activity increases with repeated use. It is not effective against many gram-negative bacteria. Use of hexachlorophene has recently been somewhat restricted since infants repeatedly immersed in soaps containing this disinfectant absorbed quantities which have produced brain damage in animals. However, hexachlorophene is of proven value in preventing nursery staphylococcal infections and the risk of brain damage in infants would appear to be too small to justify its complete abandonment.

Quaternary Ammonium Compounds

These compounds may be thought of as derivatives of ammonium chloride in which the hydrogen atoms are replaced by more or less complex organic radicals (Fig. 33-4). They represent a larger group of compounds called surface active agents, since they reduce surface tension. In contrast to other surface tension active agents (such as household detergents) quaternary ammonium compounds are cationic, which means their active groups have a positive electrical charge. Like the phenolics they act against many vegetative bacteria (attacking mainly the cell membrane), and they attack some viruses in a way which is not yet clear. Quaternary ammonium compounds are widely used for disinfection and preservation purposes, such as prevention of bacterial growth in aqueous medications.

Ammonium chloride

Benzalkonium chloride
(a quaternary ammonium
disinfectant).

FIGURE 33-4
Quaternary ammonium compounds.   * n ranges from 8 to 18

Such uses have sometimes led to tragic infections in human beings because of the ability of some bacteria, notably species of *Pseudomonas,* to persist and sometimes grow in their presence. In some instances there has been failure to recognize that these substances are easily inactivated by soaps, detergents, and by organic materials such as gauze.

Antiseptics and Disinfectants of Lesser Activity

Mercury and silver are largely bacteriostatic compounds which have been in use as disinfectants and antiseptics for many years, and were formerly thought to have strong antimicrobial properties, including the ability to kill spores. However, as recently as a decade ago many people were infected by medical instruments which had been soaked in mercury compounds and thought to be sterile. Mercury acts by binding to sulfhydryl (SH) groups in microbial proteins, and silver is thought to be a protein coagulant. Trade names include Mercurochrome and Merthiolate (mercury), and Argyrol and NeoSilvol (silver). Such agents still have valid uses as preservatives (see the following) and in other situations in which inhibition of microbial growth is desired.

Oxidizing agents (such as hydrogen peroxide and potassium permanganate) likewise have little killing ability, especially in the presence of organic materials, and are no longer recommended for disinfection.

Gaseous Agents

Fumigation has been employed at least as far back as ancient Greece, where burning sulfur was apparently used for purification. Formaldehyde gas likewise has had many years of use and is still recommended for decontaminating air filters on cabinets used for culturing *Mycobacterium tuberculosis.* Recently it has been reported that heating the powdery polymer of formaldehyde (paraformaldehyde) can be a safe and effective way of generating the gas if concentration and relative humidity are controlled to prevent explosions and repolymerization. Formaldehyde generated in that manner not only can kill the endospores of *C. botulinum* but also effectively neutralize the powerful nervous system toxin produced by this organism.

Botulism
Pages 483–485

By far the most effective and useful gaseous agent has been ethylene oxide. It penetrates well into fabrics and equipment (in contrast to formaldehyde) and can be relied on for sterilization. Ethylene oxide readily reacts with amino, carboxyl, sulfhydryl, and other chemical groups of organic substances. Since it is explosive, ethylene oxide must be mixed with some inert gas such as carbon dioxide or freon. Effectiveness also depends on temperature and relative humidity, which must be carefully controlled. Use is thus restricted to a special chamber where temperature, pressure, and relative humidity can be regulated. In practice about 4 hours are generally required to sterilize, and an additional period of 4 to 24 hours to allow release of absorbed gas, depending on the object being treated. One must allow absorbed gas to dissipate because of its irritating effect on tissues or undesired persistence of antimicrobial effect. Disposable plastic dishes, syringes, and other heat-sensitive items used in laboratories or for medical purposes have often been sterilized with ethylene oxide.

Preservatives

The ingredients of soaps, medicines, deodorants, contact lens solutions, foodstuffs, and many other everyday items often include an antimicrobial agent. Apple cider may have sodium benzoate added to it; heparin (a substance used to prevent blood clots) may contain a phenol derivative; contact lens solutions may include a mercury or a quaternary ammonium preservative; and a leather belt may be treated with a mercury compound plus one or more phenol derivatives. The purpose of these agents is to prevent or retard spoilage due to microbes which are inevitably introduced from the environment. Unfortunately they do not always work, and toxic effects or hypersensitivity states have also sometimes occurred in persons repeatedly (and usually unwittingly) exposed to them.

RADIATION

Electromagnetic radiations can be thought of as waves having energy but no mass. Examples are X-rays, gamma rays, ultraviolet, and visible light. Electromagnetic rays all travel at the speed of light, and the energy they possess is proportional to the frequency of the radiation (the number of oscillations of the wave per second). Thus electromagnetic radiations of short wavelength, such as gamma rays, have a lot more killing power than those with long wavelength, such as visible light. The spectrum of electromagnetic radiations ranges from electric waves (of very low frequency), radio waves, heat rays, visible light, ultraviolet light, X-rays, gamma rays, and the very high frequency secondary cosmic rays, with wavelength of only 10^{-11} cm. Gamma rays and ultraviolet radiations have proved to be of value as tools for microbial control.

Gamma Rays

Gamma rays represent an example of ionizing radiations which cause biological damage by producing hyperreactive ions and other molecular forms when they give up their energy to the microbe. Gamma radiation from the radioisotope cobalt 60 has proved of practical value in sterilization and decontamination. For

Gastroenteritis
Pages 477, 627–629

example, gamma irradiation is used in killing pathogens such as *Salmonella* in a food product. Such applications are analogous to the use of heat in the pasteurization process, complete sterilization being impractical because undesirable changes in color, flavor, or consistency occur with higher doses of radiation. A number of more stable biological materials (such as penicillin) and numerous plastic disposable items (like hypodermic syringes) can be effectively sterilized commercially with high dose irradiation which kills contaminating organisms without altering the material.

Various substances and conditions alter the susceptibility of microbes to ionizing radiations. Oxygen and certain chemicals may decrease resistance, while vitamin C, ethyl alcohol, and glycerol may increase resistance. Protective chemicals have, of course, great interest in view of modern-day radiation hazards. Bacterial endospores are the most radiation resistant microbial forms, whereas gram-negative rod bacteria, such as species of *Salmonella* and *Pseudomonas,* are among the most sensitive. Some bacteria which do not form spores are peculiarly resistant, however. In *Micrococcus radiodurans* and some laboratory derived mutants, for example, there are enzymes which can repair moderate degrees of radiation damage.

Ultraviolet Radiation

Certain wavelengths of radiation are much more effective antimicrobial agents than those immediately shorter or longer. The most important example of this is the band of wavelengths from 200 to 310 nm lying within the ultraviolet zone. This electromagnetic zone of enhanced killing can be explained by the fact that it includes the wavelengths optimally absorbed by nucleic acids with resulting damage to their structure and function. The absorbed energy results in changes in purines and pyrimidines so that their normal function in nucleic acid is impaired. However some members of microbial populations which have apparently been killed by ultraviolet radiation will recover if irradiated with longer wavelengths of light. Also recovery of damaged viruses may occur if they infect cells possessing repair enzymes.

In practice ultraviolet light of satisfactory germicidal properties can be produced by passing an electric current through vaporized mercury in a special glass tube similar to a fluorescent light bulb. Most of the light produced has a wavelength of 253.7 nm, close to the killing optimum of 265 nm. The antimicrobial effect falls off in almost a linear fashion very close to the bulb, and faster, as the square of the distance, as one moves farther from the bulb. Ultraviolet rays penetrate very poorly, so that a film of grease on the bulb may markedly reduce effective microbial killing, as will extraneous materials covering the microbes. Therefore in practice ultraviolet lamps are of greatest value against exposed microbes in air or on clean surfaces and at close range. Acidity, growth phase of the microbe, and the presence of spores definitely alter the effectiveness of ultraviolet irradiation, while the effect of moisture is still debatable.

LIMITATIONS OF ULTRAVIOLET IRRADIATION IN STERILIZATION

It has for years been assumed that the microbial killing effect of sunlight is due entirely to ultraviolet rays. However, much of the ultraviolet radiation is absorbed by the atmosphere and the bactericidal effect of sunlight on earth is primarily due to another mechanism, *photo-oxidation.* In photo-oxidation light energy of longer wavelengths is absorbed by microbes and results in lethal oxidation in the presence of atmospheric oxygen. By contrast, microbial killing due to ultraviolet irradiation does not require the presence of oxygen.

LOGARITHMIC DEATH
CURVE

TDT
Page 650

Certain generalizations have evolved concerning killing of bacteria and presumably they apply to other microbes as well. They may be summarized as follows.

1. *Only a fraction of the microbes present die during any given time interval.* As pointed out in Chapter 5, death of microbes usually follows a logarithmic curve, which becomes a straight-line graph if the logarithm of viable microbes is plotted against time. Curves of this kind never quite reach "zero organisms." For example, if the concentration of organisms in a fluid was 100 per milliliter at the start, and 90 percent were killed every minute, we would have the seemingly strange result of one-tenth of an organism per milliliter after 3 minutes. This result does not seem so peculiar if we interpret it (as we should) as indicating one organism per 10 ml, or more precisely, the odds of finding a living organism in any milliliter is one in ten. Sterility can thus be considered as the state where the probability is extremely low of finding even a single viable organism in the total volume involved, or of finding such an organism on the object we are dealing with. Obviously, the larger the volume, the greater the probability of finding a living organism. In commercial processes the *thermal death point* (TDP) and the *thermal death time* (TDT) have general usefulness as guides for achieving reliable sterilization. The TDP is the lowest temperature at which a suspension of microbes is sterilized in 10 minutes, and the TDT is the shortest time required to kill all the

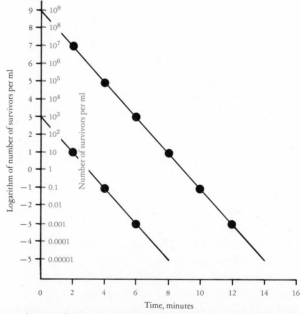

FIGURE 33-5   A bacterial killing curve. Notice that for each tenfold increase in the number of organisms at the start, it takes one additional unit of time to achieve satisfactory killing (in this example, six more minutes to kill 10^9 organisms than to kill 10^3 organisms).

microbes in a suspension at a given temperature. Because the killing curves are logarithmic, such values may show considerable variability.

589

Principles of Microbial
Killing

2. *The logarithmic death curve also indicates that the time it takes to achieve sterility depends on the number of organisms present at the start.* Thus a heavily contaminated substance would require one unit increase in time for every tenfold increase in the number of organisms (Fig. 33-5). This is one reason why prevention of contamination and mechanical methods (such as scrubbing, mopping, and washing) are useful before using disinfecting or sterilizing agents.

3. *Not all microorganisms have the same susceptibility.* There is variation among and within species. Also, there is variation with growth phase, most organisms being at peak susceptibility when in the logarithmic phase of growth. Spores of microorganisms are generally more resistant than the vegetative forms. Fungal spores are usually not as resistant as bacterial endospores, although many responsible for food spoilage easily resist pasteurization. Finally, some viruses (such as the hepatitis virus) are killed very slowly by heat and disinfectants.

COMPARISON OF AGENTS USED AGAINST MICROBES

Different means of killing microbes are compared in the laboratory by counting the number of surviving organisms after varying time intervals or by counting the survivors at a fixed time. The count is achieved by *neutralizing or removing the killing agent* and inoculating different dilutions on appropriate media. Thus the number of survivors able to grow into a visible colony are determined. Curves can then be plotted and "D" or "Z" values determined. The D value or "decimal reduction time" is the time it takes for 90 percent of the organisms to be killed at a given temperature, and the Z value represents the number of degrees one must increase the temperature to obtain 90 percent killing in a given time period. These values are used in comparing different methods of microbial killing, rather than the time required for complete killing. The latter shows wide variations due to the shape of the killing curves and is therefore of little use in scientific studies.

DECIMAL REDUCTION
TIME

Disinfectants may be evaluated by comparing the ratio of their antimicrobial activity to that of pure phenol. The larger the resulting value, the *phenol coefficient,* the more active the agent under those conditions. The phenol coefficient has general applicability but may be highly misleading since differences in killing rates depend on the species of microorganism, stage of growth, nature of the killing agent, suspending medium, pH, salt concentration, and many other factors. For these reasons any agents being compared for possible use in a given situation (such as disinfection of a hospital floor or decontamination of a spaceship) *must be evaluated under the same conditions in which they will actually be used.* Widely conflicting information on the value of disinfecting agents continually

EVALUATION OF
DISINFECTING AGENTS

arises because various and often inappropriate conditions of testing are used. For example, in counting surviving microbes it may be important to employ a medium which neutralizes the disinfectant being tested. If this is not done, traces of disinfectant carried over to the counting medium may prevent growth of the microbes and falsely indicate that they had been killed.

SUMMARY

Sterilization means killing or removing all viable organisms, while disinfection means reducing or eliminating organisms that might represent a hazard. Heating is one of the most practical methods of controlling unwanted microbes. Autoclaving and tyndallization can bring about sterilization, while pasteurization and boiling can reduce or eliminate unwanted microbes. Heat is most effective when applied under moist conditions. Filtration can be used to sterilize fluids, although with less reliability than heating. In using filters it is important to know the approximate pore size, response to solvents and pressure of filtration, and the possible affinity for biologically active substances such as enzymes. Chemical agents are unreliable for sterilizing but have wide use as disinfectants. The major classes of chemical disinfectants include alcohols, chlorine and iodine, formaldehyde, phenolics, and quaternary ammonium compounds. Gaseous agents used in sterilization include formaldehyde vapor and ethylene oxide. A variety of chemical agents may be used as preservatives, to prevent growth of destructive organisms in foodstuffs, leather goods, medications, and other materials. Electromagnetic radiations are useful in both sterilizing and in reducing the numbers of potentially harmful microbes.

Germicidal agents generally cause a logarithmic decrease in susceptible microbes, requiring time to achieve sterilization or satisfactory reduction in numbers of viable organisms. Microbes vary in their susceptibility depending on pH, temperature, growth phase, strain, and species differences.

Laboratory comparisons of germicidal agents may be made by analyzing the plotted curves of microbial killing or by determining their phenol coefficients. The most meaningful comparisons are made under conditions of actual use of the germicides.

QUESTIONS

1. Explain why not every object that has been autoclaved is sterile.
2. Explain why conditions of temperature and pressure other than those given in Figure 33–2 may be unsatisfactory for autoclaving.
3. Why do doctors and dentists no longer use alcohol to sterilize their instruments?
4. Discuss the limitations of ultraviolet light for sterilization and disinfection.
5. Why is it that laboratory comparisons of disinfectants may not always reflect their relative value in actual practice?

BROCK, T. D., *Biology of Microorganisms.* Englewood Cliffs, N.J.: Prentice-Hall, Inc., 1970.

HEDGECOCK, L. W., *Antimicrobial Agents.* Philadelphia: Lea and Febiger, 1967.

LAWRENCE, C. A. and S. S. BLOCK, *Disinfection, Sterilization and Preservation.* Philadelphia: Lea and Febiger, 1968.

PERKINS, J. J., *Principles and Methods of Sterilization.* Springfield, Ill.: Charles C Thomas, 1969.

Chapter 34

Terrestrial Microbiology

The microorganisms in soil play an indispensable role in maintaining life on this planet. The soil contains myriad microorganisms able to degrade or chemically modify organic and inorganic molecules. It also contains organisms which are indispensable to life on earth, because they transform chemical substances from forms that cannot be readily utilized to forms that serve as nutrients for a variety of living forms. For example, although man lives in a virtual ocean of air that is 79 percent nitrogen, this element cannot be utilized by most living forms unless it is converted into other forms of nitrogen such as occur in amino acids. In this way as well as in many others microorganisms contribute directly to the fertility of the soil.

The interest in soil organisms also stems from the fact that many of them produce antibiotics and other useful products. Immense screening programs have been undertaken by the drug industry with the express aim of isolating organisms that produce antibiotics. The search also continues in an effort to find organisms capable of synthesizing compounds of industrial importance. In no other habitat can one hope to find the tremendous range of synthetic capabilities as are represented by organisms in the soil.

The soil is also the habitat of a number of fungi and bacteria which are pathogenic for plants and some that are pathogenic for animals and man.

SOIL

Characterization of the Soil

Soil refers to the loose material on the outer portion of the earth's surface. Soils result from the weathering of rocks, some of which undergo many changes in the course of weathering. On a world scale soil can be divided into a number of major types associated with the major zones of climate and vegetation. The soil in forests differs from that in the desert, and there are profound differences between soil found in a coniferous forest and one found in a deciduous forest. The major differences between soils are usually related to the material from which the soil is derived. On this basis soils can be divided into two major groups. Mineral soils result from the weathering of rock and other inorganic materials, whereas organic soils result from the accumulation of organic material under conditions which do not favor its degradation such as occurs in marshes and bogs. The most common soils are of the mineral type, and these are the ones which are considered.

Formation of Soil

When rock weathers, a number of processes can take place. The action of rain dissolves water-soluble components. This may in itself be enough to cause the rock to fragment. Rapid changes in temperature result in differential expansion and contraction of the outer layers of the rock, which also lead to fragmentation. Once cracks have developed, the penetration of water and subsequent freezing aggravate the cracking. Wind-blown particles act as an abrasive and assist disintegration. In addition to the physical agents, microorganisms also play a role in weathering. Most exposed rocks have a variety of photosynthetic organisms clinging to them—lichens, algae, and mosses, in particular. When the rock is moist and sunlight available, these organisms are able to synthesize organic material which can then be used by nonphotosynthetic organisms—the heterotrophic bacteria as well as fungi. Carbon dioxide produced in the respiration of the heterotrophs becomes converted to carbonic acid by combining with water:

Lichens
Page 276

$$CO_2 + H_2O \longrightarrow H_2CO_3$$

Acid is especially important in dissolving limestone rocks. Other products of heterotrophic metabolism, particularly other acids and chelating agents which bind to inorganic substances of low solubility to form soluble materials, also help weather rocks.

Late in the weathering process plants and animals of various kinds become associated with the weathering process, and once any quantity of soil has been produced plant cover of increasing complexity begins to appear. Fresh organic matter is added to the soil following the death and decay of organisms. As the fragments of rock become smaller and smaller as a result of weathering, a layer of soil develops. During the course of weathering the interstices between the individual fragments become smaller, and water is trapped between the soil particles. This water contains dissolved gases, oxygen, and CO_2 in particular.

Components of the Soil

Soil may be divided into four major components: mineral fractions, organic material, soil moisture, and soil atmosphere. The detailed composition of the soil with respect to these four components will determine the quality and quantity of plant and animal life growing in that particular soil. The continuous interaction of all the weathering processes leads to the gradual formation of more or less distinct horizontal layers termed horizons. Mineral soils have three main horizons, commonly termed A, B, and C. The A horizon includes the surface layer or top soil, darkened by the presence of organic matter. The native organic fraction is known as humus, a mixture of complex organic materials which is resistant to microbial degradation. The B layer, the subsoil, is composed of mineral soil in which the organic compounds have been converted by the decomposers into inorganic compounds. The B horizon contains very little organic matter. Layer C is bedrock which has been only slightly weathered.

Biological Content of Soil

The soil teems with life, from the submicroscopic viruses of bacteria and plants, to the macroscopic earthworms and other large forms. Bacteria are the most numerous, ranging from a million or less to several billion per gram of soil. In terms of biomass they are not as important as the fungi, but are believed to exceed that of the algae and protozoa combined.

The most abundant soil bacteria are small coccoid rods of variable morphology. They are highly pleomorphic. Depending on the locality, 5 to 35 percent of the bacteria are of the genus *Arthrobacter.* The vast majority of the remaining bacteria are included in the following genera: *Pseudomonas, Clostridium, Achromobacter, Bacillus, Micrococcus,* and *Flavobacterium. Chromobacterium, Sarcina,* and *Mycobacterium* are less common. Members of the actinomycetes such as the antibiotic-producing *Streptomyces* and *Nocardia* are widespread and abundant. The characteristic odor of soil results from the presence of members of the *Streptomyces.* The most extensive microbial growth takes place on soil crumbs in the form of microcolonies (Fig. 34-1). Bacteria are not uniformly distributed throughout the soil profile, but their distribution usually follows closely that of organic matter. Accordingly, bacteria occur predominently in the A horizon.

The composition of the fungal flora of the soil has been difficult to determine, because the media routinely used in microbiology laboratories select for specific nutritional types and only certain fungi appear. Furthermore, the generic composition and size vary with the type of soil and with its physical and chemical characteristics. The class Fungi Imperfecti consistently contribute the greatest number of genera in soil. These genera commonly include *Aspergillus, Trichoderma,* and *Penicillium.* Phycometes are not as numerous, although members of the Mucorales are very common. Only a few genera of ascomycetes are found and the large basidiomycetes, the mushrooms, are present, but their quantity is difficult to assess because their large size is not compatible with their being cultured easily on agar medium. Larger forms of life are present in vast numbers. They range in size from moles and gophers to the earthworms and to the barely visible mites, nematodes, and insects. The numbers of these barely visible forms can be staggering. A single acre of soil may contain half a billion insect eggs and larvae, and up to a billion mites.

Arthrobacter
Pages 258–259

Streptomyces and *Nocardia*
Pages 260–262

Fungi Imperfecti
Page 289

FIGURE 34-1
A microcolony of rod shaped bacteria separated out of the soil and prepared by the freeze-etching technique. There are approximately 10 cells which is the average number of cells in a microcolony. Note that one cell is undergoing division. (Courtesy of Dr. L. E. Casida, Jr. and D. L. Balkwill)

RELATIONSHIP BETWEEN
MEMBERS OF SOIL FLORA

There are numerous interacting food chains in the soil, and one organism that preys upon another is itself preyed upon by other organisms. Several unique situations within this food chain deserve mention because of their interaction with microorganisms. The soil nematodes are a group of roundworms about the size of the letter S which resemble tiny eels. Many prey on bacteria, fungi, and protozoa as well as on other nematodes. Their major importance to agriculture is based on the fact that their mouth parts are essentially a hollow spear which can penetrate the toughest root and draw out the juices of the plant. Furthermore, they serve as very important vectors for transmitting disease-causing viruses to plants. It is estimated that they cause over half a billion dollars worth of crop damage each year in this country. Nematodes have many natural enemies, one of the most interesting being one of the soil fungi, which is much smaller than the nematode. In the presence of these nematodes the cells of the fungus form loops which are just large enough for the nematode to stick its head into. As soon as this happens, the fungal cells suddenly expand, thereby closing the opening and trapping the nematode (Fig. 34-2).

Another interesting and important relationship involves termites and the protozoan which inhabit their stomachs. Termites are major agents of wood decay in the warmer parts of the world. The termite, however, is only half the team. The gut of this insect is tightly packed with a certain type of protozoa which is able to digest the cellulose of the wood that the termite bites off. The termite can ingest but not digest the cellulose. Neither the insect nor its protozoan partner can exist without the other.

A commonly observed important plant–fungus symbiosis is the mycorrhiza (plural, mycorrhizae). These are discussed in Chapter 14.

Mycorrhiza
Pages 290–291

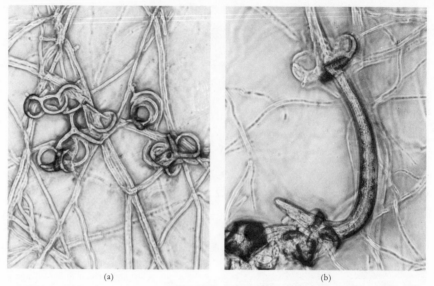

(a)   (b)

FIGURE 34-2   Predatory fungi. (a) Clusters of adhesive hyphal loops formed by the nematode-trapping fungus, *Arthrobotrys conoides*. (b) A captured nematode. (Courtesy of Dr. D. Pramer)

ENVIRONMENTAL INFLUENCES ON THE BACTERIAL AND FUNGAL FLORA

Environmental conditions affect the density and composition of the bacterial and fungal flora. The primary environmental influences include moisture, temperature, acidity, organic matter, and inorganic nutrient supply.

Moisture and Oxygen Supply

Moisture and the availability of oxygen are intimately interrelated. Wet soils are unfavorable for most bacteria simply because filling up pore space with water diminishes soil aeration. The major effect is therefore a reduction in the supply of oxygen. Continued waterlogging of the soil changes the flora from one consisting primarily of aerobes to mostly anaerobic species. When the moisture content drops to a low level, as in a desert environment, the metabolic activity and number of bacteria are markedly reduced. Although it is commonly believed that spore-formers might predominate under these conditions, it appears that many soil nonspore-forming bacteria are resistant to drying and representatives of many major groups survive for years.

Since the filamentous fungi are strict aerobes, they are found close to the surface of the soil. In waterlogged soils the numbers of fungi are also markedly reduced. Their spores, however, often persist and once the water diminishes and the level of oxygen increases, the molds recover and become an important part of the flora.

Fungi and oxygen
Pages 283–284

Acidity

Highly acid or alkaline conditions tend to inhibit the growth of many common bacteria. Many agricultural practices which adjust soils to neutrality, such as liming an acid soil, promote the growth of most bacteria. If fertilizers containing ammonium are added in excess, they can inhibit bacterial growth. This results from the fact that microorganisms can oxidize the ammonium to nitric acid, thereby creating a highly acid environment.

Many species of molds can multiply over a broad pH range—in the alkaline region up to above pH 9, and in the acidic region to pH values almost as low as 2. Because most bacteria do not multiply in highly acidic environments, the fungi dominate the population in areas of low pH. For this reason adding lime to a soil reduces the abundance of fungi and treating the soil with acid-forming fertilizers such as those containing ammonium increases their abundance. The increased abundance of fungi at low pH results primarily from the fact that the bacteria are not present and competing for food.

Temperature

Temperature governs the rates of biochemical processes through its effect on rates of enzyme reactions. A warming trend favors the biochemical changes brought about by the microbial population. Mesophiles constitute the bulk of the soil bacteria, and true psychrophilic bacteria are rare or absent. In winter the bacteria in soil are cold-tolerant mesophiles, rather than psychrophiles. Thermophiles, on the other hand, are ubiquitous.

Temperature and enzymes
Page 145

Most fungi are mesophilic and thermophiles are uncommon. The thermophiles that have been studied will grow at 50 degrees but not at 65 degrees. A few thermophilic strains can be demonstrated in soil, but they are only abundant in compost heaps which do not reach high temperatures.

Temperature and growth
Pages 105–106

Organic Matter

Nutritionally the great majority of soil bacteria are heterotrophic. The organic matter which serves as their source of energy is produced largely by the photosynthetic activity of higher plants. Although a tremendous amount of organic material is added to soil every year, there is generally not enough to supply the energy requirements of soil bacteria. Therefore the size of the bacterial population is limited by the organic matter available. The addition of organic material to soil, such as occurs when manure or crop residues are plowed into the soil, results in a dramatic increase in the number of bacteria and of fungi, since the latter are also heterotrophic.

In soils in which plants are growing a variety of organic materials exudes from the roots. The zone which surrounds the roots and contains the exudate is termed the *rhizosphere*. These exudates provide an abundant source of material for energy and therefore bacteria tend to concentrate on or near the root surface.

Nutrition of fungi
Pages 283–285

Other soil organisms, such as fungi and protozoa, are also abundant in the rhizosphere.

Growth of vegetation depends on the cycling of the major chemical elements. These elements reach the soil, as well as aquatic environments, in organic material. This material is then degraded and the elements are restored to forms which can be reutilized by plants. Oxygen and hydrogen are required elements usually abundant in the biosphere. However, unless carbon and nitrogen are continuously recirculated, plant growth would soon be limited. Phosphorus and sulfur, although needed in smaller amounts, are just as essential and also pass through complex cycles.

CARBON CYCLE

Carbon Cycle
Pages 127–128

The carbon cycle revolves around CO_2, its fixation into organic compounds by green plants, and its regeneration into CO_2 primarily by microorganisms. The overall cycle was considered in Chapter 6. The discussion here focuses on the decomposition of organic material.

One of the most important decomposition products of both animal and plant material is the heterogeneous organic humus. Humus is composed of very complex molecules whose exact structures are not known. They contain benzene rings, nitrogen side chains, and nitrogen-containing ring compounds. Because humus contains many very complex ring molecules, such as the benzene ring, it is not readily decomposed, but tends to persist for long periods. Little is known about the organisms and pathway by which humus is decomposed, but it appears

Actinomycetes
Pages 260–261

that actinomycetes may be largely responsible for its degradation to CO_2.

The decomposition of organic materials is carried out by the "decomposers," chiefly bacteria, fungi, protozoa, and small animals. The decomposition of plant residues which represent a large source of organic debris in terrestrial environments involves the activity of a number of organisms. Fungi and the true bacteria utilize the more readily decomposable organic substances such as sugars, amino acids, and proteins. The myxobacteria are capable of degrading cellulose,

Myxobacteria
Pages 241–243

a polymer highly resistant to degradation by most bacteria. Other major plant constituents such as lignin and pectins are generally decomposed by fungi. As a general rule bacteria appear to play the dominant role in the decomposition of animal flesh; fungi appear to be the most important in the degradation of wood. Protozoa, mites, and nematodes may play a more important role in decomposition than was previously suspected. If these organisms in forest litter are killed by treatments that have no effect on bacteria and fungi, the decomposition of fallen leaves and dead twigs is greatly slowed.

In the aerobic decomposition of organic matter the main gas evolved is CO_2. However, when the level of oxygen is low, as is the case with wet rice paddy

Methane bacteria
Page 256

soil, marshes, swamps, and manure piles, methane (CH_4) production may be great. The biosynthesis of methane is limited to a specialized group of bacteria, the methane bacteria, all of which are strict anaerobes which live in environments which are poorly aerated. Three genera have been described: *Methanobacterium, Methanococcus,* and *Methanosarcina.* All three genera have the ability to gain energy from the oxidation of hydrogen gas and reduction of CO_2, according to the following reaction:

$$4H_2 + CO_2 \longrightarrow CH_4 + 2H_2O$$

This is an excellent example of anaerobic respiration in which CO_2 is reduced. The methane-forming organisms cannot metabolize common organic compounds as can most heterotrophs. Indeed it appears that the largest compound that can be metabolized is the two carbon compound, acetic acid.

Anaerobic respiration
Page 155

Microbiological Decomposition of Pesticides

Most organic compounds of natural origin are capable of being degraded by one or more species of soil organism. However, this is not the case with a large number of man-made chemicals. In recent years man has added to the number of slowly degraded or nonbiodegradable compounds in the soil by adding thousands of tons of pesticides in many parts of the world. Another series of compounds which may be toxic for a variety of birds and fish are the polychlorinated biphenyls, compounds used in a large number of manufacturing operations. These aromatic molecules having chlorine atoms attached to the benzene ring are not biodegradable and their concentration in the environment is steadily increasing. The chemical nature of *herbicides* and *insecticides* covers an extremely broad range of organic compounds—organic acids, nitrophenols, chlorinated organic acids, and other organic compounds. Many of them have chemical structures similar to humus, which is also degraded very slowly. Certain of these compounds disappear from the soil within a reasonably short time, a few days or weeks. However, *many remain undegraded for years.* A great deal of research is now under way to determine how long various compounds persist in the soil, what determines how long they persist, and by what biochemical pathways they are degraded.

AROMATIC COMPOUNDS CONTAINING CHLORINE ATOMS OFTEN ARE NONBIODEGRADABLE

It is highly desirable for toxic compounds to be degradable since it is becoming apparent that most herbicides and insecticides are not only toxic to target weeds or insects but may also have far-ranging deleterious effects on a variety of birds and other animals. There is well-documented evidence that the nonbiodegradable pesticide DDT accumulates in the fat of predatory birds (Table 34-1). The data in Table 34-1 illustrate the *phenomenon of biological magnification.* The continuing ingestion-reingestion of DDT which is concentrated in fat results in the magnification of the DDT concentration as the food chain approaches its more complex members. DDT as well as other chlorinated hydrocarbon insecticides interfere with the reproductive process of birds by interfering with egg shell formation so that fragile eggs are produced which break before the young can hatch. For such reasons as this, although a degradable compound must be continually applied to maintain its effectiveness, the fact that it is biodegradable recommends its use.

TOXIC MATERIALS MAY ACCUMULATE

TABLE 34-1
Food Chain Concentration of DDT

	Parts per Million DDT Residues
Water	0.00005
Plankton	0.04
Silverside minnow	0.23
Heron (feeds on small animals)	3.57
Herring gull (scavenger)	6.00
Fish Hawk (Osprey) egg	13.8
Merganser (fish-eating duck)	22.8

From E. P. Odum, *Fundamentals of Ecology,* Saunders, Philadelphia, 1971, p. 74.

COMPOUNDS DEGRADED IN LAB MAY NOT BE DEGRADED IN NATURE

It is now apparent that *DDT is biodegradable in the laboratory,* and yet it is extremely persistent in nature. Why this is true remains unanswered. It may relate to the fact that the initial degradative reactions on the molecule do not yield energy and therefore the organism gains little by degrading the compound. In the laboratory however, energy sources are provided in addition to the DDT. Therefore organisms will be able to multiply and attack the DDT. These observations point up an important caution in the study of waste disposal by microorganisms. Biodegradability of a material under laboratory conditions does not ensure its destruction by these same organisms in their natural environment.

SLIGHT DIFFERENCES IN CHEMICAL STRUCTURE AFFECT DEGRADABILITY

Relatively slight molecular changes markedly alter the biodegradability of a compound. Perhaps the best studied example involves the herbicides 2,4-dichlorophenoxyacetic acid (2,4-D) and 2,4,5-trichlorophenoxyacetic acid (2,4,5-T). The only difference between these two compounds is the additional Cl atom on 2,4,5-T (Fig. 34-3). When 2,4-D is applied to the soil, it completely disappears within a period of several months, as a result of its degradation by a segment of the microbial population in the soil. However, when 2,4,5-T is applied, it is often still present for more than a year (Fig. 34-4). The persistence of a variety of herbicides is shown in Table 34-2. These estimates, however, are only approximations since many factors influence the rate of degradation of compounds by the soil flora. As a general rule any practice which favors the multiplication of microorganisms will increase the rate of degradation. Thus raising the temperature, maintaining the pH near neutrality, and providing an optimal amount of moisture are all likely to increase the rate of degradation of most materials added to the soil.

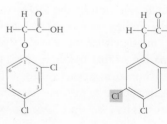

FIGURE 34-3
Formulae of 2,4-D (left) and 2,4,5-T (right). Note that 2,4,5-T has an additional chlorine atom attached to carbon atom 5.

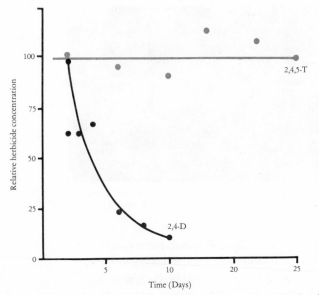

FIGURE 34-4   A comparison of the rates of disappearance of two structurally related herbicides, 2,4-D and 3,4,5-T. (From Alexander, M., *Introduction to Soil Microbiology*, John Wiley & Sons, New York, 1961)

The disappearance of a compound may result from a variety of other factors besides microbial activity. Chemical reactions, destruction by light, volatilization, and leaching from the soil may all contribute.

Some toxic compounds may be degraded to other compounds which are just as toxic as their sources. For example, the insecticide dieldrin can be degraded by both sunlight and microorganisms to another compound, photodieldrin. This compound appears to be more toxic to many biological systems than is dieldrin. Photodieldrin is termed a *stable terminal residue* because it is apparently not readily degraded and accumulates in the environment.

TABLE 34-2
Decomposition and Period of Persistence of Several Herbicides

Name of Compound	Abbreviation	Persistence in Soil	Organisms Which Degrade
3-(p-chlorophenyl)-1, 1-dimethyl urea	Monuron	4–12 months	*Pseudomonas*
2,4 dichlorophenoxyacetic acid	2,4-D	2–8 weeks	*Achromobacter Corynebacterium Flavobacterium*
2,2 dichloropropionic acid	Dalapon	2–4 weeks	*Agrobacterium Pseudomonas*
2,3,6-trichlorobenzoic acid	2,3,6-TBA	greater than 2 years	
2,4,5-trichlorophenoxyacetic acid	2,4,5-T	greater than 1 year	

From M. Alexander, *Soil Microbiology*, Wiley, New York, 1961, p. 241.

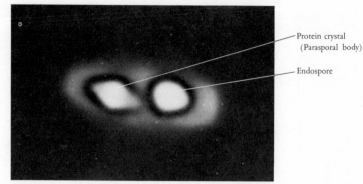

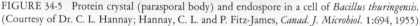

FIGURE 34-5   Protein crystal (parasporal body) and endospore in a cell of *Bacillus thuringensis*. (Courtesy of Dr. C. L. Hannay; Hannay, C. L. and P. Fitz-James, *Canad. J. Microbiol.* 1:694, 1955)

Some cases are known in which inactive compounds can be converted to active herbicides by soil organisms. Thus one compound which is itself inactive is readily converted to 2,4-D in soil by *Bacillus cereus*.

BIOLOGICAL AGENTS DISPLAY SPECIFICITY

The concern over the nonspecific effects that insecticides and herbicides exert has encouraged efforts to develop agents which are highly specific for the target organism, but harmless to everything else. It seems unlikely that organic chemical compounds can be developed which will display such specificity. Therefore efforts of many laboratories are being directed to studying biological agents which will affect only certain groups of insects. Some success has already been achieved. A spore-forming bacillus, *Bacillus thuringensis*, synthesizes a protein crystal (Fig. 34-5) in the course of sporulation which is highly toxic to the caterpillar forms of a group of insects which damage plant crops. The protein is soluble only in alkaline solutions such as are found in the gut of many insects. The ingested protein dissolves in the gut and apparently destroys the integrity of the walls. The gut's contents leak into the body fluid of the insect, increasing its pH markedly. This results in a general paralysis of the caterpillar and death ensues. The crystals are only toxic for insects with an alkaline gut, and are completely harmless to other animals and plants. These crystals are actually being used as an insecticide to control about ten different insect pests. The bacilli are grown in large fermentation vats under conditions leading to sporulation. The spores and crystals are collected, concentrated, and converted into a spray or dust that can then be applied to plants that are subject to these insects.

THE NITROGEN CYCLE

The mineral which plants require in greatest quantity is nitrogen, an important element in the makeup of proteins and nucleic acids. Although the atmosphere consists of 80 percent nitrogen, relatively few organisms can utilize this gaseous form. Inorganic nitrogen is assimilated almost entirely either as nitrate (NO_3^-) or ammonium (NH_4^+). However, the bulk of the nitrogen added to soil is in the form of organic materials, largely in the form of crop residues. Thus the organic form must be converted to ammonium ions (ammonification) and then to nitrate (nitrification). Some microorganisms convert nitrogen gas into ammonium ion, which in turn is converted into the amino group of amino acids in plant protein (*nitrogen fixation*). On the other hand, other bacteria convert nitrate to gaseous nitrogen by using nitrate as an electron acceptor in place of oxygen (*denitrification*).

All forms of nitrogen undergo a number of transformations. The entire series of reactions comprises the nitrogen cycle (Fig. 34-6). Microorganisms are essential at each stage of the cycle. The cycle consists of several individual transformations of nitrogen in which the nitrogen passes through several oxidation states from its highly reduced form (NH_4^+) to its highly oxidized form (NO_3^-). The salient features of each individual step are now considered.

Ammonification

Virtually all of the nitrogen found in the upper surface of the soil exists in organic molecules, primarily as the amino group of amino acids in protein molecules and some in covalent linkage in purines and pyrimidines.

A wide variety of microorganisms, including aerobic and anaerobic bacte-

Anaerobic respiration
Page 155

Amino Acids
Page 34

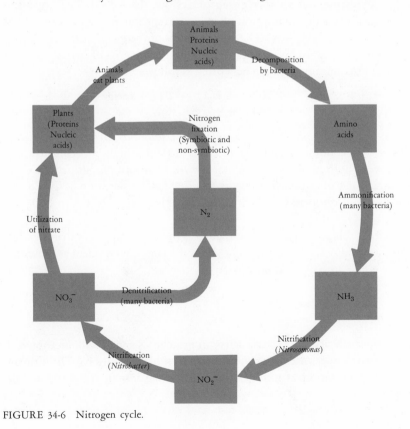

FIGURE 34-6   Nitrogen cycle.

ria, and fungi, are capable of decomposing protein initially through the action of extracellular proteolytic enzymes which convert protein into long chains of amino acids of varying lengths. Other enzymes then proceed to degrade these long chains into smaller chains and, finally, into amino acids, which are in turn degraded by a series of enzymes into CO_2, ammonium, and sulfate (derived from the sulfur-containing amino acids), and water.

Nitrification

Oxidation levels
Page 149

Nitrosomonas and Nitrobacter
Page 255

During the process of ammonification, ammonia may be released into the atmosphere, but this is likely only in alkaline soils in which very large amounts of organic nitrogen are being metabolized. The more usual sequence of events is the conversion of the ammonium to nitrite, then to nitrate. These processes are termed *nitrification*. Nitrate is readily utilized by plants. It will be helpful to consider the conversion of ammonium to nitrate in terms of the oxidation levels of the various elements involved. The ammonium ion (NH_4^+) is the most reduced form of nitrogen. As such it can be oxidized by bacteria with the release of energy. This oxidation occurs in two steps: the oxidation of NH_4^+ to NO_2^- and the further oxidation of NO_2^- to NO_3^-. One group of organisms carries out the first oxidation and another group carries out the second. The organisms which actually function in nitrification in nature represent only a few genera. The bacteria oxidizing NH_4^+ to NO_2^- are members of the genus *Nitrosomonas;* the conversion of NO_2^- to NO_3^- is carried out by a few species of *Nitrobacter*.

The reactions involved can be written as follows:

$$NH_4^+ + 1\frac{1}{2} O_2 \longrightarrow NO_2^- + H_2O + 2H^+ + energy$$
$$NO_2^- + \frac{1}{2}O_2 \longrightarrow NO_3^- + energy$$

Autotrophy
Page 110

As one would expect by looking at these reactions, they will only occur under aerobic conditions, and therefore very little nitrification occurs in waterlogged soils. Both *Nitrosomonas* and *Nitrobacter* are autotrophs, utilizing CO_2 as their sole source of carbon. The energy required for the reduction of CO_2 to the constituents of cytoplasm is obtained by the oxidation of NH_4^+ to NO_2^- or NO_2^- to NO_3^-.

Nitrite is highly toxic to plants, and it is therefore important that the organisms that oxidize the NO_2^- to NO_3^- be in the same environment as those that convert NH_4^+ to NO_2^-. Almost invariably they are.

Denitrification

Certain normally aerobic bacteria active in proteolysis and ammonification will reduce nitrate under anaerobic conditions (anaerobic respiration). The ultimate result is the liberation of nitrogen gas and the net loss of the element from the soil.

Immobilization

Nitrogen may be temporarily lost to plants by a process termed *immobilization*. This results from the assimilation of inorganic nutrients into the organic macromolecules of microorganisms in the soil. As bacteria and fungi grow they also utilize ammonium and nitrate salts for the synthesis of cell material, thereby converting nitrogen from a utilizable form into forms unusable by plants. The nitrogen is "sequestered" in the microorganisms and is unavailable until they die and are lysed. Only then can the nitrogen be changed into forms utilizable by plants. Immobilization is most likely to occur when materials which have a high carbon content, but a low nitrogen content, are added to soils. This stimulates the growth of the microbial flora, resulting in their utilization of available nitrogen. Plants cannot compete well with microorganisms when the inorganic nitrogen level is inadequate for the maximum development of either. If an inorganic source of nitrogen is added to the soil along with the carbon, there is no net immobilization of nitrogen. Thus the addition of straw to soil before planting time, which supplies carbon, should be accompanied by the addition of an ammonium salt.

Utilization of nitrogen
Page 161

Nitrogen Fixation

Nitrogen is continually being removed from the soil through the leaching action of water which removes nitrate, through its incorporation into plants which are harvested, and by denitrification. Utilizable nitrogen must therefore continually be added to the soil to maintain its nitrogen status. None of the processes described thus far explain how combined or *fixed nitrogen* (nitrogen in nongaseous form) is added to the soil. Small amounts of fixed nitrogen in the form of ammonium and nitrate ions occur in rain water. Also, fertilization of soil adds fixed nitrogen. However, the combination of these two processes often may not compensate for the amount of nitrogen lost by denitrification. Therefore nitrogen must be fixed in other ways, and microorganisms represent the most important means for this fixation.

Nitrogen fixation is brought about by two types of organisms: the free-living and those that live in symbiotic association with a group of plants, the *legumes*. A variety of bacteria are capable of reducing nitrogen gas to the amino group of amino acids in proteins. This process apparently occurs in several steps, first the nitrogen gas is activated and then reduced to the ammonium ion. The overall reaction requires an expenditure of energy. The ammonium formed inside the cell is then incorporated into other amino acids by transamination.

Transamination
Page 161

A variety of free-living microorganisms are capable of fixing nitrogen. The best known among the true bacteria, although usually not the most abundant, are members of the genus *Azotobacter*. These heterotrophic aerobic, gram-negative rods may be the chief suppliers of fixed nitrogen in grasslands and other ecosystems lacking plants with nitrogen-fixing symbionts. These organisms are rare in soils with a pH below 6.0, and members of the genus *Beijerinckia,* another

Azotobacter and *Beijerinckia*
Pages 250–251

heterotrophic nitrogen fixer, are found in highly acid soil. For reasons that are not clear, members of the genus *Beijerinckia* occur principally in the tropics and rarely in the temperate zone.

The dominant anaerobic nitrogen-fixers are members of the genus *Clostridium*. Blue-greens also fix nitrogen aerobically, especially in flooded soils. Indeed rice has been cultivated successfully for centuries without the addition of nitrogen-containing fertilizer probably because of the blue-greens developing in the water of the paddy field. Other nitrogen-fixing bacteria are found in the genera *Achromobacter, Enterobacter, Chlorobium,* and *Rhodospirillum,* the latter two being photosynthetic.

Although free-living bacteria are potentially capable of adding a considerable amount of fixed nitrogen to the soil, the *symbiotic nitrogen fixers are far more significant in benefiting plant growth and crop production.* The input of nitrogen from microbial symbionts of leguminous plants such as alfalfa may be roughly ten times the annual rate of fixation attainable by nonsymbiotic organisms in a natural ecosystem.

Members of the genus *Rhizobium* are the most important symbiotic nitrogen fixers. The plants with which they associate are all leguminous plants, which include alfalfa, clover, peas, beans, peanuts, and vetch. The symbiotic association of bacteria with the plant results in the formation of nodules on the roots of the plants. Both the plants and the bacteria can grow perfectly well without each other, although the bacteria may disappear from the soil if leguminous plants are not grown for long periods of time. However, the bacteria are only capable of fixing nitrogen when they are growing in a root nodule.

The stages in the infection process of the legumes by *Rhizobium* is understood at the descriptive level, but relatively little is understood at the biochemical level. The growth of *Rhizobium* in the rhizosphere is probably stimulated by the excretion of a number of organic compounds. The organisms multiply until they have reached a high cell density. This growth takes place primarily within a membranous layer around the outside of the root of the plant, and probably prevents the root secretions from escaping from the vicinity. The primary invasion site is the root hairs of the plant (Fig. 34-7). Several steps occur in the course of the invasion. Breakdown in the cellulose fibers which surround the root hairs may allow the bacteria to reach the hairs. Enzymes excreted by the bacteria which dissolve the cement holding the cellulose fibers together may be responsible for this breakdown. The rhizobia change their morphology from rod-shaped cells to spherical, highly flagellated cells, called *swarmer cells.* Also, rhizobia may synthesize the plant growth hormone indoleacetic acid from tryptophan excreted by the root, and this hormone may induce a deformation or curling of some of the root hairs.

Swarmer cells enter the curled root hair, the bacteria proliferate, and a hyphal-like infection thread is synthesized by the plant. The bacteria move within this thread which branches into a number of root cells. In this manner the bacteria gain entrance to the root cells. If the root cell happens to be a normal diploid cell, it is usually destroyed by the infection and degenerates. However, if it is a

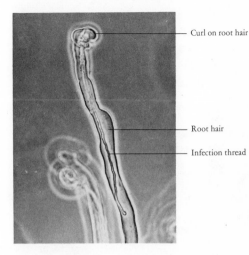

— Curl on root hair

— Root hair

— Infection thread

FIGURE 34-7
Infection thread inside a root hair. The bacteria enter the root cells via the infection thread. ($\times$310) (Courtesy of Dr. D. Munns)

rare tetraploid cell (a cell having four sets of chromosomes) it may develop into a nodule. Only about 5 percent of the infections ultimately result in nodules. By some unexplained mechanism the infection of the tetraploid cell stimulates it to divide, and the continued division of these cells results in the formation of the nodule (Fig. 34-8). The bacteria multiply rapidly within the tetraploid cells and become surrounded by a membrane synthesized by the host. They then undergo further morphological changes into swollen and irregular shapes (Fig. 34-9).

FIGURE 34-8
Nodules formed on the root system of a soybean plant. (The scale represents 1 cm.) (Courtesy of Dr. F. J. Bergersen; Bergersen, F. J., *9th Internat. Congress of Soil Sci. Trans.*, Vol. II, Paper 6, p. 49, 1968)

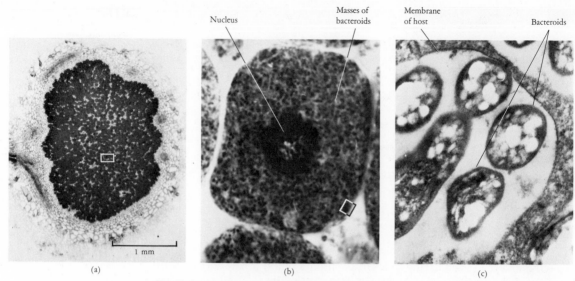

Nucleus Masses of bacteroids Membrane of host Bacteroids

(a) (b) (c)

FIGURE 34-9   Cross sections of root nodules and their components. (a) Cross section of a soybean nodule; (b) the "cut out" from (a) showing a degenerate host nucleus and masses of bacteroids; (c) bacteroids enclosed in a membrane of the host cell. (Courtesy of Dr. F. J. Bergersen; Bergersen, F. J., *9th Internat. Congress of Soil Sci. Trans.,* Vol II, Paper 6, p. 49, 1960)

These oddly shaped bacteria are termed *bacteroids,* and their formation is apparently induced by several chemicals in the plant tissue.

Because bacteroids and nitrogen fixation are both associated with nodules, numerous attempts have been made to induce the bacteroids to fix nitrogen. It has recently been shown that bacteroids removed from the nodule can indeed fix nitrogen for a limited time. However, if cultivated in the laboratory, they rapidly lose this ability. Therefore the bacteroids must contain an enzyme or enzymes involved in nitrogen fixation which they synthesize only in association with the root nodule, or else the root nodule provides environmental conditions necessary for nitrogen fixation. In this respect it was discovered in early 1900 that healthy nodules are pigmented. More recently this red pigmentation has been shown to be a hemoglobin, which is synthesized only by healthy nodules containing bacteroids. Since hemoglobin was first identified in nodules it has always been intuitively implicated in nitrogen fixation, and its role may now be clear. Several enzymes concerned with nitrogen fixation are inactivated by oxygen. Because hemoglobin can combine with oxygen, its presence may provide the nodule with the proper low level of oxygen required for nitrogen fixation.

Nodules are not found on all species of legumes, but the reason for this is not clear. Perhaps the legumes that lack nodules have not developed the capacity for symbiosis, or it is possible that the proper conditions to achieve nodulation have not yet been discovered. Likewise not all species of *Rhizobium* are capable of invading leguminous plants. Moreover the ability to develop a symbiotic relationship which results in nitrogen fixation varies markedly among infective strains. Once established the capacity of the symbiotic system to fix nitrogen is

NITROGEN FIXATION
REQUIRES ANAEROBIC
CONDITIONS

608

referred to as its *effectiveness*. It is entirely possible for a strain to be infective but not effective. There is considerable specificity between *Rhizobium* species and the particular legumes they will infect. Indeed the speciation within the genus *Rhizobium* is based on the ability of the organism to infect certain plants (Table 34-3). A single bacterial species is generally able to infect several different plants species, but the basis for this specificity is not known. It appears to reside in a single gene or in a number of closely linked genes, since the ability to infect a particular group can be transferred from one species of *Rhizobium* to another by the process of DNA-mediated transformation.

The continuous cultivation of nonlegumes may decrease the fertility of the soil in which they are planted because these plants remove fixed nitrogen. However, the cultivation of effectively nodulated legumes results in no such depletion, and succeeding nonleguminous crops frequently benefit from the nitrogen fixed by the legume. When the legume crop is plowed back into the soil, the full nitrogen gain is realized, but even when the crop is removed, there may still be a slight increase in the fixed nitrogen in the soil. The fixation of nitrogen by legumes that are effectively nodulated can be appreciable, the amount varying with the particular legume and other environmental conditions.

One cannot usually rely on effective nodulation by indigenous soil rhizobia. Therefore it is common practice today to mix a suspension of rhizobia grown in liquid culture in large vats with a carrier of moist humus or peat. The farmer mixes the seed with this rhizobial-treated peat prior to sowing the seeds, thereby ensuring effective nodulation of the plant.

Species of *Rhizobium* are notoriously susceptible to attack and lysis by bacteriophages. Bacteriophages specific for rhizobia which infect clover or alfalfa are found in the roots and nodules of the respective plants and also in soil in which these legumes have grown. Continuous harvest of alfalfa and clover sometimes results in poor plant vigor and low yields of succeeding crops. The conditions, termed alfalfa and clover sickness, were once thought to result from the lysis of rhizobia by their phages, but it is also conceivable that the plant roots excrete a toxic substance into the soil or else a plant pathogen is present.

Several genera of nonleguminous trees, including such trees as alder and ginkgo, possess nitrogen-fixing root nodules at some stages in their life cycle. The most intensively studied member of this group is the alder tree, considered to be a pioneer tree since it is able to grow in nitrogen-poor soils. This ability stems

Transformation
Pages 194–197

Virulent phage infection
Pages 314–317

TABLE 34-3
The Legumes Infected by Various Species of *Rhizobium*

Legume	Species of Rhizobium
Alfalfa group	R. meliloti
Clover group	R. trifolii
Pea group	R. leguminosarum
Bean group	R. phaseoli
Lupini group	R. lupini
Soybean group	R. japonicum

TABLE 34-4

Quantities of Nitrogen Fixed by Microorganisms

Group	Species or Habitat	N_2 *Fixed per Acre per Year (lb.)*
Nodulated legumes	Alfalfa	113–297
	Soybean	57–105
	Red clover	75–171
Nodulated nonlegumes	Alder tree	200
Blue-green algae	Arid soil in Australia	3
	Paddy field in India	30
Free-living heterotrophs	Soil under wheat	14
	Soil under grass	22
	Rain forest in Nigeria	65

From M. Alexander, *Microbial Ecology*, Wiley, New York, 1971, p. 428.

from its ability to fix nitrogen. Although many attempts have been made to isolate the microorganism(s) involved in symbiosis with the alder, the organisms have yet to be isolated in pure culture. This suggests that there may be an obligatory symbiotic relationship between the plant and the microorganism. No known species of *Rhizobium* will cause nodulation in a nonleguminous plant.

The amount of fixed nitrogen depends primarily on two factors, the organism and their habitat. Some idea of the amount of nitrogen fixed by the major groups of organisms can be gained from Table 34-4.

In the late nineteenth century there was concern that denitrifying bacteria were exhausting the nitrogen in the soil. Today, however, there are some who think that the amount of fixed nitrogen being added to the soil is excessive. Nitrogen is fixed by industrial processes as well as by microorganisms, and the amount of nitrogen fixed industrially has been doubling about every six years. It has been estimated that when this amount is added to that fixed by legumes, the total exceeds the amount that is lost through denitrification. The problem which arises is that the fixed nitrogen is being carried by water into lakes and rivers, resulting in an increase in algal growth in some cases. Furthermore, in many cases, the nitrogen in the organic wastes resulting from an ever-increasing human and animal population contributes to even more fixed nitrogen being added to our waters.

OXYGEN CYCLE

About one-fourth of all atoms making up living matter are oxygen, and this element constitutes about 20 percent of the air by weight. All of the oxygen in the air today probably originated from the decomposition of water molecules as a result of photosynthesis. The 1.5 billion cubic kilometers of water on the earth is continually being split in photosynthesis and reconstituted by respiration, the

Photosynthesis
Pages 157–158

entire cycle spanning about 2000 years. A simplified diagram of the oxygen cycle is given in Figure 34-10. It is also well substantiated that multicellular forms of life require oxygen in order to gain the maximum yield of energy from the breakdown of organic compounds and it is clear that the oxygen cycle is an absolutely essential part of life on our planet.

TRANSFORMATIONS—PHOSPHORUS

Occurrence of Phosphorus in Soil

Phosphorus occurs in the soil in both organic and inorganic forms. The organic compounds occur almost exclusively in the A horizon and generally near the surface. They are derived from surface vegetation and accordingly occur as the constituents of plant tissue, animals, and microorganisms. The inorganic forms of phosphorus occur largely as insoluble salts of calcium, iron, and aluminum. Plants cannot use these inorganic and organic forms readily. Orthophosphate (PO_4^{-3}) is readily used by most plants and microorganisms. Microorganisms play three major roles in phosphorus transformation: they mineralize organic phosphorus; they convert insoluble forms of inorganic phosphorus to soluble forms; and they immobilize inorganic phosphorus.

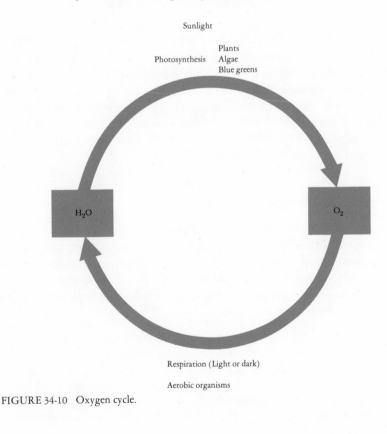

FIGURE 34-10   Oxygen cycle.

Overall Transformations of Phosphorus

Organic phosphate is converted to orthophosphate, which can then be utilized by the plants or immobilized by bacteria into organic forms of phosphorus. Some minerals containing phosphate may be converted into orthophosphate by microorganisms.

Mineralization of Organic Phosphates

Bacteria and fungi are largely responsible for rendering organic phosphorus available to succeeding generations of microorganisms as well as to plants. Most microorganisms are capable of synthesizing an enzyme, phosphatase, which cleaves the phosphate from a wide variety of compounds in which it is found. The breakdown of nucleic acids is initiated by the action of nucleases which degrade DNA and RNA into the individual subunits (nucleotides). The phosphate is then cleaved from the nucleotides by a phosphatase:

Structure of nucleotides
Page 41, Fig. 2-16

$$\text{Nucleo}\textit{tide} \longrightarrow \text{nucleo}\textit{side} + PO_4^{-3}.$$

Dissolution of Minerals

A wide variety of commonly occurring soil organisms are capable of dissolving calcium phosphate [$Ca_3(PO_4)_2$]. These include species of the genera *Pseudomonas, Mycobacterium, Micrococcus, Flavobacterium, Penicillium,* and *Aspergillus.* These organisms can grow in media with calcium phosphate as the sole source of phosphorus. Organic acids, produced by organisms in the course of metabolism, can convert calcium phosphate to the di- and monobasic phosphates [$Ca_2(HPO_4)_2$ and $CaHPO_4$, respectively], which are much more soluble than the tribasic salt [$Ca_3(PO_4)_2$]. The amount of phosphate which dissolves depends primarily on the amount of acid produced.

Immobilization of Inorganic Phosphate

Since microbial growth involves the assimilation of available forms of phosphorus into a variety of phosphorus-containing macromolecules, phosphate, like nitrogen, is immobilized by soil microorganisms. The same principles discussed for nitrogen hold for phosphorus. The addition to the soil of carbonaceous residues deficient in phosphorus may result in the net immobilization of phosphorus and the consequent decrease in crop yields.

MINERAL CYCLE—SULFUR

Formulae of amino acids
Page 34

Sulfur occurs in all living matter chiefly as a component of certain amino acids—methionine, cystine, and cysteine. The sulfur cycle bears many similarities

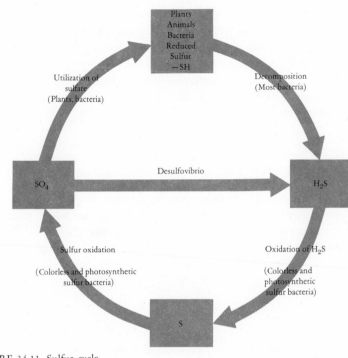

FIGURE 34-11 Sulfur cycle.

to the nitrogen cycle (Fig. 34-11). Like nitrogen, sulfur is present in the soil chiefly as a part of proteins, and it is taken up by plants in its oxidized form, sulfate (SO_4^{-2}). These proteins are degraded into their constituent amino acids by proteolytic enzymes excreted by a large variety of soil organisms. The sulfur is generally converted to hydrogen sulfide (H_2S) by a variety of soil microorganisms, a process analogous to NH_4^+ formation. The H_2S is highly toxic to biological systems and generally does not accumulate. Under aerobic conditions H_2S is oxidized spontaneously to sulfur and then to its most readily utilized form, sulfate, by sulfur bacteria. This process is analogous to nitrification. Under anaerobic conditions the counterparts of the denitrifiers operate: the sulfate-reducing bacteria reduce SO_4^{-2} to H_2S.

The microbial flora concerned with the sulfur cycle has a number of unique features. The oxidation of H_2S to SO_4^{-2} is carried out principally by two major groups of organisms, *Beggiatoa* and *Thiothrix,* and also by members of the genus *Thiobacillus.*

Another group of nonphotosynthetic bacteria which oxidize H_2S for energy are members of the genus *Thiobacillus,* the most intensely studied species being *Thiobacillus thiooxidans.*

A number of photosynthetic bacteria, the green and purple sulfur bacteria, also play a role in the sulfur cycle. The most common green sulfur bacteria are members of the genus *Chlorobium,* and *Chromatium* is a common genus of purple sulfur bacteria.

Beggiatoa and *thiothrix*
Page 239

Thiobacillus
Page 255

Photosynthetic bacteria
Pages 238–239

Desulfovibrio
Page 254

The reduction of sulfate into sulfide (sulfate reduction) is carried out by a small group of anaerobic bacteria which are capable of utilizing sulfate as the electron acceptor (anaerobic respiration). These include organisms of the genus *Desulfovibrio* as well as a few anaerobic spore-forming rods of the genus *Clostridium*. Hydrogen sulfide is thus likely to accumulate in waterlogged soils. Since this compound is toxic to higher plants, sulfate reduction is undesirable from an agricultural point of view.

PLANT PATHOGENS

Although plants are susceptible to infection by a variety of agents including bacteria, viruses, and nematodes, the fungi are by far the most important plant pathogens. One recent estimate states that in Ohio alone 1000 diseases of plants are incited by fungi, about 50 by bacteria, and 100 by viruses. Over 10 percent of the total cash yield of crops has been estimated to be lost through plant diseases.

Fungal Diseases

Fungi
Page 282, Table 14-1

The species of fungi which cause plant diseases are found among all groups: Phycomycetes, Ascomycetes, Basidiomycetes, and Deuteromycetes. Some of the more common diseases caused by members of each group are listed in Table 34-5.

Fungal diseases have virtually destroyed several plant species in the United States. An ascomycete native to the Orient, accidentally introduced into the United States, has completely wiped out the American chestnut in parts of the country. It has now invaded Europe and threatens the European chestnut with the same fate. Another ascomycete accidentally introduced into the United States from Europe is slowly but relentlessly exterminating the American elm. The fun-

TABLE 34-5
Some Fungal Diseases of Plants

Disease	Host	Group of Fungus Responsible
Brown spot	Corn	Phycomycete
Soft rot	Sweet potato	Phycomycete
Powdery scab	Potato	Phycomycete
Chestnut blight	American chestnut	Ascomycete
Ergot	Rye	Ascomycete
Scab	Apple	Ascomycete
Powdery mildew	Rose	Ascomycete
Peach leaf curl	Peaches	Ascomycete
Dutch elm	Dutch elm	Ascomycete
Scab	Peach	Deuteromycete
Late blight	Celery	Deuteromycete
White pine blister rust	White pine	Basidiomycete
Leaf rust	Wheat	Basidiomycete

TABLE 34-6
Some Bacterial Diseases of Plants

Disease	Host	Species Responsible
Crown gall tumor	Most deciduous plants	*Agrobacterium tumefaciens*
Bacterial wilt	Tobacco	*Pseudomonas solanacearum*
	Tomato	
Bacterial spot	Tomato	*Xanthomonas vesticatoria*
Bacterial canker	Tomato	*Corynebacterium michigense*
Fire blight	Apples, pears	*Erwinia amylovora*

gus is carried by a bark beetle which is susceptible to DDT. However, since this insecticide is also harmful to birds, the cities heavily planted with elms face the dilemma of whether to save the elms and kill the birds or save the birds and lose the trees.

Bacterial Plant Pathogens

The members of the genus *Erwinia,* gram-negative facultative anaerobes, represent a heterogeneous collection of organisms, often characterized by their association with plants. *Erwinia amylovora* causes a soft rot of the fleshy parts of a number of plants. The organism produces an extracellular enzyme that hydrolyzes the pectin, a "cement" that holds plant cells together. *Erwinia amylovora* also causes a wilting disease.

    Agrobacterium tumefaciens, a gram-negative, flagellated, rod-shaped organism, is closely related to the genus *Rhizobium.* The interesting feature of this relationship is that whereas *Rhizobium* infects plants and causes the formation of root nodules, *A. tumefaciens* causes a tumorlike growth in plants called crown gall tumor. In both cases the bacterial infection causes the plant cells to divide. In crown gall tumor, destruction of the bacteria does not stop the development of the tumor once it has been initiated, suggesting that the bacterial cell is not responsible for the tumor. Recent experiments have strongly suggested that bacterial DNA becomes associated—perhaps integrated—into the plant DNA and, by some unknown mechanism, causes the tumor. Conceivably a basic similarity may exist between the association that occur between various bacterial DNA's. For example, bacterial and plant DNA in crown gall tumor, bacteriophage DNA and bacterial DNA in a lysogenized bacterial cell, and the DNA of tumor viruses associated with the DNA of animal cells in certain cancer cells. Other bacterial diseases of plants are listed in Table 34-6.

Viral Disease of Plants

A large number of plant diseases are caused by viruses that are often found in soil. Some of these diseases have been discussed in Chapter 17.

Viral plant pathogens
Pages 339–344

SUMMARY

The soil is the natural habitat of a variety of organisms: viruses, bacteria, algae, fungi, protozoa, mites, and nematodes, to name the most prominent. The organisms play both useful and destructive roles. The useful roles include the decomposition of organic matter as part of the cycling process of various chemical elements, such as nitrogen, carbon, oxygen, phosphorus, and sulfur.

Most compounds generated by biological processes are degradable by living organisms. The more complex molecules such as aromatic compounds are degraded by only a very few species of organisms. Simpler structures are generally capable of being decomposed by a large number of different organisms. In general, animal flesh is degraded by bacteria and plant material by fungi. A large number of complex hydrocarbons, which often contain chlorine atoms, and are commonly found in pesticides are nonbiodegradable. Many of these compounds are fat soluble and accumulate in the fat of organisms, with harmful effects in some cases.

The elements comprising living matter must be continuously recycled. This recycling process involves a variety of microorganisms which convert the elements in organic molecules in decaying matter into inorganic forms of the elements which are utilized by plants for the synthesis of organic molecules. Fixed nitrogen is being continually lost from the soil through the action of denitrifying organisms, and so must be returned. Nitrogen-fixing organisms both symbiotic and free-living are able to convert nitrogen gas into ammonium, which is then converted into plant protein.

The most important plant pathogens are the fungi, and several species of trees in the United States have been almost completely eradicated due to fungal disease. Viral diseases are the next most important pathogen, whereas bacteria are the least important.

QUESTIONS

1. What are the major groups of organisms in the soil and what is the function of each group in the scheme of life?
2. A farmer is not pleased if heavy rains occur soon after he adds nitrate fertilizer to his soil. Why?
3. What are the advantages and disadvantages of using nonbiodegradable pesticides?
4. Compounds which are not degraded in nature may be readily degraded in the laboratory. What may account for this phenomenon?
5. Consider the oxygen and nitrogen cycle. Which steps require energy; which provide energy to the cell?
6. In the symbiotic fixation of nitrogen, what does the plant supply the bacteria, and what do the bacteria supply the plant?

ALEXANDER, M., *Introduction to Soil Microbiology*. New York: John Wiley & Sons, Inc., 1961.

CLOUD, P. and A. GIBOR, "The Oxygen Cycle," *Scientific American* (Offprint #1192).

DEEVEY, E. S., JR., "Mineral Cycles," *Scientific American* (Offprint #1195).

DELWICHE, C., "The Nitrogen Cycle," *Scientific American* (Offprint #1194).

FARB, P., *Living Earth*. New York: Harper Colophon Books, 1959.

Chapter 35

Aquatic Microbiology

And if from man's vile arts I flee
And drink pure water from the pump;
I gulp down infusoria,
And quarts of raw bacteria,
And hideous rotatorae,
And wriggling polygastricae,
And slimy diatomacae,
And various animalculae
Of middle, high and low degree.

—WILLIAM JUNIPER, *The True Drunkard's Delight*

Virtually all water contains living organisms, although not always the varieties described in the quotation above. Some are accidental contaminants, but others, the aquatic organisms, normally live in environments that are predominantly water. These unseen, often unsuspected creatures play vital roles in their ecosystems and in the geochemical cycles of the world at large. In fact human life could not survive for very long without the benefits derived from their activities. For example, the aquatic microorganisms that are photosynthetic transform energy from the sun into chemical energy, and most of them produce oxygen, essential for animal life. Others that are heterotrophic decompose wastes and participate in cycling the elements essential for life. This chapter is concerned with the nature of aquatic microorganisms and their environments, the complex interactions that occur within aquatic ecosystems, and the effects of water pollution, both on these ecosystems and on the larger environment.

618

Water has several unique properties that help to make it a necessary part of the environment for microorganisms and that also contribute to its essential functions within all living cells. For example, it can adsorb to colloids and particles by forming hydrogen bonds; it can absorb light of certain wavelengths, an important property for photosynthetic aquatic microbes; and it is an extremely efficient solvent, a property that is of paramount importance for all organisms whether they are terrestrial or aquatic.

Because water is such an excellent solvent, natural water supplies are actually dilute solutions of many substances. There is probably no truly pure water, even in laboratories where multiple distillations yield highly purified samples, since minute amounts of components of glass or other containers dissolve in the distillate. The term "pure water" is generally used to refer to water that is safe to drink but far from pure in the chemical or microbiological sense.

Rain has fewer impurities than water from other natural sources (such as rivers or lakes) since it is a distillate which becomes contaminated only from material suspended in the air. As indicated in Table 35-1 samples of rainwater were found to contain very small quantities of dissolved nutrients. As compared with rain from a nonindustrial area, rainwater collected in industrial regions may be considerably richer in dissolved materials, especially sulfate, ammonium ions, and other materials that are potential nutrients for microorganisms. It may also contain more toxic substances. When rainwater reaches the ground, it becomes further enriched with dissolved materials.

The composition of freshwater reflects its source, which may be either surface waters (such as streams, rivers, and lakes) or underground water (frequently called simply *groundwater*). The water that seeps through sandy or porous soils until it reaches hard clay or rock makes up the underground water; often it forms a layer between two rock strata. If the groundwater layer is close to the surface, it may break through in some areas to form springs; deeper groundwater, however, must be tapped by digging wells. When groundwater comes from a high level in the mountains to a lower altitude, the pressure may be sufficient to cause the water to rise above the stratum tapped for the well, or even to overflow. Such wells are known as artesian wells.

TABLE 35-1
Approximate Amounts (mg/liter) of Various Substances in Natural Water Supplies

Water Source	CO_3^{2-}	SO_4^{2-}	Cl^-	NO_3^-	NH_3	Mg^{2+}	Ca^{2+}	Na^+	K^+	P(total)
Rain	–	2.0	0.5	0.2	0.5	≥ 0.1	0.1–10.0	≥ 0.4	≥ 0.03	–
Freshwater										
Wisconsin										
soft waters	69.6	20.5	9.9	64.0	158.0	37.7	46.9	10.9	4.8	23.0
River (mean)	73.9	16.0	10.1	–	–	17.4	63.5	15.7	3.4	–
Seawater	73	2,712.	19,353.	0	0	1,294.	413	10,760	387	0.03

Data compiled from: G. E. Hutchison, *A Treatise on Limnology*, Vol. 1. Wiley, New York, 1957.
D. R. Kester, I. W. Duedall, D. N. Conners, and R. M. Ptykowicz. *Limnology and Oceanography* 12: 176–179 (1967).
T. D. Brock. *Principles of Microbial Ecology*. Prentice-Hall, Englewood Cliffs, New Jersey, 1966.

Gases dissolved in groundwater add to the pressure and may even create a fountainlike flow of water from an artesian well. If the dissolved gas is CO_2, the well water is harmless; however, if it is methane, produced during microbial decomposition of organic materials, the air around the well may become potentially explosive as the dissolved methane is freed and mixes with air. The presence of various gases dissolved in water helps to determine the microbial flora that can grow there.

The course of water through underground strata changes its properties in ways other than the addition of dissolved gases and other nutrients. Hot springs owe their increased temperatures to passage of groundwater through deep strata of the earth's crust. The water temperature increases about 1°F for every 50 feet of depth. Hot spring waters often contain large quantities of dissolved substances, including sulfur compounds such as H_2S.

The temperature of natural waters can range from approximately 0°C to nearly 100°C. It is amazing to find that various species of microorganisms can survive at any temperature within this range. Even in boiling geyser basins where the water temperature is greater than 90°C, species of bacteria adapted to grow in this environment have been selected. At the other extreme certain fungi and bacteria can grow below freezing temperatures.

The pH of natural waters, determined by their content of dissolved substances, can also vary greatly. For example, the acid springs in Yellowstone National Park generally have a pH between 2 and 4; nevertheless they support the growth of microorganisms adapted to a very low pH. Other natural waters are extremely alkaline, with a pH greater than 9; these are also the natural habitat of microorganisms. No matter how inimical the aquatic environment may seem, some kind of microbe is able to survive or even thrive there as long as nutrients sufficient for multiplication are present.

Freshwater rivers and lakes are commonly rich in nutrients (Table 35-1). Of course, there is considerable variation from one body of freshwater to another, depending on factors such as the geochemical character of the area and the nature and amount of effluents added to the waters. Thus in areas where calcium salts are plentiful in the soil, the calcium content of lakes may be considerably higher than the average figure stated in Table 35-1. The effect of added effluents is strikingly demonstrated by results of a study of Irondequoit Bay near Rochester, New York. It was found that salt used for deicing streets was washed into the bay, increasing its chloride concentration at least fivefold (from about 30 mg/liter to more than 150 mg/liter) during 1950 to 1970.

As water follows its course to the sea it becomes more and more enriched with salts, leading to marked differences in the salinity of various bodies of water. Seawater is relatively high in salt content (containing on the average about 3.5 percent as compared with less than 0.05 percent for freshwater), but it lacks certain other nutrients, such as phosphate and nitrate found in abundance in freshwater. Therefore seawater supports growth of halophilic microorganisms, but usually contains fewer organisms than freshwater. The media described in Table 35-2, used to grow marine microbes in the laboratory, was designed to approxi-

TABLE 35-2
Formula for Artificial Seawater Used for
Cultivating Marine Organisms

Salts	Grams per Liter
NaCl	19.40
$MgCl_2$	8.78
Na_2SO_4	3.25
$CaCl_2$	1.23
KCl	0.55
$NaHCO_3$	0.16
KBr	0.08
H_3BO_3	0.02
$SrCl_2$	0.02
NaF	0.0025

Adapted from D. R. Kester, I. W. Duedall, D. N. Conners, and R. M. Ptykowicz. *Limnology and Oceanography* 12: 176–179 (1967).

mate seawater. Its formula indicates the proportions of various nutrients and salts commonly found in seawater.

An unusual situation occurs in salt lakes found in hot and dry regions. Here the inflowing water with its dissolved ions is trapped, because there are no outlets. As water evaporates, more enters, keeping the water level fairly constant but increasing the content of dissolved salts. A familiar example is the Great Salt Lake in Utah, where the salinity is considerably greater than that of seawater. This sort of environment restricts the growth of all except extremely halophilic organisms.

Other examples of specialized aquatic habitats where microorganisms have become adapted to an unusual milieu include sulfur springs and mineral springs, which contain unusually high concentrations of iron, magnesium, or other minerals.

Within this wide continuum of aqueous environments, ranging from freshwater to salt lakes, a tremendous variety of microbial species exist and interact. Any body of water represents a complex ecosystem. Energy enters the system largely in the form of light, and is converted to chemical energy by the photosynthetic primary producers, either algae (in aerobic conditions) or bacteria (in anaerobic areas). Primary producers require inorganic nutrients, utilizing CO_2 as a carbon source. They serve as the first step in the food chain, providing food for protozoa and small invertebrates that in turn serve as sustenance for fishes.

In addition to dissolved inorganic nutrients, preformed organic substances also find their way into water in the form of discarded wastes, dead leaves, animal corpses, and the like. These are transformed into inorganic materials by the action of decomposers, including both bacteria and fungi. The decomposers are also part of the food chain, serving as nutrients for heterotrophic protozoa and invertebrates. A small part of the organic material in water is not decomposed, but sinks to the bottom to become part of mud sediments or to be transformed eventually into coal or oil.

Oxygen is one of the most important limiting elements within aquatic environments because of its low solubility in water. Decomposers, such as the pseudomonads and species of *Cytophaga, Caulobacter, Hyphomicrobium,* and other genera, use the available dissolved O_2 for oxidizing organic materials. Therefore the aerobic metabolism of many microbes can lead to a depletion of O_2 and consequently to conditions sufficient for anaerobic growth. Rivers are continuously aerated by the flow of water and are not as readily depleted of O_2 as are lakes. During warm weather lakes can actually become stratified into aerobic and anaerobic layers. Colder, denser water sinks to the bottom and is rapidly depleted of O_2, while warmer layers at the top are replenished with O_2 from the air. Water samples taken from various levels of such a stratified lake contain quite different populations of microorganisms. Aquatic environments can also become anaerobic following *nutrient enrichment (eutrophication)*.

THE MICROBIAL FLORA OF AQUATIC ENVIRONMENTS

Many aquatic microorganisms have unique capacities for exploiting their environments. The concentration of dissolved nutrients in water is frequently low compared with solid media; thus organisms which have special abilities have been selected. For example, many algae and aquatic bacteria can attach to stones, plants, or other solid surfaces and continually withdraw nutrients from the water flowing past them. Others float in the plankton and are moved about so that they are able to reach more nutrients than if they were sessile.

The Microbial Flora of Freshwater

The microbial flora of freshwater varies considerably with the kind of water being considered. For instance, samples of tropical rainwater were found to contain relatively few microorganisms which had been airborne in dusts and subsequently washed from the air by rain. A majority of these were actinomycetes and other soil bacteria, with fewer fungi and very few algae. After rainwater falls to the ground, it soon acquires more microorganisms of various kinds, depending on the nature of its new environment.

Algae and protozoa make up the bulk of the microbial mass of freshwater. The species of algae in lakes and their abundance at different seasons varies considerably. Algae and protozoa have been discussed in some detail in Chapters 13 and 15, respectively, and are not further considered here, except to stress that they are abundant in aquatic environments and are a predominant and an integral part of these ecosystems.

Bacteria are also essential to aquatic ecosystems. In addition to animal pathogens and other terrestrial species which are passively carried into rivers and lakes, bodies of freshwater contain species of bacteria that grow in nature only there. The population of the freshwater lake described in Figure 35-1 provides a good example of some bacteria native to freshwater habitats. Lakes can be compared

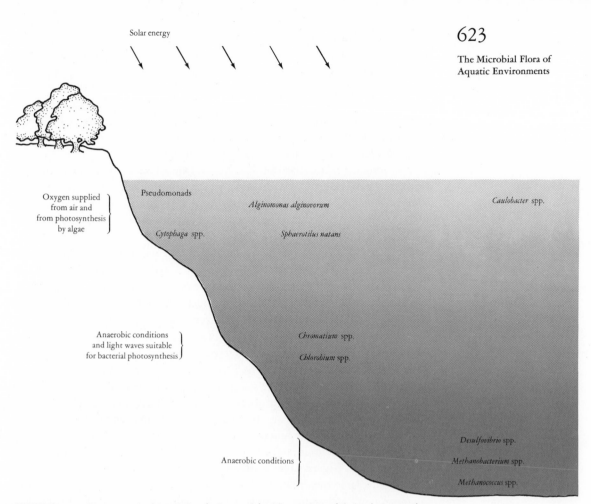

FIGURE 35-1 Representative bacteria in a freshwater lake. The species and their relative numbers varies considerably from one lake to another, and in a single lake from one time to another.

with cities, and their bacterial flora with city dwellers; various groups settle where the environment is suitable for their existence, and there is thus a constant shifting as the environment changes. To carry the analogy one step further, just as city dwellers often cluster around industrial plants that employ them, bacteria also tend to live in areas of a lake where their particular capabilities can be used to support their growth.

Thus aerobic bacteria that can degrade organic materials are usually found near the surface of lakes, where the oxygen content is highest, or in other aerated portions of the lakes; here they can act on animal wastes and plant or algal remains in the water. For example, pectins and cellulose from dead algae are broken down by the metabolic activities of heterotrophic aerobic bacteria, such as the pseudomonad *Alginomonas alginovorum*. Cellulose from dead plants and algae is also often degraded by species of *Cytophaga*. These bacteria produce a cellulose-degrading enzyme that apparently is associated with their cell surfaces, so that the bacterial cells must actually make contact with the cellulose in order to act on it.

Deeper in lake waters, where oxygen is depleted and anaerobic organisms can grow, the microbial flora is quite different from that of aerobic upper layers. In areas where light of the proper wavelength penetrates, photosynthetic bacteria prosper within the limited stratum where conditions are favorable. Here the purple and green sulfur bacteria such as *Chromatium* and *Chlorobium,* oxidize H_2S gas dissolved in the water, using photosynthesis as their system for generating energy. Sulfate ions or elemental sulfur is produced as a result.

Among the many anaerobic bacteria that are active in the sediment of lakes are members of the genus *Desulfovibrio,* which reduce sulfate ions to sulfide ions. Some of the sulfides remain in the mud, but some become dissolved in the water as H_2S, which can be oxidized by other species, leading to a cycling of sulfur from one form to another.

Other members of the flora of mud in lake sediments are active in the anaerobic decomposition of settled organic materials. For example, species of *Methanobacterium* and *Methanococcus* produce methane from organic substances, and some of the clostridia decompose nitrogen-containing organic substrates to yield amines which are subsequently oxidized to ammonia.

The prosthecate bacteria, also part of the normal flora of lakes and ponds, are uniquely adapted to exploit their environment. For example, species of stalked bacteria of the genus *Caulobacter* are able to live in water that is low in nutrients. It is thought that the stalk, an outgrowth of the bacterial cytoplasm and cell wall, serves as an area of absorption which permits the cell to take in nutrients more efficiently than nonstalked cells. The hyphae of *Hyphomicrobium* species may have the same function.

Unusual concentrations of various nutrients in natural waters leads to selection of a normal flora capable of utilizing those nutrients. Therefore iron springs that contain large quantities of ferrous ions support the growth of species of *Gallionella* and *Sphaerotilus.* Similarly, sulfur springs are the natural habitat of both the photosynthetic and nonphotosynthetic sulfur bacteria.

Although only a small portion of the actual microbial activities in a lake community has been mentioned, it is apparent that the activities of freshwater bacteria are many and complex.

The Microbial Flora of Seawater

The numbers and kinds of microorganisms in seawater are restricted by several factors. In addition to the salinity of the water and the small amounts of certain nutrients, the limiting factors include a high pH (around 8.0), lack of vitamins or other growth factors, and the intensity and wavelength of light that can penetrate the water to serve as an energy source. In addition, seawater is noted for having antibacterial properties. It has been suggested that these properties are attributable to the presence of unsaturated fatty acids, produced by algae, that inhibit many gram-positive bacteria. It is probable that other antibacterial substances are also formed by marine organisms.

Attempts to make even so gross a measurement as the total mass of living organisms (or biomass) per liter of seawater have not been very successful because of technical difficulties. One method that has been used is direct counting of organisms under the microscope, but it is tedious. Another disadvantage is that seawater contains particulate debris which sometimes cannot be distinguished visually from microorganisms. Then too not all microbes survive long enough to be counted. Attempts to grow the organisms lead to even less accurate results because of variations in growth requirements of the many species present. Probably the most accurate method used has been the measurement of ATP by the luciferin-luciferase enzyme system. This technique gives an estimate of the number of viable organisms, based on a carbon : ATP ratio of approximately 250 for most marine microbes. The measured amounts of ATP, in micrograms per liter, ranged from 0.3 or more near the surface of the ocean, to 0.05 at a depth of a few hundred meters, to 0.005 at depths below 1000 meters. Thus in general the *biomass of seawater decreases with increasing depth.*

The deep sea contains remarkably few bacteria. This may result partly from lack of energy sources and from the water pressure, which increases about one atmosphere per 10 meter depth or to as much as 1000 atmospheres (15,000 lb/sq in.) in the deepest parts of the ocean. Most microorganisms cannot survive greatly increased pressures. Those which can, called barophiles, are the only ones found deep in the sea.

Much remains to be learned about the various species of microorganisms that make up the biomass of seawater. Certainly the algae and protozoa of plankton account for a large proportion of the mass of organisms near the surface. Bacteria are found in smaller proportions, along with a few fungi and an undetermined number of viruses. Many organisms among the marine flora are probably there only temporarily as a result of having been washed from soil or air, or discarded with wastes into the saltwater.

Marine algae include much of the phytoplankton of the sea, the plantlike seaweeds, part of the flora of coral reefs, and other species. It has been estimated that the mass of phytoplankton in seawater is about ten times that of the zooplankton. Some of the genera of algae that predominate in the sea are given in Table 35-3.

Protozoa make up a large part of the zooplankton. They feed on bacteria in surface films and on submerged structures, such as masses of algae. Actively motile protozoa exist in open water where they can move to food supplies, whereas some of the ameboid or sessile ciliated species feed on bacteria and other organic materials at the bottom of relatively shallow areas of the sea.

One reason it is difficult, if not impossible, to determine the bacterial content of seawater accurately is that most marine bacteria are found associated with particles, often less than 0.1 mm, in the water near the surface. The bacteria, which utilize nutrients from the particles and from the water, are usually eaten by snails and clams.

Even though it has been estimated that bacteria account for only about 1 to 5 percent of the total biomass of the sea, these organisms are of great impor-

TABLE 35-3

Some Representative Genera of Microorganisms Commonly
Found in the Marine Environment

Genus	Characteristics
Algae	
Trichodesmium	Blue-green
Oscillatoria	Blue-green
Nostoc	Blue-green
Asterionella	Diatom
Fragilaria	Diatom
Navicula	Diatom
Gonyaulax	Dinoflagellate
Gymnodinium	Dinoflagellate
Bacteria	
Mycoplana	Gram-negative rods, polar flagella
Pseudomonas	Gram-negative rods, polar flagella
Vibrio	Gram-negative rods, polar flagella
Spirillum	Gram-negative rods, polar flagella

tance in contributing to the growth of other marine flora. Several genera of bacteria commonly found in seawater are listed in Table 35-3. Some of them oxidize organic materials to CO_2 which can be used by algae, resulting in the production of O_2 through photosynthesis. This O_2 can in turn be utilized for additional bacterial oxidations. Equally important is the production by bacteria of vitamins (especially B_{12}) which are essential for growth of many algae.

A number of bacterial species are pathogenic or nonpathogenic parasites of marine animals and plants. For example, a group of vibrios (typified by *Vibrio anguillarum*) are of major importance as fish pathogens. These halophilic bacteria have been found in about 50 species of fishes from at least 14 countries. Marine vibrios are not always pathogenic; many fishes contain nonpathogenic commensal species among their normal gastrointestinal flora. It has been postulated that some of these vibrios can act as opportunists to cause disease following injury or stress to the fish.

Some infectious diseases of fish closely parallel human disease in both the nature of the disease and the causative bacteria. One ailment resembling the human disease tularemia is caused by *Pasteurella piscicida,* a bacterial species quite similar to *Francisella tularensis* which causes tularemia. Mycobacteria are also pathogens of some marine and freshwater fish as well as man; however, the species of mycobacteria infecting fish are generally distinct from those causing human disease, and vice versa. Both *Mycobacterium marinum* and *M. salmonipilum* have been implicated in fish tuberculosis. It is probable that pathogenic *Pasteurella* and *Mycobacterium* species are transmitted as a result of fish eating portions of other infected fish. There is no evidence that either of these fish pathogens can grow in seawater outside their hosts.

Marine environments, even though they are somewhat restrictive, support complex populations of microorganisms that are essential links in food chains. In addition, marine algae are responsible for supplying a large proportion of the

O$_2$ produced by photosynthesis. These and other functions of marine microorganisms make them, as well as freshwater microbes, a vital part of the complicated web of interacting living creatures.

THE CONSEQUENCES OF WATER POLLUTION

Pollution of bodies of water often interferes with the functions of aquatic microorganisms. Trash-littered beaches, dead fish, unsightly and bad-smelling lakes, and epidemics of diseases caused by water-borne bacteria and viruses are only a few of the consequences of water pollution that have been recognized for a long time. During recent years many others have become apparent.

Water pollution is a complex problem which cannot be discussed comprehensively here. Therefore only two facets important in microbiology are considered: pollution caused by pathogenic organisms and that which leads to eutrophication.

Pollution with Pathogenic Organisms

The idea that certain diseases can be transmitted via water was not generally recognized during ancient times. Moses and the Israelites, for example, had strict rules and laws of hygiene that must have prevented the spread of many diseases transmitted by direct contact or by means of animal vectors; however, little or nothing concerning water supplies is included in these laws. There is reason to believe that Alexander the Great, about 1000 years later, may have realized the dangers of contaminated water, since it is reported that he had his troops boil their drinking water.

As the years passed more and more people began to associate certain diseases with drinking water from particular sources. In London in 1854 about 30 years before cholera vibrios were discovered, John Snow traced the source of cholera outbreak to water from a single pump on Broad Street, and suggested that the disease was caused by self-replicating agents in the water. As soon as methods for culturing bacteria became available, it was easy to prove that water can transmit a variety of pathogens. Thus for many years a high priority has been assigned to avoiding the pollution of water with human or animal wastes that may contain pathogenic organisms. Sewage disposal or treatment to eliminate or minimize this type of pollution has become prevalent in most economically advanced countries, and great progress is being made in all parts of the world. Nevertheless large water-borne epidemics occur from time to time, even in cities with water supplies that are well monitored for fecal contamination, the usual source of pathogens in water.

The 1965 epidemic of salmonellosis in Riverside, California, illustrates important points about the pollution of water with pathogenic microorganisms. This small city obtained most of its water from wells. The supply was monitored carefully, using a standard procedure (Fig. 35-2) to detect fecal contamination. This procedure is based on the fact that *Escherichia coli* is usually much more numerous than pathogens in feces and also is able to survive longer in water than

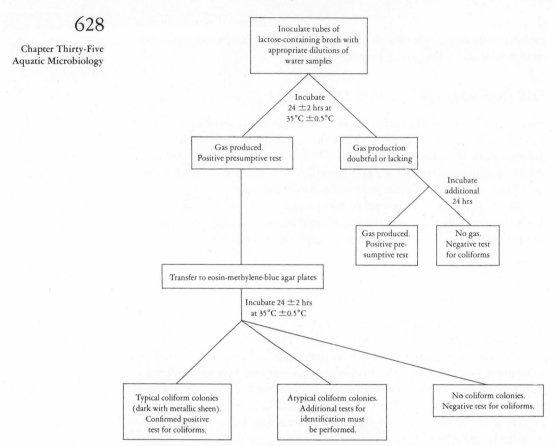

FIGURE 35-2(a)   Method recommended by the United States Public Health Service for testing for coliforms in drinking water. Coliform organisms are all aerobic and facultative, gram-negative, nonspore-forming, rod-shaped bacteria that ferment lactose with gas formation within 48 hours at 35°C. (Adapted from *Standard Methods for Examination of Water and Waste Water,* 13th ed., American Public Health Association, 1971)

most fecal pathogens. Therefore it is more practical to test for *E. coli* as an indicator of fecal contamination than to try to isolate the relatively few pathogens which may be present. In Riverside the water supply was only chlorinated when the tests indicated a significant degree of coliform contamination. At the time of the epidemic coliform counts were within acceptable limits; nevertheless more than 16,000 cases of gastroenteritis occurred. Of these more than 70 were severe enough to require hospitalization, and three patients died. The causative organism, *Salmonella typhimurium,* was shown to be ten times as numerous as *E. coli* in the water supply, accounting for the failure to detect dangerous contamination by using the standard tests.

The Riverside epidemic was unusual in that the pathogen outnumbered *E. coli* in the water, and consequently fecal contamination was not detected. Water-borne epidemics are, however, by no means unique. Accidents involving city water supplies are reported each year; for example, typhoid epidemics involving hundreds of cases have occurred in recent years at Aberdeen, Scotland,

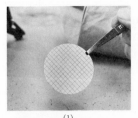

(1)   (2)   (3)   (4)

(5)   (6)   (7)   (8)

1. Sterile filter with grids for counting is handled with sterile forceps.
2. The membrane is placed on the filtering apparatus.
3. The apparatus is assembled.
4. A water sample is mixed well.
5. A portion of the sample is measured, and if necessary it is diluted with sterile diluent.
6. The portion is filtered by vacuum. Any bacteria that are present are retained on the filter.
7. The filter is placed on a dish of medium designed to select for coliforms. Nutrients can diffuse through the filter and colonies are formed on the gridded membrane where they can be counted readily.
8. The number of colonies on the filter reflects the number of coliform bacteria present in the original sample.

FIGURE 35-2(b)   Analysis of water for fecal contamination. (b) Cellulose acetate membrane (Millipore filter) method. (Courtesy of the Millipore Corporation)

and at Zermatt, Switzerland. Often the contamination can be traced to a break in a pipe which permits fecal contamination to enter the water supply. Controlled experiments have shown that bacterial pollutants can travel more than 60 meters in groundwater. In actual outbreaks of disease the pathogens of dysentery and typhoid have been shown to spread in groundwater laterally as much as 250 meters. Hence constant vigilance is required for even the best designed and operated water supplies to minimize the dangers of contamination with pathogens.

Eutrophication

Another important microbiological aspect of water pollution is the process of nutrient enrichment known as eutrophication. This process leads to increased productivity that commonly upsets the normal ecological balance and creates various kinds of problems, such as unsightly algal scums, obnoxious odors, and death of fish caused by oxygen depletion.

Algae can become a nuisance in lakes for natural reasons, but man is frequently the culprit responsible for adding nutrients and consequently causing lakes to become overproductive. Probably the material most frequently added is treated sewage which, in spite of treatment, is rich in nutrients produced by bacterial degradation of human wastes. In recent years sewage has become an even more effective fertilizer for algae because of the large quantities of phosphate-containing detergents in waste waters. Since phosphate is often a limiting

629

Algal blooms
Pages 266, 279

factor for growth of algae in unenriched lakes, increased amounts lead to tremendous increases in the numbers of algae present, resulting in algal blooms. When these algae die, they may create odor nuisances, and as they are degraded by bacteria the oxygen supply in deep water may be depleted so that fish cannot survive. In lakes that have excess phosphate because of enrichment or for natural reasons, addition of other substances, such as nitrates, can cause additional increases in algae.

Eutrophication and other forms of pollution have been in large part responsible for causing severe deterioration of rivers and lakes such as the Hudson River and Lake Erie. Concentrated efforts are being made at present to reverse the process in many areas. One of the most extensive studies of eutrophication and its successful reversal has been made by scientists at the University of Washington who studied Lake Washington. Figure 35-3 indicates that the city of Seattle, including the University, is situated between Puget Sound and Lake Washington. For many years raw sewage from the city was emptied into this large freshwater lake. During the 1930s the algal growth and odor became objectionable and untreated sewage was diverted at sea level into Puget Sound, the very large body of saltwater on the opposite side of the city. From 1940 to 1952, however, the

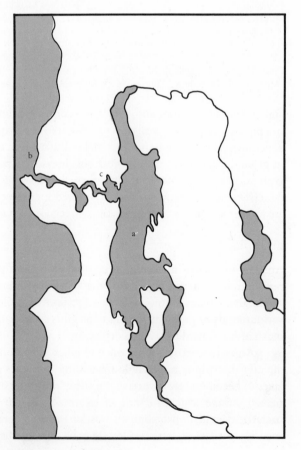

FIGURE 35-3
Lake Washington (a) and Puget Sound (b) are connected by a ship canal that intersects the City of Seattle. The University of Washington (c) is located adjacent to Lake Washington. The lake, formerly eutrophic, is now clear as a result of improved sewage disposal plans. All sewage from the city is treated and discarded into the strong currents of Puget Sound.

population around Lake Washington increased tremendously and during these years ten sewage treatment plants were built, all of which discharged treated sewage into the lake. Each year eutrophication increased and the once-clear lake waters became less transparent as the algae thrived. High concentrations of nitrates and phosphates in the water led to rapid and abundant growth of blue-green algae of the genus *Oscillatoria*.

Growing public concern about the deterioration of Lake Washington resulted in a vote in 1958 to reorganize the sewage disposal system. Over $120 million was allocated for the project. By the early 1970s, Lake Washington, which no longer received treated effluent, had regained much of its clarity and the number of fishes in the lake had increased. In addition raw sewage from the city of Seattle was no longer discarded; plants treated the sewage before discharging it deep into Puget Sound. The Sound is large enough and has enough exchange of water with the Pacific Ocean that eutrophication is not likely to result. It is, of course, necessary to maintain vigilance to be sure that Puget Sound does not become overloaded with other kinds of pollution, such as toxic metals and oil from industrial wastes.

The Lake Washington study shows that eutrophication can be successfully reversed. It also indicates the need for informed voters who are aware of the problems involved and are willing to support the efforts necessary for correction.

THE ROLE OF MICROORGANISMS
IN COMBATING POLLUTION

The interaction of aquatic microorganisms which participate in biological cycles is of the utmost importance in combating water pollution. There is a kernel of truth in the misstatement that a stream or an ocean is a natural waste-disposal unit; it is true that most natural substances can be recycled by aquatic ecosystems. However, the systems cannot be overwhelmed and remain effective. There is a limit to their capacities—a limit that is often exceeded.

Microbial degradation of pollutants is not always as effective in the deep sea as it is under certain other conditions. It was observed that food recovered from a sunken submarine after 10 months at a depth of 1540 meters was remarkably well preserved (Fig. 35-4). Subsequent experiments showed that rates of degradation by microorganisms were 10 to 100 times slower in the ocean depths than at the same temperature in the laboratory. The high pressure, lack of oxygen, and low temperature of water at great depths have a tremendous influence on microbial activities. Findings such as these must be kept in mind when materials are discarded into deep seas.

Even when aquatic microorganisms are metabolically active, they do not always operate in ways that benefit man. In the case of heavy metals and other poisons microbial activity may serve to transform or concentrate the substances to such an extent that they become harmful for human beings. For example, metallic mercury or mercury salts are often present in industrial wastes discarded

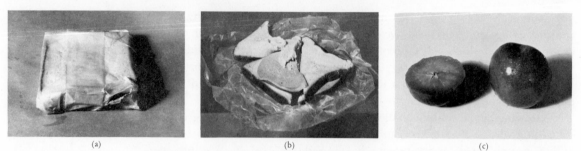

(a)   (b)   (c)

FIGURE 35-4   Lack of microbial degradation of organic materials in the deep sea. Foods from the sunken and recovered research submarine *Alvin* were strikingly well-preserved after more than 10 months at a depth of 1540 meters in the ocean 135 miles southeast of Massachusetts. (Courtesy of Dr. H. W. Jannasch; Jannasch, H. W., K. Eimhjellen, C. O. Wirsen and A. Farmanfarmaian, *Science* **171**, 672–675, 1971)

into bodies of water. In this environment the mercury compounds are converted by bacterial action into methyl mercury that is a powerful nerve toxin for man. Fish concentrate the methyl mercury in their tissues to such an extent that severe disease or death can occur in human beings who eat the fish. In Japan almost 100 people have died from eating fish caught in Minamata Bay, where an industrial plant discarded its mercury-containing wastes. Many others have suffered severe mercury poisoning, both in Japan and in other parts of the world. For this reason fishing has been banned in some areas, such as Lake St. Clair in the Great Lakes area, and swordfish have been banned as food because of their high content of mercury.

In order to combat water pollution, man must, first of all, understand the various roles played by microbes and, equally important, he must cooperate with the microorganisms and not overwhelm their capacities.

SUMMARY

Water containing dissolved nutrients serves as the natural habitat of a variety of aquatic microorganisms. Water from various sources differs considerably in its content of nutrients, offering a continuum of aqueous environments from rainwater to freshwater to saltwater. Any body of water represents an ecosystem in which the microbial flora depends on the particular environmental conditions present. Therefore the microbial flora of a freshwater lake differs considerably from the flora of rainwater or from marine flora. Within a body of water tremendous differences in microbial flora are observed, for example, between the ocean surface and the ocean depths.

Water pollution is a complex problem. Two important facets of this problem are pollution caused by pathogenic organisms and that which leads to eutrophication. Water pollution with pathogenic organisms is monitored by examining the water for *Escherichia coli* which serves as an indicator of fecal contamination. This method is usually, but not always, adequate; nevertheless constant vigilance is required for even the best water systems to minimize the dangers of contamination with pathogens. Eutrophication, by pollution of water with nutrients, leads to increased production of large numbers of algae. When these algae die and are degraded by bacteria, oxygen is depleted by bacterial metabolism. As a result fish and other oxygen-requiring organisms cannot survive and the body of water deteriorates. Eutrophication can be successfully reversed, but the reversal usually requires expensive and prolonged efforts.

Aquatic microorganisms which participate in biological cycles are of the utmost importance in combating water pollution. However, it is necessary for man to understand the various roles played by microbes so that he can cooperate with the microorganisms and not overwhelm their capacities.

QUESTIONS

1. On a trip to Yellowstone National Park and other parts of the West, the following habitats are part of the scenery; match the different kinds of organisms with the appropriate habitats (more than one kind of organism may be appropriate for a single habitat):

Habitat	*Kind of Organism*
A. Acid springs	1. Thermophiles
B. Boiling geyser basins	2. Acidophiles
C. Salt lakes	3. Psychrophiles
D. Mountain streams	4. Mesophiles
E. Freshwater lakes	5. Halophiles
F. Glaciers	6. Algae
	7. Bacteria

2. Describe the food chain in freshwater lakes, starting with the primary producers. Include the decomposers.
3. Arrange the following kinds of water samples in the order of increasing biomass:

Water sample from: Deep ocean
Surface of ocean near shore
Salt lake
Rain
Stream polluted with sewage

4. It has been said that "the solution to pollution is dilution." Discuss reasons why this is not a valid statement.
5. What is eutrophication and what can be done about it?

FURTHER READING

ALEXANDER, M., *Microbial Ecology*. New York: John Wiley & Sons, 1971.
BROCK, T. D., *Principles of Microbial Ecology*. Englewood Cliffs, N.J.: Prentice-Hall, 1966.

HEUKELEKIAN, H. and N. C. DONDERO (eds.), *Principles and Applications in Aquatic Microbiology.* New York: John Wiley & Sons, 1964.

ODUM, E. P., *Fundamentals of Ecology,* 3d ed. Philadelphia: W. B. Saunders, 1971.

SNIESZKO, S. F. (ed.), *A Symposium on Diseases of Fishes and Shellfishes.* Washington, D.C.: Special Publication No. 5, American Fisheries Society, 1970.

WOOD, E. J. F., *Marine Microbial Ecology.* New York: Reinhold, 1965.

Standard Methods for the Examination of Water and Wastewater, 13th ed., 1971. American Public Health Association, Washington, D.C. 20036.

Chapter 36

The Microbiology of Waste Disposal

During the early development of civilization, when communities were small and separated, disposal of wastes was not a major concern. Raw wastes could be emptied into rivers or lakes, or buried in the ground, where they were either destroyed by microbial activity or diluted before the contaminated waters reached other human beings. However, as men began to congregate to form large cities, the problems of waste disposal and obtaining pure water increased. Contamination of water by fecal wastes led to huge epidemics of diseases caused by water-borne pathogens. The problems have been compounded during the last few centuries by increased industrialization and the consequent addition of industrial wastes to water supplies.

At the time of the 1960 census, 70 percent of the population of the United States was urban; in fact at that time 53 percent of our population was concentrated in cities that covered only 0.7 percent of the land. Of course, the urban population has increased considerably over the last decade. To understand what this means in terms of waste disposal, consider that every day the average American uses about 150 gallons of water, 4 pounds of food, and 19 pounds of fossil fuel, which is converted into some 120 gallons of sewage, 4 pounds of trash or rubbish, and almost 2 pounds of air pollutants! This means that a city with only 1 million inhabitants is faced with the disposal of 120 million gallons of domestic sewage daily, to say nothing of industrial and other wastes.

In large American cities the collection of domestic sewage is usually carefully controlled. Disposal of the wastes, however, is often far from adequate, leading many to wonder how long the cities can continue to exist. In many smaller towns and communities the problems of waste disposal are even less efficiently handled. As late as the mid-1960s only 120 million Americans lived in areas with sewer systems, and about 70 million had to rely on cesspools or septic tanks, which more often than not are inadequate and lead to contaminated water supplies and other forms of pollution. It is clear that sewage disposal represents major problems for both the small town and the metropolis.

Clearly, every community, large or small, must provide means for transforming wastes into materials that are no longer a hazard to health and to the environment. Contaminated water, which accounts for the bulk of sewage, must be reclaimed if cities are to maintain adequate water supplies. Often the treated sewage of cities along large rivers is discharged directly into the river and becomes part of the drinking water for other cities downstream. In fact, in 1963 nearly 40 percent of Americans were using water that had already been reclaimed from domestic sewage or industrial wastes. Obviously in these instances the quality of drinking water depends on the adequacy of prior sewage or waste treatment.

Microbial activity is extremely important in the recycling of waste materials. The sections that follow consider the nature of sewage and other wastes and some of the major methods of waste treatment currently being used. It will become apparent that most of these methods depend on the *conversion,* by microorganisms, *of organic materials to inorganic forms—the process of* mineralization *or* stabilization. In general the *primary treatment* of sewage involves screening and sedimentation to remove large objects and some of the particulate material; *secondary and tertiary treatments* involve any of a number of methods for converting as much as possible of the remaining materials into odorless inorganic substances that can be reutilized.

PRINCIPLES OF MICROBIAL DEGRADATION

As indicated elsewhere life on this planet depends on a continuous recycling of wastes into materials that can be reutilized. Over millions of years a tremendous variety of organisms with diverse metabolic capabilities have evolved, so that now virtually any natural material can be degraded by microbial action, and the degradation products in turn can be reused by other organisms. Previous chapters describe some of the ways in which elements essential for life are recycled. Each of these cycles operates in the treatment of sewage.

During the *aerobic treatment of sewage,* microbial oxidations of organic compounds yield carbon dioxide and inorganic nitrogen-containing nutrients for plants. The cycle is completed when the plants are eaten, transformed into animal products, and eventually converted into human or animal organic wastes. During *anaerobic decomposition of sewage,* similar changes occur, except that anaerobic bacteria ferment organic carbon compounds to produce other organic products. These products subsequently are oxidized by the process of anaerobic respiration. The methane bacteria are particularly efficient at anaerobic degradation, almost completely converting small breakdown products formed by other bacteria from organic carbon compounds into CO_2 and CH_4 (methane or natural gas). Some of the methane bacteria can convert CO_2 into methane as well. The remaining CO_2 can be used by photosynthetic organisms and plants and the CH_4 conserved for fuel or further oxidized to CO_2 by the activities of certain aerobic bacteria.

THE NATURE OF SEWAGE

Domestic sewage contains human wastes, a source of both pathogenic and non-pathogenic microorganisms. A gram of feces may contain 100 billion bacteria, the majority belonging to anaerobic species of *Bacteroides, Lactobacillus,* and other genera. Of the less than 5 percent of fecal bacteria that are aerobic or facultative, most are *Escherichia coli;* in fact the average adult excretes approximately 200 to 400 billion *E. coli* cells per day. In addition, normal feces contain smaller numbers of *Streptococcus faecalis, Enterobacter aerogenes, Pseudomonas sp.* and a great many other bacteria, as well as a few fungi and unknown quantities of viruses. Salmonellae, shigellae, and other pathogenic bacteria are shed in the feces of infected individuals or healthy carriers, as are a number of pathogenic viruses such as poliovirus.

In addition to human wastes, domestic sewage contains other organic materials such as food wastes and a variety of cleaning compounds. When garbage disposal units are used, much more organic material is added to domestic sewage. Even so it consists of only about 1 to 2 percent organic solids and 98 to 99 percent water. Thus the microorganisms of human wastes (both pathogenic and non-pathogenic) are greatly diluted by water in raw sewage.

Large cities frequently produce sewage containing considerable amounts of industrial wastes which vary widely depending on their source. Sometimes the materials are toxic for microorganisms or plants and animals, or both. For example, studies have shown that sewage effluent from the Los Angeles area has added to the coastal waters of California considerable quantities of nonbiodegradable compounds, such as DDT and polychlorinated biphenyl compounds, that threaten marine food sources. Heavy metals, such as mercury, copper, nickel, and zinc, are also frequent industrial by-products. A comparison of parts per million of a few heavy metals in sewage sludge from highly industrialized Muskegon County, Michigan, and the relatively nonindustrial city of San Diego, California, is shown in Table 36-1. Unless special measures are taken to remove them, toxic industrial wastes may kill or inhibit waste-degrading microorganisms, consequently destroying their abilities to stabilize sewage during treatment.

Different kinds of problems arise when industries empty huge quantities of organic food wastes into the sewage. For example, it has been estimated that a single brewery may discard as much degradable organic material as a medium-sized city. Even though the material is nontoxic, it greatly increases the load of solids that must be stabilized (or mineralized) by treatment. Recent studies on waste treatment in the potato processing industry further illustrate the problem. They show that whereas in past years potatoes were consumed without proc-

TABLE 36-1

Heavy Metals in Sewage Sludge from Industrialized and
Nonindustrial Areas

Elements	Concentration (ppm) in sludge from:	
	Highly Industrialized Muskegon County, Mich.	Nonindustrial San Diego, Calif.
Cadmium	1–110	0–2.5
Chromium	26–580	0
Copper	24–690	20–33
Lead	6–510	3–11
Nickel	Trace–150	0–0.75
Zinc	90–2280	67–200

essing (and the peelings were, perhaps, fed to the livestock), at present between a third and a half of the 15 million tons of potatoes produced annually in the United States are processed to frozen French fries, potato chips, instant potato flakes, and so on. The untreated wastes produced account for 20 percent to more than 50 percent of the potatoes, and contribute a biologically degradable waste load equivalent to that of 5.5 million people. Furthermore these wastes are produced in a relatively small area of the nation, notably in Idaho and Maine.

Clearly the increases in convenience and comfort that result from industrial progress have all too often resulted in tremendous increases in both degradable and nonbiodegradable waste materials, and a consequent demand for improved methods of adequate disposal. The billions of dollars now being allocated by public and private sources for research and development in areas of waste disposal have come late; nevertheless considerable advances are being made.

SEWAGE TREATMENT

The term biochemical oxygen demand (BOD) is used to designate the oxygen-consuming property of waste water samples or, indirectly, the amount of degradable organic materials present in the water. To measure the BOD, a sample is well aerated and then incubated in a sealed container under standard conditions of time and temperature (usually five days at 20°C) before determining the amount of oxygen consumed by growth of microorganisms in the sample. The higher the BOD, the more oxygen used in biological degradation processes during incubation. Thus high BOD values reflect large amounts of degradable organic materials. For example, rich nutrient media used for laboratory cultivation of bacteria have BOD values of about 2000 to 7000 mg/liter, as compared with 100 to 300 mg/liter for raw sewage, and little or no BOD for unpolluted natural waters. The decrease in BOD during treatment reflects the effectiveness of stabilization of organic to inorganic materials.

Effective sewage treatment should decrease the BOD as much as possible, and it should also remove toxic and other objectionable materials. The methods of treatment chosen depend on a number of factors, such as the amount of material, its BOD, the presence of toxic materials, and the nature of the receiving waters (that is, bodies of water into which treatment products are emptied).

Primary treatment is designed to remove materials which can be settled. During this step sewage is passed through a series of screens to remove large objects such as sticks, rags, and trash (Fig. 36-1a) and then allowed to settle for a period of 90 minutes to 2 hours. Sometimes aluminum sulfate, ferrous sulfate, or other chemicals are added to coagulate or flocculate particles so that they settle more rapidly. After the settling period, the remaining fluid is given secondary treatment and the sedimented material from the primary tanks is usually sent to a large tank called a digester for further treatment, or it may be incinerated.

Secondary treatment (Fig. 36-1b) is designed to stabilize most of the organic materials, thereby reducing the BOD. The mode of stabilization by populations of aerobic organisms, most often used as secondary treatment in larger sewage disposal plants, is called the activated sludge method. This is the most satisfactory way to treat domestic sewage, but the process can be ruined by the presence of toxic industrial wastes. During this treatment the sewage is a nutrient source for complex populations of aerobic organisms acclimated to grow there. An abundance of oxygen is supplied by mixing in an aerator. Most of the biologically degradable organic material is converted into gases or oxidized products, and a very small percentage is incorporated into cell material of the organisms.

Complicated ecological relationships occur during activated sludge treatment. This ecosystem is analogous to those of natural waters, where animal corpses and other organic materials are degraded by microbial action. Many of the same interactions occur in both sewage treatment and self-purification of natural waters, but sewage treatment is designed to speed microbial activities under controlled conditions and in a confined place. Different populations act in turn, some degrading large organic molecules into smaller ones, which are then further oxidized by other organisms. Bacteria make up a major proportion of the microorganisms present. Among the bacteria isolated in one study of activated sludge were bacteria of the genera *Acinetobacter, Alcaligenes, Brevibacterium, Caulobacter, Comomonas, Cytophaga, Flavobacterium, Hyphomicrobium, Microbacterium, Pseudomonas,* and *Sphaerotilus,* and yeasts of the genus *Debaromyces.* It is apparent that the oxidative pseudomonads were well represented.

After bacteria and fungi have utilized nutrients, these microorganisms serve as food for ciliates, rotifers, nematodes, and other larger forms of life. The end result is a *small increase in the mass* of organisms present and a *large decrease in the amount of degradable organic materials.* The amount of organisms representing the increase in microbial mass is removed to the digester, and the original mass of sewage-adapted organisms is used to act on a new load of waste materials.

Within the digester anaerobic organisms act on the solids remaining after aerobic treatment. The digester provides anaerobic stabilization and removes water so that a minimum of solids remains. Here again various populations act sequentially. The methane-forming bacteria play an important role. They convert simple organic acids into the useful end products methane (CH_4) and CO_2, and

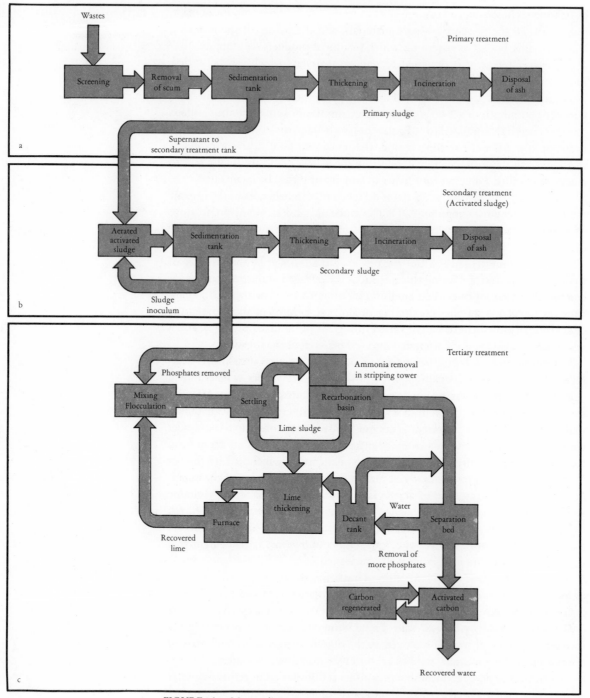

FIGURE 36-1 Metropolitan waste treatment. The most advanced waste treatment schemes utilize tertiary as well as primary and secondary treatments.

some of them can further convert CO_2 into CH_4:

$$CH_3COOH \longrightarrow CH_4 + CO_2$$
Acetic acid   Methane & carbon dioxide

$$CO_2 + 8H \longrightarrow CH_4 + 2H_2O$$

The greater efficiency of aerobic over anaerobic metabolism is reflected in more efficient removal of biologically degradable organic materials by using aerobic-activated sludge rather than other methods that utilize anaerobic organisms. The greater amount of energy provided by aerobic metabolism results in a greater cell yield. Strictly anaerobic treatment may reduce BOD 35 to 75 percent, but aerobic-activated sludge treatment usually removes about 95 percent of the BOD.

Pathogenic bacteria are generally eliminated during secondary treatment; however, disease-producing viruses may survive. Pathogens account for only a very small proportion of the total number of bacteria in feces, and they are greatly diluted by the water in sewage. Most of them soon die because they are poorly adapted for growth relative to other organisms present. During secondary treatment of sewage pathogens are forced to compete for nutrients with a huge mass of bacteria that have been adapted to grow best at the temperature and under the conditions provided. As a result most pathogenic bacteria are rapidly overgrown and eliminated by their competitors. Animal viruses, on the other hand, lack appropriate host cells in sewage and cannot replicate there, but they may survive for long periods. Therefore if large quantities of virus particles are present in raw sewage, some of them may be recovered after secondary treatment. Chlorine treatment at this point is of little or no value for killing viruses, because they are commonly secluded within small aggregates of materials, protected from the chemical.

At present many sewage treatment plants discard the residue from secondary treatment into receiving waters, often causing eutrophication. As noted in Chapter 35 large quantities of phosphates or nitrates remaining after stabilization often increase the growth of algae. These algae may gradually preclude other forms of life, as has been the case with Lake Erie, for example. Tertiary treatment of sewage to remove nitrates and phosphates can greatly alleviate the problem. At present a tremendous project is under way to restore animal life in Lake Erie, using the three-stage treatment of sewage and other methods.

Tertiary treatment to remove materials that remain after secondary treatment depends on either biological or chemical steps, or both. At South Lake Tahoe, California, over 7 million gallons of sewage can be converted within one day to water pure enough to drink. In the sophisticated, three-stage treatment system used there, both physical and chemical methods are used for tertiary treatment, part of the by-products are reclaimed, and solid wastes are burned (Fig. 36-1c).

The chemical and physical removal of materials during tertiary treatment at Lake Tahoe involves a number of steps. Following secondary treatment by the activated-sludge method, lime is added to the treated water to coagulate and precipitate phosphate-containing particles. The resulting precipitates are allowed to

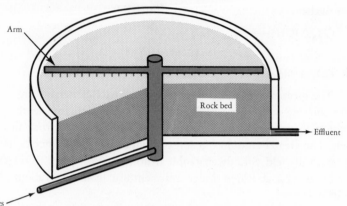

FIGURE 36-2   The trickling filter. Sewage wastes are channeled into the revolving arm and trickle through holes in the bottom of the arm onto a gravel and rock bed. The rocks are coated with microorganisms that stabilize the sewage as it trickles through the bed so that the effluent has a greatly reduced load of degradable organic materials.

settle in a clarification tank. Gaseous ammonia in the water, derived from mineralization of organic compounds, is removed by passing the water from the clarification tank through a "stripping tower." In the tower rough-textured hemlock slats break the falling water into droplets, NH_3 gas escapes and is dispersed by a fan. The water is then neutralized by adding CO_2 and is passed through a series of temporary holding ponds before being filtered through a sequence of cylinders. Some of the filters, made of coal, sand, and garnet, remove remaining particulates; others, made of activated charcoal, remove detergent, pesticides, and other toxic materials. As a final step in treatment chlorine is added to the purified water to kill or inhibit any remaining microorganisms and to oxidize any odor-producing substances that may remain.

Chemical precipitation of phosphates has been combined with biological removal of nitrates in some designs for tertiary treatment. Certain bacteria (particularly species of *Pseudomonas* and *Bacillus*) can reduce nitrates (NO_3^-) completely to N_2 (denitrification). The N_2 gas that is formed is inert, nontoxic, and easily removed.

Treatment of Small Quantities of Sewage

Considerable progress has been made in recent years in developing better methods, such as activated-sludge and tertiary treatments, for disposal of urban wastes. At present, however, these are not practical for small communities or isolated dwellings. Other means of disposal must be used, employing much the same basic principles of microbial degradation but in different ways.

Small towns sometimes depend on a process called lagooning, in which sewage is channeled into shallow ponds, or lagoons, where it remains for varying periods of time (several days to a month or more) depending on the design of the lagoon. During this time, settling occurs and materials are stabilized by anaerobic or aerobic organisms, or both. Pathogenic bacteria are usually eliminated by competition, as described previously.

Trickling filters (Fig. 36-2), frequently used for smaller sewage treatment plants, spray controlled amounts of sewage over beds of coarse sand or gravel. The pebbles and grains of sand become coated with organisms that aerobically degrade the sewage. The trickling filter is used to adjust the rate of flow of materials so that they can be maximally degraded. As is the case with activated sludge, different populations act in turn to degrade various compounds. Nematodes, rotifers, ciliates, bacteria, and other organisms cooperate during sewage stabilization, both with trickling filters and in lagoons.

Isolated dwellings or very small communities customarily rely on septic tanks for sewage disposal. In theory the septic tank is satisfactory; however, in practice, it often fails. Figure 36-3 illustrates the design of a septic tank. Sewage is collected in a large tank in which much of the solid material settles and is degraded by anaerobic microorganisms. The fluid overflow from the tank has a high BOD, and must be passed through a drainage field of sand and gravel. Theoretically stabilization should occur in the drainage field in the same manner described for the trickling filter. In fact stabilization depends on adequate aeration and sufficient action by aerobic organisms associated with sand and gravel in the drainage field—conditions which are often not met. For example, clay soil under the drainage field may prevent adequate drainage, allowing anaerobic conditions to develop, or toxic materials may inhibit microbial activity in the drain field. There is always the possibility that drainage from a septic tank contains pathogens; therefore it must never be allowed to drain where it can contaminate water supplies.

Research is being directed toward finding better ways of dealing with small quantities of sewage. One method that has proven satisfactory in some small

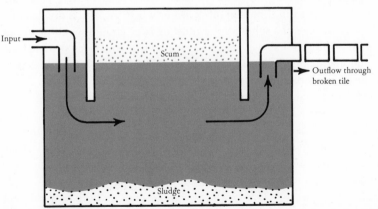

FIGURE 36-3 The septic tank. Sewage wastes enter the tank through the input pipe. Within the tank, solid materials settle and undergo anaerobic stabilization. Materials that do not settle exit through the outflow pipe through broken tile which permits seepage into the drainage field. Conditions in the drainage field must be aerobic so that materials remaining can be degraded by the activities of aerobic microorganisms. If the drainage area is not properly designed, contaminated materials readily enter adjacent surface waters.

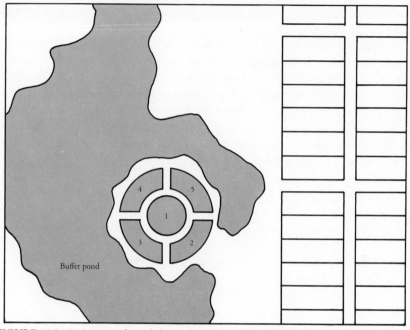

FIGURE 36-4 An integrated pond design for sewage disposal. Sewage wastes from all the lots in the subdivision (right) are pumped into the bottom of the primary pond (1) where they undergo anaerobic stabilization. Grease rises to the top and the effluent removed from near the bottom is transferred to the next pond (2). Further (aerobic) stabilization occurs in each of the other ponds. Cleaner water near the surface is transferred to the next pond in sequence (3,4,5, and the buffer pond). The buffer pond, surrounded by plantings, provides a park-like area for the subdivision. In addition, lots can be made smaller and easier to maintain because individual septic tanks and drainage fields are not necessary.

communities is the *integrated pond treatment.* Good results have been obtained in handling sewage from suburban subdivisions where the land is divided into lots. Ordinarily the size of a suburban lot must be large enough to allow installation of a septic tank and drainage field, and each homeowner is responsible for installing and maintaining his own septic tank. When the tank fails, as is often the case, health hazards and odor nuisances are created which affect the entire community. Integrated pond systems are designed for the entire subdivision or small community. Since lot sizes can be much smaller because individual septic tanks and drainage fields are not needed, the land is subdivided into smaller lots and a central portion is set aside for a series of ponds (Fig. 36-4). Sewage is channeled into successive ponds, carefully designed to carry out both aerobic and anaerobic stabilization. By the time the water reaches the final pond it is suitable for decorative or recreational use. This type of system depends on both proper design and adequate maintenance, and requires careful monitoring. However, in communities where it has been used it has provided safe and efficient waste disposal with the added bonus of a central park area around the pond system.

Progress is also being made in many other areas. For example, soil filters are being developed to remove gases (such as H_2S) that are formed during sewage treatment. If the soil is kept moist so that bacteria can grow, the malodorous and toxic H_2S is removed during passage through the soil and converted by bacteria (such as *Thiobacillus sp.*) to sulfate ions.

UTILIZATION OF TREATED WASTE RESIDUES

One area of waste treatment in which research and development are greatly needed is that of utilization of treated waste residues. Clearly, receiving waters deteriorate following the addition of treated wastes as a result of added nutrients, toxic materials, and changes in pH and temperature. Even when the best available methods of tertiary treatment are used, the quality of receiving waters is decreased. For instance, at Lake Tahoe, where the recovered water is pure enough to drink, it still contains some nutrients. For this reason the purified water is collected in reservoirs, rather than being rechanneled into the lake.

Increased efforts are being made to utilize waste residues in crop production. Treated sewage sludge has been used as fertilizer, but with limited success. For example, dried activated sludge is an effective fertilizer for shrubs and flowers and is often used by cities as fertilizer for municipal parks and golf courses; however, its cost prohibits use on commercial crops which can be fed much more cheaply with chemicals. Cost, however, is not the only factor that inhibits use of sewage residues for crops. Other factors include the presence of high concentrations of soluble salts that may interfere with germination and growth of plants; excessive nitrogen content of some wastes that leads to lower quality of crops and sometimes to increased nitrate in forage crops, thus making them unsafe for livestock; and, perhaps most important, the presence of materials such as heavy metals which are toxic for plants or hazardous to man and animals that eat the plants. In spite of these and other problems, ways are being sought (and to a limited extent are being found) to utilize waste products for the production of plants.

Industries are also increasing their efforts to find safe and useful ways to dispose of their wastes and to avoid polluting the environment. For example, sulfite waste liquors from paper manufacture have been used as nutrients for growing yeasts which can be used as food for animals.

There is also a growing awareness of the hazards involved in synthetic molecules. Even though virtually any natural organic molecule is biodegradable, unnatural configurations in materials (such as some plastics) synthesized by man may not be subject to microbial decomposition. This has certain advantages—in that plastics are not ruined by molds, mildews, and other microbes—but in terms of waste disposal it creates real problems. This was quite apparent in the case of the alkyl benzyl sulfonate detergents. These nonbiodegradable substances survive waste treatment and accumulate in treated water, leading to unsightly foaming. Their use is now banned by law in the United States. Many other types of compounds that resist microbial action during waste treatment accumulate in water, and some have potential side effects. One example is the steroid hormone being used for birth control pills. This hormone is excreted in urine and considerable quantities get into sewage. It has been found that the natural hormone is degraded by sewage microorganisms; however, the synthetic molecules used for birth control pills are resistant to degradation during sewage treatment and so can accumulate in water.

It is apparent that many problems remain to be solved. The treatment of wastes and utilization of waste residues currently offers one of the most challenging areas of research for microbiologists and other scientists.

SUMMARY

Microbial activity is extremely important in the recycling of waste materials and in reclaiming water from domestic and industrial wastes. Most methods of waste treatment involve the mineralization of organic wastes to inorganic materials by microbial action. Toxic products in the waste may interfere with treatment procedures.

Primary treatment of sewage wastes involves screening and sedimentation to remove large objects and larger particulate material; secondary and tertiary treatments involve any of a number of methods of mineralization. The activated sludge method of secondary treatment is currently the most satisfactory means of treating large quantities of domestic wastes. Tertiary treatment consists of either physical, chemical, or microbial methods for removing and sometimes reclaiming the materials left in water after secondary treatment of sewage.

Small quantities of sewage are sometimes treated by lagooning, by use of trickling filters, in septic tanks, or in a system of integrated ponds. Better methods for utilizing treated waste residues are being sought.

QUESTIONS

1. What is the nature of domestic sewage? Of industrial wastes?
2. How have modern "convenience products" such as prepared foods, detergents, and insecticides affected the problems of waste disposal?
3. Outline an ideal sewage treatment system including primary, secondary, and tertiary treatments.
4. Suppose you own 150 acres of ocean-front property that you want to develop into a community of vacation homes, retaining as much as possible of the natural setting. Of course, safe and effective sewage disposal must be a part of the plan. What are the advantages and disadvantages of each of the following for such a development:

> individual septic tanks for each building
> trickling filter
> integrated pond system

5. Plan a system for waste disposal that would be feasible for space travel.

FURTHER READING

GAUDY, A. F., JR. and E. T. GAUDY, "Microbiology of Waste Waters," *Ann. Rev. Microbiol.,* 20 (1966), 319–336.

ODUM, E. P., *Fundamentals of Ecology,* 3d ed. Philadelphia: W. B. Saunders Co., 1971.

Chapter 37

The Microbiology of Foods

Foods are part of an ecosystem in which microorganisms, insects, and a variety of other living creatures play important roles. This chapter is concerned largely with the activities of microorganisms during food spoilage and methods of preventing microbial spoilage. However, foods are spoiled also by rodents, freezer damage, and a multitude of other mechanisms beyond the scope of this discussion.

Microbial activity is necessary for the production of a number of foods, such as cheese, beer, wine, and sauerkraut. The use of yeast, molds, and bacteria in preparing these foods is considered in Chapter 38, which is concerned with commercial applications of microbiology.

SPOILAGE OF FOODS

Fruits, vegetables, dairy products, and meats all contain a variety of microorganisms on their surfaces. This is to be expected, because microbes are ubiquitous and are especially plentiful in soil and around animals. The duty of anyone who prepares food is to see that the numbers and kinds of organisms present at the time of consumption are such that the food is both safe to eat and appetizing.

Microorganisms Responsible for Food Spoilage

Microbial spoilage usually results from microbial activity following proliferation of the organisms present in and on foods. For example, potentially destructive fungi are found on virtually all apples; thus the preservation of fresh apples depends in part on controlling the number of fungi present by culling rotten apples and on adjusting the environment so as to discourage growth of fungi.

Food microbiologists have made extensive studies of the numbers of various kinds of bacteria and fungi in food samples. Representative microbial counts for meats, fruits, vegetables, and other foods may be found in most of the references at the end of this chapter. The counts may vary considerably, depending on the kind of food and its general condition. For instance, a comparison of frozen breaded shrimp samples revealed a high of 178 million organisms per gram of shrimp in poor sanitary condition and a high of 7 million per gram for samples in good sanitary condition. In other studies ground beef was found to contain from 5 million to 33 million bacteria per gram. Although no standard for ground beef has been accepted, the suggested upper limits range from 1 to 10 million organisms per gram.

The microorganisms that spoil foods are usually nonpathogenic for man. This is not surprising when the conditions of their growth are considered. Foods are most often kept well below 37°C. Pathogenic bacteria, which usually grow best near 37°C, thus prefer a warmer environment than that provided under food storage conditions. Similarly, the nutrients available in fruits, vegetables, and other foods often are not suitable for optimum growth of some human pathogens.

A number of genera of bacteria are important agents of food spoilage. *Pseudomonas* species, for example, are often involved in a wide variety of foods. The pseudomonads have extensive metabolic capabilities; hence they are able to grow on many different kinds of foods. Some of them can utilize proteins, lipids, and carbohydrates, so that they readily grow on meats as well as foods higher in carbohydrates. Psychrophilic pseudomonads that can multiply at refrigerator temperatures are especially notorious for spoiling meats and many other foods. Certain species of *Achromobacter, Alcaligenes,* and *Streptococcus* are also psychrophilic agents of food spoilage.

Members of the genus *Acetobacter* can transform ethyl alcohol to acetic acid (vinegar). Although this property is very beneficial for commercial producers of vinegar, it can be a great detriment to wine producers. In fact, *Acetobacter sp.* are one of the principal spoilage bacteria of wines. Milk products are sometimes spoiled by gram-negative rods of the genus *Alcaligenes* which cause "ropiness" of milk and make cottage cheese slimy.

Gram-positive bacteria frequently spoil foods also. In addition to some streptococci, gram-positive, food-spoilage bacteria include species of *Leuconostoc, Lactobacillus, Micrococcus, Staphylococcus, Bacillus,* and *Clostridium. Streptococcus, Leuconostoc,* and *Lactobacillus* are all lactic acid-producing bacteria. Certain strains of *Staphylococcus aureus* grow readily in high carbohydrate foods and secrete enterotoxins that cause staphylococcal food poisoning, discussed in Chapter 26. Gram-positive rods, both aerobic of the genus *Bacillus* and anaerobic of the genus *Clostridium,* are particularly troublesome during the high temperature preparation of foods because their spores are heat resistant. *Bacillus coagulans* and

Food poisonings
Pages 482–485

B. stearothermophilus, a species of flat-sour bacteria, cause sour spoiling of corn and certain other canned vegetables. Many of the clostridia putrefy and spoil foods held under anaerobic conditions; however, *Cl. botulinum* is of primary importance because of its ability to produce an extremely potent lethal exotoxin.

Requirements for Growth of Spoilage Organisms in Foods

All microorganisms require water in order to grow; however, the amount of moisture that permits growth can vary with different kinds of microbes. The water activity (a_w) is a term used by food microbiologists to designate the amount of moisture available in foods. The a_w is defined as the ratio of the water vapor pressure of food substrate to the vapor pressure of pure water at the same temperature. The relative humidity is a related measure, being 100 times a_w. Thus a typical fresh food might have an a_w of 0.995 or a relative humidity of 99.5. Salted or dried foods have lower a_w. Most bacteria that spoil foods require an a_w of 0.91 or higher for growth, although there are some notable exceptions including some of the halophilic bacteria. As a rule fungi can grow over a much wider range of a_w values than bacteria, growing at an a_w as low as 0.60. Thus yeasts and molds, rather than bacteria, are found in spoiled pickles and other brine-preserved foods, on dried foods, and on other foods which have a low a_w; acidity, however, is also a factor here.

In addition to its a_w, the pH of a food is of primary importance in determining the nature of microbial life capable of growing in or on that food. It has long been known that highly acid foods, such as vinegars, wines, and some fruits keep for long periods of time without spoiling. Furthermore, when they are spoiled, fungi are the usual spoilage agents, because the fungi grow readily in high acid media. The pH of a food may also determine whether toxins are formed by bacterial contaminants, as in the case of *Cl. botulinum.*

Clostridium botulinum
Pages 483–485

Any medium, including food, represents a complex interaction of factors affecting microbial growth. In addition to a_w and pH, other factors, such as osmotic pressure, oxidation-reduction potential, and temperature, have a great influence on the numbers and kinds of microorganisms that will propagate. Also, the kinds and amounts of nutrients present have a great deal to do with the nature of microbial growth. For example, if a certain food is deficient in a given vitamin, any organisms requiring that vitamin cannot normally grow there. The hardy pseudomonads or other bacteria capable of synthesizing the vitamin would then be able to take over and grow, providing other conditions were favorable. Thus the pseudomonads often spoil foods because they have extensive capabilities for synthesizing essential nutrients.

Rinds, shells, and similar coverings aid in protecting some foods from the invasion of spoilage organisms. Even so sooner or later a microorganism with just the right properties for living on that food will breach the defenses and cause spoilage.

Some foods contain antimicrobial substances which may help to prevent spoilage and are sometimes useful in protecting man and other animals. For example, egg white is rich in lysozyme. Thus even if lysozyme-susceptible bacteria breach the protective shell of the egg, they are destroyed by lysozyme before they can cause spoilage. Other examples include cranberries, which are rich in antimicrobial benzoic acid, and fresh milk, which has been reported to contain several antimicrobial substances. It has recently been shown that cows' milk contains an antibacterial peroxidase system, analogous to that found in human phagocytes. These peroxidase systems provide important defense mechanisms in both cattle and man. Onions, garlic, radishes, and many other foods also contain substances that kill or inhibit microorganisms. There is evidence that the Egyptians who built the pyramids consumed large quantities of garlic and radishes, and it has been postulated that these foods were used to prevent or combat infection; however, there is no scientific evidence that the antimicrobial substances they contain are active in the human body.

FOOD PRESERVATION

Some methods of food preservation, such as drying and salting, have been known throughout the ages, whereas others have been discovered or developed more recently. The major methods of preserving foods are by high-temperature treatment, low-temperature storage, desiccation, chemical treatment, and irradiation. Each of these methods is discussed separately.

High-Temperature Treatment

In 1809 a Frenchman named Appert gained fame and fortune by developing the method for canning food that, with little modification, is still used today. Although he had never heard of bacteria and other microorganisms, Appert empirically worked out a procedure for preserving foods in stoppered jars that were heated for hours in a boiling water bath.

Several general rules hold true for high-temperature preservation methods. First, as the temperature is increased, the time that the food must be heated in order to preserve it decreases. This fact reflects the thermal death times (TDT) of contaminating microorganisms. Table 37-1 indicates how the TDT of bacterial spores decreases as the temperature increases. Second, the concentration of microorganisms affects the conditions necessary for killing; the higher the concentration of bacteria, the more intensive the heat treatment required, if all other factors remain constant. For example, in tests carried out under exactly the same conditions, a given volume containing 72 billion spores of *Cl. botulinum* had a TDT at 100°C of 240 minutes, with 650,000 spores the TDT was 85 minutes, but it was only 40 minutes when 328 spores were heated at 100°C. It has been

TABLE 37-1
Some Representative Thermal Death Times of Bacterial Spores[a]

Bacterial Species	Thermal Death Time at 100°C (minutes)	Thermal Death Time at 120°C (minutes)
Bacillus subtilis	15 to 20	–
Clostridium botulinum	360	5
Bacillus coagulans	1140	17

[a]Times required to kill a given number of spores *under conditions of testing*. It should be emphasized that the times may vary under different conditions.

suggested that cells secrete proteins or other substances that protect against thermal death, and that large populations of cells secrete enough of these substances to cause the effects cited above; however, there is not much evidence to support this suggestion.

A third general rule concerning high-temperature preservation of food is that *the condition of the vegetative cells or spores influences their heat resistance.* Even within a given strain of bacteria, heat resistance varies depending on the way in which the culture has been grown, the phase of growth of the cells, the age of spores if these are present, and other factors. For example, spores become more heat resistant for a time as they age, a factor which could significantly alter the TDT for a bacterial strain. A fourth general rule is that *the environment or material in which microorganisms are heated has a major influence on heat resistance.* Moist heat kills much more efficiently than dry heat, as evidenced by the efficacy of autoclaving as a means of sterilization. Generally, organisms are most resistant to heat at the pH which supports optimum growth; therefore extremes in pH often cause increased killing. Organisms are killed by heat more readily in high concentrations of salt, and they are less heat sensitive in media containing considerable quantities of proteins and fats (protein or fat protection). These principles are illustrated by the data presented in Table 37-2.

Baked beans are alkaline, and their low water content does not permit heat to penetrate quickly; they must therefore be heated for a long period of time to kill any spores that may be present. Organisms in cream-style corn are more resistant to heat than in corn packed in brine, because of the high salt content of the brine and the protection of fats and proteins in the cream sauce. Tomatoes are highly acid and will not readily support the germination of spores; also, any

TABLE 37-2
Recommended Time and Temperature Necessary for Processing
Several Foods[a]

Food	Temperature (°F)	Time (minutes)
Baked beans	240	105
Corn, cream style	240	90
Corn, in brine	240	50
Tomatoes	212	34

[a]Recommendations based on bulletins issued by the National Canners Association.

spore-forming clostridia that may be present cannot form toxin at this low pH. Thus tomatoes can be processed safely at a lower temperature and for a shorter time than for the other foods listed.

Another means of high-temperature processing of foods is *pasteurization*. In this process food preservation is usually not the primary aim; rather, pasteurization is a heat treatment which destroys all nonsporing pathogenic organisms without significantly altering the quality of the food. Thus pasteurization of milk or other foods does not sterilize them, but only makes them free of nonspore-forming disease-producing organisms. Vinegar and a few other foods, on the other hand, are pasteurized not primarily to kill pathogens (which are not likely to survive and grow at the low pH) but to destroy certain spoilage organisms. The factors influencing high-temperature canning, discussed previously, are also applicable to pasteurization. The two methods of pasteurization in general use are the high-temperature-short time method (HTST) and the low-temperature-long time (LTLT) or holding method. For example, with milk the minimum requirements for the HTST method are 71.7°C (161°F) for 15 seconds and for the holding method 61.7°C (145°F) for 30 minutes. On the other hand, ice-cream mix, which is richer in fats than milk, requires 82.2°C (180°F) for about 20 seconds, or 71.1°C (160°F) for 30 minutes. As is the case with canning, heat treatment for pasteurization must be adjusted to the individual situation.

Low-Temperature Treatment

The growth of microorganisms is temperature dependent and low temperatures are useful in food preservation.

The so-called cellar storage of some fruits and vegetables (for example, apples and potatoes) can preserve these foods for months. In this method the food is stored at temperatures near 10°C in the dark and at the appropriate humidity.

Refrigeration can be adjusted to various temperatures, depending on the food in question. Most household refrigerators maintain a temperature around 4°C, although the temperature may vary from 0 to 10°C in different parts of the refrigerator. The proper relative humidity and ventilation in the refrigerator are important in maintaining good quality of the foods. Sometimes other methods of preservation, such as ultraviolet irradiation, are used along with chilling for better results.

During recent decades freezing has become a major means of preserving foods. This method permits foods to be stored for many months, or even longer, with little loss of color, flavor, and other criteria of quality. Various methods of freezing and a wide range of freezer temperatures are used, depending on the food being processed.

Desiccation

For many centuries drying has been used to preserve foods. This process decreases the a_w below the limits that allow propagation of most microorganisms. Often drying by natural means (sun drying) or artificial means is supplemented by other methods, such as salting or adding high concentrations of sugar or small amounts of preservative chemicals.

Recently, lyophilization (freeze-drying) has beome rather widely used for some foods. In this process the frozen food is dried in a vacuum, and the quality of the reconstituted product is much better than with ordinary freezing or drying methods.

Although drying stops growth, it may not kill all bacteria in or on foods. For example, eggs are often dried, and a number of cases of salmonellosis have been traced to dried eggs. This occurs because egg shells may be heavily contaminated with species of pathogenic salmonellae from the gastrointestinal tract of the hen. When the eggs are cracked, many of the bacteria enter and subsequently survive drying; these bacteria can cause disease in susceptible individuals who eat the eggs. Some states now have laws stating that only pasteurized dried eggs can be sold in order to prevent the transmission of such pathogens.

Chemical Preservatives

Reference has been made to methods of adding chemicals to preserve foods. Salting involves the addition of large amounts of salt that increase the osmotic strength and actually result in decreasing the a_w of the food by removal of water molecules. High sugar content of foods (as in jams, jellies, honey, preserves, sweetened condensed milk) leads to a similar result. Pickling combines salt and a very acid medium, and is very effective in controlling growth of most microorganisms. Sodium benzoate, sorbic acid, propionate, and other chemicals are often added to certain foods to retard microbial growth.

Radiation

Irradiation can be lethal to microorganisms and thus it is useful as a means of preserving foods. Adequate irradiation has been obtained from ultraviolet rays, gamma rays, and X-rays. Ultraviolet rays are restricted to surface applications, whereas gamma rays and X-rays penetrate the foods. Microwave energy is limited in use for food preservation because of the heat it generates.

Methods have been developed for irradiating foods to give essentially the same results as heat treatments that is, sterilization (as in canning) and pasteurization. It is probable that these methods will become more widely used in coming years. However, their disadvantage is that enzymes in the food are not destroyed by irradiation; therefore enzyme degradation can continue to occur and the foods may not last as long or be as appetizing as heat-treated foods unless natural enzymes are destroyed by other means such as mild heating.

CONTROL OF MICROBIAL HAZARDS IN FOODS

Standards for Control

Food production is very carefully controlled in the United States. Federal, state, and local agencies cooperate in inspections to see that protective laws are enforced. The Federal Food, Drug, and Cosmetic Acts set standards for foods that are shipped between states or territories and require accurate labeling. Other federal laws provide for the inspection of meats and poultry produced for interstate shipment.

State and local agencies enforce many of the laws that protect food consumers. Of course, food manufacturers usually act as their own watchdogs. As a result the microbiological quality of most foods sold in the United States is very good.

Microbial Food Poisoning

Food poisonings
Pages 482–485

The United States is not free of microbial food poisonings. Many cases occur each year, primarily from foods prepared at home or in large quantities in institutions. The major kinds of bacterial food poisonings, caused by salmonellae, staphylococci, and clostridia, are discussed in Chapter 26.

Additional Microbial Hazards of Foods

In addition to bacterial food poisonings, certain other hazards may result from ingesting foods. Pathogens are sometimes transmitted via contaminated foods or water. Examples are the causative organisms of brucellosis, listeriosis, leptospirosis, tularemia, tuberculosis, Q-fever, streptococcal infections, poliomyelitis, and hepatitis. Adequate cooking or pasteurization will prevent transmission of pathogens in this manner.

"Red tides"
Page 275

An unusual and often lethal form of food contamination caused by algal toxins is called *paralytic shellfish poisoning*. Shellfish, such as mussels, concentrate large quantities of toxin produced during blooms of certain algae (sometimes called a "red tide"). For several days after the red tide the shellfish contain dangerous quantities of toxin, and if eaten during this period can cause severe paralysis or death. Similarly, fungi in food can produce toxins (called mycotoxins) that may be subsequently ingested.

SUMMARY

Microbial activity plays a large role in spoilage of foods. Other microbial activities, essential in the preparation of various food products, are considered in Chapter 38. Spoilage commonly results from the activities of bacteria and fungi normally found on foods, following extensive growth of the organisms. The species involved vary greatly, depending on the type of food and on environmental conditions.

The water activity (a_w) of a food designates the amount of moisture available, and is important in determining the varieties of microbes that can thrive in the food. As a rule fungi can grow over a much wider range of a_w values than bacteria, and consequently are common causes of food spoilage even in preserved or pickled foods which have a low a_w. Other factors, such as pH, osmotic pressure, oxidation-reduction potential, temperature, and nutrient supply, also have a great influence on the numbers and kinds of microorganisms that can propagate in foods. Some foods contain antimicrobial substances that discourage growth of certain organisms.

Food preservation can be achieved by high-temperature treatment, low-temperature storage, dessication, chemical treatment, and irradiation.

Microbial hazards of foods include salmonella, staphylococcal, and clostridial food poisonings, the transmission of various pathogens by contaminated foods, and poisoning by certain algal or fungal toxins.

QUESTIONS

1. Why are species of *Pseudomonas* frequently implicated in food spoilage?
2. What is meant by the a_w of a food? Which would you expect to spoil faster, a food with an a_w of 0.994 or one with an a_w of 0.905?
3. Which is more likely to be recovered from a jar of spoiled pickles, yeast or bacteria? Why?
4. If you owned a large food-processing company, outline some general plans for microbiological research and development that you might follow.
5. What kinds of microbiological hazards might be encountered in the following foods and how can they be overcome?

> Oysters grown in fecally polluted waters
> Eggs with shells that were cracked at time of purchase
> Frozen turkey
> Fresh raw milk

FURTHER READING

FRAZIER, W. C., *Food Microbiology*. New York: McGraw-Hill, Inc., 1958.
JAY, J. M., *Modern Food Microbiology*. New York: Van Nostrand Reinhold, 1970.

Chapter 38

Commercial Applications of Microbiology

Several preceding chapters have been devoted largely to a discussion of microorganisms that cause disease in man, as well as those that destroy crops and cause deterioration of materials. Superficially it might seem that microorganisms are generally harmful—but this is far from the truth. Their tremendous benefits offered to man greatly outweigh their harmful effects. Not only are microbial activities essential in recycling nutrients and energy, but also microbiological techniques are applied commercially to provide foods, beverages, pharmaceutical products, chemicals, and even fibers.

Throughout history man has utilized microorganisms, but only during the last century has he begun to learn about the organisms involved. With increasing knowledge he has been able to improve ancient methods that were discovered empirically, such as bread- and wine-making, and to devise new practical and commercial applications of microbial activities. For example, the discovery that some organisms produce substances which kill or inhibit other microbes has opened the vital area of antibiotic production.

Continued research and increased understanding of microbial activities undoubtedly will lead to still further benefits. Environmental control, food supply, and space exploration are only a few of the areas which are expected to utilize microorganisms to an even greater degree in the future than at present. In this chapter some important commercial applications of microbiology are considered.

Ancient methods of preparing foods and beverages by utilizing microbial activities have been adapted for industry and have often been improved by scientific research. However, some of these processes—including the production of fine cheeses and wines—are far from being fully understood.

Milk Products

Milk is normally sterile until secreted; however, by the time raw cows' milk reaches the market it contains a multitude of microorganisms introduced from the udder and during handling. Along with other ubiquitous contaminants, the lactic-acid bacteria *Streptococcus lactis* and *Lactobacillus* species are commonly present. These gram-positive bacteria, described in Chapter 12, grow especially well in milk, which is rich in lactose. *Streptococcus lactis* causes the souring of unpasteurized milk by rapidly producing enough lactic acid to lower the pH to 4.8 or below, which coagulates the milk proteins. Pasteurization is sufficient to kill most spoilage bacteria as well as all pathogenic contaminants. Therefore pasteurized milk does not sour as rapidly as unpasteurized milk and remains palatable for many days at refrigerator temperature.

Lactic acid bacteria
Pages 257, 258

In the United States commercially marketed milk is usually pasteurized, and its bacterial content is monitored. The U. S. Public Health Service has developed standards for grading milk on the basis of bacterial counts as Grade A or Grade B (Table 38-1). Direct microscopic counts can be made of the numbers of visible organisms (both living and dead) in samples of raw milk; plate counts of the number of colonies grown from samples enumerate only living organisms which can grow under the conditions of culture. Both counts are used in grading the

TABLE 38-1
Recommended Standards[a] for Dairy Products

Product	Maximum Number of Microorganisms per Milliliter or Gram
Grade A raw milk prior to commingling with other milk	100,000
Grade A raw milk commingled, prior to pasteurization	200,000–300,000
Grade A pasteurized milk and milk products (except cultured products)	20,000 (5–10 coliforms)
Grade A pasteurized cultured products	10 coliforms
Ice cream and other frozen desserts	50,000 (10 coliforms)
Liquid ice-cream mix for machines	25,000 (5 coliforms)
Grade B raw milk	600,000
Grade B pasteurized milk	40,000

[a] The recommended standards vary slightly from one community to another.

milk. Grade B milk, which has a higher bacterial count, is usually made into cheese, ice cream, or other products.

Although many of the lactic acid bacteria can ferment lactose in milk, the end products vary considerably with the species or even the strain used. Whereas the growth of *S. lactis* causes souring and spoilage, certain lactobacilli (for example, *Lactobacillus bulgaricus*), often with the aid of some streptococci and yeasts, can ferment milk to yogurt. Russians, Bulgarians, and other central Europeans have eaten yogurt for hundreds of years, but it first assumed commercial importance about the beginning of the century because of the work of Elie Metchnikoff. This scientist, famous for his discovery of the importance of phagocytosis as a defense mechanism, became interested in yogurt when he observed that some of the populations which eat it regularly are exceptionally long-lived. He attempted to show that yogurt was responsible for longevity, and even wrote a book about the subject which influenced many to eat yogurt. Metchnikoff thought that lactobacilli replaced the normal bowel flora, many of which produce endotoxins, and thereby prevented exposure of the host to a variety of toxins. Unfortunately the lifespans of Metchnikoff's followers were not noticeably extended and the fad subsided. In recent years yogurt has once again become popular, both in Europe and in the United States. Certainly it is a valuable food, whether or not it contributes to longevity, because it is high in proteins and vitamins and low in calories.

YOGURT

Fresh pasteurized milk and yogurt account for about 45 percent of all the milk consumed in the United States; approximately 20 percent of the output is used for butter production. Butter can be kept much longer than fresh milk because between 80 and 85 percent consists of the fats of milk, butterfat, emulsified with 12 to 16 percent water, along with small amounts of salt and milk proteins. Butterfat is a unique combination of glycerides (Fig. 38-1), compounds that contain some of the fatty acids. Most of the bacteria and fungi present in butter after churning has removed the fats from the emulsion do not thrive on such a high-lipid diet. In addition the low water content of butter discourages growth of most microorganisms. Consequently butter can be kept for long periods without spoiling. *Pseudomonas putrefaciens* or other pseudomonads are the most common spoilage organisms of butter. They cause rancidity by breaking down the glycerides in butterfat to yield free fatty acids, as shown in Figure 38-1. Butyric acid is particularly prominent in rancid butter.

BUTTER

Although most commercially produced butter is made from pasteurized sweet cream without bacteria being added, it is possible to use mixed "starter cultures" of microorganisms to make a more highly flavored product. The special

FIGURE 38-1

The fats of butter. Butterfat includes many glycerides (fats composed of glycerol molecules with different fatty acid side chains indicated by R. in the diagram). During the spoilage of butter, the side chains are enzymatically removed by bacteria of the genus *Pseudomonas*. The fatty acids that are removed give the butter a rancid taste.

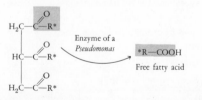

*R can represent any one of a number of different fatty acids

flavor and aroma of this "starter" butter derive principally from diacetyl, a substance produced by *S. lactis, S. cremoris,* or species of *Leuconostoc,* growing in the cream before it is churned. These bacteria produce acetylmethyl carbinol (acetoin), which is oxidized to diacetyl.

The manufacture of cheese accounts for another 10 to 15 percent of the nation's output of milk. In general cheese-making involves coagulation of milk proteins to form solid curds which can be separated from the milk liquid (whey) and treated in various ways to yield characteristic cheeses. Either lactic acid bacteria produce the proper acidity to clot milk into a firm curd, enzymes (rennet or pepsin) are added to clot the milk proteins, or both are used together.

Cottage cheese is the fresh curd of whole or skim milk (soured with *S. lactis*) removed from the whey without further treatment. Cream cheese is a similar curd made from cream. Sometimes lactobacilli are added which slightly decompose the protein and increase the flavor, but usually these cheeses are simply concentrated milk solids. Cottage cheese is high in protein, and if skim milk is used rather than whole milk, the resulting product is low in fat content and in calories. Cream cheese, on the other hand, is rich in fats and high in calories. Neither keeps very well because the fresh curds retain a high content of water and readily support growth of spoilage organisms. These cheeses have little flavor because they have not been ripened by microbial action that results in the production of substances responsible for the flavors and aromas of other cheeses.

Hard cheeses are made by heating fresh curds at low temperature for an hour or two and then removing the whey. During the curing or ripening period which may last months or years, the action of nonproteolytic bacteria or molds transforms the curd to hard cheese. The distinctive flavors of various cheeses result from the activities of different microorganisms. For example, very hard Parmesan and Romano cheeses are ripened by bacteria, as are the less hard Cheddar and Swiss cheeses. If the bacteria produce gas during the ripening process, holes are formed in the cheese (in much the same way holes are formed in bread when CO_2 is produced by the action of yeasts).

Semisoft and soft cheeses are ripened by bacteria or molds, or both. The microbial degradation of fats and proteins in the curds results in the soft final product. Limburger and Liederkrantz are soft cheeses that are ripened by bacteria; Camembert and Brie are ripened by molds. For example, *Penicillium camemberti* grows on the surface of Camembert cheese and degradation of the curd proceeds inward. The more extensive the activities of the mold, the riper and softer the cheese will be. Roquefort cheese is also ripened by mold.

The infinite variety of microorganisms which can ripen curds and the complexity of their metabolic processes account for the fact that the production of fine cheeses remains an art, even under scientific control.

Wine

A discussion of cheeses leads naturally to a consideration of wines, for the two have been partners for many centuries. It is not too surprising that early man learned to make wine, because grapes and other fruits naturally contain the yeasts needed to transform their juice to wine. These yeasts and other microbial flora collect on the waxy film, or bloom, that covers the grape and gives it a dull, matted appearance. As many as 10 million yeast cells may be found on the surface of a single grape; perhaps 1 percent of these are *Saccharomyces cerevisiae,* variety *ellipsoideus,* the strain largely responsible for fermentation to wine. However, some of the others may represent a disadvantage for winemakers by producing unwanted products.

Wine production begins with the selection of the fruit (usually grapes), because the quality of the final product depends, to a large extent, on the initial content of sugars, acids, water, and other chemicals. Selection of the appropriate starting materials is of the utmost importance to the wine producer, but the microbiologist is primarily concerned with selection of the proper yeasts to act on the raw materials.

In general wine is made commercially in the United States by crushing carefully selected grapes in a machine which removes the stems and collects the resulting solids and juices, called *must* (Fig. 38-2). If red wine is being made, the entire must of red grapes is put into the fermentation vat, the red color and rich flavors of red wines being derived from components of the grape skin and seeds. For the production of white wines only the juice of either red or white grapes is fermented. Rosé wines get their light pink color from the entire crushed red grape being fermented for about one day, after which the juice is removed and fermented alone. In the fermentation vat, sulfur dioxide (SO_2) is added to the must to inhibit growth of the natural microbial flora of the grape, and cultures of specially selected strains of *S. cerevisiae* are then added. These strains are usually more resistant than most yeasts to the antimicrobial action of SO_2. Fermentation proceeds at a carefully controlled temperature, which varies with the type of wine, for a period ranging from a few days to several weeks. During this time most of the sugar is fermented into ethyl alcohol, resulting in a final alcohol content up to 14 percent (except under special conditions). At the proper time during fermentation of red wines the juice is separated from the pulp and skins by a wine press and fermentation is allowed to continue until it is completed. The particulate debris settles and the wine is cleared by filtration through diatomaceous earth or other filters prior to aging.

An amazingly complex variety of microbial transformations occurs during the fermentation and aging of wines. Enzymes of the yeasts and other microorganisms catalyze the conversion of sugars (glucose, fructose, and others), organic acids, amino acids, pigments, and many other components of the fruit into numerous different substances. Only a few of these reactions have been identified. The fermentation of glucose to ethyl alcohol, which has been most extensively studied, was discussed in Chapter 7.

Ethyl alcohol
fermentation
Page 150

The presence of lactobacilli in wine is not always advantageous. These bacteria, along with other acid-tolerant lactic acid and acetic acid bacteria, are the most prevalent causes of wine spoilage. They are usually controlled by adding SO_2 or other inhibitory chemicals to the final product.

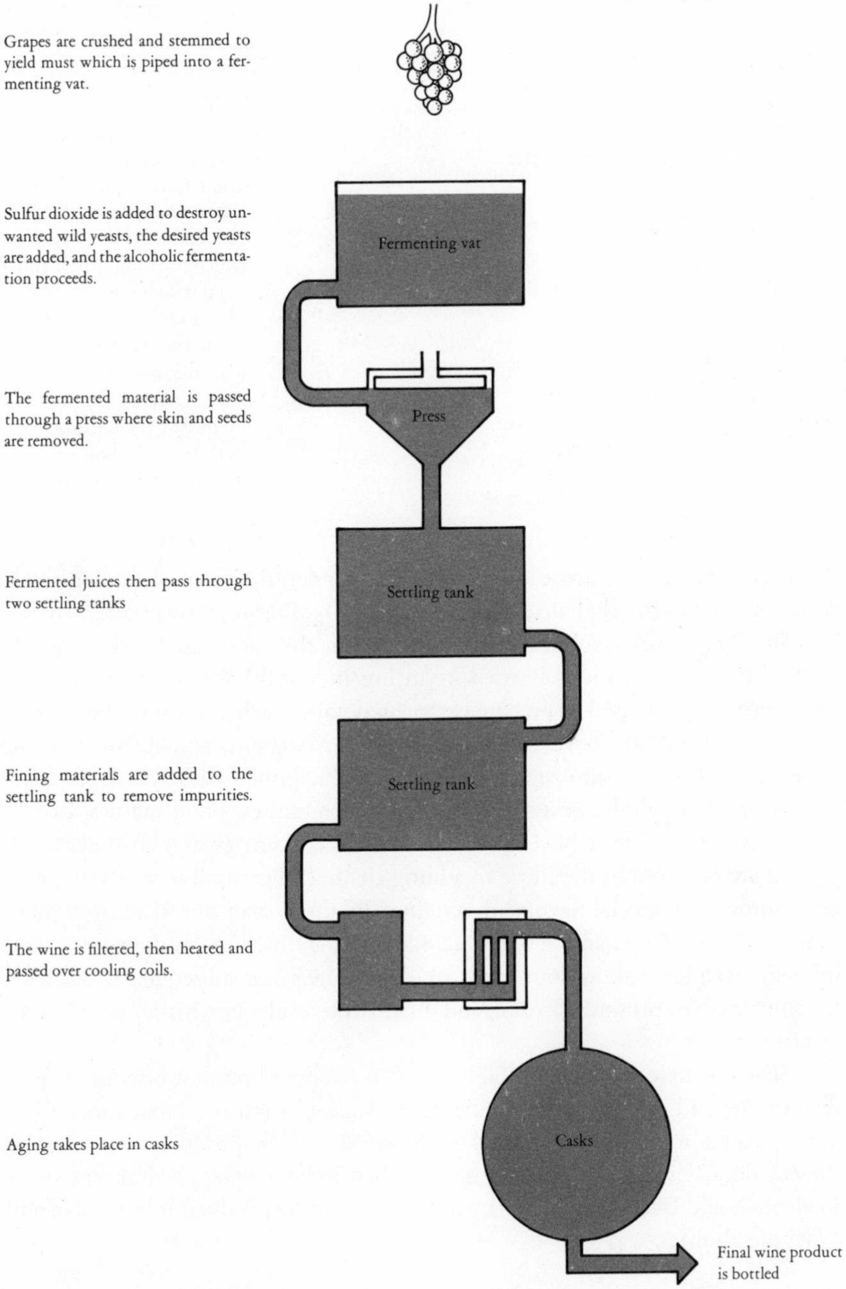

Grapes are crushed and stemmed to yield must which is piped into a fermenting vat.

Sulfur dioxide is added to destroy unwanted wild yeasts, the desired yeasts are added, and the alcoholic fermentation proceeds.

Fermenting vat

The fermented material is passed through a press where skin and seeds are removed.

Press

Fermented juices then pass through two settling tanks

Settling tank

Fining materials are added to the settling tank to remove impurities.

Settling tank

The wine is filtered, then heated and passed over cooling coils.

Aging takes place in casks

Casks

Final wine product is bottled

FIGURE 38-2 The production of red wine. Commercial wine production in the United States follows this general scheme.

TABLE 38-2
Some Alcoholic Beverages Produced by *Saccharomyces*

Beverage	Starting Material	Procedure
Beer	Germinated grain (malt)	Natural fermentation
Table wine	Fruit juice	Natural fermentation
Sake	Rice	Amylase from mold (*Aspergillus oryzae*) converts starch to sugar which is naturally fermented
Fortified wine (e.g., sherry, port)	Fruit juice	Natural fermentation plus addition of brandy to increase alcohol content to about 20%
Brandy	Fruit juice	Natural fermentation followed by distillation to increase alcohol content to about 40 to 50%
Whiskey	Grain mash	
Rum	Molasses	
Tequila	Cactus	
Vodka	Potatoes	

Every mole of glucose or fructose that undergoes alcohol fermentation yields two moles of ethyl alcohol and two of CO_2. During fermentation the release of CO_2 causes a bubbling or "working" in the vat. Usually all the gas is released before the wine is bottled, resulting in a still (noncarbonated) wine. Other processes are used to prepare carbonated wines such as champagne.

A number of the beverages produced by *Saccharomyces* activity are listed in Table 38-2. Note that almost anything that can be fermented by yeasts has been used to make alcoholic beverages—fruit juices, molasses, grain mashes, cactus, and many others. Those beverages with alcohol content greater than about 15 percent are prepared by distilling or adding alcohol to naturally fermented products. Sometimes special flavors are obtained by more than one kind of fermentation reaction. For example, lactic acid bacteria are used to produce lactic acid in grain mash for making sour-mash whiskey; *S. cerevisiae* subsequently ferments the sour-mash to produce alcohol, and distillation results in whiskey with a characteristic flavor.

The Japanese wine, sake (Table 38-2), is of special interest because its production depends on several microbial fermentation reactions. First, cooked rice is inoculated with the mold *Aspergillus oryzae* which produces amylase and thereby degrades the rice starch to sugar. Then *Saccharomyces* converts the sugar to alcohol, and lactic acid bacteria add to the flavor by producing lactic acid and other metabolites.

Beer

Beer is made by the fermentation of germinated grain (malt), usually barley, by species of *Saccharomyces*. Some 5000 years ago underbaked bread made from germinated barley was used to make beer. Small pieces of bread were soaked in water; after a few days the resulting mash fermented and was sieved or filtered and sometimes flavored with herbs or honey. Later hops came into use, being added to give a characteristic bitter flavor to the beer.

The carbohydrates of grains occur largely in the form of starch (Chapter 2) and must be converted to simple sugars before they can serve as substrates for yeast fermentation. This conversion of starch to sugars takes place soon after germination of the grain and is carried out by enzymes which develop in the germinating grain. The germinated grain, or malt, is soaked in water to yield an extract called malt wort that contains the sugars and other nutrients needed for growth of the yeast. Wort is boiled before yeast is added, partly to destroy the enzymes of the grain as well as most microorganisms that may be present, but mainly to concentrate the wort and extract the flavor. The flowers of hops are added primarily for their bitterness, but they also contain antibacterial substances that inhibit the growth of unwanted bacteria. Special brewers' yeasts are used, most often *S. carlsbergensis,* a yeast especially selected for making beer. The yeasts ferment the wort mixture for about a week. The beer is allowed to settle and is then filtered to remove microorganisms and other particulate materials before aging, carbonating, and bottling, canning, or other packaging. The process is shown in Figure 38-3. It differs from wine-making in a number of ways, but principally in that the sugars for fermentation are derived from the breakdown of starch in the grain rather than from natural sugars. The flavor of beer results from the various combinations of grain and hops used, and the alcohol content varies from 3.2 to 6 percent.

Bread

Strains of *S. cerevisiae* have also been selected especially for making bread. The yeasts utilize sugar in bread dough and produce alcohol and CO_2, but the fermentation proceeds only a few hours so small quantities of alcohol are made. During baking the volatile alcohol is lost and CO_2 bubbles are trapped in the bread, producing holes and resulting in the "lightness" of raised breads.

"Active dry yeast" (lyophilized) has become more popular than yeast cakes, especially for home baking, because it can be kept much longer and does not have to be refrigerated. The dried yeast cells retain their viability better when they are vacuum packed, or when CO_2 or nitrogen is substituted for air in the package.

Microorganisms other than yeasts can also contribute to better commercial bread products. For example, amylase produced by some of the filamentous molds is sometimes added to degrade starch in the flour, resulting in more gas production during bread-making.

Vinegar

Commercial vinegar is an aqueous solution containing at least 4 percent acetic acid, obtained from the oxidation of ethyl alcohol by species of acetic acid bacteria. The alcohol of apple cider is the substrate for cider vinegar; wine vinegar is prepared by the action of acetic acid bacteria on the alcohol in wine. Vinegar can

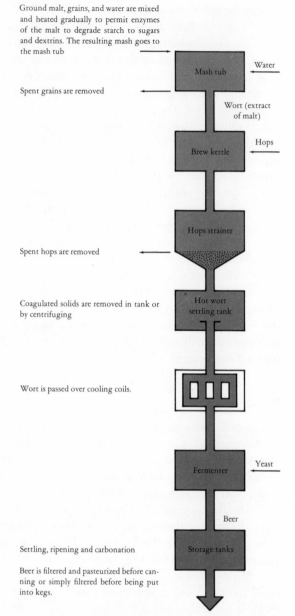

Ground malt, grains, and water are mixed and heated gradually to permit enzymes of the malt to degrade starch to sugars and dextrins. The resulting mash goes to the mash tub

Spent grains are removed

Coagulated solids are removed in tank or by centrifuging

Wort is passed over cooling coils.

Settling, ripening and carbonation

Beer is filtered and pasteurized before canning or simply filtered before being put into kegs.

FIGURE 38-3
The commercial production of beer.

be made from many alcoholic beverages; however, wine, malt, and cider vinegars are most important industrially.

Acetic acid bacteria are strictly aerobic, gram-negative rods, characterized by their ability to carry out a number of incomplete oxidations. They can tolerate high concentrations of acid and thus can grow in acidic wines or ciders, oxidizing alcohol to acetic acid, by the following reactions:

$$2CH_3CH_2OH + O_2 \longrightarrow 2CH_3CHO + 2H_2O$$
Ethyl alcohol + oxygen $\longrightarrow$ acetaldehyde + water

$$2CH_3CHO + O_2 \longrightarrow 2CH_3COOH$$

Acetaldehyde + oxygen $\longrightarrow$ acetic acid

The intermediate product, acetaldehyde, is quickly converted into acetic acid and does not accumulate during the process. *Acetobacter aceti, A. orleanse,* and *A. schuezenbachi* are species important in the manufacture of vinegar.

Transformation of alcohol to acetic acid (acetification) may be slow or rapid. The best quality (and most expensive) vinegars are obtained by the old French or Orleans process of slow acetification. In this process oak barrels half full of wine are placed on their side; the barrels have holes drilled in the upper part to provide the aerobic environment necessary for growth of *Acetobacter*. A "starter" of vinegar containing acetic acid bacteria is added to the wine and the bacteria develop as a film on the surface of the liquid. Approximately one to three months are required for maximum oxidation of ethyl alcohol to acetic acid.

Rapid acetification in a vinegar generator is widely used to produce most vinegars. The generator consists of a tank (usually wooden, to resist the highly acid vinegar) containing wood shavings or other materials which provide a large surface area. Acetic acid bacteria develop aerobically as a film on the wood shavings. The alcoholic substrate is sprayed on the shavings and trickles through, being oxidized en route by the abundance of *Acetobacter* organisms. In principle the vinegar generator operates much as the trickling filter does during degradation of wastes.

Sauerkraut

Sauerkraut is the product of a natural lactic acid fermentation carried out by lactobacilli and certain other bacteria that are normally present on cabbage. The cabbage is shredded and layered, with salt between the layers. The salt content, 2.0 to 2.5 percent of the total weight, inhibits the growth of many bacteria and also dehydrates the cabbage. The layers are firmly packed and pressed down to provide an anaerobic environment. Acid produced as a result of fermentation soon inhibits growth of most microorganisms, but favors growth of *Leuconostoc, Lactobacillus plantarum, L. brevis,* and similar species which grow and produce large amounts of lactic and acetic acids, alcohol, and CO_2, and smaller amounts of other compounds. When the desired flavor has been attained, usually after two to four weeks at warm room temperatures, the sauerkraut is canned.

Lactic acid fermentations also assume great commercial importance in the production of pickles. The same species of lactic acid bacteria that ferment sauerkraut are also of primary importance in the production of pickles.

PICKLES

PHARMACEUTICALS AND FOOD SUPPLEMENTS

Microorganisms are of great value in the manufacture of a number of pharmaceuticals and food supplements.

Man can synthesize twelve of the twenty major amino acids. The other eight, the essential amino acids, must be ingested. Some microorganisms can synthesize these amino acids. Derepressed mutant strains of bacteria have been selected which produce unusually large quantities of a particular amino acid. For example, the essential amino acid L-lysine has in the past been produced commercially by the joint activities of selected strains of two common species of bacteria, *Escherichia coli* and *Enterobacter aerogenes*. During the first part of the process a mutant of *E. coli* which cannot produce lysine supplies enzymes which catalyze the incorporation of nitrogen from inorganic ammonium ions into amino groups of the organic compound diaminopimelic acid. The second part of the process involves enzymes of *E. aerogenes*, which catalyze the removal of CO_2 from diaminopimelic acid, to yield L-lysine:

Ammonium ions + glycerol + corn-steep liquor $\xrightarrow{\text{enzymes of } E. coli}$ Diaminopimelic acid

$$\text{COOH—CH—NH}_2\text{—CH}_2\text{—CH}_2\text{—CH}_2\text{—CH—NH}_2\text{—COOH} \xrightarrow{\text{enzymes of } E. aerogenes}$$
Diaminopimelic acid

$$\xrightarrow{\hspace{2cm}} CO_2 + \text{CH}_2\text{—NH}_2\text{—(CH}_2)_3\text{—CH—NH}_2\text{—COOH}$$
L-lysine

In almost all restaurants—and many home kitchens as well—monosodium glutamate (MSG) is added to foods to enhance their flavor. Commercially, MSG is prepared from a nonessential amino acid, L-glutamic acid, which is synthesized from molasses by *Micrococcus glutamicus* (a gram-positive coccus) or from corn-steep liquor by *Brevibacterium flavum* (a gram-positive rod). Although MSG in moderate amounts enhances food flavor without any adverse effects, it has been found that excessive quantities added to food can cause a curious set of symptoms in rare individuals, described as the "Chinese restaurant syndrome," caused by the high sodium concentration resulting from added MSG.

Organic acids (for example, lactic, citric, gluconic, and butyric acids) are also produced industrially by microbial activities. Lactic acid, for example, can be synthesized by *Lactobacillus bulgaricus* from the lactose in whey that is a waste product from cheese manufacture. More often glucose or other sugars are used as a substrate for fermentation by *L. delbruckii*. The product, lactic acid, is used in the manufacture of food products, for treating hides to make leather, in pharmaceuticals, paints, and plastics. Butyric acid, produced by species of *Clostridium*, is an important organic solvent often used in paints and plastics.

Molds have been used commercially to produce citric, oxalic, and gluconic acids. Special strains of *Aspergillus niger* ferment beet molasses (or sugars from other sources) to citric acid. Gluconic acid can also be made commercially by *A. niger*, but other strains must be selected which produce gluconic acid but not other organic acids. A simple glucose-salts medium provides the substrate for gluconic acid production by *A. niger*.

Vitamins

Yeasts synthesize vitamins very efficiently and certain strains of yeast are often used as food supplements. An ascomycete (*Ashbya gossypii*) is widely used to produce riboflavin industrially. A particularly important commercial process is the production of vitamin B_{12} by actinomycetes and some other bacteria, including species of *Streptomyces* and also certain strains of *Bacillus megaterium*.

Steroids

Although it is difficult to recover large quantities of *steroid hormones* from animal sources, it is possible to manufacture them from sterols or related compounds obtained from animals or plants. Special strains of the mold *Rhizopus* and of the bacteria *Corynebacterium* and *Streptomyces* are used to change particular chemical groups on the molecules by enzyme activity thereby producing the steroids (Fig. 38-4).

ANTIBIOTICS

The manufacture of antibiotics, begun more than 30 years ago by growing *Penicillium notatum* in flasks, has developed into today's efficient antibiotic industry. The tremendous growth of the industry in the United States is illustrated by statistics which show that in 1945 about 12,000 pounds of antibiotics were produced, as compared with well over 3 million pounds in 1961.

In the American antibiotic industry, penicillin is currently being produced by special strains of *P. chrysogenum*. The natural penicillin molecule can be altered chemically to give a variety of semisynthetic penicillins with special properties (Fig. 32-1).

Figure 32-1
Page 557

Antibiotics represent only a small proportion of the metabolic output of the microorganisms which synthesize them. In fact they seem to be minor by-products of metabolism, not essential to the cells which produce them, and it is not clear how or why they are made. In order to harvest sufficient quantities of antibiotics, mutant strains of microorganisms have been selected which pro-

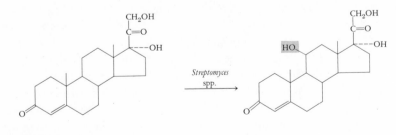

Compound S

FIGURE 38-4 Steroid transformation. Special strains of bacteria or molds can enzymatically convert compound S, an inactive substance widely distributed in nature, to valuable steroids such as hydrocortisone.

(a)   (b)

FIGURE 38-5   The effects of gibberellins on plant growth. Untreated plants are pictured on the left, and plants treated with gibberellins are on the right. (a) Bush beans. (b) Head lettuce. (Courtesy of Dr. M. J. Bukovac)

duce large amounts of a particular antibiotic. Even so, the large residues of microorganisms left after antibiotics have been produced present tremendous disposal problems. In the production of penicillin, for example, only a tiny part of the final culture fluid is recovered as penicillin, and huge amounts of mold mycelium and expended culture medium must be discarded. These wastes are an environmental hazard unless properly treated because they are rich in nutrients.

VACCINES, TOXOIDS, AND ANTITOXINS

Vaccines, toxoids, and other materials used to immunize must be prepared under the most rigid conditions. Exotoxins, such as diphtheria and tetanus toxins, that are prepared commercially must be transformed into innocuous toxoids and carefully tested before use. Similarly, vaccines made of infectious agents, whether living (for example, the Sabin polio vaccine) or inactivated (such as the Salk polio vaccine) must be produced industrially and yet scrupulously monitored to ensure their safety. There are several recorded instances where mistakes have been made in the preparation of vaccines, such as the incident some years ago when a batch of the polio vaccine was not completely inactivated. As a result a number of children who received the vaccine developed paralytic poliomyelitis.

Some commercially produced immunizing agents of major importance are listed in the table in Appendix IV. Many of these agents are grown in chick embryos or cells; thus they present a real danger to those few persons who are allergic to eggs. Even the tiny amounts of egg antigens present in an immunizing dose of these vaccines can cause an anaphylactic reaction and be fatal to an individual who is highly sensitive to eggs.

In toxoid preparation from exotoxins fairly mild treatment with formalin or heat can be employed to destroy toxicity while retaining the ability to incite antibody formation. Infectious agents (such as bacteria or viruses) may be killed or inactivated and injected to stimulate the production of antibodies without causing disease. Alternatively, living, attenuated agents are administered in order to immunize. For example, the Salk polio vaccine is a mixture of inactivated polioviruses of all three antigenic types, whereas the Sabin vaccine in common use now consists of living mutants of the polioviruses which cause an infection but are not able to invade the central nervous system or to lead to severe disease. Protection results following either of these immunizing procedures.

Antitoxins, also prepared commercially, sometimes are the only effective treatment against powerful exotoxins. Often horses or other large animals are immunized with toxoid and the gamma globulin portion of blood serum is prepared. When possible, gamma globulin from immunized human volunteers is prepared to decrease the risk of hypersensitivity reactions in people who receive the antitoxin. Antitoxins are used primarily in botulism, tetanus, diphtheria, and in cases of snake venom or certain other kinds of intoxications.

Vaccines and antitoxins must be standardized during preparation to ensure that each dose has the proper amount of activity. Reference standards have been established by federal health agencies and are carefully enforced. In a number of cases international standards, established by the World Health Organization, are used.

ENZYMES

It is obvious from previous discussions that yeasts, molds, and bacteria produce an immense number of enzymes. Fungi are used commercially in the manufacture of invertase (which converts sucrose to glucose and fructose), proteases (which degrade protein to constituent peptides and amino acids), and a large variety of other enzymes, as outlined in Table 38-3.

TABLE 38-3
Some Enzymes That Are Produced Industrially by Microorganisms

Enzyme	Reaction Catalyzed	Organisms Frequently Used for Production
Amylase	Starch to maltose	*Aspergillus sp.*
Maltase	Maltose to glucose	*Rhizopus sp.*
Proteases	Proteins to polypeptides or amino acids	*Bacillus subtilis* *Aspergillus sp.*
Pectin-degrading	Degradation of pectin	*Aspergillus sp.* *Rhizopus sp.*
Invertase	Sucrose to glucose and fructose	*Saccharomyces cerevisiae*
Penicillinase	Degradation of penicillin	*B. subtilis* *B. cereus*
Streptodornase	Degrades DNA	*Streptococcus pyogenes*
Streptokinase	Degrades fibrin clots	*S. pyogenes*

The major types of
hypersensitivities
Table 21-1, Page 394

In recent years detergents containing enzymes of *Bacillus subtilis* have gained popularity because the enzymes remove many kinds of stains from fabrics. Thermophilic strains of *Bacillus* may be used because their enzymes are often more heat resistant than those of mesophilic strains. The detergents often are in powder form and generate dusts which can be inhaled, sometimes resulting in allergic sensitization to the bacterial enzyme proteins. This kind of sensitization has led to some serious respiratory problems in industrial workers exposed to large amounts of detergent dusts (probably a mixture of atopic and immune-complex-mediated allergic reactions, Chapter 21). At present intensive investigations are in progress to determine whether there is a significant danger for consumers exposed to very small amounts of the enzymes.

GIBBERELLINS

The gibberellins comprise a group of substances which have remarkable capacities to stimulate plant growth and flowering. They include gibberellic acid and closely related compounds with a complex cyclic organic acid structure. The fungi *Gibberella* and *Fusarium* are used commercially to produce gibberellins. Figure 38-5 illustrates the effects of gibberellins on plant growth.

DEXTRANS

Some species of *Leuconostoc* can convert sucrose into fructose and large polymers of glucose called dextrans. Solutions of dextran isotonic with blood can be used to supplement plasma in patients who have had extensive blood loss.

OTHER APPLICATIONS

In addition to those chemicals already mentioned, alcohols and acetones can be manufactured by microbial activities. Industrial ethyl alcohol may be produced from molasses by *S. cerevisiae* and other yeasts, and distilled. The lactose in whey, starch in grains, and sugars in molasses are all readily converted by yeasts to ethyl alcohol suitable for industrial uses. Butyl alcohol and acetone are produced by the anaerobic metabolism of some species of *Clostridium;* however, fermentation is no longer used commercially, because it is now cheaper to produce acetone and butanol synthetically.

Linen, hemp, and other fibers are made by retting, a process in which microbial enzymes break down pectin in plant stems and release the natural fibrous materials. For example, in preparing linen, bundles of flax are submerged in water and anaerobic conditions develop which permit clostridia to grow in the material. During a period of about two weeks clostridial enzymes attack the pectin and free the fibers, which are subsequently cleaned and made into linen cloth. Other bacteria can carry out aerobic retting processes.

Microorganisms are frequently used to assay the amounts of certain substances present in a mixture. Some organisms can detect traces of materials too small to be demonstrated by chemical assays. Microbes can also distinguish between D- and L-isomers of amino acids which cannot be distinguished by ordinary chemical analyses.

Vitamins are usually measured by microbiological assay. Strains of bacteria which require the vitamin are grown on medium lacking it except for the quantities present in the added test material. Growth of the bacteria occurs only if the substance is present and growth is proportional to the amount of vitamin present.

Microbiological assays
Pages 111–112

SUMMARY

Microbial activities have many commercial applications. Milk products which result, at least in part, from microbial activity include yogurt, butter, and cheeses. Wine results from the fermentation of fruit juices by *Saccharomyces cerevisiae,* and other strains of this yeast ferment grains to yield beer. These fermentation products can be distilled to produce whiskies and other beverages with high alcohol content. *Saccharomyces* yeast is also used for manufacturing bread. Vinegar, which is a solution of acetic acid, is produced by the oxidation of ethyl alcohol in fermented cider, wine, or malt products by acetic acid bacteria. Lactic acid bacteria are responsible for the fermentation reactions necessary for producing sauerkraut and pickles. Other commercial applications of microbiology include the manufacture of amino acids and a number of other organic acids, vitamins, antibiotics, vaccines, toxoids, antitoxins, enzymes, and a host of other useful products. Frequently microorganisms are also used to assay the amount of certain substances, such as a vitamin, present in a sample.

QUESTIONS

1. List at least five industries that depend on the alcoholic fermentation carried out by yeast of the genus *Saccharomyces.*
2. What are some of the foods produced by lactic acid fermentations? What organisms are involved in preparing each of these foods?
3. What are the principles involved in producing toxoids, antitoxins, and vaccines?
4. What is the source of enzymes in "enzyme detergents"? Are there any possible hazards involved in using these products?
5. Strains of *Penicillium* that produce penicillin can be isolated from fresh fruits. Therefore it might be supposed that penicillin is frequently ingested in fresh foods, but in fact it is not. Why?

AMERINE, M. A., "Wine," *Scientific American,* 211 (Aug., 1964), 53.

KLEYN, J. and J. HOUGH, "The Microbiology of Brewing," *Ann. Rev. Microbiol.,* 25 (1971), 583–608.

PRESCOTT, S. C. and C. G. DUNN, *Industrial Microbiology,* 3d ed. New York: McGraw-Hill, Inc., 1959.

Progr. Indust. Micro., Vol. 1 (1959) through Vol. 10 (1971).

UNDERKOFLER, L. A. and R. J. HICKEY (eds.), *Industrial Fermentations,* Vols. I and II. New York: Chemical Publishing Co., 1954.

Medically Important Bacteria

Bacterial Group	Medically Important Genera
1. Phototrophic	None recognized
2. Gliding	None recognized
3. Sheathed	None recognized
4. Appendaged and budding	None recognized
5. Spirochetes	*Treponema, Borrelia, Leptospira*
6. Spiral and curved	*Vibrio, Spirillum*
7. Strictly aerobic gram negative rods	*Pseudomonas, Brucella, Bordetella, Francisella, Moraxella, Acinetobacter, Bartonella.*
8. Facultatively anaerobic gram negative rods	*Escherichia, Shigella, Enterobacter, Klebsella, Serratia, Proteus, Salmonella, Yersinia, Flavobacterium, Pasteurella, Aeromonas, Haemophilus*
9. Strictly anaerobic gram negative rods	*Bacteroides, Fusobacterium*
10. Nonphotosynthetic autotrophs	None recognized
11. Gram negative cocci	*Neisseria*
12. Gram positive cocci	*Micrococcus, Staphylococcus, Streptococcus, Peptococcus, Peptostreptococcus*
13. Endospore-forming	*Bacillus, Clostridium*
14. Nonspore-forming gram positive rods	*Lactobacillus, Erysipelothrix, Listeria, Corynebacterium, Propionibacterium*
15. Branching bacteria	*Mycobacterium* (pathogenic species do not usually branch), *Nocardia, Actinomyces, Streptomyces, Micromonospora, Thermoactinomyces*
16. Mycoplasmas	*Mycoplasma*
17. Obligate intracellular	*Rickettsia, Coxsiella, Chlamydia*

Appendix II

Classification of Bacteria

Some of the principal groups of bacteria as classified in the seventh edition of Bergey's Manual,* 1957, are given below. Many of the genera mentioned in this book are placed in parentheses after the family names. Newer classification schemes are in limited use, but that below is widely known.

Order I: **Pseudomonadales**

Thiorhodaceae (sulfur bacteria)
Athiorhodaceae (non-sulfur bacteria)
Nitrobacteriaceae (*Nitrosomonas, Nitrobacter*)

Order II: **Chlamydobacteriales**

Chlamydobacteriaceae (*Sphaerotilus*)

Order III: **Hyphomicrobiales**

Hyphomicrobiaceae (*Hyphomicrobium*)

Order IV: **Eubacteriales**

Azotobacteriaceae (*Azotobacter*)
Rhizobiaceae (*Rhizobium, Agrobacterium*)
Achromobacteriaceae (*Flavobacterium*)
Enterobacteriaceae (*Escherichia, Aerobacter, Klebsella, Erwinia, Serratia, Proteus, Salmonella, Shigella*)

* *Bergey's Manual of Determinative Bacteriology*, Robert S. Breed et al, editor. Baltimore: Williams and Wilkins Co., 1957.

674

Brucellaceae (*Pasteurella, Bordetella, Brucella, Haemophilus, Moraxella*)

Bacteroidaceae (*Bacteroides, Fusobacterium, Dialister*)

Micrococcaceae (*Micrococcus, Staphylococcus, Methanococcus, Peptococcus*)

Neisseriaceae (*Neisseria, Veillonella*)

Lactobacillaceae (*Diplococcus, Streptococcus, Peptostreptococcus, Lactobacillus*)

Propionibacteriaceae (*Propionibacterium*)

Corynebacteriaceae (*Corynebacterium, Listeria, Erysipelothrix, Arthrobacter*)

Bacillaceae (*Bacillus, Clostridium*)

Order V: **Actinomycetales**

Mycobacteriaceae (*Mycobacterium*)

Actinomycetaceae (*Actinomyces, Nocardia*)

Streptomycetaceae (*Streptomyces, Micromonospora, Thermoactinomyces*)

Order VI: **Caryophanales**

Caryophanaceae (*Simonsiella*)

Order VII: **Beggiatoales**

Beggiatoaceae (*Beggiatoa, Thiothrix, Leucothrix*)

Order VIII: **Myxobacteriales**

Cytophagaceae (*Cytophaga*)

Myxococcaceae (*Chondrococcus*)

Order IX: **Spirochaetales**

Spirochaetaceae (*Cristispira*)

Treponemataceae (*Treponema, Borrelia, Leptospira*)

Order X: **Mycoplasmatales**

Mycoplasmataceae (*Mycoplasma*)

The *Rickettsiaceae* (*Rickettsia, Coxsiella*), *Chlamydaceae* (*Chlamydia*), *Bartonellaceae* (*Bartonella*), and the viruses, were placed in a separate class called the *Microtatobiotes,* consisting of two additional orders.

Appendix **III**

Survey of Groups
of Animal Viruses

Although no universal classification of animal viruses has been generally agreed upon, for purposes of discussion the various groups of viruses infecting man and other animals will be divided into the following eleven categories, plus a twelfth unclassified group:

DNA Viruses		RNA Viruses
1. Poxviruses		6. Picornaviruses
2. Herpesviruses		7. Arboviruses
3. Adenoviruses		8. Myxoviruses
4. Papovaviruses		9. Paramyxoviruses
5. Parvoviruses or picodnaviruses		10. Reoviruses
		11. Rhabdoviruses

1. Poxviruses

The poxviruses are the most complex in structure as well as the largest viruses of vertebrates. Several poxviruses cause human disease; among these are the viruses of smallpox, cowpox, and viruses causing local skin infections which may be acquired by contact with sheep or cows. Cow, sheep, swine, and fowl pox are all economically important diseases of domestic animals. The myxoma virus of rabbits was once used as a deliberate means of controlling these animal pests both in Australia and Europe. The poxviruses have a complex morphology with a DNA-containing core and an outer membrane. Poxviruses are unusual among the DNA viruses because they replicate in the cytoplasm of host cells rather than in the nucleus.

2. Herpesviruses

These are large enveloped DNA viruses which replicate within cell nuclei. A distinguishing characteristic is their ability to establish long term latent infections. The virion contains DNA within an icosahedral capsid surrounded by a lipoprotein envelope. Herpesviruses are known to cause disease in a wide variety of vertebrates, birds, and amphibia as well as mammals.

The commonest human herpesvirus infection results in the "cold sore." This is due to herpes simplex virus carried as a latent infection which can become activated and cause local lesions on the skin, particularly the face, or on mucous membranes, following some stress or change in fitness of the carrier. Occasionally more serious infections occur, particularly in the cornea, which can lead to blindness. This is one of the few viral infections where drug treatment can be used successfully. Certain inhibitors of DNA synthesis (such as iododeoxyuridine) locally administered can effectively prevent viral replication. Genital lesions are produced by one type of herpes simplex virus. Recently, a new treatment for genital herpes has been reported which seems promising. In this treatment, a light-sensitive dye is introduced into the lesion, and the area is exposed to visible light; the dye absorbs light energy which breaks strands of viral DNA, resulting in death of the virus. Results have been favorable in the limited number of patients treated in this way to date.

Another common disease caused by a member of the herpesvirus group is chickenpox. The virus responsible, the varicella-zoster virus, also causes shingles, a disease which represents a reactivation of latent virus. Cytomegalovirus is another common herpesvirus. It can cause severe disease or death in newborns but frequently causes an inapparent infection in children.

3. Adenoviruses

The adenoviruses occur as latent infections of the tonsils, adenoids and respiratory tract in man and indeed are responsible for a proportion of respiratory conditions, especially in military recruits and young children. Adenoviruses were, in fact, discovered during attempts to isolate the common cold virus. Some human adenoviruses have been shown to be oncogenic; that is, capable of producing tumors in experimental animals under certain conditions. However, up to the present time, no evidence exists that they are oncogenic in their natural human hosts.

Some thirty or forty known types of human adenovirus may be distinguished; only a few of these are oncogenic. Other adenoviruses have various animal species as their natural hosts. Clearly, this is a fairly diverse group of viruses; although the structure is similar, the base composition of the viral DNA is quite variable.

4. Papovaviruses (*Papilloma, Polyoma, Vacuolating Viruses*)

These are relatively small DNA viruses chiefly of interest as a result of their oncogenic character. However, no papovaviruses are known to be associated with human cancer; the only member of the group infecting man is the causative agent of warts, a kind of papilloma. On the other hand, several members of this group have been extensively studied since they induce tumors in a variety of animals, particularly rodents. The papilloma viruses of rabbits first produce benign tumors some of which may later become malignant. The relationship of these viruses to cancer is considered in Chapter 18.

The common human wart occurs in localized groups, particularly in exposed areas such as hands and feet. Virus is found in the external layers of the skin. Transmission of virus from one location to another is common, occurring mostly in areas subject to skin abrasion such as the fingers or knees.

A monkey virus, SV40 (Simian Vacuolating Virus 40), classified in this group undertakes a rather strange interaction with various members of the adenovirus group. When simultaneous infection of a cell with SV40 and an adenovirus occurs, the cell yields particles consisting of adenovirus capsids surrounding an SV40 genome or a mixture of SV40 and adenovirus DNA. These hybrid viruses may contain all or part of the two genomes. In some cases SV40 and adenovirus DNA are covalently linked and a new hybrid genome is evident. These observations are of great theoretical interest, since they may present some insight into the mechanism of evolution of new viruses.

5. Parvoviruses

This is a group of very small viruses some of which contain single-stranded DNA. An alternate name, picodnaviruses (pico = small-dna-viruses) is often used. The category probably consists of two separate groups since some of the viruses appear to have double-stranded DNA. All are of very small size, and some are unable to propagate except with the assistance of other "helper" viruses. For example, some parvoviruses with double-stranded DNA are completely dependent upon associated adenoviruses for replication. This presumably results from the fact that such virus genomes contain only the information necessary to specify the capsid protein but not other functions essential to replication. Other parvoviruses such as mouse minute virus contain single-stranded DNA and are structurally similar to the well-studied *E. coli* bacteriophage ϕX174. No known pathogenic conditions in man are attributable to these small viruses; however, some of them are pathogenic for animals.

6. Picornaviruses

These are one of the largest groups of pathogenic human viruses and are among the smallest of the RNA viruses. In fact, the designation itself spells that out (pico = small-rna-viruses). So diverse are the viruses within this group that they are divided into subgroupings as follows:

I. Enteroviruses—predominantly in the gastrointestinal tract

   a. Polioviruses

   b. Coxsackieviruses

   c. Echoviruses

II. Rhinoviruses—predominantly in the nasal passage

The virions are small and stable to inactivation. Enteroviruses are especially acid tolerant. All are naked icosahedrons of approximately the same dimensions (20-30 nm diameter). They grow mainly in primate cells and often are highly destructive to them.

a. *Polioviruses.* Three antigenic types are known. The mode of intracellular replication is as well understood as that of any animal virus. Poliomyelitis is now under control as a result of the development and application of vaccines.

b. *Coxsackieviruses.* These are also intestinal viruses, similar in structure to poliovirus. Coxsackievirus infections may involve a remarkably wide range of the major organ systems and may occasionally cause severe disease. These viruses were discovered by accident in the course of studies on poliomyelitis and were named for the village in New York where this occurred.

c. *Echoviruses.* These viruses were also discovered in the course of studies on poliomyelitis. They were isolated from normal individuals and designated ECHO viruses (Enteric Cytopathogenic Human Orphan). The term orphan was coined originally because they were not recognized to be associated with any specific disease ("virus in search of disease"). Nevertheless, echoviruses have now been associated with many human afflictions, ranging from respiratory to central nervous system infections. Their distinction from coxsackieviruses is somewhat arbitrary, being based partially on the susceptibility of suckling mice to infection by coxsackie but not to echoviruses.

II. *Rhinoviruses.* Although colds are among the most common illnesses of man, early attempts to identify specific viruses as the causative infectious agents met with difficulty for a number of reasons. For example, the symptoms of a cold can be caused by many different means, including allergy, and bacterial as well as viral infection. Then too, the transfer of viruses isolated from cold victims to volunteers often gave inconsistent results. Often some recipients, but not all, became sick. It was hard to interpret whether failure of some volunteers to get cold symptoms when exposed to a particular virus was due to their immunity, state of physical fitness, or other factors. In spite of all these difficulties, a large number of different rhinoviruses are now known to be causes of this tiresome human ailment. Rhinoviruses differ from enteroviruses in that the RNA is of somewhat greater molecular weight (approximately 2.5×10^6), the slightly larger virion is more stable to thermal inactivation but less resistant to acid, and rhinoviruses replicate better at $33°C$, the temperature of the nasopharynx. Other than that, these viruses are remarkably similar to enteroviruses in form and mode of replication.

7. Arboviruses

Unlike all other classes of viruses, the arbovirus group is defined on the basis of their being arthropod-borne, rather than on their physical-chemical structure. In fact, this is a heterogeneous group. An arbovirus multiplies in an insect vector and is transmitted by a bite to the bloodstream of a vertebrate host. More than 200 arboviruses are known, some of which fall into other accepted categories, such as rhabdoviruses. However, the majority are medium-sized, enveloped, single-stranded RNA viruses, roughly spherical in shape (40–100 nm diameter).

Most arboviruses are capable of multiplication in an insect vector and any one of several vertebrate hosts; however, yellow fever and dengue are caused by viruses which have only man and other primates as their natural hosts. In the laboratory many animals may be infected with arboviruses, although very young mice are favored. In general, encephalitis and hemorrhagic fevers are the most serious associated diseases both in the natural human host and laboratory animals.

8. Myxoviruses

This group is represented by the influenza viruses, of vital importance to man. Virions are irregular in shape, but roughly spherical, about 100 nm in diameter. They contain five pieces of single-stranded RNA. A prominent viral envelope is apparent, consisting of lipoproteins derived from the plasma membrane or membrane precursors of cells in which the virus was grown. However, two important proteins coded for by the viral genome also occur in the envelope. One is responsible for hemagglutination, a characteristic of this and many other viruses, and the other a neuraminidase. This latter enzyme degrades neuraminic acid in the mucus which covers cells in the respiratory tract and allows the virus to reach and attach to host cell receptors. This enzyme thereby promotes entry and spread of the virus, as well as having other effects.

The myxovirus or influenza group is divided into several antigenically defined types, subtypes, and strains. These cause the different forms of "flu", a febrile disease that is well-known. In fact, a distinguishing characteristic of influenza virus is its propensity for antigenic change which prevents the prospective human victim from becoming immune to all influenza. This is called "antigenic drift." Every generation or so, a major change in antigenic structure seems to occur resulting in a world-wide epidemic. It is postulated that these changes may come about as a result of genetic recombination between the five strands of RNA in the virion in mixed infections.

Genetic instability
of myxoviruses
Page 459

9. Paramyxovirus and Rubella

As the name implies, these viruses are structurally similar to myxoviruses. However, they differ in being somewhat larger, more irregular in shape and in their mode of RNA replication. All paramyxoviruses appear to contain a single strand of RNA of about 7×10^6 daltons and in most cases the envelope has hemagglutinating activity.

Measles and mumps viruses are the most important human viruses classified in this group. Both are typical paramyxoviruses. Vaccines have been developed for protecting against both of these paramyxoviruses.

Rubella virus has been considered as part of the paramyxovirus group by virtue of size, lipid content, and other physical properties. However, electron microscopic studies show rather irregular particles which preclude accurate classification.

Rubella is usually a trivial disease of children or adults, involving mild fever and rash and followed by immunity. The importance of the disease is based upon the teratogenic effects of infection in a woman during the first three months of pregnancy. The risk of damage to the fetus is considerable if infection occurs during early development. Little is known of the basis of this effect, although it may be related to the mildness and noncytocidal character of the infection which allows it to persist. Apparently the virus retards normal cell division.

10. Reoviruses

These viruses, distinguishable as the only group containing double-stranded RNA, are readily isolated from the feces and the respiratory tract of man, other mammals, and birds. The wound tumor virus of plants is almost identical in structure and nucleic acid composition. Despite their ubiquitous distribution and their frequent occurrence in patients with mild respiratory conditions, and occasionally with fatal pneumonia or with rashes, they have not been consistently associated with human disease, thus they are called respiratory enteric orphan, or *REO*-viruses. Judged by the incidence of antibodies to reoviruses, they are among the most widely distributed of all viruses both in human and animal populations.

11. Rhabdoviruses

This is a group of RNA viruses distinguished by their bullet-like shape. The group is chiefly of interest by virtue of the inclusion of rabies virus. A single-stranded RNA is enclosed in an enveloped virion about 70 nm wide and 180 nm long. Spikes with hemagglutinating activity project from the envelope, similar to the spikes of influenza virus.

Rabies virus is unusual in its wide host range. It is apparently able to infect all mammals; for most species including man it is nearly always lethal although other species such as bats can be carriers exhibiting no symptoms.

12. The Unclassified Slow Viruses

Not enough is known about their nature to classify the so-called "slow viruses." These include the virus of kuru, a disease of man, and scrapie and visna viruses of sheep. Visna virus causes a "slow" neurological infection in sheep. Its importance lies in the analogy with certain central nervous system diseases of man. These infections, which are also observed in other species, involve slow progressive destruction of the brain and spinal cord accompanied by loss of the myelin sheath which normally surrounds nerves. The first symptoms are abnormal posture and gait, later followed by complete paralysis, a sequence of events reminiscent of multiple sclerosis in man. Recently it has been shown that visna virus contains an RNA-dependent DNA polymerase, similar to that found in oncogenic viruses. This finding raises the exciting possibility that "slow virus infections" involve mechanisms similar to those in oncogenesis.

Reverse transcriptase
Pages 165, 352

Appendix **IV**

Some Commercially Produced Immunizing Agents of Major Importance

Agent	Active Against	Production Method
Diphtheria toxoid	Exotoxin of *Corynebacterium diphtheriae* (cause of diphtheria)	Exotoxin produced by strains of *C. diphtheriae*; converted into toxoid by formalin treatment
Tetanus toxoid	Exotoxin of *Clostridium tetani* (cause of tetanus)	Exotoxin produced by *C. tetani*; converted into toxoid by formalin treatment
Pertussis vaccine	*Bordetella pertussis* (cause of whooping cough)	Killed suspension of *B. pertussis*
Typhoid vaccine	*Salmonella typhi* (cause of typhoid fever)	Killed suspension of *S. typhi*
Cholera vaccine	*Vibrio cholerae* (cause of cholera)	Killed suspension of *V. cholerae*
Plague vaccine	*Yersinia pestis* (cause of plague)	Killed suspension of *Y. pestis*
Tularemia vaccine	*Francisella tularensis* (cause of tularemia)	Attenuated or killed suspension of *F. tularensis*
Bacille Calmette-Guerin (BCG)	*Mycobacterium tuberculosis* (cause of tuberculosis)	Avirulent BCG tubercle bacilli, living in suspension

Agent	Active Against	Production Method
Typhus vaccine	*Rickettsia prowazekii* (cause of typhus fever)	Suspension of formalin-killed bacteria, grown in the yolk sac of eggs
Rocky Mountain spotted fever vaccine	*Rickettsia rickettsii* (cause of Rocky Mountain spotted fever)	Suspension of formalin-killed bacteria, grown in the yolk sac of eggs
Vaccinia virus vaccine	Variola virus (cause of smallpox)	Living virus, harvested from lymph of calves or sheep
Poliovirus vaccine	Poliovirus (cause of paralytic poliomyelitis)	Living or killed viruses of all three types, grown in monkey kidney cell cultures
Rubella vaccine	Rubella virus (cause of German measles)	Living, attenuated virus grown in rabbit kidney or other cells
Rubeola vaccine	Rubeola virus (cause of measles)	Live, attenuated virus grown in chick embryo cell cultures
Mumps vaccine	Mumps virus	Live, attenuated virus grown in chick embryo cell cultures
Yellow fever vaccine	Yellow fever virus	Live, attenuated virus grown in cell cultures and eggs
Rabies vaccine	Rabies virus	Phenol- or ultraviolet-killed rabies virus, grown in embryonated duck eggs
Influenza vaccine	Influenza viruses	Prevalent types of influenza viruses, grown in egg allantoic fluid and killed with formalin

Glossary

In addition to terms and their definitions, the following list includes the names of organisms referred to in the text. Where the pronunciation of a term or name is not obvious, the authors have attempted to show the commonly used pronunciation within parentheses following the term.

abcess (ab′-sess)   A localized collection of pus surrounded by inflamed tissue.

activated sludge method   A method of sewage treatment in which wastes are aerated and degraded by complex populations of aerobic microorganisms.

activation energy   The energy required to elevate molecules from one energy level where they are non-reactive to a higher energy level where they can react spontaneously.

active site   The sequence of amino acids on an enzyme molecule which is concerned with binding the substrate. Also termed catalytic site.

active transport   The energy requiring process by which molecules are carried across cell boundaries. Often the molecules are moved from a lower to a higher concentration.

adenosine triphosphate (ATP) (a-den′-o-seen tri-fos′-fate)   A chemical compound that is the major storage form of energy in cells. The chemical energy is stored in the form of two high energy phosphate bonds, which when hydrolyzed release energy.

adsorption (viral)   The process of the interaction of a virus with specific receptors on host cells, resulting in retention of the virus on the cell surface.

aerobic organisms (aerobes)   Organisms that can utilize oxygen as the final electron acceptor during metabolism.

agar (ah′-grr)   A polysaccharide obtained from various species of seaweeds, used to gel materials (especially microbiological media). It cannot be broken down by most bacteria.

akinetes (ack′kineats)   Nonmotile resting cells produced by some algae; characterized by a thickened cell wall and food reserves.

alginates (al′-jin-ates)   Polysaccharides derived from some of the brown algae used to aid in emulsifying materials (such as causing suspension of small globules of one liquid in a second liquid with which the first will not mix).

allergy (hypersensitivity)   An immunological response that results in damage to the responder, usually manifested by impaired function of a tissue.

allosteric protein (al-o-steer′-ic pro′-teen)   A protein, commonly an enzyme, which changes its shape or conformation as a result of binding small molecules. The change in shape affects its ability to carry out its catalytic or other function.

allosteric site   The sequence of amino acids on an allosteric protein which binds the small molecules. It is distinct from the active or catalytic site.

ameba (pl. amebae)   A unicellular organism with an indefinite changeable form. Also spelled amoeba.

amino acid (ah-mean′-o acid)   An organic compound containing both a carboxyl (COOH) and an amino ($-NH_2$) group, bonded to the same carbon atom. The 20 amino acids which are the subunits of proteins vary in the structure of their side chains.

aminoglycoside (ah-meé-no-gly′-ko-side)   A group of chemical compounds derived from sugars such as glucose and containing amino groups. Several antibiotics (including streptomycin) are aminoglycosides.

ammonification (a-moan′-i-fi-ka-shun)   The reactions which result in the release of ammonia (NH_3) from organic nitrogen containing molecules.

anaerobic jar (an-air-ro′-bic)   A glass jar from which oxygen can be evacuated to provide an oxygen-free environment. Used in the growth of anaerobic organisms.

anaerobic organisms (anaerobes)   Organisms that do not use oxygen as the final electron acceptor during metabolism; organisms that grow in the absence of air.

anaerobic respiration The metabolic process in which electrons are transferred from one compound to an inorganic acceptor molecule other than oxygen. The most common acceptor molecules are carbonate, sulfate and nitrate.

anamnestic (memory) response (an'-am-nes'-tick) The immunological response to a second or subsequent dose of antigen; characterized by a more rapid, efficient and long-lasting response than resulted following the first (primary) dose of antigen.

anemia (an-eem'-ee-a) A condition characterized by having less than the normal amount of hemoglobin.

anopheline (a-noff'-fa-leen) Refers to mosquitos of the genus *Anopheles*.

antibiotic A chemical substance produced by certain molds and bacteria which inhibits the growth or kills other microorganisms. Some antibiotics can now be synthesized by man and others can be altered by chemical methods so as to enhance their usefullness.

antibody (an'-ti-body) Protein produced by the body in response to the presence of an antigen; can combine specifically with that antigen.

antibody reaction site The part of an antibody molecule that combines with the specific antigen.

anticodon (anti-ko'-don) The three nucleotides in transfer RNA which are complementary to a sequence of three nucleotides in messenger RNA (a codon).

antigen (an'ti-jen) A substance that can incite the production of specific antibodies and can combine with those antibodies.

antigenic determinant The part of an antigen molecule that combines with the specific antibody.

antiseptic A substance that will inhibit or kill microbes. As often used, this term implies that the substance is sufficiently non-toxic to apply to skin or other tissue.

antiserum A serum useful in treatment, or in identification of antigens, because of its content of specific antibody or group of antibodies.

arthrospores (arth'-ro-spores) Fungal spores formed by breaking apart of the cells composing a septate hyphae.

ascus (ask'-us), pl. asci (ass-sigh) Sac-like structure in which fungal spores (ascospores) are borne.

asporogenous (ay-spore-ah'-jen-us) Not producing spores.

asymptomatic (ay-sim-to-mat'-ik) Without symptoms; when used to describe infections, the infected person notices no abnormalities.

atopic (ay-tow'-pick or ay-topic) allergic reactions Allergic (or hypersensitivity) reactions caused by antibodies of the class IgE.

attenuated Modified or weakened; used to describe microorganisms that are modified in such a way as to be nonvirulent but still retain the ability to incite immunity (e.g. live attenuated polio virus is used to immunize against polio).

autoclave A device employing steam under pressure, used for sterilizing materials stable to heat and moisture.

autoimmune Immune or hypersensitive to some constituents of one's own body.

autotroph (aw'-tow-trof) An organism that can utilize CO_2 as its main source of carbon.

auxotroph (ox'-o-trof) Term applied to laboratory derived mutants of a microorganism that requires an organic supplement, other than a carbon and energy source, for growth. For example, an amino acid, purine, pyrimidine or vitamin.

Acetobacter (a-see'-toe-bak'-ter)

Acinetobacter (a-sin-et'oh-bak-ter)

Actinomyces (ak-tin-oh-my'-seez)

Actinoplanes (ak-tin-oh-plane'-eez)

adenoviruses (ad'-e-no viruses)

Aedes (ah-ee'-deez)

Aeromonas (air-oh-moan'-az)

Agrobacterium (ag-roh-bak-teer'-ee-um)

Allomyces (al-oh-my'-seez)

Ancylostoma duodenale (an-kil-lahst'-toh-mah duo-deen-ah'-lee)

Arthrobacter (are-throh-bak'-ter)

Ascaris lumbricoides (ass'-kah-riss lum-brik-koid'-eez)

Aspergillus (ass-per-jill'-us)

autolyze The breaking up or disintegration of a cell, usually by means of its own enzymes.

avirulent (ay-veer'-u-lent) Lacking virulence; when used to describe microbes, means properties which normally promote ability of an agent to cause disease are lacking.

axial filament The structure responsible for motility of spirochaetes. It resembles a flagellum that lies within the layers of cell wall.

axilla (ak-sil'-a) Armpit

Azotobacter (ah-zoh-toh-bak'-ter)

Bacillus subtilis (bah-sil'-us sut'-till-lis)

    B.megaterium (B. meg-a-teer'-e-um)

    B.polymyxa (B. poly-mix'-a)

bacteriocin (back-tear'-e-o-sin) Protein molecules, produced by bacteria, which kill other strains of the same or closely related species. Bacteriocins are named after the organisms which produce them, i.e. enteric bacteria such as *E. coli* produce colicins; *B. megaterium* synthesizes megacins. The chemical structure and mode of action differ for different bacteriocins.

bacteriocin factor (back-tear'-e-o-sin factor) A DNA plasmid or episome which carries the genetic information for the synthesis of a bacteriocin.

bacteriophage (back-tear'-e-o-fayge) A virus that infects bacteria; often abbreviated "phage."

Bacteroides fragilis (back-ter-oid'-eez frah-jil'-us)

bacteroids (back'-tear-oids) Bizarre, irregularly shaped forms which certain bacteria can assume under certain environments; for example, *Rhizobium* in root nodules.

Balantidium coli (bal-an-tid'-ee-um coh'-lee)

barophiles (bear'-oh-files) Organisms that grow only under conditions of high pressure, such as those adapted to grow in ocean depths (where the pressure is as high as 1,000 times the normal at sea level).

Bartonella bacilliformis (bar-ton-ell'-ah bah-sil-ah-for'-miss)

basal body The part of the bacterial flagellum which attaches the flagellum to the cell.

base analogs A chemical whose structure is similar (analogous) to one of the purine or pyrimidine bases found in DNA or RNA.

base pair A purine with its complementary pyrimidine in DNA and RNA. Adenine pairs with thymine, and guanine with cytosine, for example.

basidium (bah-sid'-e-um) (pl. basidia) (bah-sid'-e-ah) A club-like structure on which certain fungal spores (basidiospores) are borne.

Bdellovibrio bacteriovorus (dell-oh-vib'-ree-oh back-ter-oh-vor'-us)

Beggiatoa (bej-ee-a-toe'-ah)

Beijerinckia (by-yer-ink'-ee-ah)

benign (be-nine') Non-malignant. Favorable or neutral rather than serious.

Bifidobacterium (by-fid-oh-back-teer'-ee-um)

binary fission (bye'-ner-e' fish'-en) An asexual reproductive process in which one cell splits into two independent daughter cells.

biochemical mutants A general term applied to mutants which have defects in biochemical pathways; generally applied to mutants defective in the synthesis or degradation of small molecules.

Biochemical Oxygen Demand (BOD) The oxygen-consuming property of water.

686

biological magnification The phenomenon of the increasing concentration of a chemical substance, such as a pesticide, in organisms the higher they are in the food chain.

biomass (bye′-o-mass) The total mass of living organisms in a given volume (e.g. in seawater—the total mass of living organisms per liter). The total weight of all organisms in any particular environment.

biopsy (bye′-opsee) The surgical removal of small pieces of tissue for microscopic examination to aid in diagnosis.

biosphere (bye′-o-sphere) The portion of the earth in which life exists.

Blastomyces dermatitidis (blast-oh-mye′seez der-mah-tit′-id-us)

blastospore (blast′-oh-spore) Reproductive cell formed by budding, as in the yeasts.

bloom (of algae) Overabundant growth of algae, often as a result of enrichment with a nutrient that would otherwise be scarce enough to limit growth.

Bordetella pertussis (bor-deh-tell′-ah per-tuss′-siss)

Borrelia recurrentis (Bor-rell′-e-ah ree-cur-ren′-tiss)

Brucella (bru-sell′-ah)

budding An asexual form of cell multiplication in which protuberances (buds) form at various sites on the parent, enlarge, and develop into new individuals.

calcareous (kal-care′-e-ous) Containing calcium oxide (lime).

Candida albicans (can′-did-ah al′-bik-anz)

capsid (kap′-sid) Protein coat that surrounds and protects the nucleic acid of a virus.

capsomere (kap′-so-mere) A subunit of the protein capsid of a virus.

carbohydrate An organic compound consisting of many hydroxyl

(—OH) groups and containing either a ketone (—$\overset{\overset{\textstyle O}{\|}}{C}$—) or aldehyde

(—$\overset{\overset{\textstyle O}{\|}}{C}$—H) group. Examples include sugars, cellulose, glycogen, starch.

carcinogen (car-sin′-o-jen) A cancer-producing substance.

carcinogenesis (car-sin-oh-jen′-uh-sis) The induction of cancer.

carrier An individual who has pathogenic microbes in or on his body without showing any signs of illness. The carrier state occurs during incubation and convalescence of infectious disease, and with asymptomatic infection, colonization or contamination. As usually used, the term implies that the microbes have access to the exterior of the body and thus potentially to other people.

catabolism (ka-tab′-o-lism) The metabolic degradation of organic compounds.

catabolite repression (ka-tab′-o-lite repression) The inhibition of genes governing a pathway of degradation by substances that can be efficiently utilized as sources of energy.

catalase (kat′-a-laze) An enzyme, found in man and many microorganisms, which degrades hydrogen peroxide to oxygen and water.

Caulobacter (caw-loh-back′-ter)

cell-bound antibodies Antibodies that are attached to cells, either because they were not released from the cells that formed them or because they have attached to other cells.

cellulose A polysaccharide, composed of glucose subunits; the most abundant organic compound in the world.

central dogma The colloquial phrase given to a major theme of molecular biology—that the flow of information follows the pattern: DNA to RNA to protein.

Cephalosporium (sef-alloh-spore′-ee-um)

cercaria (ser-care′-ee-ah) The last developmental stage of the free larvae of flukes, distinguished by possessing a tail.

chelate (key′-late) To combine with a metal in weakly dissociated complexes in which the metal is part of a ring.

chelicera (kel-iss′-serah) Appendages close to the mouth of arachnids; evolutionary equivalent of pinchers.

chemostat (keme′-o-stat) A growth chamber with an inlet and outlet which allows a controlled continuous addition of nutrients and outflow of cells.

chert A dull colored, flintlike quartz often found in limestone.

chitin (kite′-in) A polysaccharide found in the cell walls of fungi and in the external covering of arthropods and crustaceans.

Chlamydomonas (klah-mid′-o-moan′-as)

chlamydospore (klah-mid′-oh-spore) A form of asexual spores of some fungi, characterized by a thickened wall and by resistance to adverse environmental conditions.

chocolate agar Blood agar which has been heated under carefully controlled conditions so as to release hemin and destroy an enzyme which breaks down NAD. The heating causes the blood to assume a chocolate color.

Chlamydia (klah-mid′-ee-ah)

chloramphenicol (klor-am-fen′-i-kol) A bacteriostatic antibiotic produced by a species of *Streptomyces* which inhibits protein synthesis in a variety of gram positive and gram negative organisms.

chorioallantoic membrane (kore′-e-oh-allan-tow′-ick membrane) The outer (or) one of the membranes surrounding an embryo such as in an embryonated chicken egg, often used to grow viruses.

chlorophylls (klor′-o-fills) Green pigments necessary for photosynthesis.

Chondrococcus columnaris (kon-droh-kock′-us cahl-um-nare′-iss)

chromatographic column (crow-mat′-oh-graf′-ick column) An absorbent material packed into glass columns through which mixtures of substances are passed; separation of components of the mixture may be achieved by selective absorption to different parts of the absorbent material or by differences in the ease by which the substances are removed (eluted) from the absorbent material.

cilia (sill′-e-a) Hairlike appendages on certain eucaryotic cells, which may serve to move the cell.

ciliates (sill′-e-ates) Protozoa that bear multiple cilia, for example the paramecia.

cilium (cilia) A short, projecting hairlike organelle of locomotion, similar to a flagellum but thicker and having different internal structure.

clone (klohne) A family of cells derived from a single parental cell by repeated divisions. The offspring of a single cell.

Clostridium botulinum (Kloss-trid′-ee-um bot-u-line′-um)

   C. perfringens (C. per-fringe′-enz)

   C. tetani (tet′-an-ee)

Coccidioides immitis (Kock-sid-ee-oid′-eez im′-mit-iss)

coenocytic (seen′-oh-sit′-ick) Having many nuclei within a single-celled organism.

coenzyme (co-en′-zyme) A small, organic molecule, that transfers small molecules from one enzyme to another.

colicin (kole′-i-sin) One kind of bacteriocine; a protein produced by one strain of enteric bacteria which kills other strains of enteric bacteria.

coliform (cole′-e-form) Gram-negative rods, including *Escherichia coli* and similar species, which normally inhabit the colon; commonly included in the coliform group are *Enterobacter aerogenes, Klebsiella* sp., and other related bacteria.

colloid (koll′-oid) A suspension of finely divided particles, approximately 5 to 5,000 Å in size, which do not settle rapidly and are not readily removed by filtration.

colonization Establishment of a site of reproduction of microbes on a material, animal or person without necessarily resulting in tissue invasion or damage.

combined therapy The term applied to the simultaneous use of more than one technique in the treatment of a disease; often refers to the use of several antibiotics simultaneously.

commensalism (koh-men'-sa-lism) State of living on or in an organism without causing harm or giving benefit to the organism.

community All of the organisms inhabiting a common habitat and interacting with one another.

competent cell A bacterial cell which is capable of taking up and integrating high molecular weight DNA into its own chromosome in DNA-mediated transformation.

competitive inhibition The inhibition of enzyme activity caused by the competition of the inhibitor with the substrate for the active (catalytic) site on the enzyme. Impairment of function of an enzyme due to its reaction with a substance chemically related to its normal substrate.

complement (C) A system of at least 11 serum proteins that act in sequence producing certain biological effects concerned with inflammation and the immune response.

complex lipids Lipids which contain other elements in addition to hydrogen, oxygen and carbon.

compost heap A pile of organic plant materials, commonly leaves and grass, which undergoes decomposition by microorganisms.

congener (kon'-jen-er) A thing closely related to another; in a chemical sense, a structure closely related in form to another chemical structure.

conjugation A mechanism of gene transfer in bacteria which involves cell to cell contact.

conjunctiva (kon-junk-tye'-vah) The mucous membrane that lines the inner surface of the eyelids and exposed surfaces of the eyeball.

conidia (sing., conidium) (cone-id'-e-ah) The reproductive spores of fungi.

conidiospores (cone-id'-e-oh-spores) Fungal spores borne on specialized structures called conidiophores.

constitutive enzyme (kun-stit'-u-tive) An enzyme whose synthesis is not altered in response to changes in the environment.

contagious (kun-tay'-jeeus) Infections transmissible by contact; catching.

coralline (kor'-rahl-lean) Coral-like; refers to a class of coral-like algae with hard, calcareous cell walls (the class Corallinaceae).

corals (kor'-rahls) Primitive, calcareous plants that grow principally in tropical waters.

core enzyme of RNA polymerase The portion of the enzyme which is specifically concerned with catalyzing the formation of ester bonds between nucleotides. It has no specificity as to where it initiates or stops transcribing.

corepressor (ko'-ree-press-sore) The small molecule which activates the repressor (regulator) protein to regulate gene function.

corticosteroid (kor-tik-oh-stee'-roid) A medication or hormone which resembles in structure and action corticosterone, a hormone produced by the outer portion of the adrenal gland.

Corynebacterium diphtheriae (koh-ryne-nee-bak-teer'-ee-um dif-theer'-ee-ee)

C. hoffmani (C. hoff'-man-ee)

C. xerosis (C. zeer-oh'-siss)

covalent bond (ko-va'-lent) A strong chemical bond formed by the sharing of electrons.

Coxsiella burnetii (cox-see-ell'-ah bur-net'-ee-ee)

crustacean (krus-tay'-sheean) A member of the class Crustaceae; which includes water fleas, shrimp, crabs and similar forms.

Culex (ku'-lex)

cysts (sists) Dormant resting cells characterized by a thickened cell wall.

cytochromes (site'-o-chromes) Iron-containing substances that help convey electrons to molecular oxygen in the electron transport chain; they function in the generation of ATP.

cytopathic effects (CPE) (site-oh-path'-ick) Observable changes in cells *in vitro* produced by viral action; for example, lysis of cells or fusion of cells.

Cytophaga (sye-toff'-ah-gah)

cytoplasmic membrane The flexible structure immediately surrounding the cytoplasm in all cells.

dalton (doll'-ton) A unit of weight equal to the weight of a single hydrogen atom.

darkfield microscopy A microscopic technique in which no light is visible except that which is reflected from an object; this results in the object being light and the background dark.

DDT An insecticide, dichlorodiphenyl trichloroethane, that is toxic to man and other animals when swallowed or absorbed through the skin.

death phase The stage in which the number of viable bacteria in a population decreases at an exponential rate.

deciduous Having parts, such as leaves, that fall off at a certain season.

decomposers Heterotrophic microorganisms that decompose organic substances, with the production of inorganic materials.

decontamination Removing or inactivating pathogenic microbes and their toxic products.

defective virus A virus that lacks a part of its nucleic acid which consequently cannot direct the replication and release of complete, infective virions.

Demodex folliculorum (dem'-oh-dex foh-lik-u-lore'-um)

denature To destroy the native configuration of a macromolecule, such as protein or nucleic acid.

denitrification (de-nite'-trif-fi-ka-shun) The process by which nitrate is reduced to nitrogen gas.

density The ratio of the mass of a substance to its volume.

deoxyadenosine monophosphate (dAMP) One of the subunits of DNA, a mononucleotide in which the sugar is deoxyribose. The other subunits of DNA are abbreviated dTMP, dGMP, and dCMP.

deoxyribose (de-ox'-e-rye-bose) A five carbon sugar found in DNA.

dextran (dex'-tran) Branched polysaccharide of D-glucose subunits; often found as a storage product in bacteria and yeast.

deoxynucleoside triphosphates (de-ox'-e-new-klee-o-side try-fos'-fates) Purine and pyrimidine bases, bonded to the sugar deoxyribose which in turn is bonded to a string of three phosphate molecules.

dermis The portion of the skin lying just beneath the surface layers and containing nerves, blood vessels, and connective tissue.

diabetes Condition characterized by excessive urine production. Sugar diabetes is the colloquial term for diabetes mellitus, a condition marked by deficiency of the hormone insulin and consequent elevation of blood and urine glucose.

diaminopimelic acid (dye'-uh-mean'-oh-pie-meal'-ick acid) A chemical found in nature only in procaryotic organisms, particularly in the cell-wall mucocomplex of bacteria.

diatomaceous earth (dye-uh-tow-may'-see-ous) (diatomite, kieselguhr) Deposits of empty cell walls of diatoms; this extremely hard silica-containing material has many practical uses.

Dermacenter andersoni (der-mah-sen'-ter an-der-soh'-nee)-

Desulfovibrio (dee-sull-foh-vib'-ree-oh)

diatoms (dye'-uh-toms) Algae with hard silica-containing cell walls; members of the class Bacillariophyceae, phylum Chrysophyta.

diauxic growth (dye-ox'-ik) The two-stage growth curve displayed by certain bacteria when grown in a medium which contains two different compounds which can serve as sources of energy.

diffusion The movement of molecules from a region of high concentration to a region of low concentration.

digester A large tank in which sedimented wastes are degraded by anaerobic microorganisms.

dimer (dye′-mer) The molecule resulting from the association of two identical subunits.

dimorphic (dye-more′-fick) Having two forms, as in the fungi that may exist either as molds or in yeast-like forms.

dinoflagellates (dye-no-fla′-jell-ates) Algae of the class Pyrrhophyta, primarily unicellular marine organisms.

Diphyllobothrium latum (dye-fil-oh-bah′-three-um lah′-tum)

diploid (dip′-loyd) Refers to a cell in which there are two copies of each type of chromosome (2N); each chromosome of a pair originates from a chromosome of one of the parent sexual cells that fused to begin the line of diploid cells.

direct selection The technique of selecting mutants by plating organisms on a medium on which the desired mutants will grow but the parent will not.

disaccharide (dye-sack′-a-ride) A carbohydrate molecule consisting of two monosaccharide molecules joined together.

disease A process resulting in tissue damage or alteration of function, producing symptoms, or noticeable by laboratory or physical examination.

disinfection Killing pathogenic microbes on or in a material without necessarily sterilizing it. Use of this term usually implies that a liquid or gaseous chemical agent is employed for the microbial killing.

DNA polymerase (DNA poll-lim′-er-race) The enzyme which catalyzes the formation of DNA. A number of different DNA polymerase enzymes exist, and the function of each *in vivo* is not known.

donor cell The bacterial cell which transfers its chromosome, or a portion of its DNA, to another cell (the recipient cell).

dorsoventrally In the direction from back to front; e.g., in vertebrates, from vertebral column to the abdomen.

doubling time The time required for the number of cells in a population to double in number; essentially the same as the generation time.

DNA hybridization A technique for assessing the relatedness between DNA molecules. The strands of DNA are pulled apart (denatured) and their ability to associate (renature) with the single stranded DNA from another organism is measured. Only DNA strands with complementary base sequences will associate.

eclipse The period during which virus exists within the host cell as free nucleic acid.

ecological niche The particular role that a species plays in the activities of the ecosystem, as well as the physical space occupied by an organism.

ecology The study of the relationships of organisms to one another and to their environment.

ecosystem (ee′-ko-system) An environment and the organisms which inhabit it; for example, a lake with all its inhabitants (from bacteria and algae to fish) make up a particular ecosystem. All of the organisms in a habitat plus all of the factors in the environment with which the organisms interact.

ectoparasite A parasite that lives or feeds on the outer surface of the host's body.

eczema A condition characterized by a blistery skin rash, with weeping of fluid and formation of crusts, usually due to an allergy. Eczema vaccinatum is the disease resulting from infection of a person with eczema by vaccinia virus, the agent used to immunize against small pox.

eczema vaccinatum (eks-e′-mah vak′-sin-ah-tum) A sometimes fatal disease characterized by a generalized pox eruption that results from vaccinia virus infection of a person with eczema; a common consequence of exposure of a patient with eczema to someone recently vaccinated vs. smallpox.

electron A subatomic particle of negative electrical charge that orbits the positively charged nucleus of an atom. For maximum stability, an atom must have a certain number of electrons in its outermost orbit.

electrophoresis (e-lek′-tro-for-ee′-sis) A technique for separating proteins based on the fact that different proteins have different electrical charges and so will move at different rates in an electrical field.

encephalitis (en-sef-ah-lye′-tis) Inflammation of the brain.

endemic Present in a community at all times, but only in small numbers of cases.

endospore (en′-dough-spore) A very resistant form of a resting bacterial cell which develops from a vegetative cell by a series of biochemical reactions termed sporulation.

end product The chemical compound which is the final product in the sequence of a metabolic pathway.

end product inhibition Also termed "allosteric inhibition" or feedback inhibition, the inhibition of the first enzyme of a biosynthetic pathway by the end product which combines with the allosteric site of the enzyme, thereby changing its shape and altering its enzymatic activity.

end product repression The regulation of gene activity by a process in which the end product of a biosynthetic pathway activates the repressor molecule which then combines with the operator region and prevents the functioning of RNA polymerase.

energy The ability to do work.

enrichment culture A technique used for isolating an organism from a mixed culture by manipulating conditions so as to favor growth of the organism sought while minimizing growth of the other organisms present.

Entamoeba histolytica (en-tah-mee′-bah his-toh-lit′-ik-ah)

Enterobacter aerogenes (enter-oh-bak′-ter air-ah′-jen-eez)

Enterobius vermicularis (enter-oh′-bee-us ver-mik-u-lah′-ris)

Epidermophyton (epee-der-moff′-fit-ton)

envelope (viral) The outer lipid-containing layer possessed by some virions; obtained from modified host cell membranes when the virus leaves the cell.

envelope (of bacterial cell) The portion of the cell that encloses the cytoplasm, including the cytoplasmic membrane, cell wall, and capsule if present.

enzyme An organic catalyst; a protein molecule which lowers the activation energy of substrates allowing them to react at temperatures compatible with life.

epidemic The presence of many cases of a disease within a region.

epidermis (epee-der′-miss) The outermost skin layers.

episome (ep′-pi-soam) A particle of functional DNA which can exist either covalently bonded to the bacterial chromosome or in an extrachromosomal state.

equine Pertaining to horses.

Erwinia (er-win′-ee-a)

Erysipelothrix (air-ree-sip′-ell-oh-thrix)

erythrasma (air-ree-thraz′-mah) A painless condition involving moist areas of the skin, characterized by redness or brown discoloration, and large numbers of the bacterium *Corynebacterium minutissimum*.

erythrocytes (e-rith′-row-sites) Red blood cells.

erythromycin (e-rit-throw-my′-sin) An antibiotic, produced by a strain of *Streptomyces,* which inhibits protein synthesis primarily of gram-positive organisms.

Escherichia coli (esh-er-ik′-ee-a coh′-lee)

ester bond   A bond, found in lipids and nucleic acids, in which a hydroxyl group (—OH) is bonded to a carboxyl group (—COOH) with the removal of HOH.

eucaryotic cell (you-carry'-aw'-tick)   A complex cell type, characterized by having a nuclear membrane, mitochondria and numerous chromosomes. All living cells except bacteria and blue greens are eucaryotic.

eutrophication (you'-trow-fi-kay'-shun)   Nutrient enrichment leading to over-production of algae.

excise   To cut out.

extrachromosomal   Not associated with the chromosome, applied to episomes and plasmids in bacteria, and to DNA in eucaryotic organelles.

exudate   A material such as pus composed of fluid from the vascular system, cells and sometimes products of tissue breakdown.

F prime cell (F')   A bacterial cell which carries an extrachromosomal F particle to which is attached a fragment of chromosomal DNA.

facultative organisms   Those organisms able to carry out both aerobic and anaerobic metabolism; may also be used in terms of other properties such as photosynthesis (e.g. facultatively photosynthetic organisms can use either photosynthesis or other means of gaining energy).

fastidious   Exacting; used to refer to organisms that require many growth factors.

fats   Simple lipids consisting of esters of glycerol with fatty acids.

F^+ cell   A bacterial cell that can transfer the extrachromosomal F (fertility) particle to a recipient (F^-) cell.

F^- cell   A bacterial cell that does not contain an F particle, but can act as a recipient and receive one from an F^+ cell.

feedback inhibition   Another term for allosteric or end product inhibition. The inhibition of the first enzyme of a biosynthetic pathway by the end product of the pathway.

fermentation   The metabolic process in which the final electron acceptor is an organic compound.

fission (fish'-en)   An asexual reproductive process in which one cell splits into two or more independent daughter cells; in binary fission two daughter cells are produced.

fixation   The combination of an element with other atoms in covalent linkage; for example, the combination of nitrogen gas with hydrogen to form NH_3, or of carbon into organic compounds.

fixed nitrogen   The form of nitrogen when it is combined with other elements, i.e. NH_3, $—NO_2^-$, NO_3^-.

flagellates (fla'-jell-ates)   Organisms that bear flagella; the term is usually applied to protozoa or algae.

flagellum, pl. flagella (fla-jell'-um)   A long whiplike organelle of locomotion made up of intertwined molecules of protein.

Flavobacterium meningosepticum (flay-voh-bak-teer'-ee-um men-injoh-sep'-tik-um)

flavoproteins (flay-vo-pro'-teens)   Enzymes which form a part of the electron transport scheme. The flavin portion of the molecule is a coenzyme which is a derivative of the vitamin riboflavin.

fluke   A short flattened parasitic worm of the class *Trematoda*.

follicle   A very small excretory sac such as the ones from which body hair protrudes.

fomite (foh'-mite)   An inanimate object, such as a book, tool or towel which can act as a transmitter of pathogenic microbes even though not supporting their growth.

forespore   One of the more readily identified stages in the process of sporulation, identified as a refractile body which is not yet resistant to heat.

Francisella tularensis (fran-siss-sell'-ah tu-lah-ren'-siss)

free energy (G)   Energy that is available to do work, particularly in causing chemical reactions.

fruiting body   Specialized reproductive structures containing many spores; examples are mushrooms (the fruiting bodies of certain basidiomycetes) and the fruiting bodies of slime molds and some myxobacteria.

frustules (frus'-tewles)   Hard, silica-containing cell walls of diatoms.

Fusobacterium (fuzoh-bak-teer'-ee-um)

galactose (ga-lak'-toes)   A six carbon sugar; together with glucose it forms the disaccharide lactose (milk sugar).

gametes (gam'-eats)   Haploid cells which fuse with other gametes to form the diploid zygote in sexual reproduction.

gametocyte (gam-ee'-tow-site)   Cell that produces or becomes a mature sexual cell.

gastroenteritis (gas'-trow-en-ter-ite'-is)   Inflammation of the gastrointestinal tract, usually characterized by diarrhea, nausea, and vomiting.

gene   A portion of a chromosome which carries the genetic information for the synthesis of one polypeptide chain.

generation time   The time required for the population of cells to double in number; also termed division time.

genus (gene'-us)   A category of related organisms, usually containing several species; the first name of an organism in the Binomial System of Classification.

genetic code   The composition and sequence of all of the sets of three nucleotides which code for the amino acids. The sequence of three nucleotides which codes for one amino acid is termed a code word.

genetic recombination   The process by which a segment of DNA of one chromosome is integrated into the DNA of another chromosome.

genome (gene'-ome)   The total genetic information needed to reproduce a cell or a virus. In bacteria, it is equivalent to one chromosome.

genotype (gene'-o-type)   The sum total of the genetic constitution of an organism. Much of the genotype is not expressed at any one time.

germicide   An agent which kills microbes.

germinate   To begin to grow.

germination   The sum total of the biochemical and morphological changes that an endospore or other resting cell undergoes, prior to its becoming a vegetative cell.

Giardia lamblia (jee-are'-dee-ah lamb'-lee-ah)

glomerulonephritis (glom-er-ulo-nef-rye'-tiss)   A kidney disease characterized by inflammation of the capillary loops through which portions of blood plasma are normally filtered.

glucose   A six carbon sugar; also called dextrose.

glucose effect   The inhibitory effect of glucose on the synthesis of degradative enzymes concerned with energy metabolism in a variety of pathways.

glycosidic bond (gly-ko-sid'-ik)   A bond in carbohydrates which links a hydroxyl group (—OH) of one sugar with the aldehyde group

$$\overset{O}{\overset{\|}{(—C—H)}}$$ of another.

Gonyaulax (gone-ee-aw'-lax)

growing point   The site on the DNA molecule at which replication is taking place.

growth curve   The graphic representation of the change in numbers of organisms in a culture with time.

growth factor   An organic compound other than the carbon and energy source which an organism requires and cannot synthesize, e.g. a vitamin, amino acid, FAD, etc.

Haemophilus influenzae (hee-moff'-ill-us in-flew-en'-zee)

half-life ($T_{1/2}$)   Time required for the disappearance of half of a sam-

690

ple; with reference to immunoglobulins, the time required for the disappearance from the blood of half of the molecules in a sample injected into the bloodstream.

Halobacterium (hah-loh-bak-teer'-ee-um)

halophile (hale-oh-file)

halophilic organism Literally, means "salt-loving" organism which will grow in a medium to which has been added a concentration of NaCl which inhibits the growth of other bacteria; obligate facultative halophiles will grow in NaCl but do not require it; obligate halophiles require NaCl; extreme halophiles will grow only in media where the NaCl concentration is very high.

haploid (hap'-loyd) The chromosome state in which each type of chromosome is represented only once (1N), although there may be several copies of the same chromosome. All procaryotic cells are haploid, as are the sexual cells of eucaryotic forms.

hapten (hap'-ten or hap'-teen) Substance that can combine with specific antibodies but cannot incite the production of those antibodies unless it is attached to a large carrier molecule.

haustorium (pl. haustoria) (house-tore'-ee-um) Tiny root-like projections of fungi; function to increase the surface area and aid in uptake of nutrients by absorption.

heat shock The short heat treatment, usually 60–70°, given to bacterial endospores to induce the process of germination.

hemadsorption (heem-ad-sorp'-shun) The adsorption of red blood cells to host cells infected with certain viruses. Hemadsorption results from incorporation into the host cell membrane of viral proteins which combine with the red blood cells.

hemagglutination (heem'-uh-gloot-uh-nay'shun) Agglutination, or clumping together, of red blood cells.

hemolysin (hee-mah'-liss-in) A substance which disrupts the membrane of red blood cells, causing them to release their hemoglobin.

herbicide A chemical which is used to kill weeds. However, many herbicides are not selectively toxic, and therefore they will kill most plants and animals if not carefully applied.

hermaphroditic (her-maff-roh-dit'-ik) Possessing both male and female sex organs.

herpes simplex (her'-peez sim'-pleks) A blistery localized skin rash caused by herpes simplex virus, usually located on the lip (cold sore, fever blister) or the genitalia.

herpes viruses (herp'-eez viruses)

herpes zoster (her'-peez zos'-ter) A localized blistery infection of the skin supplied by one or a group of sensory nerves, sensory nerve ganglia, and sometimes other tissues, often painful, by the varicella-zoster virus.

heterotroph (organotroph); heterotrophic organism (het-uh-row-trowf) An organism which obtains energy from organic compounds. An organism that utilizes an organic compound as its main source of carbon.

hexose The general term for a six carbon sugar.

Hfr cell (high frequency of recombination) A strain of bacterium that has the ability to transfer its chromosome to an F⁻ cell. The F particle is integrated into the chromosome of the Hfr cell.

high energy bond A bond which yields a large amount of energy when the bond is broken. This energy can be harnessed to do work.

histamine (hiss'-ta-meen) A low molecular weight substance found in an inactive form in certain animal and plant tissues that upon release may cause relaxation of the walls of blood vessels, increased permeability of the vessels, passages of the lung, and other effects. It is one of the substances playing a role in the inflammatory process.

Histoplasma capsulatum (hiss-toh-plaz'-mah cap-su-lah'-tum)

homologous Corresponding in structure, position and origin; refers to DNA molecules which have identical base sequences and therefore arise from the same species.

hook One of the structural components of the bacterial flagellum.

hormogones (horm-oh-gohnes) or hormogonia Short rows of filamentous cells that function as reproductive elements; the hormogones break off from old filaments and glide away to form new filaments.

horizons The layers of soil, generally three, that are distinguishable when soil is examined in cross section.

host An organism on or in which smaller organisms or viruses live, feed or reproduce. When dealing with parasites having complex life cycles, the host in which the adult lives, or the one in which sexual reproduction takes place, is called the *definitive host*. A host harboring larval or asexually reproducing forms is called an intermediate host.

humoral antibodies (hume'-uh-ral) Antibodies in body fluids, such as blood.

humus (hew'-mus) A dark colored, highly complex organic material in soil, that is not readily degraded by microorganisms.

hyaluronic acid (hye-al-you-ron'-ik) A mucopolysaccharide gel which fills intercellular spaces in tissue, and also occurs as a capsular substance of some bacteria.

hypertonic (hye-per-ton'-ik) A fluid having an osmotic pressure greater than another fluid with which it is compared.

hydrocarbon Any compound composed of only hydrogen and carbon.

hydrogen atom This atom consists of one proton, and one electron which moves in orbit around the electron. Weight = 1 dalton.

hydrogen bond A weak attraction (bond) between an atom which has a strong attraction for electrons and a hydrogen atom which is covalently bonded to another atom that attracts the electron of the hydrogen atom.

hydrolytic reaction A reaction in which a molecule reacts with water and is broken down into two or more smaller molecules. AB + HOH → AOH + HB.

hydrophobic A substance which lacks an affinity for water.

hydrophobic bonds Bonds which arise as a result of the tendency of polar water molecules to exclude nonpolar molecules. The nonpolar molecules associate with one another (by van der Waals forces).

hypersensitivity reaction (allergic reaction) An immunological reaction that causes impaired functioning of the host making the immunological response.

hypha (pl. hyphae) (high'-fah, high'-fee). Filament of fungal cells. A single filament of a fungus. A large number of hyphal filaments (hyphae) constitute a mycelium.

icosahedral symmetry (eye-coh-suh-heed'ral) A symmetry characterized by a structure which has 20 sides, and each side is an equilateral triangle.

imino The group N—H; the amino acid proline has this group and is an imino acid.

"immune" (lysogenic bacteria) Bacteria which are not susceptible to a particular temperate phage because they carry that phage in prophage form. The "immunity" depends on the presence of a repressor whose synthesis is directed by the prophage.

immune cells Lymphoid cells (usually lymphocytes) capable of participating in cell-mediated immunity.

immune response Specific response to an antigen by production of humoral antibodies or immune cells. Primary—the response which occurs after first exposure to an antigen. Secondary—the response which occurs after second or subsequent exposure to an antigen (see anamnestic response).

immunity State of protection; for example, state of protection against the mumps virus. Natural or innate immunity—protection that results from the genetic make up of the host; for example, domestic animals have an innate immunity to mumps virus. Acquired immunity—immunity gained as a result of exposure to an agent; for example, immunity to mumps virus is usually acquired (actively) by response to infection with the virus, but it may also be acquired (passively) by the administration of specific antibodies formed by another host.

immunofluorescence (im-mune′-oh-flewo-res′-ence) Fluorescence resulting from a reaction between a substance and specific antibodies that are bound to a fluorescent dye.

immunoglobulins (im-mune-oh-glob′-yew-lins) Antibodies or antibody-like protein molecules, usually part of the gamma-globulin portion of blood.

immunological enhancement (im-mune-oh-lodge′-e-kul) Enhanced survival of incompatible grafts (tumors or normal tissue) caused by specific humoral antibodies.

immunological tolerance (immunological unresponsiveness) Failure to respond to a potential antigen; lack of response to a specific substance that would normally cause a response.

immunosuppression Nonspecific inhibition (suppression) of the ability to make an immune response, resulting in an overall depression of the response to all or most antigens.

impetigo (im-pa-tye′-go) A bacterial disease of the skin, often highly contagious, characterized by small blisters, weeping of fluid and formation of crusts.

inclusion body Round, oval and sometimes irregular bodies occurring in the cytoplasm or nuclei of cells infected with certain viruses or other intracellular parasites.

indicator agar (differential medium) Agar containing components that can be modified in distinctive and readily recognized ways by specific microorganisms. The medium serves to differentiate between various organisms.

indirect selection The selection of mutants by determining which organisms do not grow on a certain medium; the mutants do not grow, the parents do grow.

inducer A substance responsible for activating certain genes, thereby resulting in the synthesis of new proteins.

inducible enzyme An enzyme that is synthesized only in response to a particular substance in the environment, the inducer.

induction (viral) The process by which a prophage is excised from the host cell DNA.

induration (in′-dew-ray′-shun) Thickening.

inert (noble) elements An element whose outer electron shell is filled; therefore it does not form bonds with other elements.

infection Invasion of tissues (including skin or mucous membranes) by microbes with or without the production of disease.

infective dose (ID_{50}) The dilution of virus at which 50% of inoculated hosts are infected.

inflammatory response A nonspecific response to injury, characterized by redness, heat, swelling, and pain in the affected area.

infusion The liquid extract of any material soaked in water.

integration With reference to DNA, the covalent bonding of DNA from a donor organism with the recipient chromosome.

interferons (in-ter-fear′-ohns) Proteins that are produced by virus-infected cells (or by cells in response to certain other stimuli) that prevent the replication of viruses in other cells.

in vitro Literally means in glass; in culture; outside of the body.

in vivo In the body.

insecticide Any substance that kills insects; if applied improperly it may kill other organisms as well.

inspissation (in-spiss-say′-shun) Making a fluid thicker by evaporating volatile components.

iodophore (aye-o′-do-fore) A complex of iodine with a carrier molecule that liberates free iodine in solution. Such compounds are used as antiseptics.

ion An atom or group of atoms which has gained or lost an electron and therefore carries a negative or positive charge.

isomers Two compounds, which have identical chemical compositions, but differ in configuration, in that one is the mirror image of the other.

Isospora belli (eye-soss′-pore-ah bell′-ee)

isotope Two forms of the same element which have the same atomic number but different atomic weights. One of the forms is unstable and decomposes spontaneously, so that its presence can be readily detected.

jaundice (jawn′-diss) A condition characterized by abnormally large amounts of certain hemoglobin breakdown products in the blood, skin and mucous membranes, giving them a yellow color.

kinetic energy Energy which results from motion.

kinetics (kin-et′-icks) Used to describe the dynamics of ongoing processes, such as growth of bacteria or enzyme reactions.

kinin (kye′-nin) A group of peptide substances found in blood and tissue in inactive form, which upon activation play a role in causing inflammation, blood clotting and other effects.

Klebsiella (kleb-see-ell′-ah)

label A marker, commonly a radioactive atom which allows the location of the molecule to be monitored.

Lactobacillus bulgaricus (lack-toe-ba-sil′-lus bull-gar′-e-kus) *L. casei* (L. case′-e-eye)

lactose A sugar (disaccharide), consisting of one molecule of glucose and one of galactose.

lag phase The stage in the growth of a bacterial culture characterized by extensive synthesis of nucleic acid and protein; but no increase in the number of viable cells.

lamella (la-mel′-la) A layer of internal cell membranes as organelles of photosynthesis.

larva (lar′-va) An immature form of certain animals which differs morphologically from the adult form.

larynx (la′-rinks) Area between the upper end of the windpipe and the throat and containing the vocal cords; "voice box."

lecithin (less′-sith-in) A phospholipid found in animal tissues.

legume (leg′-ume) A group of plants (including peas, beans and clover) with pods that split in two when mature; some can develop a symbiotic relationship with nitrogen-fixing bacteria.

Leptospira (lep-toh-spire′-ah)

lethal dose (LD_{50}) The dilution of virus or other test agent which kills 50% of inoculated hosts.

leukocytes (lou′-ko-sites) White blood cells. There are several morphological types: polymorphonuclears (polys), lymphocytes, eosinophiles, monocytes and basophiles.

lichen (like′-en) A fungus and an alga growing together to form a single organism. An alga and a fungus living symbiotically.

lignin (lig′-nin) A hard, complex carbohydrate which is found in plant cell walls.

lime A basic material, the oxide of calcium.

lincomycin (link-o-my′-sin) A bacteriostatic antibiotic produced by *Streptomyces* which inhibits protein synthesis.

lipid Any of a diverse group of organic substances which are relatively insoluble in water but soluble in alcohol, ether, chloroform, or other fat solvents.

lipopolysaccharide (lip′-o-poly-sak′-a-ride) The macromolecule formed by the covalent attachment of lipid with polysaccharide.

lipoproteins (lip-o-pro′-teens) A weak combination of lipid and protein.

Listeria (liss-teer′-ee-ah)

692

log phase   The stage of growth of a bacterial culture when the cells are multiplying exponentially.

lumen (lou'-men)   The cavity within a tubular structure such as the intestine, blood vessel or windpipe.

lymphocyte (limf'-oh-site)   Small, round or oval lymphoid cell with a large nucleus and scanty cytoplasm; lymphocytes are inactive cells but when stimulated they can transform into active, dividing cells; found in blood and lymph, and also in many body tissues.

lymphoid (lim'-foid)   Pertaining to lymphocytes or the lymphatic system, as lymphoid tissue = tissue composed largely of lymphocytes, and lymphoid cells = cells characteristic of lymphoid tissue.

lymphoma (lim-fo'-ma)   A tumor or neoplastic disorder of lymphoid tissue.

lysogenic (bacteria) (lye-so-jen'-ick)   Bacteria which are carrying a prophage of a temperate virus.

lysogenic conversion   The change in the properties of bacteria as a result of their carrying a prophage; for example, the conversion of non-toxin producing bacteria to toxin-producing bacteria by the action of a prophage.

lytic (virus) (lit'-ick)   A virus that replicates within a host cell and causes lysis (death and disruption) of the host cell.

macrolides (mak'-roh-lydes)   A group of antibiotics such as erythromycin, composed of compounds having a large cyclical sturcture.

macromolecule (mak-roh-mol'-e-kule)   A large molecule made up of subunits bonded together in a characteristic fashion.

macrophage (mak'-roh-fayge or mack'-row-fahge)   A large mononuclear phagocyte; macrophages are found in virtually all parts of the body.

malaise (ma-layze')   A vague feeling of being ill.

malignancy (ma-lig'-nan-cy)   A progressive increase in virulence or seriousness of any pathologic condition leading eventually to death.

mannose (man'-nose)   A six carbon sugar.

maturation (viral)   The assembly of complete, infective virions from viral components replicated separately within a host cell.

meiosis (my-o'-sis)   The process by which the chromosome number is reduced from diploid (2N) to haploid (1N). Segregation and reassortment of the genes occur in this process and gametes or spores may be produced as the end product of meiosis.

meningitis (men-in-jye'-tiss)   Inflammation of the meninges, membranes that cover the brain and spinal cord.

merozoite (mare-o-zo'-ite)   The youngest and smallest of the developmental forms of malarial parasites growing in the mammalian host.

messenger RNA (mRNA)   A type of RNA which transcribes the genetic information of the DNA of the cell; mRNA serves as a template for protein synthesis.

metachromatic   Staining different colors with the same dye. For example, *C. diphtheriae* may stain blue with red inclusions when the dye methylene blue is applied.

metachromatic granules   Granules inside certain bacteria which stain a different color than the dye used for staining, as red with a blue dye. Some are known to consist of long chains of phosphate molecules termed volutin.

metazoan   Refers to members of the metazoa, multi-cellular animals which show tissue differentiation.

Methanobacterium (meth-ain-oh-bak-teer'-ee-um)

Methanococcus (meth-ain-oh-kok'-us)

Methanosarcina (meth-ain-oh-sahr'-sin-ah)

microaerophile (my-kroh-air'-oh-file); **microaerophilic organism**   Organism that requires low concentrations of oxygen for growth (lower than the concentration in air). Sometimes used to indicate organisms that can grow in air but grow much better under anaerobic conditions.

Micrococcus (my-kroh-kock'-us)

microgram   One millionth of a gram (one thousandth of a milligram). Abbreviated μg.

Micromonospora (my-kroh-moan-ah'-spora)

Microsporum (my-kroh-spore'-um)

mineralization (stabilization)   Conversion of organic to inorganic materials.

mineralize   To convert organic substances into inorganic molecules.

miracidium (meer-a-sid'-ee-um)   The free swimming larva which emerges from the eggs of flukes.

mitosis (my-toe'-sis)   A process of chromosome duplication. The chromosomes divide longitudinally and the daughter chromosomes then separate to form two identical daughter nuclei.

molt   Shedding of the outer covering of arthropods and other forms.

monolayer cell culture   An *in vitro* culture in which the cells form a single layer on the container.

monosaccharide (mono-sak'-kar-ride)   A sugar, a simple carbohydrate, generally having the formula $C_nH_{2n}O_n$ where n can vary from 3 to 8. The most common are 5 and 6.

Moraxella (more-ax-ell'-ah)

motor nerve   Nerve responsible for controlling muscles (as opposed to nerves transmitting sensations).

mucilage   A sticky substance; in algae, secreted polysaccharide materials that aid in attachment to surfaces and in the motility of some forms.

mucoprotein (mu-ko-pro'-teen)   A protein associated with a carbohydrate.

mutagen (mutagenic agent) (mu'-ta-jen)   Any agent which increases the frequency at which DNA is altered, thereby resulting in an increase in the frequency of mutations.

mutant (mu'-tent)   An organism whose base sequence in DNA has been modified resulting in an altered protein which gives the cell properties which differ from the parent.

mutation (mu-ta'-shon)   A modification in the base sequence of DNA in a gene resulting in an alteration in the protein coded by the gene.

mutualism   Both host and organism benefit. A living together of two organisms whereby both benefit from the association.

mycelium (my-seal'-e-um, my-seal'-e-ah) (pl. mycelia)   A tangled, mat-like mass of fungal hyphae, exemplified by the cottony growth of molds. The collection of hyphae composing a fungus.

Mycobacterium leprae (my-koh-bak-teer'-ee-um lepp'-ree) *M. tuberculosis* (M. tuberk-you-loh'-siss)

Mycoplasma pneumoniae (my-koh-plazz'-mah nu-moan'-ee-ee)

mycorrhiza (my-koh-rize'-ah)   A symbiotic relationship between certain fungi and the roots of plants.

myxoviruses (mix'-oh-viruses)

nares (nair'-eez)   Nostrils.

nasopharynx (nay-zo-fair'-inks)   Area where the nasal passages join the throat.

Necator americanus (neck-kay'-tor ah-mair-ik-ahn'-us)

Neisseria catarrhalis (nye-seer'-ee-ah cat-arr-al'-iss)

   N. gonorrheae (N. gahn-oh-ree'-ee)

   N. meningitidis (N. men-in-jit'-id-iss)

neoplasm (nee'-o-plasim)   See tumor.

Nicotinamide adenine dinucleotide   NAD (nik-o-tin'-a-mid add'-den-neen di-new'-klee-o-tide)   A coenzyme which transfers H atoms. It is a derivative of the vitamin, niacin. When NAD is carrying the H atom, it is indicated as NAD_{red}.

natural amino acid   The L-isomer of the amino acid. Only L amino acids are found in proteins.

nitrification (nye-tri-fi-kay'-shun)   The conversion of NH_3 or NH_4^+ to nitrite (NO_2^-) and nitrite to nitrate (NO_3^-).

Nitrobacter (nye-troh-bak′-ter)

nitrogen fixation   The combination of nitrogen gas with other atoms to form compounds such as NO_2^-, NO_3^-, NH_3.

Nitrosomonas (nye-troh-soh-moan′-ass)

Nocardia asteroides (noh-kar′-dee-ah ass-ter-oid′-eez)

nonsense codon   A sequence of three purines and pyrimidines in messenger RNA which does not have a corresponding complementary sequence of three purines and pyrimidines in tRNA; therefore, the presence of a nonsense codon terminates the synthesis of a polypeptide chain.

nonpolar bond   A bond in which the charge is evenly distributed on each of the atoms which form the bond.

niacin (nye′-a-sin)   A B vitamin; it can be transformed into the coenzyme nicotinamide adenine dinucleotide (NAD).

numerical taxonomy   A technique for determining the relationship between organisms by determining the number of characteristics that the organisms have in common.

oncogenesis (on-ko-jen′-i-sis)   The process of tumor formation.

oncogenic (on-ko-jen′-ick)   Tumor inducing.

operator region   The gene which combines with an active repressor molecule.

opsonins (op′-soh-nins)   Substances that cause increased phagocytosis; for example, antibodies and complement.

opsonize   Increase the susceptibility of microorganisms to phagocytosis.

organ   A structure composed of different tissues, coordinated to perform a special function. Absent in members of the Protista.

organelle (or-gan-ell′)   Any structure which performs a specific function in a cell.

orifice (or′-ih-fiss)   Entrance or outlet of a body cavity.

origin   Refers to the site on the DNA molecule at which DNA synthesis is initiated in the process of chromosome duplication.

orthophosphate (or-tho-fos′-fate)   The inorganic form of phosphate; sometimes abbreviated P_i.

osmosis (os-moh′-sis)   The passage of a solvent through a membrane from a dilute solution into a more concentrated one.

osteomyelitis (oss′-tee-oh-my-e-light′-iss)   Inflammation of the bone.

otitis (oh-tye′-tiss)   Inflammation of the ear. Inflammation of the chamber just beneath the eardrum is called otitis media.

outgrowth   The stage in the development of endospores into vegetative cells which follows endospore germination.

oxidative phosphorylation (oxidate′-tiv fos-for-i-lation)   The generation of energy in the form of ATP which results from the passage of electrons through the electron transport chain to a final electron acceptor.

papovaviruses (pap-oh′-vah-viruses)

parvoviruses (par′-voh-viruses)

Pasteurella multocida (pass-ture-ell′-ah mull-toss′-id-ah)

pasteurization (pass′-ture-yze-ay-shun)   The processes of heating food or other substances under controlled conditions of time and temperature (for example 63 °C for 30 minutes) in order to kill pathogens and reduce the numbers of other microbes.

pathogenesis (path-oh-jen′-isiss)   The sequence of changes which culminates in disease.

pathogenic   Causing disease. "Pathogenic" is also used to designate microbes which commonly cause infectious diseases, as opposed to those which do so uncommonly or never.

para amino benzoic acid (PABA)   A growth factor for certain bacteria; a precursor of the vitamin folic acid.

parasite   An organism that lives in or on another organism (the host) and gains benefit at the expense of the host.

peat   Partially decayed plant material found in swamps.

pectins (pek′-tins)   Polysaccharide substances found in the cell walls of some algae; also found in ripe fruits and used to make jellies. A polysaccharide which serves as an important component of many plant cell walls.

Pediculus humanus (ped-ik′-u-luss hu-man′-us)

pellicle (pell′-e-kull)   Thickened, elastic cell covering; differ from cell walls, in that pellicles are not rigid.

penetration (viral)   The process of introducing viral nucleic acid into a host cell.

Penicillium (pen-e-sill′-e-um)

pentose   A general term for any five carbon sugar.

peptide bond   The covalent bond which joins adjacent amino acids together in proteins—an amino group of one amino acid is bonded to the carboxyl group of the next amino acid, with the formation of water.

Peptococcus (pep-toh-kok′-us)

Peptostreptococcus (pep-toh-strep-toh-kok′-us)

peristalsis (pay-riss-tall′-siss)   Wavelike contractions of the digestive tract responsible for propulsion of contents through the tract.

permeability   The capacity of a membrane or other structure to allow passage of substances through it.

permease (per′-me-ace)   A system of enzymes which are concerned with the transport of nutrients into the cell.

pesticide   A general term for any agent that kills pests, either plant or animal. If applied indiscriminately, it may kill other forms of life.

petrolatum (pet-tro-lay′-tum)   A greasy, jelly-like substance made from petroleum, often referred to by the trade name vaseline.

phagocytosis (fag′-oh-sigh-toe′-sis)   Cellular ingestion of particulate materials within membrane-bound vesicles; the particles are engulfed by pseudopods of the cell.

phenotype (feen′-o-type)   The sum total of the observable properties of an organism, resulting from the interaction of the environment with the genotype.

phospholipid (fos-fo-lip′-id)   A lipid which has a phosphate molecule as part of its structure.

Photobacterium (foh-toh-bak-teer′-ee-um)

photosynthesis (foh-toh-sin′-the-sis)   The sum total of the metabolic processes by which light energy is utilized to convert CO_2 and a reduced inorganic compound to cytoplasm. $6H_2X + 6CO_2 \rightarrow C_6H_{12}O_6 + 6X$.

Phthirus pubis (theer′-us pu′-biss)

phylum (phyla) (fie′-lum, fie′-lah)   A large division of related families in classifying living organisms. The classification is subdivided progressively from Kingdom → Subkingdoms → Phyla → Classes → Orders → Families → Genera → Species.

phytoplankton (fie-tow-plank′-ton)   The floating and swimming algae and procaryotic organisms of lakes and oceans.

Picornaviruses (pick-orn′-ah-viruses)

pigments   Coloring substances; in algae, substances such as chlorophylls, carotenoids, and phycobilins that absorb light energy and give color to the algae.

pilosebaceous gland (pile-oh-seh-bay′-see-us)   The tiny gland (sebaceous gland) that secretes an oily substance into the hair follicle.

pinocytosis (pine′-oh-sigh-toe′-sis)   Cellular ingestion of liquids; process essentially similar to phagocytosis.

Pityrosporum orbiculare (pit-ee-rah′-spore-um orb-ik-u-lah′-ree) *P. ovale* (P. oh-val′-lee)

plankton (plank′-ton)   Organisms that float in water.

plaque (plack)   A clear area in a lawn or monolayer of cells; viral plaques are created by viral lysis of infected cells within the clear area.

plaque (dental)   A collection of bacteria tightly adhering to a tooth surface; responsible for dental caries.

plasma (plaz′-ma)   The fluid portion of blood before clotting.

plasma cells   Lymphoid cells specialized to produce immunoglobulins in large quantities.

plasmodium (pl. plasmodia) (plaz-mode-e-um, plaz-mod-e-ah)   A mass of cytoplasm, enclosed by only a cytoplasmic membrane (but lacking a cell wall) and therefore free to ooze over a substrate; an example is found in the acellular slime molds.

Plasmodium vivax (plazz-moh′-dee-um vee′-vax) *P. falciparum* (P. fahl-sip′-ahr-um)

pleomorphic (plee-oh-more′-fik)   Occurring in various shapes.

photooxidation   A chemical reaction occurring as a result of absorption of light energy in the presence of oxygen; is sometimes responsible for death of microbes.

polychlorinated biphenyl compounds (PCBS)   Nonbiodegradable, toxic organic chemicals that are widely used in plastics and paint manufacture. (When discarded as wastes into bodies of water, these compounds are toxic for marine organisms).

polyene (polly-een′)   A chemical compound containing alternating single and double bonds; a group of antibiotics having such a property.

polymer   A high molecular weight compound which results from the chemical combination of large numbers of subunits.

polymorphic (polly-more′-fick)   Having different distinct forms at various stages of the life cycle.

PMN, polymorphonuclear leukocyte   (polly-morf-oh-new′-kle-er lew′-koh-site)   A phagocyte with a lobed nucleus; also called neutrophilic granulocyte, poly, or neutrophil.

polysaccharide (polly-sak′-karride)   Long chains, branched or unbranched, of monosaccharide subunits.

poly-β-hydroxybutyric acid granules   A storage reserve of carbon and energy which occurs as long chains of individual subunits forming readily visible granules.

polysome (polly′-soam)   A number of ribosomes attached to the same molecule of messenger RNA.

population   A group of organisms of the same species; i.e. an interbreeding group.

potential energy   The energy that is available in a substance as a result of its position in relation to its surroundings.

primary structure   The sequence of amino acids in a protein molecule.

procaryotic cell (pro-karry-ot-tik)   The simple cell type, characterized by the lack of a nuclear membrane, the absence of mitochondria, and a haploid state at all times.

polypeptide   A chain of amino acids bonded together by peptide bonds; the length varies from several amino acids to the length coded for by one gene.

primary producers   In the food chains, the organisms that carry out photosynthesis and thus produce organic materials from CO_2 and H_2O, using sunlight as an energy source.

progeny   Offspring; descendants.

prophage (pro-fayge or pro-fahge)   The nucleic acid of a temperate phage when it is integrated with the host cell DNA.

Propionibacterium shermanii (proh-pee-ah-ee-bak-teer′-ee-um shermahn′-ee) *P. acnes* (P. ak′-neez)

prostaglandin (pross-tah-gland′-in)   A group of substances found in various tissues and body fluids which cause relaxation of the walls of certain blood vessels and other effects. Like kinins and histamine, probably play a role in inflammation.

protease (pro′-tee-ace)   An enzyme that breaks down proteins. Proteases differ markedly in their activity at different pH. Some proteases degrade proteins to amino acids, others degrade only to polypeptides.

protein (pro′-teen)   A macromolecule containing one or more polypeptide chains.

Proteus mirabilis (proh′-tee-us mee-rab′-il-us)

Protista (pro-tis′-ta)   The kingdom in which the members of the microbial world are classified—lack of extensive tissue differentiation characterizes its members.

protoplast (pro′-tow-plast)   A cell which has its rigid cell wall removed.

prototroph (pro′-toh-trof)   An organism which has no organic growth requirements other than a source of carbon and energy.

pseudohyphae (sue-dough-high′-fee)   A cellular filament resembling hyphae, but formed by adhering elongated yeast cells which have formed by budding.

Pseudomonas aeruginosa (sue-dough-moan′-ass a-rue′-gin-o′-sa)

pseudomycelium (sue-dough-mye see′-lee-um)   A mass of fungal filaments resembling hyphae but composed of loosely united elongated cells formed by budding.

pseudopod (sue′-dough-pod, sue-dough-pohd′e-uh) (pseudopods or pseudopodia)   Extension of cytoplasm that aids in engulfing particles and functions in the motility of ameboid cells.

pupa (pew′-pah)   A stage in the development of many insects in which most of the changes from worm-like form to adult occur.

pure culture   A culture which contains only a single strain of organism.

pus cells   Cells in pus, principally dead PMNs.

putrefaction (pew-tre-fak′-shun)   The decomposition of proteins with the production of foul smelling products.

pyogenic (pye′-oh-jen′-ik)   Producing pus.

pyrophosphate (pie-row-fos′-fate)   Two phosphate molecules bonded together.

quarternary ammonium compound (quart-ter′-nary)   A substance with an N atom bonded to four organic groups, thereby giving a positive charge to the N atom. A chloride or bromide atom commonly makes up the rest of the molecule.

R factor   A transferable plasmid, found in many enteric bacteria, which carries genetic information for resistance to one or more chemotherapeutic agents. Often associated with a transfer factor that is responsible for conjugation with another bacterium and transfer of the R factor.

R genes   The genes on the R factor which are specifically concerned with resistance to chemotherapeutic agents.

radioactive   A substance emitting an ionizing radiation which can be detected and thereby used to locate the radioactive compound.

raphe (ray′-fee)   Open, elongated slit in the frustule of some diatoms, through which mucilage can extrude; involved in the motility of diatoms.

receiving waters   Bodies of water into which waste treatment products are emptied.

recipient cell   The cell which receives DNA from a donor cell.

refractile (ree-frak′-tile)   A body which bends a ray of light which passes through it so that the body often appears brighter than the surrounding objects.

regulatory genes   Genes whose function is to control the rate of synthesis of other gene products; regulatory genes often code for repressor proteins.

replica plating   A technique for transferring organisms from a large number of separated colonies from one medium to others. Used in the indirect selection of bacterial mutants.

replication (viral)   The process of synthesis of viral components within a host cell, culminating in the formation of complete virions.

replication fork   The Y shaped structure in the replicating DNA molecule; the site of DNA replication.

replicative form   A double-stranded form of nucleic acid, produced during multiplication of single-stranded viruses and consisting of the single strand of nucleic acid and a complementary strand; the replicative form serves as the intermediate in the synthesis of new single strands of viral nucleic acid.

repressor A regulatory substance (usually a protein) which prevents the function of a certain gene or genes.

repressor protein (regulatory protein) A protein capable of regulating the function of a gene or group of genes.

reproduction The initial process through which propagation of species is carried on; may be asexual in which a single cell divides, or sexual in which two cells unite before daughter cells are formed.

reservoir The sum of all the sources of an infectious agent.

resistance transfer factors (RTF) A set of genes associated with R factors which carry the genetic information for their transfer.

resolving power The distance two objects must be from one another before they can be distinguished with the microscope as two separate objects.

respiration (res-spir-ray′-shun) The sum total of metabolic steps in the degradation of foodstuffs when the electron (hydrogen) acceptor is an inorganic compound.

reticuloendothelial system (re-tick′-yew-low-end-oh-theal-e-al) The distribution of mononuclear phagocytes throughout the body in such a way as to clear foreign materials from the circulation; particularly abundant in liver, spleen, lung, lymph nodes, and bone marrow. Certain mononuclear phagocytic cells of the body located especially in liver, spleen, lung, lymph nodes and bone marrow.

reverse mutation or reversion The modification of the base sequence in a mutant gene to its original (parent) sequence.

reverse transcriptase An enzyme which synthesizes DNA complementary to an RNA template.

rhabdoviruses (rab′-do-viruses)

Rhizobium (rise-oh′-bee-um)

rhizopodial (rise-oh-poh′-dee-al) Algae that are motile by means of ameba-like pseudopods.

Rhizopus (rise′-oh-puss)

rhizosphere (rise′-o-sphere) The region immediately surrounding the roots of plants.

Rhodomicrobium vannielii (road-oh-my kroh′-bee-um van-neel′-ee-ee)

ribose (rye′-bose) A five carbon sugar found in ribonucleic acid.

ribosomal RNA (rye-bow-soam′-mal) A type of RNA which is found in ribosomes.

rich medium A culture medium containing many different kinds of nutrients; the nature and quantity of each nutrient is usually undetermined.

Rickettsia prowazekii (rik-ett′-see-ah prow-ah-zek′-ee-ee)

RNA polymerase (poh-lim′-mer-ace) An enzyme that catalyzes the synthesis of RNA from ribonucleoside triphosphates. The enzyme which links together nucleotide subunits during the transcription of DNA to form RNA. Its action represents the first step in protein synthesis.

Saccharomyces cerevisiae (sac′-a-row-my′-seas sara-vis′-i-ee)

Salmonella typhi (sall-moh-nell′-ah tye′-fee)

salmonellosis (sall-moh-nell-oh′-sis) Infection with pathogenic gram-negative rods of the genus *Salmonella*; well-known forms of salmonellosis include the common gastroenteritis caused by many strains of *S. enteritidis* and the less common typhoid fever caused by *S. typhi*.

saprophyte (sap′-roh-fyte) A microorganism which lives on dead organic material.

sarcoma (sar-ko′-ma) A tumor of connective tissue. Usually fibrillar and highly malignant.

Sarcoptes scabiei (sar-kop′-teez scabe′-ee-ee)

Schistosoma mansoni (shiss-toh-soh′-mah man-soh′-nee)

schistisomiasis (shiss-toh-soh-mye′-assiss) The disease resulting from infection with flukes of the genus *Schistosoma*.

secondary bonding forces Weak chemical bonds, such as hydrogen bonds, van der Waals forces.

selective enrichment A technique for specifically encouraging the growth of a particular group or even a single species of organism.

selective medium A medium which has components which restrict growth to organisms of a particular type.

semi-conservative replication The type of replication of DNA in which new daughter strands complementary to each of the original strands are synthesized in each round of replication.

sensitized lymphoid cells Lymphoid cells, usually lymphocytes, that have the capacity to respond to specific antigen in such a way as to cause reactions of delayed-type hypersensitivity and cellular immunity or production of antibody in the anamnestic response.

septate Divided by membrane; for example, fungal filaments may be septate or may lack septa (aseptate or nonseptate).

septum (sep′-tum) A membrane separating two cavities or two cells in a filament of cells.

serological Having to do with serum.

Serratia marcescens (ser-ray′-shee-ah mar-sess′-senz)

serum (see′-rum) The fluid portion of blood which remains after the blood clots.

sessile (sess′-ile) Fixed or attached.

sexual recombination The exchange of portions of DNA molecules, from each parent, in a cell; provides for the cell having some properties of each of the parental cells.

Shigella dysenteriae (shigg-ell′-ah diss-en-tair′-ee-ee)

Siderocapsa (sid-er-oh-cap′-sah)

Siderococcus (sid-er-oh-kok′-us)

sigma (σ) **factor** Subunit of RNA polymerase that recognizes specific sites on DNA for initiation of RNA synthesis. A subunit of the enzyme RNA polymerase which is responsible for determining where the enzyme begins transcribing.

simple lipid A lipid that contains only the elements C, H and O.

Simonsiella (see-moan-see-ell′-ah)

soap A substance resulting from reaction of a fatty acid and an alkali; a salt of a fatty acid.

Sphaerotilus (sfeer-aht′-il-us)

spherule (sphere′-ule) Large, round, thick-walled structure containing many fungal spores; characteristic of the tissue phase of *Coccidioides immitis*.

sphincter (sfingk′-ter) A ringlike muscle that normally constricts a passage or orifice of the body and relaxes as required for normal function.

spicule (spick′-ule) A needle-like structure.

Spirillum (spye-rill′-um)

Spirochaeta (spyro-kee′-tah)

spontaneous mutants Mutants which arise without the addition of recognized mutagenic agents.

sporangium (spore-ran′-gee-um) Sac-like structure within which asexual spores are formed.

spores Specialized reproductive cells; asexual spores germinate without uniting with other cells, whereas sexual spores of opposite mating types unite to form a zygote before germination occurs.

sporocyst (spore′-oh-sist) A stage in the life cycle of certain protozoa in which two or more of the parasites are enclosed within a common wall.

Sporotrichum schenckii (spore-oh-trick′-um shenk-ee-ee)

sporozoan (spore′-oh-zoe′-uhn) A protozoan of the class Sporozoa.

sporozoite (spore-oh-zoh′-yte)   A developmental form in the life cycle of certain protozoan parasites such as *Plasmodium* species. The sporozoite is the form in which the malarial parasite is transmitted to man by the infected mosquito.

sporulation   Spore formation; also used to mean a process of multiple fission by which some protozoa divide.

stable terminal residue   The last in a series of soil degradation products which is not capable of being further degraded by natural means, and therefore remains.

stabilization (mineralization)   Conversion of organic materials to inorganic substances.

Staphylococcus aureus (staff-il-oh-kok′-us aw′-ree-us) *S. epidermidis* (S. epi-der′-mid-iss)

starches   Branched polysaccharides consisting of glucose subunits; serve as nutritional reserves in plants and some eucaryotic protists.

stasis (stay′-sis)   A staying in one place, or a stoppage of flow; for example of blood or urine.

stationary phase   The stage of growth of a culture in which the number of viable cells remains constant.

sterilization   Rendering an object or substance free of all viable microbes. Practically speaking, to sterilize an object is to make it extremely improbable that a single living organism or virus remains.

Streptococcus pyogenes (strepto-kok′-us pye-ah′-jen-eez)

Streptomyces griseus (strepto-mye′-seez gree′-see-us) *S. mediterranei* (S. med-it-ter-an′-ee-ee) *S. venezuelae* (S. ven-ez-u-ayl′-ee)

substrate   The substance on which an enzyme acts to form the product.

succession (ecological)   The predictable and orderly sequence of change in a community.

sucrose   A disaccharide consisting of a molecule of glucose and one of fructose.

sugar   A common term for the low molecular weight carbohydrates. Common table sugar is sucrose. There are D and L forms of sugars; the D forms are found in nature.

sulfanilamide (sul-fa-nil′-a-mide)   A drug whose chemical structures is very similar to that of para-amino benzoic acid. It competes with PABA for the enzyme which converts PABA to folic acid.

suture (su′-chore)   Stitch used by a surgeon to close a wound.

swarmer cells   The cells of *Rhizobium* which penetrate the root hairs to set up the symbiotic association with legumes. They are more nearly spherical than the rod shaped cells of *Rhizobium*.

symbiosis (sim-bye-oh′-siss)   Living together.

synapsis (sin-nap′-sis)   The pairing of identical chromosomes that occurs prior to the first division in meiosis.

synthetic medium   A medium in which every component has been identified with respect to its chemical composition and quantity.

taxonomy   The science of the classification of organisms.

temperate phage   A phage which can either become integrated into the host cell DNA as a prophage, or replicate outside the host chromosome, and lyse the cell.

template   The molecule which serves as the mold for the synthesis of another molecule. Macromolecular synthesis involves templates.

termites   A type of insects which harbor cellulose digesting protozoa in their gut.

tetracycline (tet-ra-sigh′-clean)   An antibiotic produced by a member of the genus *Streptomyces;* acts by inhibition of protein synthesis.

therapeutic index   The ratio of minimum toxic dose to the effective dose of a medication.

Thermoactinomyces (ther-moh-ak-tin-oh-mye′-seez)

Thiobacillus (thy-oh-bah-sill′-us)

Thiothrix (thy′-oh-thrix)

thorax   The chest.

tincture   An alcoholic solution of a medication or other chemical substance.

tissue   A group of cells, of one or a few similar types which associate into a structural subunit and perform a specific function.

titer (tight′-er)   The concentration of a substance in solution; for example, the amount of a specific antibody in serum, usually measured as the highest dilution of serum which will give a positive test for that antibody. The titer is often expressed as the reciprocal of dilution; thus, a serum which gives a positive test when diluted $1:256$, but not at $1:512$, is said to have a titer of 256.

toxin   A poisonous substance originating from microbes, plants or animals.

toxoid   A substance derived from a toxin in which toxicity has been destroyed while maintaining its ability to induce immunity to the toxin.

Toxoplasma gondii (tox-oh-plazz′-mah gahn′-dee-ee)

transamination (trans-am-in-a′-shun)   A reaction in which keto $acid_1$ and amino $acid_2$ react to form keto $acid_2$ and amino $acid_1$; there is an exchange of the keto and amino groups.

transcription   The process of transferring genetic information coded in DNA into informational messenger RNA. The copying of the message of DNA into RNA, by a process which involves base pairing of complementary RNA with the DNA and polymerization of the bases so ordered.

transduction   Transfer of genetic information by temperate phages. *Generalized transduction*   Transduction of any part of a bacterial chromosome. *Restricted or specialized transduction*   Transduction of a particular portion of the bacterial chromosome adjacent to the site of prophage integration.

transfer RNA (tRNA)   A type of RNA which binds and carries amino acids to the ribosome.

transformation   A conversion from one form to another. In the context of animal virology, the process by which viruses change normal cells to malignant cells. In bacterial genetics, the process of transfer of DNA as "naked" DNA.

translation   The process by which genetic information in the messenger RNA directs the order of the amino acids in protein during protein synthesis.

Treponema pallidum (trep-oh-nee′-mah pal′-ee-dum)

Trichinella spiralis (trick-in-ell′-ah speer-ahl′-iss)

trichocysts (trick′o-cysts)   Small dartlike structures found in some ciliates; can be ejected as a defense mechanism.

Trichomonas vaginalis (trick-oh-moan′-us vaj′-in-alice)

Trichophyton (trick-oh-phye′-ton)

trophozoites (trowf′-oh-zoe′-ites)   The vegetative forms of some protozoa, as opposed to the resting forms called cysts.

tumor   A swelling involving a mass of new tissue which grows independently of surrounding tissue and may invade it. Synonym, neoplasm.

tyndallization (tin-dal-i-za′-shun)   The technique of killing endospores by treating a material with heat sufficient to kill vegetative organisms, incubating to allow any spores to germinate and develop into vegetative cells, and then reheating to kill the new vegetative cells. Tyndallization requires that the substance allow germination of endospores if they are present.

typhus   An infectious disease caused by *Rickettsia prowazekii*.

ultracentrifuge   A high-velocity centrifuge used to separate colloidal or submicroscopic particles, such as proteins or viruses.

ultraviolet light (UV)   Electromagnetic radiation with a wavelength between 175–350 nm (shorter than visible light). Certain wavelengths absorbed by nucleic acids, result in mutation and death.

valency Capacity to combine with something else; for example, of atoms to combine in specific proportions with other atoms, and of antibodies to combine with a certain number of antigenic determinants.

van der Waals interactions (van-der-walls) Weak chemical bonds which result from the nonspecific attraction of atoms when they come near each other.

vascular Pertaining to blood or lymphatic vessels.

vectors An animal, usually an insect, which carries an infectious agent from one host to another.

vegetative cell The growing or feeding form of a microbial cell, as opposed to resistant resting forms.

vegetative phage Phage that is replicating within a host cell and producing mature virions.

vehicle An inanimate carrier of an infectious agent from one host to another.

Veillonella (vah'-yon-ell-ah)

vesicle A small sac.

vestigial (ves-tijh'-e'-al) Occurring as a visible trace or sign of something that no longer exists.

Vibrio cholerae (vib'-ree-oh kahl-er-ee)

virion (veer'-e-ohn) A viral particle consisting of nucleic acid surrounded by a protein capsid and, in some viruses, an outer envelope.

virulence The sum of properties of an organism which enhances pathogenicity.

virulent phage A bacteriophage that causes lysis of the host cell following phage replication in the cell. It cannot be integrated into the chromosome as opposed to temperate phage of the host.

virus An obligate intracellular parasite consisting of a bit of nucleic acid, surrounded by a protein coat and sometimes enclosed by an envelope. Different viruses are capable of infecting animals, plants and bacteria.

vitamin A substance, required in very small quantities, which is converted into a coenzyme.

volutin (vol'-u-tin) Another name for polyphosphate compound; occurs as metachromatic granules; long chains of phosphate molecules joined together to form a storage material in certain cells.

Xenopsylla cheopis (zeen-oh-sill'-ah kee-op'-iss)

Yersinia pestis (yer-sin'-ee-ah pess'-tiss)

zooplankton (zoe-oh-plank'-ton) The floating and swimming small animals and protozoa of open waters.

zoospores (zoe'-oh-spores) Motile, asexual spores.

zygote (zigh'-goat) Diploid cell formed by the sexual fusion of two haploid gametes. A cell formed by the sexual union of a male and female gamete.

Index

702

Mitochondria, 91–93
Mitomycin C, 563
Mitosis, 205–206
Mixed cultures, 100
Molds, 281–294
    commercial uses, 662, 666, 667, 669, 670
    in food spoilage, 647–649
    in lichens, 276–277
    in soil, 594
    source of antibiotics, 667
Molecular hybridization, DNA, 231–234
    in studies of viral oncogenesis, 354
Molecules, small, 31 (*see also* Macromolecules)
Monocystis sp., 302, 305
Monocytes, 376, 377
Monosaccharide, 41
Mosquito, 537–541
    anatomy, 537
    groups, 537
    as vector of equine encephalitis viruses, 541
        F. tularensis, 538
        Plasmodium sp., 538–541
Mouse mammary tumor virus, 349
Mucocomplex, structure and component of bacterial cell wall, 77–78
Multicellular parasites and human diseases, arachnids, 544–545
    flatworms, 548–552
    insects, 537–544
    roundworms, 546–548
Mumps, 475
Mushrooms, 281–288
Mutagenic agents (mutagens), 183–187
    base analogs, 183
    mode of action, 183
    nitrous acid, 183
    ultraviolet light, 185
    tumor induction by, 348–349
    X-rays, 185
Mutation, 182–192
Mutualism, 359
Mycelia, actinomycetes, 521–522
    fungi, 282, 283
    streptomycetes, 261
Mycelial bacteria, aquatic, 262
    characteristics, 57, 63, 259
Mycobacterium sp., characteristics, 259
    M. bovis as cause of tuberculosis, 463–464
    M. leprae, cause of leprosy, 508
        laboratory studies, 508–509
    M. smegmatis, 260
    M. tuberculosis, cause of tuberculosis, 462
        characteristics, 462–463
        immunity to, 386, 388, 682 (BCG)
        latent infections, 366
Mycoplasma pneumonia, 458
    cold agglutinins, 458

Mycoplasma sp., characteristics, 57, 64, 262
    M. bovirhinus, 262
    M. felis, 262
    M. gallisepticum, 262
    otitis media, 451
    M. pneumoniae, cause of pneumonia, 458
        characteristics, 262–263
        cold agglutinins, 458
        laboratory identification, 458
    sterols in cytoplasmic membrane, 84
Mycorrhizas, 290–291
Mycoses (fungal infections), 291
    caused by, *Blastomyces dermatitidis,* 467
        Candida albicans, 441, 499–500
        Coccidioides immitis, 464–466
        Cryptococcus neoformans, 291
        Epidermophyton sp., 441
        Histoplasma capsulatum, 466–467
        Microsporum sp., 441
        Sporotrichum (*Sporothrix*) *schenckii,* 523
        Trichophyton sp., 441
Myocarditis, viral, 530
Myxobacteria, characteristics, 241–243
    degradation of cellulose, 598
Myxomycetes (slime molds), 291–293
Myxoviruses, causing colds, 446
    causing influenza, 458
    characteristics, 680

N

Naked virus, 334, 335
Nasal chamber, normal flora, 445–446
Necator americanus, 546
Negri bodies, 512
Neisseria sp., characteristics, 256
    N. catarrhalis, normal flora of throat, 358
    N. gonorrhoeae (gonococcus), as agent of gonorrhea, 492–495
    N. meningitidis, among normal flora of nasopharynx, 446
        bacteriocin typing, 505
        cause of meningitis, 504–507
        endotoxin, 506
        epidemiology, 505
        transforming DNA, 505–506
        types, 505
Nematodes, 546–548
    Ancylostoma duodenale, 546
    Ascaris lumbricoides, 547
    Enterobius vermicularis, 546
    Necator americanus, 546
    Trichinella spiralis, 548
Neomycin, 177
    mode of action, 177
    structure, 560
    use in therapy, 561

S

718

of insects, 344
methods to study, 326–331
of plants, 339–344
tumor, 349–356
Vitamins, microbial synthesis of, 626, 667
produced by diatoms, 279
Volutin, of *Corynebacterium* sp., 258

W

Warts, 677–678
Waste disposal, principals of microbial degradation, 636
nature of sewage, 637
sewage treatment, 638, 642
utilization of treated waste residues, 645
Water, as vehicle of infectious agents, 419–420
Water activity (a_w) of foods, 649
Water ecosystems, 621–625
Water molds, 286
Waxes, degradation by mycobacteria, 259
Weil-Felix test, 544
White blood cells (*see* Leukocytes)
Whooping cough, 460–461
Wine, 659–663
Work of cells, 154
World Health Organization, and control of epidemics, 423
role in control of malaria, 540
role in control of smallpox, 437
role in establishing standardized antibiotic sensitivity
testing methods, 571

Wound infections, 516–524
aerobic infections, 517
anaerobic infections, 518–521
animal bites, 522
burns, 517
fungal, 523
gas gangrene, 521
organisms responsible, 516, 518
surgical, 517
tetanus, 518–521

X

Xanthomonas sp., characteristics, 250
Xenopsylla cheopis, 543

Y

Yeasts, commercial uses, 291, 660–663, 669, 670
definition, 283
electron micrograph of, 284
life cycle of, 290
Yersinia pestis, cause of plague, 542
classification, 252
toxin production by, 364
Yogurt, commercial production of, 658

Z

Zygote, in sexual recombination, 205
Z value, 589